Control Handbook

Other McGraw-Hill Handbooks of Interest

AMERICAN SOCIETY OF MECHANICAL ENGINEERS · ASME Handbook—Metals Properties
AMERICAN SOCIETY OF MECHANICAL ENGINEERS · ASME Handbook—Engineering Tables
BAUMEISTER AND MARKS · Standard Handbook for Mechanical Engineers
BRADY · Materials Handbook
CALLENDER · Time-Saver Standards
CARRIER AIR CONDITIONING COMPANY · Handbook of Air Conditioning System Design
CONOVER · Grounds Maintenance Handbook
CROCKER AND KING · Piping Handbook
CROFT, CARR, AND WATT · American Electricians' Handbook
DAVIS AND SORENSEN · Handbook of Applied Hydraulics
EMERICK · Handbook of Mechanical Specifications for Buildings and Plants
EMERICK · Heating Handbook
EMERICK · Troubleshooters' Handbook for Mechanical Systems
FACTORY MUTUAL ENGINEERING DIVISION · Handbook of Industrial Loss Prevention
FINK AND CARROLL · Standard Handbook for Electrical Engineers
FRICK · Petroleum Production Handbook
GAYLORD AND GAYLORD · Structural Engineering Handbook
GUTHRIE · Petroleum Products Handbook
HARRIS · Handbook of Noise Control
HARRIS AND CREDE · Shock and Vibration Handbook
HAVERS AND STUBBS · Handbook of Heavy Construction
HEYEL · The Foreman's Handbook
HUSKEY AND KORN · Computer Handbook
KATZ · Handbook of Natural Gas Engineering
KING AND BRATER · Handbook of Hydraulics
KLERER AND KORN · Digital Computer User's Handbook
KORN AND KORN · Mathematical Handbook for Scientists and Engineers
LA LONDE AND JANES · Concrete Engineering Handbook
MAGILL, HOLDEN, AND ACKLEY · Air Pollution Handbook
MANAS · National Plumbing Code Handbook
MANTELL · Engineering Materials Handbook
MERRITT · Building Construction Handbook
MERRITT · Standard Handbook for Civil Engineers
MOODY · Petroleum Exploration Handbook
MORROW · Maintenance Engineering Handbook
MULLIGAN · Handbook of Brick Masonry Construction
MYERS · Handbook of Ocean and Underwater Engineering
PERRY · Engineering Manual
ROSSNAGEL · Handbook of Rigging
STANIAR · Plant Engineering Handbook
STETKA · NFPA Handbook of the National Electrical Code
STREETER · Handbook of Fluid Dynamics
TIMBER ENGINEERING CO. · Timber Design and Construction Handbook
URQUHART · Civil Engineering Handbook
WADDELL · Concrete Construction Handbook
WOODS · Highway Engineering Handbook

Industrial Pollution Control Handbook

Edited by

HERBERT F. LUND

President, Leadership Plus, Inc.
Former Editor-in-Chief, *Factory* magazine

McGRAW-HILL BOOK COMPANY

New York St. Louis San Francisco Düsseldorf Johannesburg
Kuala Lumpur London Mexico Montreal New Delhi Panama
Rio de Janeiro Singapore Sydney Toronto

Sponsoring Editors Harold B. Crawford/Daniel N. Fischel
Director of Production Stephen J. Boldish
Editing Supervisor Lila M. Gardner
Designer Naomi Auerbach
Editing and Production Staff Gretlyn Blau,
Teresa F. Leaden, George E. Oechsner

INDUSTRIAL POLLUTION CONTROL HANDBOOK

 Library of Congress Catalog Card Number 70-101164

07-039095-9

4567890 MAMM 7543

UC

To my dear, patient wife, Belle,
and for Lisa, Ilene, Holly, and Donna

Contributors

ORLAN M. ARNOLD *Vice President, Ajem Laboratories, Inc., Livonia, Michigan (Environmental Research Consultant, Grosse Pointe Park, Michigan)*

R. P. BARZLER *Senior Mechanical Engineer, Giffels Associates, Inc., Architects, Engineers, Planners, Detroit, Michigan*

B. BASKIN *Senior Mechanical Engineer, Giffels Associates, Inc., Architects, Engineers, Planners, Detroit, Michigan*

R. M. BILLINGS *Director of Environmental Control, Kimberly-Clark Corporation, Neenah, Wisconsin*

JOHN A. BLATNIK *Congressman from Minnesota, House of Representatives, Rayburn House Office Building, Washington, D.C.*

RALPH R. CALACETO *Vice President, Airetron Engineering Division, Pulverizing Machinery, A Unit of the Slick Corporation, Midland Park, New Jersey*

W. ALLEN DARBY *Consultant, Water Management Systems, Dorr-Oliver Inc., Stamford, Connecticut*

G. G. DEHAAS *Manager, Chemicals Development Department, Weyerhauser Company, Longview, Washington*

EMIL F. DUL *Engineering Specialist, Pollution Control Division, Lockwood Greene Engineers, Inc., New York, New York*

THOMAS A. FRIDY, JR. *Project Director, Water and Waste Treatment, Civil and Marine Engineering Department, Lockwood Greene Engineers, Inc., Spartanburg, South Carolina*

NORMAN FRISCH *Manager, Chemical Process Development, Research and Development Division, Research-Cottrell, Inc., Bound Brook, New Jersey*

D. J. GIFFELS *Project Director, Giffels Associates, Inc., Architects, Engineers, Planners, Detroit, Michigan*

L. C. GILDE *Manager–Engineering, Water Administration, Campbell Soup Company, Camden, New Jersey*

HUGH P. GRUBB *WED Enterprises, Inc., Glendale, California*

FRANK N. KEMMER *Manager, Pollution Control Department, Nalco Chemical Company, Chicago, Illinois*

JOHN E. KINNEY *Sanitary Engineering Consultant, Ann Arbor, Michigan*

LESLIE E. LANCY *President, Lancy Laboratories Inc., Metal Finishing and Waste Treatment Engineers, Zelienople, Pennsylvania*

BRUCE S. LANE *Director of Engineering and Maintenance, The Upjohn Company, Kalamazoo, Michigan*

HERBERT F. LUND *President, Leadership Plus, Inc., Environmental Engineering, Manufacturing Analysis, and Motivation Training Consultants, One Wellington, Stamford, Connecticut. (Former Editor-in-Chief,* Factory *magazine)*

WILLIAM W. MOORE *Vice President and General Manager, Air Pollution Control Division, Research-Cottrell, Inc., Bound Brook, New Jersey*

EDMUND S. MUSKIE *United States Senator from Maine, United States Senate, Washington, D.C.*

EDGAR G. PAULSON *Manager, Process and Waste Water Engineering, Calgon Corporation, Pittsburgh, Pennsylvania*

DAVID H. REEVE *President, Effluent Controls, Inc., Lakeland, Florida*

ROBERT G. REICHMANN *Plant Engineer, Electronic Division, Aerojet-General Corporation, Azusa, California*

R. D. ROSS *Vice President, Thermal Research & Engineering Corporation, Conshohocken, Pennsylvania*

ROBERT M. SANTANIELLO *Vice President and General Manager, Gulf Degremont, Inc., Liberty Corner, New Jersey*

HERBERT A. SCHLESINGER *Manager, Pollution Control Division, Lockwood Greene Engineers, Inc., New York, New York*

JOSEPH SCRUGGS *Project Engineer, D. J. Stark and Associates, Inc., New York, New York*

STANLEY K. SMITH *Supervisor of Facilities Engineering, Honeywell Inc., St. Petersburg, Florida*

EMMET SPENCER, JR. *Pollution Control Engineer, FMC Corporation, Chemical Divisions, New York, New York*

ARTHUR C. STERN *Professor, Department of Environmental Sciences and Engineering, The University of North Carolina, Chapel Hill, North Carolina*

KENNETH WATSON *Director of Environmental Control, Research and Development Division, Kraftco Corporation, Glenview, Illinois*

JEROME WILKENFELD *Director, Environmental Health, Hooker Chemical Corporation, Stamford, Connecticut*

EDWARD WILLOUGHBY *Chief Civil Engineer, Giffels Associates, Inc., Architects, Engineers, Planners, Detroit, Michigan*

JOSEPH B. ZUZIK *Supervisor, Industrial Hygiene and Safety, American Cyanamid Company, Stamford, Connecticut*

Foreword

by Senator Edmund S. Muskie

Public concern over environmental pollution is increasing at a geometric rate. The concern stems from an awareness of the threats to health and welfare from the wastes of our society; it leads inevitably to pressure on industry to reduce the discharge of contaminants into the air and public waterways.

The central question in pollution abatement is no longer whether; it is how. The "Industrial Pollution Control Handbook" is designed to answer the question of how best to reduce waste discharges with the latest technology. It should provide a major contribution to the dissemination of control techniques and the encouragement of sound corporate decisions in waste management.

The implementation of an effective environmental improvement program depends on the coordinated efforts of the general public, government, and private industry. I am encouraged by the publication of this Handbook as evidence of private industry's growing awareness of the need to reduce waste discharges as a contribution to societal health and as a sound business practice.

Preface

One of the contributors to this Handbook, Jerome Wilkenfeld, informs us that pollution has a long history. The problems of water pollution have been recorded during the Roman Empire. In the medieval period, King Edward of England signed the first air pollution law, the Sea Coals Act. There are even references to pollution in the Bible. So pollution problems are not new. They have just come on stronger, much stronger during these last two decades.

Basically, there are two reasons for the accelerated concern for the deterioration of our environment: the rapid increase in pollution and the dynamics of television and newspaper coverage. Although there isn't conclusive evidence, there is a parallel relationship between the increasing incidence of pollution and population growth. Simply, the more people, the more cars; the more people, the greater the need for power from generating plants; and the more concentrated the population near urban centers, the greater the likelihood of industries settling there. And, given the dynamics of the communications media, the competition for hot news has brought the dirt of pollution to the forefront. The public has definitely been aroused during the last ten years, but sometimes through questionable reporting. For the most part, the general good control practices by industry are played down while the infrequent poor practices are embellished in bold headlines. Let's set the record straight concerning industry and its social responsibility.

The United States probably suffered from a lack of industrial pollution control as far back as the Civil War, but nobody told us about it. Then, as now, our waterways served as the source of cooling water and also as the fluid conveyor of industrial wastes. Similarly, the sky above was the conveyor to disperse gaseous wastes. For years, industry considered it had the right to dump wastes into the sky and streams.

For the last ten years, we have been telling industry that the old methods of getting rid of air and liquid wastes must be completely changed. But we must also be aware that it will cost industry extra funds in capital equipment and operating costs with little or no return on the investment. In relation to the hundred years of industrial history, ten years are relatively short for a complete reversal of disposal procedures. Nevertheless, I do feel industry has been given sufficient warnings and time to set aside both money and technical knowledge to prevent pollutants from reaching the surrounding population and plant life.

In my opinion, an industrial pollutant is a liability. A liability in the sense that a company or a plant owes the population surrounding it breathable air and drinkable water. It can be considered as a long, long term loan of the sky and waterways granted to industry. The pollutant liability that industry has built up most of these hundred years now should be paid up. Now, the needs of the growing population demand the calling in of that loan in the form of good breathable air and water of drinking condition. Industry's problem is to accomplish this feat practically, economically, without excessive burden to itself.

The "Industrial Pollution Control Handbook" is intended to help industry meet its obligation. It is a communication bridge between the best practical experts on industrial pollution problems and all of the responsible industrial managers. This Handbook was conceived as a practical, easy-to-understand guide to solving pollution problems.

From the beginning, we looked at pollution as a total system problem. The various interactions between air, water, and solid wastes must be evaluated by a plant or even an entire industry. The treatment of a water pollutant may create an air pollutant, for instance, an objectionable odor. And of course, the collection and handling of the dust particles from a bag house or cyclone will produce a solid waste disposal situation. This particular problem is described in depth by R. D. Ross in Chapter 7. Since pollution control is evolving as a total system, possibly a closed-loop phenomenon within a plant or complex, the primary section of this Handbook focuses on pollution control by industry types.

Every major industry where pollution is a difficult problem has been described in a separate chapter: foundries, plating, pulp and paper, foods, chemicals, textiles, pharmaceuticals, aerospace, and even a separate review of European plant practices. Each chapter has been written by an expert practical contributor or team of contributors.

In front of these bread basket-type chapters, we trace the evolution of industrial pollution control. This includes a statement of the problem by Jack Kinney, the history of federal legislation by Congressman John Blatnik, and a similar treatise on state and local legislation by Jerry

Wilkenfeld. Air and water quality standards and criteria, always a difficult point of discussion, has been clearly described by Art Stern, now of The University of North Carolina and R. M. Santaniello of Gulf Degremont, Inc. Also in this evolution section, other contributors present: air and water pollution control systems, W. W. Moore and Dr. Norman Frisch of Research-Cottrell and Ed Paulson, Calgon; disposal of collected wastes, R. D. Ross, Thermal Research & Engineering; joint municipal-industrial control programs by Ken Watson, Kraftco Corporation; and finally, what practical areas of research are needed for pollution control, Joe Zuzik, American Cyanamid.

There have been many problems related to the design, installation, operation, and costs of control equipment. For the last five years, I have always stated that claimed operating efficiency, whether 78 or 99.9 percent, is only possible when the equipment is running. But if the equipment is stopped for any reason, the efficiency becomes zero. The last section on pollution control equipment is dedicated to help you avoid, minimize, or hopefully eliminate the zero efficiency of your control equipment. First, you should organize a plant department. Then the design details and fabrication materials of air and water pollution control equipment are thoroughly described by Mr. Calaceto and Mr. Darby, respectively. Probably the most often neglected aspects of control equipment are the proper training of operators, equipment start-up, and the specifics of performance guarantees. The latter by Dr. O. M. Arnold, warns plant managers to demand protective contracts which will assure that claimed equipment efficiencies become actual efficiencies.

To help you become more conversant with technical terms, a Glossary of nearly forty pages has been amassed and placed after the equipment section. I have been told that many people judge the value of a technical book or handbook on the quality of its index. My wife, Belle, and I have devoted many hours and tired sets of eyes to produce a useful, descriptive, source index. For the busy manager, this index will be a time saver.

We did not avoid cost information. In fact, contributors were encouraged to furnish capital, installation, and operating costs. Each of these cost categories are broken down separately in the index.

And of course, both the McGraw-Hill Book Company and I are indeed grateful to Senator Edmund S. Muskie for his eloquent introductory remarks.

Herbert F. Lund

Contents

Section 1

Evolution of Industrial Pollution Control

Chapter 1

The Industrial Pollution Problem

JOHN E. KINNEY

Sanitary Engineering Consultant, Ann Arbor, Michigan

The challenge of a handbook is to provide adequate guidance in what should be done and how it can be done. Many either ignore the first issue or presuppose general knowledge and hence consider it superfluous. These books proceed with varying degree of detail to guide the doing of the job.

No such assumption is valid for pollution problems. Furthermore, the question of what should be done demands a predetermination of why it should be done if there is to be assurance that how it is done is adequate.

DeMadariaga's observation is apropos. He noted that Europeans facing a crisis divide into two groups: those who ask "what is to be done" and those who ask "who shall do it"—the peoples-what and the peoples-who. Americans are no different in that respect, but in addition, Americans characteristically act energetically with but little consideration as to future consequences. Progress is measured in terms of speed rather than direction. Little time is invested in a prior determination of why. Fundamentals can be and often are ignored.

Certainly this has been our history in water pollution abatement, and every indication is that there will be a repetition of these errors in air pollution control.

The spontaneous answer as to why there should be a control program would be simply to eliminate pollution and prevent further pollution. But such an answer has no substance, for we have no definition of pollution! We have no specific objective!

There are many varied and even conflicting opinions of pollution. Twenty years of federal enactments have failed to be specific. The closest to congressional guidance was contained in the House Public Works Committee Report 2021 (89th Congress, 2d Session). In a chapter "What Is Water Pollution," the committee report enumerated the various sources and types of wastes and concluded:

> It it obvious there are many potential sources of waste which may cause pollution. Whether pollution exists, however, depends on the actual water quality, for pollution is an impairment of quality such that it interferes with the intended uses.

This sense of Congress which was contained in the House Report accompanying the Clean Streams Act of 1966 has been given little heed by the administration—so little in fact that the House Appropriations Committee expressed its concern in the report of July, 1967, on the Public Works Appropriation for 1968 and specifically stated what is expected.

> As the power to control water quality and quantity is not only the power to make or break business and agriculture but is a power over the life of the nation itself, it is essential that the Federal Water Pollution Control Administration not only closely coordinate its plans and activities with all the Federal agencies involved, but also with each of the States, local jurisdictions, and private interests affected by the program.
>
> The imposition of restrictions and controls without full and equitable consideration of the essential and varied interests involved in water supply, including priority of use and riparian rights, could have a most serious adverse effect on the various segments of the economy dependent upon water for their existence. The Committee wishes to emphasize the importance, therefore, of the new Administration undertaking its most difficult and essential program of water pollution control with a sense of balance and caution to assure any disruptive or adverse effects on the economy are minimized.
>
> Although the Committee appreciates the need for adequate planning, it questions the soundness of the comprehensive planning that has been done in the past. Much of it appears to have consisted largely of extensive and expensive surveys of sources of pollution. . . . [This] is only one of many ingredients needed to develop an effective program for the control of pollution. There is an urgent need to expedite the planning necessary to achieve pollution control on a comprehensive river basin approach. This essential objective cannot be accomplished at the pace or in the manner pursued in the past.

But again congressional views caused no immediately noticeable change in the procedures of the establishment.

From the variety of pollution abatement projects underway one might presume the questions on how abatement can be achieved have been answered. However, from the tenor of questions being asked at federal, state, and local levels, it is obvious there is no general agreement on either policy or mechanics. This conclusion is strengthened when the research program is appraised. Rather than be directed toward solution of problems selected on basis of priority of importance, the grant monies have been held as inducement for ideas of any kind on any facet of individual interest.

This introduction to this handbook is intended to trace the development of the present situation as a basis for recommending such decisions as are prerequisite to a sound program and then to calm the waves of hysteria so application of knowledge compiled by experts in particular areas can be applied with assurance. Knowing limitations permits employing gages of adequacy in effectively abating and preventing pollution.

The "industrial pollution problem" cannot be completely isolated. Its very definition is dependent on decisions in other facets of water quality control. Its cost, which is a cost of doing business, bears directly on industrial location and operation.

So this appraisal must necessarily be more an environmental pollution assessment, exploring the political-social-economic-public relations aspects as well as the technical. Pollution in its broadest sense is an impairment, and most certainly activities in these areas are impairing the effective and efficient protection and use of our air and water resources.

Necessarily such an appraisal is critical. The near hysteria evidenced at times deserves positive therapy, and if the hysteria is created by distortions or untruths, the only medicine is baring the truth. If a new program is required, its structure should correct present errors, and only a critical examination can delineate deficiencies. Short of that the effort at best would simply be another reorganization.

> We tend to meet any new situation by reorganizing,
> and a wonderful method it can be for creating the
> illusion of progress while producing confusion,
> inefficiency and demoralization.
> —PETRONIUS ARBITER, A.D. 65

LEGISLATIVE HISTORY

Congress establishes broad policy by making the big decisions via legislation with respect to responsibility, scope of action, and money.

But the administrators can construe and implement policy by making the small decisions via regulations and interpretations and thus control policy, responsibility, scope of action, and money.

Rhetoric and reality unfortunately are often poles apart. This has been the story in the much publicized and, at times, highly overstressed "crisis" of water pollution. Since the air pollution story is a parallel, the details of what has happened in water will serve both.

A brief recounting of legislative activity may be helpful. The first federal enactment was in 1948, after a long struggle by persons dedicated to cleaning up the rivers and lakes, but many of whom incidentally never promoted strengthening state programs. These persons or others representing their organizations appeared periodically before congressional committees loudly endorsing proposals for greater federal authority and larger federal expenditures. Successes in 1956 and 1961 were evidenced by amendments to the federal law.

Each chorus before the Congress had two notes in common: the situation was worsening; more policing authority was essential. Apparently no one noticed the indicated correlation—every time more authority was granted the problem got worse.

Meanwhile, there was concurrent activity within the states. At the end of World War II few states had effective laws. Action was predicated on a complaint of nuisance or activated by a dedicated official who could sell a reason, responsibility, and a program. Budgets were literally nonexistent. Happily, the public awakening to the threats of pollution which promoted the federal legislation was also responsible for enactment of state laws establishing specific agencies charged with pollution control.

On the day President Truman signed the first federal control law establishing the program in the Public Health Service, he also signed a second document initiating a program for the Ohio River Basin. June 30, 1948, was the birth date of the Ohio River Valley Water Sanitation Commission, ORSANCO, an interstate agency of eight states draining into the Ohio River. This compact was approved by Congress.

The development and achievements of this interstate compact have been detailed in easy-to-read format by E. J. Cleary in "The ORSANCO Story" (Johns Hopkins Press).

A few references to ORSANCO should suffice to make the point in this appraisal of our national situation. ORSANCO maintained a small professional staff of five—engineers, a biologist, and a public relations expert. It had no laboratory. The budget amounted to less than 1 cent per person per year for the valley residents over its first 18 years. Court cases were nil—two cities were threatened.

Yet things were done. In 1949, when the staff was assembled, only 1 percent of the sewered population in the basin had sewage treatment, and industrial waste treatment was practically nonexistent. Within the first 5 years, and prior to the availablity of the first federal grants for municipal plant construction, 56 percent of the population had financed treatment plants.

During this period ORSANCO enlisted the active assistance of industrial management and technical representatives from companies located in the basin. This voluntary contribution could not possibly have been matched by any budget.

There was also a volunteer team of municipal and industrial water plant operators, a team of federal and state agency laboratory directors and technicians from industry, a team of aquatic biologists, and a team of specialists from industry and private toxicology laboratories. In addition, there was active liaison between teams.

This feat was all the more remarkable because in 1949–1950 few industries would confide in each other, to say nothing of confiding in a regulatory agency. State programs were not well established, and states were reluctant to admit weaknesses. Central repositories for information were unavailable. In fact, information was scarce on industrial wastes and on effects of various constituents on man, animals, and fish. There was no agreement, even among states, on analytical methods. The unknowns greatly outweighed the knowns.

ORSANCO's success in transforming this polyglot group into a fountainhead of strength and achievement in a relatively short time was due to a belief, shared by the drafters of the compact as well as by the commission and staff it created, that the commission should counsel with those it would legislate and jointly determine the objectives which must be achieved if the purpose of the compact was to be fulfilled.

The spirit was one of "balance and caution," with each party recognizing and accepting personal responsibility; with each goal justified; with the means of implementation jointly investigated and fully appraised as to merits.

Quite simply there was a sharing of what was known. Those discharging wastes were invited to assess sources, methods of measurement, and possible controls. A joint assessment by all water users of the quality criteria necessary to protect specific uses encouraged government and industry contributions to search the literature and summarize and evaluate what was known and outline deficiencies. And there was an assessment of technical and economic feasibility in meeting different quality objectives. ORSANCO assumed the added responsibility of correlating and reconciling data and analytical methods and provided interpretation of the data.

All parties knew why they were involved and what they were attempting to accomplish before they got involved in how to do it. When they did, adequacy of approach was the control.

The path was not smooth. But arguments were settled by data before a court of peers. The dogmatist and the generalist got little sympathy from anyone.

Nor were the findings kept secret. Publications authored by the experts and reviewed by those who would apply the information found a worldwide demand.

The underlying philosophy was that agreement on objectives was sought only after there was an adequate technical basis for sound judgment.

ORSANCO represents the "peoples-what"; they are interested in the thing and deal with it effectively.

The federal agency represents the "peoples-who"; they are interested in who the person is to run the show and neglect the thing.

As DeMadariaga noted, the peeoples-what are republican; the peoples-who are monarchist. Public affairs run a smooth course with the peoples-what; public affairs do not work with the peoples-who.

His conclusions are fully supported by an analysis of the American water pollution control program. Neither ORSANCO's pattern nor its success has been evidenced in the federal approach. Rather the federal agency program has appeared as a series of concurrent activities characterized by lack of interdependency. Emphasis has been on the individual person and on his title of autonomy in each activity.

For example, an enforcement program was initiated with stress on large public gatherings wherein water quality conditions were generalized but sources of waste were particularized. Meanwhile the research program wavered between grant applications from persons with particular areas of interest to programs within the agency which reflected entirely different personal interests. The program planning was a third entity and in theory was to provide comprehensive plans for each major drainage basin. Basin inventories on waste load discharged were exercises in minutiae, and efforts to relate them to water quality were nonexistent. Many of the comprehensive plans are still in the "secret" stage; at least the progress, scope, and objectives are not commonly available. And neither enforcement decisions nor research programs have awaited or been correlated with the comprehensive plans.

Senator A. Ellender (D. La.), of the Committee on Appropriations, submitted a report in connection with the 1964 Public Works Appropriations Bill on "Comprehensive River Basin Studies." In it he pointed up the duplication by federal agencies in four departments (Army, Interior, HEW, and Agriculture), each of which is preparing comprehensive plans for major drainage basins. The report noted, "The basic question involved here is the cost of providing the degree of refinement to which these studies are to be carried at this time for future development."

The report concluded with the committee belief that:

> Comprehensive planning should mean giving appropriate attention and evaluation to the broad needs of an area now and in the foreseeable future and only broad-scale planning for the development of water resources to achieve these goals. It does not believe that comprehensive planning need embrace the expenditures of large sums of money to conduct detailed and costly field investigations and designs of specific projects that obviously will not be built for a great many years.

Senator Ellender was concerned about lack of teamwork among the agencies in the four departments and the wastage of money on minutiae. The same indictment can be made against the various activities in the federal pollution control agency.

The objective of this several-pronged attack has been to convince the public the pollution problem is cancerous in character, malevolent, and increasing. The graphic presentations, the radio and TV commercials, and the emotional utterances at meetings of civic groups have carried this conviction.

Results have been as expected. Congress has responded to the public pressures without determining how much is generated by the agency itself. Parkinson's law is correct. With over 1,000 employees the agency had become self-perpetuating. Even more, the huge budget controlled; the tail wagged the dog; the agency no longer was satisfied to be within and part of the Public Health Service. The PHS was blamed for the failure of this rapidly growing behemoth to coordinate its movements and show constructive tendencies. Demands to take it from "under the dead hand of the Public Health Service" were answered with the establishment of a new and separate agency in the Department of Health, Education and Welfare (HEW) by the Water Quality Act of 1965, with the program under the responsibility of an assistant secretary.

The other demand by this new agency was for authority to decide to what uses America's waters could be put. The agency spokesmen asked, and the Senate magnanimously concurred, to give the Secretary of HEW power to determine quality standards for the nation's waters. Some senators strenuously objected to so much authority being given any individual, but the objections with few exceptions were in private. As one senator complained, "To be against a water pollution control bill is to be against motherhood." He missed the point. Decisions on water can be as irrevocable as motherhood, and the increasing number of illegitimate children indicates not all efforts toward motherhood are in the right direction. Neither motherhood nor water purity is free from complications. Only Senator John Sherman Cooper (R. Ky.) publicly and on the floor of the Senate denounced the enactment as giving an agency head more authority than the President possesses.

And the House of Representatives agreed with him. Establishment of water quality standards requires a predetermination of the uses of the water. So after a more thoughtful reappraisal of the situation, the House Public Works Committee refused to make such concessions of authority, and the House provided a unanimous vote of agreement.

However, the route of legislation is one of compromise. The insistence by the House that the states set standards was modified to permit the states to hold public hearings, set standards, and submit them to the Secretary of HEW for approval. With his agreement, the standards would provide guidance for all concerned to know what is required and thus permit orderly accomplishment.

The states were given about 18 months to do the job. This required separate public hearings by the state on those portions of interstate waters within its jurisdiction. Since the states were accepting responsibility for decisions and the hearings were public, emphasis generally was on data to substantiate both criteria of quality and a time schedule for implementation.

Incidentally, this combination of quality criteria and plan of implementation was defined in the 1965 enactment as standards of quality. Including the plan was a new twist, but it had merit in that it was intended to permit orderly achievement of an objective without harassment.

BUT THE FATES INTERVENE

No amount of planning ahead overcomes the effect of plain dumb luck. So although this was the plan and expectation of the architects of the law, there was an unanticipated factor which changed the mix and thus the final product. The administrators insisted the new program conform to past practice. And a further isolation of this agency from direct congressional control under a specific committee was provided by an executive order of Feb. 28, 1966, which transferred the agency from the Department of HEW to Interior.

The Public Works Committee has been the congressional sponsor of both the Corps of Engineers and water pollution control authorizations. The Interior Committee has sponsored Interior Department programs. Western senators on the Interior Committee have been agitating for a reorganization of the executive agencies which would transform the Department of Interior into a Department of Natural Resources. To this new agency would go the Corps of Engineers, Soil Conservation Service, and water pollution control. The proposed reorganization legislation would also delegate authority to the Secretary to control the water resources planning which would be submitted to the Congress for action.

So the executive order transfer was in line with this. Arguments to coordinate agencies dealing with water have merit. As Congressman John Saylor (R. Pa.) has detailed for Congress, there are 38 federal agencies dealing with water, and many have overlapping functions and responsibilities. Interior had responsibility for fish and wildlife, geological survey, mines, recreation, desalination, and reclamation; so there was reason to support the transfer to Interior of the pollution control program.

The transfer started speculation. It was heightened by an administration proposal in 1966 to once more amend the pollution control law to give the Secretary of Interior authority to propose river basin development programs. The Senate bought this concept, but the House did not. With Congress granting such authority, the House realized that the Secretary would effectively control resources, for only those plans approved by the Secretary would be submitted to Congress. It was the authority the western senators had been proposing. If granted, a further reorganization by executive order would complete the new agency.

Speculation was also spurred by the fact that the much larger budget of the water pollution control agency set it apart from other Interior Department agencies. Moreover, there was a well-developed public relations section and a history of personnel talking directly with, and even working intimately in, the offices of members of Congress and committee staffs. Other Interior groups were actually forbidden to do this. For an agency spawned by fishermen in 1948, it had indeed assumed major proportions.

Congressional committees cherish their appropriations authorizations; so the Public Works Committee was not interested in losing control of the pollution program. Yet the Interior Committee was not intrigued by the prospects of a cell within the Department with a budget which could exert control over the other activities. And this was what happened. The research budget of $60 million over 3 years was second only to the construction budget in influence. With Vietnam causing restrictions on research considered politically unessential (this exempted restrictions on water pollution control), the other agencies sought financial assistance from water pollution control to keep projects going. Concessions were made in jurisdiction and policy. But the most marked effect was the silencing of federal agencies, once proud of their knowledge and responsibility, after pollution control spokesmen made their pronouncements.

With its transfer to Interior in May, 1966, the control administration published "Guidelines" for the states to follow in determining whether standards would be acceptable. The law had become effective in October, 1965, and standards were to be submitted by June 30, 1967.

Actually, the guidelines extended the authority of the Secretary beyond that provided by law. For example, it required that standards should incorporate all conclusions issued by the Secretary after previous conferences. Also, it called for a minimum of secondary treatment for municipalities and a comparable degree of treatment for industries unless a lesser degree of treatment could be shown necessary. With this provision the guidelines transformed the program from one of stream standards to one of effluent standards. The degree of treatment was the important criterion, not the protection of the stream. The guidelines further emphasized this by declaring that all wastes amenable to treatment are to be treated even if they do not adversely affect water quality standards.

These guidelines had been issued when the House held its hearings in 1966 on the administration bill proposing authority for the Secretary in basin planning. The House Committee Report 2021 prepared by the Public Works Committee under the joint leadership of John Blatnik (D. Minn.) and William Cramer (R. Fla.) attermpted to place things in focus. It offered not only the definition of pollution noted earlier, but an assessment of the objectives as well as the deficiencies in knowledge deserving research attention. It provided an approach for control of pollution.

The sense of this report was endorsed by a series of reports by a Subcommittee of the House Science and Astronautics Committee chaired by Emilio Daddario (D. Conn.). These reports on the state of knowledge in air and water control and solid waste disposal outlined deficiencies and still promoted action where it would be effective. The reports of this subcommittee deserve more attention than they have received.

CRITERIA AND STANDARDS

The Water Quality Act of 1965 offered a hope, made a promise, and exacted a monumental effort.

The hope: There would be a reasonable program capable of orderly achievement.

The promise: The states could determine the uses of their interstate waters if they would, after holding public hearings, adopt reasonable criteria of quality and plan of implementation.

The effort: The states must expend the energies required to hold responsible hearings and present objectives able to be substantiated.

Cynics to the contrary, the states with few exceptions took up the challenge and did jobs of which they were proud. For the first time in many watersheds nebulous objectives were reduced to specifics. There were the expected exceptions, but for the most part the desires of special interest groups were documented or, through lack of substantiation, made to appear as less than in the public interest.

The states completed their tasks and sent the standards to the Secretary for review. Then the difficulties started. The standards for two states were accepted and copies sent to the other states as examples of the manner in which the Secretary expected their submissions to be modified. On Aug. 9, 1967, in a report on progress to Senate Public Works Committee members, the Secretary of Interior noted:

> The most significant single thing about the standards that I have approved is that they call for a minimum of secondary treatment for all municipal wastes and a comparable degree of treatment of industrial wastes.

Thus the modification in traditionally accepted definitions of criteria and standards effected by the federal law had been changed again by the Secretary. Prior to the 1965 amendment criteria of quality were considered as yardsticks to permit best judgment on the levels required or desired for specific use conditions. Standards existed when the specific quality level was selected for a given water. The 1965 enactment defined standard as a combination of quality criteria and plan of implementation. But the announcement of the Secretary changed standard to mean a type and degree of treatment.

In other words, the federal law set up stream quality standards and included the program to achieve them; the Secretary's pronouncement set up effluent standards irrespective of stream requirements.

A look at the law points up the tenuous position of the Secretary and the states which follow his lead. Section 10(c)(5) makes it specific and certain that the federal law deals with stream standards and limits action against discharges to those which exceed those limits.

> The discharge of matter . . . which reduces the quality of such waters below the water quality standards established under this subsection . . . is subject to abatement. . . .

In other words, there is no action against a discharge which does not reduce the quality below the standard even if it does not have the degree of treatment specified as mandatory.

Furthermore, Sec. 10(c)(1) provides in part:

> . . . if such criteria and plan are established in accordance with the letter of intent and *if the Secretary determines that such State criteria and plan are consistent with paragraph (3) of this subsection,* such State criteria and plan shall thereafter be the water quality standards applicable to such interstate waters or portions thereof. [Italics added.]

Paragraph (3) reads thus:

> Standards of quality established pursuant to this subsection shall be such as to protect the public health or welfare, enhance the quality of water and serve the purposes of this Act. In establishing such standards the Secretary, the Hearing Board or the appropriate State authority shall take into consideration their use and value for public water supplies, propagation of fish and wildlife, recreational purposes, and agricultural, industrial and other legitimate use.

Plainly, the Secretary's review was intended to make certain the state had, in public hearing and in adopting standards, given attention to "use and values" of the waters and that all "legitimate uses" were evaluated. Nothing in the law gives the Secretary authority to either change or force changes in the state-adopted standards. Also, the phrase "enhance the quality of water" is only one of three objectives, and the second sentence defines how the achievement of these objectives is to be determined.

The law permits the Secretary to promulgate standards if the states do not do so or if the Secretary desires a revision in such standards as the states adopt. But he must first have a conference of all interested parties (defined in the law as states, interstate agencies, municipalities, and industries) before preparing regulations setting forth standards, and then he must allow the state 6 months to adopt its own standards consistent with the act, and can promulgate his standards only if the state does not do so. However, if the governor of the state objects to the Secretary's standards within 30 days, the Secretary must then convene a formal hearing board which would have authority to adopt or revise the Secretary's standards. The hearing board composition of federal and state representatives is specified in the law.

Obviously if the federal agency can convince the state agency to modify its standards "voluntarily" and thus have the state submit the amended standards, the work of the Secretary is vastly less than if he has to (1) prove the state-adopted standards are inconsistent with paragraph (3); (2) convene a conference and promulgate regulations setting forth standards; and (3) be prepared to justify the standards if challenged before a hearing board or later before a court.

Moreover, if the state does the modifying or amending of its standards to meet the Secretary's desires, and the proffered changes are made in writing by the state, the state then, for the record, assumes responsibility for validity of the changed standards. State changes have been instigated by oral suggestions from the federal agency. The record is not one of the federal agency proposing changes and substantiating them in writing.

This point is important. If a municipality or industry is convinced a standard is not reasonable, the simplest course of action is to do nothing. Enforcement would bring the issue to a head, for the court action is not merely an administrative procedure review but actually a review of the total situation. The law is specific for the guidance of the court:

> In any suit brought under the provisions of this subsection *the court shall receive in evidence* a transcript of the proceeding of the conference and hearing provided for in this subsection, together with the recommendations of the conference and Hearing Board and the recommendations and standards promulgated by the Secretary, and *such additional evidence,* including that relating to the alleged violation of the standards, *as it deems necessary to a complete review of the standards and to a determination of all other issues relating to the alleged violation. The court, giving due consideration to the practicability and to the physical and economic feasibility of* complying with such standards, *shall have jurisdiction to enter such judgment and orders* enforcing such judgment *as the public interest and the equities* of the case may require. [Sec. 10(c)(5).] [Italics added.]

Two points of note: (1) there is a long-term legal procedure if the federal agency seeks litigation (and this invariably stops progress in abating pollution); and (2) a hearing board and a court must review the Secretary's standards—or standards adopted by the state and approved by the Secretary.

Thus, if the Secretary assumes the task of changing state-adopted standards, the time element is against progress in abating pollution. The burden of proof in declaring state standards inadequate must pass two formal hearings. During this time there is very good reason for cities and industries to halt activity and wait for a decision. If the Secretary can get the state to adopt modified standards, the state assumes responsibility for achieving court support or else the state will stand in the unenviable light of an incompetent agency which set standards it could not enforce. This can then be offered as evidence more federal authority is required.

ASSIMILATIVE CAPACITY AND ECONOMICS

Underlying this situation is the difference in philosophy of two schools of thought:

Whether water and air resource utilization will be directed by engineering studies of fact and economic studies of costs and benefits or whether direction will be by administrative fiat and legal contract

Whether assimilative capacity of the resource (air or water) is a physical and economic capacity to be utilized or whether the requirement shall simply demand the maximum treatment available

Actually, the second is the more often stated argument, but it really acts as a screen for the first. The decision determines whether engineering or law, data or compromise, specific objectives or constantly changing goals will be our way of life. Doubt about this is dispelled by comparison of progress in achieving control in areas strictly under state dominance with areas where federal conferences have been adjusting objectives.

Viewpoint makes a difference, but exaggeration often excites vocal support. For example, one argument frequently offered by proponents of clean streams is that sanitary engineers advocate using assimilative capacity in order to keep a stream as dirty as possible while those interested in clean streams propose keeping it as clean as possible. Making it either black or white is an easy way to confuse the issue.

Utilizing the assimilative capacity means to treat wastes before discharges and/or to control the discharge so as to remove any material or polluting property which would cause an impairment of quality in the receiving water such as to interfere with the other uses determined in the public interest. But treatment beyond that actually benefits no one, adds greatly to the costs, and really is unwarranted. It is definitely not a case of allowing the water to be as dirty as possible. That would mean no treatment at all.

But in another sense it does mean determining how much can be left in without causing harm. People who argue water should be as clean as possible have no concept of natural water quality. Going to the extreme—if the clean waters advocates' statement is to be taken literally—would be to require distillation of all waste waters. It can be done. However, sterile water devoid of dissolved solids and organics would kill fish. It does not exist in nature.

However, treatment is only one factor in utilization of assimilative capacity. The other is dispersal. The load discharged is not the control; concentration in the receiving water is. In order to utilize the capacity of the receiving water, proper dispersal is a must. Furthermore, if there is adequate dispersal, the visible or aesthetic effect will be minimized. When sewers discharge to the surface of receiving waters, the warmer water tends to float and from the air is often visible. If the discharge waste water is a different color, either cleaner or discolored, the effect is pronounced.

When sewers carry high concentrations of nutrients and/or triggering salts, algal blooms are promoted by the warmer water if the discharge stratifies in the upper receiving water where sunlight is available. This also occurs in harbors. The same effect occurs when warmer tributary streams reach lakes. The need for better dispersal control is not limited to sewer outfalls.

The House Public Works Committee (Report 2021) addressed itself to the issue of assimilative capacity:

> The committee also understands that comprehensive studies have demonstrated the desirability of having more effective techniques to appraise the assimilative capacity of streams in order to better assess effects of increasing populations and industrial and agricultural usage. The committee feels that since the proper usage of such capacity is both an economic resource and an integral component in assessing water quality standards, demonstration projects to determine practical measuring control techniques would be most desirable.

The question then is whether there is justification for some arbitrary limit between the absurdity of no use and the level defined by sanitary engineers and biologists as acceptable quality.

How important this decision is can be shown in the costs of treatment for process waste waters. Too often not enough consideration is given to operating and maintenance—the continuing costs established by the kind of treatment plant installed. These costs for treatment have been averaged as:

$0.10 per 1,000 gal for primary treatment
$0.20 per 1,000 gal for secondary treatment
$0.40 per 1,000 gal for advanced treatment

(These figures include amortization. Another estimate sets the initial capital investment at one-fourth to one-third the total cost of the facility over its lifetime. If the unit costs $1 million, its operation and maintenance will be over $2 million.) These figures are averages for industry as a whole, and for any particular industry the cost, particularly for advanced treatment, may be two or three times as much.

If boiler blowdown must be treated to meet suggested dissolved solids concentration limits, it would require a desalination process at about $1 per 1,000 gal. Assuming 5 percent of the industrial plant intake as boiler blowdown, this would mean annually $48 million to American industry.

Cooling is the principal water usage in industry. The 1964 usage was reported at 9,759 billion gallons. If arbitrary limits are set on temperature at discharges, cooling is required. This means recirculated systems with treatment of blowdown. Using $0.30 per 1,000 gal for the latter and $0.10 for recycling, the annual expense would be over $1 billion.

For American industry the annual budget for treating process water has been estimated by Bramer at:

$ 370 million for primary treatment
740 million for secondary treatment
1,480 million for advanced treatment

Thus the annual price tag will vary from $370 million if treatment is a minimum to $1.74 billion if secondary treatment and cooling are required. This jumps to $2.5 billion a year if industry is to remove everything possible—"to make the waters as clean as possible."

Since in many instances primary treatment will not be sufficient, the actual annual tariff will vary somewhere between $370 million and $1.8 billion depending on the standards selected.

Operating and maintenance expenditures are directly deductible. If excess treatment is required, i.e., treatment which provides no measurable benefit in the stream, then the added operating expense causes a direct reduction in taxable income and a loss of revenue to the federal government and to states.

Once the facilities are installed, the operating expense continues during boom or recession. Also, adding 1 to 6 percent to the cost of consumer goods results in increased cost of living. The cost of secondary treatment for each phase of production such as foods, paper, textiles, steel, chemicals, and fabrication is estimated at 1.5 to 2 percent, and this then is cumulative in the price of consumer goods.

TAX RELIEF

The magnitude of the expenditure for waste control has excited members of Congress to propose extending the federal doctrine of financial assistance to municipalities to include industrial waste control facilities. Their argument is that pollution control facilities will benefit all. Since the federal government is participating and setting accelerated time requirements, they feel it consistent to have the federal government participate in financing the program.

Different approaches have been proposed:

1. *Investment credit.* A credit of a percentage of the capital cost in the year of construction. Cost to government is the actual credit allowed.

2. *Accelerated amortization.* A write-off in 1 to 5 years. Cost to government is the interest on the money over the shortened period. This is the least cost to government approach.

3. *Grants.* Costs the same as investment credit but does not set a precedent

in the tax regulations and so is preferred by government economists if tax relief must be adopted. Permits selectivity in recipients. However, just as grant program now controls rate of municipal plant construction (cities await their turn), the same effect can be anticipated with industry.

4. *Low-interest loans.* Of little real help to industry. Raises question of ownership and responsibility of operation of facilities. Would increase cost of administration.

5. *Effluent fees.* This is a punitive approach. As an "incentive" it is totally negative. Reports that effluent fees are effective in the German Ruhr Basin are a misstatement of the situation. Ruhr does not have effluent fees. Rather there is a charge for water purchased and a sewer service charge just as in metropolitan areas of this country for handling wastes.

Tax relief has been mistakenly called a tax incentive. It is an incentive in the same manner as the tax deduction allowed the individual for gifts to charity. The incentive is the law. Tax relief helps finance the new facilities and more quickly get the capital investment funds back so as to invest in revenue producing, and thus tax producing, facilities.

A number of states have recognized the necessity to encourage industrial growth as a means of expanding the tax base if services are to be provided without tax increases. These states have also affirmed the public benefits of waste treatment and have provided tax relief—ad valorem, franchise and sales.

RESEARCH

Second to the construction grant program is the research grant program.

And second only to the demand more be done faster and expenditures be increased is the lamentation on the dearth of answers and inadequate talent.

The consoling thought in an atmosphere which tends to promote hysteria is that the American people are extending their life span, and apparently proportionately to the news media propaganda that pollution of air and water is imperiling health. It is also helpful to remember one can safely drink water anywhere in this country. Health is not a crisis despite the innuendoes and direct allegations by federal officials, politicians, and do-gooders.

And this inclination to make allegations rather than determine fact is considered by many as the fundamental fault of the federal agency research program in both air and water.

Conclusions have been used, in fact are still being used, to promote stronger federal legislation and increased grants for construction and research. Research monies are directed toward mechanics to achieve what the conclusions say should be done rather than first establishing whether these conclusions are correct.

Two illustrations, one for air and the other for water, will show how this research policy error influences criteria of quality.

At present the whipping boy in air pollution is sulfur. It can be measured, although there are questions on what is measured and the accuracy attained. Sulfur compounds in the air are credited with causing deaths by respiratory ailments as well as affecting the heart. Limiting concentrations are being set. But the crisis is claimed to be so critical that ambient air standards (concentrations in the air where it is utilized) are being bypassed and limits are being set on permissible concentrations of sulfur in the fuel burned.

In New York City this limit has been set at 0.3 percent. In the Ohio Valley at a federal hearing for the Ohio–West Virginia airshed, the limit was set at 1 percent. Such a difference raises a question as to the relative importance of health in the two regions if sulfur is the culprit.

However, in either case the immediate effect is an economic one, for the decisions eliminated a large segment of the coal industry and set higher costs for oil. Moreover, if the pattern continues for other areas of the country, the economy of neighboring oil-producing countries will be adversely affected.

The effort is limited to means of reducing sulfur, and the justification deserves attention. During the 1967 Senate hearings of the Public Works Committee, G. N. Stone and A. J. Clarke of the Central Electricity Generating Board, London, England, reported on the British study underway since 1952 which had attempted in every way possible to prove a correlation between sulfur dioxide and adverse health effects. The conclusion was that sulfur is not the responsible agent but rather that particulate matter is primarily responsible for trouble. The 1952 smog incident with its notorious inversion had a high sulfur dioxide reading which approached 2 ppm, and during that smog there were some 4,000 more deaths than would have been expected with the normal London mortality rates. In 1962 there was a second episode with sulfur dioxide levels slightly higher than in 1952, but there were only 400 deaths. Clarke noted that the difference was in a 40 to 50 percent reduction in smog content and attributed it to the construction of high stacks which, although they did not cause a reduction in the emission load, did cause a reduction in concentration of solids below the inversion level. Stone noted that the medical evidence was totally against sulfur as the culprit, and hence the British were not seeking to control sulfur by legislation. Describing the emphasis on sulfur as a witch hunt, Clarke reported that efforts now were to determine the true cause of trouble.

Controlling particulate matter is much easier and less costly than eliminating sulfur compounds. However, unless history is made by a change in practice, sulfur will continue to be the whipping boy in the United States and people with respiratory ailments will continue to suffer until our research is reoriented.

Nor does the charge withstand scrutiny that sulfur in the air is the factor influencing heart disease death rates. Past studies have documented the influence of traces of vanadium in water supplies in those areas where heart disease incidence is lowest. In New York State, where the rate is highest, vanadium is absent. There was no correlation with urbanization, but there was with the presence or absence of vanadium.

For the illustration on water consider the classic "dying Lake Erie." A little data in the hands of an opportunist can create havoc if there is no compunction about using innuendoes and speculation.

When the Federal Water Pollution Control Administration (FWPCA) made its survey of Lake Erie, the agency reported a 2,800 sq mi area in the central lake devoid of oxygen. The map of the lake was wreathed in black. There was a pronouncement the lake was dying and fish could not survive. Described as a chemical tank and an open cesspool by eloquent but scientifically illiterate members of Congress as well as by personnel in the federal agency, this lake became the symbol of a despairing nation.

Lake Erie produces about one-half the fish catch of all the Great Lakes—about 50 million lb a year. This weight has varied little since 1875. Fish species have changed, but they changed in Lake Michigan at the same time. The last commercial catch of sturgeon in both lakes was in 1895. Commercial fishermen had deliberately eradicated the big fish because they were damaging whitefish nets. Whitefish disappeared from the St. Clair in 1878, and the hatchery for western Lake Erie was closed in 1890. Selective fishing by commercial fishermen and changing migration habits were the cause, not pollution. In addition, smelt was introduced in 1912 and the lamprey and alewife made their entrance from the ocean. These are efficient feeders and, coupled with selective fishing, helped displace the desired species. Fish management has been almost totally nonexistent.

When lakes stratify because of temperature density effects, there is a zone below the thermocline where the water does not move. Organic material will deplete the oxygen in this zone if the zone is relatively shallow. This is a natural phenomenon occurring in many lakes. Above the thermocline oxygen is plentiful.

When the FWPCA made its survey, the lake level was at a record low; so the thermocline was low and hence the oxygen was low in the bottom 3 to 10 ft.

Instead of identifying the bottom material which depleted the oxygen, the investigators published the conclusions that algae caused the trouble, that phosphorus in detergents was the primary source of nutrient, and that this should be eliminated. Shortly thereafter an International Symposium on Eutrophication was held at the University of Wisconsin. These sessions, which developed all the facets involved in algal blooms, evolved the conclusion that the federal approach is a naïve exercise in futility. What is required is a water management program.

These illustrations are typical of the lack of definition of the problem. The research is directed toward how to do things, rather than why. Another complaint is common. Applications for research projects are encouraged. Then the application is given a preliminary screening and a reply statement indicating desirability, but further information or modifications are recommended. This correspondence continues over a period of months—12 or 15 months is not uncommon. The applications submitted exceed the amount of money available. This permits the plea that greater appropriations are required.

From the practical point of view financing construction is as important as developing technology. The House Public Works Committee report pointed this up:

> Industrial research should not be limited to the technology of waste treatment. It should also include an investigation of possible financial methods of providing this treatment, including methods of providing treatment works for smaller industries on an installment basis.

ANALYTICAL CONSIDERATIONS

Partial responsibility for lack of knowledge on actual effects of waste discharges, for both air and water, can be assigned to present data collection philosophy. In too many instances the analytical techniques have been devised to measure constituents rather than the single or combined effect of several constituents. A relationship is then assumed between the constituent measured and an observed effect. After a while the assumption is forgotten and the specific analysis becomes the control.

For example, phenol could be measured and odor was considered subjective; so phenol was measured and the cause of odor was assumed to be defined by phenol concentration. Eventually limits were set on phenol and the objective became a control over phenol concentrations. Other causes of odor were ignored; so was odor. Even worse, phenol was measured only on occasions of odor trouble, and whatever concentration of phenol was present was considered as causing the trouble.

Dow Chemical Company developed a technique to measure odor because Dow learned control over phenol offered no assurance there would be no odor troubles. Test work by steel and oil companies in laboratory panels and river surveys confirmed the Dow findings. The American Society for Testing and Materials adopted and published the method as a standard. So did the industrial waste section of Standard Methods.

But a limit for phenols remains in many regulations although it has no validity.

Other analyses have a similarly undistinguished history. Many are run routinely only because of tradition. The data have little significance, but since the data are not used, this bothers no one. Routine continues, often because some regulatory agency set up a tabulation and it has become the regulation.

Cost of analytical work is increasingly important, and unless data are minimal, accurate, interpreted, and utilized, much money is wasted.

Obtaining the most benefit at least cost is most easily done by:

1. Analysis of river above and below the industrial plant, or of surrounding area for air quality control, to determine whether there is impairment of quality and whether it originates or is aggravated by plant discharges
2. Determination of sources within the plant of the constituents contributing to the problem and then an in-depth survey by sections of the plant to determine the waste loads, the variation with production, the areas where controls can be provided to minimize losses, and the concentrations and flows at points of discharge
3. Routine monitoring of the river, and/or surrounding area, coupled with periodic checks on sewer outfalls and stack discharges to be assured concentrations are within the range established during the in-depth survey

Experience with both municipalities and industries highlights the importance of bypassing facilities due to improper maintenance. With increasing argument for more federal policing of facilities, the significance of bypassing is obvious. However, to keep analytical costs to a minimum, work should be limited to essentials, adequate to prove lack of liability.

Data gathering and reporting is not an exact science. Lack of agreement continues on methods of sampling, preservation of samples, analytical techniques, and reporting of results. The range of results among qualified laboratories checking the same sample can be amazing. Some limiting concentrations set by regulatory agencies are at or below the sensitivity of analytical techniques. At those levels accuracy is questionable.

Also of concern is the reporting of discharges as either concentrations or loads unless correlated to effect in the receiving air or waters. Large flows with low concentrations mean large poundage. Whether it is significant depends on local conditions. Measuring outfalls without making allowance for the intake concentration causes trouble. For example, a sewer carrying 70,000 gpm with a concentration of solids of 50 parts per million (ppm) would have a load of 41,700 lb a day of solids. If the intake water solids concentration is 40 ppm, the net, which is what is added by the plant, would be 8,300 lb. Whether this is significant would depend on its effect—as measured by particle size, settling rate, concentration and size of particles in the stream, velocity in the stream, etc. Weight in itself means little, and yet it is a commonly accepted standard.

The Corps of Engineers is now proposing contracts with industry whereby the company would pay for dredging caused by solids discharged from the plant. The assessment is to be based on the weight of solids, but to be equitable, there must be a determination as to whether the solids will settle, whether the settled solids in fact impede navigation, the density of the solids, and the bulking factor (conversion of weight to volume) to be employed.

The increasing administrative attention to waste discharge measurements in terms of proposing effluent standards sets the stage for increasingly more stringent regulations on sewer outfalls as a measure of improved pollution abatement. The alternative is a positive definition of effect and a court action to prove or disprove its validity as control.

Developing the mechanics for a reporting of waste loads is an essential element in proposals to levy an effluent fee on industrial water users. Accuracy is not so important as having an approximation of waste loads and volumes of discharge. The proposition is being offered that waste loads above some agreed-upon level would be considered as pollution and the fee levied would be punitive to encourage abatement. Discharge of loads below this figure would be assessed a fee as a rental of the assimilative capacity of the receiving water. The same principle would be applied to air.

The reasoning is quite simple. It assumes air and water are publicly owned resources and as such should be used for public benefit, and if used by industries, such industries should pay for the privilege. Once the precedent is set, the extension of the principle to require payment (taxes) from individuals to use water and breathe air would be logical. Such assessments are proposed as alternatives to excise taxes or as additional income for public agency disbursement. Pollution abatement is obviously only an excuse for application of the fee system.

Primarily an economic issue, this is discussed under analytical considerations because it points up not only the importance of accuracy now in establishing records of discharge and resultant water quality, but also the potential increase in cost of analytical controls as well as cost of using the air and water resources.

One way to reduce costs of analytical work is to consolidate sewers into a few outfalls. This can have the added advantage of providing greater initial mixing and possibly dilution before discharging to the receiving water.

This theme should be continued to its logical conclusion. The objective is stream protection, and since concentration is the important factor, not load as is often supposed, the dispersal of the discharge in the receiving stream is most desirable. When warmer water is discharged at the surface of a water, it tends to float. Sampling downstream from such discharges permits great range in results according to the sampling site selected. At the surface the concentration is high; below surface it may not be picked up.

With regulations setting limits "after initial discharge" or "after initial mixing," this factor of design can be most important. Obviously, as mentioned earlier, unless proper dispersal is provided, the potential assimilative capacity is not utilized. Also, dispersal devices below water surface achieve cooling as well as dilution and reduce the visual effect, the principal public indicator of pollution.

ADMINISTRATIVE MECHANICS

A major deficiency in the American emphasis on pollution control lies in the absence of established administrative techniques which can be appraised for effectiveness. And yet, unless the programs are administered effectively, efforts to provide reasonable standards are in large measure wasted.

The major categories of administrative agencies are federal, interstate, state, and regional. With the often heard argument that pollution effects are not confined to political boundaries, the advocate for fast action claims the higher echelons of government would be most effective. Added support is provided by arguing the large costs involved require big government.

But 20 years of experience belies this thesis. Pollution abatement is achieved by installation of facilities and controls. The federal government has not distinguished itself in providing treatment for federal installations, let alone convincing cities and industries of what should be constructed. Federal dicta are generalizations on types of treatment rather than tailored proposals. With the exception of the Ohio River Valley Water Sanitation Commission, interstate compacts have been less than spectacular in achieving results. What has been achieved can be largely attributed to state agencies. Even in the Ohio Basin the interstate agency did not usurp or assume the responsibilities of the states.

However, control efforts to date have been limited to construction of treatment facilities at ends of sewers rather than to least cost, most effective programs. In other words, no team endeavors or in-stream treatment. Even the low-flow augmentation benefits incorporated in large dam construction by the Corps of Engineers have been unused because no administrative arrangements have been devised to permit decisions on the use of the water.

In part this is due to the unwillingness of the federal water pollution control agency to agree that release of augmentation flows can be utilized as cheaper than secondary treatment. An interagency memorandum with the Corps of Engineers bypassed the intent of Congress and placed the Corps in the position of justifying projects with benefit monies which the public cannot use because of an opinion in the pollution control agency that secondary treatment must first be supplied to all waste discharges. In the Ohio Basin, for example, some 37 reservoirs have been financed in part by low-flow augmentation benefits and as yet no water has been released from any of them for this purpose.

Another example: Release of additional flows on the Delaware River during the few weeks when fish go upstream to spawn would achieve the same result and at much less cost than installation of higher degrees of municipal and industrial waste treatment.

This arbitrary ruling by the federal agency also contradicts the expressed intent of Congress as reported in the House Public Works Committee Report 2021:

> Rather than broadbrush recommendations, the committee urges the possibilities in attaining the objective at the optimum low-cost solution. The committee has reorganized the research and demonstration section to permit direct Federal financing in industrial as well as municipal waste control endeavors in order to excite new techniques on shore and in the river.
>
> Since the objective of the pollution control program is to provide water of acceptable quality, research is needed to explore supplemental treatment techniques, and methods for treating the residual water quality problems which remain in areas where treatment facilities have been constructed.

Another obstacle to the development of less cost, more effective control is the reluctance of counties, cities, or industries to assume the responsibility entailed in accepting leadership to promote regional efforts.

An answer lies in the formation of conservancy foundations in river basins or portions of basins. It should consist of counties, cities, and industries and could be financed by voluntary contributions from the government and industry members. If established for a specified period of time (say 3 to 5 years), it would provide the mechanism by which a unified effort could be initiated and, concurrently, develop the mechanics for a continuing organization.

Such an organization can get action started and places the burden of proof on others that the action is not in the best public interest. It permits joint effort and, if capably directed, at least cost. A local joint determination of alternatives, costs, and benefits can do well in any court review.

Such a local agency can be formed without prior determination of taxing powers or complicated review and control procedures. These can be developed during the trial period if shown desirable.

The obvious restriction on such an approach is the shortage of technically competent persons who are qualified to define and willing to defend a public stand against emotion arousers.

However, if the people in a region can understand that alternative approaches to solution of resource management problems are available, and this includes pollution control, and if they appreciate how large the financial stakes are on the choice of solution, the chances for local area decision are enhanced.

If industry appreciates how intimately these decisions affect its market position and realizes that an active role in achieving resource protection can help assure other area benefits, industrial leadership participation in resource management can be positive and will be effective.

Chapter 2

History of Federal Pollution Control Legislation

JOHN A. BLATNIK

**Congressman from Minnesota,
House of Representatives,
Rayburn House Office Building,
Washington, D.C.**

Air and water pollution, two modern concerns, have been with us since the United States underwent its metamorphosis from an agricultural nation to an industrial one over one hundred years ago. At that time, no one was particularly alarmed to see plumes of smoke rising into the air or to see effluents flowing into a clear stream—they were the price of progress. Moreover, to the 31 million people who inhabited the country at that time, the air was infinite, as were the waters, and a little dirt would hardly be noticed. So we thought.

Yet, as early as 1899 with the River and Harbor Act, 1912 with the Public Health Service Act, and 1924 with the Oil Pollution Act, Congress was giving some attention to the increasing evidence of water pollution. Concern over air contamination began in 1907, with the formation of the Smoke Prevention Association, and in 1912 with the first governmental gesture toward control of air pollutants with the Bureau of Mines studies of smoke control. Admittedly, these were feeble gestures.

Increasingly after 1912, the gravity of air and water pollution received attention from succeeding congresses, along with the delicate ancillary question of states' rights. Obviously, neither waterways nor air currents recognized political boundaries of states, counties, and cities, and this elementary fact was a formidable obstacle in the enactment of sound, effective legislation. Again and again proposals for federal controls were introduced and defeated in the House or vetoed by the President, in a discouraging effort to stimulate state action on what was later to be recognized as a national issue.

The outbreak of World War II not only halted congressional engagement with water and air pollution problems but complicated the issue. The demands of wartime meant hasty construction of new war production plants, increased use of waterways for sewage and production wastes, and skies darkened by industrial smoke. Among the major pollutants of this period was the federal government, with its military supply factories and over twenty separate federal agencies, each contributing its own kind of dirt to the atmosphere and the water.

WATER POLLUTION CONTROL LEGISLATION

By 1947, when I came to Congress, there was ample experience of the ravages and wastes of dirty water and air, and widespread recognition that both were increasing at staggering rates. In 1949, the Water Pollution Control Act was passed, a step which was frankly experimental and limited to a trial period of 5 years. Nevertheless, its enactment represented a major breakthrough in terms of federal recognition of a national problem. The Act declared that pollution was best dealt with at the local level but recognized that this could not be done without the assistance of the federal government. In summary, the Water Pollution Control Act of 1948 had these provisions:

1. The U.S. Surgeon General of the Public Health Service would administer its provisions.
2. Research with emphasis on industrial waste.
3. Technical assistance to state and interstate agencies.
4. Federal financial aid to states and interstate agencies in the form of grants for industrial waste studies. One million dollars per year appropriations for 5 years.
5. Federal financial assistance to states, interstate agencies, and municipalities in the form of grants for planning pollution abatement works. Annual appropriation of $1 million for 5 years.
6. Federal financial aid to municipalities in the form of 2 percent loans for construction of abatement works, but restricted to one-third of the project cost, or not more than $250,000. The annual appropriations authorized were $22.5 million for 5 years.
7. Development by watershed comprehensive plans for pollution control in cooperation with states and interstate agencies, municipalities, and industries.
8. Promotion of interstate cooperation through encouragement of interstate compacts and uniform state laws.
9. Enforcement restricted to pollution affecting a state, other than a state in which it originates, through federal court action, subsequent to public hearings and contingent on the consent of the state in which the pollution starts.
10. Establishment of a National Water Pollution Control Board.
11. Construction of a sanitary engineering research center.

Despite its frankly experimental probings, the 1948 Act was good legislation for its time. But neither the authors of the first federal public works legislation of the early past century nor those writing the 1948 Act could envision the startling expansion of our nation. Literally, most of our public facilities and services suddenly became inadequate for the burgeoning needs and demands of a soaring population and the millions returning from defense careers into civilian life. Whole new cities appeared, with new housing, highways, public buildings, and shopping areas adding contamination anew to waterways. War-born industries poured new chemicals and toxics into streams too fast for technology to keep abreast of them. They added to the air pollution caused by the vastly increased motor vehicles which jammed our cities and highways.

The 1948 Act was strengthened and extended by the Congress in 1952, but it was due to expire on June 30, 1956. We knew that the pollution problem was running far ahead of our knowledge of what clean water really meant, that we were not sure how to correct new sources of contamination, and that the federal role as overseer and coordinator of state and local efforts would need much more emphasis. We knew also that higher appropriations were needed everywhere for antipollution systems and construction.

In answer to these challenges, the Congress passed the Federal Water Pollution Act of 1956, a major step in the fight against dirty water, especially in its enforcement provisions, which gave the federal government substantially more authority to deal with polluters of interstate waters through conferences, hearings, and lawsuits. This section laid the basis for future establishment of federal sanctions, so essential for an effective control program. The 1956 Act:

1. Reaffirmed the policy of the Congress to recognize, preserve, and protect the primary responsibilities and rights of the states in preventing and controlling water pollution.
2. Authorized continued federal-state cooperation in the development of comprehensive programs for the control of water pollution.
3. Authorized increased technical assistance to states and broadened and intensified research by using the research potential of universities and other institutions outside of government.
4. Authorized collection and dissemination of basic data on water quality relating to water pollution prevention and control.
5. Directed the Surgeon General to continue to encourage interstate compacts and uniform state laws.
6. Authorized grants to states and interstate agencies up to $3 million a year for the ensuing 5 years for water pollution control activities.
7. Authorized federal grants of $50 million (up to an aggregate of $500 million) for the construction of municipal treatment works, the amount for any one project not to exceed 30 percent of cost or $250,000, whichever was smaller.
8. Modified and simplified procedures governing federal abatement actions against interstate pollution.
9. Authorized the appointment of a Water Pollution Control Advisory Board.
10. Authorized a cooperative program to control pollution from federal installations.

A comparison of the summaries of the 1948 and 1956 Acts reveals first that any project for control of environmental contamination cannot be static. While legislation can and must delineate the guidelines, there must be a constant reappraisal of the current state of water and air conditions, which can vary from region to region rapidly, principally because of our population, industrial, and military mobility.

These bills and subsequent legislation show also that while pollution controls are popular in a generalized sense, they can cause conflicts of interest and misunderstandings in specific details. The federal backup of state and local programs had been developed at a studied pace but evidently ran ahead of general popular acceptance. The 1956 legislation demanded that water quality standards be set by the states, but it was opposed vigorously. By 1965, it became law.

Although the 1956 Act struck the first formidable blow at national water pollution, it became evident to me that it needed revisions and more liberalized provisions and appropriations. With this objective, early in 1959, I introduced H.R. 3610. The purpose of this bill was to expand the construction grant program by increasing federal participation and relaxing some of the requirements for communities to become eligible under the 1956 legislation. My bill as sent to the White House would have increased to $90 million the yearly authorization for construction grants. It also would have retained the 30 percent of project cost limitation but would have increased the individual federal grant ceiling from $250,000 to $450,000. It also asked permission for several communities to unite in constructing a centralized water treatment facility. President Dwight D. Eisenhower vetoed the bill.

In retrospect, Congress and those groups within the academic, scientific, engineering, and public health fields had reasonable confidence that the new legislation being sponsored was meeting the challenge of water and air quality decay. By hindsight, we know now that we were explorers in an unknown land and that the problem of pollution was not one, but many problems.

From hearings in House and Senate committees and investigations by many private and semipublic groups, we learned much in the late 1950s.

Among the complexities were some basic truths. The old approach to water pollution as essentially a public health issue, while vital and necessary, was inadequate. We had to control water from the first drop that fell on a moutain, follow its course through every human and industrial use until it eventually went back to the sea. We learned that reaction to crisis and emergency had to be replaced by preventive measures.

Encouraged by President John F. Kennedy's "Message on Natural Resources" of Feb. 23, 1961, my colleague in the Senate, Edmund S. Muskie (D. Maine), and I sponsored the Federal Water Pollution Control Act Amendments of 1961, which were enacted into law and signed by the President on July 20, 1961. The law was expanded and strengthened by these main provisions:

1. Extending federal authority to enforce abatement of intrastate as well as interstate pollution of interstate or navigable waters and strengthening enforcement procedures.
2. Increasing the authorized annual $50 million federal financial assistance to municipalities for construction of waste treatment works to $80 million in 1962, $90 million in 1963, and $100 million for each of the four following fiscal years 1964–1967; raising the single grant limitation from $250,000 to $600,000; and providing for grants to communities combining in a joint project up to a limit of $2,400,000.
3. Intensifying research toward more effective methods of pollution control; authorizing for this purpose annual appropriations of $5 million up to an aggregate of $25 million and authorizing the establishment of field laboratory and research facilities in seven specified major areas of the nation.
4. Extending for 7 years until June 30, 1968, and increasing federal financial support of state and interstate water pollution control programs by raising the annual appropriation authorization from $3 million to $5 million.
5. Authorizing the inclusion of storage for regulating streamflow for the purpose of water quality control in the survey or planning of federal reservoirs and impoundments.
6. Designating the Secretary of Health, Education and Welfare to administer the Act.

The amendments to the Act in 1961 received widespread publicity and comment and at the time doubled the annual appropriations of $50 million in ascending amounts to $100 million.

My estimate now is that our broadening base, from concern with the public health aspects of pollution to the overall view of water as a natural resource, was the paramount important change.

The amended Act assigned to the Department of Health, Education and Welfare the responsibility for cleaning up our waters and keeping them clean, propagation of fish and wildlife, recreational uses, and industrial and agricultural supplies, including irrigation.

The fivefold increase in funds for research, from $5 million to $25 million, proved to be a blue-chip investment, It stimulated investigations by states, cities, universities, and industry, and made water and air pollution an urgent consideration in other state and federal programs relating to our American life in general.

The public began to see that all our progress would be fruitless if we could not get rid of smoggy air and sewage-laden waters. This new involvement was reflected in the Public Works and Economic Development Act of 1965; the Appalachian Regional Development Act of 1965; the Housing and Urban Development Act of 1965; the Consolidated Farmers Home Administration Act of 1965, and the Demonstration Cities Act of 1966.

Statistics updated to June 30, 1965, revealed what was accomplished by federal construction grants beginning with 1957. A total of 6,200 projects, with a combined

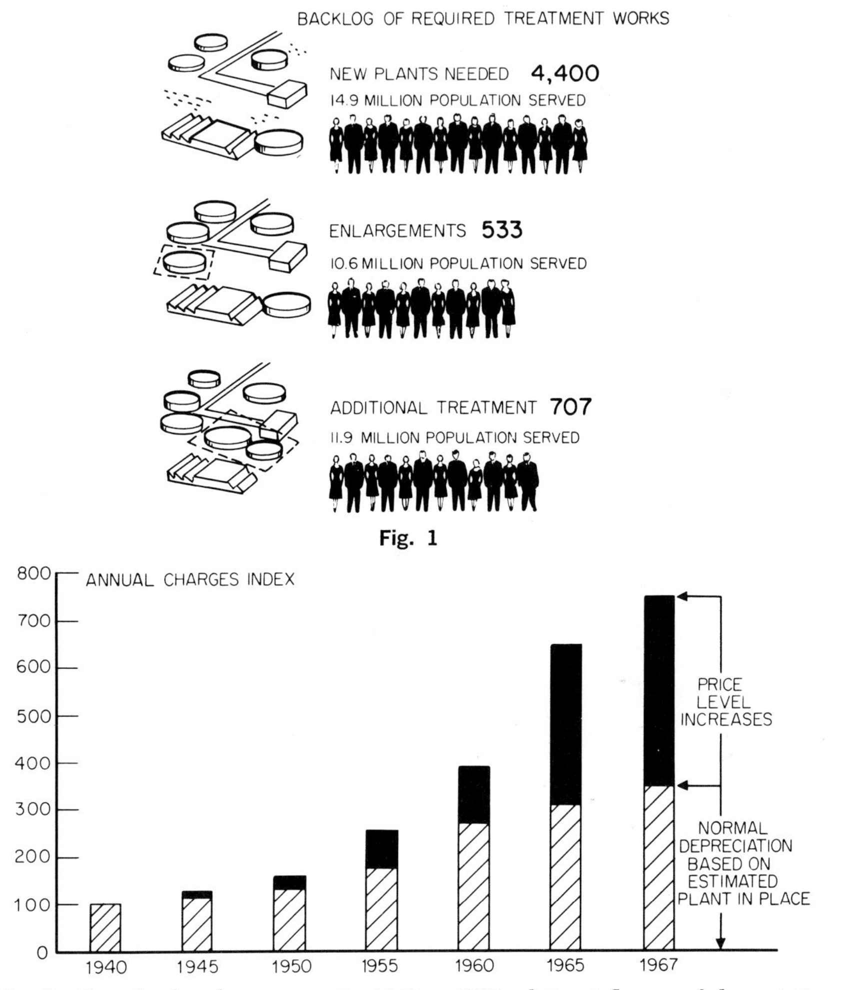

Fig. 1

Fig. 2 Growth of replacement costs, 1940 to 1967 relative influence of depreciation and price levels over time. (*Source: The Cost of Clean Water and Its Economic Impact, vol. 1, U.S. Dept. of Interior, FWPCA, January, 1969.*)

cost of about $2.6 billion, received $658 million in federal aid, with $108 million provided under the Accelerated Public Works Act. While commendable, new disposal plants provided by the Act merely kept pace with the demands created by new population growth and increased urbanization, or replaced old facilities which technological advances and population relocation had made obsolete.

Yet even with this progess, the backlog of unmet needs everywhere in the nation was formidable, and federal assistance was still as inadequate as some of the early pollution control installations.

Data on the needs of the nation's 100 largest cities were compiled. These showed that 80 of these communities needed 367 projects, to cost an estimated $1.3 billion. They revealed also that annual surveys of backlogs were falling short by at least $1 billion. To this were added the needs of communities with populations under 100,000. The primary incentive aspects of the federal construction grants, while encouragingly successful, had not been realized in the most critical pollution centers, the large metropolitan areas.

Coping with staggering financial burdens, the largest cities could not find the federal ceiling of $250,000 in aid much of an incentive to launch programs costing billions. As an example, St. Louis had voted a $95 million bond issue to build new sewage and treatment plants. It used its $250,000 of federal aid well, but obviously that sum could not have been the primary incentive. A new federal aid ceiling of $600,000 in 1961 received encouraging response in the form of renewed effort by our cities, but this still was not commensurate with what had to be spent by communities in financial straits.

The response to the increased federal ceiling by the cities was indeed encouraging, but perhaps more gratifying was the flood of books, articles, newspaper editorials, and television and radio documentaries about air and water pollution. At last people were really waking up to the magnitude of the problem before us. Scores of business, fraternal, service, and educational organizations endorsed the efforts of the Congress and demanded greater ones.

The Water Quality Act of 1965

It was in this favorable climate that the Congress began its deliberations on what was to become the Water Quality Act of 1965. The legislation received overwhelming support in both houses and was signed by President Johnson on Oct. 2, 1965.

At the signing he remarked: "Today, we proclaim our refusal to be strangled by the wastes of civilization. Today, we begin to master our environment. But, we must act, and act swiftly. The hour is late, the damage is large."

In 1965, President Johnson stressed the need for clean water in his State of the Union message and in his message on natural beauty and in both asked Congress to enact new and stronger legislation. The bill passed the House unanimously and was approved by the Senate by a big majority, after some months of discussion and clarification of the water quality standards provisions. A landmark in federal efforts in antipollution, the Water Quality Act of 1965 further strengthened preceding laws with these provisions:

1. Established a national policy for the prevention, abatement, and control of water pollution.
2. Created the Federal Water Pollution Control Administration (a new agency, originally within the Department of Health, Education and Welfare but now within the Department of the Interior).
3. Authorized a new 4-year program of grants and contracts for the development of new or improved methods of controlling waste discharges from storm sewers and combined storm and sanitary sewers. It authorized annual appropriations of $20 million for these projects, designed mostly to assist older cities.
4. Doubled the dollar limitations on construction grants.
5. Increased from $100 million to $150 million the annual appropriation to encourage cities and states to embark on new controls.
6. Provided for establishment of water quality standards for interstate waters.
7. Required that grant recipients keep records, open to audit.

Similar to previous legislation in the area of antipollution, the Water Quality Act of 1965 not only embodied growing knowledge and techniques, it revealed again the enormity of the task and the demand for a bolder attack and, especially, higher funding.

The value of federal incentive grants as an instrument for encouraging municipalities to build, enlarge, or modernize their waste treatment plants had become apparent in 1956. During the next 5 years, these federal grants stimulated a 62 percent increase in this construction, with contract awards rising from an annual average of $266 million to $432 million. The 1961 amendments had further stimulating effects. Construction reached $539 million in 1961, $654 million in 1962, and $815 million in 1963 but dropped to $600 million in 1964 because the $815 million figure for 1963 reflected the additional $108 million granted under the Accelerated Public Works Program.

From the beginning of the program to Sept. 30, 1965, a total of 6,351 projects were approved for federal grants of $678 million. The local communities contributed an additional $2.5 billion to meet the total cost of $3.2 billion. Approved projects in this 9-year period improved the quality of water in 55,000 miles of waterways, serving about 53 million of our national population.

Nevertheless, construction did not reach the necessary level to bring the municipal waste problem under control within the decade. A survey showed that 1,342 communities were still discharging raw sewage into waterways, an additional 1,337 cities and towns needed enlarged facilities, and 2,598 communities did not have sewage systems, much less treatment plants. Almost 33 million Americans, before 1965, had either inadequate waste treatment plants or none.

The Water Quality Act of 1965 took these circumstances into consideration by increasing from $100 million to $150 million the annual authorization for appropriations for the fiscal years 1966 and 1967, the remaining two years for which federal waste treatment works construction grants were authorized.

During the hearings for the Act spirited pleas were made by governors, mayors, and other representatives of the most highly populated areas, for a greater share of the construction grants. They were told that the grants did not provide incentive when measured against prospective costs.

Congress responded by doubling the maximum amount for a single project from $600,000 to $1.2 million, doubling the maximum amount for a multimunicipal project from $2.4 million to $4.8 million, and making an additional $50 million available to states on a straight population basis, rather than a population per capita income basis. With those additional funds the federal government promised to pay up to 30 percent of the project cost, if the state matched the federal grant.

Two other actions were taken to meet primarily metropolitan needs. The Congress voted that grants could be increased by 10 percent if a project was certified by a metropolitan or regional planning agency as conforming with a comprehensive metropolitan area plan. The thinking of the Senate Committee (Senate Report 556, 88th Congress, 1st Session) was that such planning "had become no less than essential in view of our rapid urbanization. . . . Orderly development of municipal areas must be planned and followed to eliminate factors which lead to the breeding of slum and blight-impacted areas and to effect those sizable economies and efficiencies ordinarily made possible through the coordination of common interests and needs."

The congressional committees were impressed by the necessity for reducing pollution caused by the existence of combined storm and sanitary sewers in many of the nation's older municipalities. During periods of storm runoff, even in small amounts, the sewers discharge flows of storm water and sanitary sewage in excess of the capacity of treatment plants. As a result, much untreated waste is bypassed into receiving waters, creating a situation that is worse than if all storm water runoff were discharged into receiving waters in an untreated state.

Congress acted on information that there were more than 1,100 communities whose entire waste collection systems were of the combined sewer type. These sewers were serving a population of almost 21 million people. Another 810 cities of 37.8 million population had systems which consisted in part of combined sewers.

Complete separation of combined systems would entail expenditures of up to $30 billion. Many witnesses suggested that research and development grants be authorized to demonstrate new or improved methods to eradicate this problem. Congress agreed and established a 4-year program of grants and contract authority, with annual appropriations of $20 million for fiscal years 1966 through 1969.

As the law is written, the grants and contracts may be made with public and private agencies, with institutions, and with individuals, but the total amount appropriated for the contracts may not exceed 25 percent of the funds appropriated for these purposes in any fiscal year. Furthermore, the federal government may not pay more than half of the cost of any single project, and no single grant may be more than 5 percent of the funds authorized to be appropriated in any one fiscal year.

Research in other aspects of water pollution control, which is conducted directly in federal water laboratories and, by means of grants, in universities and other institutions, will continue as provided in the Federal Water Pollution Control Act of 1956 (as amended in 1961) and in other legislation.

When the new bill was drafted, the provisions concerning water quality standards created the most controversy. There was broad agreement that constant effort to improve the quality of the water supply was necessary to make water available for more uses, that water quality standards provided reliable and sound guidelines and provided a basis for preventive action, and that such standards enabled municipalities and industries to develop realistic plans for new plants or expanded facilities without uncertainties about waste disposal requirements on interstate waters. The issue was: Who would establish the standards and enforce them?

As worked out, the new law provided authority for the Secretary of Health, Education and Welfare to establish standards of water quality to be applicable to interstate waters, in the absence of state action. The standards were to be formulated on democratic procedures with hearings, public notices, and consultations with public and private interests. Voluntary compliance was stressed and court action avoided.

By June 30, 1967, all of the 50 states had established programs to clean their waterways, the Federal Water Pollution Control Administration reported.

Clean Water Restoration Act of 1969

The major change in the 1966 Act, in contrast to previous acts, was the clearly stated view of the Congress that water pollution control will require vast sums of money. It said frankly that the federal role had been weak and hesitant in approaching solutions to a federal concern and problem.

It was a milestone on the road to solution, chiefly because it removed hesitation in 1966, when the Clean Water Restoration Act was enacted. It was outstanding also, because the Act was "very popular legislation," in and out of the Congress.

The Clean Water Restoration Act was approved unanimously by the Congress and was signed by President Johnson on Nov. 3.

This landmark measure authorized appropriation of $3.9 billion for construction grants to help build sewage treatment plants for research and for grants to state water pollution control programs. In addition, the Act, without setting a fixed dollar figure, authorized the appropriation of such additional funds as might be necessary for enforcement, comprehensive planning, and other functions.

Construction grants totaling $3.5 billion were authorized for a 5-year period beginning in fiscal 1967.

Funds for construction grants were increased from $150 million in fiscal 1967 to $450 million in fiscal 1968, $700 million in fiscal 1969, $1 billion in fiscal 1970, and $1.25 billion in fiscal 1971.

The Act removed dollar ceiling limits on construction grants, effective July 1, 1967. The previous maximum was $1.2 million for single municipal projects and $4.8 million for multimunicipal grants. Lifting the dollar ceiling helped big cities because every municipality was able to receive at least a 30 percent federal grant and, under certain conditions, as much as a 55 percent federal grant.

If a state was willing to put up 30 percent of the cost of all projects for which federal sewage treatment plant construction funds were available, the federal share was raised to 40 percent, leaving the city only 30 percent to pay.

If a state adopted enforceable water quality standards for the stream into which a proposed sewage treatment plant discharged, the federal government provided

50 percent of the cost of a project, and the state's share was reduced to 25 percent. Under these conditions, a city paid only the remaining 25 percent of a project cost.

The federal government paid an additional 10 percent of the amount of a grant if a project was certified by a metropolitan or regional planning agency as conforming with the comprehensive plan for a metropolitan area.

The new Act continued the $20 million a year program for research and demonstration projects by states and interstate and local jurisdictions to provide improved ways of handling the problem of combined sanitary and storm water sewage. Also, the federal government could then pay 75 percent of project costs.

The Act authorized $20 million a year for state, interstate, and local advanced waste treatment projects, water purification, and industrial waste treatment. The federal share would be 75 percent of project costs.

The Act also authorized $20 million a year for 3 years to help improve ways of treating industrial wastes. The federal share of a project could be 70 percent, up to $1 million. Projects were to have industry-wide application.

General research, training, investigations, and information activities were authorized at $60 million in fiscal 1968 and $65 million in fiscal 1969.

Three million dollars for a special study of pollution in estuaries was authorized.

The act provided strong support for the development of comprehensive water quality programs for entire river basins.

The new Act authorized $5 million in fiscal year 1967 and then $10 million annually for fiscal years 1968–1971 for grants to states and to interstate agencies to enable them to expand and improve their water pollution control programs, including the training of needed personnel.

A new section provided that at the request of the majority of conferees in any enforcement conference or public hearing, the Secretary of the Interior could ask an alleged polluter to file a report with him on the kind and quantity of discharges he was sending into a river or other body of water.

Enforcement conference machinery was expanded to include pollution problems involving boundary waters or rivers the United States shares with Canada and Mexico.

The Secretary of the Interior was given the responsibility of administering the Oil Pollution Act. The Act was strengthened and expanded to include the Great Lakes and other inland waters not previously covered. Under the Act, anyone dumping oil must remove it or reimburse the federal government for the cost of removal.

A special study was ordered on the cost of carrying out the federal water pollution control program and on the national requirements to obtain clean water.

The new Act also ordered a study of the need for additional trained state and local government personnel to carry out water pollution control programs.

The Act called for a special study of pollution from boats and what should be done about it.

A SUMMING UP

By early 1967, the Congress wanted to know of the progress that had been made during the past 12 years and especially since enactment of the Water Quality Act of 1965 and the Clean Waters Restoration Act of 1966. Consequently, I asked Secretary of the Interior Stewart D. Udall a series of questions about the status of the antipollution program. The following reply to Chairman Fallon's inquiry, I believe, is a good summary of what had been done by April, 1967:

U.S. Department of the Interior,
Office of the Secretary,
Washington, D.C., April 24, 1967.

Dear Mr. Fallon: Congressman Blatnik's letter of April 13, 1967, set forth a series of questions relative to this Department's administration of the Federal water pollution control program. While I intend to discuss generally each of the points raised in my testimony before your Committee on April 25, 1967, I believe a more detailed response would also be most helpful to the Committee. Accordingly, there follows your questions and comments thereon:

Question. To what degree is the Department geared up to carry out the Water Quality Act of 1965?

Answer. Since the transfer of the Federal Water Pollution Control Administration last May 10, the Department has successfully sought qualified personnel to conduct the entire program. To date, we have a little more than 1,900 people with about 100 vacancies to be filled. A number of these people are now actively engaged in providing technical assistance and guidance to the States in their efforts to develop approval water quality standards by June 30 of this year.

Once the State standards are received, some have already been received, our immediate task will be to analyze and correlate them definitely in determining whether they may be approved.

This will be a formidable task, but one for which we believe we are ready. But, let no one think that the approval of these standards will be a "pro forma" operation. It will take considerable effort to do a thorough and prompt job. Now is the time to do such a job.

If we permit ourselves to be hurried into acting on any State's standards, we will be sorry later, and more importantly, we will not have carried out the spirit and intent of the 1965 act. We must balance our need for haste in establishing standards with the need for being thorough in establishing adequate, realistic, and enforceable standards. We are determined and ready to maintain this balance.

Question. In particular, what is the status of work toward and the schedule for completing the adoption of water quality standards for interstate and coastal waters of the States?

Answer. On the day after the water pollution control program was transferred to this Department, I issued guidelines for the States to follow in adopting approval water quality standards. Since then, we have, as I have stated, been providing assistance to the States in preparing standards for our approval.

We are now ready to receive and review them, but we have not, for the reasons I just indicated, established any schedule for completing our review. It is both premature and unrealistic to establish a schedule for completion of our review. Possibly, when all the standards are received, we will be able to give some estimates of when our work will be completed. But, I doubt that we could consider, even then, that those estimates would be firm.

Again let me emphasize, that we must do the job right the first time, because every thing we do thereafter will be built on this initial effort. If the foundation is weak, what chance has the building of withstanding the onslaughts of time? We will act as rapidly as good judgment dictates.

Question. In what ways, and to what extent, will the establishment of water quality standards increase the workload of the Administration?

Answer. Once State or Federal standards, including a plan for enforcement, are adopted and approved, our next task will be to see that these plans are effectively carried out. In many cases, this will require a considerable amount of monitoring of the plan of implementation and surveillance by our technicians to assure accurate data. We believe that this work will result in some workload increases, but we are not ready to say now what those increases will be.

Question. How far has the Department gone in implementing the Clean Water Restoration Act of 1966?

Answer. In the few months since the 1966 Act was passed by the Congress and approved by the President, we have done much towards full implementation of all of its provisions.

In the area of construction grants where, with the exception of the reimbursement provisions, the new authority is not effective until July 1, 1967, we have nearly completed our regulations governing the making of the increased grants and of the reimbursement grants. In this regard, we estimate that about twelve States will have passed legislation enabling them to take advantage of the increased grant provisions of the 1966 Act.

We have made grants totaling over $10 million in the area of waste treatment and industrial waste and are currently considering proposals totaling about $48 million for these two programs and the combined sewer program.

All the studies authorized by the 1966 Act are underway and on schedule.

In March of this year, we issued new guidelines for our State program grants which represented a marked departure from previous guidelines. The major thrust of these guidelines is, consistent with the objectives of the 1966 Act, to help the States upgrade programs and, through effective and careful administration on our part, to make the plans which they submit to us more meaningful and useful documents.

Question. What is being done to accelerate its river basin planning projects?

Answer. The guidelines or requirements to carry out this vital part of the program as it was strengthened by the 1966 Act are nearly ready and should be available very soon. We are also bolstering our staff to provide technical assistance to State and river basin planning groups to help them in obtaining grants. I believe this effort, plus other efforts by the Department to move the program toward cleaning up entire river basins, is of high priority.

Question. To what extent, and what areas, are the Administration's operations handicapped by personnel shortages?

Answer. As I have indicated, when the water pollution control program was transferred to us last year we did not have adequate personnel to get the job done. Since then, we have established a new organizational structure and recruited many qualified people. We have not had any difficulty in recruiting these people and are continuing to do so until we reach our present estimated personnel requirements. With this increased personnel plus careful utilization of their talents, we have been able to speed up our efforts and we believe we are meeting our commitments.

Question. What is the present and projected scope of the Administration's enforcement efforts?

Answer. With the approval of water quality standards, I expect that there will be some need to increase our enforcement efforts, although I believe that, at least initially, this increase should not be substantial. As I have said, the plans for enforcement, to be meaningful and realistic, must give the States, municipalities, and industry a reasonable time to take voluntary corrective measures necessary to insure that the waters will meet the levels of quality set by the Standards. Litigation is not the first step.

The 1967 fiscal year enforcement budget is $3,799,000. The President's budget for fiscal year 1968 requests $3,409,000. Although the 1968 estimate is $390,000 less than fiscal year 1967, this should not be construed as a deceleration of our enforcement efforts. As a matter of fact, we are proposing program increases for enforcing water quality standards where necessary. This action should, however, only be necessary where the State fails in its regulatory responsibilities, or, at the request of a Governor. One of the most important features of the standards is the State enforcement plan. We want the States to act and we will cooperate with them.

Additional funds are also proposed to enable the Department to implement and effectively carry out the provisions of the Oil Pollution Act of 1924.

Question. What will be the effects of the proposal to cutback the construction grants appropriation for fiscal year 1968 from the authorized sum of $450 million to about $200 million?

Answer. We do not view the President's request for $200 million for these grants as a cutback. It represents a one-third increase in the sum authorized in fiscal year 1967. Further, there are important factors to consider in arriving at this figure. Some States will have carry-over funds from fiscal year 1967 estimated to be about $25 to $40 million. Also, many States, although not all, will not be geared up to handle a large influx of projects because they have just spent much of their efforts on the adoption of standards. Lastly, the reimbursement provisions of the 1966 Act may enable the States to proceed with their construction projects if they are ready.

In summary, we believe that the $200 million figure in the President's budget is a good indication that we mean business and it is in keeping within the fiscal constraints which all must recognize.

Question. What will be the effects of the proposed cutback in research and development?

Answer. New funds for research and development have increased by about

40 percent from 1967 to 1968. Although the authorization was not funded at the levels authorized in the 1966 Act, the overall increases in research and development will allow for significant strides in the development of technology to control pollution. We are heartened with the response from industries and communities in developing proposals to carry out these new provisions.

Question. When does the Administration expect to have completed, and at what total cost, the research and development programs it is now authorized to conduct?

Answer. We have outlined a broad-scale research effort, which would enable us to develop the necessary technology for effective and economical control of pollution from most sources. As you know, research results are particularly hard to predict, but our scientists have set a target date of 1976. They expect to have a substantial and increasing output of new technology continuously between now and the 1976 target date. The total research budget requirements estimated at this time are $450 million during the 1968–1976 period.

Question. What is the status, and schedule time for complete establishment of the proposed national inventory of industrial wastes?

Answer. A questionnaire to develop the necessary information has been designed. This is being sent to the Bureau of the Budget for clearance. Once cleared, the questionnaire will be distributed to that segment of manufacturing industry which comprises over 90 percent of the manufacturing wastes discharged. We expect the report stemming from this effort to be available by about the end of fiscal year 1968. Thereafter, we expect to expand and update the inventory, probably biennially.

Question. Is the Oil Pollution Act adequate to deal with oil spillage from tankers such as recently occurred near Cornwall, England, and near Yorktown, Virginia, on March 26?

Answer. The Oil Pollution Act, 1924, as amended, makes it unlawful to discharge oil into the navigable waters of the United States from a boat or vessel, except in cases of "emergency imperiling life or property, or unavoidable accident, collision, or stranding," or except as permitted by Secretarial regulations. As your Committee knows, the Torrey Canyon incident occurred about fifteen miles from the shores of England—that is, beyond that Country's territorial waters—and it was apparently due to an accident. Thus, if the same situation occurred near the United States, we could not have dealt with the problem under the 1924 Act. Even under the Oil Pollution Act of 1961, as amended, which is not administered by this Department, there would not be any method of dealing with the Torrey Canyon problem because the leakage resulted from damage to the vessel.

In regard to the Yorktown incident, it is reported that there was accidental discharge due to someone turning the wrong valve on a passing tanker. If, and details about this incident are still sketchy, the discharge was caused by a grossly negligent or willful spilling, leaking, pumping, pouring, emitting, or emptying of the oil, the 1924 Act could be used to invoke criminal penalties and to make the offending vessel owners remove the oil. We are looking into the incident, but, as recent articles have pointed out, proof of gross negligence or willful spilling, etc., is quite difficult.

We believe that these two situations serve to point up the need for a further strengthening of our authorities to cope with these and similar situations adequately. It is our intention to recommend soon remedial and strengthening legislation based on studies now being conducted, which would be designed to meet these problems in a reasonable manner.

Question. Does the Federal Water Pollution Control Act adequately define the waters covered by the requirements of the Act?

Answer. While that Act does not define in precise terms all the waters covered by the Act, we are administering the program without any real difficulty. We therefore believe that the present definitions are adequate.

Question. What progress is being made under the new research grant programs, advanced waste treatment and joint industrial municipal waste treatment systems, and prevention of pollution by industry?

Answer. Good progress is being made under these new programs. In 1967, although no funds were appropriated for these purposes, the program utilized the new provisions of the Clean Water Restoration Act of 1966 which permit the utilization of storm and combined sewer funds for this purpose in the event activity in the storms and combined sewer area was not reaching expectations. This proved to be the case. Early in November, 1966, it did not appear that an adequate number of proposals would be forthcoming to utilize available funds or combined sewer and stormwater projects.

Therefore the Administration selected proposals for projects in the two new areas. The response was great and in December, grants and contracts amounting to over $10 million were awarded out of 1966 funds in the advanced waste treatment and industrial waste areas.

Since then, there has been considerably more interest shown in stormwater and combined sewer projects. Currently under review are proposals amounting to nearly $48 million for all three program areas. Thus, the interest shown and progress being made in this program is great.

In summary, we are very pleased with the first year's progress of the program in this Department. We are looking forward to even better results in future years and are working expeditiously to that end.

Sincerely Yours,

Stewart W. Udall,
Secretary of the Interior.

Under present regulations, the states have begun submission of receiving water standards based on the criteria furnished by the Water Quality Criteria Advisory Committee established by the Secretary of the Interior. This sequence of criteria—standards—enforcement is the scientific, factual basis for both air and water legislation. It was carefully developed to avoid arbitrary designation of environmental quality objectives and to reserve to the states the choice of ambient standards.

Criteria are scientific statements of cause and effect. For example, a lowering of the dissolved oxygen in a stream to x parts per million will result in the death of 50 percent of a certain species of fish within 24 hours. A concentration in air of x parts per million sulfur dioxide for more than Y days each year will cause upper respiratory complications to a significant fraction of the population. Criteria are descriptive statements of the ramifications of environmental changes or contaminations. The laws call for the federal government to sponsor the research necessary and to issue criteria statements, because these relationships should apply nationwide.

After publication of the criteria by the federal government, states and local government choose their own standards which they submit to the federal government for approval. This assures that every locality in the United States achieves some minimum environmental quality. Obviously, federal review is also necessary to bring conformity of state standards in interstate river basins or air regions.

Enforcement involves relating the ambient air or water standards to the sources of pollutant emission. Local monitoring programs identify sewage outfalls or gas stacks. Each source is assigned a maximum output in order to keep the quality of the shared receiving environment up to the agreed upon standard. If a source exceeds its allowance, enforcement action can proceed ranging from notification of violation, to fines, to an injunction which may shut down the polluting operation.

As the state standards for water quality were filed, some gaps in the law became apparent. The problem of oil spillage from tankers had not been covered adequately by the Oil Pollution Act of 1924. The drainage of polluted waters from mines, especially abandoned excavations, was a serious problem in many states. The difficulty of correcting facilities operations after construction led to consideration of preconstruction certification that government and private facilities discharging into surface waters would meet quality standards.

A comprehensive Water Quality Improvement Act of 1969 passed the House in April. The report accompanying the bill stated its major purposes:

Control of Pollution by Oil and Other Matter

The water pollution problem is a large and complex one. Previous reports submitted by this committee in support of the Federal Water Pollution Control Act of 1948, the amendments enacted in 1961, the Water Quality Act of 1965, and the Clean Water Restoration Act of 1966, have dealt in detail with what is happening to our water, what permitting the damage to continue will do to the Nation's health, safety, and development, and what must be done to stop the continuing damage and reverse the pollution process.

Repeatedly, increasing scientific and technical expertise have brought to the fore aspects of the broad problem of which we were not aware, or with which we had dealt inadequately in the past. Events of catastrophic proportions have confronted us with dramatic evidence of the need for new or better preventive or control laws and procedures.

It is in this context that the committee recommends—urges—the enactment of H.R. 4148, the Water Quality Improvement Act of 1969.

Recognition that oil is a potentially serious water pollutant is not new. The Congress recognized the need for control in the Oil Pollution Act of 1924 reported by this committee. But by almost any relevant yardstick, 1924 was an altogether different life. The breakup of the tanker *Torrey Canyon,* with its incalculable damage to the coast of England and its nearly $8 million cleanup cost, warned us that existing Federal oil pollution control programs would be inadequate to handle a similar catastrophe if it occurred here. Devastation of California's beaches 2 months ago by oil from an offshore drilling rig made the lesson loud and clear. It may not even be possible to assess the vast damage to marine life and recreation. This committee made an on-the-site investigation at Santa Barbara and found the physical situation appalling.

The Oil Pollution Act of 1924 is simply not sufficient to cope with such problems. It applies only to discharges and spills that are grossly negligent or willful; limited to vessels, it does not apply at all to spills from fixed installations such as pipelines, oil deposits, refineries, or manufacturing plants or other types of industrial activity using and storing large quantities of oil. Confined to oil, the 1924 act provides no protection against dozens of other potentially hazardous substances.

In addition to its contamination of water, shoreline, and beaches, oil often has severe effects on fish and wildlife, shellfish, and recreation. Untold ecological damage can result not only from the oil itself but also from chemicals used in attempting to deal with the oil. We must be able to combat this type pollution and prevent, wherever possible, catastrophes like these. It is in large part to that need that H.R. 4148 is addressed.

H.R. 4148 is not simple legislation. A section-by-section analysis is essential to its adequate presentation. That analysis appears as part II of this report. Part I of the report, here, is therefore confined to elaboration not appropriate in the bill itself.

Oil Pollution

Section 17(a) sets forth the definitions of the terms used in the bill. Subsection 17(a)(2) is a general definition of matter which would present an imminent and substantial hazard. The term could extend to more than 200 substances. The Secretary of the Interior is now reviewing a list of over 200 substances to determine what should in fact be held to be hazardous. Before this subsection can become meaningful, the Secretary will have to issue regulations, following the usual administrative procedures governing such issuance, identifying hazardous matter. The committee expects that the Secretary will proceed as rapidly as possible in this regard. Specific note should be made of the fact that the definition of matter does not include by-product material, source material, and special nuclear material as defined in the Atomic Energy Act of 1954.

The requirement that notice of discharge of oil or matter be given to appropriate authority, contained in subsection 17(b), is essential to expeditious and efficient cleanup action. It is a requirement placed upon the individual who is operationally *responsible* for the vessel or facility involved. It is not intended to include seamen, in the case of a vessel, for example, or to a night watchman or janitor in the case of a facility. By this clarification, however, we do not mean that the requirement is limited to the president of a large corporation or the owner of a vessel. The emphasis is on *operationally* responsible at the time of discharge.

This provision does not extend to private waters such as landlocked ponds specifically built, for instance, to receive drilling refuse, or for similar purposes.

Public vessels are exempt from the penalty for failure to give notice but the committee expects that public agencies will by appropriate regulation or instruction require that operational personnel give the notice of discharge the bill demands.

The committee is aware that the term "substantial" as it appears in subsection 17(b) and in subsequent subsections of the bill is subject to varying interpretation and judgments. It is, as a practical matter, impossible to substitute a more specific term. What is a "substantial" discharge into a river or a harbor, for example, might be insignificant if discharged into the Atlantic when turbulent water would result in rapid dispersal. By the same token, an insignificant discharge at sea might well be overwhelming in another area.

With respect to the cleanup authority vested in the Federal Government under subsection 17(d)(1), the committee calls attention to the fact that the Federal Government is responsible for discharge cleanup without regard to the cause of the discharge (including acts of God) or the location of the waters of the United States into which it occurs. The committee urges that to the extent feasible in the particular situation the State and local groups already formed for continuing cleanup programs will be called upon for assistance.

The authority of the United States to remove or destroy a vessel where a marine disaster situation has created a substantial threat of pollution follows closely both the philosophy and the language of the same authority vested in the Corps of Engineers, by law in 1899, institutions where the vessel constitutes a threat to navigation. It should be noted that liability for the cost of such removal is limited to vessels negligently operated and thereby the cause or contributory to the cause of the disaster involved.

Subsection 17(e)(1) requires that the owner or operator immediately remove any discharge. If he fails to do so and the United States performs the cleanup, the willful or negligent owner or operator and vessel is liable for up to $10 million of the cleanup cost, or a sum equal to $100 per gross registered ton, whichever is less. This liability is a liability per vessel, per owner, except that in the case of a barge tow, each barge in the tow is considered to be a separate vessel. Under subsection 17(e)(1), the United States may proceed against the owner or operator of any vessel that causes or contributes to the cause of the discharge, even though the discharge actually comes from another vessel.

The cleanup liability provision with respect to onshore and offshore facilities, contained in subsection 17(f)(3), sets a maximum possible liability of $8 million. As to onshore facilities, it requires that the Secretary of the Interior, through full public hearing procedure and in consultation with interested Federal agencies, including the Small Business Administration, establish classifications of onshore facilities and activities and set differing limits of liability for each classification, none of which may be in excess of $8 million. This subsection is so written as to provide a high liability for a large discharge from a major facility and at the same time insure that reasonable low liability will be set for the hundreds of small businesses and other facilities along our waters whose potential discharge would be small and upon whom a large liability could very well impose a ruinous burden. The subsection does not apply to any onshore facility until the Secretary establishes its classification, and the Secretary must submit intended classifications and liability limitations to the Congress at least 60 days before they are to become effective.

Subsection 17(g) requires that the Secretary of the Interior issue regulations "establishing environmental quality" criteria relating to the methods and procedures for the removal of oil and matter. The words "establishing environmental quality" refer solely to ecological protection and the chemicals, substances, or conceivable devices that may or may not, because of their efficacy on the one hand or deleterious effects on the other, be used in cleanup. These words do not extend, in any sense, to the subject of environmental quality generally.

Subsection 17(g)(2) prescribes the civil penalty for violation of cleanup regulations. It protects the vessel owner or operator, and the person who owns or operates an onshore or offshore facility by requiring that notice be given and an administrative hearing be held before the penalty may be assessed. If the individual concerned disagrees with the assessment of the penalty, he has full de novo judicial protection in that the United States would have to bring an action in an attempt to collect the penalty.

Subsection 17(h)(1) authorizes a revolving fund to cover Federal cleanup costs. The "other funds" referred to in the subsection would be derived from reimbursements of cleanup costs and penalties. The fund may be used, under subsection 17(h)(2), to reimburse the States for their assistance in cleanup operations.

The reference to the "national contingency plan" in subsection 17(h)(2) includes the regional contingency plans provided for in the national plan.

Barges are specifically included in subsection 17(k)(1), requiring evidence of financial responsibility, because many barges are not registered.

The study of requirements for financial responsibility and limits of liability called for in subsection 17(k)(3) is necessary because neither the affected industries nor international underwriters have had previous experience in this area of discharge cleanup, and they were unable to supply the committee with adequate factual information in this regard. It is hoped that the results of the study, plus any experience gained in the interim, will disclose any need for amendment that may exist.

Control of Sewage from Vessels

Wastes from ships and boats are still another cause of pollution. It is most severe in bays, inlets, lakes, harbors, and marinas. These pollutants include sewage, oils, chemicals, and other wastes. Most vessels are not equipped to provide even minimal treatment. The growing popularity of recreation craft is almost certain to increase this source of pollution to significant proportions if corrective and preventive action is not set in motion now.

Section 18, contained in H.R. 4148, seeks to provide the corrective and preventive potential.

The definition of "marine sanitation device" includes chemicals, biochemicals, etc., that may be found to be sufficient for the necessary sewage treatment.

The committee wishes to make it clear that it expects the Secretary of the Interior to hold full administrative hearings before he issues the regulations with respect to standards of performance for marine sanitation devices required by subsection 18(b).

The committee also wishes to make it clear that in the application of the standards and regulations for existing vessels, the most careful consideration should be given to the problem of economic costs. The American-flag merchant marine is already in a critical position. More than 80 percent of the nonpassenger vessels are more than 20 years old, and if they are still in operation when the regulations become effective, the cost of their refitting would almost certainly be prohibitive. American-flag passenger vessels are few and are a marginal operation at best. It is obvious that a reasonable approach in these circumstances is called for.

In enforcing the prohibitions contained in subsection 18(b)(4) the committee expects reason to prevail. If the marine sanitation device on a given vessel stops operating while the vessel is out on the water, for example, enforcement action would certainly be inappropriate.

It should be emphasized that the research authority granted the Secretary of the Interior in H.R. 4148 is not intended to eliminate or otherwise affect research activities in similar areas being conducted by the Coast Guard, the enforcement arm for much of the control contained in this bill.

Area Acid and Other Mine Water Pollution Control Demonstrations

Acid mine drainage is another longstanding source of water pollution. The chemical quality of water is altered in streams receiving such drainage. The acid flow destroys fish and fish food organisms, damages recreational and esthetic values, corrodes transport equipment, bridges, and other structures exposed to the water. Waters so polluted require extra and expensive treatment when utilized for municipal and industrial water supplies. Acid drainage is associated with active as well as abandoned coal mining operations; the latter continue to produce acid for indefinite periods after mining is discontinued. An estimated 3.5 million tons of acid mine waters drain into the streams of the United States annually, damaging approximately 4,000 miles of streams.

Although we know what acid mine drainage does, we do not yet know what to do about it. The problem is substantial and possible solutions must be found. Section 10 contained in H.R. 4148 is a straightforward, self-explanatory authorization of a $15 million demonstration research program to seek those solutions.

Training Grants and Contracts

Sections 20 through 23 of the bill expand the training grants program already authorized in the Water Quality Act in an effort to alleviate a critical shortage of skilled engineering aides, scientific technicians, and treatment plant operators. The need for this expansion is clearly supported by the report on the study of manpower and training needs in water pollution control. The recommended expansion is clearly set forth and the committee sees no interpretation problems.

Cooperation by All Federal Agencies in the Control of Pollution

Section 3 of H.R. 4148, in subsections 11(a) and 11(b), requires maximum feasible cooperation by all Federal agencies in the control and prevention of water pollution.

Subsection 11(a) deals directly with procedures for control of pollution caused by the administration or actual operation, either directly or by contract, of federally held real property or facilities. In attempting to insure that Federal facilities will be in compliance with the applicable water quality standards, the problems to be considered, the priorities to be assessed, and the relative values and public interests to be weighed, are very much akin to the problems, priorities, and interests which must be taken into account by a State when it is establishing water quality standards for a given area, by industries when they are making decisions on how and where they will expand capital investment, and by local governments in attempting to achieve a balance among health and welfare, economic development potential, and supportable tax structure.

The Federal Government obviously must balance much the same factors, within the limits of the funds available and within the broad and complex context of the national interest.

The disposition of dredged spoil is currently the most highly publicized of the possible sources of pollution from a Federal activity. Research is underway seeking to determine whether dredged spoil is in actuality an active pollutant, or if it is, to what extent its introduction into any given body of water does in fact lower the quality of that water. Whatever the answers are, the continuing viability of the rivers and harbors that produce the spoil are essential to the economies of the regions they serve and hence to the total national interest.

Transporting the spoil to available land disposition sites is extremely expensive; the land is costly and the transportation is costly. The States and localities would much prefer to preserve their land for more economically productive use. The Federal Government must allocate its available tax revenues among a great many equally clamorous public demands. The dilemma is clear; the attainable solutions are dimly seen at this point. Section 11(a) sets forth its present requirements accordingly, subject to future amendment as technology and available money increases and as more definitive local determinations, supportable in the national interest, are made.

Subsection 11(b) requires, first, that any applicant for a Federal license or permit to conduct an activity which may discharge into the navigable waters of the United States provide the Federal agency issuing the license or permit with a certification from the affected State or States or interstate water pollution control agency that the activity will be conducted in a manner that will not reduce the quality of the water below applicable water quality standards; second, that where water quality standards are issued by the Secretary under the Water Quality Act of 1965 or where a State or interstate agency lacks authority to issue certification, the Secretary shall provide the certification; third, that in the case of multiple licenses or permits by one or more Federal agencies for the same activity, if the applicant receives a certification for one agency, it need not obtain a certification for the other agency or for succeeding permits or licenses unless the Secretary or the State, upon receipt of notice, objects, except that this provision does not apply to an application for an operating license or permit; and fourth, that no Federal license or permit may be issued until certification is received, except that in any case where actual physical construction of the facility itself has been lawfully commenced prior to the enactment of this act (and by this is meant actual excavation or building; site acquisition, construction of access roads, or similar preliminary or collateral activity would not satisfy the requirement) no certification shall be required under this subsection for a license or permit for the activity after the date of enactment and except, further, that any such license or permit issued without certification shall terminate at the end of the 2-year period beginning on the date of enactment of the Water Quality Improvement Act of 1969 unless prior to such termination date the person having the license or permit submits to the licensing Federal agency a certification which otherwise meets the requirements of this subsection.

A wide variety of licenses and permits (construction, operating and otherwise) are issued by various Federal agencies. Many of them involve activities or operations potentially affecting water quality. The purpose of subsection 11(b) is to provide reasonable assurance (as determined by the affected State, States, or the Secretary of the Interior) that no license or permit will be issued by a Federal agency for an activity that through inadequate planning or otherwise could in fact become a source of pollution.

The language of the legislation is intended to eliminate duplicating certification requirements, and to afford a safeguard against too broad a use of the single certification.

On March 11 last, our esteemed colleague, the chairman of the Joint Committee on Atomic Energy, expressed his concern about possible conflict between this proposed legislation and the regulatory authority and responsibilities of the Atomic Energy Commission. The legislation here reported, as it relates to nuclear-generating facilities, concerns itself with thermal pollution. Thirty-four of the Nation's 54 jurisdictions now have approved thermal standards for water quality. Heat pollution from industrial and powerplant sources can be expected to increase at a very substantial rate, based on projections of industrial growth and electric power demands. Increased water temperatures affect a stream's capacity to assimilate wastes. Temperature changes also can ruin water for fishing and recreation.

The chairman of the Joint Committee raised, essentially, six points. First, he was fearful that an undesirable competitive factor would develop by virtue of the possibility that a significant fraction of all new electrical generating capacity (other than nuclear) would not be covered by subsection 11(b). The committee believes this concern is met by the fact that a Federal license or permit of some kind is required for almost all electric generating plants, and any Federal agency granting the relevant license can and should condition the grant upon compliance with applicable water quality standards.

Second, the chairman questioned the need for certification for both the construction license and the operating license which the Commission grants. Based on testimony by the Commission, the committee has concluded that the very different character of the two applications, the long period of time that elapses between their issuance, and the uncertainty as to the finality of plans at the construction license stage, all support the requirements for certification with respect to both applications.

Third, the chairman of the Joint Committee recommended that where construction licenses have already been issued, those facilities be exempt from any certification requirement. The problem of a construction permit already issued when the act is passed is of course not peculiar to nuclear facilities, and the committee recognizes that some relief must be accorded in such cases. The subsection therefore provides, as has been outlined above, that where actual physical construction of a plant is already underway, certification is postponed for a period of 2 years, which the committee believes is sufficient time to permit whatever action may be necessary to comply with the water quality standards and obtain the certification.

Fourth, the chairman recommended that judicial authority to suspend a permit or license should be discretionary rather than mandatory, and that recommendation has been followed.

Fifth, the chairman recommended that the certifying agency be required to state "reasonable assurance" of compliance with water quality standards, rather than guaranteeing compliance, and that recommendation has been followed.

And finally, the chairman was concerned that the subsection might in some way, through the water quality standards, alter the Commission's preemptive authority as to radiological health and safety standards, as it is contained in the Atomic Energy Act of 1954. The committee is informed that nothing in subsection 11(b) could be construed as an amendment to the Atomic Energy Act of 1954.

The Atomic Energy Commission has informed the committee that doctors, hospitals, universities, and research institutions are licensed by the Commission to possess and use limited quantities of nuclear materials that might, in minute quantities, be disposed of through a waste treatment system. It is not intended that subsection 11(b) apply to these specific types of licenses or permits.

Section 8 of H.R. 4148 changes the name of the Federal Water Pollution Control Administration to the National Water Quality Administration. The committee believes that the agency should bear a designation that bespeaks its positive goals.

H.R. 4148 authorizes the following appropriations:

Item	Section identification	1970	1971	1972	Total amount
Cleanup revolving fund	17(h)(1)	$ 20,000,000			$ 20,000,000
Acid mine drainage reserve	19(d)(1)	15,000,000			15,000,000
Training grants and contracts	23(c)(3)	12,000,000	$ 25,000,000	$25,000,000	62,000,000
Estuary research extension	5(k)(4)	1,000,000			1,000,000
General research, investigation, and training extension	5(l)	65,000,000	65,000,000		130,000,000
Project research extension	6(e)	60,000,000	60,000,000		120,000,000
Total		$173,000,000	$150,000,000	$25,000,000	$348,000,000

In the water pollution control laws enacted in earlier years, the Congress has authorized the expenditure of approximately $3 billion for all phases of the program. Thus far about two-thirds of that has actually been appropriated at the Federal level. Progress in development of the technology for dealing effectively with water pollution has made tremendous strides. Progress in applying the technology creeps along, and achieving clean water becomes more and more a life and death matter with each passing year. The committee, therefore, respectfully urges that the authorizations contained in this bill be fully funded, and that as to the water pollution control program as a whole, Federal, State, and local governments and the Nation's industries carefully reexamine their present positions in terms of the national jeopardy inherent in the failure to act affirmatively on a large scale.

A companion bill (S. 7) is, at the time of this writing, being considered by the Senate.

A major difficulty in the water pollution abatement effort has been the shortfall in federal funds for sewage treatment plant construction grants. Whereas the Congresss had authorized amounts of $450 million for 1968, growing to $1 billion in 1970, the administration requested only $214 million for the fiscal year 1969 and the same amount for 1970. This deficiency has delayed many municipalities in initiating construction and is a major setback to the goal of restoring the quality of heavily populated river basins. At the time of this writing, over half of the members of the House of Representatives have united in a bipartisan drive to fund in full the $1 billion authorized for 1970 in the Clean Water Restoration Act. It is only through realistic funding that our cities can even begin to approach adequate waste treatment—and even $1 billion is not enough.

Much of our pioneering attack on pollution was confined to stopping the discharge of community sewage into our streams. It appeared to be the chief pollutant because it was so obvious. Further investigations, however, revealed that industry, either directly or indirectly, is responsible for well over half of our water and air pollution.

If the trend toward pollution is not reversed, we are warned, the expense of cleaning water and air for industrial use—let alone for our personal use—will increase enormously and could reach the point of noneconomic return. Continuing to pour carbons into the atmosphere could create a greenhouse effect over the earth. The trapped heat could melt the polar icecap and inundate coastal cities around the world. The grim reality is that man is gradually destroying the environment that makes it possible for him to live.

M. A. Wright, President of the Chamber of Commerce of the United States, in December, 1966, said,

> Today we in business and industry still have the freedom to make a reasoned and resolute response to the problem [of air and water pollution]. Tomorrow our actions may be tightly controlled by government regulations. If our efforts in this area are made mandatory, not only will we be forced to take more costly and less efficient actions, but we will also forfeit our claim to being a responsible segment of society. To those who say they cannot afford to take effective anti-pollution measures, I can only respond that they can't afford not to.

Mr. Wright's advice that industry cannot afford to ignore antipollution controls has been demonstrated in several fields. One is in the area of public relations. A recent Harris poll revealed that by 2 to 1 margins, Americans demand that industry, the automotive industry, communities, and county, state, and federal authorities multiply their antipollution programs. In some portions of the country, the majority expressed willingness to pay an additional $15 a year in taxes to help finance the campaigns.

The court of public opinion can deal harshly with industries, whose products have day-to-day contact with the public, if the producers are shown to be air or water polluters. Such enterprises spend millions to create a good public image, which could be destroyed by a photograph showing what it is doing to water or air. The swift remedies the detergent manufacturers evolved to stop detergent foam in our waterways are the handiest illustration of what outraged shoppers can accomplish. A company selling its paper products in the supermarkets flirts with economic reprisals if it fouls streams, and so do oil and gasoline producers who sell automotive fuel to hunters and fishermen.

There is solid evidence that public demand is having its effect on industries with close public contacts, but it is evident also that manufacturers with only indirect contact do not feel such pressure.

There has been a holdback by many individual industries, understandably, to assure that they are not alone in curbing air and water pollution to the detriment of their competitive position. Why install expensive equipment if the plant just down the river, a few blocks away, does not?

Generally, however, American industry has seen the handwriting on the wall. There is now an almost universal recognition that it must stop polluting air and water or face public reprisals and police action from government.

Industrial moves toward cooperation have been delayed for two main reasons. The first is cost. The second is lack of knowledge and techniques to reform its sins.

The cost to industry and the taxpayer to reverse pollution is staggering. One of the reasons, said former Commissioner James Quigley of the Federal Water Pollution Control Commission, "is the big lag in doing the do-able. Treatment plants that are being built now could have been constructed fifty years ago. There is no reasonable choice but to get it done as quickly as possible and as inexpensively as we can."

Outlays by industry and governmental agencies now are estimated at $5 billion annually, with rising expenditures to a possible $10 billion by 1970. These expenditures can be divided thus: 40 perecent for controlling air pollution, 40 percent for controlling water pollution, and 20 percent for disposal of rubbish, garbage, and other solid waste materials.

It is a demonstrated fact that most industries can afford waste treatment without hurting their competitive position. It is equally true that some older, marginal plants could be forced to close if they have to spend vast sums to stop pollution, but these can get help in working out a reasonable program. The federal government does not want to force anyone out of business. However, industry must accept pollution controls as a necessary expenditure in construction and day-by-day operations. Fortunately, such acceptance increases.

Smaller manufacturers, in antique but adequate buildings, say that they cannot afford to install water pollution controls, because the expense could bankrupt them. Many of these have a solution, however, which apparently has been missed by many. Under the federal construction grant program, some industries deliver their wastes to the municipal treatment plant. Where this can be done, it is the cheapest way for an industry to conform with the regulations.

Fifteen states by mid-1967 have offered various forms of tax relief to industries which install water pollution equipment. Eleven of these also give tax deductions to industries which take measures to cease air pollution. All of the 15 states give quite liberal incentives. Some do not collect property taxes on the new installations; others give as much as 60-month amortizations on income tax, increased by the expenditures; while a few permit income tax deductions for the full cost of antipollution control measures.

Giving industry incentives to build antipollution facilities in older buildings through tax write-offs is a lively proposal before the Congress. When present inflationary trends are brought under control, however, further incentives in the form of tax relief certainly will be reconsidered. Because antipollution is partly a public health measure, without immediate profit or reward to industry, it is proper that the federal government consider what encouragements it can give to industry in tax relief.

An indirect subsidy to industry is the research being financed by the Interior Department. Initially, contracts totaling $239.500 were let to six research firms and the State University of Kansas in late 1967 to study the waste problems of 10 leading industries.

For study purposes, industries were divided into groups according to the amount of waste water they discharged in 1964. Investigations determined the costs and effectiveness of alternative methods of reducing wastes by changing processing methods, by using various treatments, and by recovering wastes. These studies have been of great importance, not only to the Congress, which asked for them in the Clean Water Restoration Act, but for industry.

The industries studied first were blast furnace and steel mills, motor vehicle and motor parts, paper mills, petroleum refining, textile mills, canned fruits and vegetables, leather tanning and industry, the dairy industry, plastics and resins, and meat products. This research undoubtedly will be continuous and on an increasing scale; such continuance will be asked of each new Congress. There can be no legitimate complaints by smaller industrial plants that they cannot afford research, because it will be free—public information for those who seek it.

It is often said that the waste treatment techniques of today are essentially the same as they were 50 years ago. Then the question is asked why we are spending billions for plants that may soon be obsolete, when a technology breakthrough arrives.

The best present judgment is that plants built now will serve indefinitely. Technologies are being developed in every field, but we do not have time to await the answers. If we do not stop pollution, the time will come when it cannot be reversed. Some waters such as Lake Erie already cannot be restored.

We can be optimistic about what the new regional water study laboratories in every region of the country will do, not only in evolving new methods of antipollution abatement, but also in developing the professionals who will specialize in this task. Generally, there has been excellent response from industry in cooperation with government on all levels. Reports of industries recovering the costs of antipollution systems by reuse of water or extraction of usable materials from what was formerly sent into the atmosphere are heartening.

AIR POLLUTION CONTROL

There are 8,400 places in the United States which have air pollution problems, ranging from minor to major to grave, and ranging from a rural community of about 700 persons made miserable by smells from a tannery, to big cities, where health authorities have found high cancer incidence in the wake of air contamination. By the end of this century about 82 percent of all Americans will live in urban areas, where pollution is most rampant.

The publicity about smog, caused by air inversions in a few big metropolitan areas, gives a misleading picture of where smog afflicts the country. It is a nationwide problem. The annual percentage of hours of inversion in various regions is as follows:

Region	Percentage
Rocky Mountains	35–50
West Coast	35–40
Appalachian Mountains	30–45
Central Plains	25–40
Northwest Pacific Area	25–30
Great Lakes	20–30
Atlantic Coast	10–35

The insidious effects of air pollution are by no means entirely urban. The economic significance of air pollution on crops and forests was not extensively known until 1958, when studies in California and New Jersey and by the Southeastern Forest Experiment Station of the U.S. Department of Agriculture were completed and published. By 1961, these were followed by investigative reports in 19 states. Every cash crop, including the most popular flowers grown for the market, was affected in varying degrees by industrial and automotive pollution, very often from sources far away from the farmlands and the forests. Within a few years, other reports on crop damage became almost worldwide.

The annual economic losses to agriculture from air pollution are yet to be measured precisely, but the 1962 National Conference on Air Pollution said it "amounts to hundreds of millions of dollars a year." President Kennedy in his "Special Message on Natural Resources," in 1961, estimated annual damage to all enterprises, including vegetation, at $7.5 billion as a current figure. Undoubtedly, the estimate is greater today.

There was considerable dispute even in the early 1960s about the effects of air pollution on the public health. The U.S. Public Health Service asserted that there was then "strong evidence" that contaminated air contributed to respiratory ills, including the common cold, bronchitis, pulmonary emphysema, bronchial asthma, and lung cancer.

Smog in London killed 4,000 persons in 1952 and 340 in 1962. The Donora, Pa., air inversion in 1948 made one-third of the population sick and killed 20. Later evidence revealed that a smog concentration in New York City in 1953 had killed about 200 persons.

What was speculative in the early 1960s now is positive. Studies made in the United States, Great Britain, and Japan reveal that air pollution contributes directly to pulmonary and upper respiratory ailments and causes susceptibility to a host of other debilitating or fatal afflictions.

All of the bad effects of inhaling carbon monoxide, the hydrocarbons, and lead, plus the gases caused by additives in all types of motor fuels, will not be known for some time. However, there is hardly a sizable community in the United States in which the Public Health authorities have not conducted studies of the effect of motor emissions on human health. The conclusion is that exhaust contaminants are killers over long exposure.

The remarkable proliferation of all types of automotive equipment after the end of World War II was a decisive factor in the inauguration of air pollution control as a national aim. With wartime restrictions on production and buying of automobiles lifted, federal and state authorities, anticipating the wholesale purchase of new cars, built about 100,000 miles of new highways. But when they were constructed, there were 300,000 new cars to roll over them. Traffic jams in urban areas, immobilizing cars, poured motor exhaust gases in unprecedented amounts into narrow streets. The grave effects these emissions had on drivers, pedestrians, businesses, buildings, and city vegetation quickly convinced many that the old belief was wrong, that smoke was not the only air pollution.

In December, 1949, President Truman asked Secretary of the Interior Julius A. Krug to organize an interdepartmental committee, which would call the first United States Technical Conference on Air Pollution.

The first concerted congressional activity in the air pollution situation came in 1950. Two resolutions passed the House, asking for research. They died in the Committee on Interstate and Foreign Commerce. That year, the American Municipal Association adopted its first resolutions on air pollution, starting the so-called "urban lobby."

The Technical Conference, requested by President Truman, met in May, 1950. More than 750 persons, representatives of government, industry, official and unofficial civic groups, and citizens' organizations interested in conservation and beautification projects had spirited discussions. The outstanding resolution asked that the federal government take the leadership in identification of air pollution, give determined proof of its causes, and develop the techniques to stop it.

PIONEERING IN AIR POLLUTION

After its investigations, the House Interstate and Foreign Commerce Committee in 1952 passed a resolution now regarded as landmark legislation for air pollution. It directed the Surgeon General, the Secretary of the Interior, and the Secretary of Agriculture "to intensify their respective activities within the scope of their existing authority," to begin research. The U.S. Public Health Service had authority from the Public Health Service Act of 1944.

The House passed the resolution of July 2, 1952. It failed Senate passage by one vote, when one conservative senator objected because there was no definite indication of what was to be spent. Congress adjourned three days later.

Attempts to do something about air pollution failed again in 1954. But late in that year the Secretary of Health, Education and Welfare established an Ad Hoc Committee with representatives of nearly all government departments and agencies concerned and the National Science Foundation. This committee did yeoman work on every facet of the air pollution situation, telling President Eisenhower that in some areas the air could no longer absorb further pollution.

The President, in his State of the Union message in 1955, proposed strengthening programs for air controls and in January issued a special message, pointing out that air pollution was a serious health problem. He asked higher appropriations for the Public Health Service to gather more scientific data and find more effective controls.

In April, 1955, the Senate Flood Control subcommittee of the Rivers and Harbors Committee considered both water and air pollution, but emphasized that it was a state and local problem, not a federal concern, and did not recommend federal enforcement.

By a voice vote, the Senate accepted the Committee's recommendations, including a $3 million annual appropriation for 5 years for the air program, on May 31. On July 5, the House passed a bill given to it by its Interstate and Foreign Commerce Committee, which raised annual funding to $5 million and differed with the Senate about the federal role. The next day, the Senate accepted the House version.

On July 14, 1955, President Eisenhower signed the legislation, which became Public Law 84-159. This remained as the basic statute for federal air pollution involvement, until passage of the 1963 Clean Air Act.

From 1955 through 1962, there were growing discussions, hearings, seminars, and demands for more and stronger legislation in the Congress and state legislatures to curb air pollution, but there was little legislation anywhere. The so-called "urban lobby," composed of the United States Conference of Mayors, the American Municipal Association, and the National Association of Counties, persisted in their demands for more and better financed federal actions.

The National Association of Manufacturers opposed any increase in the air pollution programs, as did the automotive, chemical, iron and steel, coal, and other industries. The American Medical Association approved of limited federal expenditures for research but, in concert with the majority of industries, believed that the primary state and local responsibilities for pollution controls should be maintained. The Association of State and Territorial Health Officers agreed. All believed the federal role should be in research, but not in strong enforcement.

Over several years the debate among various federal agencies and in the Congress about jurisdiction over enforcement of anti-air-pollution controls had excellent results. Universities, industries, health authorities in cities and states, and scientific societies were spurred into making independent studies about what constituted pollution and countermeasures.

During the height of this debate, Abraham A. Ribicoff was Secretary of Health, Education and Welfare. Ribicoff had been Governor of Connecticut and knew the problems that caused air contamination firsthand. In November, 1962, he was elected to the U.S. Senate from Connecticut. His first act was to introduce "The Clean Air Act" with 19 cosponsors. Senator Ribicoff asked for assignment to the Senate Public Works Committee, which would oversee his legislation, but was offered a seat on the Finance Committee. The Clean Air Act then was managed by Senators Maurine Neuberger (D. Ore.), Clare Engle (D. Calif.), and Thomas Kuchel (R. Calif.). Both S. 432, the Ribicoff measure, and S. 444, sponsored by the Oregon and California senators, were debated through the year.

In the House, Representative Ken Roberts (D. Ala.) introduced H.R. 4415, on Feb. 28, 1962, which was considered by the Committee on Interstate and Foreign Commerce. Roberts was Chairman of the subcommittee on Health and Safety and had been immersed in anti-air-pollution studies for several years. Other bills identical with the Ribicoff bill in the Senate were companion measures with the Roberts proposal.

After extensive hearings, suggestions from many witnesses were incorporated into a new bill, H.R. 6518. With remarkably little opposition, the legislation passed the House on July 24, 1963.

With the adoption of the Clean Air Act in December, 1963, federal policy in the field of air pollution control underwent significant evolution. Although there was no change in the view that responsibility for the control of air pollution rests primarily with state and local governments, the federal government responded to a very real need by equipping itself to aid state and local control programs more effectively and to stimulate them to the increased level of activity considered necessary. Thus, the preamble adds a new dimension to the federal role when it states that "Federal financial assistance and leadership is essential for the development of cooperative Federal, State, regional, and local programs to prevent and control air pollution." The preamble points out that most of the nation's people now live in urban areas, including metropolises which sprawl across municipal, county, and state boundary lines, and it specifically mentions motor vehicles as one of the major contributors to the mounting air pollution problem.

The Congress has instructed the federal government to assume the responsibility for directly aiding in the development of state, regional, and local control programs sufficiently equipped and empowered to reverse the trend toward ever more polluted air. To begin with, the Act continues and expands the authority for the ongoing research, development, and technical assistance programs of the Division of Air Pollution. It places considerable emphasis on the fact that there is still much to be learned in the technical and scientific spheres and that the federal government has a responsibility for seeing that this knowledge is developed. The same is true of the need for additional trained personnel to work in the fields of air pollution research and control.

Among the important new authorities provided by the Clean Air Act is that for program grants. Briefly, these grant funds may be made available to state and local agencies for the purpose of developing, establishing, or improving air pollution control programs. Federal funds will be available on a matching basis—$2 for every $1 for single jurisdictional programs, and $3 for every $1 for programs operating on a regional basis. The objective of this provision is not to participate in the maintenance cost of ongoing air pollution control programs throughout the nation, but rather to stimulate state and local agencies to develop new programs or to expand existing control efforts.

The new Clean Air Act also includes for the first time limited legal regulatory authority on the federal level for abatement of specific air pollution problems. This limited regulatory power is clearly intended to supplement the abatement powers of state and local governments with respect to two types of situations:

First, with respect to an interstate problem in which pollution arising in one state may be endangering the health or welfare of persons in another state, the Secretary of Health, Education and Welfare may, on his own initiative or on official request as specified in the Act, initiate formal proceedings for the abatement of the pollution as found to be necessary.

Second, with respect to a similar air pollution problem, which is purely intrastate in nature, the Secretary may invoke such formal abatement proceedings only on official request from designated officials in the state involved.

These federal regulatory powers are intended to supplement the state and local authorities by (1) providing a means of dealing with interstate problems which are not easy and sometimes are impossible to reach by the remedies available to a single state, and (2) providing technical and other assistance from the federal government in cases with which, although intrastate in character, it is difficult for state or local authorities to deal.

The regulatory abatement procedures authorized in the Act are similar to those in use for several years under the provisions of the Water Pollution Control Act—involving the steps of conference with the cognizant official agencies, public hearing, and finally court action. The procedure may, of course, terminate at the initial or second step of the process if the problem is resolved.

Several other provisions of the Clean Air Act reflect the new and evolving federal air pollution control policy. For example, the Act directs the Department of Health, Education and Welfare to develop and promulgate criteria of air quality for the guidance of state and local authorities in establishing standards for source emissions and ambient air. In addition, specific directives are included to give particular research attention to the removal of sulfur from fuels and to the development of effective and practical devices for controlling air pollution. The Act also calls for the formation of a technical committee on motor vehicle pollution, composed of representatives from the Department of Health, Education and Welfare and the automotive industry, the manufacturers of motor vehicle pollution control devices, and the producers of motor fuels. This committee will review progress toward effective control of vehicular pollution and indicate specific areas in which additional research and development are needed. The Secretary of Health, Education and Welfare is required to report to Congress periodically on this aspect of the air pollution problem and recommend any new legislation that he determines is warranted. Thus, the Congress has initiated a process of almost continuous review of the motor vehicle pollution problem.

Finally, the Act retains the previous directive that federal facilities should, to the fullest extent possible, seek to minimize or eliminate air pollution for which they are responsible. In addition, the new act authorizes the Secretary of Health, Education and Welfare to designate classes of potential pollution—sources for which federal agencies would be directed to obtain permits from him, subject to such conditions as he may prescribe. The Secretary is required to report to the Congress each January on the status of these permits and the compliance with their terms.

The deaths of two senators had a substantial effect on passage of a strong air pollution measure in 1963. Dennis Chavez (D. N.Mex.), Chairman of the Public Works Committee, died on Nov. 18, 1962. Robert Kerr (D. Okla.), second-ranking member of the Committee and a foe of federal control in this field, died on Jan. 1, 1963.

Succeeding to chairmanship of the Public Works Committee was Senator Pat McNamara (D. Mich.), urban-oriented and impressively familiar with the problems of highly industrialized areas. He appointed Senator Edmund Muskie (D. Maine) as chairman of a special subcommittee to handle both water and air pollution. It was established on Apr. 30, 1963. Muskie had sponsored antipollution bills and was regarded as an expert on pollution problems, because of three terms in the Maine legislature and as Governor of his state from 1955 to 1959. Aware of the wealth and attractions of Maine's natural resources, Muskie was as alert to what such resources meant to the country at large. Muskie's Republican colleague on this committee was Senator Caleb Boggs (R. Del.), who had been Governor of his state from 1953 until 1961. This special committee achieved an impressive record of bipartisan cooperation.

The number of expert witnesses who appeared before the committees in both House and Senate gave valuable insights into the air pollution situation, offering the results of newer research with improved methods and techniques. Previous opposition to federal controls either had disappeared or was greatly diminished, although there were some diehards who simply did not want any controls.

The Air Pollution Control Act of 1955 can be regarded rightly as the first action by the federal government to recognize the problem and to become engaged in seeking remedies for it. It reflected the general thinking in the Congress at the time that, given the tools of research and technological assistance, plus federal incentive grants, the states and local governments could do the job that demanded to be done.

After 12 years of close surveillance by the Congress, HEW, state, and municipal health authorities and research by academic and scientific institutions, one salient fact emerged: the job was bigger than anyone knew. Air pollution was increasing at jet speed while remedial measures were proceeding at mule-and-cart pace. Science showed that all efforts were doing little to slow air contamination. It had become a threat not only to the United States but to the earth.

After 18 days of hearings in five regional centers, the Air Quality Act of 1967, S. 780, sponsored by Senator Muskie, with bipartisan support, was sent to the Senate on July 18, 1967. The following description of the Air Quality Act was given by Senator Muskie to the Senate, before debate:

> Mr. President, the Senate has demonstrated its recognition of air pollution as a serious national problem. Beginning with the Clean Air Act of 1963 the Senate has given unanimous approval to legislation designed to expand Federal support for the battle to preserve the quality of our air resources.
>
> We realize there are no panaceas, no overnight cures for the complex problems of air pollution, but we are pledged to protect the health and welfare of every citizen of this Nation whether that person be healthy, or suffering from a bronchial disorder.
>
> Mr. President, there is an abundance of compelling evidence to indicate that air pollution is a hazard to health. There is more compelling evidence to indicate that the public welfare is adversely affected by indiscriminate pumping of waste into the air. We know this as individuals who have experienced discomfort from foul odors, had our eyes burn from smog or looked at the color of a white shirt after a day in any of our industrial cities.
>
> At the same time popular concern for air pollution control has risen dramatically as the result of increased leisuretime, greater publicity, increased awareness of health problems and a variety of other reasons. There is a demand for action, and all the evidence received by the Public Works Committee this year in 18 days of hearings, in consultations and in research supports that demand.
>
> Mr. President, S. 780, as amended by the committee, is complex, as are the problems of environmental control. The problem of air pollution is neither local nor temporary. It is a universal problem, and, so long as our standard of living continues to increase, it will be a permanent threat to human well-being.

HISTORY OF THE AIR QUALITY ACT OF 1967

The committee's recommendations provide far-reaching opportunities for a comprehensive, broad-based attack on the Nation's air pollution problem while expanding the potential of control technology and identifying the health and welfare effects of air pollution. The objective of S. 780 is the enhancement of air quality and the reduction of harmful emissions consistent with maximum utilization of an expanding capacity to deal with them effectively. At the same time, it provides authority to abate any pollution source which is an imminent danger to health, by whatever means necessary.

The Air Quality Act of 1967, therefore, serves notice that no one has the right to use the atmosphere as a garbage dump and that there will be no haven for polluters anywhere in the country.

The committee believes that, to date, public and private efforts to accomplish air quality objectives have been inadequate. Research has been insufficient, with little significant development of new and improved methods for controlling or eliminating air pollution. As each day passes there is a greater urgency for closer cooperation between government and industry in an effort to make substantial inroads on air pollution control and abatement.

I would like to point out, for clarification, that the bill as reported by the committee is the entire Clean Air Act as amended by this year's action. Because the amendments are complex and because they recast the force effect of the Federal air pollution control effort the committee thought it wise to present to the Senate a complete act containing both the amendments and previously adopted language which was not changed. The Cordon print beginning on page 63 of the report indicates the changes in existing law.

In order to facilitate the objective of a national abatement program which will enhance the quality of our Nation's air, the proposed amendments to the Clean Air Act provide the Secretary of Health, Education, and Welfare with the following authority:

First. To request an immediate injunction to abate the emission of contaminants which present "an imminent and substantial endangerment to the health of persons," anywhere in the country.

Second. To designate "air quality control regions" for the purpose of implementing air quality standards, whenever and wherever he deems it necessary to protect the public health and welfare.

Third. In the absence of effective State action in accordance with the provisions of the act, to establish ambient air quality standards for such regions.

Fourth. In the absence of effective State action in accordance with the provisions of the act, to enforce such standards.

Fifth. In the absence of action by the affected States, to establish Federal interstate air quality planning commissions.

It should be emphasized that it is the intent of the committee to enhance air quality and to reduce harmful pollution emissions anywhere in the country, and to give the Secretary authority to implement that objective in the absence of effective State and local control. It is believed that the Air Quality Act of 1967 carries out that intent.

The committee recognizes the potential economic impact and therefore economic risk, associated with major social legislative measures of this type. But this risk was assumed when the Congress enacted social security, fair labor standards, and a host of other legislation designed to protect the public welfare. Such a risk must again be assumed if the Nation's air resources are to be conserved, and enhanced to the point that generations yet to come will be able to breathe without fear of impairment of health.

S. 780 is a logical expansion of the Clean Air Act of 1963 as amended. In the basic act the Congress provided for development by the Public Health Service of "air quality criteria" to identify the effects of pollutants on health and welfare. To date, one such set of criteria relating to oxides of sulfur has been issued and the committee understands that criteria on several other contaminants including carbon monoxide, particulates, and oxidants will be released within the next 6 months.

The committee recognizes this activity and has encouraged its expansion under S. 780. In addition to continuing the analysis of the quantitative and qualitative effects of air pollution under strengthened and refined procedures of evaluation, the committee bill authorizes two new areas of activity. A first and important step will be the identification of those areas of the Nation which have significant air pollution problems. The designation by the Secretary of these problem areas will trigger the setting of air quality standards related to those pollutants for which criteria have been developed.

In concert with expanded criteria development will be the compilation of information of methods of pollution control, which will be a publication of the technology and economically feasible methods of control of pollutants subject to criteria. This information will be designed to assist the States in carrying out their responsibility to control air pollution within their respective boundaries.

These three steps, designation of air quality control regions, criteria development, and publication of control technology information are the tools for development of air quality standards and an early warning system to those industries and others in problem areas who will be required to control their emissions.

The committee does not suggest that, with one fell swoop, the air pollution problems of the Nation will be solved. Development of the above-mentioned information will require a time equivalent to that considered in the initial administration proposal for uniform national emission standards. And after the information is available and the States adopt standards, there will be the necessary time to achieve desired emissions control.

But we will have a national program of air quality. The States will retain the primary responsibility to determine the quality of air they desire. In no event, however, will the Federal Government approve any air quality standard or plan for implementation of that standard which does not provide for protection of the public health and welfare of all citizens within the air quality region.

Moreover, in the absence of concertive State action the bill gives the Secretary of Health, Education, and Welfare the authority to set standards in such regions.

In addition, new industries, wherever they locate, will know that control is inevitable and plan for it. The fact that an area is not now a problem area will not mean that controls will never be required. When the air quality of any region deteriorates below the level required to protect public health and welfare, the Secretary is required to designate that region for the establishment of air quality standards, enforceable by the Federal Government if the States fail to act. It should be pointed out in this connection that the Public Health Service has expressed the view that every urban area of 50,000 or more population now has an air pollution problem. There are also population areas under that size which clearly have problems related to particular pollution sources and conditions.

Considerable attention was given, in the hearings and also in informal conferences and executive sessions, to the concept of national emission standards. Such standards were urged by the administration first, as a means of eliminating the economic disadvantage of complying with air pollution controls as a local requirement and the temptation for industry to leave or avoid areas where such controls are presently necessary; and, second, on the ground that some industries, by their nature, are a danger to health and welfare wherever they are located.

In the judgment of the committee, these arguments were offset by the following considerations:

First. The administration itself did not propose uniform national emission standards but rather minimal national standards. Clearly, therefore, there would be local variations which would not eliminate economic disadvantages. Dr. John T. Middleton, Director, National Center for Air Pollution Control, Health, Education, and Welfare, said:

"Our intention is to get minimum national standards to help insure that no single pollution source would, in itself, be a threat to public health and welfare. These standards would be based on scientific criteria of the effects of air pollutants on man, animals, vegetation, and the air resources itself. The criteria we would use would be those which we are authorized to publish under the provisions of the Clean Air Act."

He said later:

"The setting of such standards at the Federal level would not relieve States and communities of the responsibility of insuring that pollution sources located within their jurisdictions are controlled to the full extent necessary. States could adopt emission standards more stringent than those set at the Federal level."

Second. Administration witnesses testified that PHS has made no findings with respect to industries which, in and of themselves, constitute a danger to public health and welfare.

Third. Under the bill approved by the committee, the Secretary's authority has been extended so that he can deal effectively with any situation which, by its nature, is a danger to health and welfare, in any location.

Fourth. National emission standards would eliminate some control options—relocation of pollution sources, fuel substitutes, and so forth—which may be essential in serious problem areas in the absence of effective technology.

Fifth. Wise use of capital resources dictates that the first priority for the pollution control dollar is in those areas where the problem is most critical. National emission standards would give equal priority to critical areas and areas where no problem presently exists.

Sixth. The program authorized in the committee bill will lead to control of the industries described on a national basis, with the kind of local variations envisioned by administration witnesses.

The committee does recognize the need for national action on sources of pollution which move in interstate commerce. For the purpose of further consideration of national emission standards, for moving and stationary sources, the committee bill directs the Secretary to undertake a 2-year study of the concept and the full range of its implications.

Other areas to which the committee gave serious attention were the questions of incentive assistance to industry and alternatives to internal combustion. These matters are discussed later in the report under "Pending Questions."

Summary of Provisions of S. 780

S. 780, as ordered reported, includes the following provisions:

First. Authority for the Secretary to go immediately to court in the event that he finds a particular pollution source or combination of sources, wherever such source or sources may be located, is presenting an "imminent and substantial endangerment to the health of persons" to seek an injunction against the emission of such contaminants as may be necessary to protect public health.

Second. Provision for establishment of the Federal interstate air quality planning agencies if the States do not request designation of a planning agency for an interstate air quality control region.

Third. Provision for the Secretary of Health, Education, and Welfare to set ambient air quality standards in any designated air quality control region, if the States fail within 15 months after receiving a criteria and recommended control techniques, to adopt such standards and an acceptable plan for implementation.

Fourth. Provision for the Secretary to go to court, after 180 days notice, to enforce any violation of standards in any designated air quality control region.

Fifth. Specific directive to the Secretary to continue to use existing enforcement procedures as may be necessary to protect public health or welfare during the standards development period; and provision for participation by interested parties in an abatement conference.

Sixth. A three-step approach to development of air quality standards including (1) designation, by the Secretary of Health, Education, and Welfare, of air quality control regions based on the need for pollution control and protection of health and welfare; (2) expansion of the existing provision for development and issuance of criteria as to the health and welfare effects of pollutants or combinations of pollutants; and (3) publication of information on the control technology required to achieve various levels of air quality.

Seventh. An expanded research and demonstration program to advance the technology for controlling pollution from fuels and vehicles, including specific authorization of $375 million for 3 years—through 1970.

Eighth. Federal preemption of the right to set standards on automobile exhaust emissions, with waiver of application of preemption to any State (California) which had adopted standards precedent to promulgation of Federal standards.

Ninth. Expanded State and local program grants provision to encourage comprehensive planning for air quality standards.

Tenth. Establishment of a statutory President's Air Quality Advisory Board and such other advisory committees as may be necessary to assist the Secretary in performing the functions authorized.

Eleventh. A study of the concept of national emission standards. including an analysis of the health benefits to be derived, as well as the economic impact and costs.

Twelfth. Federal assistance to the States to develop motor vehicle emission and device inspection and testing systems.

Thirteenth. Federal registration of fuel additives.

Fourteenth. Comprehensive cost analyses of the economic effect on the Nation, industries, and communities of air pollution control, and a report thereon to Congress and the President.

Fifteenth. Comprehensive reports to the Congress.

Sixteenth. Three-year authorization of $325 million for programs other than research on control of pollution from fuels and vehicles—total authorization, including research, $700 million.

Improvement of man's environment centers on the enhancement of the quality of human life. This was appropriately stated by Dr. William H. Stewart, Surgeon General of the United States:

"Thanks to many advances in protecting people against disease, we are able in the health professions to think about the positive face of health—the quality of individual living. The healthy man or woman is not merely free of specific disability and safe from specific hazard. Being healthy is not just being unsick. Good health implies, to me, the full and enthusiastic use by the individual of his powers of self-fulfillment.

"Therefore in controlling air pollution for the benefit of health we are working toward an environment that is not only safe but conducive to good living. I know that you and the members of this committee share this aspiration."

The Air Quality Act of 1967 was signed into law by President Johnson in November of that year. Its implementation has been steadily making inroads on air pollution. Again the sequence established by law is one of criteria—standards—enforcement.

The Secretary of Health, Education and Welfare has established air quality regions—that is, natural geographical areas which share a common air mass and where airflows between emission sources and affected populations are consistent. This is a key provision because it unites arbitrary political jurisdictions in abatement activities.

Simultaneously, criteria have been issued by the federal government for sulfur oxides and particulate matter. These are two major pollutants encountered in most metropolitan areas. The National Air Pollution Control Administration has also issued compilations of abatement technology for each pollutant. Criteria and technology documents are being prepared for a number of other individual contaminants such as carbon monoxide, oxidants, lead, nitrogen oxides, and fluorides.

Armed with this authoritative information, the states are now required to develop ambient air quality standards and plans for their implementation. Where an air quality region involves more than one state, the standards and plans for each state must be compatible. The Secretary of Health, Education and Welfare has the authority to approve standards as being consistent with the aims of the federal law.

Automobile pollution control is a special issue because of the mobility of this emisson source. Following 1965 amendments, standards have been promulgated on a nationwide basis for hydrocarbon and carbon monoxide emissions. Each year the standards are reviewed and tightened when necessary to meet ambient air quality goals, and as improved automotive technology becomes available. Federal research funds are going into programs to develop alternative power sources such as steam or electricity for cars.

Several heavily polluted areas have not waited for the full sequence of federal law steps to begin abatement. California, of course, has been a leader in air pollution control because of the intense problem in its coastal population centers. New York, Chicago, Pittsburgh, and St. Louis are other cities which have produced noticeable improvements in air quality despite increases in industrialization and population. These and many other municipalities have benefited from the provisions of the federal Act which makes grants to initiate and sustain local control programs. Personnel may be trained, equipment acquired, and monitoring measurements made with the matching funds which reaffirm the policy that control is a local responsibility.

Other provisions of the 1967 Act are described in this language from a Department of Health, Education and Welfare analysis:

> The Air Quality Act of 1967 is a blueprint for a systematic effort to deal with air pollution problems on a regional basis. It calls for coordinated action at all levels of government and among all segments of industry.
>
> The system which this new legislation develops hinges on the designation of regions where two or more communities—either in the same or different States—share a common air pollution problem and on the development and implementation of air quality standards for such regions.
>
> The Department of Health, Education, and Welfare will define the broad atmospheric areas of the Nation and will designate specific air quality control regions.
>
> The Department will develop and publish air pollution criteria indicating the extent to which air pollution is harmful to health and damaging to property and detailed information on the cost and effectiveness of techniques for preventing and controlling air pollution.
>
> As soon as air quality criteria and data on control technology are made available for a pollutant or class of pollutants, States will be expected to begin developing air quality standards and plans for implementation of the standards. They will have 90 days to submit a letter indicating that they intend to set standards, 180 days to set the standards, and 180 days to develop plans for implementing them.
>
> Air quality standards will be developed and applied on a regional basis. Wherever an air quality control region includes parts of two or more States, each State will be expected to develop standards for its portion of the region.
>
> If the Secretary of Health, Education, and Welfare finds that the air quality standards and plans for implementation of the standards in an air quality control region are consistent with the provisions of the Air Quality Act, then those standards and plans will take effect.
>
> If a State fails to establish standards, or if the Secretary finds that the standards are not consistent with the Act, he can initiate action to insure that appropriate standards are set. States may request a hearing on any standards developed by the Secretary; the hearing board's decision will be binding.
>
> States will be expected to assume the primary responsibility for application of the air quality standards. If a State's efforts prove inadequate, the Secretary is empowered to initiate abatement action.
>
> OTHER PROVISIONS
>
> The Air Quality Act of 1967 also provides for: Expansion of the Federal Government's air pollution research and development activities.
>
> Continuation of grants to States and communities to assist them in their efforts.
>
> Financial aid for planning activities in interstate air quality control regions.
>
> Retention of authority for Federal action to abate interstate air pollution problems, and, on request from States, intrastate problems.
>
> Action by the Secretary to obtain court orders to curtail pollution during emergencies.
>
> Continuation of Federal standard-setting to control motor vehicle pollution.
>
> Awarding of grants to States to assist them in developing programs for inspection of motor vehicle pollution control systems.
>
> Continued efforts to control pollution at Federal installations.
>
> Creation of a 15-member Presidential Air Quality Advisory Board.
>
> Establishment of advisory groups to assist the Department of Health, Education, and Welfare.
>
> Registration of fuel additives.
>
> A study of the need for and effect of national emission standards for stationary sources of air pollution.

A study of the feasibility of controlling pollution from jet and conventional aircraft.

Comprehensive economic studies of the cost of controlling air pollution.

An investigation of manpower and training needs in the air pollution field.

CONCLUSION

The Congress has led the way in applying a national policy to restore and maintain the quality of our environment. In the past decade a series of laws have been enacted and refined which blend the powers and resources of the federal government with the responsibilities and prerogatives of local government and the free enterprise economy. The steps to clean air and water are a logical, science-based sequence in which the needs of a progressive, highly technical society can be balanced against the long-term preservation of environmental quality.

The United States with its wealth of resources and technical skills has refused to sacrifice the values of air, water, and landscape in the name of economic gain. Our laws depend upon the will power and the purse power of the electorate to make them work. Although the job will never be finished, the encouragement of pollution abatement successes just beginning to be visible is a clear sign that the legislation is proving to be effective. We will continue to improve it as the practical results of implementation become available.

APPENDIX

In its deliberations and writing of air pollution legislation the Congress uses the following definitions as standard, although others may be added as new techniques and technologies evolve.

Sources and Control of Emissions

I. Automotive
- A. Emissions
  1. Automobile: hydrocarbons, carbon monoxide, and oxides of nitrogen
  2. Diesels: smoke, odor, oxides of nitrogen and benzopyrene
- B. Controls
  1. Automobiles: blowby, exhaust control devices, and engine modification
  2. Diesels: better operational control, improved fuel, and possibly exhaust control device

II. Stationary
- A. Dust, smoke, and mist
  1. Emissions: fly ash, soot, smoke, iron oxides, particles suspended in moisture, particles suspended in gaseous substances, etc.
  2. Controls: settling chambers, separators, packed beds, collectors (such as baghouses), scrubbers, precipitators, and air filters
- B. Gas and vapor
  1. Emissions: sulfur dioxide, oxides of nitrogen, fluorides, hydrocarbons, and hydrogen sulfide
  2. Controls: stacks (for dispersion), absorbers or scrubbers, incinerators, catalytic combustion, and absorption
- C. Odor
  1. Emissions: general offensive odors from chemical plants, pulp and papermills, stockyards, slaughterhouses, etc.
  2. Controls: dispersal or dilution, combustion, absorption, and modification

Prior Legislation for Air Pollution Control

The prior authority of the Department of Health, Education and Welfare with respect to air pollution is derived primarily from the Air Pollution Control Act, Public Law 159, 84th Congress, approved July 14, 1955, as amended.

This act authorized a program of research and technical assistance to obtain data and to devise and develop methods for control and abatement of air pollution by the Secretary of Health, Education and Welfare and the Surgeon General of the Public Health Service. The act recognized the primary responsibilities and rights

Source and Effect of Pollutants

Pollutant	Major sources	Principal effects
Sulfur dioxide	Fuel combustion (coal, oil, cellulosic material), industrial processes	Sensory and respiratory irritation, plant damage, corrosion, possible adverse effects on health
Oxidants	Atmospheric photochemical reactions involving nitrogen oxides, organic gases, vapors, and solar radiation	Sensory and respiratory irritation, plant damage. Provides, indirectly, an index of visibility reduction due to photochemical aerosols. Possible adverse effects on health
Carbon monoxide	Gasoline-powered vehicles, fuel combustion, industrial processes	Reduction in the oxygen-carrying capacity of blood
Total gaseous hydrocarbons	Fuel combustion, industrial processes, evaporation of hydrocarbons	Visibility reduction, plant damage, and sensory irritation are effects produced in photochemical reactions involving reactive hydrocarbons; ethylene itself causes plant damage
Nitrogen oxides (nitric oxide and nitrogenic dioxide)	Fuel combustion, industrial processes	Visibility reduction, plant damage, and sensory irritation are produced in photochemical reactions involving nitrogen oxides; these gases may also cause adverse health effects, and nitrogen dioxide can cause decreased visibility
Total aliphatic aldehydes, formaldehydes, and acrolein	Fuel combustion, incineration of wastes, atmospheric photochemical reactions	Sensory irritation, plant damage, visibility reduction, and possible adverse effects on health
Carbon dioxide	Combustion processes	Used as an index of pollution from combustion operations
Suspended particulate matter	Combustion, and industrial and natural processes	Visibility reduction, soiling
Hydrogen sulfide	Coke, distillation of tar, petroleum and natural gas refining, manufacture of viscose rayon, and in certain chemical processes	Odor nuisances, caused deaths in Poza Rica, Mexico, when large quantity escaped from units of a natural gas refining plant
Hydrogen fluoride	Heating to high temperatures of ores, clays, or fluxes containing fluorine. Generally from steel mills, ceramic works, aluminum reduction plants and superphosphate factories	Damage to citrus and certain other agriculture plants, flowers; affects teeth and bones of cattle when forage crops have been consumed
Lead	Internal-combustion engines, industrial emissions, open burning of lead paint coated wood	Lead poisoning

of the states and local governments in controlling air pollution, but authorized federal grants in aid to air pollution control agencies to assist them in the formulation and execution of their research programs directed toward abatement of air pollution.

Under the provisions of the act, the Surgeon General was authorized to prepare or recommend research programs, encourage cooperative activities by state and local governments; conduct studies and research and make recommendations with respect to any specific problems of air pollution, if requested; conduct research and make grants for research, training, and demonstration projects; and make available to all agencies the results of surveys, studies, investigations, research, and experiments relating to air pollution and abatement.

Public Law 86–493, approved June 8, 1960, directed the Surgeon General of the Public Health Service to conduct a thorough study of motor vehicle exhaust as it affects human health through the pollution of air. A report on this study was published as House Document 489. In 1962, the Air Pollution Control Act was amended by Public Law 87–761 so as to make permanent the requirement that the Surgeon General conduct studies relating to motor vehicle exhaust. The act was further amended so as to authorize appropriations to carry out the act until June 30, 1966.

Although the Air Pollution Control Act, as amended, constituted the basic authority for the Department's activities in the field of air pollution, sections 301 and 311 of the Public Health Service Act have also been utilized as a basis for appropriations to support these activities. Section 301 is the basic section of the Public Health Service Act with respect to the Surgeon General's authority relative to research, research training, and related functions; section 311 is the basic section authorizing federal-state cooperation and technical assistance. In addition, section 314(c) of the Public Health Service Act authorizes grants to states, counties, etc., to assist in establishing and maintaining adequate public health services, including grants for demonstrations and for training of personnel for state and local health work.

The program that developed under authority of the 1955 air pollution legislation was primarily focused on research and technical assistance. In shaping this program, it was felt that effective control would depend upon greatly increased knowledge of the types and amounts of pollutants being discharged to the atmosphere; better understanding of the meteorological and climatological factors that influence the dispersion of pollutants in the atmosphere; more sophisticated knowledge of the physical and biological effects of pollutants, especially in the relatively low concentrations in which they are usually encountered in community air; a fuller awareness of the importance of certain specific air pollution problems, such as the motor vehicle; and improved information on the administrative, legal, social, and economic factors involved in the control of air pollution.

The technical assistance aspect of the program was primarily centered on efforts to define and characterize the air pollution problems existing in various cities and states, and some interstate metropolitan areas. The goals of this technical assistance activity was to help in the establishment or strengthening of State or local control programs by helping to identify and clarify the air pollution problems in certain areas and to plan effective programs to achieve better control of these problems.

Two other major areas in which the Federal program has been active relate to (1) the training of technical personnel, and (2) the dissemination of information. Training activities were undertaken in recognition of the fact that there are not enough trained personnel to staff the control programs that are needed now in cities and States throughout the country.

Providing authoritative and comprehensive information about air pollution to the many official groups, professional organizations, and other segments of the population who have a direct interest in the problem has been a major element of the Federal air pollution program since its inception.

The information accumulated in the years 1955–63, concerning the magnitude of the air pollution problem and the general inadequacy of State and local control programs, contributed to the recent reshaping of Federal policy in this field. The committee became convinced that "control programs must be accelerated" and that "the nationwide character of the air pollution problem requires an adequate Federal program to lend assistance, support, and stimulus to State and community programs."

The Clean Air Act of 1963

With the adoption of the Clean Air Act in December, 1963, Federal policy in the field of air pollution control underwent significant evolution. Although there was no change in the view that responsibility for the control of air pollution rests primarily with State and local governments, the federal government responded to a very real need by equipping itself to aid State and local control programs more effectively and to stimulate them to the increased level of activity considered necessary. Thus, the preamble adds a new dimension to the federal role when it states that "Federal financial assistance and leadership is essential for the development of cooperative Federal, State, regional, and local programs to prevent and control air pollution." The preamble points out that most of the nation's people now live in urban areas, including metropolises which sprawl across municipal, county, and state boundary lines, and it specifically mentions motor vehicles as one of the major contributors to the mounting air pollution problem.

The Congress has instructed the federal government to assume the responsibility for directly aiding in the development of state, regional, and local control programs sufficiently equipped and empowered to reverse the trend toward ever more polluted air. To begin with, the act continues and expands the authority for the ongoing research, development, and technical assistance programs of the Division of Air Pollution. It places considerable emphasis on the fact that there is still much to be learned in the technical and scientific spheres and that the federal government has a responsibility for seeing that this knowledge is developed. The same is true of the need for additional trained personnel to work in the fields of air pollution research and control.

Among the important new authorities provided by the Clean Air Act is that for program grants. Briefly, these grant funds may be made available to State and local agencies for the purpose of developing, establishing, or improving air pollution control programs. Federal funds will be available on a matching basis—$2 for every $1 for single jurisdictional programs, and $3 for every $1 for programs operating on a regional basis. The objective of this provison is not to participate in the maintenance cost of ongoing air pollution control programs throughout the nation, but rather to stimulate state and local agencies to develop new programs or to expand existing control efforts.

The new Clean Air Act also includes for the first time a limited legal regulatory authority on the federal level for abatement of specific air pollution problems. This limted regulatory power is clearly intended to supplement the abatement powers of state and local governments with respect to two types of situations:

First, with respect to an interstate problem in which pollution arising in one state may be endangering the health or welfare of persons in another State, the Secretary of Health, Education and Welfare, may, on his own initiative or on official request as specified in the act, initiate formal proceedings for the abatement of the pollution as found to be necessary.

Second, with respect to a similar air pollution problem, but which is purely intrastate in nature, the Secretary may invoke such formal abatement proceedings only on official request from designated officials in the state involved.

These federal regulatory powers are intended to supplement the state and local authorities by (1) providing a means of dealing with interstate problems which are not easy and sometimes are impossible to reach by the remedies available to a single state, and (2) providing technical and other assistance from the federal government in cases with which, although intrastate in character, it is difficult for state or local authorities to deal.

The regulatory abatement procedures authorized in the act are very similar to those in use for several years under the provisions of the Water Pollution Control Act—involving the steps of conference with the cognizant official agencies, public hearing, and finally court action. The procedure may, of course, terminate at the initial or second step of the process if the problem is resolved.

Several other provisions of the Clean Air Act reflect the new and evolving federal air pollution control policy. For example, the Act directs the Department of Health, Education and Welfare to develop and promulgate criteria of air quality for the guidance of state and local authorities in establishing standards for source emissions and ambient air. In addition, specific directives are included to give particular research attention to the removal of sulfur from fuels and to the development of effective and practical devices for controlling air pollution. The Act also calls for the formation of a technical committee on motor vehicle pollution, composed of representatives from the Department of Health, Education and Welfare and

TABLE 1 Summary of the Cost of Carrying Out Provisions of the Federal Water Pollution Control Act of 1966

(In thousands of dollars)

Section	1970[a]	1971	1972	1973	1974	Total 1970–1974
1. Total	305,972	[b]	[b]	[b]	[b]	[b]
Construction grants and construction grants administration	217,050	[b]	[b]	[b]	[b]	[b]
Other costs	88,922	181,907	191,403	191,108	183,913	837,253
3. Comprehensive programs for water pollution control:						
(a) Comprehensive pollution control basin studies	7,935	7,750	8,200	8,800	9,100	47,785
(b) Regulation of streamflow studies		1,400	1,500	1,500	1,600	
(c) Grants to nonfederal water pollution control agencies	2,000	4,000	5,000	6,000	6,000	23,000
Grants administration	[c]	275	275	275	275	1,100[d]
4. (a and b) Uniform state laws and interstate cooperation	[e]	165	165	165	165	660[d]
5. Conduct and coordinate research and training:						
(a) Conduct and coordinate research:[f]						
1. Collect research material						
2. Grants to institutions and individuals for research, training, and demonstration projects	25,494	47,000	51,000	48,000	42,000	213,494
3. Advice and assistance to consultants						
4. Grant research fellowships (training)	4,580	8,500	9,000	11,000	15,000	48,080
5. Manpower development and training	1,264	2,650	2,600	2,400	2,500	11,414
(b) Technical assistance to state and interstate agencies	3,750	7,500	9,500	11,000	12,000	43,750
(c) Collect and disseminate basin data on water quality	2,600	7,000	11,000	14,000	14,000	48,600
(d) Conduct research and demonstrations:[f]						
1. Municipal waste treatment and water reuse						
2. Identification and measurements of effects of pollution						
3. Augmented streamflows to control pollution						
(e) Field laboratories, establish, equip, and maintain[g]	....	14,000	5,000	1,500	1,500	22,000
(f) Great Lakes—research and technical development[h]						
(g) Estuarine pollution	300[i]	....	....	....	....	300[i]
6. Grants for research and development:						
(a) To states and municipalities for:						
1. Controlling storm water discharges						
2. Demonstrating waste treatment and water purification methods and joint treatment systems	20,000	58,000	64,000	62,000	55,000	259,000
(b) To individuals for prevention of industry pollution						
7. Grants for state and interstate water pollution control programs	10,000	10,000	10,000	10,000	10,000	50,000
Grants administration	[c]	530	530	530	530	2,120[d]
8. Grants for construction of treatment works	214,000	[b]	[b]	[b]	[b]	[b]
Grants administration	3,050	[b]	[b]	[b]	[b]	[b]
9. Water Pollution Control Advisory Board[j]						
10. Enforcement and water quality standards:						
(c) Maintenance of water quality standards	752	762	768	773	778	3,833
(d) Enforcement conferences						
(e) Remedial actions	3,700	5,075	5,075	5,075	5,075	24,000
(f) Public hearings						
11. Federal interagency cooperation to control pollution	900	1,250	1,400	1,400	1,400	6,350
12. Administration (executive direction and support)	5,217	5,600	5,900	6,200	6,500	29,417
16. (a) Cost studies	430	450	490	490	490	2,350

SOURCE: *Water Pollution Control 1970–1974: The Federal Costs*, U.S. Department of Interior, FWPCA, January, 1969.

NOTE: Sections 1, 2, 13, 14, and 15 do not require authorizations. Sections 16(b), 17, and 18 have been carried out in previous fiscal years.

[a] Derived from the President's budget.

[b] Total costs and cost of carrying out Section 8 are not shown for fiscal years 1971–1974 because of proposed legislation affecting Section 8.

[c] Grants administration costs for fiscal year 1970 are included with Section 3 costs.

[d] Four-year total fiscal years 1971–1974.

[e] Included in Section 10(d), (e), and (f).

[f] Costs shown for Section 5(a) are the combined costs of carrying out Sections 5(a) and 5(d). A portion of these costs also assist in carrying out activities authorized by Subsections 5(b), (c), (e), and (f).

[g] Includes only planning, construction, equipment, and repair and improvement costs.

[h] Included in Sections 3, 5(b), (c), (d), and 10(c).

[i] Includes $120,000 for the National Estuarine Pollution Study to be completed in fiscal year 1970. Remaining $180,000 is for cost of activities related to the marine environment. Fiscal years 1971–1974 marine environment costs are included with Section 3.

[j] Included in Section 12.

the automotive industry, the manufacturers of motor vehicle pollution control devices, and the producers of motor fuels. This committee will review progress toward effective control of vehicular pollution and indicate specific areas in which additional research and development are needed. The Secretary of Health, Education and Welfare is required to report to Congress periodically on this aspect of the air pollution problem and recommend any new legislation that he determines is warranted. Thus, the Congress has initiated a process of almost continuous review of the motor vehicle pollution problem.

Finally, the Act retains the previous directive that federal facilities should, to the fullest extent possible, seek to minimize or eliminate air pollution for which they are responsible. In addition, the new act authorizes the Secretary of Health, Education and Welfare to designate classes of potential pollution—sources for which federal agencies would be directed to obtain permits from him, subject to such conditions as he may prescribe. The Secretary is required to report to the Congress each January on the status of these permits and the compliance with their terms.

TABLE 2 Federal Grants to State and Interstate Water Pollution Control Programs and State and Interstate Expenditures

Fiscal year	Federal grants	State agency expenditures	Interstate agency expenditures	Total grants and expenditures
1957	$ 1,800,000	$ 4,004,501	$181,132	$ 5,985,633
1958	2,700,000	*	*	*
1959	2,700,000	6,514,980	370,628	9,585,608
1960	2,700,000	6,755,822	345,232	9,801,054
1961	2,700,000	7,606,088	717,801	11,023,889
1962	4,500,000	8,162,954	835,840	13,498,794
1963	5,000,000	9,277,135	636,021	14,913,156
1964	5,000,000	9,530,490	615,542	15,146,036
1965	5,000,000	11,204,986	681,778	16,886,764
1966	5,000,000	12,271,400	702,103	17,973,503
1967	5,000,000	17,642,924	506,333	23,149,257
1968	10,000,000	†	†	†

SOURCE: *Water Pollution Control 1970–1974: The Federal Costs,* U.S. Department of Interior, FWPCA, January, 1969.

* Not available.

† Data have not been received from all state and interstate agencies.

Sewage Treatment Needs

A recent survey by the Conference of State Sanitary Engineers disclosed that the backlog of municipal waste treatment demand now involves about 5,640 communities, with total populations of 37.4 million. Constructing new, enlarged, and improved sewage treatment plants and their ancillary facilties to meet these needs will cost an estimated $2.6 billion. Included in the survey of communities are 2,661 with populations totaling 6.1 million who need sewage collection systems, as well as sewage treatment plants. Many other communities need extensions of sewage interceptors to provide treatment for new developments in their jurisdictions.

The chart illustrates the approximate needs now.

Sewage Treatment Needs of the 100 Largest Cities in the United States—Present Needs and Future Needs through 1972[a]

State[b]	Present needs			Additional needs through 1972		
	Estimated number of projects	Estimated total cost	Federal share, 30%	Estimated number of projects	Estimated total cost	Federal share, 30%
Alabama:						
Birmingham	4	$ 28,000,000	$ 8,400,000	4	$ 32,000,000	$ 9,600,000
Mobile	1	2,500,000	750,000	1	3,000,000	900,000
Montgomery	3	11,000,000	3,300,000	3	12,500,000	3,750,000
Total, Alabama	....	$ 41,500,000	$ 12,450,000	....	$ 47,500,000	$ 14,250,000
Arizona:						
Phoenix	3	$ 1,750,000	$ 525,000	10	$ 15,000,000	$ 4,500,000
Tucson	2	3,500,000	1,050,000	4	5,000,000	1,500,000
Total, Arizona	....	$ 5,250,000	$ 1,575,000	....	$ 20,000,000	$ 6,000,000
California:						
Los Angeles	2	$ 9,500,000	$ 2,850,000	21	$ 75,000,000	$ 22,500,000
San Francisco	8	2,500,000	750,000	25	15,600,000	4,680,000
San Diego	3	200,000	60,000	6	17,650,000	5,295,000
Oakland	None	....	....	1	20,000,000	6,000,000
Long Beach	6	990,000	297,000	12	4,200,000	1,260,000
San Jose	3	2,580,000	774,000	1	12,000,000	3,600,000
Sacramento	None	....	....	1	10,000,000	3,000,000
Fresno	1	290,000	87,000	1	750,000	225,000
Total, California	....	$ 16,060,000	$ 4,818,000	....	$ 155,200,000	$ 46,560,000
Colorado, Denver	5	$ 35,667,000	$ 10,700,100	4	$ 8,500,000	$ 2,550,000
Connecticut:						
Hartford	2	$ 2,500,000	$ 750,000	23	$ 15,800,000	$ 4,740,000
Bridgeport	1	7,000,000	2,100,000	1	3,700,000	1,110,000
New Haven	1	1,800,000	540,000	2	3,600,000	1,080,000
Total, Connecticut	....	$ 11,300,000	$ 3,390,000	....	$ 23,100,000	$ 6,930,000
District of Columbia, Washington	11	$ 14,104,000	$ 4,231,200	13	$ 20,277,000	$ 6,083,100
Florida:						
Miami	c	c	c	c	c	c
Tampa	c	c	c	c	c	c
Jacksonville	c	c	c	c	c	c
St. Petersburg	4	$ 6,300,000	$ 1,890,000	None		
Total, Florida	....	$ 6,300,000	$ 1,890,000			
Georgia:						
Atlanta	6	$ 70,000,000	$ 21,000,000	6	$ 100,000,000	$ 30,000,000
Savannah	3	10,000,000	3,000,000	6	12,000,000	3,600,000
Total, Georgia	....	$ 80,000,000	$ 24,000,000	....	$ 112,000,000	$ 33,600,000
Hawaii, Honolulu	4	$ 3,050,000	$ 915,000	14	$ 12,150,000	$ 3,645,000
Illinois:						
Chicago	50	$ 196,585,400	$ 58,975,620	69	$ 375,989,900	$112,796,970
Rockford	1	7,031,000	2,109,300	None		
Total, Illinois	....	$ 203,616,400	$ 61,084,920			
Indiana:						
Indianapolis	7	$ 12,575,000	$ 3,772,500	11	$ 49,300,000	$ 14,790,000
Gary	2	4,400,000	1,320,000	3	11,100,000	3,330,000
Fort Wayne	2	1,500,000	450,000	5	10,000,000	3,000,000
Evansville	2	1,500,000	450,000	6	10,000,000	3,000,000
South Bend	2	1,750,000	525,000	5	5,000,000	1,500,000
Total, Indiana	....	$ 21,725,000	$ 6,517,500	....	$ 85,400,000	$ 25,620,000
Iowa, Des Moines	2	$ 3,430,000	$ 1,029,000	8	$ 3,495,000	$ 1,048,500
Kansas:						
Wichita	2	$ 4,750,000	$ 1,425,000	2	$ 550,000	$ 165,000
Kansas City	2	6,472,800	1,941,840	2	3,000,000	900,000
Topeka	2	685,000	205,500	7	2,490,000	747,000
Total, Kansas	....	$ 11,907,800	$ 3,572,340	....	$ 6,040,000	$ 1,812,000

Sewage Treatment Needs of the 100 Largest Cities in the United States—Present Needs and Future Needs through 1972[a] (Continued)

State[b]	Present needs			Additional needs through 1972		
	Estimated number of projects	Estimated total cost	Federal share, 30%	Estimated number of projects	Estimated total cost	Federal share, 30%
Kentucky, Louisville	13	$ 6,673,000	$ 2,001,900	22	$ 32,320,000	$ 9,696,000
Louisiana:						
New Orleans	25	$ 27,879,000	$ 8,363,700	15	$ 69,122,000	$ 20,736,600
Shreveport	3	13,500,000	4,050,000	4	15,500,000	4,650,000
Baton Rouge	None	...	...	8	12,000,000	3,600,000
Total, Louisiana	....	$ 41,379,000	$ 12,413,700	....	$ 96,622,000	$ 28,986,600
Maryland, Baltimore	3	$ 6,000,000	$ 1,800,000	5	$ 42,150,000	$ 12,645,000
Massachusetts:						
Boston	9	$ 15,000,000	$ 4,500,000	2	$ 4,000,000	$ 1,200,000
Worcester	2	15,000,000	4,500,000	None		
Springfield	3	6,000,000	1,800,000	None		
Total, Massachusetts	....	$ 36,000,000	$ 10,800,000	....	$ 4,000,000	$ 1,200,000
Michigan:						
Detroit	4	$ 68,480,000	$ 20,544,000	2	$ 151,000,000	$ 45,300,000
Flint	2	13,200,000	3,960,000	1	5,000,000	1,500,000
Grand Rapids	2	16,500,000	4,950,000	1	2,500,000	750,000
Total, Michigan	....	$ 98,180,000	$ 29,454,000	....	$ 158,500,000	$ 47,550,000
Minnesota, Minneapolis–St. Paul	3	$ 43,046,000	$ 12,014,070	4	$ 32,475,000	$ 9,742,500
Mississippi, Jackson	5	$ 20,000,000	$ 6,000,000	None		
Missouri:						
St. Louis	4	$ 67,471,000	$ 20,241,300	1	$ 300,000	$ 90,000
Kansas City	14	6,830,000	2,040,000	10	25,753,000	7,725,900
Total, Missouri	....	$ 74,301,000	$ 22,290,300	....	$ 26,053,000	$ 7,815,900
Nebraska:						
Omaha	1	$ 658,000	$ 197,400	2	$ 4,000,000	$ 1,200,000
Lincoln	2	1,700,000	510,000	1	1,250,000	375,000
Total, Nebraska	....	$ 2,358,000	$ 707,400	....	$ 5,250,000	$ 1,575,000
New Jersey:						
Newark and Paterson[d]	3	$ 2,800,000	$ 840,000	2	$ 35,000,000	$ 10,500,000
Jersey City	2	2,700,000	810,000	3	43,000,000	13,500,000
Total, New Jersey	....	$ 5,500,000	$ 1,650,000	....	$ 80,000,000	$ 24,000,000
New Mexico, Albuquerque	5	$ 4,787,000	$ 1,450,100	10	$ 10,443,629	$ 3,133,089
New York:[c]						
New York	....	...	...	12	$ 780,000,000	$234,000,000
Buffalo	....	...	...	2	36,000,000	10,800,000
Rochester	....	...	...	3	33,000,000	9,900,000
Syracuse	....	...	...	4	30,000,000	9,000,000
Yonkers	....	...	...	1	13,000,000	3,900,000
Albany	....	...	...	2	12,000,000	3,600,000
Total, New York	....	...	...	....	$ 904,000,000	$271,200,000
North Carolina:						
Charlotte	5	$ 2,115,000	$ 634,500	18	$ 10,954,000	$ 3,286,380
Greensboro	6	2,593,000	777,900	0	2,350,000	705,000
Total, North Carolina	....	$ 4,708,000	$ 1,412,400	....	$ 13,304,600	$ 3,901,380
Ohio:						
Cleveland	3	$ 28,500,000	$ 8,550,000	None		
Cincinnati	3	40,000,000	12,000,000	None		
Columbus	5	9,150,000	2,745,000	3	$ 1,300,000	$ 390,000
Toledo	3	3,332,000	999,000	1	3,900,000	1,170,000
Akron	2	6,205,000	1,861,500	1	1,000,000	300,000
Dayton	2	3,000,000	900,000	4	5,140,000	1,542,000
Youngstown	None	...	...	1	10,000,000	3,000,000
Total, Ohio	....	$ 90,187,000	$ 27,056,100	....	$ 21,340,000	$ 6,400,200

Sewage Treatment Needs of the 100 Largest Cities in the United States—Present Needs and Future Needs through 1972[a] (Continued)

State[b]	Present needs			Additional needs through 1972		
	Estimated number of projects	Estimated total cost	Federal share, 30%	Estimated number of projects	Estimated total cost	Federal share, 30%
Oklahoma:						
Oklahoma City	6	$ 20,700,000	$ 6,228,000	8	$ 30,536,000	$ 9,160,800
Tulsa	1	300,000	90,000	8	15,680,000	4,704,000
Total, Oklahoma	....	$ 21,000,000	$ 6,318,000	....	$ 46,216,000	$ 13,864,800
Oregon, Portland	8	$ 4,080,000	$ 1,224,000	4	$ 7,560,000	$ 2,268,000
Pennsylvania:						
Philadelphia	3	$ 1,300,000	$ 392,700	5	$ 6,985,000	$ 2,095,500
Pittsburgh	2	5,155,000	1,546,500	2	32,000,000	9,600,000
Erie	None	....	....	5	5,600,000	1,680,000
Total, Pennsylvania	....	$ 6,484,000	$ 1,939,200	....	$ 44,585,000	$ 13,375,500
Rhode Island, Providence	None	....	....	2	$ 100,000	$ 30,000
Tennessee:						
Memphis	3	$ 70,000,000	$ 21,000,000	None		
Nashville	4	120,000,000	36,000,000	None		
Chattanooga	1	25,000,000	7,500,000	None		
Total, Tennessee	....	$ 215,000,000	$ 64,500,000			
Texas:						
Houston	12	$ 9,700,000	$ 2,910,000	98	$ 43,850,000	$ 13,155,000
Dallas	1	1,100,000	300,000	4	13,200,000	3,960,000
San Antonio	6	11,100,000	3,330,000	2	10,000,000	3,000,000
El Paso	6	10,000,000	3,000,000	3	7,100,000	2,130,000
Austin	None	....	....	9	9,200,000	2,760,000
Corpus Christi	3	8,300,000	990,000	6	6,700,000	2,010,000
Amarillo	None	....	....	3	2,800,000	840,000
Lubbock	None	....	....	3	3,950,000	1,185,000
Fort Worth	None	....	....	9	11,300,000	3,390,000
Total, Texas	....	$ 85,100,000	$ 10,530,000	....	$ 108,100,000	$ 32,430,000
Utah, Salt Lake City	1	$ 12,000,000	$ 3,600,000	None		
Virginia:						
Norfolk	3	$ 10,500,000	$ 3,150,000	11	$ 29,225,000	$ 8,767,500
Richmond	10	6,825,000	2,067,500	3	11,025,000	3,307,500
Total, Virginia	....	$ 17,325,000	$ 5,197,500	....	$ 40,250,000	$ 12,075,000
Washington:						
Seattle	0	$ 6,220,000	$ 1,865,000	13	$ 23,870,000	$ 7,161,000
Spokane	7	1,000,000	300,000	25	14,000,000	4,200,000
Tacoma	3	3,000,000	900,000	4	3,150,000	945,000
Total, Washington	....	$ 10,220,000	$ 3,036,000	....	$ 41,020,000	$ 12,306,000
Wisconsin:						
Milwaukee	5	$ 115,000,000	$ 34,500,000	None		
Madison	None	....	....	None		
Total, Wisconsin	....	$ 115,000,000	$ 34,500,000			
Grand total	367	$1,323,279,100	$396,983,730	617	$2,583,941,129	$775,182,339

[a] Division of Water Supply and Pollution Control.
[b] U.S. Census Population, 1960, pp. 66 and 67, U.S. Department of Commerce, Bureau of the Census.
[c] Information not available.
[d] Passaic Valley sewerage commissioners include Newark and Paterson plus other communities.
[e] Needs through 1972 include present needs; not able to distinguish present needs.

The Water Pollution Control Funding Gap
Authorizations vs. Allocations under
the 1966 Clean Waters Restoration Act
(in millions)

States	1968		1969		1970		1968–1970 funding gap totals‡			
	Authorized	Allocated	Authorized	Allocated	Authorized	Allocated*	Total $ authorized	Total $ allocated	Total $ gap	% not funded
Totals	$450.0	$203.0	$700.0	$214.0	$1,000.0	$214.0	$2,150.0	$631.0	$1,519.0	70.7%
Alabama	8.4	4.1	12.9	4.1	18.3	4.1	39.6	12.3	27.3	68.9
Alaska	1.2	0.9	1.5	0.9	1.9	0.9	4.6	2.7	1.9	41.3
Arizona	3.8	2.0	5.6	2.1	7.8	2.1	17.2	6.2	11.0	64.0
Arkansas	5.2	2.9	7.6	2.8	10.6	2.8	23.4	8.5	14.9	63.7
California	35.3	14.6	56.9	14.9	82.8	14.9	175.0	44.4	130.6	74.6
Colorado	4.7	2.4	7.1	2.4	10.0	2.4	21.8	7.2	14.6	67.0
Connecticut	6.2	2.9	9.7	2.9	13.9	2.9	29.8	8.7	21.1	70.8
Delaware	1.6	1.1	2.3	1.1	3.0	1.1	6.9	3.3	3.6	52.2
District of Columbia	2.3	1.3	3.3	1.3	4.6	1.3	10.2	3.9	6.3	61.8
Florida	11.8	5.3	18.6	5.4	26.8	5.4	57.2	16.1	41.1	71.9
Georgia	9.8	4.6	15.1	4.6	21.6	4.6	46.5	13.8	32.7	70.3
Hawaii	2.2	1.4	3.0	1.3	4.1	1.4	9.3	4.1	5.2	55.9
Idaho	2.5	0.5†	3.4	1.6	4.5	1.6	10.4	3.7	6.7	64.4
Illinois	22.9	9.6	36.7	9.8	53.4	9.8	113.0	29.2	83.8	74.2
Indiana	11.1	4.9	17.5	5.0	25.2	5.0	53.8	14.9	38.9	72.3
Iowa	6.9	3.3	10.7	3.3	15.3	3.3	32.9	9.9	23.0	69.9
Kansas	5.7	2.8	8.6	2.8	12.2	2.8	26.5	8.4	18.1	68.3
Kentucky	7.8	3.7	12.0	3.8	17.0	3.8	36.8	11.3	25.5	69.3
Louisiana	8.3	4.0	12.7	4.0	18.1	4.0	39.1	12.0	27.1	69.3
Maine	3.1	1.9	4.5	1.9	6.1	1.9	13.7	5.7	8.0	58.4
Maryland	7.5	3.5	11.8	3.6	17.0	3.6	36.3	10.7	25.6	70.5
Massachusetts	12.0	5.3	19.1	5.4	27.6	5.4	58.7	16.1	42.6	72.6
Michigan	18.0	7.7	28.7	7.8	41.6	7.8	88.3	23.3	65.0	73.6
Minnesota	8.4	3.9	13.1	3.9	18.7	3.9	40.2	11.7	28.5	70.9
Mississippi	6.2	3.4	9.2	3.4	12.8	3.4	28.2	10.2	18.0	63.8
Missouri	10.3	4.7	16.3	4.8	23.4	4.8	50.0	14.3	35.7	71.4
Montana	2.4	0.7†	3.3	1.5	4.5	1.5	10.2	3.7	6.5	63.7
Nebraska	4.0	2.2	5.9	2.1	8.2	2.1	18.1	6.4	11.7	64.6
Nevada	1.2	0.9	1.7	0.9	2.2	1.0	5.1	2.8	2.3	45.1
New Hampshire	2.2	1.4	3.0	1.4	4.0	1.4	9.2	4.2	5.0	54.3
New Jersey	14.0	6.1	22.4	6.2	32.4	6.2	68.8	18.5	50.3	73.1
New Mexico	3.1	0.7†	4.4	1.9	6.0	2.1	13.5	4.7	8.8	65.2
New York	37.6	15.5	60.7	15.8	88.4	15.8	186.7	47.1	139.6	74.8
North Carolina	11.1	5.2	17.4	5.2	24.9	5.1	53.4	15.5	37.9	71.0
North Dakota	2.3	0.3†	3.3	1.6	4.3	1.6	9.9	3.5	6.4	64.6
Ohio	22.1	9.4	35.5	9.6	51.5	9.6	109.1	28.6	80.5	73.8
Oklahoma	6.1	3.1	9.3	3.1	13.2	3.1	28.6	9.3	19.3	67.5
Oregon	4.7	2.4	7.1	2.4	10.1	2.4	21.9	7.2	14.7	67.1
Pennsylvania	25.7	10.8	41.3	11.0	60.0	11.0	127.0	32.8	94.2	74.2
Rhode Island	2.7	1.6	3.9	1.6	5.3	1.6	11.9	4.8	7.1	59.7
South Carolina	6.5	3.4	9.7	3.4	13.7	3.3	29.9	10.1	19.8	66.2
South Dakota	2.5	0.3†	3.5	1.7	4.6	1.8	10.6	3.8	6.8	64.2
Tennessee	9.0	4.3	13.9	4.3	19.8	4.3	42.7	12.9	29.8	69.8
Texas	22.0	9.4	35.2	9.6	51.0	9.6	108.2	28.6	79.6	73.6
Utah	2.9	0.7†	4.1	1.8	5.6	1.8	12.6	4.3	8.3	65.9
Vermont	1.8	1.4	2.4	1.3	3.0	1.3	7.2	4.0	3.2	44.4
Virginia	9.7	4.5	15.1	4.5	21.7	4.5	46.5	13.5	33.0	71.0
Washington	7.0	3.3	11.0	3.3	15.7	3.3	33.7	9.9	23.8	70.6
West Virginia	5.2	2.7	7.8	2.8	10.8	2.8	23.8	8.3	15.5	65.1
Wisconsin	9.5	4.4	15.0	4.4	21.5	4.4	46.0	13.2	32.8	71.3
Wyoming	1.5	0.005†	2.1	1.2	2.6	1.2	6.2	2.4	3.8	61.3
Guam	1.6	0.8†	1.6	1.5	1.7	1.4	4.9	3.7	1.2	24.5
Puerto Rico	6.6	3.5	9.8	3.5	13.7	3.5	30.1	10.5	19.6	65.1
Virgin Islands	1.5	1.5	1.5	1.4	1.6	1.4	4.6	4.3	0.3	6.5

* From *The Nation's Cities, Magazine of the National League of Cities*, September, 1969.

† Actual amounts used by these eight states, although they were entitled to use more. Unused amount totaling $8.3 million from these eight reallocated to other states.

‡ 1970 appropriations still pending.

History of State and Local Pollution Laws

JEROME WILKENFELD

Director, Environmental Health,
Hooker Chemical Corporation,
New York, New York

INTRODUCTION

While recent legislative activity in the field of pollution control appears to be rising at an exponential rate, references to legal efforts to control pollution extend back as far as recorded history. It is probable that the rate of passage of pollution legislation will taper off and that in the future legislative changes will occur at a slower rate so that in the overall a plot of frequency of legislative change vs. time will follow a gaussian distribution.

Ancient History

The earliest commonly available record of air pollution can be found in the old testament, Exodus 9:10, where it is stated, "And they took ashes of the furnace, and stood before Pharaoh; and Moses sprinkled it up toward heaven; and it became a boil breaking forth with blains upon man, and upon beast." Apparently this incident of pollution indirectly resulting from combustion processes but with directly related health effects was not sufficiently severe to cause public arousement because the regulatory agency, in this case the Pharaoh, did not legislate against it. It took several more plagues of increasing severity to cause him to act.

Biblical references to water pollution also exist. However, the first real effort to deal with problems of water borne wastes is seen in Rome, where the municipal sanitary waste disposal problem of a large number of people living in a small area forced the construction of the Cloaca Maxima, or great sewer, for the collection of sanitary and commercial wastes for discharge to the Tiber River. One can be sure that regulations existed for the use of such a sewer. This system depended on dilution and dispersion, with biological activity in the stream handling any treatment required. Apparently this natural assimilation resulted in acceptable stream recovery within a reasonable distance, since the system was used for many years. Of course, the problems of contagious disease transmission also existed but were not recognized.

Medieval History

These two instances of pollution and its control may have affected commercial operations, but the first legislative effort to control pollution with a direct, demonstrable effect on industry appears to have been the Sea Coals Act promulgated by King Edward of England almost 650 years ago. This act prohibited the use of sea coal in London, where air pollution by smoke and particulate matter coupled with sulfur oxide fumes had become intolerable to the local populace. This local regulation, applicable only to London, affected not only the resident who burned sea coal for home heating and cooking but any commercial user. At that time this included only small shop owners or craftsmen, since industry as we know it did not exist. It should be noted that this was a process specific, source control type of regulation, which for a limited problem area is still found to be an effective means of control.

The Victorian Age

On the other hand, the first and strongest effort to correct water pollution is found in programs started at the end of the nineteenth century aimed at control of water borne infectious diseases. Since this is directly related to human wastes, primarily from domestic sources either singly or in municipal collection systems, standardized effluent treatment definition was the most common type of regulation developed. For example, requirements for solids removal and disinfection or these plus removal of oxygen consuming substances biologically to the extent possible by biologic oxidation were normal requirements on all wastes until very recently. Attempts to enforce regulations of this type on industrial sources during the last 25 years have caused much difficulty in determining what constitutes reasonable and effective control. Enforcement was further hampered because control agencies were normally staffed with sanitary engineers in health oriented agencies, resulting in more time being spent on communicating and educating than in correcting problems. Fortunately this is largely overcome now with a substantial cross migration of trained personnel between industry and government agencies.

The Modern Age

In the past cities and counties with significant industrial or commercial waste loads in combined systems usually regulated to avoid problems with the collection or sewer systems rather than by definition of treatment methods. This is exemplified by the City of Niagara Falls Air Pollution Ordinance of 1949, which flatly prohibited the discharge of any odorous, corrosive, or toxic substances in any quantiy into the city sewer system. Of course, this attempt to avoid any problem caused considerable difficulty in definition of control requirements. However, during the life of the ordinance the responsible agency personnel and the industrial contributors usually were able to work out reasonable controls.

As for receiving water quality, except for a few isolated areas of the country, reasonable use as considered under the *doctrine of riparian rights* controlled until the last decade. This doctrine, which has its roots in the common law, requires that every downstream riparian owner is entitled to receive the flow of the stream at his property undiminished in quantity or quality by upstream usage. This doctrine, if rigidly enforced, would make current water practices unsupportable.

In practice it has been subject to continual reinterpretation, and presently in some Eastern states consideration is being given to modification by legislation to allow for appropriation of all flow above that normally used by riparian owners. This excess flow would be made available for diversion to uses in the common interest, such as impoundment, irrigation, or diversion into other watersheds, and would be a partial appropriation of water rights by and for public purposes.

In contrast to the riparian rights doctrine in the Eastern or water rich states the arid Western states developed the doctrine now known as *prior appropriation,* in which a first user has the right to remove the water and use it for his own purposes even if it is to the detriment of a user who arrived on the scene at a later time. Application of this doctrine does not seem to have pollution implications beyond those of flow, since it applies to the removal or nonremoval of water for irrigation purposes rather than to the contamination of water remaining in the stream.

From these factors and the historical development of air and water pollution control regulations, it can be concluded that emphasis in early air pollution regulation was primarily on a local basis and directed at sources which were obvious or of a nuisance nature. They were therefore usually of local or limited area coverage. Water pollution control, on the other hand, once it included communicable disease aspects capable of affecting users quite some distance downstream and beyond the regulatory preview of the local receptor, required action by a higher governmental body, such as the state.

Early Local Regulations

The local air pollution regulations were of the type that are typified by smoke control ordinances such as that of the village of Ossining, N.Y., which adopted its ordinance in 1936 with amendments in 1952. This ordinance forbids the burning of coal or fuel oil in any boiler, furnace, or burner within the village in such manner as to permit smoke, ashes, soot, noxious vapors, and gases to pass into the air from the combustion thereof, in such volume as to be offensive to persons, or damaging to real or personal property of the inhabitants of the village except in the portions of the village designated in the zoning ordinance of the village as "industrial district." Another early interesting ordinance is the 1921 ordinance adopted in the village of Hastings on the Hudson, N.Y., which provides that: "Every factory or business established after August 16, 1921, which causes more smoke to be generated than by the private residences now in existence shall be equipped with smoke stacks at least 400 feet in height and the smoke thereof shall under no circumstances come within 250 feet of any private residence within the village."

On the other hand, a much more sophisticated approach can be found in early water legislation, where effects on navigation and health were the subject of action by the United States Congress as early as the "Laws for Protection and Preservation of the Navigable Waters of the United States," approved Mar. 3, 1899, which prohibits the deposit of "any refuse matter of any kind or description whatever other than that flowing from streets and sewers and passing therefrom in a liquid state into any navigable water of the United States or into any tributary of any water from which the same shall float or shall be washed into any tributary of any water from which the same shall float or shall be washed into any such navigable water." The "Oil Pollution Act of 1924" deals specifically with "the deposition of oils from vessels into coastal or tidal navigable waters." On a state level the state of Michigan in Act 245, C.A. 1929, as amended by Act 17, P.A. 1949, created a water resources commission to "protect and conserve the water resources of the state, to have control over pollution of any waters of this state and the Great Lakes, with power to make rules and regulations governing the same, and to prescribe the powers and duties of such commission; to designate the commission as the state agency to cooperate in this act."

As another example, prior to 1949 when New York State expanded and upgraded its water pollution control law, Article V of the Public Health Law governed the prevention of pollution of the waters of the state by requiring the approval of plants and a permit from the State Commissioner of Health for the discharge of sewage or industrial wastes. The Commissioner of Health was also empowered to issue orders requiring the abatement of pollution when sewage waste discharges were in quantities injurious to public health or when conditions constituted a public nuisance. However, to be valid such an order had to be countersigned by the governor and attorney general.

WATER POLLUTION CONTROL

During the late 1940s and early 1950s most states enacted more direct water pollution legislation than had been available. Until that time there were various types of regulations from the sophisticated Michigan law to an assortment of single purpose controls such as those over the deposition of oil or sawdust in streams which might affect fish and other aquatic life administered by state conservation agencies. The advent of laws of comprehensive water pollution control in the late 1940s and the early 1950s generally assigned authority to a water pollution control board which had the function of establishing rules and regulations, conducting investigations, surveying the state's watercourses for pollution, reviewing and approving new and modified pollution control installations, issuing permits on discharges, and in general managing the amount of the materials which could be discharged into the watercourses without adverse effect. Discharge quality standards were established in several ways. Some were as specific as Louisiana's criteria for sugar mill wastes, which covered many industry-specific problems and wastes, while others such as those in Colorado defined the effluents from industrial operations in such chemical and physical terms as settleable and suspended solids, color, and BOD. On the other hand, many more established water quality criteria, or stream standards, for receiving waters and required emitters either to receive prior approval for their effluents or to submit plans for proposed treatment systems. The evaluation of the impact of these proposed discharges on receiving streams was left to the state agency staff. As one might expect in the development of such complex legislation and regulations with broad-scale impact, attempts to limit the powers of regulatory agencies and to assure reasonableness in enforcement resulted in the development of loopholes. One example is in the New York State law of 1949, which allowed economic infeasibility as a defense against requirement of installation of control equipment. This was absolute, without any requirement that equities be weighed. As a result of this the famous Long Island duck farms cases as well as the municipal discharges on the Mohawk River were ruled exempt from the State Water Pollution Control Board jurisdiction on the basis of economic infeasibility. The end result of these exclusions from reasonable control was the degradation of public waters out of balance with the economic benefit, and added pressure on the legislature to pass more restrictive legislation. This occurred as part of the revision to the state's total water resources program effective Jan. 1, 1962, wherein only a temporary delay can be obtained on the basis of technological or economic feasibility. In both areas corrective installations are being constructed.

In any event the best law is no better than the weakest enforcement, and this in part is dependent on the funding of the program. Many state laws in fact resulted in nothing more than a program of negotiation on what would be required in the way of control of existing pollution or the avoidance of new problems, since state agencies during the 1950s quite often operated on such limited budgets and staff that they could not undertake broad-scale investigations and enforcement actions.

Current Ordinances

Many local ordinances now, in addition to the previously mentioned specific limitations for the protection of the collection system, limit high concentrations of wastes and toxic materials which might adversely affect treatment processes. They may seem reasonable and logical, but this recognition that industrial and domestic wastes can and should be treated jointly when practical is still not accepted without question in many places. This is especially true where the industrial portion of the treatment plant load is large. Where combined treatment of industrial with municipal wastes can effectively be carried out, this is strongly urged by federal as well as state regulatory agencies. The limitation on the number of points of discharge to public waters and the sharing of the treatment costs as well as enlargement of treatment plant size make this combining attractive. Coupled with this is the fact that many municipalities or sewage districts have sewer taxes under various names. It might be classified as a sewer service charge, sewer rental, or levy included as part of the price of water sold by the municipality, but in any event it is in many cases an effort to raise revenues, although in some it is an effort to share the cost of treatment equitably. Rarely are sewer charges installed as a means of forcing control of the discharge of waste materials, since this can result in a license to pollute if the cost of treatment is high.

The methods employed for assessing the charges vary quite considerably, the simplest being use of a multiplier on the standard water sales rate. Of intermediate complexity are charges on the basis of the number of commodes in an installation, or based on the type of business or operation such as residence, hotel, office building, or commercial establishment. The most sophisticated system assesses charges on the basis of quantity of water discharged multiplied by factors for the oxygen consuming constituents as well as suspended and settleable solid materials. As would be expected, choice among these various methods is frequently dependent on the size of the community and mix of sewage sources in the locality assessing the charge. By and large, however, the vast majority are on a simple rate basis and are therefore primarily aimed at adding to the ad valorum tax base. No cases are found where these charges are assessed on direct discharges to public waters and where no restriction on waste water quality or effect exists. That is, effluent charges alone are not considered a practical means for pollution control.

What Federal Water Quality Act Means

With the passage of the several modifications of the Federal Water Pollution Control Act culminating in the law passed in 1966, and based on the factors outlined above, it is apparent that the state level must be the implementation and enforcement level in water pollution programs. The level of technical competence required to mount an effective program and the relationship of various communities as one moves downstream in watercourses make it imperative that local considerations be included while freedom to apply pressure on upstream sources to correct problems makes state level involvement mandatory. Even a cursory review of potential pollutants such as that outlined in "Water Quality Criteria," published by the State Water Quality Control Board of the State of California, reveals the high degree of professional competence required to assess adequately the possible impact of materials that can be discharged from industrial operations. In addition a strong need is developing for judgment on the potential effects of new materials in a technologically advancing society such as ours, clearly demonstrating that the staffing for avoidance of pollution and control of watercourse quality will cost a great deal more than most municipal bodies can afford. The high degree of specialized technical competence required will find sufficient use only at the state level. In some cases the frequency of use and broad applicability make federal involvement more appropriate than state.

During the last several years there has been a considerable acceleration of effort to expand the state programs along with a quantification of standards as a result of Public Law 89–234, The Federal Water Quality Act of 1965. Among the provisions of this act of direct interest to industry is one calling for the establishment of water quality standards for interstate waters. These are to be adopted by the states and will become applicable if the governors file letters of intent and adopt standards on a predetermined timetable. All states filed such letters prior to the Oct. 2, 1966, deadline, held hearings, and adopted water quality criteria prior to the June 30, 1967, deadline.

Legal Aspects of Water Quality Criteria

The law requires that these water quality criteria be acceptable to the Secretary of Interior along with standards which must include programs for implementing and enforcing the criteria consistent with the purposes of the Act, which is "to enhance the quality and value of our water resources and to establish a national policy for the prevention, control, and abatement of water pollution." The Secretary, in guidelines issued in May, 1966, further noted that the standards of quality shall take into consideration the water's "use and value for public water supplies, propagation of fish and wild life, recreational purposes, agricultural, industrial, and other legitimate uses." (Sec. 10 C3 of the Act.) In addition the standards proposed by the states should provide for the "upgrading and enhancement of water quality and use or uses of streams or portions thereof that are presently affected by pollution" along with "the maintenance and protection of quality and use or uses of the water now of a high quality or of a quality suitable for present and potential future uses." These last two statements should be read very critically, since they can be the basis for extremely stringent control requirements. Before this is discussed, however, note that the resulting widespread definition of water quality requirements for various uses has a salutary effect, since engineers will now be in a better position to determine just what degree of treatment is required to maintain and protect classified waters. Typical of the quality level desired are the standards developed by the Ohio River Valley Water Sanitation Commission:

Minimum Conditions Applicable to All Waters at All Places and at All Times

1. Free from substances attributable to municipal, industrial or other discharges or agricultural practices that will settle to form putrescent or otherwise objectionable sludge deposits;
2. Free from floating debris, oil, scum and other floating materials attributable to municipal, industrial or other discharges or agricultural practices in amounts sufficient to be unsightly or deleterious;
3. Free from materials attributable to municipal, industrial or other discharges or agricultural practices producing color, odor or other conditions in such degree as to create a nuisance;
4. Free from substances attributable to municipal, industrial or other discharges or agricultural practices in concentrations or combinations which are toxic or harmful to human, animal, plant or aquatic life.

Stream-quality Criteria

for public water supply and food processing industry

The following criteria are for evaluation of stream quality at the point at which water is withdrawn for treatment and distribution as a potable supply:

1. *Bacteria:* Coliform group not to exceed 5,000 per 100 ml as a monthly average value (either MPN or MF count); nor exceed this number in more than 20 percent of the samples examined during any month; nor exceed 20,000 per 100 ml in more than five percent of such samples.
2. *Threshold-odor number:* Not to exceed 24 (at 60 deg. C.) as a daily average.
3. *Dissolved solids:* Not to exceed 500 mg/l as a monthly average value, nor exceed 750 mg/l at any time. (For Ohio River water, values of specific conductance of 800 and 1,200 micromhos/cm (at 25 deg. C.) may be considered equivalent to dissolved-solids concentrations of 500 and 750 mg./l.)
4. *Radioactive substances:* Gross beta activity not to exceed 1,000 picocuries per liter (pCi/l), nor shall activity from dissolved strontium-90 exceed 10 pCi/l, nor shall activity from dissolved alpha emitters exceed 3 pCi/l.

5. *Chemical constituents:* Not to exceed the following specified cencentrations at any time:

Constituent	*Concentration (mg/1)*
Arsenic	0.05
Barium	1.0
Cadmium	0.01
Chromium (hexavalent)	0.05
Cyanide	0.025
Fluoride	1.0
Lead	0.05
Selenium	0.01
Silver	0.05

FOR INDUSTRIAL WATER SUPPLY

The following criteria are applicable to stream water at the point at which the water is withdrawn for use (either with or without treatment) for industrial cooling and processing:

1. *Dissolved oxygen:* Not less than 2.0 mg/l as a daily-average value, nor less than 1.0 mg/l at any time.
2. *pH:* Not less than 5.0 nor greater than 9.0 at any time.
3. *Temperature:* Not to exceed 95 deg. F. at any time.
4. *Dissolved solids:* Not to exceed 750 mg/l as a monthly average value, nor exceed 1,000 mg/l at any time. (For Ohio River water, values of specific conductance of 1,200 and 1,600 micromhos/cm (at 25 deg. C.) may be considered equivalent to dissolved-solids concentrations of 750 and 1,000 mg/l.)

FOR AQUATIC LIFE

The following criteria are for evaluation of conditions for the maintenance of a well-balanced, warm-water fish population. They are applicable at any point in the stream except for areas immediately adjacent to outfalls. In such areas cognizance will be given to opportunities for the admixture of waste effluents with river water.

1. *Dissolved oxygen:* Not less than 5.0 mg/l during at least 16 hours of any 24-hour period, nor less than 3.0 mg/l at any time.
2. *pH:* No values below 5.0 nor above 9.0, and daily average (or median) values preferably between 6.5 and 8.5.
3. *Temperature:* Not to exceed 93 deg. F. at any time during the months of May through November, and not to exceed 73 deg. F. at any time during the months of December through April.
4. *Toxic substances:* Not to exceed one-tenth of the 48-hr median tolerance limit, except that other limiting concentrations may be used in specific cases when justified on the basis of available evidence and approved by the appropriate regulatory agency.

FOR RECREATION

The following criterion is for evaluation of conditions at any point in waters designated to be used for recreational purposes, including such water-contact activities as swimming and water skiing:

Bacteria: Coliform group not to exceed 1,000 per 100 ml as a monthly average value (either MPN or MF count); nor exceed this number in more than 20 percent of the samples examined during any month; nor exceed 2,400 per 100 ml (MPN or MF count) on any day.

FOR AGRICULTURAL USE AND STOCK WATERING

Criteria are the same as those shown for minimum conditions applicable to all waters at all places and at all times.

Also note that all these levels include substantial safety factors for the protection of the various uses, since in many cases complete and accurate scientific evidence has not been developed to confirm with precision the limits set. In such circumstances it is customary to err on the side of safety.

Federal Guidelines on Water Treatment

Despite the fact that the federal law calls for basing control on the best usage and the establishment of watercourse quality, the Secretary of Interior has in his guidelines stated,

> No standard will be approved which allows any waste amenable to treatment or control regardless of the water quality and water use or uses adopted. Further, no standard will be approved which does not require all wastes, prior to discharge into any interstate water, to receive the best practicable treatment or control that will provide for water quality enhancement commensurate with proposed present and future water uses.

This administrative interpretation of the law in short has led to a blanket requirement of secondary treatment for all industrial wastes regardless of stream need or receiving water quality. In addition, in a release dated Feb. 8, 1968, the Secretary of Interior stated that his interpretation of the Water Quality Act requires:

> Waters whose existing quality is better than established standards as of the date on which such standards become effective will be maintained at their existing high quality. These and other waters of the state will not be lowered in quality unless or until it has been affirmatively demonstrated to the State Water Pollution Control Agency and the Department of Interior that such change is justifiable as a result of necessary economic or social development and will not interfere with or become injurious to any assigned uses made of, or presently possible in, such waters. This will require that any industrial, public, or private project or development which would constitute a new source of pollution or an increased source of pollution to high quality waters will be required, as part of the initial project design to provide the highest and best degree of waste treatment available under existing technology, and since these are also Federal standards, these wastes treatment requirements will be developed cooperatively.

He further stated that "no standards will be approved from here on that do not contain a satisfactory Anti-Degradation Provision, and those states whose standards have been previously approved will be asked to revise them to include such provisions where needed."

No comment is made on the possible incompatibility of a secondary treatment or equivalent requirement with this antidegradation proposal. This arises in the tacit recognition that secondary treatment gives only 80 to 90 percent removal of treatable waste constituents and thereby results in detectable changes in receiving water contaminant levels. It is assumed this is what will be allowed in the federal review.

It has been the experience of many states that the standards submitted have required extensive negotiation and modification prior to their acceptance by the Federal Water Pollution Control Administration. Much of this has been handled in informal conferences. For example, several state criteria on thermal pollution and the levels at which this must be controlled in different climatological sections of the country have been modified by this procedure.

Industrial Equivalent of Secondary Treatment

From the industrial viewpoint there is considerable uncertainty as to what consitutes "equivalent to secondary treatment" for materials which are insoluble and removable by the various filtration processes. For organic materials readily degradable biologically, equivalency can be reasonably developed. However, when materials are refractory or of a different order of biological degradability, there is little in the way of appropriate guidelines. It is apparent from the federal guidelines and statements issued and from actions taken by the Corps of Engineers in processing applications that direct approvals of some discharges by the FWPCA are being required even though the law does not call for this.

It therefore appears that, despite the mandate of the federal act that stream criteria and standards with state primacy should be the mechanism for control in the United States, there still is debate on the details of the program. This results from the apparent advantage from the viewpoint of control agencies of discharge and process regulations as a means of enforcement.

Administratively they are much more convenient to enforce, as they provide a means for relating legal requirements directly to a source. They also offer an opportunity for limitation of the number of discharge points by pressing for joint treatment with municipalities wherever practical. Surveillance problems are considerably reduced with effluent controls, since sampling and testing of outfalls is much more easily conducted than stream surveillance and in many cases this can be made the responsibility of the emitter.

Interpretation of Legislation

As of this writing there have been no court tests of the interpretations by the Secretary. In the entire field of water pollution control it is rare that any of the regulations are carried through ultimate court tests, so that a somewhat unique situation exists where untried laws are being enforced on a wide scale without the benefit of impartial review. Apparently this results from the adverse public relations implications of court testing, to the highest judicial levels, of pollution control regulations. Several people have noted facetiously that pollution is unique in that it, among all social ills, is the only one without a defender. The maximum extent of objection usually ends with debate on the method and degree of control. On this point there are many people who will speak up pointing out that emission limitations, unless they are clearly and logically related to receiving water quality, can only result in excessive and extensive overcontrol requirements high in cost without concomitant benefits.

Based on developments to date, future patterns are clear.

1. There will, as previously noted, probably be a slowing up in the rate of passage and of new modifications of existing water pollution control legislation.
2. It is now fairly well defined that the lowest reasonable regulatory level for water pollution control will be the state government. Federal involvement, however, will increase.
3. The types of control regulations which will ultimately be enforced will be a blend of receptor and emission requirements.

Some variation in emission requirements from place to place will continue but will be in keeping with variations in receiving watercourse size and quality, local climatology, and usage.

4. Procedure will be developed for the detection of potential problems prior to their becoming troublesome. This may include prior approval of new materials before use or the development of methods and procedures within industries for such evaluation without regulatory requirements.
5. A rollback is starting on existing sources where levels of contaminants exceed desirable quantities in watersheds. These will go beyond just achieving safe limits to allow for the future growth of population and industry.
6. Better systems of weighing and balancing equities of situations on new and existing sources will be developed. These will take into account in quantitative terms not only economic but social value, priority, and impact.

AIR POLLUTION CONTROL

Until approximately 1960, most states and local communities had ordinances or laws of the type mentioned in the opening paragraphs of this chapter. That is, they were primarily of the local nuisiance type and were rather vague on the definition of pollution and what constituted an actionable emission. Many were nothing more than smoke control regulations, and in some cases where clearly defined programs were mandated in the law, the power of the state to enforce control was severely limited.

Early Local Legislation

A good example of such an act was the State of Washington Air Pollution Control Act of 1957, which recognized the problems of air pollution control and prevention to be regional in nature. The purpose of the act was to "provide for creation of separate districts to control and prevent air pollution in each area where it may exist or is likely to occur." Air pollution was defined as "the presence in the outdoor atmosphere of substances put there by man in concentration sufficient to cause an unreasonable interference with the comfort, safety, or health of man for the reasonable use and enjoyment of this property." Further the act gave cities, towns, counties, or specially created air pollution control districts considerable power in preventing or controlling air pollution. These political units had the power to take unilateral action or to join with other cities, towns, or counties to form a district for the control of pollution. The district was deemed a political corporate body with the power to levy district taxes against real and personal property. The district in addition could hire staff, conduct investigations, issue regulations, and carry out a full-scale air pollution control program. However, there was no means by which the state could force action where pollution problems existed if the local political entities, for whatever reason, did not wish to act. As a result, while there was some action taken and correction of some major problems occurred, the program could not keep pace with increasing control requirements, and the law was modified considerably in 1967.

The New Jersey Control Act

In a different case, the New Jersey Air Pollution Control Act of 1954 created a strong state level program based on an air pollution commission which included cabinet rank ex officio members as well as citizen representation. The law gave this statewide body the power to formulate and promulgate, amend and repeal codes, rules, and regulations as well as the power to conduct research, investigate sources, require registration of sources, receive and initiate complaints, and take action against polluters. This broad-scale, detailed program was effective but was abandoned in 1967 in favor of a different approach, apparently because control action was taken too cautiously and the program did not move rapidly enough to satisfy the governor and legislature.

The New York State Control Act

New York State, as a third example, passed its comprehensive air pollution control law in 1957, and while it was somewhat different in form from the New Jersey law, in substance it was essentially the same. This program too has been modified by the legislature but only to the extent that enforcement procedures were simplified and broader classes of controls allowed to permit accelerated action. If any overall conclusion can be drawn from the history of state level regulations, it is apparent that there has been an underestimation of the speed with which corrective action has to be instituted and of the willingness of the public to submit to costly and detailed control of their activities. There has been no major dispute with the rationale that air pollution control problems need to be considered in the context of the local area needs and political mores. Where the local political entity is large or forceful enough to handle the problem, it can best be done at that level. Any mention of large, forceful local programs must consider the development of the Los Angeles Air Pollution Control Program. This agency's history provides an example of the problems which can arise and the effect on industry when a large-scale air pollution control program is undertaken with extreme pressure for precipitous action.

Evolution of Los Angeles Air Pollution Control Program

Shortly after World War II, it became very apparent that with the rapid growth of population in the Los Angeles basin, air quality was deteriorating badly and the visibility cutting smoke and fog which occasionally occurred in the sunny basin were becoming more than a simple nuisance because of increasing frequency and intensity. The highly vocal, outdoor loving population of that area was not willing to accept this. In hopes of correcting it, the County Board of Supervisors passed a strong air pollution control ordinance in 1947, which included extensive police powers. The program called for broad-scale enforcement action with very complete control over all stationary sources including industrial sources. The immediate desire for action and correction prior to development of knowledge of the cause of the problem led to widespread control of anything that was visible or that the control authorities thought might contribute to the problem.

The public at large as individuals first felt the impact of their desire for rigid control in the requirement that all backyard rubbish incinerators be abandoned. However, since the Los Angeles area did not have public garbage collection, public pressures and the time required for the municipality to gear up and take action resulted in a 2-year delay on this phase of the control program. Where public effort was not required, the pressure was greater and more precipitous control action was required from industry. Stringent regulations on particulates, sulfur dioxide, and any industrial plume that was visible were promulgated. It is estimated that in less than 2 years $12 million was spent in Los Angeles County for reduction and control of air pollution. While this herculean effort did result in substantial reductions in particulate SO_2 emissions, the primary problem of smog intensity and frequency was not improved but rather continued to get worse. Finally basic research into the chemistry of organics in the atmosphere was completed by Dr. A. Haagen-Schmidt, indicating the sources of the problem to be the sunlight catalyzed reaction of some olefinic organics with oxides of nitrogen and ozone to form higher molecular weight compounds. These compounds condense to form aerosol sized droplets which are irritating to the eyes and throat and light obscuring.

This relatively high concentration of organics in the atmosphere results primarily from the massive use of automobiles for transportation by the public, coupled with an unusually low ventilation rate.

Currently there are some who feel that the control of sulfur dioxide has hindered rather than helped the problem, since this reducing substance should act as a reaction chain stopper for the oxidation reactions which occur in the Los Angeles atmosphere. With the advent of controls on new automobiles starting during 1964, it is expected that the rate of deterioration of the Los Angeles atmosphere can be slowed or stopped, and when coupled with the long-range planning for control of other sources of photochemically reactive materials the future growth pattern of the area is being anticipated, to avoid future problems.

It is therefore apparent that much of the added expense necessitated by the hasty application of controls was not germane to the basic problem. In fact automobile controls are the province of a separate statewide board rather than the county agency because of the mobile nature of the sources.

The 1967 changes in the California State Air Pollution Control Law have vested more of the power in the state and less in the local community, which should help correct past differences in emphasis on the types of sources.

San Francisco Bay Area Regulations

The magnitude of the Los Angeles area problem should under no circumstances be underestimated or the need for control minimized. There are, however, a substantial number of knowledgeable people who feel that the rate of correction could have been greater and at lower cost had the program been more soundly based on technical evidence. The San Francisco Bay area, for example, has passed three regulations compared with the Los Angeles 1967 rules, yet has what appears to be a very effective program maintaining air quality without extensive public dispute. This is said in full recognition of the differences in magnitude of problems between Los Angeles and San Francisco. In San Francisco the primary method of control is limiting the allowable emission rates from sources rather than detailing the allowable control procedure for processes as is done in Los Angeles. In both these cases the ultimate point of control is the quality of the atmosphere at the point of use as determined by monitoring. In neither case is the relationship of ambient air quality to emission level very clearly defined.

Trend toward Ambient Air Quality Regulations

Alternatively some areas are attempting to base control regulations more directly on ambient air quality requirements at the point of use. This allows a maximum degree of flexibility in control procedures yet utilizes the natural advantages of an area to its maximum economic and social benefit. Of necessity these regulations are more complex and are somewhat more difficult to understand and enforce. They also require a more extensive control agency organization but result in lower overall cost to the community for the desired degree of control. This must be understood by the industrial segment of the public as well as the public at large, since it means that for effective control under such programs higher agency budgets must be supported.

The alternative is to adopt emission limitations based on what is technologically feasible without consideration of actual need. This has resulted in the adoption of Los Angeles developed emission limitations in wind swept areas because the numbers are low and political prestige can be enhanced. Frequently these are also never enforced.

In order to implement an effective program of ambient air based regulations, it is first necessary to determine what the problems and needs of the area are. To this end determinations must be made of the type of raw materials available to the location, the climatological, meteorological, and topographical factors and their interrelationship. It is also necessary to determine what the areas, type, and distribution of industry are, along with projections for the future. Other sources must also be investigated, including types of household heating used, transportation patterns, the distribution of residences and their relationship to roads and factories, and the general long-range projections for the area including zoning. Coupled with this source inventory and demographic study is the need for an extensive sampling and monitoring program to establish the relationship of area effect to sources, since this can vary considerably from place to place. With these factors in hand, along with an understanding of the political structure of the area, regulations can be derived which may have some commonality with those in other regions yet are significantly different in other respects. Regulations of this type recognize the relationship between cause and effect and also recognize the extent of control desired by the public at large. In all cases to date, the agreed on levels to be achieved are, and should be, beyond minimal requirements for the protection of health.

The following tabulation of the range of values found in typical state ambient air standards gives some idea of the variation resulting from differing needs. It is by no means exhaustive, and a study of the actual standards shows a much

Typical State Ambient Air Quality Standards

Total suspended particulates, mg/cu m, 30-day avg	100–250
Soiling index, COH, M ft	0.4–0.6
Smoke, Ringelmann No.	1–2
SO_2, ppm, 8-hr avg	0.1–0.25
Reactive SO_4, mg SO_3/100 cm^2, 24-hr avg	0.5–30
Suspended SO_4 or H_2SO_4 mist, mg/cu m, annual avg	4
H_2S, ppm, 1-hr avg	0.1
Oxidants, ppm, 1-hr avg	0.1–0.15
CO, ppm, 8-hr avg	15–30
NO_x, ppm, 1-hr avg	0.1–0.25
Fluorine as HF, ppb, 24-hr avg	1–5
F in forage crops, ppm	35
Beryllium, mg/cu m, 30-day avg	0.01
Lead, mg/cu m, 30-day avg	5.0

broader range of effects and limitations as the details of the standards are included. For example, some sampling periods required vary from 10 through 30 min, 1 hr, 8 hr, 24 hr, 30 days, to annual averages. In addition some allow standards to be exceeded for specified periods or at a specified frequency.

South Carolina's Stack Emissions Control

An example of this type of regulation using the current knowledge of meteorology and dispersion to relate source to receptor can be found in South Carolina's Regulation 3 covering particulate emissions from fuel burning operations. This regulation recognizes that ash carried out of a stack from fuel burning operations can create undesirable conditions as a result of excessive dust deposition of large visible particles. To this end all particles larger than 60 microns in diameter must be collected at the source. Since the smallest suspended dust fraction can also create undesirable concentrations beyond the property line, a limit is placed on the maximum average concentration of such particles of 100 micrograms/cu m in residential communities and 200 micrograms/cu m in nonresidential and industrial areas. To meet these ground level goals, atmospheric dispersion calculations based on published studies are used to relate the ground level concentration to source stack, with individual limits placed on the mass emission rate of particulates from specific operations in an area. Methods for deriving these individual limits are included in the regulation recognizing thermal buoyancy, exit gas velocity, and physical stack height as factors affecting the allowable emission rate. In the regulation itself the following values are listed as maxima:

South Carolina Maximum Allowable Ground Level Concentrations for Particulates

Duration of occurrence	Predominantly residential, micrograms/cu m	Predominantly nonresidential and industrial, micrograms/cu m
3–15 min.	100	200
30 min to 1 hr.	50	100
24 hr.	25	50

The regulation also makes a distinction between new and existing sources with the understanding that it is practical to control new sources to a greater extent than existing sources.

This approach is also followed in regulations promulgated in other states and is considered logical, workable, and meaningful by experts in the field.

Texas and Compliance for Existing Facilities

Texas regulations on particulate matter and sulfur compounds (regulations 1 and 3, respectively) follow a similar pattern and recognize that, while the effective dates of the regulation can be set to occur shortly after promulgation for new or modified equipment or facilities, existing facilities may require a longer period of time to be brought into compliance. To this end provision is made for the granting of variances on the presentation of justification. Provision is also made for simplified filing on short-term overlapping of corrective action with the required compliance date. Texas also spells out in considerable detail, in appendixes to the regulations, sampling procedures and exact methods for determining compliance or noncompliance including procedures for property line measurement of ground level concentrations.

New York State's Ambient Air and Emissions Control

An alternate approach to the type of regulation described above is New York State's Part 187 controlling contaminant emissions from process, and exhaust and ventilation systems. This rule is a hybrid in that it is part ambient air regulation and part emission control limitation and results from the wide range of density of sources in New York State from that found in the large metropolitan and industrialized areas to the occasional sources in the rural and recreational areas where emissions may or may not have to be restricted. The rule includes an environmental rating system for classification of different types of emissions based on effects on the community. The rating takes into account the properties and quantities of contaminants emitted, effect on human, plant, and animal life or property, meteorological parameters, stack height, characteristics of the community, and the ambient air quality classification of the area in which the sources are located or which it affects. Procedures are spelled out for the determination of the environmental rating, and reduction requirements are then related to the potential emission rate of the process. The potential emission rate is considered to be the rate at which contaminants would be emitted to the outdoor air in the absence of air pollution control facilities or other control measures, and a method for considering cyclic operations is included recognizing that instantaneous emission potential, as well as total emission potential over the period of the cycle, is of importance.

While this rule has a broad spectrum of control requirements from extremely restrictive (over 99 percent removal of contaminants) to a category D where "the degree of air cleaning may be specified by the Commissioner provided satisfactory dispersion is achieved," one overriding factor in the rule is the requirement that the rule not be construed to allow or permit any person to emit air contaminants in quantities which alone or in combination with other sources would contravene any established air quality standard. This in effect means in polluted areas it may be necessary to control to a greater degree than required by the formulas in this regulation.

These several regulations cited indicate that control of a complex problem such as air pollution can take different forms and meet differing needs in different areas.

Basically the overriding argument in favor of the approach of controlling on the basis of need rather than cost has been the ability to minimize costs to the community and economy even at an increase in regulatory expense. The flexibility of basing control on need is also resulting in the development of unique and economic procedures for reduction of emissions rather than freezing on specific control techniques.

This is showing up frequently in the area of process modification, where an effort to avoid the generation of pollutants rather than the installation of control equipment is the basis for control.

Effect of Enforcement Inequity on Competitive Status

One problem which receives frequent discussion in industry is the question of equality vs. equity arising in the enforcement of these rules. Members of the industrial community who have not been closely involved in the development of air pollution regulations are now realizing that direct dollar costs for control installations by competitors may not be exactly equal when they are located in different air basins. In addition when one is planning new construction the cost may be higher than for existing competitors in the same area. This results in complaints by the new installer of unequal treatment. To date it has been possible to justify these differences on the basis that in overall equity consideration competitive advantages of a different nature in other cost areas also vary. These include probable better efficiency for new installations as well as labor and tax rate variations.

Priorities for Control

Currently it appears that in most areas surveyed the primary pollutants to be controlled are particulate matter, sulfur dioxide, oxidants, carbon monoxide, oxides of nitrogen, and reactive organics. These are not given in any priority order, although the strongest effort seems to be on particulates and sulfur dioxide, with the others in a secondary category. Smoke and other visible plumes along with odors are also mentioned as problems. Visible plumes are frequently regulated on the basis of equivalence with a Ringelmann number. Regulations on odor control are usually vague, with little more than a subjective statement aimed at avoidance of a nuisance being used.

The Local Community Function

As can be readily seen from the types of regulations described, local communities, unless they are of substantial size, have considerable difficulty in mounting effective broad-scale programs. As a minimum a relatively small community must spend upward of $100,000 a year for staff and expenses for any air pollution program that is to be effective. Even at this level it would include only surveillance and very limited monitoring to determine contaminant levels. The more sophisticated techniques required for stack sampling, ambient air quality determination, and development of new techniques or analytical methods of necessity are being carried out by the more extensive staffs of federal, state, and major metropolitan agencies. The local community in most cases can and should act as a monitoring arm on the more readily determinable types of pollution such as visible sources, particulate measurements, and the operation of monitoring stations in area wide networks. They also can act as interpreters of programs and regulations to the local community as well as a vehicle for conveying the needs and desires of the community back to the higher levels of government. This appears to be the procedure developing in several of the states such as New York, where outside of the city of New York few have staffs large enough to mount an effective independent program. Other local jurisdictions which have varying types of regulations are Erie County in New York State, Chicago, Illinois, the Puget Sound area in Washington, and Detroit, Mich. These ordinances cover the broad scale of emission regulation types as well as ambient air quality based regulations. In some instances these inappropriately include the Los Angeles process weight tables, which were developed for their specific particulate control needs and based on process size as it exists there.

The cost of these local control programs varies considerably, with the most expensive being Los Angeles. It is generally felt that a minimal program will cost upward of $0.45 per capita in a modest sized city.

Impact of Federal Act on State and Local Control

As in water the federal legislation in this field has an expanding impact on state and local control requirements. The passage of the Federal Air Quality Act of 1967 will have an effect on air control regulations similar to that of the Water Pollution Control Act of 1965 on water programming. Of direct interest is the fact that the law calls for the following program under the Secretary of Health, Education and Welfare:

1. Define the broad atmospheric areas of the nation and designate specific air quality control regions.
2. Develop and publish air pollution criteria indicating the extent to which air pollution is harmful to health and damaging to property and detailed information on the cost and effectiveness of techniques of preventing and controlling air pollution. This includes an updating of the air quality criteria for sulfur dioxide published in 1967.
3. Following the publication of these criteria and the data on control technology available, the states have to begin developing air quality standards and plans for implementation.
4. The standards are to be set within 1 year and 6 months of the publication of the criteria. These air quality standards are developed and applied on a regional basis, and where two or more states are involved, each state is expected to develop standards for its portion of the region.
5. The Secretary further has the power of review and acceptance or rejection of these standards if they are not consistent with the act.
6. Advisory groups have been established to develop the criteria and to investigate the economics and practicality of existing control technology.

Educating Industrial Management

This presentation of timetables and quantification of objectives and programs should aid considerably in the expediting of air pollution control and the avoidance of hazardous situations. To industry it means that every plant operator has to expand his knowledge of materials handled and emitted and increase his testing requirements and control reliability. Every plant manager has to know in quantitative terms how much and what is coming out of each vent in his plant and what the adverse environmental effects of these materials are, the levels at which no adverse effect occurs, the makeup of the surrounding community, the control measures needed to meet these levels, and last, but by no means least, where his plant stands in meeting these requirements. Just the analytical effort required to determine the quantitative status of any substantial installation is a large and expensive undertaking.

For this reason it is necessary to plan the program and timetable for implementation in some detail, with preliminary estimates of probable installation needs along with capital and operating forecasts made early in the program, upgrading as the program proceeds. The new federal law calls for the National Center of Air Pollution Control to investigate national emission standards. However, the pressure for consideration of such standards is bound to be tempered quite considerably if the state programs are effectively mounted under the mandate of the act.

Predictions on Future Regulation

Based on the developments to date, it appears that the future pattern will be one where the lowest broad-scale regulatory level will be the state and region, with heavy federal pressures and monitoring. Local control will be of a limited nature in most areas.

There will be a blend of emission requirements tempered considerably by their effect on the receptor. Procedures will have to be developed for the detection of potential problems before they occur, along with improvement in techniques for rolling back the quantity emitted from existing sources to allow for future population as well as technologic growth. Members of the industrial community will have to understand why dual standards on new as opposed to existing sources exist; however, this is expected to be acceptable. One additional factor that is not faced up to often enough is the fact that, as additional knowledge of cause and effect develops, there may be times when regulations in effect will have to be changed, sometimes to more restrictive levels, though occasionally loosened. All interested parties will have to review the basis of these changes critically and accept them when there is a demonstrable need.

Concern over Total Environmental Control

In the overall it is apparent that the future program in air is not too different from that described for water pollution control, and the similarities of the programs and their procedures are understandable. The interrelationship between air and water pollution control along with solid wastes disposal is clear, and while substantial differences in control techniques do exist, close coordination is necessary. This is particularly necessary since the field of solid wastes disposal is only now on the threshold of broadening regulatory requirement, as are the other environmental areas of in-plant and consumer protection, all of which are subject to a fair amount of regulation at present. Some state and local governments, as well as private agencies such as large corporations, are developing coordinated programs which include all environmental factors. A paraphrase of the factors used in the determination of an environmental rating in New York State's Part 187 can be helpful in assessing overall impact. This includes properties, quantities, and rate of emission, physical surrounding of the emission source, population density, anticipated future growth, dispersion characteristics, geographic relationships, as well as consideration of receptor handling capability and in addition the latest findings on effect, both toxicologically and aesthetically, on the receptor, possible hazardous side effects of mixing with other contaminants already present in the environment, and the ultimate fate of materials handled and distributed, along with potential hazards in use. It is obvious that in the entire field of pollution control or environmental management, more restrictive demands and rules will occur in the future. Man and his ecology are rapidly being treated in a systematic manner, but this is not to be interpreted to mean that such management must require uniformity of regulation. The maintenance of flexibility in mode of action will result in continued competition to gain advantage, and by this means we can assure the maximum use of native ingenuity in the development and maintenance of a desirable environment.

Chapter 4

Air and Water Pollution Quality Standards

Part 1: Air Pollution Standards

ARTHUR C. STERN

Professor, Department of Environmental Sciences and Engineering, The University of North Carolina, Chapel Hill, North Carolina
and
President and Board Chairman of Triangle Universities Consortium on Air Pollution (Duke, North Carolina State, and University of North Carolina)

INTRODUCTION

An air quality standard says, "The concentration of a pollutant in the atmosphere at the point of measurement shall not be greater than some specified amount."

An emission standard says, "The amount of a pollutant emitted from a specific source shall not be greater than some specified amount." The amount may be expressed as a concentration in the effluent gases or as a total amount regardless of concentration in the effluent gases.

In general an air quality standard is developed by formally or informally considering air quality criteria, which are compilations of effects associated with various concentrations and durations of exposure of pollutants, plus other factors, such as cost and technological feasibility of emission control, and social questions.

An air quality standard is generally not enforceable as such. Standards which attribute to a source the decrease in air quality at the boundary of the land surrounding the source, and limit the emission from the source on this basis, are emission standards, even though stated in terms of air quality. Air quality is brought to the air quality standard by the enforcement of emission standards, by land use planning and zoning standards, and by limitations on minimum stack height and fuel composition.

Although land use planning and zoning standards will not be further discussed in this chapter, they are useful means of air pollution control. Included in this category are standards relating to streets, roads, parks, open space, transportation, structures, etc.

AIR QUALITY STANDARDS

United States

In the United States the Clean Air Act requires that the Secretary of Health, Education and Welfare designate air quality control regions, which may be intra- or interstate, and that the individual states adopt air quality standards for each air quality control region so designated by the Secretary. The procedure involved in the adoption of such standards by the states is outlined in Fig. 1, which charts

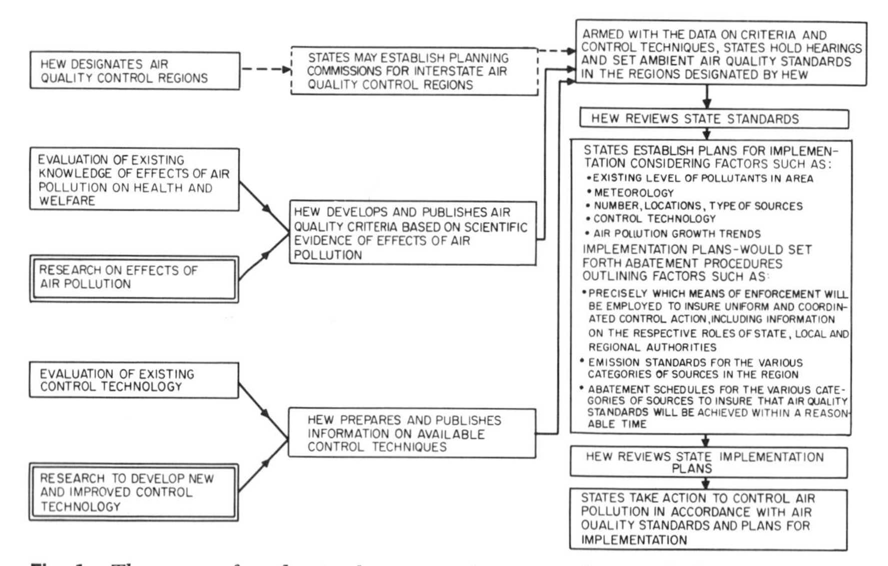

Fig. 1 The process for adoption by a state of an air quality standard for a designated air quality region which is acceptable to the Secretary of HEW.

the process for the most straightforward route contemplated by the act, i.e., the adoption by a state of a standard, for a designated air quality region, which is acceptable to the Secretary. When a state fails to adopt an air quality standard for a designated air quality control region within the prescribed time interval after receipt from the Secretary of Health, Education and Welfare of air quality criteria for a pollutant and documentation of the control technology applicable thereto, or adopts a standard or a plan for its implementation which the Secretary finds to be inadequate, the Secretary may propose the standard for the designated air quality control region in the state. When a state finds the standard proposed by the Secretary unacceptable, approval or modification of the standard is then recommended to the Secretary by a board composed of five or more persons appointed by the Secretary, after such board has held appropriate hearings.

Although no state is required to adopt an air quality standard for any pollutant for which the Secretary has not yet issued air quality criteria and documentation of applicable control technology or for any area of the state not designated by the Secretary as an air quality control region, there is no prohibition in the act which would prevent a state from adopting air quality standards for substances on which the Secretary had not yet so acted and for areas of the state not so designated. There is, however, the presumption in the law that an air quality standard for a pollutant, adopted by a state, as applicable to the state as a whole, or to an area of the state including all or part of a designated air quality control region, prior to its receipt from the Secretary of air quality criteria and documentation of applicable control technology for that pollutant, must be submitted to the Secretary for his approval within the prescribed time period after the Secretary has made such issuance.

As of the effective date of the latest amendments to the Clean Air Act, Nov. 21, 1967, the Secretary had previously (March, 1967) published air quality criteria for only one class of pollutants—the sulfur oxides. The law required that this document be reevaluated before formal issuance to the states. The air quality standards which have been adopted by the several States to date are given in Tables 1A, 1B, and 1C.

U.S.S.R.

In the U.S.S.R., air quality standards are promulgated by the Federal Ministry of Public Health on the basis of recommendations made to it by a permanent expert committee which also has the responsibility for seeing to it that the research necessary for the development of recommendations is undertaken in appropriate institutes. In general, U.S.S.R. air quality standards represent the lowest pollutant concentrations at which a response to a standardized battery of tests was evoked in man or in test animals. There is an a priori assumption that such a response is adverse and therefore is to be avoided. This procedure and philosophy lead to air quality standards considerably lower than in other countries where (1) a larger number of criteria are considered than in the U.S.S.R. and (2) the definition of an adverse response differs. U.S.S.R. air quality standards are for 24-hr and instantaneous exposure, the latter generally interpreted as the shortest averaging time of the air sampling and analysis techniques utilized, which, for most substances, is about 20 min (Table 2). An important feature of the U.S.S.R. standards are the rules for the combined effect of several pollutants. Some of the other eastern European member states of the Council for Mutual Economic Aid make official use of these U.S.S.R. air quality standards.

Poland

The air quality standards officially adopted in Poland are listed in Table 3.

West Germany

In West Germany there are both official (Table 4) and quasi-official (Table 5) air quality standards. The quasi-official standards are those adopted by the Kommission Reinhaltung der Luft (Clean Air Commission) of the Verein Deutscher Ingenieure (VDI—German Society of Engineers), which comprises task groups representing industry, government, and academic institutions.

Other Countries

Several countries and provinces other than those listed above have adopted air quality standards (e.g., Czechoslovakia, Table 6, and Ontario, Table 7). Other countries have adopted standards for just one or two pollutants—particularly sulfur

TABLE 1A State Air Quality Standards for Gases (U.S.A.) (ppm)

Substance	State	Averaging time				Note
		24 hr	1 hr	Other	Other	
Carbon monoxide	California	......	120	30	8 hr	Serious level
	New York	......	60	15	8 hr	[a]
		......	......	30	8 hr	
	Pennsylvania	25	......	.....	......	
Ethylene........	California	......	0.5	0.1	8 hr	Adverse level
Fluorides (as HF)	Montana	0.001	......	.....	......	Averaging time assumed
	New York:					
	Rural	0.001	......	.....	......	Also fluorine
	Urban	0.002	......	.....	......	Also fluorine
	Industrial	0.004	......	.....	......	Also fluorine
Soluble........	Pennsylvania	0.007	......	.....	......	
Hydrogen sulfide	California	......	0.1	.....	......	Adverse level
	Missouri[b]	......	......	0.05	30 min	[c]
		......	......	0.03	30 min	[d]
	Montana	......	......	0.05	30 min	[c]
		......	......	0.03	30 min	[d]
	New York	......	0.1	.....	......	
	Pennsylvania	0.005	0.1	.....	......	
Nitrogen dioxide	California	......	0.25	.....	......	Adverse level
		......	3	.....	......	Serious level
Nitrogen oxides	Colorado	......	0.1	.....	......	[e]
Oxidant (by KI)	California	......	0.15	.....	......	Adverse level
	Colorado	......	0.1	.....	......	[e]
	Missouri[b]	......	0.15	.....	......	
	New York:					
	Rural	0.05	......	0.1	4 hr	
	Urban	0.1	......	.....	......	
	All	......	0.15	.....	......	
Sulfur dioxide....	California	......	10	.....	......	Emergency level
		......	5	.....	......	Serious level
		......	1	0.3	8 hr	Adverse level
	Colorado	0.1	0.5	.....	......	[e]
	Delaware:					
	Rural	0.08	0.21	.....	......	[f]
		0.02	0.05	.....	......	[g]
	Residential	0.1	0.25	.....	......	[f]
		0.03	0.07	.....	......	[g]
	Commercial	0.12	0.28	.....	......	[f]
		0.04	0.09	.....	......	[g]
	Industrial	0.15	0.32	.....	......	[f]
		0.06	0.12	.....	......	[g]
	Florida (Dade County)	......	......	0.1	8 hr	
		......	......	1	20 min	
	Missouri[b]	0.1[h]	......	0.02	1 year	Tentative
		0.05[i]	0.1[j]	0.25[k]	5 min	
	Montana	0.1	......	0.02	1 year	[e]
		......	0.25	.....	......	[l]
	New York	0.1	0.25	.....	......	[m]
		0.15	0.4	.....	......	[m]
	Pennsylvania	0.25	0.5	0.05	30 days	
	South Carolina	0.1	0.3	.....	......	Nonindustrial areas
		0.17	0.5	.....	......	Industrial areas
	Texas:					
	Industrial	0.3	......	0.5	30 min	
	Other	0.2	......	0.4	30 min	

[a] 8 hr average time value not to be exceeded more than 15 percent of the time in a year. 1 hr average time value not to be exceeded more than 1 percent of the time in a year.
[b] St. Louis metropolitan area.
[c] Not to be exceeded over twice a year.
[d] Not to be exceeded more than twice in any 5 consecutive days.
[e] Not to be exceeded more than 1 percent of the time in any 3 months.
[f] 95th percentile (annual basis).
[g] Geometric mean (50th percentile) (annual basis).
[h] Not to be exceeded over 1 day in any 3-month period.
[i] Not to be exceeded more than once in any 90 days.
[j] Not to be exceeded more than once in any 4 days.
[k] Not to be exceeded more than once in any 8 hr.
[l] Not to be exceeded more than 1 hr in any 4 consecutive days.
[m] Not to be exceeded more than 1 percent of the time in any year in certain specified geographic regions.

TABLE 1B State Air Quality Standards for Dusts and Mists (U.S.A.) (mg/cu m)

Substance	State	Averaging time				Note
		24 hr	1 hr	Other	Other	
Beryllium....	New York	0.00001	.....	....	....	
	Montana	....	.....	0.00001	30 days	
	Pennsylvania	....	.....	0.00001	30 days	
	Texas	0.00001	.....	....	....	a
Calcium oxide	Oregon	....	.....	0.02	....	b
Lead (as Pb)	Montana	....	.....	0.005	30 days	
	Pennsylvania	....	.....	0.005	30 days	c
Sulfates:						
(as H_2SO_4).	Pennsylvania	....	.....	0.01	30 days	d
(suspended)	Missouri	....	.....	0.004	1 year	d,e
		....	.....	0.012	....	e
	Montana	....	.....	0.004	1 year	a
				0.012	....	e
Sulfuric acid	Missouri	0.01[f]	.....	0.004	1 year	g
				0.012	....	
				0.03	30 min	e
	Montana	....	0.03[e]	0.004	1 year	
				0.012	....	
	New York	0.1				
Suspended particulate matter	Colorado	....	.....	0.12	3 months	
	Delaware:					
	Rural	0.06	.....	....	....	n
		0.13	.....	....	....	o
	Residential	0.075	.....	....	....	n
		0.15	.....	....	....	o
	Commercial	0.095	.....	....	....	n
		0.17	.....	....	....	o
	Industrial	0.125	.....	....	....	n
		0.20	.....	....	....	o
	All	....	0.5	....	....	
	Missouri	0.2[e]	.....	0.075	1 year	d
	Montana	0.2[h]	.....	0.075	1 year	
	Oregon:					
	Industrial	....	.....	0.25	....	b,i
		....	.....	0.30	....	i
	Other	....	.....	0.15	....	b,i
		....	.....	0.20	....	i
	Pennsylvania:	0.5	.....	0.15	30 days	
				0.15	30 days	a
	Air basin	....	.....	0.1	30 days	j
	Fugitive dust	....	.....	2	10 min	a,k
	South Carolina	0.025	0.05	0.1	15 min	Residential
		0.05	0.1	0.2	15 min	Nonresidential
	Texas:					
	Residential	0.125	.....	....	....	l
	Commercial	0.15	.....	....	....	l
	Industrial	0.175	.....	....	....	l
	Other	0.2	.....	....	....	l,m

[a] At any point beyond the property on which the source thereof is located.
[b] Above normal background concentration.
[c] Tentative.
[d] St. Louis metropolitan area.
[e] Not to be exceeded over 1 percent of the time.
[f] Not more than once in any 90 days.
[g] Not more than once in any 48 hr. This value appears elsewhere as the hourly average not to be exceeded over 1 percent of the time.
[h] Not to be exceeded more than 1 percent of the days in a year.
[i] The regulations are intended as a rate, and consequently 24 hr, 1 hr, or 1 min averaging time can be used.
[j] Based on results from geographically uniformly spaced sampling stations.
[k] From a source near ground level other than a flue.
[l] Not to be exceeded more than 10 percent of the days in a month.
[m] Vacant, range, or agricultural land, except within 100 ft of a residence.
[n] Geometric mean (50th percentile) (annual basis).
[o] 95th percentile (annual basis).

TABLE 1C State Air Quality Standards for Deposited Particulate Matter (U.S.A.)

State	Area	Tons/sq mi/month	Note
Missouri	Nonindustrial	10	[a]
	Heavy industrial	25	[a]
Montana	Residential	15	[b]
	Heavy industrial	30	[b]
Oregon	Residential-commercial	15	[c]
	Industrial	30	[c]
Virginia	Residential-commercial	15	[c]
	Industrial	30	[c]
		Mg/sq cm/month	
Pennsylvania	Any	1.5	[d]
	Air basin	1.0	[e]
	Emission standard	0.6	[f]

[a] In St. Louis metropolitan area, 3 months average above 5 tons/sq mi/month background.
[b] Includes 5 tons/sq mi/month background.
[c] Above normal background concentration.
[d] Not to be exceeded as the average of three successive sampling periods.
[e] Based on results from geographically uniformly spaced sampling stations.
[f] At any point beyond the property on which the source thereof is located.

dioxide (Table 8). In many of the world's nations, the situation is analogous to that in the United States; i.e., enabling or authorizing legislation has been but recently adopted and the resulting air quality standards are, as of this writing, being developed, but have not yet been promulgated.

EMISSION STANDARDS

United States

The Clean Air Act authorizes the promulgation of federal emission standards for motor vehicle exhaust. It does not authorize the setting of federal emission standards for any other class of source. The Act requires that within a specified time after the states submit state air quality standards to the Secretary of Health, Education and Welfare for his approval, they must indicate the means by which these air quality standards are to be enforced by the state and its subordinate jurisdictions. Although the law does not say so, in so many words, the most obvious means of implementation will have to be the adoption of state and local emission standards, the enforcement of which should achieve the air quality standard adopted (Fig. 1).

There is no prohibition in the Act of states and their jurisdictions having, in their regulations, emission standards (whether adequate or inadequate) for pollutants and processes for which no state air quality standard has been adopted; nor for areas of the state outside designated air quality control regions; nor for having in their regulations both air quality and emission standards (whether adequate or inadequate) for pollutants for which the Secretary has not yet sent to the states air quality criteria and documentation on applicable control technology. However, it appears implicit in the Clean Air Act that, for those pollutants for which the Secretary has issued air quality criteria and documentation of applicable control technology, the state and its jurisdiction cannot have in their regulations emission standards applicable to designated air quality control regions which are inadequate to achieve within these regions the air quality standards approved by the Secretary for the state. It would therefore seem that emission standards now on the books which are inadequate for this purpose will have to be changed. Tables 9 and 10 list state and local emission standards presently existent in the United States.

TABLE 2 U.S.S.R. Air Quality Standards

Substance	Single exposure (approx 20-min averaging time)		24-hr averaging time	
	mg/cu m stp	ppm	mg/cu m stp	ppm
Acetaldehyde	0.01	0.005	......	......
Acetic acid	0.2	0.08	......	......
Acetic anhydride	0.1	0.02	......	......
Acetone	0.35	0.15	0.35	0.15
Acetophenone	0.003	0.0006	0.003	0.0006
Acrolein	0.3	0.12	0.1	0.04
Ammonia	0.2	0.28	0.2	0.28
Amyl acetate	0.1	0.019	0.1	0.019
Amylene	1.5	0.5	1.5	0.5
Aniline	0.05	0.013	0.03	0.008
Arsenic (as As)	......	......	0.003	......
Benzene	1.5	0.5	0.8	0.25
Butane	200	85	......	......
Butanol	0.3	0.1	......	......
Butyl acetate (-*n*)	0.1	0.021	0.1	0.021
Butylene	3	1.3	3	1.3
Butyric acid	0.015	0.004	0.01	0.003
Caprolactum	0.06	0.013	0.06	0.013
Caprylic acid	0.01	0.002	0.005	0.001
Carbon black (soot)	0.15	......	0.05	......
Carbon disulfide	0.03	0.01	0.01	0.0033
Carbon monoxide	3	2.7	1	0.9
Carbon tetrachloride	4	0.7	......	......
Chlorine	0.1	0.033	0.03	0.01
Chloroaniline (-*p*)	0.04	0.008	......	......
Chlorobenzene	0.1	0.02	0.1	0.02
Chloroprene	0.1	0.028	0.1	0.028
Chlorophenyl isocyanate (-*m*)	0.005	0.001	0.005	0.001
Chlorophenyl isocyanate (-*p*)	0.0015	0.0002	0.0015	0.0002
Chromium, hexavalent (as CrO_3)	0.0015	......	0.0015	......
Cyclohexanol	0.06	0.015	0.06	0.015
Cyclohexanone	0.04	0.008	0.04	0.008
Dichloroethane	3	0.75	1	0.25
2,3-dichloro 1,4-naphthaquinone	0.05	......	0.05	......
Diethylamine	0.05	0.02	0.05	0.02
Diketene	0.007	0.002	......	......
Dimethylaniline	0.0055	0.001	......	......
Dimethyl disulfide	0.7	0.18	......	......
Dimethylformamide	0.03	0.01	0.03	0.01
Dimethyl sulfide	0.08	0.03	......	......
Dinyl (diphenyl + its oxides)	0.01	0.0015	0.01	0.0015
Divinyl	3	1.2	1	0.4
Epichlorohydrin	0.2	0.05	0.2	0.05
Ethanol	5	2.5	5	2.5
Ethyl acetate	0.1	0.029	0.1	0.029
Ethylene	3	2.3	3	2.3
Ethylene oxide	0.3	0.15	0.03	0.015
Formaldehyde	0.035	0.029	0.012	0.01
Fluorides (as F)	0.03	......	0.01	......
Fluorides (insoluble)	0.2	......	0.03	......
Furfural	0.05	0.013	0.05	0.013
Gasoline (as C) (from crude oil)	5	1.25	1.5	0.38
Gasoline (as C) (from shale)	0.05	0.01	0.05	0.01

TABLE 2 U.S.S.R. Air Quality Standards (Continued)

Substance	Single exposure (approx 20-min averaging time)		24-hr averaging time	
	mg/cu m stp	ppm	mg/cu m stp	ppm
Hexamethylene diamine	0.01	0.002	0.01	0.002
Hydrochloric acid (as H+)	0.006	....	0.006	......
Hydrochloric acid (as HCl)	0.2	0.15	......	......
Hydrogen fluoride	0.02	0.03	0.005	0.008
Hydrogen sulfide	0.008	0.005	0.008	0.005
Isopropyl benzene	0.014	0.003	0.014	0.003
Isopropyl benzene hydroperoxide	0.007	0.001	0.007	0.001
Lead (as Pb)	....	....	0.0007	......
Lead sulfide (as Pb)	....	....	0.0017	......
Malathion	0.015	....	......	......
Maleic anhydride	0.2	0.05	0.05	0.01
Manganese (as Mn)	0.03	....	0.01	......
Mercury (as Hg)	....	....	0.0003	......
Mesidine (2-amino-1,3,5-trimethyl benzene)	0.003	....	......	......
Methanol	1	0.75	0.5	0.38
Methyl acetate	0.07	0.023	0.07	0.023
Methyl acrylate	0.01	0.003	......	......
Methyl aniline	0.04	0.01	......	......
Methyl mercaptan	9×10^{-6}	....	......	......
Methyl methacrylate	0.1	0.025	0.1	0.025
Methyl parathion	0.008	....	......	......
Methyl styrene ($-\alpha$)	0.04	0.01	0.04	0.01
Naphthaquinone ($-\alpha$)	0.005	0.001	0.005	0.001
Nitric acid (as HNO_3)	0.4	0.15	......	......
Nitric acid (as H+)	0.006	....	0.006	......
Nitrobenzol	0.008	0.001	0.008	0.001
Nitrogen dioxide	0.085	0.045	0.085	0.045
Phenol	0.01	0.0026	0.01	0.0026
Phosphoric anhydride	0.15	0.026	0.05	0.0085
Phthalic anhydride	0.2	0.03	0.2	0.03
Propanol	0.3	0.12	......	......
Propylene	3	1.5	3	1.5
Pyridine	0.08	0.023	0.08	0.023
Styrene	0.003	0.0007	0.003	0.0007
Sulfur dioxide	0.5	0.19	0.15	0.058
Sulfuric acid (as H+)	0.006	....	0.006	......
Sulfuric acid (as H_2SO_4)	0.3	....	0.1	......
Suspended particulate matter (dust)	0.5	....	0.15	......
Thiophene	0.6	0.17	......	......
Toluene	0.6	0.15	0.6	0.15
Toluene di-isocyanate	0.05	0.0071	0.02	0.0029
Tributyl phosphate	0.01	....	......	......
Trichloroethylene	4	0.67	1	0.17
Valeric acid (-*n*)	0.03	0.008	0.01	0.003
Vanadium pentoxide	....	....	0.002	......
Vinyl acetate	0.2	0.06	0.2	0.06
Xylene	0.2	0.05	0.2	0.05

TABLE 3 Polish Air Quality Standards

Substance	Single exposure 20-min averaging time		24-hr averaging time	
	mg/cu m stp	ppm	mg/cu m stp	ppm
Carbon disulfide*	0.045	0.015	0.015	0.005
Carbon monoxide†	3	2.7	0.5	0.45
Gasoline†	2.5	0.6	0.75	0.2
Hydrogen sulfide*	0.06	0.04	0.02	0.013
Hydrogen sulfide†	0.008	0.005	0.008	0.005
Nitrogen oxides*	0.6	0.33	0.2	0.11
Nitrogen oxides†	0.15	0.08	0.05	0.03
Sulfur dioxide*	0.9	0.35	0.35	0.13
Sulfur dioxide†	0.25	0.1	0.075	0.03
Sulfuric acid*	0.3	.....	0.1	.....
Sulfuric acid†	0.15	.....	0.05	.....
Suspended particulate matter (dust)*	0.6	.....	0.2	.....
Suspended particulate matter (dust)†	0.2	.....	0.075	.....

NOTE: There are also deposited particulate matter standards of 250 tons/sq km/year* and 40 tons/sq km/year† and 6.5 tons/sq km/month.

* For protected areas.

† For specially protected areas.

TABLE 4 Federal Republic of Germany Air Quality Standards

Substance	30-min averaging time			
	Long-term exposure		Not to be exceeded more than once in any 8 hr	
	mg/cu m stp	ppm	mg/cu m stp	ppm
Chlorine	0.3	0.1	0.6	0.2
Hydrogen sulfide	0.15	0.1	0.3	0.2
Nitrogen oxides	1	0.5	2	1
	Long-term exposure		Not to be exceeded more than once in any 2 hr	
Sulfur dioxide	0.4	0.15	0.75	0.3

NOTE: There are also deposited particulate matter standards (in g/sq m/month) of 0.65 and 1.3 for general and industrial areas, respectively, and for the average of 12-month averages of 0.42 and 0.85, respectively.

TABLE 5 VDI (German Engineering Society) Air Quality Standards

Substance	30-min averaging time			
	Long-term exposure		Not to be exceeded more than once in any 4 hr	
	mg/cu m stp	ppm	mg/cu m stp	ppm
Acetaldehyde	4	2	12	6
Acetic acid	5	2	15	6
Acetone	120	50	360	150
Acrolein	0.01	0.005	0.025	0.01
Amyl acetate	30	5	90	15
Amyl alcohol	20	5	60	15
Aniline	0.8	0.2	2.4	0.6
Benzene	3	1	10	3
Butanol	15	5	45	15
Butyl acetate	25	5	75	15
Carbon tetrachloride	3	0.5	10	1.5
Chlorobenzene	5	1	15	3
Chloroform	10	2	30	6
Cresol	0.2	0.05	0.6	0.15
Cyclohexanone	10	2	30	6
Dichloroethane	8	2	25	6
Diethylamine	0.03	0.01	0.09	0.03
Diethyl ether	65	20	195	60
Dimethylamine	0.02	0.01	0.06	0.03
Dinitrobenzene	0.035	0.005	0.1	0.015
Dioxane	20	5	60	15
Ethanol	100	50	300	150
Ethyl acetate	75	20	225	60
Ethylene oxide	4	2	12	6
Formaldehyde	0.03	0.02	0.07	0.06
Furfural	0.08	0.02	0.25	0.06
Gasoline	80	20	240	60
Higher alkyl benzols	....	5	....	15
Methanol	15	10	40	30
Methyl acetate	15	5	45	15
Methyl ethyl ketone	30	10	90	30
Methyl isobutyl ketone	20	5	65	15
Methylene chloride	20	5	55	15
Monoethyl amine	0.02	0.01	0.06	0.03
Monomethyl amine	0.02	0.01	0.06	0.03
Naphthalene	2.5	0.5	7.5	1.5
Nitrobenzol	0.3	0.05	0.85	0.15
Perchlorethylene	35	5	110	15
Phenol	0.2	0.05	0.6	0.15
Propanol	50	20	150	60
Pyridine	0.7	0.2	2.1	0.6
Tetrahydrofuran	30	10	90	30
Toluene	20	5	60	15
Toluene di-isocyanate	0.007	0.001	0.021	0.003
Trichloroethane	30	5	90	15
Trichloroethylene	30	5	90	15
Triethylamine	0.04	0.01	0.12	0.03
Trimethyl amine	0.02	0.01	0.06	0.03
Turpentine	25	5	75	15
Vinyl acetate	20	5	60	15
Xylene	20	5	60	15
	Long-term exposure		Not to be exceeded more than once in any 8 hr	
Chlorine	0.3	0.1	0.6	0.2
Nitrogen oxides	1	0.5	2	1
	Long-term exposure		Not to be exceeded more than once in any 2 hr	
Hydrochloric acid (as HCl)	0.7	0.5	1.4	1
Nitric acid	1.3	0.5	2.6	1

TABLE 6 Czechoslovak Air Quality Standards

Substance	Single exposure (approx 30-min averaging time)		24-hr averaging time	
	mg/cu m stp	ppm	mg/cu m stp	ppm
Ammonia	0.3	0.42	0.1	0.14
Arsenic (as As)*	.....	.....	0.003	......
Benzene	2.4	0.75	0.8	0.25
Carbon black (soot)	0.15	.....	0.05	......
Carbon disulfide	0.03	0.01	0.01	0.0033
Carbon monoxide	6	5.4	1	0.9
Chlorine	0.1	0.033	0.03	0.01
Formaldehyde	0.05	0.033	0.015	0.01
Fluorides (inorganic, gaseous)	0.03	.....	0.01	......
Hydrochloric acid (as H+)	0.01	.....	......	......
Hydrogen sulfide	0.008	0.005	0.008	0.005
Lead (as Pb)†	.....	.....	0.0007	......
Manganese (as MnO_2)	.....	.....	0.01	......
Nitric acid (as H+)	0.01	.....	......	......
Nitrogen oxides (as NO_2)	0.3	0.16	0.1	0.06
Phenol	0.3	0.075	0.1	0.026
Sulfur dioxide	0.5	0.19	0.15	0.06
Sulfuric acid (as H+)	0.01	.....	......	......
Suspended particulate matter (dust)	0.5	.....	0.15	......

* Also its inorganic compounds, except arsine.
† Also its compounds, except tetraethyl lead.

In addition to the standards listed in these two tables, a number of American jurisdictions have adopted emission standards for total solid particulate pollutants in effluent air or gases from sources in general, and from combustion sources in particular. A wide variety of units have been used to specify these standards. When converted to the common units of mg/cu m stp (in the case of combustion gases, adjusted to either 12 percent CO_2, 50 percent excess air, or 6 percent O_2), the standards for sources in general range from 120 to 1,210 mg/cu m and those for combustion sources from 690 to 3,150 mg/cu m. Of the latter, 60 percent are at the value of 1,030 mg/cu m, and 25 percent are less than that value. The most recent trend is for values lower than 840 mg/cu m.

Tables 9 and 10 list only the single-valued emission standards. A number of jurisdictions use multivalued limits which require use of a chart or table. Most of these charts and tables can be found in either "A Compilation of Selected Air Pollution Emission Control Regulations and Ordinances," Public Health Service Publication 999-AP-43, U.S. Department of Health, Education and Welfare, Washington, D.C., 1968; or A. C. Stern, "Air Pollution Standards," in "Air Pollution," vol. III, 2d ed., pp. 601–718, Academic Press Inc., New York, 1968.

Smoke emission standards (Table 11) limit such emissions, in terms of smoke density on the Ringelmann scale, which varies as follows:

Scale Number	*Percent Blackness*
0	0
1	20
2	40
3	60
4	80
5	100

TABLE 7 Ontario (Canada) Air Quality Standards

Substance	Single exposure (30-min averaging time)*		24-hr averaging time	
	mg/cu m stp	ppm	mg/cu m stp	ppm
Ammonia	3.5	5		
Beryllium	0.00001	.....	0.00001	
Calcium oxide (lime)	0.02			
Carbon disulfide	0.45	0.15		
Carbon monoxide	5.5	5		
Chlorine	0.3	0.1		
Fluorides (as HF):				
Industrial—commercial	....	.....	0.0026	0.004
Residential—rural	....	.....	0.007	0.001
Fluorides	....	0.01		
Hydrogen chloride	0.05	0.04		
Hydrogen sulfide	0.045	0.03		
Lead (as Pb)	0.02			
Nitrogen oxides	....	.....	0.18	0.1
Oxidant (by KI)	....	.....	....	0.1
Sulfur dioxide:				
Industrial—commercial	0.78	0.3	0.52	0.2
Residential—rural	0.78	0.3	0.26	0.1
Suspended particulate matter (dust):				
Industrial—commercial	0.2	.....	0.175†	
Residential—rural	0.1	.....	0.09†	
	1-hr averaging time		8-hr averaging time	
Carbon monoxide	66	60	16.5	15
Nitrogen oxides	0.36	0.2		
Oxidant (by KI)‡	....	0.15		
			1-year averaging time	
Sulfur dioxide:				
Industrial—commercial	1.04	0.4	0.13	0.05
Residential—rural	0.65	0.25	0.05	0.02
Suspended particulate matter (dust):				
Industrial—commercial§	....	.....	0.11	
Residential—rural§	....	.....	0.06	

NOTE: There are also deposited particulate matter standards (in tons/sq mile/month) of 20 and 40 for residential-rural and industrial-commercial areas, respectively; for the average of 12-month averages of 13 and 25, respectively; and a single-exposure* value of 15.

* Maximum ground level concentration from a single source measured on the centerline downwind from the stack.

† 90 percent of 24-hr samplings.

‡ 90 percent of samplings in 1 month must be less than 0.07 ppm.

§ Geometric mean.

TABLE 8 Air Quality Standards for Sulfur Dioxide in Other Countries

Country	Single exposure (approx 30-min averaging time)		24-hr averaging time	
	mg/cu m stp	ppm	mg/cu m stp	ppm
The Netherlands[a] (avg over 1 yr)	....	....	0.15	0.05
[a,b]	....	....	0.3	0.1
[a,c]	....	....	0.5	0.18
Rumania	0.68	0.26	0.23	0.09
Sweden[d]	0.65	0.25	0.26	0.1
Switzerland:				
Mar. 1–Oct. 31[e]	0.75	0.3	0.52	0.2
Nov. 1–Feb. 28/29[e]	1.3	0.5	0.75	0.3
Valois Canton[f]	....	....	0.52	0.2

[a] Proposed standard (not yet adopted).
[b] Not to be exceeded more than 2 percent of the time.
[c] Not to be exceeded more than 0.3 percent of the time, or 1 day per year.
[d] 24-hr mean not to be exceeded more than once in any 30 days; 30-min mean not to be exceeded more than 1 percent of the time in any 30 days. Also a monthly standard of 0.05 ppm (0.13 mg/cu m).
[e] Single exposure standard not to be exceeded more than once in any 2 hr.
[f] Outside area of source.

Although there are no federal emission standards universally applicable in the United States for other than motor vehicles (Table 12), there are standards adopted by the federal government for application solely to federal installations (Table 13).

Great Britain

In Great Britain, no air quality standards have been adopted. However, under the provisions of the Alkali Act, a number of emission standards have been adopted (Tables 14 and 15).

Federal Republic of Germany

In addition to the air quality standards adopted officially and by the VDI Kommission Reinhaltung der Luft previously referred to, there are also both official and quasi-official (VDI) emission standards in the Federal Republic of Germany (Tables 16 and 17).

Japan

The Japanese have adopted several emission standards (Table 18).

Other Countries

In the U.S.S.R. and the other eastern European member states of the Council for Mutual Economic Aid, there has been a tendency until now to rely solely on air quality standards and not to adopt official emission standards. Czechoslovakia is an exception (Table 19). A few of the countries other than those already discussed (Tables 20 and 21) have adopted emission standards for specific pollutants. A more widespread coverage may be expected in the future.

TABLE 9 State and Local Emission Standards for Specific Pollutants (U.S.A.)

Substance	Jurisdiction	Source of emission	Original units	mg/cu m stp	ppm
Carbonyls[a,b]	Bay Area APCD	Incineration	ppm	.....	50
Fluorine	Florida	Primarily from fertilizer mfg.	0.4 lb/ton P_2O_5	.....	......
Hydrocarbons[b]	Bay Area APCD	Incineration	ppm	.....	50
Nitrogen oxides[c]	California	Automobiles	ppm	630	350
Organic compounds	Bay Area APCD[d]	As hexane	ppm	.....	50
		As total carbon	ppm	.....	300
		....	20 lb/day	.....	......
	Los Angeles APCD[e]	Heated or baked	15 lb/day	.....	......
		Unheated	40 lb/day	.....	......
Reactive compounds	Bay Area APCD[d]	All olefins, substituted aromatics and aldehydes	10 lb/day	.....	......
	Los Angeles APCD[f]	Dispose or evaporate	1.5 gal/day	.....	......
Sulfur dioxide	Bay Area APCD	....	ppm	5,200	2,000
	Dade County, Fla.	....	ppm	5,200	2,000
	Eugene, Ore.	Combustion	0.2 % by vol	5,200	2,000
	Los Angeles APCD	....	0.2 % by vol	5,200	2,000
	Missouri[g]	Existing nonfuel sources	ppm	5,200	2,000
		New nonfuel sources	ppm	1,300	500
		Fuel burning sources[h]	2.3 lb/million Btu heat input	.....	......
	Natick, Mass.	Fuel burning sources	0.4 % by vol	10,400	4,000
	New Jersey[i]	Other than fuel burning	ppm	5,200	2,000
		Sulfur removal from H_2S	ppm	39,000	15,000
	New York, N.Y.	....	ppm	5,200	2,000
	Riverside County, Calif.	....	0.2 % by vol	5,200	2,000
	Sarasota County, Fla.	....	0.2 % by vol	5,200	2,000
Sulfuric acid	Bay Area APCD	Acid mfg. from S or pyrites	0.08 grain/cu ft	184	......
		Acid mfg. from other materials	0.3 grain/cu ft	690	......
	Missouri	Existing nonfuel sources	mg/cu m	70	......
		New nonfuel sources	mg/cu m	35	......
	New Jersey	....	10 mg/cu m	350	......

[a] As formaldehyde.
[b] Corrected to 5 percent O_2 in the flue gas.
[c] Average for specified test cycle.
[d] Where less than 5 percent by volume of the organic compounds is "reactive" or "reactive" compound emissions are reduced at least 85 percent, these limits do not apply. There are other exemptions in Regulation 3.
[e] Except where there is 85 percent or more solvent removal from effluent.
[f] As defined by Rule 66 of Los Angeles APCD.
[g] St. Louis metropolitan area.
[h] Different dates of applicability for installations larger and smaller than 2,000,000,000 Btu/hr.
[i] Not applicable with under 3,000 cfm of gas, and 50 lb/hr SO_2 emission or 100 lb/hr instantaneous emission rate.

TABLE 10 Emission Standards for Total Particulate Matter in Effluent Air or Gases from Specific Processes (U.S.A.)

Source of emission	Jurisdiction or code	Standard—original units	mg/cu m stp
Incineration of refuse	Allegheny County, Pa.	0.2 lb/1,000 lb gas	240
	Bay Area APCD	0.2 grain/cu ft	460[a]
	Cleveland, Ohio	0.2 lb/1,000 lb gas	240
	Cincinnati, Ohio	0.4 lb/1,000 lb gas	490[a]
	Dearborn, Mich.	0.3 grain/cu ft at 500°F	1,280[a]
	Florida	0.2 grain/cu ft	460[a]
	New York, N.Y.	0.65 lb/1,000 lb gas	790[a,b]
Existing residential and commercial	Delaware	0.75 lb/million Btu	....
Existing industrial and rural	Delaware	1 lb/million Btu	....
New industrial and rural	Delaware	0.75 lb/million Btu	....
New residential and commercial	Delaware	0.5 lb/million Btu	....
200 lb/hr and over	Missouri, Maryland	0.2 grain/cu ft	460[a,c]
	Falls Church, Va.	0.3 lb/1,000 lb gas	370[a,b]
	Montgomery County, Md.	0.3 lb/1,000 lb gas	370[a,b]
	Prince Georges County, Md.	0.3 lb/1,000 lb gas	370[a,b]
	Rockville, Md.	0.3 lb/1,000 lb gas	370[a,b]
Less than 200 lb/hr	Missouri, Maryland	0.3 grain/cu ft	690[a,c]
	Falls Church, Va.	0.65 lb/1,000 lb gas	790[a,b]
	Montgomery County, Md.	0.65 lb/1,000 lb gas	790[a,b]
	Prince Georges County, Md.	0.65 lb/1,000 lb gas	790[a,b]
	Rockville, Md.	0.65 lb/1,000 lb gas	790[a,b]
Less than 1,000 lb/hr	Illinois	0.35 grain/cu ft	805[a]
1,000 lb/hr and more	Illinois	0.2 grain/cu ft	460[a]
Up to 10,000 lb/hr	Philadelphia, Pa.	0.2 lb/1,000 lb gas	240[a]
Over 10,000 lb/hr	Philadelphia, Pa.	0.6 lb/1,000 lb gas	740[a]
Municipal	Michigan	0.3 lb/1,000 lb gas	370[a,d]
Industrial, less than 400 lb/hr	Michigan	0.65 lb/1,000 lb gas	790[a,d]
Industrial, more than 400 lb/hr	Michigan	0.3 lb/1,000 lb gas	370[a,d]
Residential, less than 200 lb/hr	Michigan	0.65 lb/1,000 lb gas	790[a,d,e]
Residential, more than 200 lb/hr	Michigan	0.3 lb/1,000 lb gas	370[a,d]
Industrial processes:			
Asphaltic concrete plants	Florida	0.3 grain/cu ft	690[f]
	Wayne County, Mich.	0.2 lb/1,000 lb gas	240[d]
Stationary	Michigan	0.3 lb/1,000 lb gas	370[g]
Portable:			
0–100 tons/hr	Michigan	0.6 lb/1,000 lb gas	740[g]
101–150 tons/hr	Michigan	0.5 lb/1,000 lb gas	620[g]
151–200 tons/hr	Michigan	0.45 lb/1,000 lb gas	550[g]
Over 200 tons/hr	Michigan	0.35 lb/1,000 lb gas	430[g]
Cement, grinding, crushing, etc.	Michigan	0.15 lb/1,000 lb gas	180[h]
Cement, clinker coolers	Michigan	0.3 lb/1,000 lb gas	370[g]
Corn milling (wet)	Illinois	0.75 grain/cu ft	1,725[i]
Kilns, cement:	Illinois	0.1 grain/cu ft	230[j]
Lime	Michigan	0.25 lb/1,000 lb gas	310[h]
	Michigan	0.2 lb/1,000 lb gas	240
Design	Wayne County, Mich.	0.2 lb/1,000 lb gas	240[d,k]
Operation	Wayne County, Mich.	0.3 lb/1,000 lb gas	370[d,l]
Metallurgical processes:			
Ferrous:			
Bessemer converters	Allegheny County, Pa.	0.65 lb/1,000 lb gas	790[a,m]
	Cleveland, Ohio	0.2 lb/1,000 lb gas	240[n]
Blast furnace gas, bled	Allegheny County, Pa.	0.5 lb/1,000 lb gas	610
	Detroit, Mich.	0.5 lb/1,000 lb gas	610[d]
	Illinois	0.1 grain/cu ft	230
	Wayne County, Mich.	0.5 lb/1,000 lb gas	610[d]
Blast furnace gas, burned	Allegheny County, Pa.	0.35 lb/1,000 lb gas	430
	Detroit, Mich.	0.2 lb/1,000 lb gas	240[d]
	Illinois	0.5 grain/cu ft	115
	Wayne County, Mich.	0.2 lb/1,000 lb gas	240[d]
Cupolas	Chicago, Ill.	0.35 grain/cu ft	805
	Cleveland, Ohio	0.5 lb/1,000 lb gas	610
New	Allegheny County, Pa.	0.5 lb/1,000 lb gas	610
Existing	Allegheny County, Pa.	0.65 lb/1,000 lb gas	790
	Missouri	0.4 grain/cu ft	920[c,o]

(footnotes on page 4-16)

TABLE 10 Emission Standards for Total Particulate Matter in Effluent Air or Gases from Specific Processes (U.S.A.) (Continued)

Source of emission	Jurisdiction or code	Standard—original units	mg/cu m stp
Metallurgical processes:			
Ferrous:			
Cupolas (continued) :			
Jobbing, operation	Detroit and Wayne Counties, Mich.	0.4 lb/1,000 lb gas	490[d,i]
	Michigan	0.4 lb/1,000 lb gas	490[i]
Production, operation	Detroit and Wayne Counties, Mich.	0.25 lb/1,000 lb gas	310[d]
0–10 tons/hr	Michigan	0.4 lb/1,000 lb gas	490
11–20 tons/hr	Michigan	0.25 lb/1,000 lb gas	310
21 and over tons/hr	Michigan	0.1 lb/1,000 lb gas	120
Production-design	Detroit and Wayne Counties, Mich.	0.1 lb/1,000 lb gas	120[d,i]
Furnaces:			
Electric, basic oxygen, and open hearth	Allegheny County, Pa.	0.2 lb/1,000 lb gas	240
	Cleveland, Ohio	0.2 lb/1,000 lb gas	240
	Illinois	0.1 grain/cu ft	230
Operation	Detroit and Wayne Counties, Mich.	0.2 lb/1,000 lb gas	240[d]
	Michigan	0.15 lb/1,000 lb gas	180
Design	Detroit and Wayne Counties, Mich.	0.1 lb/1,000 lb gas	120[d]
	Michigan	0.1 lb/1,000 lb gas	120
Heating and reheating	Allegheny County, Pa.	0.3 lb/1,000 lb gas	370
	Michigan	0.3 lb/1,000 lb gas	370
	Detroit and Wayne Counties, Mich.	0.3 lb/1,000 lb gas	370[d]
Sintering plants	Allegheny County, Pa.	0.2 lb/1,000 lb gas	240
	Cleveland, Ohio	0.2 lb/1,000 lb gas	240
	Illinois	0.1 grain/cu ft	230
Operation	Detroit and Wayne Counties, Mich.	0.2 lb/1,000 lb gas	240[d]
	Michigan	0.2 lb/1,000 lb gas	240
Design	Detroit and Wayne Counties, Mich.	0.15 lb/1,000 lb gas	180[d]
Nonferrous:			
Furnace or smelter, general	Cleveland, Ohio	0.2 lb/1,000 lb gas	240

[a] Adjusted to 6 percent O_2, 12 percent CO_2, or 50 percent excess air.
[b] Maximum emission, 250 lb/hr.
[c] St. Louis and Washington metropolitan areas.
[d] In Detroit and Wayne County, Mich., less water vapor from wet collector, if employed.
[e] Does not apply to domestic incinerators having less than 5 cu ft storage capacity.
[f] Except for portable plants with fewer than three inhabited residences within 1 mile.
[g] In remote locations; if no water is available and emission limit of 0.3 lb/1,000 lb gas cannot otherwise be satisfied, plant may be located in 1 mile radius of uninhabited buffer zone.
[h] Up to 15,000 bbl/day kiln capacity. Special regulations for over 15,000 bbl/day total plant kiln capacity.
[i] Also supplemental regulations.
[j] Or 99.7 percent collector removal efficiency.
[k] Rotary kilns must have 99 percent weight collection efficiency, if more restrictive.
[l] Rotary kilns must have 98.5 percent weight collection efficiency, if more restrictive.
[m] Installations after July 5, 1960, only.
[n] Future date of applicability.
[o] Or 85 percent collector removal efficiency.

TABLE 11 Smoke Emission Standards (U.S.A.)

Emission greater than this Ringelmann number prohibited*	Date of enactment				Status as of end of 1965
	Before 1940	1940–1949	1950–1959	1960–1965	
3	61	(−1)	(−2)	(−7)	51
2	14	13	57	20	104
1	0	0	0	2	2
Total (net)	75	12	55	15	157

* Except in some cases, for special circumstances such as soot blowing or starting cold fires.

TABLE 12 Federal* Emission Standards for Motor Vehicles (U.S.A.)
Average for Specified Test Cycle

Substance	Original units	mg/cu m stp	ppm
Carbon monoxide:			
Over 140 cu in. displacement	1.5% by vol	16,500	15,000
100–140 cu in. displacement	2% by vol	22,000	20,000
50–100 cu in. displacement	2.3% by vol	25,500	23,000
Hydrocarbons:			
Crankcase	None permitted		
Over 140 cu in. displacement	ppm	......	275
100–140 cu in. displacement	ppm	......	350
50–100 cu in. displacement	ppm	......	410

* Also California.

TABLE 13 Federal Emission Standards for Federal Installations (U.S.A.)

Source of emission	Original units	mg/cu m stp
Particulate Matter		
Incineration of refuse:		
200 lb/hr and over*	0.2 grain/cu ft	460
Less than 200 lb/hr*	0.3 grain/cu ft	690
Sulfur Dioxide		
Fuel burning installations:		
In New York, N.Y.	0.35 lb/million Btu heat input	...
In Philadelphia	0.65 lb/million Btu heat input	...
In Chicago	0.65 lb/million Btu heat input	...

* Adjusted to 12 percent CO_2.

TABLE 14 British Emission Standards for Specific Pollutants

Substance	Source of emission	Original units	mg/cu m stp	ppm
Antimony[a,b]	Less than 5,000 cfm	0.05 grain/cu ft	115	......
Antimony[a,b]	More than 5,000 cfm	0.02 grain/cu ft	46	......
Arsenic[a,b]	Less than 5,000 cfm	0.05 grain/cu ft	115	......
Arsenic[a,b]	More than 5,000 cfm	0.02 grain/cu ft	46	......
Cadmium[a,c]	Max 30 lb/168 hr	0.017 grain/cu ft	39	......
Chlorine[a]	......	0.1 grain/cu ft	230	77
Hydrogen chloride	Alkali (salt cake) works	0.2 grain/cu ft	460	328
Hydrogen chloride	Hydrochloric acid works	0.2 grain/cu ft	460	328
Hydrogen fluoride[d]	......	0.1 grain/cu ft	230	......
Hydrogen sulfide	......	ppm	7.5	5
Lead[c,e]	Up to 3,000 cfm of exhaust gas	0.05 grain/cu ft	115	......
Lead[c,f]	3,000–10,000 cfm of exhaust gas	0.05 grain/cu ft	115	......
Lead[c,g]	10,000–140,000 cfm of exhaust gas	0.01 grain/cu ft	23	......
Lead[c]	Over 140,000 cfm of exhaust gas	0.005 grain/cu ft	12	......
Miscellaneous[h]	Superphosphate fertilizer manufacture	0.1 grain/cu ft total acidity as SO_3	230	......
Nitrogen oxides[d]	......	1 grain/cu ft	2,300	1,280
Nitrogen oxides[d]	Nitric acid plants	2 grains/cu ft	4,600	2,560
Sulfur dioxide	Chamber sulfuric acid plants	4 grains/cu ft (as SO_3)	9,200	......
Sulfur dioxide	Sulfuric acid concentration	1.5 grains/cu ft (as SO_3)	3,450	......
Sulfur dioxide	Contact sulfuric acid plants, sulfur burning	2% of the sulfur burned		......
Sulfur dioxide	Contact sulfuric acid plants other than sulfur burning	4 grains/cu ft (as SO_3)	9,200	......
Total solid particulate matter:				
Greater than 10 microns	......	0.2 grain/cu ft	460	......
Less than 10 microns	......	0.05 grain/cu ft	115	......

[a] Also compounds of the element.
[b] As the trioxide.
[c] As the element.
[d] As SO_3 equivalent in original units.
[e] 100 lb/week mass emission limit.
[f] 400 lb/week mass emission limit.
[g] 1,000 lb/week mass emission limit.
[h] Or efficiency of condensation of acid gases greater than 99 percent.

TABLE 15 British Emission Standards for Particulate Matter from Specific Processes

Source of emission	Original units	mg/cu m stp
Blast furnace gas, bled:		
Integrated works	0.2 grain/cu ft	460
Merchant furnaces	0.5 grain/cu ft	1,150
Cement, grinding and crushing:		
New plants	0.1 grain/cu ft	230
Existing plants	0.2 grain/cu ft	460
Kilns:*		
Up to 1,500 tons/day clinker	0.2 grain/cu ft	460
1,500–3,000 tons/day clinker	Sliding scale from 0.2 to 0.1	.....
3,000 and over tons/day clinker	0.1 grain/cu ft	230
Rock crushing, dust handling	0.2 grain/cu ft	460
Cupolas, hot blast:		
Grit and dust†	0.2 grain/cu ft	460
Fume†	0.05 grain/cu ft	115
Electric generating plants:		
Old	0.2 grain/cu ft	460
New‡	0.05 grain/cu ft	115
Furnaces, steel, all oxygen refining processes producing fume†	0.05 grain/cu ft	115
Nonfume processes†	0.2 grain/cu ft	460
Sintering plants	0.2 grain/cu ft	460

* Older kilns, up to 0.5 grain/cu ft.

† Grit and dust, over 10 microns; fume, less than 10 microns.

‡ New 2,000- to 4,000-Mw plants must have 99.3 percent fly ash removal efficiency after 12 months' operation.

TABLE 16 Federal Republic of Germany Emission Standards for Particulate Matter from Specific Processes

Source of Emissions	*mg/cu m stp*
Blast furnace gas:	
Bled	20
Burned	50
Cement grinding and crushing	150
Copper smelting	500
Incineration of refuse:	
Less than 20 tons/day	200*
More than 20 tons/day	150*
Lead smelting:	
Reducing furnaces	400
Refining furnaces	200
Slag blowing	100
Oxygen-using steelmaking furnaces	150†
Sintering plants:	
Continuous operation	150
Special cases	300‡
Zinc smelting:	
Distillation process	200
Electrothermic process	100
Rotary process	500
Stationary retorts	400
Miscellaneous§	150

* Adjusted to 7 percent CO_2.

† Particles smaller than 10 microns.

‡ Where, for instance, raw material is to be used in the form of fine dust and the applicant is able to show that, although the present state of technical development does not permit keeping within the 150 mg/cu m limit, no objectionable effects need be feared in the neighborhood.

§ Exhaust from screening, crushing, or filling plants and from other similar sources of emission.

TABLE 17 VDI (German Engineering Society) Emission Standards for Particulate Matter from Specific Processes

Standard No.	Source of emission	mg/cu m stp
2094	Cement kilns; cement grinding and crushing	150
2095	Sintering plants	300
2099	Blast furnace gas, burned	50
2100	Coke crushing and screening	150
2101	Primary copper smelting, refining, reverberatory and shaft furnaces	300
2102	Secondary copper smelting, blast and refining furnaces and converters	300
2112	Oxygen-using steelmaking furnaces	150[a]
2284	Zinc smelting:	
	Distillation process	200
	Electrothermic process	100
	Rotary process	500
	Stationary retorts	400
2285	Lead smelting:	
	Reducing furnaces	400
	Refining furnaces	200
	Slag blowing	100
2286	Aluminum reduction:	
	Alumina grinding	150
	Alumina calcining	100
	Primary reduction	100
2287	Copper smelting:	
	Hydrometallurgical:[b]	
	Cobalt calcination	1,000
	Inhibition plant	500
	Roasting plant	100
	Zinc calcination with scrubbing	3,000
	Zinc calcination with hydroelectric dust extraction	500[c]
2292	Coal briquetting	300[d]
2293	Coal preparation plants	300[d]
2301	Incineration of refuse:	
	Less than 20 tons/day	200[e]
	More than 20 tons/day	150[e]
2441	Aluminum reduction, secondary recovery furnaces	300[f]

[a] Particles smaller than 10 microns.
[b] Recovery after chloridizing roasting.
[c] Output over 30 tons/day of zinc.
[d] Limit for industrial exhaust ventilating system effluent, 150 mg/cu m.
[e] Adjusted to 7 percent CO_2.
[f] Total dust emission not to exceed 1 percent of aluminum production.

TABLE 18 Japanese Emission Standards for Particulate Matter from Specific Processes

Source of Emission	*mg/cu m stp*
Cupolas	200
Electric generating plants:	
Water wall	1,000
Other	1,200
Furnaces:	
Blast	500
Calcining	100
Electric	900
Gas generating	1,000
Glass melting, tank	700
Metal heating	700
Metal refining	700
Open hearth	600
Oxygen-using	1,000
Petroleum refinery	700
Reactor*	1,200
Reverberatory	700
Other	1,000
Incineration of refuse	700
Kilns:	
Cement	600
Calcining	
Continuous	700
Other than continuous	2,000
Drying	1,200
Lime	1,500
Sintering plants	1,000

* Including direct burning carbon black furnaces.

TABLE 19 Czechoslovak Emission Standards for Specific Pollutants

Substance	*kg/hr**
Acrolein	3
Ammonia	3
Arsenic (inorganic compounds, except arsine)	0.03
Benzene	24
Carbon black (amorphous carbon)	1.5
Carbon disulfide	0.3
Carbon monoxide	60
Chlorine	1
Fluorine (gaseous inorganic compounds)	0.3
Formaldehyde	0.5
Hydrochloric acid (as hydrogen ion)	0.1
Hydrogen sulfide	0.08
Lead (except tetraethyl lead)	0.007
Manganese (as MnO_2)	0.1
Mercury (metallic)	0.003
Nitric acid (as hydrogen ion)	0.1
Nitrogen oxides (as NO_2)	3
Phenol	3
Sulfuric acid (as hydrogen ion)	0.1
Total solid particulate matter†	5

* Emission rate above which it is necessary to submit a report to the government; where discharge is for less than 1 hr, there is a proportionate reduction in emission permissible without such reporting.

† Maximum SiO_2 content 20 percent.

TABLE 20 New South Wales and Queensland Emission Standards for Specific Pollutants

Substance	Source of emission	Original units	mg/cu m stp	ppm
Antimony	As the element	0.01 grain/cu ft	23	.....
Arsenic	As the trioxide	0.01 grain/cu ft	23	.....
Cadmium	As the element	0.01 grain/cu ft	23	.....
Chlorine	As the element	0.1 grain/cu ft	230	77
Hydrogen fluoride	....	0.05 grain/cu ft	115	.....
Hydrogen sulfide	....	ppm	7.5	5
Lead	As the element	0.01 grain/cu ft	23	.....
Mercury	As the element	0.01 grain/cu ft	23	.....
Nitrogen oxides	....	1 grain/cu ft	2,300	1,280
Nitrogen oxides	Nitric acid plants	2 grains/cu ft	4,600	2,560
Sulfur dioxide*	Metallurgical plant SO_2 recovery	4 grains/cu ft	9,200	3,540
†	Sulfuric acid manufacture	3 grains/cu ft	6,900	2,650
Sulfuric acid	....	0.1 grain cu ft	230	.....
Total solid particulate matter greater than 10 microns	....	0.2 grain/cu ft	460	.....

* New South Wales only.
† Or efficiency of condensation of acid gases greater than 99 percent.

TABLE 21 Emission Standards for Sulfur Dioxide from Specific Processes in Other Countries

Country	Source of emission	Original units	mg/cu m stp	ppm
Belgium	Lead and zinc smelting	1 part/1,000 parts	2,600	1,000
Paris, France	Combustion:			
	Less than 1 million kcal/hr	1 g/1,000 kcal	.....	.....
	More than 1 million kcal/hr	4 g/1,000 kcal	.....	.....

STACK HEIGHT STANDARDS

The principal other type of air pollution standard is that which requires certain minimum stack heights.

FUEL STANDARDS

Emission standards limiting the emission of sulfur dioxide and smoke can be complied with by controls on either process or fuel. However, some jurisdictions have chosen to limit directly the sulfur content of fuel or the volatile matter content of coal that may be burned in the jurisdiction.

Part 2: Water Quality Criteria and Standards for Industrial Effluents

ROBERT M. SANTANIELLO

Vice President and General Manager, Gulf Degremont, Inc., Liberty Corner, New Jersey

This year, approximately fifty thousand billion gallons of water will be fed into United States industrial plants and power generating facilities. Much of this water must first be treated in order to qualify it for the intended application. Such treatment may be quite simple or relatively complex but in most cases is amply justified by engineering or economic facts of life. Not every gallon reentering the free water domain as effluent matches its quality as an influent. A considerable quantity of effluent fails to meet quality requirements, still ill-defined, but nonetheless real, needed to prevent pollution. This pollution may be grossly evident as oil slick, foam, floating debris, or marked turbidity. It may, on the other hand, be manifest only in diminution or alteration of aquatic life, altered taste, or high concentrations of relatively nontoxic chemicals. Naturally, attention is directed most at those losses of water quality in which the industrial effluents are rendered unfit for their intended use as water for public consumption; maintenance of fish and wildlife; agriculture; recreation; or as influent, after standard treatment, in a downstream manufacturing facility.

Generally applicable control criteria and standards have been difficult to establish because of the variability of plant location, water sources, and processes and the practical need for controls. Agreement among even the foremost authorities in the field has been difficult to achieve.

The federal government has acted through the Federal Water Pollution Control Act as amended by the Water Quality Act of 1965 to initiate the establishment of standards and criteria for water pollution control. Under the terms of the act such controls were to originate at the state level whenever possible. Only in those cases in which the states did not specify criteria and standards in time to meet the June 30, 1968, deadline was the government to take it upon itself to establish "permissible" and "desirable" criteria. These criteria have recently been proposed by the National Technical Advisory Committee on Water Quality Criteria, reporting to the Federal Water Pollution Control Administration.

Industry must now take a hard look at these criteria in an effort to find the most effective and expeditious manner in which to comply. In its own best interests, industry should remain alert to the possibility that criteria[1] have been established which cannot be met practically and for which justification may not be entirely clear. However, in reviewing these criteria, it must be kept in mind that pollution problems require a 100-year bookkeeping outlook rather than the more customary 3- to 5-year payout. Because this long-range view is necessary, the problem is one which cannot be borne by industry alone. Society as a whole must carry its share of this responsibility.

The pollutant capability of a manufacturing facility or power generating station effluent is a function of the nature of that effluent and of the intended use of the body of water which receives it. That is, industrial pollution may be quantitated in each instance in terms of the way in which an effluent affects the evaluated use of the receiving body of water. For example, waters used downstream for cattle farming may safely contain up to 1 g/l of SO_4^{--}. If these same waters were to be used also as public waters, the recommended upper limit for SO_4^{--} is 0.25 g/l. A logical analysis preliminary to the establishment of water pollution control standards can be based upon the division of water sources into major end uses. The FWPCA Committee divides water end uses into five major categories. These are (1) public supply, (2) recreation and aesthetic value, (3) preservation of fish and wildlife, (4) agricultural use, and (5) industrial use.

The intended end use of water determines the degree of pollution permissible by certain specific constituents or characteristics. The concentration at which the designation "pollutant" is earned varies for each constituent from relatively high levels to *any* detectable quantity.

Pollutants directly attributable to industrial processes may be categorized as physical, chemical, and radioactive. Current criteria and standards for pollution control, based upon the recommendations of the Committee on Water Quality Criteria acting through its subcommittees, may be described by breaking down these large categories of pollutants into their major individual constituents.

PHYSICAL POLLUTANTS

Color

The presence of color in water may not necessarily be due to waste discharge from upstream facilities. Color in many instances is added to the watercourse because of the presence of decaying vegetation and bacteria.

Among the industries which commonly contribute to water color are the pulp and paper, textile, petrochemical, and chemical industries.

In general, the standards for color in waters designated for recreation and aesthetic use are not specified other than to contain the recommendation that surface waters should be free from all substances attributable to discharges or waste which produce objectionable color. This includes waters classified as primary contact[2] waters and those designated for secondary contact such as boating and fishing.

The criterion now designated as "permissible criterion" for color in waters used for public water supply is 75 color units (platinum cobalt—standard). A value

[1] There has recently been considerable controversy concerning the interpretation by the FWPCA and Mr. Udall that water supplies in any of the categories covered under the Act may not be subjected to any form of degradation, even if such degradation yields a water quality above that permitted under the criteria and standards suggested by the Committee on Water Quality Criteria.

[2] Primary contact recreation is defined as that recreational use in which there is such intimate contact with the water that ingestion of significant amounts is probable. Secondary contact recreation refers to the use of water in which significant ingestion is unlikely.

less than 10 color units is the recommended "desirable criterion." The criterion for color, as for other pollutants, of public water supplies is, in many cases, established on the basis of water quality attained by specific standard treatment methods. The water under evaluation is not to have been treated more vigorously than by coagulation employing 50 ppm alum, ferric sulfate, or copperas with alkali addition as necessary but without the use of coagulant aids or activated carbon; 6 hr of sedimentation; rapid sand filtration at 3 gal/sq ft/min; and disinfection with chlorine.

Colors which result from dyes and from other industrial and processing sources cannot be measured by the platinum cobalt test standard. They should be present only in concentrations which can be removed by the standard water treatment plant. In addition to color removals obtained through processes for treatment of other waste constituents, removals are obtained by standard chemical precipitation methods with associated absorption by the precipitant of flocculant. Carbon adsorption can be applied to the more difficult color removal problems.

Waters designated for maintenance of fish, other aquatic life, and wildlife require color standards. This is required because 10 percent of the light striking the surface of a body of water must arrive at the bottom of any photosynthetic zone in which it is desired to maintain adequate dissolved oxygen concentrations. Color in excess of 50 units may limit photosynthesis and by reducing dissolved oxygen levels disturb the balance between aquatic life forms.

In the case of marine and estuary waters, color additions should be limited to those which have been shown conclusively not to be deleterious to aquatic life.

For agricultural waters, it has been recommended that color not exceed 15 color units on the basis that such excess becomes aesthetically objectionable although it may not represent a health hazard.

Table 22 lists, among other constituents, allowable color concentrations for point of use waters in the power generating, textile, paper, petrochemical, and chemical, food, and cement industries. These values then are tolerable levels after recommended water treatment. Success in achieving these values with only the basic treatment system depends largely on the nature of upstream effluents, if any.

Temperature

In some instances, the most economic solution to a thermal pollution problem is more effective utilization of heat in the process systems or better heat recovery by higher-efficiency heat exchangers. For the thermal pollution problem associated with large volumes of waste cooling water and minimal temperature difference, the appplication of cooling towers and recirculation of cooling water through the system is a most practical answer.

It is recommended that for primary contact recreation waters, water temperature should not exceed 85°F (30°C), except where higher temperatures are caused by natural phenomena. It has been found that prolonged exposure to waters warmer than 85°F may produce undesirable physiological effects.

For public water supplies it is again recommended that water temperature not exceed 85°F and that there not be more than a 5°F water temperature increase or variations greater than 1°F/hr above that caused by ambient conditions. Any water temperature change adversely affecting the aquatic life, taste, odor, or chemistry of the water is to be avoided. Likewise any variation which adversely affects water treatment facilities or which decreases the acceptance of water for drinking or cooling purposes is to be avoided.

For waters in which a well-rounded population of warm and cold water fishes is to be preserved, it is recommended that heat not be added to a body of water at any time in amounts greater than that which will raise the temperature of the water flowing at the minimum daily gpm more than 5°F. For certain surface water zones in lakes this recommended temperature rise should not exceed 3°F. These increases should be calculated on the monthly average of maximum daily temperatures.

In general, cooling water should not be pumped from the constant temperature

TABLE 22 Point of Use Water Requirements for Selected Industries

Characteristic	Power generation	Textile	Pulp and paper	Chemical process	Petroleum	Food	Soft drink	Cement
Hardness ($CaCO_3$)	0–20*	25	100	250–900	350	250	‡	‡
pH, units	8–10*	2.5–10.5†	6–10	6.2–8.7	6–9	6.5–8.5	‡	7†
Calcium, mg/l	0	....	20	60–100	75	100	100	....
Chlorides, mg/l	No problem	....	200–1,000	500	300	250	500	250
Manganese	0.01–0.3	0.01–0.05	0.05–0.1	0.1–0.2	....	0.2	0.05	0.5
Iron, mg/l	0.01–1*	0.1–0.3	0.1–0.3	0.1–0.3	1	0.2	0.3	25
Color units	No problem	5	10–30	20	No problem	5	10	No problem
Alkaline ($CaCO_3$)	40–140	....	....	125–200	....	250	85	400
Suspended solids	0–10	5	10	5–30	10	10	0	500

* A function of boiler pressure.
† Depending upon operation (sizing, bleaching, scouring, etc.).
‡ Controlled by treatment for other constituents.

zone near the bottom of a lake to be discharged into the same body. Discharge of a heated effluent into this zone is not recommended. Normal daily and seasonal temperature variations present before the addition of artificially provided heat should be maintained. Table 23 lists the currently recommended maximum temperatures compatible with spawning and growth requirements for various fish species.

TABLE 23 Maximum Recommended Temperatures for Development of Various Fishes

Max temp, °F	Spawning and/or egg development	Growth
48	Lake trout, walleye, northern pike, Atlantic salmon	
55	Salmon, trout	
68	Perch, smallmouth bass	Salmon, trout
75	Largemouth and other bass	
80	Catfish, shad	
84	....	Pike, perch, smallmouth bass, walleye
90	....	Largemouth bass, bluegill, crappie
93	....	Catfish, gar, other bass, shad

Preservation of marine and estuarine organisms requires that there not be more than a 4°F rise in temperature associated with the addition of heat of artificial origin to coastal and estuarine waters during the months September through May. There should not be more than a 1.5°F rise during the summer months.

Temperature variations are not considered critical in waters destined for agricultural use, except in those cases where large quantities of water are used for hydrocooling farm products.

Odor and Taste

In general, odor in water is due to the presence of dissolved gases, such as hydrogen sulfide, and the presence of volatile organic compounds. When the threshold odor value exceeds three units, based on *n*-butyl alcohol calibration tests, it is generally considered objectionable. The criterion for water quality with respect to odor is simply that it not be objectionable.

Taste is an important factor in that it may be a direct indication of the presence of dissolved inorganic salts of iron, zinc, manganese, copper, sodium, potassium, et al. Certain inorganic and organic constituents may taint fish and other marine organisms, imparting characteristic tastes. The hydrocarbons, phenolic compounds, sodium pentachlorophenate, coal tar wastes, and the wastes from sewage, coal coking, kraft paper, and petroleum processes among others may contribute to the production of objectionable tastes in water or aquatic life. Chlorophenol, for example, has been found to produce an unpleasant taste in fish at concentrations of only 0.0001 mg/l.

Copper concentrations of 0.019 mg/l seem quickly to impart a green color and characteristic unpleasant taste to oysters. These organisms seem to have special capacity to absorb, store, and concentrate certain metals and nonmetals, demonstrated dramatically by the so-called superiodized oysters produced in France before World War II.

Taste and odor in water supplies may also result from natural phenomena such as decaying vegetation, algae, or bacterial slime. As in the case of color removal, taste and odor may also be removed through processes for treatment of or control of other waste constituents. Processes employed for taste and odor removal are coagulation, carbon adsorption, aeration, and oxidation with chlorine or other oxidizing agents.

Turbidity

In most cases turbidity is caused by suspended clay or silt, dispersed organics, and microorganisms.

The desirable criterion for public water supplies is that there be virtually no turbidity, that is, that there be low levels and that it not be objectionable. Moreover, the basic water treatment plant must be able to remove it adequately, continuously, and at reasonable cost. Since water treatment plants for turbidity removal are designed for the specific application, increases or fluctuations in turbidity loading above the capacity of the plant are considered excessive.

The common operations employed for the removal of turbidity are coagulation or flocculation, sedimentation, and filtration.

Methods used for measuring turbidity customarily may not adequately measure characteristics harmful to a public water supply. Since water with a given turbidity may coagulate more rapidly than one with less turbidity with or without the addition of coagulation aids, the establishment of turbidity criteria in terms of standard units has not been possible.

For waters designated for aquatic life preservation, turbidities due to the discharge of waste should not exceed 50 Jackson units[1] in warm water streams or 10 Jackson units in cold water streams.

Turbidities in warm water lakes should not exceed 25 Jackson units, and cold water lake turbidity should not exceed 10 units. For agricultural waters, it should not exceed 5 units, mainly on the basis of aesthetics.

CHEMICAL POLLUTANTS

pH

Control of effluent pH obviously requires adjustment of the hydrogen or hydroxyl ion concentrations. This can be effected by the addition of alkalies or acids as required. Adjustment of the hydrogen or hydroxyl ion can also be accomplished by the indirect application of acids and bases through use of ion exchange systems operated on a hydrogen or hydroxide cycle. Fixed bed or continuous systems may be used. Continuous systems, when applicable, offer higher efficiencies.

For primary contact recreation waters, it is recommended that the pH be in the range 6.5 to 8.3, in no case falling below 5 or rising above 9. Such criteria are based largely upon the physiology of the human eye and specifically upon the buffering capacity of tears.

For public water supplies the permissible criterion is 6.0 to 8.5.

Waters in which aquatic life is to be maintained should not be exposed to a pH below 6.0 or above 9.0. In the case of saline waters, materials which alter the pH by more than 0.1 unit should not be introduced. At no time in this case should the pH be less than 6.7 or greater than 8.5. It has been noted that aquatic plants of greatest value as waterfowl foods grow optimally in waters in which the pH ranges from 7.0 to 9.2.

Agricultural waters are by and large immune to nondrastic changes in pH, tolerating values between 4.5 and 9.0.

Alkalinity

Any of the standard techniques for neutralization may be used for control of alkalinity.

Alkalinity in public water supplies should be sufficient to cause effective floc formation but not high enough to be toxic or to produce a corrosive or encrusting water.

Alkalinity is determined by the relative amounts of bicarbonate, carbonate, and hydroxide ions. It is also related to pH and calcium content. In general, alkalinity

[1] Based on the Jackson candle turbidimeter, commonly read with a direct reading colorimeter calibrated in Jackson units.

should not be less than 30 mg/l, based upon 500 mg/l dissolved solids and a pH range of 6.0 to 8.5. Values higher than 400 to 500 mg/l are considered too high for public water supplies, except when normal for the water source. Since processing control is impaired by frequent variations from normal values, such variations should be minimized.

In order to maintain the proper chemical balance for preservation of aquatic life, it is recommended that no highly dissociated materials be added to a water supply in quantities sufficient to induce pH changes above or below the range of 6.0 to 9.0. In order to protect the carbonate buffering system, acid should not be added in quantities which lower the total alkalinity below 20 mg/l, expressed as $CaCO_3$. Some weakly ionized materials may exert their prime effect upon aquatic life not in terms of pH changes but as the result of the formation of toxic substances. For example, such weakly dissociated compounds as HCN, HClO, and H_2S may be toxic because the anion formed or the undissociated molecule itself is harmful.

Total Dissolved Solids

Removal of total dissolved solids (TDS) from waste waters per se is one of the more difficult and, in many cases, more expensive waste treatment procedures.

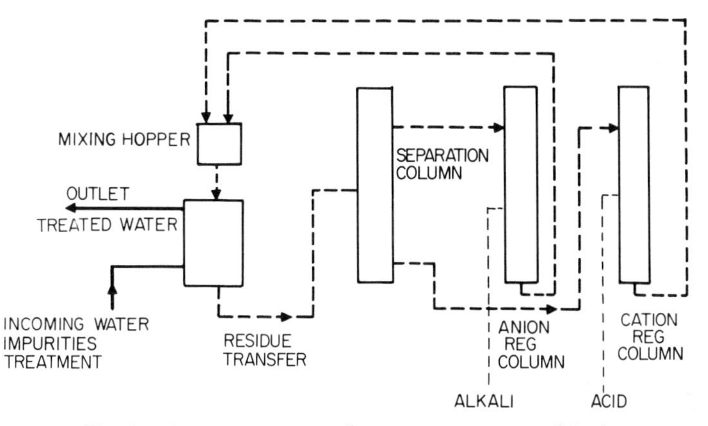

Fig. 2 Continuous ion exchange process—mixed bed.

Where high total dissolved solids are associated with the heavy metals or hardness, reduction may be accomplished by precipitation operations. Where the total dissolved solids are present as sodium or potassium compounds, total dissolved solids reduction may require distillation or ion exchange (Fig. 2) and then an effective method for disposal of the ion exchange regenerant wastes.

Deep-well disposal of brines from oil field operations is common. Most often TDS disposal requires controlled discharge to the receiving body.

It is recommended that total dissolved solids not exceed 500 mg/l for public water supplies. High values for TDS may exert adverse economic effects and harmful physiological effects, primarily osmotic, and may impart taste or odor. High levels of sulfate and chloride are often associated with corrosion of water system components.

In general, it may be said that the concentration of dissolved materials in natural fresh waters may be somewhat below the optimum levels for preservation and promotion in aquatic life. Thus, it may be that the addition of dissolved solids to a water supply benefits aquatic life in the receiving body. However, there exist upper tolerable limits beyond which osmotic effects cause dehydration of tissue cells.

It is recommended that for maintenance of fauna, total dissolved materials in a watercourse not exceed 50 millimoles, equivalent to 1.5 g/l of NaCl. Such a recommendation assumes that the dissolved constituents are fundamentally innocuous. In those cases in which there is toxicity associated with any constituent, the level must be lowered accordingly. Further, it is recommended that total dissolved materials not exceed more than one-third the concentration normal for that water source. This latter prohibition is based upon the fact that aquatic life food sources, such as diatoms, may be more sensitive to dissolved materials than the organisms they feed. Destruction of the food source may then result not only in kill-off of desirable organisms but in the overgrowth of nuisances.

For marine waters, it is recommended that industrial effluents not alter the salinity of the water strata more than ±10 percent of the natural variation.

TDS levels in waters destined for agricultural uses may be as high as 2,000 to 5,000 mg/l when carefully used for tolerant plants in permeable soil. Usually, concentrations below 500 mg/l are preferred since there are few detrimental effects at this level.

Ammonia

Ammonia is a by-product of many industrial processes. It is produced in greatest quantity by the distillation of coal for the production of gas, coke, and compounds used in the manufacture of chemicals for the textile and chemical process industries. For the most part ammonia recovery processes are employed in these operations, but often the level of ammonia in the process effluent rises to 1.0 mg/l as NH_3. The waste products are principally free ammonia, cyanide, and thiocyanate salts and a variety of aromatic compounds.

Past practice for ammonia removal from waste effluents has been by stripping or lime addition to decompose ammonia salts followed by stripping. Ammonia is effectively reduced by biological treatment. Lesser concentrations may be discharged to a secondary treatment plant where ammonia is consumed in the biological process. In cases where the nutrient balance is not favorable, it may be necessary to supplement phosphorus for effective ammonia reduction.

The permissible criterion for ammonia levels has been established as 0.5 mg/l in waters designated as public water supplies. The desirable criterion is less than 0.01 mg/l. Excessive ammonia requires higher feed of chlorine for effective disinfection.

Aquatic life exposed to levels of 1 mg/l may suffocate because of the significantly reduced oxygen combining capacity of blood. For preservation of aquatic life, ammonia pollution must be evaluated in terms of the most sensitive organism in receiving waters considered to be of economic or ecologic importance. It is necessary to determine the life stage in which this most sensitive organism is most vulnerable to ammonia and then to use an application factor[1] of 1/20 for tests carried out on this reference organism in order to determine the safe concentration.

Arsenic

Arsenic pollution is often associated with the manufacture or use of herbicides and pesticides. It may be a by-product of mining operations.

Where arsenic is present as a pollutant with heavy metals, treatment by precipitation of the heavy metals will lead to a reduction of approximately 90 percent of the initial arsenic. The mechanism is not well understood but may be due to precipitation of arsenic as a complex with the heavy metal ions.

The permissible criterion for presence of arsenic in public water supplies is 0.05 mg/l. The desirable criterion is complete absence. The basic treatment plant exerts little or no effect upon the concentration of arsenic in influent waters.

Arsenic hydride and the trioxide are especially toxic. Arsenate acts by tying

[1] The ratio of the safe concentration of a given pollutant under prescribed conditions to the 96-hr median tolerance limit (TL_m) is called the application factor for that pollutant. It permits rapid calculation of the safe concentration, having determined the 96-hr TL_m. The 96-hr TL_m is that concentration of the pollutant which is lethal for 50 percent of the test organisms exposed to it for 96 hr. TL_m should not be confused with the safe limit.

up active sites in cellular substituents. The situation is not always so clear-cut. For while arsenic is a notorious poison, arsenical "dips" have been used recently for livestock and arsenic compounds in low dosages are added to poultry feed as coccidiostatics. Sodium arsenate, when present in livestock water supplies at levels of 5 mg/l, functions in some unknown way to reduce the toxicity of selenium in those same waters. A tolerance level of 1 mg/l has been established for irrigation waters.

Toxic doses of arsenic vary widely with the animal species, as indicated in Table 24.

TABLE 24 Relative Resistance of Some Animal Species to Arsenic Intoxication

Animal	*Index of Resistance to the Toxic Effects of Arsenic*
Poultry	1
Dogs	2
Swine	10
Sheep, goats, horses	200
Cattle	300

The tissues of many organisms accumulate arsenic. Its harmful effects may therefore be delayed when the water concentration is low—but are lethal nonetheless.

Barium

The permissible criterion for barium in public waters is 1.0 mg/l.

Barium may exert its detrimental physiological effect by forming a barium sulfate precipitate, thus effectively diminishing sulfate ion concentration below body requirements.

Boron

Industrially, boron pollution occurs as the result of the manufacture or use of synthetic boranes. Boron may be present naturally in concentrations as high as 15 mg/l, particularly in the Western United States. The permissible criterion for public water supply is 1.0 mg/l. Ideally, it should be absent.

Boron is toxic for many organisms in concentrations as low as 1 mg/l, while other organisms tolerate levels above 15 mg/l. Boron is an essential plant nutrient in concentrations up to 0.5 mg/l. On the other hand irrigation waters containing more than 0.5 mg/l have caused destruction of such sensitive crops as apples, citrus fruits, and nuts. Generally, water containing more than 4 mg/l of boron is unsatisfactory for all crops.

Boric acid concentrations up to 2,500 mg/l are tolerated by many organisms, although inhibition of growth has been noted.

Cadmium

Cadmium is used widely by industry in the production of copper, lead, silver, and aluminum alloys. It also is used in metal finishing, ceramic manufacture, and photography and is a by-product of nuclear reactor operation. Salts of cadmium are used as insecticides and as antiparasitic agents.

Cadmium may be removed from waste water by precipitation as the metal hydroxide. Recovery of the metal by ion exchange can be economically attractive in some instances.

The permissible criterion for cadmium in public water supplies is 0.01 mg/l. Cadmium is especially dangerous, since it may combine synergistically with other toxic substances. Its principal effect upon aquatic life is the inhibition of bivalve shell production in mollusks.

Cadmium has been reported in concentrations as high as 3.2 mg/l in the waters off Long Island, New York. Such concentrations are the result of electroplating, textile, and chemical process operations.

Cadmium toxicity has been implicated in serious cardiovascular diseases in man. Therefore, based upon current knowledge, agricultural waters should not contain more than 0.005 mg/l in order to prevent cumulative toxicity in crops headed for market.

Chloride

Chloride and sulfate wastes are more closely associated with high total dissolved solids, and control of chloride waste is accomplished in much the same way as control of total dissolved solids.

It is recommended that chloride concentration not exceed 250 mg/l and that ideally it be less than 25 mg/l.

High chloride concentrations may result from oil field operations or from industrial effluents associated with papermaking, galvanizing, and water conditioning. Fortunately, this ion is not toxic in less than extreme concentrations and so does not present a major pollution problem.

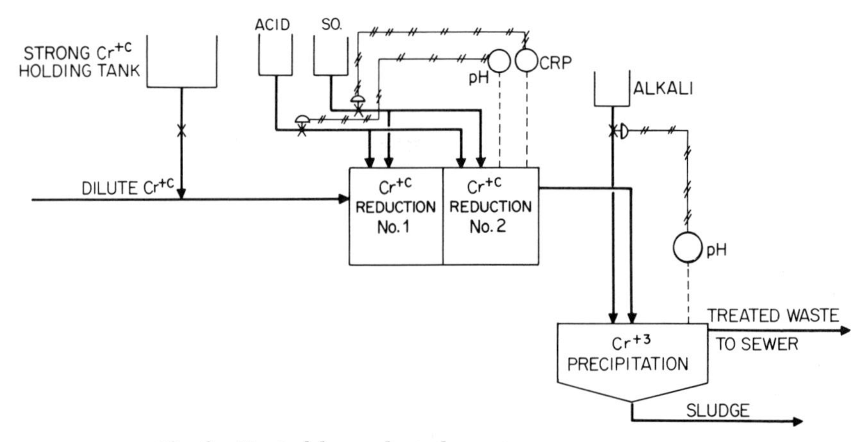

Fig. 3 Typical hexavalent chromate treatment system.

Chromium

Chromium compounds may be present in industrial wastes from a wide variety of processes including tanning, electroplating, and cooling tower effluents.

Chromium wastes lend themselves to chromate recovery (Fig. 3) as commonly practiced in the metal plating industry. For final disposal, chromium is precipitated as the metal hydroxide after reduction of chromium to the trivalent state. Another less common practice is precipitation of chromium by addition of soluble barium salts.

For public water supplies chromium in the hexavalent form should not exceed 0.05 mg/l and preferably should be absent entirely.

The toxicity of chromium varies with pH, temperature, the organism exposed, and the valence form. In any given form, toxicity may vary with the presence of other compounds.

For agricultural waters, tolerance of chromic and chromate ions varies with the plant species. More sensitive plants are adversely affected by concentrations of 5 mg/l.

Copper

Control of copper wastes and iron, lead, manganese, and zinc wastes is most commonly effected by precipitation as the metal hydroxides. Precipitation is usually effected at a pH between 7 and 9.

Public water supplies may tolerate as a permissible criterion 1.0 mg/l. A desirable criterion of virtual absence has been expressed. In general, excess copper is highly toxic to algae, sea plants, and invertebrates but is only moderately toxic to mammals. The relationships between concentrations of different heavy metals in a water supply are extremely important in the determination of ultimate copper toxicity. Positive and negative synergistic effects make necessary the individual evaluation of each situation.

Kills of 35 to 100 percent of primitive fresh water plant forms were obtained with copper concentrations as low as 0.5 mg/l. Concentrations of copper above 0.1 mg/l were found to be toxic for oysters. Lobsters transferred to tanks lined with copper, after living in aluminum, stainless steel, and iron tanks for 2 months, died within a day.

For agricultural waters, it appears that copper toxicities are more feed related than water related. It should be noted, however, that copper toxicities have been observed at copper concentrations as low as 0.1 mg/l in plant nutrient waters. Table 25 lists maximum safe concentrations for some trace ions.

Iron

The permissible criterion for iron in the filterable form for public water use is 0.3 mg/l. Again, as with the pollutants described above, the defined water treatment plant is relatively ineffective for iron removal.

TABLE 25 Maximum Safe Concentrations for Trace Ions

Trace Ion	*Allowable Concentration in Farmstead Waters, mg/l*
Arsenic	0.05†
Barium	1.00
Cadmium	0.01†
Chromium (Cr^{6+})	0.05†
Copper	1.0*
Cyanides	0.20
Iron	0.30*
Lead	0.05†
Manganese	0.05*
Selenium	0.01†
Silver	0.05
Zinc	5.0*

* Recommended limits.

† Also the maximum allowable concentration in farm animals' water supplies.

In general, iron is one of the less toxic pollutants. Criteria for waters designated for the preservation of aquatic species do not specifically define upper limits for iron. Statements on standards relating to the use of iron bearing waters for agriculture are confined either to the lack of available data or the assertion that iron is not likely to be a problem.

Lead

A permissible criterion of 0.05 mg/l and complete absence as a desirable criterion are the FWPCA's Committee on Water Quality Criteria recommendations for public waters. Lead may arise as a contaminant of groundwaters both from natural sources and in the form of various industrial and mining effluents. A major problem with lead pollution is that the element is a cumulative poison. There is considerable variation in the toxicity of lead, depending upon the form of the element. Drinking water supplies for animals should not contain concentrations of lead exceeding 0.5 mg/l. For irrigation waters an upper limit of 5 mg/l is proposed.

Manganese

A permissible criterion of 0.05 mg/l has been proposed for public water supplies. In general, toxicity data on manganese for aquatic life and agricultural waters are not readily available. Cattle fed up to 600 mg/kg for 20 to 45 days suffered no serious effects.

Phosphorus

Unnaturally high phosphorus concentrations may be the result of animal and plant processing or fertilizer and chemical manufacturing operations.

The ultimate effect of high phosphorus effluents depends upon a highly complex set of factors which have made establishment of generally acceptable limits impossible.

However, it can be said generally that total phosphorus concentrations above 50 mg/l contribute to the overgrowth of objectionable plant forms.

When the concentration of complex phosphates is greater than 100 mg/l, coagulation processes may be adversely affected. Another effect of high phosphorus is that it may bring about oxygen depletion in a body of water. This results from the requirement for 160 mg of O_2 to oxidize completely 1 mg of phosphorus from an organic source.

Selenium

Selenium, when present with heavy metals in a waste stream, will be removed in much the same way as arsenic is removed in heavy metal precipitation.

It is recommended that the value 0.01 mg/l be established as the permissible criterion for the presence of selenium in public water supplies. As a desirable criterion complete absence is recommended.

Selenium is toxic for animals when it accumulates in tissue at the level of 5 mg/kg. On the other hand, it is an essential trace mineral and is commonly fed in quantities sufficient to cause accumulations of 1 to 2 mg/kg.

The tolerance limits for selenium should be based upon animal toxicity rather than toxicity in plants. It is not uncommon for plants to accumulate concentrations of selenium well above the levels toxic for animals if irrigating waters permit these accumulations. It has been suggested, therefore, that the concentration of selenium in waters used for irrigation not exceed 0.05 mg/l.

Uranyl ion

The permissible criterion for uranyl ion is 5 mg/l. Preferably, it should be absent. The standard for uranyl ion (UO_2^{--}) is drawn on the basis of its chemical properties rather than on the basis of its radioactivity. The ion is of concern in public water supplies because of the possibility of renal damage.

Taste and color appear in water at a level of 10 mg/l—a level considerably lower than the concentration at which physiological damage is manifested. The standard of 5 mg/l is established rather arbitrarily at one-half that at which taste and color appear, providing a considerable safety margin.

Zinc

Zinc bearing effluents may be the result of primary metal and chemical process operations, among others.

It is recommended that zinc be present in public water supplies in concentrations no higher than 5 mg/l. Ideally, it would be virtually absent.

A complex relationship exists between zinc concentration, dissolved oxygen, pH, temperature, and calcium and magnesium concentrations. Prediction of toxicities has been less than reliable, and controlled studies on separate effects have been little documented.

For any given sensitive organism, a concentration of $\frac{1}{100}$ of the 96-hr TL_m is considered a safe level.

Zinc is normally found in seawaters at a concentration of 0.01 mg/l. Marine life may contain zinc in concentrations up to 1,500 mg/l. It has been determined, however, that levels of 10 mg/l cause a 50 percent inactivation of photosynthesis in certain kelp.

A variety of fresh water plants tested manifested toxic symptoms at concentrations of 10 mg/l. For this reason a tolerance limit of 5 mg/l is proposed for irrigation waters.

Nitrates and Nitrites

The permissible criterion for nitrates and nitrites (determined as N) in public water supplies is 10 mg/l. It is desirable that they be virtually absent. Nitrite is included in this recommendation because it reacts with the oxygen-carrying pigment in blood, hemoglobin, to produce a compound which is a less effective oxygen transporter and may produce serious physiological effects.

Conversion of nitrates to nitrites in the stomachs of ruminant animals may produce this effect (at levels above 2,800 mg/l of nitrate). High nitrate concentration may also favor the growth of undesirable plants.

Sulfate and Sulfide

Sulfides and their oxidation products, sulfates, are found in water as the result of natural processes and as the by-product of oil refinery, tannery, pulp and paper mill, textile mill, chemical plant, and gas manufacturing operations.

The recommended permissible criterion is not more than 250 mg/l of sulfate in public water supplies. Recommended desirable criterion is less than 50 mg/l.

Concentrations in the range of less than 1.0 to 25.0 mg/l of sulfides may be lethal in 1 to 3 days to a variety of fresh water fishes.

Sulfate levels of 2,000 mg/l have been found to cause progressive weakening and death in cattle.

ORGANIC CHEMICALS

Carbon Chloroform Extract (CCE)

Current recommendations subject to change with additional low-flow rate data are that 0.15 mg/l be regarded as the permissible criterion for the presence of organic materials, as determined by carbon chloroform extraction. New analytical techniques indicate that the 0.2 mg/l level specified in "Drinking Water Standards[1] as determined by older, less reliable methods is actually 0.15 mg/l as determined by newer techniques.

In general, CCE determinations are considered too complex for application to farmstead water supplies. Instead, an estimate of the CCE may be based upon values in nearby municipalities using the same or smaller water sources.

Methylene Blue Active Substances[1]

The permissible criterion for methylene blue active substances is 0.5 mg/l. They should be virtually absent.

Cyanide

The cyanides, represented by hydrocyanic acid and its salts, are important and ubiquitous industrial chemicals. They are extremely toxic, especially at low pH. It is thought that cyanide acts by inhibiting the phosphorylative oxidation reactions which permit cellular respiration.

Many lower animals and fishes seem to be able to convert cyanide to thiocyanate ion, which does not inhibit respiratory enzyme activity. Cyanide compounds formed by the reaction of CN^- with heavy metals may be even more toxic substances. For these reasons control of cyanide in industrial effluents is extremely important.

As with many of the water constituents described previously, cyanide ion is unaffected by the basic water treatment plant.

Cyanide removal from industrial effluents is commonly effected by the stagewise application of lime and chlorine, which progressively oxidizes cyanides (Fig. 4)

[1] U.S. Department of Health, Education and Welfare, Public Health Service Publication 956, Washington, D.C., 1962.

to cyanates and then to carbon dioxide and nitrogen. In many European operations this process goes no further than the cyanide to cyanate conversion.

A permissible criterion of 0.20 mg/l and a desirable criterion of complete absence from public waters are the recommendations of the FWPCA Water Quality Committee.

It is recommended that for preservation of aquatic life, cyanide be determined by the flow-through bioassay method with particular attention paid to standardization of dissolved oxygen, temperature, and pH.

Oil and Grease

Oil and grease pollution may be the result of bilge and ballast water; refinery and industrial plant wastes resulting from the lubrication of machinery; rolling mills, gasoline filling stations; and fat manufacture or processing.

Oil and grease should be absent from industrial effluents—virtually absent as a permissible criterion and completely absent preferably. The specification of the subcommittee that no oils or greases be present in water supplies is based upon the troublesome taste and odor problems associated with even minute quantities in a water supply. Moreover, even very small quantities produce scum lines on water treatment facilities, swimming pools, retention basins, reservoirs, and other containers.

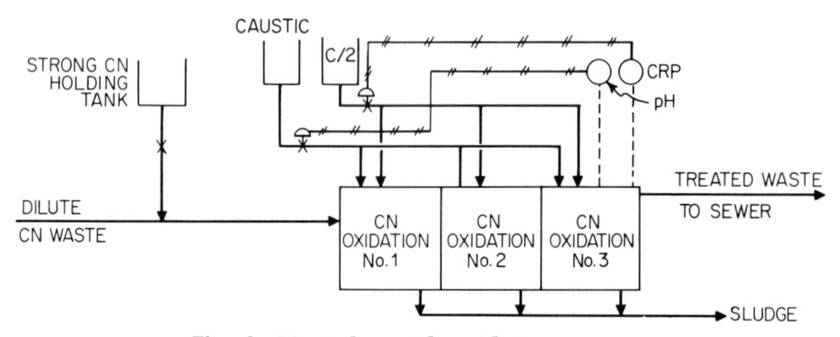

Fig. 4 Typical cyanide oxidation system.

An oil slick is just barely visible at concentration of approximately 25 gal/sq mile. When the oil concentration is 50 gal/sq mile, an oil film three-millionths of an inch thick is visible as a silvery surface sheen.

Oils coat the gill filaments of fish, producing suffocation in low concentration. Some crude oils contain a water soluble fraction which is highly toxic to fish life. Oil and greases may, moreover, coat and destroy algae, plankton, and bottom dwelling organisms. Oil films may interfere with reaeration and photosynthesis and destroy aquatic insect life. Crayfish weighing 35 to 38 g die in oil concentrations of 5 to 50 mg/l within 18 to 60 hr. Similar lethalities exist for other small marine organisms.

Oil is especially destructive to water birds and other aquatic life species. Its deleterious effects range from inhibition of egg laying and hatching to the destruction of water impermeable compounds normal to plumage and fur.

The effects of oil pollution have recently been tested extensively as the result of oil tanker wreckings. The literature is now replete with data on the toxicities, color, taste, and synergistic effects of oil pollution. For an oil removal system, see Fig. 5.

Pesticides and Herbicides

The permissible and desirable criteria for the presence of pesticides and herbicides in public water supply vary with the specific compounds and are shown in Table 26.

BIOCHEMICAL OXYGEN DEMAND

Biochemical oxygen demand (BOD) is a measure of the dissolved oxygen in a water supply needed to depose organic materials in a measured time and at a constant temperature. Adequate dissolved oxygen is a principal requisite for main-

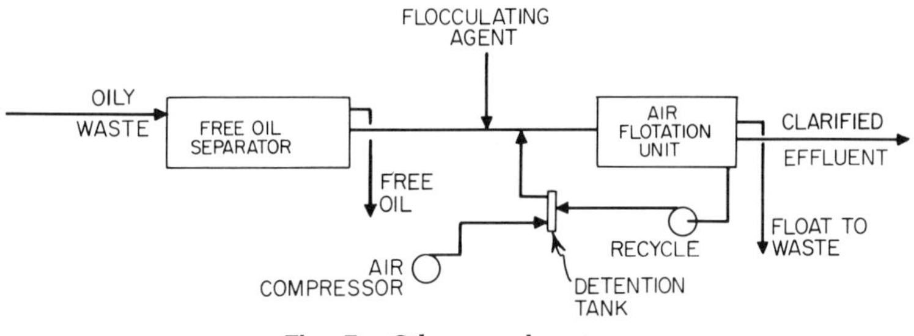

Fig. 5 Oil removal system.

tenance of aquatic life. It is also a factor in the corrosivity, septicity, and photosynthetic activity of a water supply.

To determine BOD, uptake from an independent supply is measured over a 5-day period in a system maintained at 20°C. At the end of the test the total amount of oxygen consumed is a measure of the requirements of the system over and above those supplied by the dissolved oxygen in the water source.

TABLE 26 Current Criteria for the Presence of Some Pesticides and Herbicides in Public Water Supplies

Compound	Permissible criteria, mg/l	Desirable criteria
1,2,3,4,10,10-hexachloro-1,4,4a,5,8,8a-hexahydro-1,4,5,8-dimethanonaphthalene (Aldrin)	0.017	Absent
1,2,4,5,6,7,8,8-octachloro-3a,4,7,7a-tetrahydro-4,7-methanoindane (chlordane)	0.003	Absent
1,1,1-trichloro-2,2-bis (parachlorophenyl) ethane (DDT)	0.042	Absent
1,2,3,4,10,10-hexachloro-6,7-epoxy-1,4,4a,5,6,7,8,8a-endo-exo-octahydro-1,4,5,8-dimethanonaphthalene (Dieldrin)	0.017	Absent
Same as above except endo-endo (Endrin)	0.001	Absent
1,2,3,4,5,6-hexachlorocyclohexane (Lindane)	0.056	Absent
1(or 3a),4,5,6,7,8,8-heptachloro-3a,4,7,7a-tetrahydro-4,7-methanoindane (Heptachlor)	0.018	Absent
1,1,1-trichloro-2,2-bis (paramethoxyphenyl) ethane (Methoxychlor)	0.035	Absent
Organic phosphates plus carbamates*	0.1	Absent
2,4-D plus 2,4,5-T plus 2,4,5-TP (herbicides)	0.1	Absent

Subject to expert toxicological evaluation. The specified concentrations are those deemed not to be harmful if ingested over extensive periods.

* Certain compounds or mixtures may require even lower concentrations.

Chemical oxygen demand (COD) is a part of BOD and may be determined independently in a test requiring only the use of an oxidizing agent, potassium dichromate, and a sample of the receiving water. BOD determination is an important index of industrial pollution when organic material is deposited as an industrial waste in a receiving stream. In fact, organic material is the single largest constituent of industrial waste.

The principal factors in a waste stream with high BOD are (1) carbonaceous organic material normally degraded by aerobic microorganisms and the main, if not the only, oxygen-demanding constituent of sewage; (2) oxidizable nitrogen in organic or inorganic forms; (3) certain chemical reducing compounds such as ferrous iron, sulfite, and sulfides.

Major problems may arise when high BOD industrial effluents enter municipal sewage plants. Clogging or corrosion of the sewer system, hydraulic overloading of the equipment, or overloading of the grit chambers, screens, or comminutors are just a few of the attendant problems. Industrial wastes may reduce the efficiency of settling tanks and of solids removal. They may produce scum problems associated with high oil and grease content. In addition to problems which have a mechanical origin, certain toxic industrial wastes destroy microbiological populations required to degrade sewage.

The permissible criterion for dissolved oxygen in public water supplies is that it be greater than or equal to 4 as a monthly mean with near saturation as a desirable criterion. The average BOD of domestic sewage is approximately 100 to 300 mg/1, and sewage streams receiving industrial wastes should not have BODs greater than 100 mg/l if overloading of the system is to avoided.

Table 27 lists BODs of typical industrial wastes based upon arbitrarily chosen units of production for each industry.

TABLE 27 BODs of Typical Industrial Wastes

Industry	Production unit	BOD, mg/l per production unit
Meat packing	One hog	1,048
Dairy	1,000 lb raw milk	570
Beet sugar process water	Ton of raw beets	1,230
Brewery	bbl of beer	1,000
Paper mill (drinking)	Ton of paper	300
Pulp mill (sulfite)	Ton of pulp	443
Tanning (vegetable)	100 lb of hides	1,200
Laundry	100 lb of clothes	320
Oil refinery	bbl crude	20

INTERNATIONAL WATER QUALITY CRITERIA AND STANDARDS

French Legislation

Drinking water standards French legislation establishes the maximum permissible concentrations of specific water constituents as indicated in Table 28.

Color in water must not exceed 20 units (platinum-cobalt colorimetric scale) and turbidity must not exceed 15 drops of a 1/1,000 alcohol solution of gum mastic, under normal operating conditions. The code permits turbidity up to 30 drops of mastic in 50 mm of optically clear water, for limited periods of time.

Suspended solids may not exceed 0.1 unit on the Baudrey clogging capacity test. In this test a water sample is passed through a fine fabric filter 1 sq cm in area under a constant head of 10 cm. The volume of water passing through the element is collected until the water no longer emerges in a steady stream but flows discontinuously. At this point clogging capacity and, thus, an index of suspended solids may be derived from standard curves.

The total mineral content of any water for drinking may not exceed 2 g/l, and water must be free from any unpleasant smell or taste.

Under a more recent amendment, additional criteria have been established specifying low nitrate content, noncorrosivity, and sulfates below the level at which components of the water distribution network would be affected.

TABLE 28 Maximum Permissible Concentrations of Specific Water Constituents*

Constituent	*Maximum Permissible Concentration, mg/l*
Lead (as Pb)	0.1
Selenium (as Se)	0.05
Fluorides (as F)	1.0
Arsenic (as As)	0.05
Chromium (as CR^{6+})	Undetectable
Cyanides	Undetectable
Copper (as Cu)	1.0
Iron (as Fe)	0.2
Magnesium (as Mn)	0.1
Zinc (as Zn)	5.0
Phenolics	Nil
	Desirable Criteria, ppm
Magnesium (as Mg)	125
Chlorides (as Cl)	250
Sulfates (as SO_4)	250
Nitrates:	
(As N)	2–5
(As NO_3)	9–22

* From French legislation.

TABLE 29 World Health Organization Standards for Portable Water

Constituent	Criterion	
Lead (as Pb)	0.1 mg/1	Maximum allowable concentration
Selenium (as Se)	0.05 mg/1	
Arsenic (as As)	0.2 mg/1	
Chromium (as Cr^{6+})	0.05 mg/l	
Cyanide (as CN^-)	0.01 mg/l	
	Permissible	Excessive
Total solids	500 mg/l	1,500 mg/l
Color	5 units	50 units
Turbidity	5 units	25 units
Taste	Unobjectionable	....
Odor	Unobjectionable	....
Iron (Fe)	0.3 mg/l	1.0 mg/l
Manganese (Mn)	0.1 mg/l	0.5 mg/l
Copper (Cu)	1.0 mg/l	1.5 mg/l
Zinc (Zn)	5.0 mg/l	15 mg/l
Calcium (Ca)	75 mg/l	200 mg/l
Magnesium (Mg)	50 mg/l	150 mg/l
Sulfate (SO_4)	200 mg/l	400 mg/l
Chloride (Cl)	200 mg/l	600 mg/l
pH	7.0–8.5	(6.5 or) 9.2
$(Mg + Na)SO_4$	500 mg/l	1,000 mg/l
Phenolics	0.001 mg/l	0.002 mg/l
α emitters	10^{-9} c/ml	....

Total hardness should be less than 30 French degrees (300 ppm), preferably 12 to 15 French degrees (135 ppm). Drinking water must in no circumstances contain more than 2 mg/l of phosphorus pentoxide.

French law states generally that any discharge or effluent entering a watercourse directly or indirectly and thereby harming or destroying fish or their sources of food may bring civil penalties including fines up to 5,000 French francs and terms of imprisonment up to 5 years.

Trade wastes The following regulations have been established regarding the preliminary disposal techniques for trade wastes in general. All effluents must be neutralized to a pH value between 5.5 and 8.5. When lime neutralization is employed, the upper limit may be 9.5.

The effluent must be reduced to a temperature not exceeding 30°C (86°F). Discharges of cyclic or hydroxylated compounds or their halogen derivatives are prohibited.

Any discharge which contributes to the appearance of abnormal odor, taste, or color for water to be used as public supplies is likewise prohibited. An industrial effluent may not contain any product likely to emit toxic or inflammable gases or vapors. The effluent may not contain floating material or substances which may be destructive to fishes or their food.

World Health Organization Standards and Criteria

The standards represent, in general, the consensus of world opinion. The maximum allowable concentrations and permissible criteria are described in Table 29 for the principal pollutant characteristics or constituents of water.

United Kingdom Recommendations for Water Quality

In England and Wales there are more than 800 authorities or joint bodies acting to prescribe water standards—each under obligation to ensure that water sources remain unpolluted in the individual localities. When pollution occurs, they are charged with the responsibility for tracking it down and eliminating the source.

Published British standards are especially concerned with bacterial contamination and do not speak specifically of industrial pollutants other than to specify generally that continuous analysis of water sources, surface waters particularly, is necessary in order to control pollution.

Chapter 5

Air Pollution Control Programs and Systems

WILLIAM W. MOORE

Vice President, Research-Cottrell, Inc.,
Bound Brook, New Jersey

and

NORMAN W. FRISCH

Director of Research,
Cottrell Environmental Systems, Inc.,
Bound Brook, New Jersey

Air pollution control technology is an extensive and highly complex subject involving alternate general approaches to a problem, many types of emissions, and many types of control equipment. Thus, a comprehensive and detailed treatment of the subject is beyond the scope of this chapter. A generalized treatment of the control technology available and the critical factors involved in the determination of the basic approach to the selection of a particular air pollution control system is presented. Such basics should be of value to the plant engineer in getting off to the proper start when confronted with an air pollution problem. If outside sources are consulted on air pollution problems, the plant engineer will want to have an understanding that will enable him to participate in discussions with them. It is in this context that the chapter is written.

Generally speaking, air pollution control technology can be divided into three main approaches:

1. Process change to prevent pollutant generation
2. Dispersion to reduce concentrations to acceptable levels at receptor sites
3. Removal of pollutants after generation and before dispersion

All these approaches should be examined in the order given at an early stage in the formulation of plans for achieving a given air pollution control objective. Many times, changes in process operating conditions and/or raw material inputs can either meet a given objective alone or modify it in such a way as to simplify the overall solution of the problem by facilitating the use of dispersion and/or collection. Admittedly, in a majority of cases, this approach will be found technically or economically impractical, but this does not mean that it should not be considered as part of the overall problem analysis. The fact that some current legislation dictates approaches such as these (limiting the sulfur content of fuel, for example) attests to their viability.

Dispersion is an effective technique of air pollution control where the control objective is limitation of the pollutant to maximum levels—either quantitatively or subjectively defined—at specific receptor sites. The technique involves the utilization of local meteorological factors—such as wind speed and horizontal and vertical diffusion parameters—local topography, and elevation, temperature, and velocity of the source of pollutant carrying gases. The use of tall chimneys has been found to be effective in reduction of receptor site concentrations of many pollutants by several orders of magnitude, a capability which is particularly important in many odor problems where reduction of concentrations below the threshold value by other techniques is either technologically or economically impractical.

The important factor to be remembered, however, in the consideration of tall chimneys as a control device, is that the objective must be defined in terms of some criterion involving concentration or threshold at a receptor site. Thus, if the code limits a gaseous pollutant to a specified maximum ground level concentration, an odor to a specific threshold, or particulate to a maximum allowable fallout, dispersion should always be considered as a most powerful weapon for achieving these objectives. On the other hand, there is no point in considering it as a pollution control technique if the objective is defined purely in terms of maximum quantitative emission.

When process change and/or dispersion cannot reach the desired air pollution control goal, reliance is placed on the third general method: removal of the objectionable—or pollutant—contaminants of the process effluent gas stream. This method, which involves many techniques and types of hardware, is at the core of pollution control technology. Used alone, or in conjunction with either or both of the previously discussed techniques, it is the foundation upon which present-day concepts of air resource management rest.

For the purposes of a discussion on pollution control technology, the pollutants themselves are most conveniently divided into three broad classifications: particulates, gases, and odors.

Particulates are chemical elements or compounds in either solid or condensed liquid droplet form. They are usually described in terms of those physical characteristics which affect the mechanisms for their separation. Most important of these are physical size and density, for these parameters are related to most mechanisms of separation. These parameters are also related to important properties of the aerosols of which they are a part, such as optical transmission and stability. In fact, the classification of various types of particulates into dusts, smoke, mists, and fogs is based upon their particle size and mass. A chart which conveniently presents an overview of particulate characteristics is shown in Fig. 1.

Gases, of course, are chemical elements or compounds with boiling points sufficiently low that they exist in volatilized form at ambient temperatures. They are described in terms of their chemical composition, concentration, and odor threshold. Critical parameters which are required include the concentration level to which the gas must be reduced, the equilibrium solubility of the gas, as well as heats of solution and reaction, if the latter are appreciable.

It is convenient to classify gases of concern on the basis of their key chemical element. The key elements which are involved in gaseous pollutants are, to a large extent, contained in the group sulfur, nitrogen, halogens, and carbon. Table 1 is a classification of gaseous pollutants. Some pollutants, of course, contain more than one of these elements; hence the classification is somewhat arbitrary, but nevertheless useful.

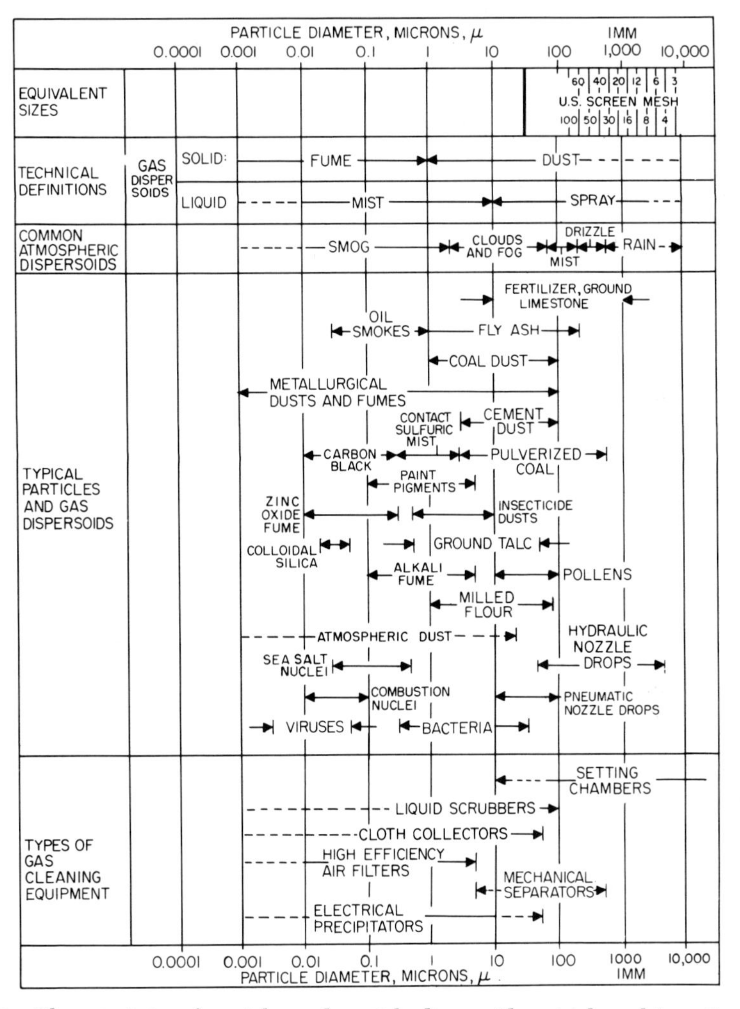

Fig. 1 Characteristics of particles and particle dispersoids. (*Adapted from Stanford Research Institute.*)

Odors are exceedingly elusive of explicit definition. They are sometimes chemically identifiable—such as in the case of R-SH (mercaptans), H_2S, and CS_2. In these cases, they can be treated in a discussion of control technology along with other chemically identifiable gaseous pollutants, as has been done in Table 1. The principal difference between chemically identifiable odors and nonodorous gaseous pollutants is the usual requirement for reduction of odorous gases to concentrations several orders of magnitude below those required for nonodorous gaseous pollutants.

Many odors—such as those from sewage, meat rendering, and food processing—

arise from complex substances which are not readily defined chemically. In these cases, it is common to attempt to define odors subjectively via use of descriptive terms such as flowery, fish, fecal, or garlicky, and to determine odor threshold levels by progressive dilution in air until the concentration is reached at which the odor is just identifiable (recognition threshold). But even here, great difficulties are encountered in precise definition because of great variations in individual receptor response as well as the extremely low recognition threshold concentrations.

The mammalian olfactory system is still the most satisfactory, though uncalibrated,

TABLE 1 Typical Gaseous Pollutants and Their Sources

Key element	Pollutant compounds	Example of source of pollutant
S	SO_2	Boiler flue gas
	SO_3	Sulfuric acid mfg
	H_2S	Natural gas processing, sewage treatment, paper and pulp industry
	R-SH (mercaptans)	Petroleum refining, paper and pulp
	CS_2	Viscose mfg
N	NO, NO_2	Nitric acid mfg, high temp oxidation processes using air, motor vehicles, nitration processes
	NH_3	Ammonia mfg
	Other basic N compounds, pyridines, etc.	Sewage, rendering, pyridine base, mfg, solvent processes
	Amines (cadaverine, etc.)	
Halogen:		
F	HF	Phosphate fertilizer, aluminum mfg
	SiF_4	Ceramic, fertilizer
Cl	HCl	HCl mfg, PVC combustion, organic chlorination processes
	Cl_2	Chlorine mfg
C	Inorganic:	
	CO	Incomplete combustion processes, motor vehicles
	CO_2	Combustion processes (not generally considered a pollutant; slow increase of CO_2 in atmosphere is viewed by some as potential source of long-range climatic change)
	Organic:	
	Hydrocarbons	
	Paraffins	
	Methane	
	Butane	Solvent operations
	Hexane	
	Olefins	
	2-butene	Gasoline, petrochemical operations
	Aromatics:	
	Ethylbenzene	Solvents
	Oxygenated hydrocarbons:	
	Aldehydes	
	Formaldehydes	Partial oxidation processes
	Acrolein	
	Ketone	
	Methylethylketone	Surface coating operations
	Alcohols	
	2-propanol	Surface coating operations
	Phenols	Petroleum processing, plastic manufacture
	Oxides	Ethylene oxide mfg
	Chlorinated solvents:	
	Perchlorethylene	Dry cleaning
	Trichlorethylene	Degreasing operations

instrument for odor measurement. In spite of these difficulties, a technology for control (not quantitative detection) of odors by a variety of techniques is available.

POLLUTION CONTROL BY PROCESS CHANGE

Pollution control by process change is really the province of the process designer or operator rather than the pollution control system designers. From the viewpoint of the plant manager or engineer confronted with a problem, the method cannot be overlooked as an effective technique. A detailed treatment of the method is obviously impossible because of the tremendous number of processes and process variables that would have to be considered. Yet a few principles should demonstrate the importance of giving this method the earliest consideration in arriving at a problem solution.

Presuming that the problem has been defined in terms of the existence of certain pollutants and the levels to which they must be reduced to reach the control objective, one should always and immediately consider possible changes to the process inputs or the process conditions as a means for achieving either partial or complete achievement of objectives. This should be done on the simple principle of "what goes in must come out." What are the process inputs which give rise to the existence of the pollutants in the emissions? Would their reduction or removal as process inputs make the process inoperable or only more expensive? In the case of the latter, what are the economic consequences of partial or complete removal of undesirable process inputs?

Analysis should also be done on the basis of any available knowledge on the effects of changes in certain operating variables on the percentage of process inputs that end up as air pollutants. Are there any changes that can be made to process operating conditions that will reduce air pollutants with the same process inputs? What are the economic consequences of these changes? Information on these factors will allow consideration of the feasibility of the process change option as a partial or complete solution when aligned against the other techniques of dispersion and removal.

An example involving change in operating conditions is the case of a medium-sized refuse incinerator plant. The plant is presently processing 50 tons of refuse per day operating on an 8 hr/day cycle with a small continuous feed incinerator equipped with underfire air. The particulate emissions are in excess of the new air pollution code by a factor of 5.

The immediate consideration appears to indicate air pollution control equipment with a minimum efficiency of 80 percent is required; costs are estimated for its installation plus an annualized cost to own and operate the new pollution control equipment. Then, an examination is made of possible process changes. It is known that particulate emission can be reduced to the desired level by reducing the underfire air supply to the furnace, but at the expense of reducing the 8-hr capacity of the furnace to about 20 tons.

A consideration given to operating the plant over a 24-hr rather than an 8-hr period suggests an overall 24-hr capacity of 60 tons, while continuing to meet the required emission code. Estimates of additional operating and maintenance costs indicate an annual figure less than that for owning and operating pollution control equipment. The decision is reached to operate the plant on a 24-hr day.

Admittedly, this is an oversimplified illustration, but it emphasizes the importance of considering the process change approach in the overall analysis of pollution control situations.

The same general approach is applicable to gaseous and odorous emission problems as well. An example is provided by the kraft pulp process in which a process liquid, black liquor, is burned to produce heat and recover a useful chemical. Existing firing methods do not always completely incinerate odors produced in the furnace. And the problem is complicated where direct contact heat exchange is used for heat recovery from the combustion gases, since hot gas volatilizes odorous components from the liquor. Process change here includes improved firing practice, and the use of recuperative heat exchange to eliminate contact between stack

TABLE 2 Classification of Emission Control Equipment

Particulates	*Gases and Odors*
Filters	Wet absorbers
Fabric*	Scrubbers
Fibrous bed	Absorption towers*
Granular bed	Dry adsorbers
Precipitators	Fixed bed*
Dry—plate type*	Dynamic bed
Dry—pipe type	Dispersed
Wet	Incinerators
Dry inertial	Direct thermal (flame)*
Cyclones*	Catalytic
Baffled chambers	
Wet inertial	
Scrubbers*	

* Most common.

TABLE 3 Typical Industry Usage of Particulate Collectors

Electrostatic precipitators:
- Coal, oil, lignite, fluid coke-fired furnaces
- Kraft and soda papermaking process recovery boilers
- Lime kilns
- Cement, phosphate, gypsum, alumina, bauxite processing
- Blast furnaces, open hearths, basic oxygen furnaces, coke ovens, ore roasters, pyrites roasters, hot scarfing, sinter kilns
- Nonferrous roasters, smelters, converters
- Elemental phosphate furnaces
- Sulfuric acid production
- Petroleum cracking catalyst recovery
- Municipal incinerators

Mechanical collectors:
- Stoker-fed coal, wood, bark, bagasse-fired furnaces
- Cement, phosphate, gypsum, alumina, lime, bauxite processing
- Blast furnace sinter processing
- Taconite processing
- Roasters
- Refinery catalyst recovery
- Coal drying
- Spray drying
- Machining operations
- Flour, lumbering, woodworking operations

Fabric filters:
- Cement, phosphate, gypsum, alumina processing
- Electric furnaces
- Insecticide and fertilizer production
- Lead furnaces
- Titanium dioxide processing
- Spray drying
- Machining operations
- Feed and flour milling
- Woodworking
- Carbon black production

Wet scrubbers:
- Paper mill recovery boilers
- Lime kilns
- Phosphate and gypsum processing
- Blast furnaces, open hearths, electric furnaces
- Foundry cupolas
- Pyrites roasting
- Lead furnaces
- Aluminum production
- Sulfuric, phosphoric, nitric acid production
- Municipal and apartment house incinerators
- Hot coating processes

gas and black liquor. In a so-called "dry" system only combustion air is heated in direct contact evaporators; so the odors it picks up will be subject to incineration.

POLLUTION CONTROL BY REMOVAL

Classification of control technology for the removal of pollutants after their formation is again most conveniently expressed in terms of the two types of pollutants: particulates; gases and odors. A general outline of the mechanisms and types of equipment in general use today for removal of pollutants is shown in Tables 2 and 3.

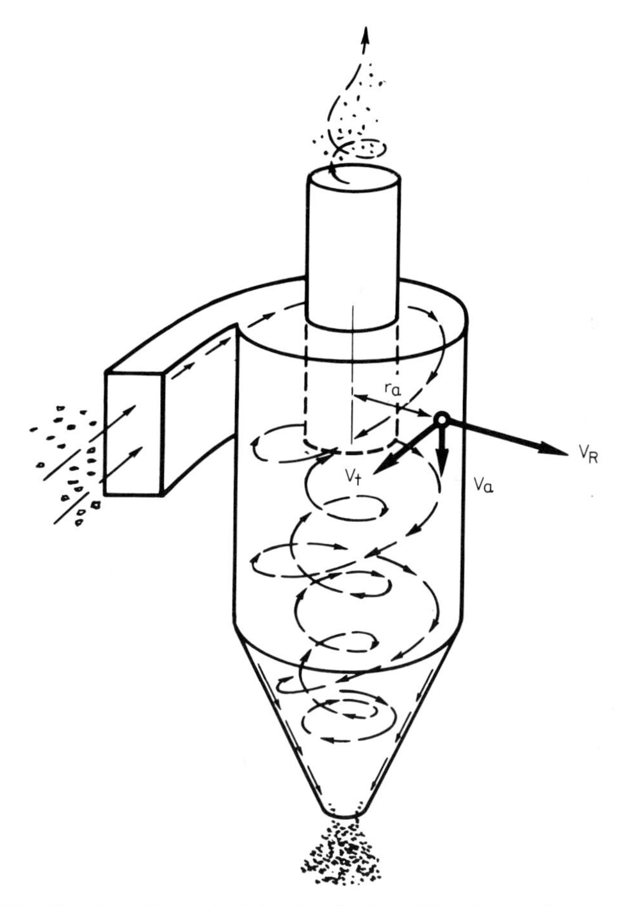

Fig. 2 Operating principle of a *dry* inertial cyclone collector.

Particulates

Cyclones The simplest form of particulate collector is the dry inertial collector. The most prevalent type in use today for air pollution control is the cyclone.

Figure 2 is a simple schematic of the basic cyclone collector. The entire mass of the gas stream with the entrained particulates is forced into a constrained vortex in the cylindrical portion of the cyclone. By virtue of their rotation V_t with the carrier gas around the axis of the tube and their higher density with respect to the gas, the entrained particulates are forced toward the wall at velocity V_R by centrifugal force. Here they are carried by gravity and/or secondary eddies toward the dust outlet at the bottom of the tube. The flow vortex is reversed in the lower portion of the tube, leaving most of the entrained particulates behind. The cleaned gases then pass through the central, or exit, tube and out of the collector.

There are many variations of the basic cyclone. In many, the vortex is achieved with a vane cascade rather than a simple tangential entry—these are usually called vane-axial or multitube types. Many utilize energy recovery devices in the gas outlet tube, after the separation of entrained particulates. These convert some of the rotational kinetic energy back into static pressure and thus reduce the total energy input required to establish desired separation velocities. This is shown in Fig. 3.

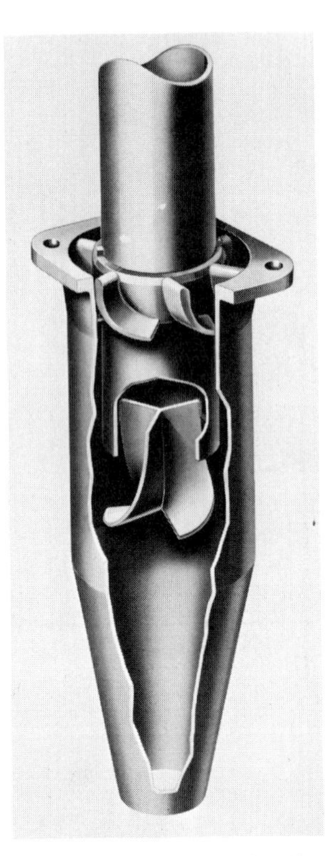

Fig. 3 Typical multitube mechanical collector tube, showing inlet vanes and outlet recovery vanes. (*Research-Cottrell, Inc.*)

Simplified equations which govern the performance of all cyclone collectors are shown in Fig. 4.

The first requirement is to accelerate the gas in the cyclone to get a high V_t. The relationship between the operating variables, cyclone design, and tangential velocity is shown in Eq. (1) of Fig. 4, which is derived from Bernoulli's equation. Acceleration of the gas and entrained particles to V_t is achieved by expansion of gas through the pressure change ΔP. The pressure change depends on the operating variables such as gas pressure, temperature, and molecular weight—and on the design of the tube, which essentially determines how efficiently static pressure is converted to velocity. Rotation of the particles around the tube axis at V_t results in centrifugal force, which causes the particles to migrate outward at a velocity relative to the gas of V_R.

The critical variables involved in V_R are identified in Eq. (2). They are particle radius, particle density, tangential velocity which is a function of power input ΔP, gas viscosity, and the tube radius. Generally speaking the efficiency is an increasing function of V_R. This relationship cannot be defined explicitly, principally because of factors such as reentrainment, bounce, and particle interactions. But with these equations, those factors which affect efficiency, and their direction, can be identified. It is clear, for example, that as particle radius increases, V_R will increase appreciably, resulting in an increase in efficiency. This is illustrated in Fig. 5, which shows the basic performance characteristic of a typical cyclone collector. This is known as the collector fractional efficiency curve. It is usually experimentally established by the supplier of such cyclones. It is the basic engineering tool, along with the flow–pressure drop formula, for design of cyclones. With these curves and a knowledge of the particle size distribution and specific gravity of material to be removed, one can calculate, by simple numerical integration, the overall efficiency to be expected.

It is interesting to note how rapidly the efficiency of cyclone collectors falls off on particles below about 10 microns in diameter. This is the outstanding characteristic of cyclone collectors—their inability to collect very fine materials at high efficiency and the rapid drop-off in efficiency below a certain critical particle size, a factor which often leads to their use as classifiers.

The attempt to minimize the radius of rotation to maximize V_R, and thus efficiency, has led to the extensive use of small-diameter cyclone tubes used in parallel to

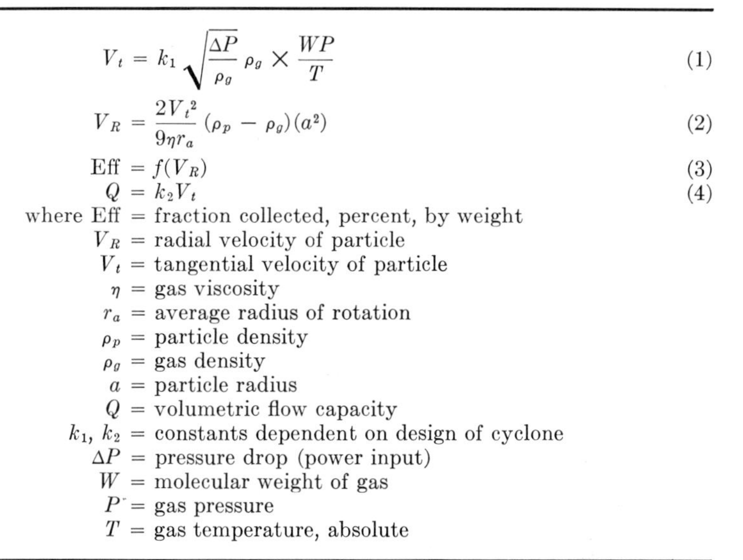

$$V_t = k_1 \sqrt{\frac{\Delta P}{\rho_g} \rho_g \times \frac{WP}{T}} \tag{1}$$

$$V_R = \frac{2V_t^2}{9\eta r_a} (\rho_p - \rho_g)(a^2) \tag{2}$$

$$\text{Eff} = f(V_R) \tag{3}$$

$$Q = k_2 V_t \tag{4}$$

where Eff = fraction collected, percent, by weight
V_R = radial velocity of particle
V_t = tangential velocity of particle
η = gas viscosity
r_a = average radius of rotation
ρ_p = particle density
ρ_g = gas density
a = particle radius
Q = volumetric flow capacity
k_1, k_2 = constants dependent on design of cyclone
ΔP = pressure drop (power input)
W = molecular weight of gas
P = gas pressure
T = gas temperature, absolute

Fig. 4 Cyclone collector formulas.

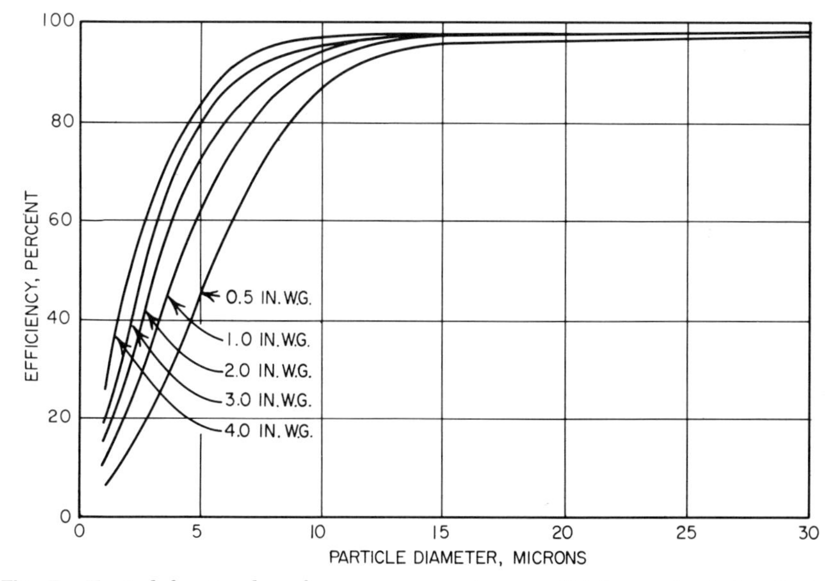

Fig. 5 Typical fractional performance curves for a multitube mechanical collector.

provide sufficient flow capacity in large installations without excessive pressure drop. A typical multitube array of vane-axial cyclone collectors is shown in Fig. 6.

Mechanical collectors can be used to clean gases with significant moisture content if the temperature remains above the dew point and operation is not cyclic. Dips below the dew point, however, may cause plugging problems, especially when very small diameter tubes are used.

Because of their simplicity of construction and lack of any moving parts, cyclones are available in a variety of materials, including ceramic (or ceramic lined), and are thus applicable over a broad range of temperatures, from below ambient to 2000°F+. They can be used at high gas pressures, the only critical factor being the requirement for proportionate increase in power input to maintain required separation velocities. They are not sensitive to inlet concentration—in fact, their efficiency increases with increasing particle concentration, probably because of particle interactions.

Fig. 6 A modern multitube mechanical collector with top inlet (from right) and side outlet (to left). (*Research-Cottrell, Inc.*)

Cyclones alone find most prevalent use under the following general situations:

1. Where dust is coarse
2. Where concentrations are fairly high—above 3 grains/scf
3. Where classification of entrained material is desired
4. Where very high efficiency is not a critical requirement
5. Where reduced stack opacity—which is primarily dependent on finer particle size fractions—is not a criterion
6. As precollectors in conjunction with collectors that are more efficient with fine particulates
7. As "scalpers" to remove coarse particles in tandem with other equipment which removes fine particles

Cyclones are the lowest in initial cost of the four generic classes of particulate collectors. For small-diameter off-the-shelf multitube collectors, the initial cost for collectors in the 100,000 acfm (actual cubic foot per minute) capacity class ranges from about $0.08 to $0.10 per acfm of gas treated. Special custom designed, lined cyclones may have an initial cost as high as $0.25 per acfm. Installation costs usually add another 25 percent. Operating cost for most cyclone collectors (which involve draft loss in the range of 2 to 4 in. water gage) runs about $0.05 per year per acfm design capacity. Maintenance cost—principally tube or lining replacement due to dust erosion—generally runs in the range of $0.03 to $0.04 per year per acfm design capacity.

Scrubbers The next particulate collector in terms of relative complexity and initial cost is the wet scrubber. As in the case of dry inertial collectors, there is a great variety of individual scrubber types and configurations. Again, they all operate on the same basic principle. In a wet scrubber, the primary aerosol particles are confronted with so-called impaction targets—which can be wetted surfaces or individual droplets. In most high-performance scrubbers, these are almost invariably the latter, with solid or liquid surfaces functioning as demisters. Demisters or mist eliminators perform the function of separating the scrubbing liquid from the gas stream. Performance capability requirement is less rigorous in the demister, since the liquid droplets with their captured particles have sufficient mass to make inertial separation simple. Demisters may also perform a cooling function.

$$\Psi = \tfrac{2}{9}(\rho_g - \rho_p)\,\frac{V_{RV}a^2}{\mu_g D_b} \tag{1}$$

$$\text{Eff} = 1 - \exp(-\,kL\sqrt{\Psi}) \tag{2}$$

$$D_b = f(\mu_L, V_N, \sigma_L, L) \tag{3}$$

where Ψ = impaction parameter
ρ_p = particle density
ρ_g = gas density
V_{RV} = relative velocity, particle to target
V_N = atomizer nozzle velocity
a = particle radius
μ_g = gas viscosity
μ_L = liquid viscosity
D_b = target diameter
Eff = fraction collected
L = liquid to gas volume ratio
k = constant
σ_L = liquid surface tension

Fig. 7 High-energy wet scrubber formulas.

As the primary particles attempt to follow the streamlines around the target droplets, their inertia causes them to move relative to the streamlines toward the surface of the droplets. A certain percentage in the total cross section swept out by a moving droplet will be collected by the droplet. This percentage is known as the target efficiency of the droplet. Target efficiencies have been analytically and experimentally correlated with a dimensionless parameter known as the "impaction parameter" or "separation number." This impaction parameter is defined in Fig. 7, together with the other simplified basic equations governing scrubber operation. Looking at the equation for the impaction parameter, it is seen that again particle size, particle density, and gas viscosity are critical variables. In addition, there are two others: relative velocity between particle and target, and target droplet diameter.

The impaction parameter defines the target efficiency of an individual droplet. An efficiency equation for a scrubber having a large number of particles and large numbers of target droplets has been utilized and experimentally verified under controlled conditions. This takes the form shown in Fig. 7. From this equation, it may be seen that the higher the impaction parameter, and the higher the liquid-to-gas ratio, the higher the efficiency. It is evident from the impaction parameter and the efficiency equation that high efficiencies will be obtained when particle radius, particle density, and relative velocity between particle and target droplet are high, and when gas viscosity and target droplet size are low.

Information is available on atomizing nozzles which shows that for both straight hydraulic and multifluid nozzles target droplet size decreases with increasing nozzle velocity, and increases with increased liquid-to-gas ratio, liquid viscosity, and liquid surface tension.

Unfortunately, a specific relationship between system energy input, scrubber geometry, and relative velocity has not yet been established. It has been shown, however, based upon a large number of individual tests, that for practical engineering design purposes the ratio V_{RV}/D is a function of the total power dissipated in turbulence in the system—regardless of the geometry of the device. This is a very important fact and essentially provides the basis for a practical design methodology in the absence of a complete understanding of the process.

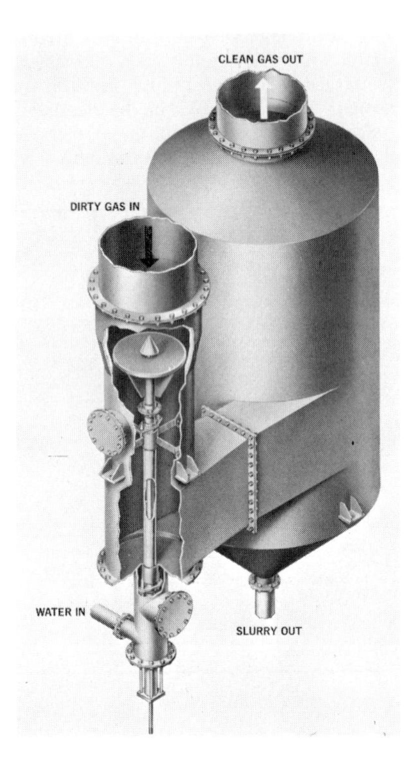

Fig. 8 Flooded-disk scrubber, a high-energy scrubber adustable to maintain optimum pressure drop despite variations in gas flow. (*Research-Cottrell, Inc.*)

What this methodology says is that for a given total energy loss in turbulence per unit volume of gas, the efficiency of the scrubber depends only upon the other variables in the efficiency equation—particle density, particle size, liquid-to-gas ratio, and physical properties of the liquid. It says if these are fixed, then in theory any scrubbing device—regardless of its geometry—will do the same job if it is operated at the same total turbulence per unit volume of gas. This turbulence can be introduced by losses in the main gas stream itself, by use of atomizing nozzles which use an external source of energy such as hydraulic or pneumatic pressure, by the use of mechanical agitators, or by any combination of the three. But the end result is the same.

Thus the critical variables in wet scrubbers are particle size and specfic gravity, total power input per unit volume of gas, viscosity, physical properties of the liquid such as viscosity and surface tension, and liquid-to-gas ratio.

Since the turbulence required for scrubbing can be produced in so many ways, there is a tremendous number of scrubber designs varying from simple spray chambers to highly complex and expensive mechanical devices. Wet scrubbers of the high-energy input type have the capability to collect particulates in the submicron range. Figure 8 shows an example. Further, since scrubbers utilize liquid media, they can be used simultaneously as chemical mass transfer devices. And because of the large gas-liquid interface, they are efficient heat-transfer devices and can cool the gases to the adiabatic saturation temperature of the inlet gas stream.

High-energy scrubbers generally find use where:

1. Fine particles must be removed at high efficiency.
2. Cooling is desired and moisture addition is not objectionable.

3. Gaseous contaminants as well as particulates are involved.
4. Gases are combustible.
5. Volumes are not extremely high (because of relatively high operating cost per cfm).
6. Large variations in process flows must be accommodated (variable orifice type only).
7. Relatively high pressure drop is tolerable.
8. The problem of scrubbing liquid polluted with the material(s) removed from the gas can be handled.

Initial cost for wet scrubbers in the 100,000 acfm capacity range may range from $0.25 to $0.35 per acfm in mild steel to $0.65 in stainless. Erection is usually about 25 percent more. Operating cost is relatively high, because of the energy input required to collect low-mass particles inertially. Operating cost ranges

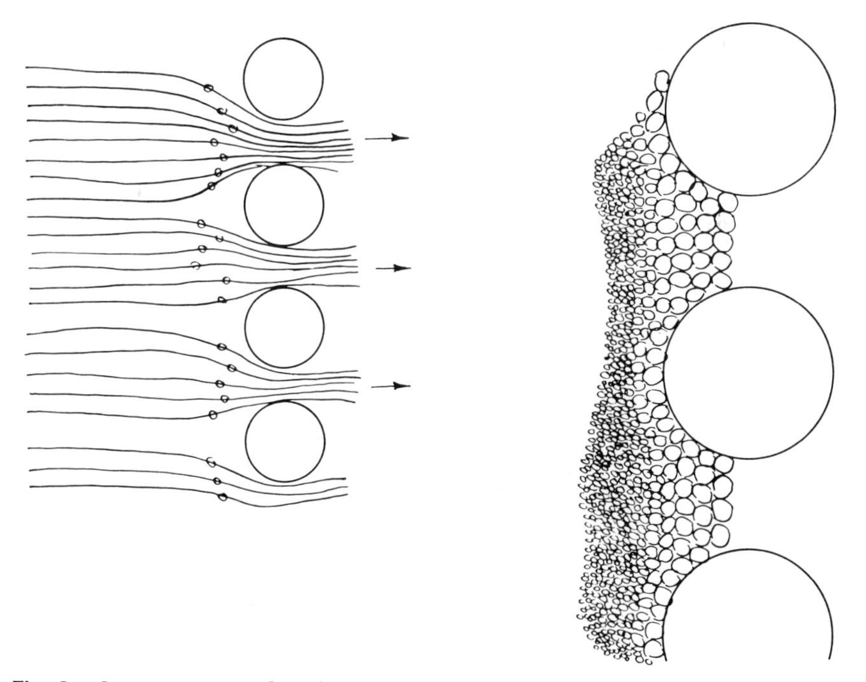

Fig. 9 Operating principles of a surface type fabric filter. Initial condition at left; filtering effect of "cake" indicated at right.

from $0.25 to $0.75 per year per acfm capacity depending on the problem. Maintenance cost is low, in the range of $0.01 to $0.02 per year per acfm capacity.

Fabric filters In essence, fabric filters are very large vacuum cleaners, with filters of various configurations made of porous fabrics which can stand the thermal, chemical, and mechanical rigors of individual applications. The most usual form of fabric filter comprises a number of cylindrical bags, usually inflated by the gases to be cleaned with the gas passing through the fabric filter medium from the inside. Particles suspended in the gas stream impinge on, and adhere to, the filter medium and are thus removed from the gas stream. In so-called "surface" filters, the deposit of dust so collected becomes, in turn, the filtering medium for succeeding particles. In the type known as "depth" filters the filter medium is a felted fabric into which particles may penetrate but do not pass through because of the constricted tortuous path they would have to follow.

This type of filter usually requires substantially less cloth area for a given loading than a surface type fabric filter. But it cannot be used with certain dusts, and where volumes are large the economic advantage of its high air-to-cloth ratio tends to be counterbalanced by other cost factors. Accumulated particles also aid the filtration operation in this type of filter, and both types can achieve very high efficiency in collection of particles down to submicron size.

This basic mechanism of surface filtration is illustrated in Fig. 9. Shown on the left is the filter medium and flow conditions for the initial condition. The mechanism of initial collection is identical to that described for wet collectors—namely, inertial impaction—although in fabric filtration direct interception, diffusion, and electrical effects may play an important role as well. As the filter cake is built up with time, the situation shown on the right develops. The role of the cake itself gives insight into what is probably the outstanding characteristic of fabric filters: their capability for very high efficiency on even the finest particles.

In considering fabric filters from an engineering point of view, one is rarely concerned with prediction of efficiency, for fabric filters have the capability of 99 percent + almost automatically—provided they are maintained in proper operating condition. The engineering problem is the selection of size, fabric, cleaning method and cycle, and operating pressure drop which will yield operational reliability for long periods at minimum overall cost. These factors are illustrated in Fig. 10, which gives typical operating characteristics for a typical bag or compartment of a fabric filter. Starting at time zero with a clean filter, pressure drop is low. Pressure drop rises according to the equation shown.

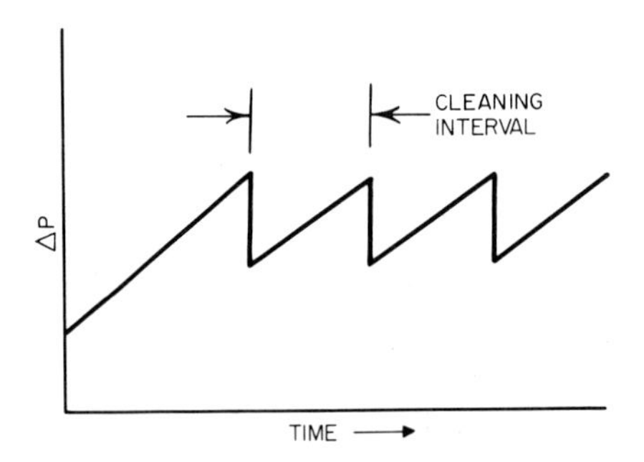

Fig. 10 Fabric filter formulas and cycle diagram. $\Delta P = \mu V_s(K_D + K_1 w)$; ΔP = pressure drop; V_s = superficial velocity; K_D = weave resistance coefficient (0.01–0.04); K_1 = cake resistance coefficient (shape, humidity); w = cake wt/sq ft surface; μ = gas viscosity.

The principal variables can be identified as:

1. Superficial velocity (also known as air-to-cloth ratio).
2. Fabric resistance coefficient (permeability).
3. Cake resistance coefficient—which, in turn, is dependent on particle size and shape, humidity, etc. Humidity is a particularly important factor—operation should remain above the dew point at all times to prevent excessive residual deposits and so-called "blinding" of fabrics.
4. Weight of cake per unit area—which is related to the concentration of particulates.
5. Gas viscosity.

As time increases, pressure drop increases to the point where it must be reduced lest either the flow capacity of the system be reduced or the bag rupture. At this point, the filter is cleaned by one of a variety of means. The cleaning operation may be destructive to the fabric—particularly fiber glass, which is presently the principal high-temperature medium—and will eventually, after a repeated number of cycles, cause it to fail and necessitate its replacement. Thus, a great deal of effort has been devoted to the evolution of many cleaning methods to increase the bag life—mechanical shaking, collapsing, reverse flow, shock wave, pressure pulse, etc. In fact, since all fabric filter suppliers generally utilize the same fabric filter media, the principal distinction between the fabric filters of different manufacturers is the method of bag suspension and cleaning.

From Fig. 10 the relationship between superficial velocity, operating power input, and bag life for a given problem can be seen. Increasing superficial velocity to reduce initial size and capital cost of the fabric filter will increase the rate at which pressure drop builds and thus the cleaning frequency required to hold a given pressure drop. But increasing cleaning frequency may reduce fabric life and increase maintenance costs. The state of the art of engineering design of fabric filters is selection of filter medium, superficial velocity, and cleaning method that will give the best economic compromise.

Figure 11 shows a schematic of a type of multicompartment fabric filter which

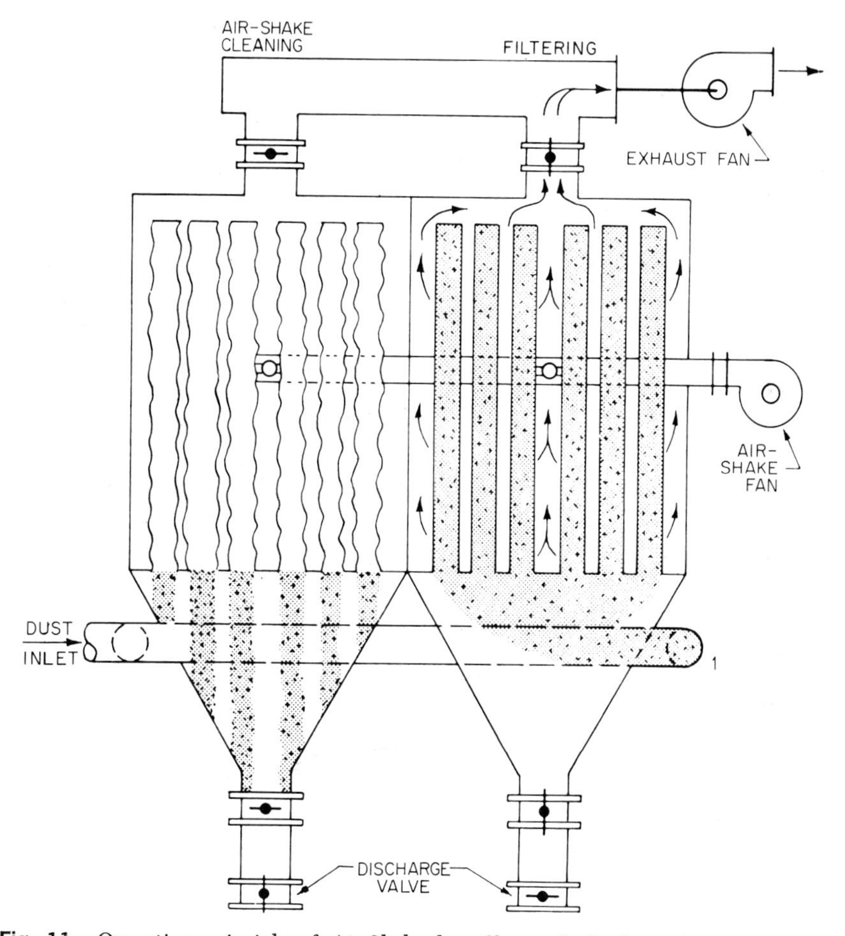

Fig. 11 Operating principle of Air-Shake bag filter, which cleans bags gently but thoroughly by air rather than mechanical shaking. (*Research-Cottrell, Inc.*)

utilizes a jet of air to effect gentle bag shaking. Note that fabric filters are multicompartment arrangements, where the compartment being cleaned is taken off the main stream of gas during the cleaning cycle. Dirty gas enters the fabric filter at point 1, passes upward into the bags on the right, through the filter medium, and out the exhaust at the top. After a period of time, the compartment on the left assumes the operating mode of the compartment on the right by automatic operation of inlet valves. At this time, the fan is operated, forcing air in a series of jets between the rows of right-hand compartment bags. This shakes and cleans them, causing the collected material to fall into the storage hopper. At the end of this operation, the compartment is brought back on stream. Figure 12 shows a mechanically shaken filter designed for intermittent operation. It is not compartmented; typically the entire unit would be shut down at a noon hour or shift end for cleaning. Gas inlet openings below floor level, and one or more outlet openings in the roof or end, are cut in the field to suit the installation layout.

Naturally, selection of fabric to stand thermal and chemical conditions is critical. Table 4 presents the characteristics of some of the natural and synthetic fibers used in the manufacture of filter media. Note particularly that at the present state of the art highest continuous temperature capability, offered by glass fiber fabric, is about 550 to 600°F. Ceramic fibers show promise for high-temperature fabric, but more development is needed.

In view of the foregoing characteristics, fabric filters are usually utilized:

1. Where very high efficiencies are desired (99 percent +).
2. Where operation is generally above dew point, i.e., low humidity in gases.

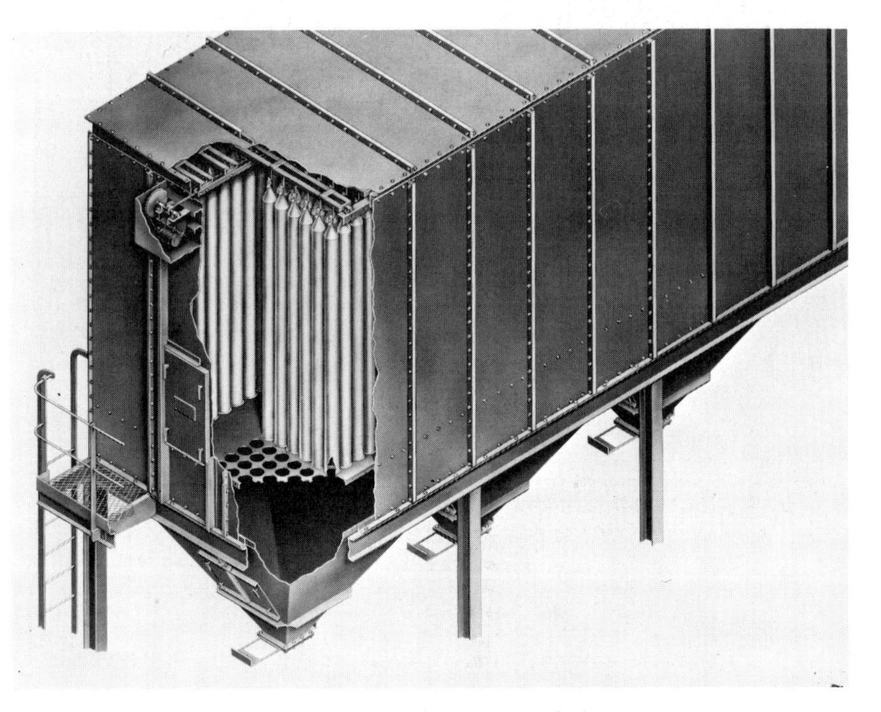

Fig. 12 Mechanically shaken fabric filter designed for intermittent operation.

3. Temperatures are not excessive.
4. Volumes are relatively low and therefore scrubbers or precipitators are not normally used; fabric filters can be and have been built for very large volumes.
5. Valuable material is to be collected dry.
6. Water availability and disposal is a problem.

Initial cost for fabric filters ranges from $0.50 to $1.20 per acfm capacity depending principally on the filter medium. Erected cost is approximately 1.30 × f.o.b. cost. Operating cost runs about $0.07 per year per acfm. Maintenance cost—principally bag replacement—runs 10 to 25 percent of original installation cost per year, or about $0.05 to $0.25 per year per acfm depending on the filter medium.

Electrostatic precipitators The basic elements of an electrostatic precipitator are a source of unidirectional voltage, preferably with some ripple on it; a corona or discharge electrode; a passive or collecting electrode; and a hopper to receive collected material. Three unit operations occur in electrical precipitation:

1. Particle charging
2. Particle collection
3. Transport of collected material

These unit operations may be conducted separately or simultaneously. In most industrial electrostatic precipitators they occur simultaneously. Such precipitators are often called single stage or Cottrell type. In some special applications—notably cleaning air for recirculation in occupied spaces—particle charging, particle collection, and material removal are conducted in distinct steps in separate sections. In either case the fundamental operating principles are the same.

TABLE 4 Properties of Some Common Fabric Filter Fibers

Fiber	Nominal temp limit, dry, °F	Resistance to moist heat	Resistance to abrasion	Resistance to mineral acids	Resistance to alkalies
Glass	550–600	E	P	E (except HF)	P
Teflon	400	E	F	E	E
Dacron	275	F	E	G	G
Acrylics	240–260	G	G	G	F
Rayon	200	G	G	P	F
Polypropylene	200	F	E	E	E
Nylon	200	G	E	P	G
Cotton	180	G	G	P	G
Acetate	Under 160	F	G	P	P
Polyethylene	Under 160	F	G	G	G

E = excellent, G = good, F = fair, P = poor.

Particle charging is achieved by the formation of an electrical plasma of positive ion-electron pairs in the region very close to the discharge electrode. This discharge is commonly called the corona discharge. This corona discharge provides a source of unipolar gas ions which migrate out toward the grounded passive electrode—filling the interelectrode space with a unipolar space charge. Particles entrained in the gases passing through this space charge collect these gas ions and themselves become highly charged. The same electric field that has produced the space charge then becomes the collecting field in which the charged particle is caused to migrate toward the passive or collecting electrode. Once the particle reaches the collecting electrode, it adheres to it until removed by mechanical means into the hopper below for storage and ultimate disposal.

Figure 13 gives simplified basic equations which govern the electrostatic precipitation process. The efficiency is seen to be related primarily to the total surface area of collecting electrode per unit volume of gas treated, and the so-called "particle drift velocity" or precipitation rate parameter. It is important to note that collection in the electrostatic precipitator is fundamentally independent of the velocity of the main stream flow (although excessively low or high velocities can cause problems aside from the basic collection process). In all other particulate collection devices it is necessary to accelerate the entire mass of the gas to be cleaned before separating forces can be obtained. The electrostatic precipitator, on the other hand, applies the separating forces directly to the particles without the necessity for accelerating the gas. This unique feature results in an extremely low power input requirement for the collection of fine particulates compared with any other form of collector.

It will be noted that the precipitation rate parameter is critically dependent on the field strength: it appears as the square. Note also that the rate parameter is dependent only on the first power of the particle radius; in contrast, inertial devices are more strongly affected by particle size. In contrast, a precipitator is extremely sensitive to those factors which affect the maximum voltage at which it can operate. These are principally the gas density and the electrical conductivity of the material being collected. The higher the gas density, the higher the w and the higher the efficiency—all other things being equal.

Field strengths in high-performance electrostatic precipitators usually run in the range of 15,000 volts/in., or about 60,000 to over 100,000 volts in industrial electrode geometries.

Most economical design and operation of dry electrostatic precipitators is obtained when the electrical conductivity (or its reciprocal, the resistivity, which is the commoner term) falls within certain desired limits. If the resistivity is too low, particles on reaching the collecting electrode rapidly lose their charge and can easily become reentrained in the gases flowing past the collecting electrode. If the resistivity is too high, charged particles reaching the collecting electrode cannot lose their charge because of the low conductivity of the material deposited earlier. As a result, a large voltage gradient is built up across the deposited layer, subtracting from both the charging and collecting fields and causing the values of w to fall.

Eventually, as this process proceeds longer, or with very highly resistive materials, the point is reached where the voltage gradient across the dust layer is sufficient

$$\text{Eff} = 1 - \exp\left(\frac{-Aw}{V}\right) \tag{1}$$

$$w = \frac{E_0 E_p a}{2\pi\eta} \tag{2}$$

$$E_0, E_p = f(\rho_g, R) \tag{3}$$

where Eff = fraction collected
A = surface area of collecting electrodes
V = volumetric flow rate
w = particle drift velocity or precipitation rate parameter
E_0 = charging field, volts/distance
E_p = collecting field, volts/distance
a = particle radius
η = gas viscosity
ρ_g = gas density
R = particle(s) bulk resistivity

Fig. 13 Electrostatic precipitator formulas.

to cause a dielectric breakdown and a phenomenon called "back ionization" or "back corona" which provides a source of ions of charge opposite to those being generated at the discharge electrode. These effectively neutralize the unipolar space charge and have a disastrous effect on the whole process. The electrical resistivity is a critical operating factor in the application of electrostatic precipitators. Fortunately, resistivity can be controlled in many cases by selection of proper operating temperature of the precipitator or use of simple conditioning agents like water vapor. But there are situations in which resistivity factors suggest that alternate equipment solutions may be more economical, at least on a first cost basis, than electrostatic precipitators.

There are many other factors—such as adequately uniform gas flow and maximum treatment velocities—which are important in the design of electrostatic precipitators, but the important thing for the purpose of this discussion is to remember that a precipitator is basically a volumetric device, as are all collectors—and that it is particularly sensitive to gas density and the conductivity of the material to be collected.

The term "sensitive" has been used several times in describing the precipitator, which is understandable in view of the very tiny difference in material removal per cubic foot of gas represented by, say, 99 percent instead of 95 percent collection efficiency. With an inlet grain loading of 7 grains/cu ft of gas, the difference is removal of an extra 0.00004 lb of material per cubic foot of gas. To achieve such delicate performance, a precipitator must be finely tuned; however, with today's modern controls this is substantially a design rather than an operating problem.

Gas density and flow rate are bound to fluctuate somewhat, and not be uniform throughout the precipitator. Automatic control is therefore provided to keep voltage continuously at the highest level allowed by changing conditions within the precipitator, and within the current and voltage ratings of the power supply. Moreover, for highest efficiency the electrical system is divided into two or more

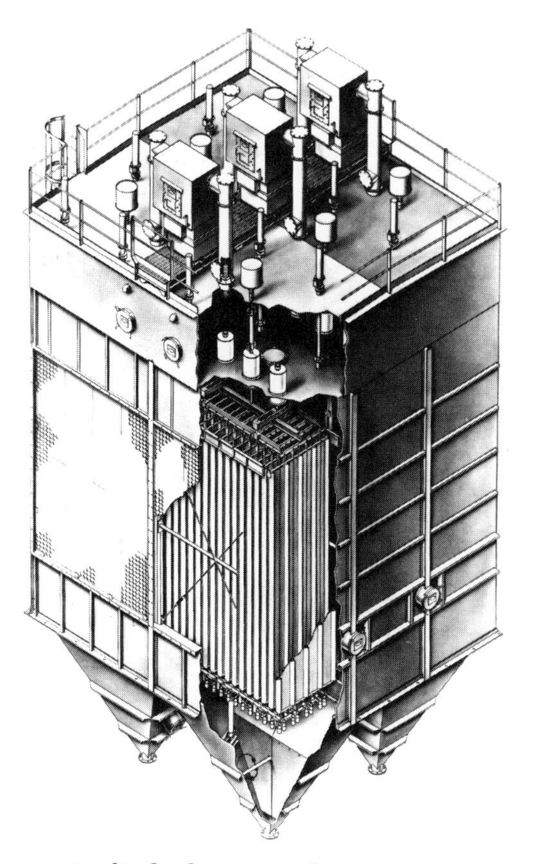

Fig. 14 Modern sectionalized plate type electrostatic precipitator. (*Research-Cottrell, Inc.*)

sections each responsive only to conditions within its area of the precipitator, not limited by conditions existing elsewhere within the unit.

Figure 14 illustrates a typical dry plate type precipitator. In some applications the collecting electrodes are vertical pipes with a discharge electrode hung in each.

Because of their relatively low power input, because they do not rely on inertial separating forces, and because they can operate completely dry, electrostatic precipitators find their most prevalent use where:

1. Very high efficiencies are required on fine materials.
2. Volumes of gas are very large.
3. Water availability and disposal are problems.

4. Valuable dry material is to be recovered.

Initial cost for electrostatic precipitators in the 100,000 acfm range is about $0.80 per acfm. Actually, 100,000 acfm is a relatively low volume for an electrostatic precipitator. In the more usual size range, of the order of 1,000,000 acfm, the initial cost is about $0.40 per acfm. Erected cost is highly variable but for grass roots plants usually runs about 1.70 × f.o.b. cost. Operating cost is very low, amounting to about $0.03 per year per acfm capacity. Maintenance cost is in the range of $0.02 to $0.03 per year per acfm capacity.

Electrostatic precipitators are also utilized, although not as extensively, for the collection of mists and fumes in saturated atmospheres. Here the principal requirement is that collected material flows freely, or can be easily flushed with intermittent sprays, from the collecting electrodes. In this type of application, pipe type precipitators, with somewhat different economics due—usually—to the need for corrosion resistant materials of construction, are commonly utilized.

Precipitators are also being considered with increasing frequency for removal of fine particulates from gas streams at substantially elevated temperatures and pressures. Feasibility has been demonstrated at pressures approaching 1,000 psig and temperatures to 1500°F.

REMOVAL OF GASEOUS AND ODOROUS POLLUTANTS

Adsorption

Adsorption is a process by which molecules from a gas or liquid stream attach themselves on the surface of a solid. The process is based on the attractive force between the solid surface and the adsorbed molecules. These forces vary considerably with the nature of the surface and molecule, and the intensity of the force is the basis of subdivision between physical and chemical (chemisorption) adsorption. In the latter, there is a distinct chemical reaction between the surface molecules and the adsorbate molecules, generally leading to a condition of irreversibility. For most industrial applications, this condition of irreversibility is not economically feasible since either the adsorbent must be used only once, or the energy requirements for regeneration are excessive. Consequently, physical adsorption is much more important as an air pollution control approach.

In order to obtain economic rates of removal, it is necessary to expose large adsorbent surfaces to the stream to be processed. Generally, highly porous solids with large internal areas due to fine capillaries are used in these processes. Finely divided materials of intermediate surface area levels find some application also.

Activated carbon is the most important adsorbent. It shows good performance with solvents and odor molecules. It is relatively nonpolar, so that the adsorptive performance for organic molecules in the presence of water is only mildly depressed.

Alumina, bauxite, molecular sieves, and silica gel are more polar adsorbents. They find utility in drying operations and in liquid phase adsorption processes. Clays and ion exchange resins find some utility in decoloring operations. Specialty carbons such as those containing inorganic additives (Ag, Cu, Zn, etc.) are used for gas masks and related applications. Other specialty adsorbents may combine an oxidizing agent on a high-surface-area support for treating low levels of oxidizable air pollutants ($KMnO_4$ on alumina, etc.).

Many adsorbents are also used in catalytic processes.

Adsorbents are manufactured in different grades of surface area, porosity, and particle size. Additionally, the source of the material (vegetable, petroleum, etc., in the case of activated carbon), the chemical composition ($SiO_2/Al_2O_3/Na_2O$ in the case of molecular sieves), the type of impurity, and especially the method of activation are important factors in influencing adsorptive performance.

Capacity Equilibrium isotherms define the equilibrium extent of adsorption as a function of solute concentration. With a given solute, the extent of adsorption increases with solute concentration, system pressure, and reduced temperature level. Large molecules of high boiling points are adsorbed to a larger extent, limited only by those pore diameters smaller than that of the molecule.

In addition to these equilibrium concepts, kinetic concepts may be important in practice. At high gas flow rates, only a small fraction of equilibrium capacity may be achieved before breakthrough of the solute occurs and insufficient removal of solute from the gas stream may take place.

In mixtures of two or more solutes, complex effects may occur. While adsorption of both solutes will occur simultaneously, as adsorption proceeds the higher-boiling components may displace the more volatile one.

Regeneration In most applications, efficient regeneration is required so that economic reuse of the adsorbent can take place. Thermal regeneration, typically using steam to raise bed temperature to about 650°F, is generally used. Steam is passed into a bed at low velocity, generally in a direction opposite to the flow during solute removal. The adsorbate molecules are loosely held at the elevated temperatures and are partitioned between the adsorbent and steam. Cooling and drying of the carbon bed are then generally practiced.

In some cases, when very low concentrations of strongly bound components are removed from a gas, it is economically reasonable to dispose of the loaded adsorbent material.

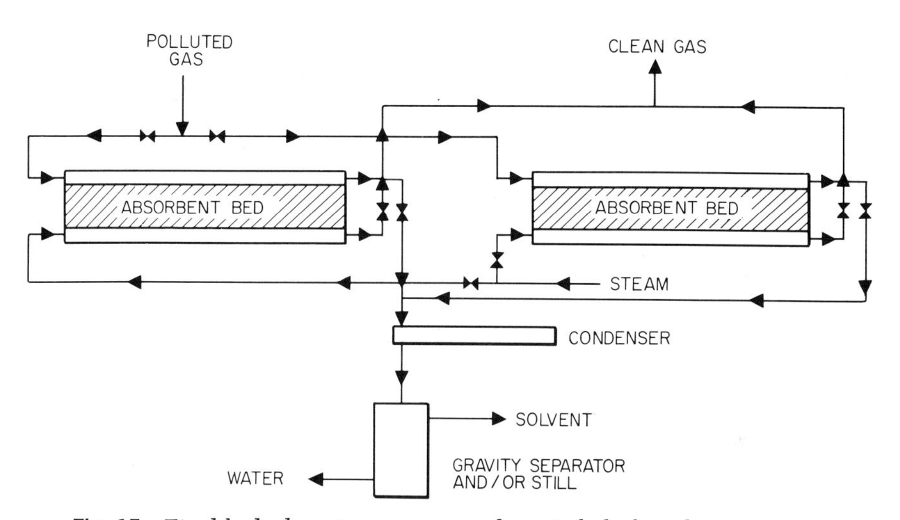

Fig. 15 Fixed bed adsorption process, used particularly for solvent recovery.

Adsorption equipment

Fixed Bed Adsorption. Generally, fixed bed processes have been used. In large installations, horizontally oriented beds 3 to 25 ft in depth are used. Ratings of standard units are roughly 2,000 to 40,000 cfm (dual units). Capacities of these units for common solvents are of the order of 200 to 3,000 lb/hr. Complete units, with blowers, steam condenser, separator (decanter), and control, are supplied. Controls initiate various phases of the process, either on a time basis or on an analysis basis. Figure 15 shows a simple schematic.

The basic technique of the fixed bed unit involves loading and alternately regenerating two or more units. Frequently, three units are used, with one on stream, a second being regenerated by steaming, and the third cooling. Cooling is frequently effected by passage of solute-free air through the previously steamed bed. Loading is usually limited to a maximum temperature of 100°F.

Typical gas flow rates are on the order of 60 to 100 fpm (based on open cross-sectional area).

While economics favor recovery of solvents from concentrated air streams, it is necessary to avoid the hazards which accompany operation in the explosive range. Generally, sufficient air must be provided to avoid exceeding 25 to 50 percent of the lower explosive limit (LEL). Typical values of LEL are given in Table 5.

Recovered solvent is generally subjected to a distillation operation, especially so if it is partially contaminated with water. Frequently, recovered solvent may be recycled to the process without difficulty, but it is wise actually to test the performance of recovered solvent obtained in a pilot installation.

Generally, above about 1,000 ppm feed concentration, recovery of solvent is advantageous. At trace levels, 1 ppm, disposal of loaded carbon is generally practi-

TABLE 5 Typical Lower Explosive Limit Values (percent by volume)

Acetone	2.5
Benzene	1.2
Carbon disulfide	1.1
n-Hexane	0.96
Isopropanol	1.8
Methane	6.1
Methyl ethyl ketone	1.8
Toluene	1.1
Xylene	0.9

cal. In the intermediate range of 1 to 1,000 ppm oxidative regeneration has been used. Air is passed into the spent carbon and in conjunction with traces of catalytic metals on the carbon effects oxidation of the pollutant.

Continuous adsorption permits use of smaller charges of adsorbent for a given duty. While large fluidized beds are useful as continuous adsorbers for process applications, for air pollution the use of a rotating fixed bed unit containing the carbon in a rotating annulus increases the efficiency of carbon utilization. Figure 16 illustrates the principle. Feed air enters the shell through ports, passes through

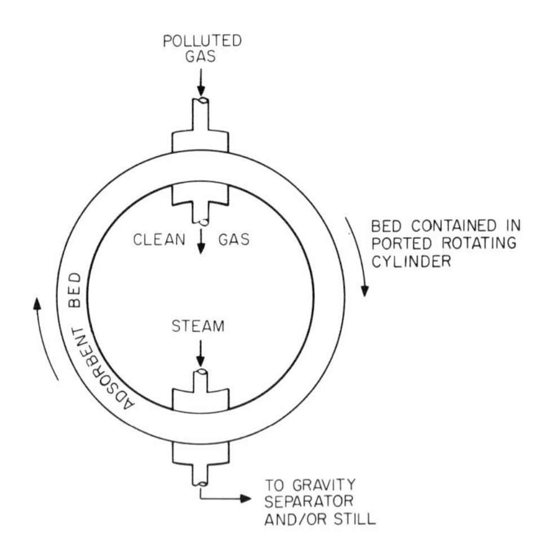

Fig. 16 Operating principle of a continuous absorption process.

the carbon bed, and exits through ports located in the inside of the cylindrical bed. The solvent-free gas travels along the cylinder axis and is exhausted to the atmosphere. A similar arrangement permits feed of steam, by use of valves, to the carbon as it reaches the lowest point on the cylinder.

Adsorption is a high-initial-cost process, frequently warranted when recovered product for reuse is involved. Capital costs for package units are \$4 to \$10 per cfm for carbon steel and \$5 to \$13 per cfm for alloy construction. Installation varies considerably because of the nature of the gas stream and contaminants;

it may range from 20 to 50 percent of equipment cost. In general, it is practical to recover a large fraction of solvents vented in printing operations, solvent use, plastic operations, and extraction operations. In other plants, low levels of odorous materials may be removed but not recovered from air streams. Installations in this category include fermentation, glue, natural product processing, and tanning operations.

Absorption

Absorption is a process in which a soluble gas is transferred from a gas stream into a liquid. The gas may become physically dissolved in the liquid or may react with a dissolved constituent in the liquid.

Absorption serves as a method both for recovering gaseous components and for purifying gas streams. It is the latter operation in which we are interested. In many air pollution control operations soluble gaseous components to be removed constitute levels of 1 percent or less of the main gas.

Gas absorption is a diffusional operation which depends upon the rate of molecular and eddy diffusion. Ultimately, the transfer must take place across a liquid-gas interface. The interface may be formed by the use of liquid films, gas bubbles, and liquid droplets. A large variety of devices have been devised to effect gas absorption based on either dispersed liquid or dispersed gas phase. A brief classification of these equipment types follows:

- Dispersed liquid
  - Liquid film
    - Packed tower
    - Wetted wall tower
  - Liquid drops
    - Spray towers
    - Cyclone spray chamber
    - Venturi scrubber
    - Flooded disk scrubber
    - Centrifugal gas absorber
- Dispersed gas
  - Gas bubbles
    - Plate (tray) tower
    - Sparging tank
    - Agitated vessel

Gaseous solutes which are to be removed by absorption must exhibit solubility in the liquid scrubbing medium. For most air pollution control applications, water is the most suitable scrubbing medium based on availability, cost, level of corrosiveness, volatility, viscosity, and ease of disposal. Frequently, the level of removal desired decrees that the scrubbing liquid rate must be quite high because of the limited solubility of the solute in water. This condition of high liquid rate results in increased pressure drop and/or increased capital costs. Consequently, the most desirable gas absorption systems are those in which the dissolved solute exerts negligible partial pressure over the solution. This is commonly achieved by means of chemical reaction. Thus, acids are absorbed in alkaline solutions to form non-volatile salts (HCl in caustic solutions, H_2S in ethanolamine, etc.).

In some cases, when the solute concentration level is sufficiently high and when it possesses economic value, recovery of solute in concentrated form is warranted. In those cases, regenerative processes are used. Typically, heat unstable compounds (bicarbonates, amine salts) are formed, and then the solution is subjected to thermal regeneration, solute recovery, cooling, and reuse. This approach has not been extensively adopted in air pollution control because of the extremely low levels of pollutants involved and the complexity of some polluted streams. However, regenerative absorption processes are under study for SO_2 removal from stack gas.

Another common means of reducing the partial pressure of a pollutant is by oxidation in solution. Oxidants include chlorine, permanganate, and oxygen (with dissolved catalyst) as well as reversible oxidants such as quinones, chelates of iron, cobalt, nickel, and the like. Thus, H_2S of limited water solubility is oxidized to (nonvolatile) SO_4^{--} or S. Volatile odorous compounds (ketones, aldehydes) are converted to less volatile acids and to CO_2.

Kinetics of absorption The basic expression for the rate of transfer of solute from the gas phase to the liquid phase is given in Fig. 17.

$$N_A = K_G A P \,\Delta Y$$

where N_A = moles transferred per hour
K_G = mass transfer coefficient, moles/hr/sq ft/atm
P = total pressure
ΔY = driving force differential, expressed in gas phase concentration units
A = interfacial area

Fig. 17 Rate of absorption formula.

The expression in this form, based on gas phase concentration units, is most useful in the field of air pollution, since a major portion of the gas adsorption situations involve the removal of fairly soluble solutes present at low concentrations in air. For these cases, the resistance to mass transfer is largely controlled by gas phase diffusion, and the use of partial pressure concentration units is most reliable.

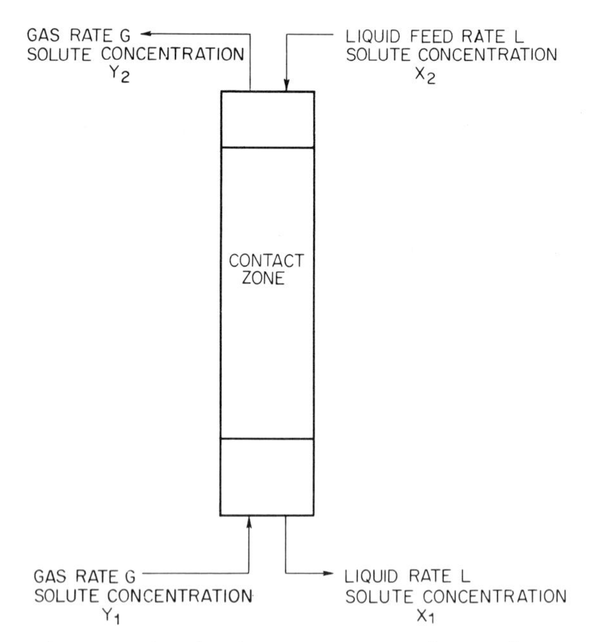

Fig. 18 Absorption operation showing parameters used in performance calculation.

The mass transfer coefficient based on gas phase concentrations is a function of solute-carrier physical properties, turbulence level, temperature, and characteristic dimensions of the system.

The area term A refers to the interfacial area of the system. In the case of liquid droplets, the term defines the total surface area of the droplets. In packed towers, the term is the totality of interfacial surface area and is most frequently expressed on a volumetric basis (as a combined term K_GA), although it is quite useful to be able to define separately the effect of liquid rate on wetted surface in the piece of equipment.

The term ΔY is the driving force available for mass transfer and is the difference between the Y, the concentration of solute in the gas, and Y^*, the equilibrium concentration of solute corresponding to the actual liquid composition. In the most favorable case, that achieved by irreversible chemical reaction, $Y^* \rightarrow 0$.

The driving force expression is best characterized by an X, Y diagram which relates concentration of solute in the gas phase and in the liquid phase by a mass balance, and represents the relationship between equilibrium liquid and gas concentration. For the countercurrent packed tower arrangement of Fig. 18, Fig. 19 gives the X, Y plot.

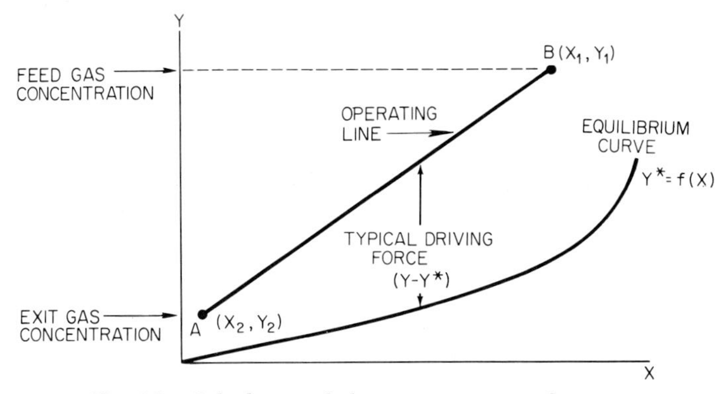

Fig. 19 Calculation of absorption tower performance.

For dilute gases, with a relatively nonvolatile solvent, the operating line AB is a straight line. The equilibrium line, representative of a partially soluble gas, is relatively straight at low concentrations, showing considerable curvature at higher concentrations. The driving force for a gas-film controlling case is shown as Y-Y^*.

A convenient approach to specifying the performance of a unit is based on the transfer unit concept. The number of transfer units required for a given duty in the dilute gas case is given by

$$N_{OG} = \int_{Y_2}^{Y_1} \frac{dY}{Y - Y^*}$$

The height of packing required for a specified performance is

$$h = \frac{G_M}{K_GAP} N_{OG}$$

where h = packed depth, ft
G_M = molar in gas (mass) velocity based on tower cross section, lb mole/(hr) (sq ft)
K_GA = mass transfer coefficient, lb mole/(hr)(cu ft)(atm)
P = total pressure, atm

For the simple case of zero back pressure (solution is reactive medium, oxidation, etc.)

$$N_{OG} = \ln \frac{Y_1}{Y_2}$$

In this case, one transfer unit results in 63 percent removal of solute. Three transfer units are required for 95 percent solute removal.

Packed towers vs. plate towers These are the commonest types of gas absorbers. The former is commoner in air pollution control situations. The plate tower finds application in large sizes where the distribution problems inherent in large packed towers become important. The plate tower is also useful when cooling is desired, or when very high liquid rates are required. The packed tower is well suited for corrosive applications, low pressure drop operations, and relatively small scale operations.

Typical Packed Tower. The unit consists of a cylindrical shell, a packed section held on a support plate, a liquid distributor, possibly a liquid redistributor, access manholes, gas inlet and outlet, and for those cases in which a recirculation of the liquid is desirable, a sump with recirculation pump and overflow. A typical unit is diagramed in Fig. 20.

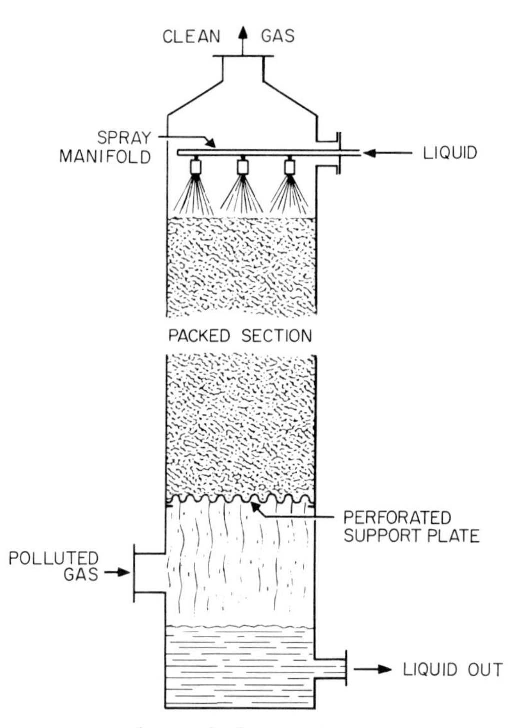

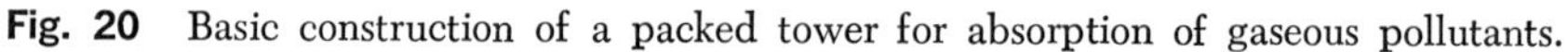

Fig. 20 Basic construction of a packed tower for absorption of gaseous pollutants.

Packing. In the past, broken rock was a typical material. Now more sophisticated materials are available which are relatively inexpensive, permit high liquid and gas rates without excessive pressure drop, and are consistent with good gas absorption rates.

A variety of materials of construction include ceramic, carbon, metal, plastic, and wood. Popular shapes include Raschig and Pall rings, Berl and Intalox saddles, and low pressure drop grids.

Combined particulate removal and gas absorption Many processes emit particulate matter and noxious gases simultaneously. An economic solution for air pollution control in this case is a combination of a high-energy scrubber with a low-energy packed tower. Such a combination is shown in Fig. 21. The high-energy flooded disk scrubber achieves a scrubbing action by the atomization of the scrubbing medium into small droplets in a variable area conical throat. The performance level can be maintained relatively fixed irrespective of gas flow rate by maintaining a fixed pressure drop across the unit.

Accompanying the particulate removal in the scrubber is considerable absorption by the finely divided solvent droplets. When additional gas absorption is required, a packed tower addition is a satisfactory solution.

Mild steel packed towers for air pollution control applications cost $1 to $5 per scfm.

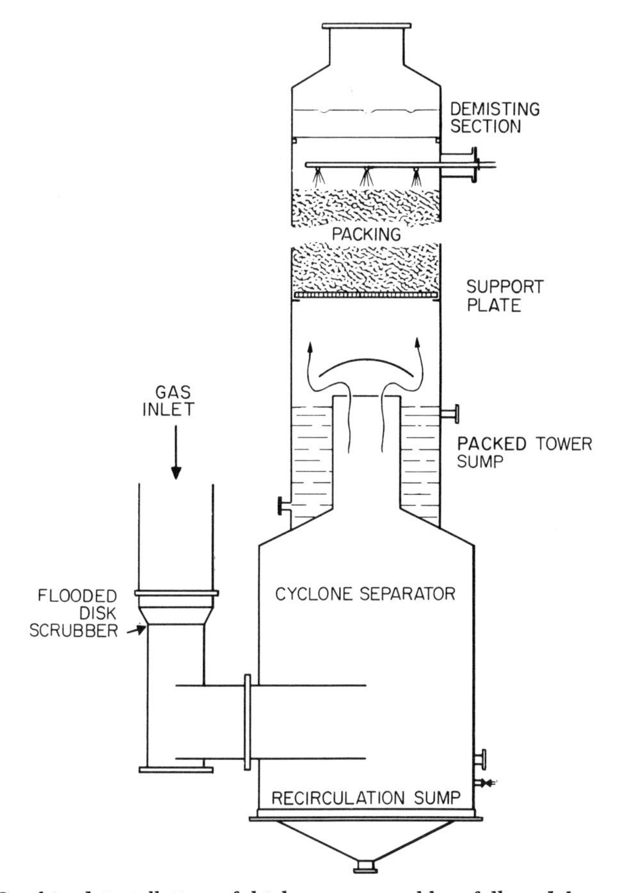

Fig. 21 Combined installation of high-energy scrubber followed by packed tower can remove both particulate and gaseous pollutants from a gas stream. (*Research-Cottrell, Inc.*)

Combustion

Flame Many processes produce fumes which have no recovery value because of either their complexity or the inherent nature of the material. Disposal by adsorption or absorption therefore may not be warranted. However, the fumes may form a combustion or health hazard or constitute a noxious odor source. When the fumes contain combustible material, one means of disposal is flame incineration by which the organic matter (both gaseous and particulate) is converted to carbon dioxide and water. Sulfur compounds are converted to sulfur dioxide.

In most air pollution situations, the concentration level of combustibles in the gas stream is below the lower limit of flammability. Auxiliary fuel must be burned to heat the fumes to a sufficiently high temperature with adequate reaction time for complete combustion, because partial oxidation products may be formed which are likewise unacceptable in the gas stream. Under proper conditions, flame incineration is effective in removing combustible aerosols and gases from waste streams, improving stack visibility, reducing odors, eliminating toxic components, and eliminating the danger of explosions. In many cases, efficiencies of 90 percent or more may be achieved.

Considerable variation in combustion zone temperature exists for complete removal of various compounds. Typically 1200 to 1500°F is the most representative temperature range for good combustion. Efficient combustion depends upon (1) adequacy of oxygen supply, (2) completeness of mixing of polluted gases with flame, (3) adequacy of combustion zone temperature, and (4) adequacy of contact time at combustion zone temperature.

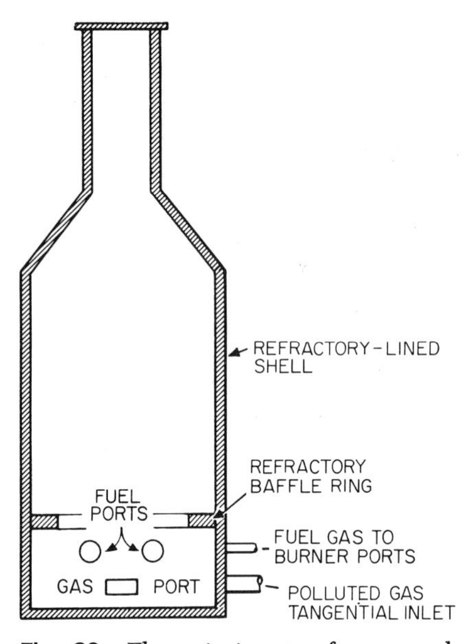

Fig. 22 Flame incinerator for removal of gaseous pollutants, with auxiliary fuel used to maintain necessary temperature.

Generally there is little problem in supplying sufficient oxygen. Typically the levels of combustible to be destroyed in the air are low and the necessary air requirements are based on fuel consumption.

Good mixing may be attained by supplying the contaminated air directly into the throat of the afterburner at velocities of the order of 25 fps. Figure 22 sketches a common arrangement using tangential entry of both fuel gas and polluted air into the base of the combustor. In some units combustion air consists solely of the polluted air. This type of unit restricts consumption of fuel. It is preferable for applications in which a relatively constant flow of air is to be treated.

The advantages of this technique for air pollution abatement include simplicity, high efficiency, and relative insensitivity to contaminant composition. With respect to the latter condition, more resistant contaminants require longer exposure or somewhat longer contact time. Generally a contact time of the combustion zone of 0.3 to 0.5 sec is supplied, with temperatures ranging to about 1500°F. Other advantages include the compact nature of the equipment and the insensitivity of the abatement approach to aging (vs. catalytic combustion).

Flame incinerators operate at low pressure drop and require little maintenance if properly constructed. The chief operating cost is that of the fuel required to attain temperature levels of ~1500°F. In most large installations considerable recovery of heat values can be achieved by use of heat exchangers or by directing the combusted gas back into the process, for example, to a drying or curing oven.

In situations in which there is considerable variation in flow of the contaminated stream, it is advisable to install modulating burner controls to maintain constant temperature in the incineration chamber. It is also customary to record the operating temperature so that malfunction is readily noted and corrected.

Baking and curing operations involving inks, paints, and floor coating generally produce solvent laden atmospheres which are conveniently treated by flame incineration. Similarly roasters, meat smokehouses, rendering, varnish cooking, blood driers, and sewage treatment operations produce odorous gases which may be controlled by flame incineration.

Costs of flame incineration units are reported to average $5 to $10 per scfm (installed).

Catalytic combustion Many objectionable emissions contain combustible components at low concentration levels, unsuitable for direct combustion without the addition of considerable heat value.

Often these components are poorly defined chemically so that direct chemical reaction by absorption is not practical. Catalytic combustion is a technique which can often convert components to harmless carbon dioxide and water under reasonable reaction conditions.

Since oxidation conducted over a catalyst requires a lower ignition temperature than the thermal combustion process, it is possible in many cases to effect high reduction of pollutant concentrations without excessive fuel charges.

The basis for the catalytic process is a metallic catalyst, generally distributed on an inert support. At reaction conditions momentary adsorption of an organic gas molecule on the active catalyst surface occurs; at an adjacent site, oxygen is adsorbed to form an activated metal-oxide complex whose nature depends upon the particular metallic catalyst. Oxygen is transferred to the organic molecule, and the molecule is released ultimately as carbon dioxide (and water).

The process occurs under dynamic conditions of oxidation and reduction of the surface, and if poisons are not present in the gas stream, catalytic activity is maintained at a high level and regeneration or replacement of the catalyst is not required for long service periods.

Process Requirements. The process requires a combustible pollutant at a concentration less than its lower explosive level (LEL). Generally, for safety, operating concentrations are maintained below 25 to 50 percent of the LEL.

In order to maintain the catalyst in an active state and supply sufficient oxygen to the process for complete combustion of organic compounds, 1 percent excess oxygen, based on stoichiometry, is required. Addition of oxygen is beneficial in terms of higher removal efficiencies at given contact time and temperature.

The kinetic nature of the catalytic reaction at fixed contaminant concentration and exposure time indicates a narrow temperature range, often 50 to 100°C, over which the extent of combustion of the contaminant increases from 10 to more than 90 percent. The temperature range varies with the reactivity of the contaminant.

Decreased contact time or reduced catalyst activity can be partially compensated for by the use of higher bed temperatures. In practice, the inlet temperature to the bed is increased as the catalyst ages to maintain efficiency at a constant level.

In the active temperature range, it is often possible to achieve high throughputs with a catalytic unit, being limited only by pressure drop resulting from operation at excessive bed velocities. Generally, contact time as such is not specified; rather it is more useful to specify space velocity, the number of gas volumes treated by a unit volume catalyst bed per hour. Thus, for many applications, the honeycomb type catalyst bed can operate efficiently at 30,000 to 125,000 (vol)/vol/hr. Some modular units (Oxy-Cats) are rated at 5 to 15 scfm/unit. Units may be rated on face velocity, cfm/sq ft, provided depth of the unit is specified.

Catalysts. Most units are based on platinum metals, which are highly active in the presence of oxygen; they exhibit, under most conditions, good thermal stability and resistance to poisoning. Other metals require higher reaction temperatures and are generally less useful.

A wide variety of designs have been used, but most of the units use either a platinum metal electrodeposited on a heat-resistant Ni-Cr alloy, platinum metal dispersed on a high area support, or a combination of the latter on a ceramic honeycomb network such as that shown in Fig. 23. The advantages of low pressure drop, good distribution of gas to the entire catalyst surface, ease of handling, high volumetric efficiency, and thermal stability are achieved in the honeycomb unit. For the honeycomb unit, an aluminum salt solution containing alumina particles may be used to coat the cell walls of the honeycomb. After drying and thermal treatment, a film of approximately 0.003 in. of strongly adherent material is built up on the honeycomb structure. This alumina is impregnated with a solution of a platinum metal salt and thermally treated. In this fashion an active surface of above 20 sq m/g is prepared, which facilitates gas treatment at high volumetric rates.

Temperature Control. In some cases, a hot gas stream requiring no preheat is to be treated and then vented. Usually, this situation will require an auxiliary burner to maintain catalyst temperature during periods in which gas flow does not occur.

If the process gas is available below the ignition temperature, fuel may be burned to preheat the gas stream. Adequate precautions must be taken to ensure mixing of the gases before passage into the catalyst bed.

It is frequently desirable to supply the necessary heat to the cool gas stream by exchange with the purified, hot exhaust gas stream. Alternatively, the hot gas stream may be recycled to the process or to other energy recovery stages (steam generation or turbine power recovery).

In some systems, the concentration of combustible vapors is limited to a predetermined concentration by the use of a gas analyzer which samples the gas and controls a damper on the entrance of the catalytic unit. An alarm may be incorporated into the system to warn when the predetermined concentration is exceeded. Automatic shutdown at still higher concentration levels is also a feature of this control system. This control is incorporated into the system to prevent overheating of the catalyst.

Fig. 23 Honeycomb catalyst. (*Oxy-Catalyst, Inc.*)

Poisoning. In mechanical poisoning the surface of the catalyst is coated with finely divided particles carried in by the gas stream. This should be avoided by proper gas cleaning techniques, as discussed earlier in this chapter.

In general, the platinum metal catalysts are relatively insensitive to poisons; however, few types of impurities can deactivate the catalyst, sometimes at a rate which decrees that noncatalytic removal technique be used.

Halogen compounds act as suppressors of catalytic activity; at sufficiently low concentrations, it is possible to operate at increased catalyst temperature to overcome this suppression.

Phosphorus, arsenic, and boron are chemical poisons for platinum which are bound to the catalyst surface. Their presence in the gas stream should be avoided. Pilot plant tests may be required to determine if catalyst deactivation will be a problem in practice.

In some cases, it is practical to clean the catalyst surface of accumulated matter. In one installation, iron particles were removed during shutdown periods by dismantling this unit and soaking each catalytic element in oxalic acid solution. Other cleaning techniques are described in the patent literature.

In some cases, it is possible to deposit carbon on the catalyst surface. This may occur during periods of insufficient preheat with gases containing high molecular weight organic compounds, or it may result from operation during periods with insufficient oxygen in the feed. Cautious oxidation is required to clean the catalyst surface without overheating.

Applications. A few of the applications of catalytic combustion are treatment of fumes from fabric finishing operations, wire enameling operations, printing ink drying, deep fat frying, nut roasting, paraffin wax treating, amine effluents, ethylene oxide tail gas from scrubbing operations, phthalic anhydride tail gas, varnish kettle fumes, sewage odors, conversion of carbon monoxide, as well as treating of fumes from petrochemical operations.

While catalytic oxidation is a most important approach to air pollution problems, the elimination of nitrogen oxide fumes requires a catalytic reduction technique using similar catalytic equipment illustrated in Fig. 24. Conversion to N_2 and O_2 can be effected. This technique is applicable when the effluent contains only low levels of oxygen since the presence of oxygen requires additional amounts of reducing gas, generally H_2 or CH_4.

In addition to the cost factor, heat-release problems become excessive with increasing oxygen content, requiring staging of catalyst beds with interstage cooling to avoid damage to the catalyst. The technique of catalytic reduction is widely practiced in nitric acid manufacturing plants on NO_x emissions.

Catalytic oxidation vs. flame incineration Flame incineration provides a technique for achieving a constant and high removal efficiency with sufficient flexibility to handle a variety of process fumes. Flame incineration suffers in a situation

Fig. 24 Honeycomb catalyst support for a unit designed to reduce NO_x pollutants. (*Oxy-Catalyst, Inc.*)

in which intermittent feeding of pollutant occurs, because of slow response of the incinerator to change in temperature and the requirement to achieve a higher temperature level than in the catalytic case. When catalyst poisons are present, it may not be feasible to use the catalytic approach.

In a normal situation, the relative advantage depends largely upon fuel costs. When the fumes are amenable to low-temperature catalytic oxidation, this technique may offer a large advantage over flame incineration.

In control of fumes from a printing press drier, the following operating costs were prescribed for 13,000 scfm units. Catalytic operating cost, including $5,710 for catalyst replacement, was $42,135 per year without heat recovery, and $11,100 with heat-exchange equipment provided. Flame incineration without heat recovery would be $119,960, which heat exchange could reduce to $54,100 per year.

These costs are based on 800°F operation of the catalytic unit and 1200°F for the flame unit. The catalyst unit is assumed to undergo a 50°F temperature rise requirement per year, and after 2 years the catalyst is replaced. In the above situation, the concentration of fumes was 12.5 percent of the lower explosive limit, a condition approximating some gravure operations. Costs were based on $0.10 per therm (100,000 Btu).

Costs. The cost of catalytic oxidation equipment, above 2,000 cfm levels, runs from $2 to $5 per scfm depending upon cfm level. A 5,000 cfm treating unit costs $15,000 to $18,500 excluding blower and heat exchanger. Installed costs are roughly $5 to $10 per scfm. Costs vary depending upon materials of construction. Above 1100°F, stainless steels are required. Above 1300°F refractory construction is used.

Masking and Counteraction of Odors

When specific odors cannot be defined chemically, or their odor thresholds cannot be established, or incineration or absorption is ineffective, two techniques have been used in limited cases with some effectiveness. Odors may be masked, or counteracted. The former technique involves a stronger and more agreeable odor to overpower the disagreeable odor. This may involve high chemical consumption and can create a new odor problem, since even a pleasant odor may be objectionable if very strong.

Counteraction utilizes more viable methods such as the selection of an additive which counteracts or partially neutralizes the offensive odor. Some investigators have found that the introduction of one of a pair of odors into each nostril, at the right concentration, can result in odor neutralization. For example, it is known that skatole is a powerful odor source in fecal matter. It also is an important characteristic in jasmine. By using a jasmine formulation lacking in the indole-skatole notes as a counteractant to sewage odor, it is possible to develop a floral odor lacking the offensiveness of the fecal odor. There is some necessity to maintain the level of counteractant relative to the malodor in a fixed concentration range to effect proper conversion of a malodor into a more agreeable one.

Other antagonistic pairs are:

- Butyric acid–oil of juniper
- Chlorine–vanillin
- Camphor–eau de cologne

One commercial supplier of counterodorants offers five families of chemical counteractants for the following classes of odors:

- Phenolic (also thiophenolic)
- Butyric
- Amine
- Acrylic
- Mercaptan

The odors at cumene-base phenol plants due to cumene, phenol, and acetone are treated with a single counterodorant. Mercaptan and hydrogen sulfide sources at a paper mill are similarly treated with a single odorant.

Searches for new odor solutions based on the art of odor neutralization are continuing.

A variety of techniques have been used for industrial application of masking and counteraction. The odorant mask may be fogged into the base of a stack. Levels from 10 ppm up to 2 percent are common. It is important to achieve proper drop size to avoid excessive chemical consumption and to achieve complete volatilization.

Metering of the liquid and feeding by use of ejectors are techniques which have been used.

Lagoons can be treated by injection of additive in waste streams. In some cases a continuous film on the surface of the water may be formed which the offensive odorous vapors must pass through to escape, thus picking up a small concentration of additive.

The offensive gas stream may be scrubbed with a solution containing a volatile additive for the dual purpose of removing some water soluble component and simultaneously volatilizing some counterodorant.

In some situations it is satisfactory to introduce the odorant into the raw material before processing.

One technique which has been used in sewage treatment plants located in the proximity of residential locations involves the generation of an air curtain around the source. A line of stovepipe 6 to 12 in. in diameter is laid along the periphery of the offending units, as close as possible to the offended receptors. Air containing the volatilized counterodorant is injected into the line. The volatilization of odorant may be controlled by passage of air over a pan of odorant held at a proper temperature. Another useful technique uses a windmill type arrangement whose rotor partially dips in the reservoir of odorant and is rotated under the impulse of flowing air. Through the use of properly spaced holes in the stovepiping a curtain of material some 20 ft high is formed which serves as a source of reodorizing chemicals to neutralize the offensive odors before they can diffuse into the neighborhood.

In a typical installation 400 ft of 6-in. piping was laid around the periphery of a unit. Holes ½ in. in diameter were drilled at 3-ft intervals along the pipe. An air flow of 300 cfm was supplied at a power consumption of ⅓ hp. Typical cost of such piping installed runs to $1.5 to $3.5 per ft. Cost of chemicals was $5 per day (24-hr basis).

Dispersion

Natural processes, in many cases, are quite efficient in removing pollutants from the atmosphere. Thus SO_2 is oxidized to SO_3 and then to sulfates, especially the calcium salt. Hydrogen sulfide is similarly oxidized chiefly to sulfates. Other chemical and biological processes are responsible for the conversion of CO (relatively slow conversion), ammonia, and amines. These natural processes are quite attractive as a solution to our air pollution control problem, provided the concentration of the pollutants at ground level can be maintained at a sufficiently low value. To achieve this effect it is necessary to vent the gas under suitable conditions from a tall chimney or stack.

A chimney has two main parts: a shaft or outer shell, and a liner. The liner is designed to withstand the temperature and chemical constituents of the gas; the shaft is designed to protect the liner from wind loads and weather and, sometimes, to support it. The shaft is usually brick or reinforced concrete, although metal is sometimes used for short stacks. The liner is usually brick, although in very tall chimneys steel is used instead.

Metal liners are subject to corrosion if temperature is not maintained continuously above the dew point, while bricks and mortars are available to resist virtually any kind of chemical attack by condensed vapors. Common mortars are furfuryl alcohol-based, which has a pH resistance range of 4 to 12 and a maximum temperature of 350°F; potassium silicate, which has a pH tolerance from 0 to 5.5 and a limit of 2000°F; and colloidal silica, which has a pH resistance range from 0 to 7 and a temperature limit of 2000°F but costs approximately 120 percent of the cost of potassium silicate. Furfuryl alcohol type is about twice as costly as potassium silicate.

Sectional liners are those supported by the shaft on corbel projections from the shaft ID. Clearance between the liner OD and shaft ID is left empty or filled with insulating material. These are generally not as resistant to chemical attack as free-standing liners built to support their own weight, with open annular space between liner and shaft which is open at the top and usually has air inlets at the bottom. A chimney design with free-standing liner is shown in Fig. 25.

An important and complex portion of chimney design is related to defining the necessary conditions to restrict maximum ground level concentrations below those specified by the control authorities.

Effect of atmospheric conditions To a large extent the temperature profile of the atmosphere is responsible for the behavior of plumes from stacks. It has been found that the rate of change of temperature for rising dry air is 5.5°F (1,000 ft). This is termed the adiabatic lapse rate. If a steeper gradient exists, a gaseous element will continue to rise from a stack, thus offering the opportunity to achieve considerable dilution in the atmosphere.

Various temperature profiles are encountered in practice. In some cases the temperature of the atmosphere increases with altitude, resulting in a condition termed an inversion. Such a condition leads to stratification and reduced mixing. In particular, inversion conditions make it difficult to disperse pollutants.

Effect of topography Buildings and hills can act to reduce both wind velocity and level of turbulence and consequently exert a detrimental effect in reducing ground level concentrations.

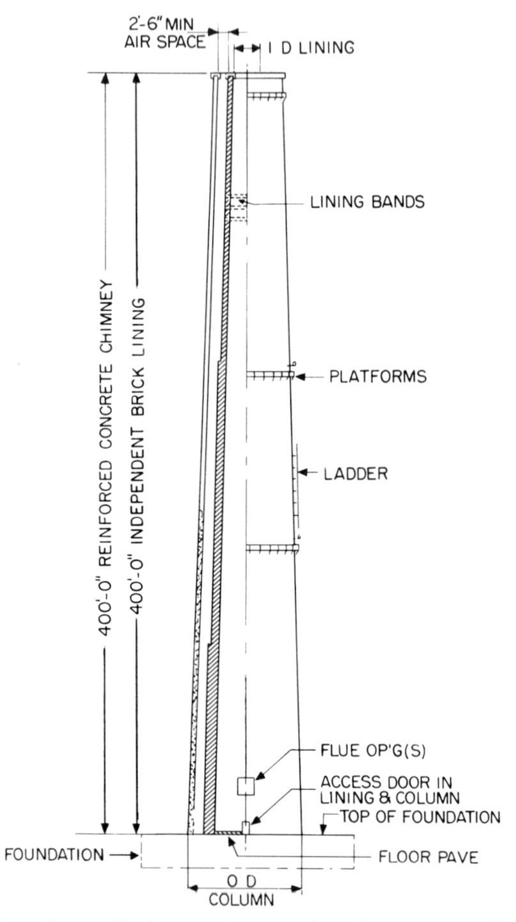

Fig. 25 Typical modern tall chimney has reinforced concrete column and—except for very tall chimneys—a free-standing brick liner. (*Custodis Construction Company.*)

Prediction of the effect of topography on dispersion is best approached from an experimental viewpoint.

Regulations are written in a variety of ways. Obviously if a code specifies a maximum emission level per pound of product, etc., dispersion is not a satisfactory approach.

In New Jersey, the code for the discharge of sulfur-containing gases in the atmosphere specifies adjusted stack height required as a function of emission levels. For 5 to 8,000 lb/hr SO_2, the stack heights are 20 to 800 ft. The adjusted stack height takes into account stack exit velocity and stack gas temperature.

Calculation of ground level concentration Obviously the maximum ground level concentration of a pollutant is an important quantity that should be predicted for a given stack. Equations have been developed, and empirical modifications have been applied to these basic approaches.

$$H_X = Z_s + (Z_v)_X + (Z_d)_X \tag{1}$$

where H_X = effective stack height, ft
X = distance in feet measured downwind, horizontally from base of stack
Z_s = actual height, ft
Z_v = velocity rise, ft
Z_d = density rise, ft

The maximum velocity rise $(Z_v)_{\max}$ is given by

$$(Z_v)_{\max} = \frac{4.77}{1 + 0.43(V_W/V_s)} \frac{(Q_1V_s)^{1/2}}{V_W} \tag{2}$$

where V_W = wind velocity, fps
V_s = gas exit velocity, fps
Q_1 = volumetric emission rate, ft³/sec

The actual velocity rise at distance X from the base of the stack is given by

$$(Z_v)_X = (Z_v)_{\max}\left[1 - 0.8\frac{(Z_v)_{\max}}{X}\right] \tag{3}$$

The density rise (Z_d) is given by

$$(Z_d)_X = \frac{6.37gQ_1(T_g - T_{ga})z}{V_W{}^3/T_{ga}} \tag{4}$$

where g = acceleration due to gravity
T_g = absolute gas temperature
T_{ga} = absolute gas temperature at which gas density would equal atmospheric air density

Values of z are related to horizontal distance (X) by the factor x from Bosanquet's experimental data. Factor x is found by

$$x = \frac{V_WX}{3.57(Q_1V_s)^{1/2}} \tag{5}$$

The ground level concentration at a horizontal distance (X feet) downwind from the stack and at a distance (Y feet) measured perpendicular to the wind direction is found by

$$(C_o)_{X,Y} = \frac{M}{2\pi pq60V_WX^2}e^{-\left(\frac{H_X}{P_X} + \frac{Y^2}{2q^2X^2}\right)} \tag{6}$$

where C_o = concentration at ground level at given point
M = volumetric rate of pollutant discharge, ft³/min
p, q = Bosanquet's vertical and horizontal diffusion coefficients

---

Fig. 26 Formulas for calculating concentration at any point at ground level of pollutants discharged from a tall chimney.

In essence as a gas leaves the chimney it possesses a given velocity, density, and temperature. The velocity and density contribute to an effective stack height. The gas begins to mix with the atmosphere and is conveyed in the downwind direction. Mixing occurs by a diffusion mechanism.

A typical analysis for ground level concentration approaches the problem in two stages. Initially it is useful to establish an effective stack height for the pollutant which takes into account stack height, a velocity rise, and a density rise.

The approach of Brink and Crocker as used in a computer program for stack design is outlined in Fig. 26. Many refinements are possible. Lapse rates other than adiabatic can be incorporated into the program.

There are rules of thumb which are useful for orientation purposes and also can be used for planning small stacks where extensive computational studies are not warranted. These rules follow:

1. Stacks should be at least 2½ times as high as the surrounding buildings or countryside so that the surroundings do not introduce significant turbulence.
2. Gas ejection velocities from the stack should be greater than 60 fps so that the stack gases will escape the turbulent wake of the stack. Velocities in excess of 90 fps may erode brick liners.
3. Gases from stacks with diameters less than 5 ft and heights less than 200 ft will hit the ground so soon, part of the time, that ground concentrations may be excessive.
4. The maximum ground concentration of stack gases subjected to atmospheric diffusion usually occurs at about 5 to 10 stack heights from the stack.
5. When stack gases are subjected to atmospheric diffusion (and building turbulence is not a factor), ground level concentrations of the order of 0.001 to 1 percent of the stack concentration are possible for a properly designed stack.
6. Ground concentrations can be reduced by the use of higher stacks. The ground concentration varies inversely as the square of the effective stack height.

Chimney costs vary widely with geographical location, gas, conditions, allowable construction time, and height (cost per foot of height increases with increasing height by a factor of 1.25 to 1.5 per unit increase). A cost approximation for chimneys with free-standing liners of 8-ft ID, 200 to 400 ft in height, would range from $500 to $1,000 per ft of height.

Determination of Requirements

From the foregoing rudimentary treatment of the operating principles of the generic types of control equipment, it is clear that there are certain basic requirements for data that the customers must supply for a rational approach to the design of the most effective control system. As an absolute minimum, for particulate collectors the following must be defined:

The volumetric flow rate of gases
The gas temperature, pressure, and humidity
The particulate form (wet, dry) and concentration
The particulate size distribution and specific gravity
The particulate bulk electrical resistivity
The required removal efficiency
Process constraints with regard to use of water

From this rather minimum quantity of basic data, it is possible to perform an initial screening on the type of system required. For example, if we have an efficiency requirement in the 99 percent range and the particulate size distribution indicates all material is below 10 microns, it is known the application cannot utilize a cyclone. Similarly, if the gas temperature is 650°F, we know we cannot use a bag filter without precooling the gases. If water vapor plumes cannot be tolerated we cannot use a wet scrubber on hot gases unless we take further steps to reduce the humidity or increase the temperature before discharge of gases to the stack.

It cannot be overemphasized that definition of these basic data requirements is absolutely mandatory. If such information is not immediately available, then it is well worth the minimal time and expense required to get it—for it will be time and money saved many times over in the final solution. There are adequate engineering services—from both consulting organizations and/or equipment suppliers—available to assist in obtaining these basic data.

The preliminary screening will define those alternative solutions which are feasible in the light of today's technology. The selection of the optimum will require a detailed analysis of many additional technical and economic factors, with the depth and complexity of the analysis being primarily related to the magnitude of the problem. Since the optimum solution is almost always the reliable achievement of a desired set of performance objectives at the lowest possible cost to own and operate, such detailed analysis will be dominated by economic factors such as:

1. Capital investment
   a. Base collector
   b. Auxiliaries such as fans, pumps, controls, flues, waste disposal, and water treatment
2. Operating costs
   a. Water
   b. Electric power
   c. Maintenance—material, labor
   d. Chemical costs
3. Cost of capital
   a. Interest
   b. Taxes
   c. Amortization

Again, competent engineering services are available from many quarters to assist in definitions of the problem and data inputs required for selection of the most effective and economical system.

Water Pollution Control Programs and Systems

EDGAR G. PAULSON

Technical Consultant, Environmental Engineering, Calgon Corporation, Pittsburgh, Pennsylvania

Management must continuously be concerned with the quality of the water being discharged from its plants. This concern should be expressed in the form of interest and stimulation of planned programs to control or resolve each individual situation. This can be accomplished only by having available facts on which true decisions can be made. Too often, people are concerned with the possibility or assumed reality that they face a major capital expenditure to resolve their waste water problem, and thus they hesitate to initiate the first stage of the planning process required to truly define their needs.

Phases involved in a full pollution control program are:

1. Definition of the problem and development of an action plan
2. Detailed engineering
3. Construction and startup

An outline of the steps involved in each of these phases is shown in Table 1. The first phase, where a problem is defined and a course of action developed, represents only a small percentage of the potential investment that may be required. It is highly recommended that this program be initiated early so that time is available to gain maximum benefit from the study phase.

The scope of the fact finding study, or "water audit," is partially dependent upon what it is desired to accomplish. For example, in some cases, the objective

TABLE 1 Water Pollution Control Program

I. Survey or feasibility study
  A. Fact finding
    1. Develop a plant water balance for average and peak operating conditions.
    2. Inventory all industrial processes using water.
    3. Determine characteristics of the receiving waterway both upstream and downstream from plant's discharge.
    4. Determine chemical characteristics of waste streams.
    5. Study all operations using water and producing wastes.
    6. Determine local requirements with respect to pollution.
  B. Analyze data to determine
    1. Sources of offending contaminants.
    2. Feasibility of segregating contaminated wastes requiring treatment from dilute wastes which would be acceptable without treatment.
    3. Availability of "natural" dilution waters, that is, waters employed for useful purposes but not contaminated.
    4. Quality of effluent required for compliance with stream standards.
    5. Whether treatment is necessary.
  C. Exploit in-plant and/or process changes to minimize the problems by
    1. Reducing wastes or waste volume at sources.
    2. Exploring the possibilities for reuse of process materials without treatment.
    3. Investigating recovery of valuable process materials.
    4. Reexamining the degrees of treatment required to meet state standards.
    5. Reevaluating to decide whether treatment is necessary.
  D. Detailed report on the engineering survey
    1. Recommend a preliminary course of action.
    2. Advise management whether a waste treatment plant is necessary.
    3. Describe the general type of plant required.
    4. Provide preliminary estimate of construction cost.
    5. Prepare preliminary estimate of operating costs.

II. Detailed engineering
  A. Process design and evaluation
    1. Assign liaison and engineering personnel as required.
    2. Evaluate bench scale or pilot plant data.
    3. Translate the total evaluated data into process flow diagrams and functional specifications for the treatment plant.
    4. Prepare plot plan showing layout on plant site.
    5. Assemble an engineering report for client review and approval.
    6. Obtain preliminary approval of regulatory agency.
  B. Definitive engineering
    1. Prepare detailed engineering flow diagrams which form the basis of final plant design.
    2. Obtain approval by client of overall plant design.
    3. Complete the definitive design.
    4. Obtain final approval and permit of regulatory agency.

III. Construction and startup
  A. Procurement and scheduling
    1. Prepare complete equipment specifications, bills of material, and preliminary timetable.
    2. Prepare item delivery and installation schedule.
    3. Use critical path scheduling when warranted.
    4. Coordinate and inspect all phases of the work performed by fabricators.
  B. Facilities erection and testing
    1. Plan, supervise, and coordinate erection of the complete waste water treatment plant.
    2. Conduct unit tests, after assembly, to assure proper functioning of all related facilities.
    3. Inspect, adjust, and calibrate instruments and controls to conform to high accuracy standards with engineers performing the work.
  C. Operator training
    1. Prepare detailed operating manuals for all unit operations in the plant.
    2. Assemble vendors' manuals for use by plant personnel in maintaining, repairing, and replacing mechanical, instrument, and electrical equipment parts.
    3. Assist with training of operating crews while construction work is in the final stages.
  D. Startup of treatment facilities
    1. Initiate a control testing program.
      a. Operational
      b. Quality of effluent
    2. Initiate an efficiency testing program.
    3. Establish conditions for operations.
    4. Initiate a development program.
    5. Establish record-keeping procedures.
  E. Supervise operation

may be only to characterize the effluent from a plant to determine its effect on the receiving stream. This would involve sampling and gaging of the plant effluent, and stream sampling upstream and downstream from the plant. These data are necessary to determine whether a problem exists, and the approximate order of magnitude of the problem. Such a program is inadequate to design a treatment plant properly, except where relatively simple processes involving only a limited number of water using operations are employed. A thorough, plant-wide water audit is essential to develop an effective program for plants utilizing more complex processes.

A comprehensive fact finding program is initiated with the collection of certain key documents. These are:

A *plot plan* of the plant property showing the location of property lines, buildings, departments, etc.

A *sewer layout* that shows the location of the existing sewer system or collection system.

A *process flow sheet* showing the manufacturing operations with particular emphasis on the processes where water is being used.

A *location map* showing the geographical relationship and location of the plant and the receiving stream as well as information on nearby water users or other waste water contributors.

The next step involves the completion of a water inventory, which includes a tabulation of all water using operations, together with other pertinent data such as

1. Identification of water using operations
2. Location of operation
3. Quantity of water used
4. Quality of water used or required
5. Quality of water discharged
6. Location of water discharge point

ELEMENTS OF A POLLUTION CONTROL PROGRAM

Flow Measurement

An integral part of this program is the need to obtain flow information on various in-plant streams as well as the plant outfalls. Waste water flows are measured for the following reasons:

1. To determine the quantity of water being discharged, as well as variations in the flow rate.
2. To determine the pounds of constituents being discharged, based on the analytical data and the determined flow rate.
3. To evaluate segregation possibilities.
4. To determine the effect of the waste water discharge on the receiving stream, if applicable.

Rates of flow can be approximated by the following methods:

Water meters on influent lines. Water consumption in the plant should be determined during a waste water survey to check on waste water flow measurements and to compute a water balance for the plant. Meters also can be installed at particular water using operations to obtain flow data.

Container and Stopwatch. The time required to obtain a given volume of water in a container is measured. Volume can be determined either by weight added or by a calibrated collection container. The weight of water added is divided by 8.3 lb/gal to determine gallons collected. The flow in gallons per minute is then determined by the formula

$$\frac{\text{Gal in container} \times 60}{\text{Time, sec, to fill container}} = \text{gpm}$$

If the container fills in less than 10 sec, the accuracy of this method is questionable.

Time to fill a tank, pit, or sump when its volume is known. The time needed to fill the tank is determined, and the flow is calculated using this formula:

$$\frac{\text{Tank volume, cu ft} \times 7.5}{\text{Time, min, to fill tank}} = \text{gpm}$$

(1 cu ft = 7.5 gal.)

Pump Capacity and Running Time. If the waste water is pumped, the flow can be calculated using the capacity of the pump and the amount of time it is in operation during the sampling period. It is desirable to calibrate the pump, if possible, rather than just depend on curves provided by the manufacturer.

Depth and Velocity of Flow in Partially Filled Sewers. A given length of sewer between two points is measured, and the velocity of flow between the points is obtained. The velocity can be determined using cork or wood floats, dyes, or

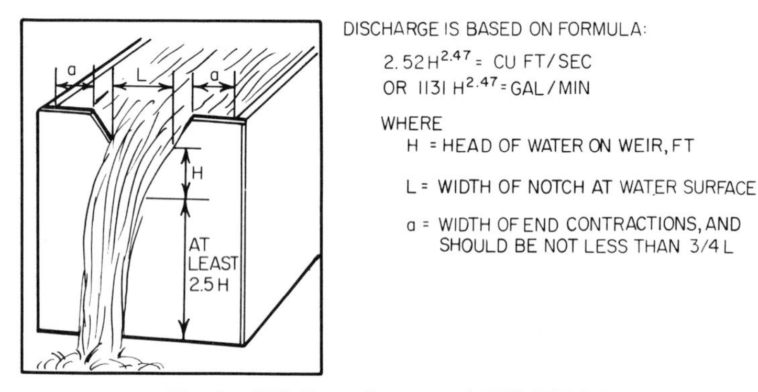

Fig. 1 90° V-notch weir. (*ORSANCO.*)

current meters. Depth of flow is converted to cross-sectional area of liquid. Flow is calculated by

$$\frac{\text{Distance, ft,} \times \text{area, sq ft,} \times 7.5}{\text{Time, min}} = \text{gpm}$$

This method is not extremely accurate and should be used only for estimating purposes or when other methods are not practical.

Salt Concentration. A known strength sodium chloride solution (or other applicable salt) is added at a constant measured rate to the flow in the sewer. The chloride concentration at a downstream point in the sewer is then determined. The chloride concentration of the water blank is determined first. Flow is calculated as follows:

$$\frac{\text{lb/hr salt added} \times 2{,}000}{\text{ppm measured} - \text{ppm in blank}} = \text{gpm}$$

Weirs. A weir acts like a dam or obstruction, with the water flowing through the notch, which is usually rectangular or V shaped.

To ensure accurate weir measurements:

1. The weir crest must be sharp or at least square-edged. Steel is the best material of construction, but tempered wood is also used.
2. The weir must be ventilated. There must be air on the underside of the falling water.
3. Leaks around the weir plate must be sealed.
4. The weir must be exactly level.
5. Weirs should be kept clean.

6. The head on the weir should be measured at a distance of 2.5 times the head upstream from the weir.

7. The channel upstream from the weir should be straight, level, and free from disturbing influences. A stilling box may be used to quiet the water flow.

8. The weir should be sized after the flow is estimated using other methods. The head on any weir should probably be greater than 3 in. but not more than 2 ft.

90° V-notch Weir. This type of weir is limited to measuring low flows because of its geometry. It is especially valuable when there is wide variation in flow. An example of this type of weir is shown in Fig. 1.

Rectangular Weir. With this type of weir, the maximum head possible is desired, but the sides of the weir must not overflow. If the flow over the weir is too low, the reading will be inaccurate.

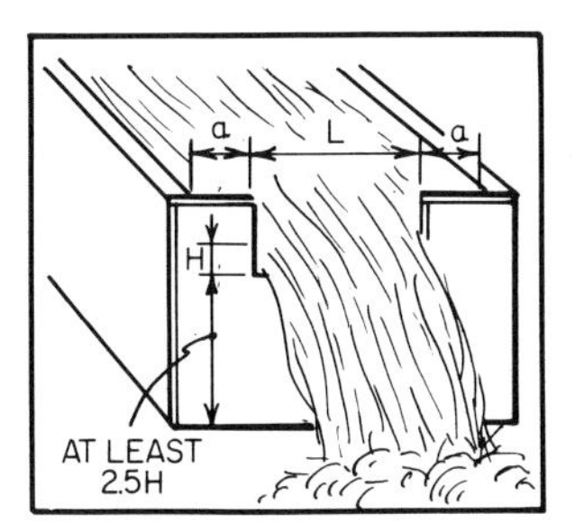

DISCHARGE IS BASED ON FRANCIS FORMULA:

$3.33\,LH^{3/2}$ = CU FT/SEC

OR $1495\,LH^{3/2}$ = GAL/MIN

WHERE

L = LENGTH OF WEIR CREST, FT

H = HEAD OF WATER ON WEIR, FT

a = WIDTH OF END CONTRACTIONS AND SHOULD NOT BE LESS THAN 2.5 H

(a)

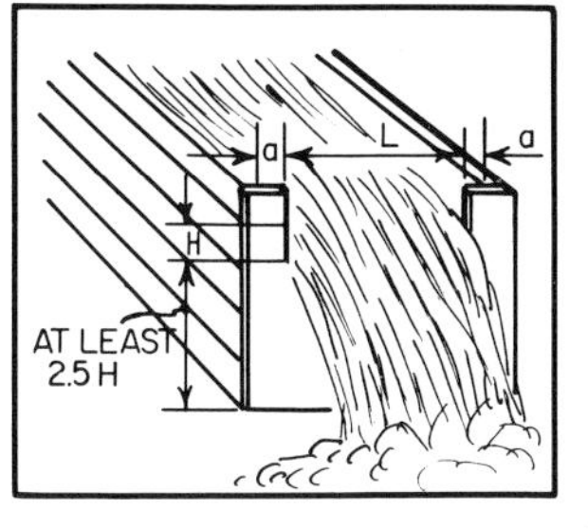

DISCHARGE IS BASED ON FORMULA BY SCHODER:

$3.00\,LH^{3/2}$ = CU FT/SEC

OR $1346\,LH^{3/2}$ = GAL/MIN

WHERE

L = LENGTH OF WEIR CREST, FT

H = HEAD OF WATER ON WEIR, FT

a = WIDTH OF END CONTRACTIONS LARGE TO PERMIT FREE PASSAGE OF AIR DOWN SIDE OF WEIR PLATE

(b)

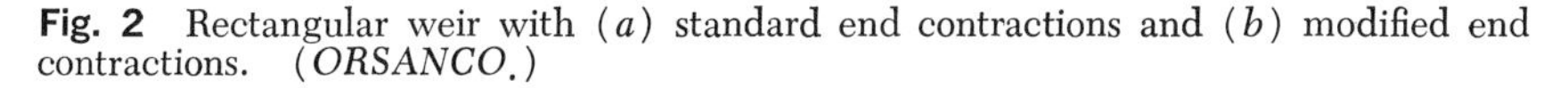

Fig. 2 Rectangular weir with (*a*) standard end contractions and (*b*) modified end contractions. (*ORSANCO.*)

The width of the available channel and the required weir opening size dictate whether a weir with standard end contractions or modified end contractions must be used. The relationship of head on the weir vs. flow is somewhat different in the two cases. Examples of rectangular weirs are shown in Fig. 2*a* and *b*.

Tables are available which provide flow data in gpm at given heads over the weir for given lengths of weirs.

Parshall flume A Parshall flume can be used to measure flows in open channels at or near ground surface. This device is valuable when it is not possible to dam the water. It is also advisable for a permanent installation because it is self-cleaning. A typical installation is shown in Fig. 3.

Flow under submerged conditions can be calculated from readings taken at gages. If the water surface downstream from the flume is high enough to retard the rate of discharge, submerged flow exists.

When there is no backwater effect, water passing through the throat and diverging section assumes a level which corresponds to the floor of the channel. This pattern demonstrates free flow.

The flow of a free discharge from a Parshall flume is calculated by

$$Q = 4WH^n$$

where $n = 1.522\ W^{0.026}$
Q = flow, cfs
W = width of throat, ft
H = head of water above level floor, ft

Flow under submerged conditions can be calculated from readings taken at gages, one located at a point two-thirds the length of the converging section measured

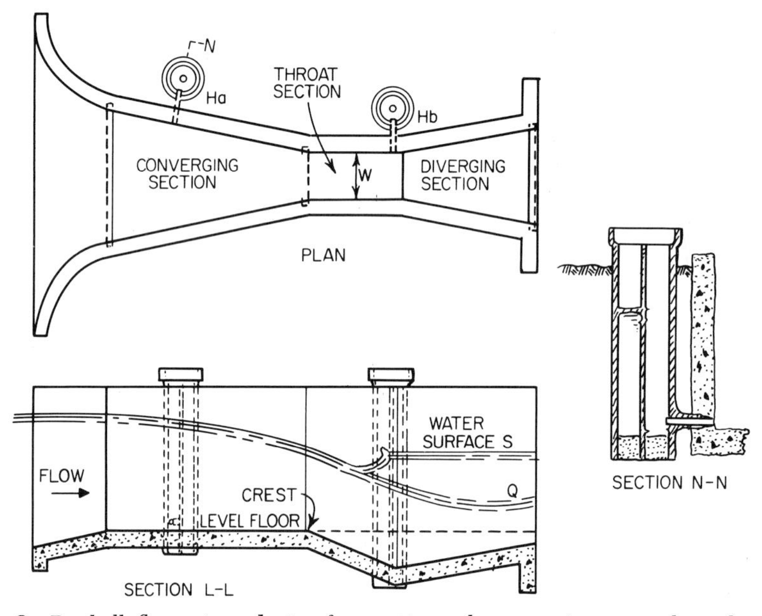

Fig. 3 Parshall flume is a device for continuously measuring open-channel waste water flow at or near ground surface. Parshall flume is a better measuring device where solids are present in waste water because of its self-cleaning characteristics. It is generally used in permanent installations. (*ORSANCO.*)

back from the crest of the flume H_a and one located near the downstream end of the throat section H_b. Degree of submergence is given by the ratio H_b/H_a. Tables are available for obtaining the flow under these conditions.

Velocity meter This type of device is usually used to measure stream flows. It consists of a rotating vane and cup arrangement with a sounding component by which velocity is measured according to the revolutions of the submerged cups. With this method the stream is broken into imaginary sections. The depth of the water and its velocity are measured at depths proportional to the total water depth, and the average velocity is obtained for each section. Velocities are averaged to determine the amount of water in the stream and its rate of flow in cubic feet per second.

Head measurement The head produced on a weir or a Parshall flume can be measured with a ruler. For inconvenient locations or to record the head, the following methods may be used.

Manometer or Recorder. A method for measuring head is by passing air through a valve and bubbler and then a Y with one branch running to the weir and the other to a manometer or other pressure recording device. A typical drawing showing this system is Fig. 4.

Water Level Recorders. These instruments provide convenient records of flow data by measuring the head of water flowing over the lip of a weir by means of a float. The rise and fall of the float with changing water levels is recorded against time.

Waste Water and Stream Sampling

Waste waters are sampled and analyzed to determine whether the water is contaminated, to identify those contaminants that require treatment, and to aid in selecting the proper treatment process.

Streams are sampled and analyzed to determine current condition and the condition expected after installation of a treatment process. (In certain watersheds, stream sampling has been made unnecessary by regulatory agencies which have

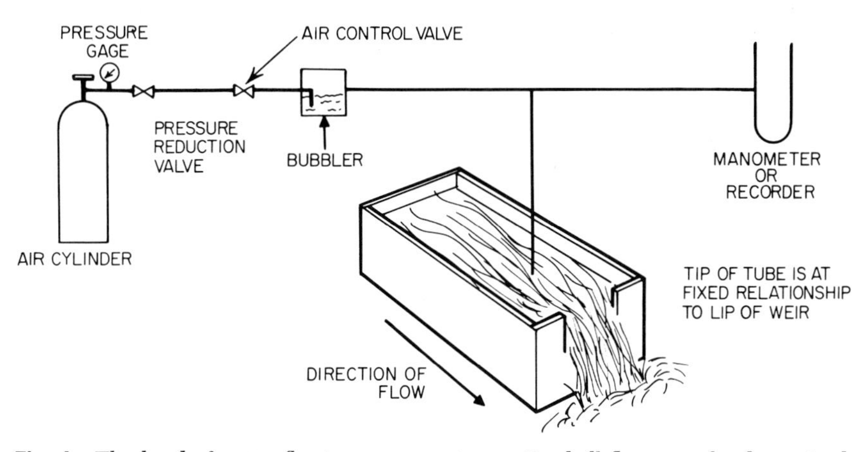

Fig. 4 The head of water flowing over a weir or a Parshall flume can be determined and recorded by the use of an air bubbler system. For a permanent installation, the recorder can be set to read directly in gallons per minute and can also totalize flow.

established allowable effluent concentrations and in other situations have allocated contaminant loadings for individual plants on a pounds per day basis.)

Composite samples Since a waste water's characteristics can vary considerably, composite samples are collected to obtain a truer representation of the waste.

Small samples are collected at frequent intervals during the sampling period. They are mixed together to form the composite sample.

Depending on plant operation, 8-, 16-, or 24-hr composites can be collected. Daily sampling for 3 days generally constitutes a sampling program. Samples can be composited on the basis of:

Flow. The amount of sample collected at any time during the sampling period is proportional to the flow of waste water at that time.

Time. The same amount of sample is collected at every interval during the sampling period regardless of variations in waste water flows.

The sample size collected at any one time should be at least 200 ml. Composite samples can be collected either *manually* or with *automatic samplers.* Automatic, battery-operated samplers are available for collecting composite samples on the basis of flow or time. One such sampler is capable of collecting approximately 2 gal per 24 hr (minimum). (However, if less than 24-hr composite samples are collected, sufficient samples may not be available to satisfy laboratory requirements.)

Other automatic samplers are available which collect one large composite sample or small samples in individual bottles.

One disadvantage of composite sampling is that indications of extreme conditions are lost. For example, if the pH of waste varies from 3 to 11 during the sampling period, the pH of the composite sample could be 7.0.

Grab samples are collected when:

1. Water to be sampled does not flow on a continuous basis, such as an intermittently dry discharge outlet or a contaminated process tank dumped periodically.
2. Appearance of a discharge changes rapidly. Grab samples are collected to determine the cause of the change.
3. It is desired to determine if a composite sample is camouflaging extreme conditions of the waste.

Amount of sample to be collected depends upon the laboratory tests to be run. The amount of sample required for each test to be performed should be determined before the sampling program is begun to ensure that sufficient sample is collected.

Determinations that may be conducted on a waste water sample are:

pH
Alkalinity or acidity
Total hardness
Chloride
Sulfate
Suspended solids
Volatile suspended solids
Total solids
Volatile total solids
Settleable solids
Total nitrogen
Ammonia nitrogen
Total phosphate
Copper
Nickel
Zinc
Chromium, hexavalent
Chromium, total
Iron
Manganese
Solvent soluble (oil)
Phenol
Biochemical oxygen demand (BOD)
Chemical oxygen demand (COD)
Total organic carbon (TOC)
Cyanide

Naturally, the analysis run for any specific sample will depend on the processing operation involved and the type of material added to the water, as shown in Table 2. Standard methods are available for conducting these determinations.

Handling of Samples

Some analytical tests require that the sample be specially fixed immediately after it is collected. This is done so the amount of constituent present in the sample will not change during shipment. Common analytical determinations that require specially fixed samples are:

Cyanide A separate, 1-qt sample is needed for this test. The pH of the sample is raised to 11.0 or above with sodium hydroxide. The sample should be kept cool and shipped to the laboratory by the fastest means.

Phenol A separate, 1-qt sample is needed. The pH of the sample is lowered

TABLE 2 Waste Water Analysis

	Analysis for inorganic waste water	Partial analysis for inorganic waste water	Analysis for organic waste water	Partial analysis for organic waste water
pH	X	X	X	X
Alkalinity or acidity	X	X	X	X
Total hardness	X		X	
Chloride	X			
Sulfate	X			
Suspended solids	X	X	X	
Volatile suspended solids			X	
Total solids	X		X	
Volatile total solids			X	
Settleable solids	X		X	
Total nitrogen			X	
Ammonia nitrogen			X	
Total phosphate	X			
Copper	X	X	X	
Nickel	X			
Zinc	X		X	
Chromium, hexavalent	X			
Chromium, total	X			
Iron	X		X	
Manganese	X			
Solvent soluble organic	X			
Phenol			X	
Biochemical oxygen demand			X	X
Chemical oxygen demand			X	X
Total organic carbon			X	X
Cyanide	X	X		

Wastes are classified as organic or inorganic. This table includes various analytical determinations usually performed to characterize a waste water sample properly.

to 4.0, and 1 g of copper sulfate should be added. The sample should be kept cool and shipped by the fastest means.

Nitrogen A separate, 1-qt sample is needed; 0.8 ml of concentrated sulfuric acid should be added to the sample. The sample should be kept cool and shipped to the laboratory by the fastest means.

While the following tests do not require that the sample be specially fixed in the field, they do require a significant amount of sample or special handling.

Heavy metals A separate container is usually required for determination of heavy metals. While no special fixing requirements in the field are necessary, the contents of this container are acidified in the laboratory before the specific tests are run. This is to ensure that the entire concentrations of metals are in solution and that no metals are clinging to the walls of the container.

Solvent soluble (oil) It is prudent to collect a separate sample for this determination since a large amount of sample is usually required.

Biochemical oxygen demand A relatively large sample is required for this test. In addition, it is necessary that any samples for BOD analysis be kept cool and shipped to the laboratory by the fastest means to prevent biological activity which will change the characteristics of the sample. Shipment by air express has been found to be satisfactory where cold samples can be left unrefrigerated for approximately 8 to 16 hr without significant changes taking place. Samples can be shipped in dry ice. Lowering the pH of samples has also been used to slow down biological activity when long shipment times are required.

Reporting of Analytical Results

When a water sample is analyzed, the analyst is asked to determine the amount of various contaminants present. Since the amounts determined are a few milligrams per liter and sometimes even fractions of a milligram, it would be impractical to report results by percentage, which is used in many other types of analysis. Ordinarily, samples are measured by volume; therefore, it is current practice to report results in terms of milligrams per liter (mg/l).

The term parts per million, which has been used historically in this field, is a measure of proportion by weight. In terms of percentage, 1 part per million is equal to 1 ten-thousandth of 1 percent. Frequently, a sample is analyzed and reported in ppm when in essence, because of the procedure used in its analysis, it should have been reported in mg/l. Since a liter of water weighs approximately 1,000 g or 1,000,000 mg, 1 mg/l was thought of as being equal to 1 ppm. This relation is correct for the analysis of most waste waters where the specific gravity is essentially the same as that of tap water. However, for some industrial wastes the close corollary between ppm and mg/l does not hold since the specific gravities vary from that of water itself.

Although most water analyses are reported in mg/l, they are also expressed as equivalents per million, or epm. An equivalent per million is a unit chemical equivalent per million unit weights of solution. Concentration in this unit can be obtained by dividing concentration in ppm by the equivalent weight of the ion or substance. Results may also be reported as milligrams/liter or milligram equivalents/kilogram, which means epm.

Waste Load Calculations

At this point, the waste water engineer has the data needed to determine contaminant loading. The following formulas are most frequently used:

$$\text{Waste load, lb/day} = \frac{(\text{mg/l}) \times \text{flow, total gal/day}}{120{,}000}$$

Frequently, the survey period does not extend over 24 hr. When this is so, the following formula can be used:

$$\frac{\text{Load, lb}}{\text{Time, hr}} = \frac{(\text{mg/l})(\text{gpm})(\text{time, hr})}{2{,}000}$$

Batch system Contaminant loading is determined thusly:

$$\text{Load, lb/batch} = \frac{(\text{mg/l})(\text{gal/batch})}{120{,}000}$$

Analysis of Data

The major consideration is a comparison of the plant effluent characteristics with established regulatory agency guidelines to confirm if a waste water problem exists and to determine the severity of the problem.

Then the effluent data and in-plant data must be organized into a usable form.

It is usual to use a material balance. This may be a line diagram showing sources, quantity and quality of water, poundage of various contaminants, etc. These data can be organized by tabulating the source of waste, the volume of water involved, and the pounds of contaminant contributed. It is useful to tabulate data on the basis of contaminant added by each process.

In addition, it is necessary to know the locations of various waste producing operations and how the wastes are being collected and discharged from the plant. This involves combining the plot plan, the sewer layout plan, and the process location plan. This becomes particularly important when segregated collection systems must be considered.

In completing the material balance on any given outfall, the route from an in-plant source of waste water may become questionable. Tracer techniques must then be employed. One technique is to follow a known constituent used in a selected process. This system can be used only if there are a limited number of sources for this particular constituent, and their presence will not complicate interpretation of the developed data.

Another technique is the use of tracer dyes. A water-soluble dye is injected into the water leaving the process, and the route is followed by stationing personnel at various manholes and outfalls to watch for the colored water. It is desirable to have available a field communication system, such as walkie-talkies, to follow the route of the dye, to speed up communication, and to coordinate the program more completely.

Reducing Waste Load by In-plant Changes

If after data analysis it is determined that certain effluent streams are not acceptable for discharge, the next logical step is to review the processes contributing the particular contaminant to determine whether the amount of contaminant can be reduced or recovered for reuse. Process water use practices also should be reviewed to determine if water use can be reduced.

Here is an area where original thinking is mandatory. Some illustrative examples of approaches that have proved successful are:

Still rinses This involves using a rinse water tank which does not have any flowing water passing through it. This can serve as a preliminary concentrator. The concentration of the tanks may be set so that treatment of downstream flowing rinses is not necessary. Another approach is to set the concentration of the tank so that this water can be returned to the process tank for reuse. It can be brought up to solution strength either by the addition of fresh process chemicals or by evaporation.

Fog rinses These are very fine spray rinses. They result in a higher concentration of process material in the resultant rinse water solution. This solution may be of such strength that it can be returned to the process tank for reuse with or without concentration.

Countercurrent rinses This involves reusing the rinse water from a downstream tank in another or several tanks. The reduction in water usage can be calculated by the following formula:

$$F = \frac{S}{T}\left(\frac{C_0}{C+}\right)^{1/n}$$

where $C+$ = final concentration of contaminant in rinse, expressed as oz/gal, mg/l, etc.

C_0 = concentration of contaminant in process solution, expressed as oz/gal, mg/l, etc. ($C+$ and C_0 must be expressed in consistent units)

F = flow of fresh water, gpm

n = number of rinse tanks

S = dragout, gal

T = time interval between rinse operations, min

Ion exchange recovery This is used to concentrate contaminants on ion exchange resins, from which they are reclaimed for reuse in the process.

Reduction in dragout This reduces the amount of contaminant entering the rinse water by use of squeezer rolls, air blowing, drip stations, etc. This undiluted solution is returned to the process for reuse.

After completion of in-plant changes, the effluent quality should be reassessed to determine if treatment is still necessary. In some cases, this may require a sampling and gaging program of the plant effluent.

Segregation of Waste Streams

As soon as terminal treatment is deemed necessary, an area that must be considered immediately and reconsidered as treatment plant plans are being made is the desirability or necessity of segregating various waste water streams so the collection system will be developed concurrently with the treatment plans.

As a general rule, all plants should be provided with three basic systems, namely:

1. Storm water
2. Sanitary waste
3. Industrial waste

The *storm water system* should receive all surface and storm runoff. This system also can be used for discharging uncontaminated waters, such as cooling waters, that require no treatment prior to discharge. For the system to be effective, it is necessary to monitor the quality of these discharges periodically to make sure that no contaminated waters are inadvertently tied into this system. While it is desirable to keep uncontaminated waste water out of the treatment plant, cost of installing separate collection systems for small, isolated streams may be so high that bypassing the treatment plant is uneconomic.

The *sanitary system* should collect the waste waters from all washrooms, shower rooms, etc. For most industrial plants, it is usually desirable to send these wastes to a municipal plant for treatment, rather than treat them individually.

The *industrial system,* then, should receive all other waste streams. While this is listed as a single system, a multiplicity of collection systems may be necessary. For example, where continuous treatment is to be used, it may be necessary to collect concentrated batch dumps of spent process solutions separately so they can be bled into the continuous treatment system at a controlled rate. The classic example of separate collection systems is a large metals finishing operation where separate collection systems are required for (1) cyanide wastes, (2) hexavalent chromium, (3) oil, and (4) acid-alkali wastes. In addition, there is also the possibility that separate collection systems for concentrated spent process solutions may be required for one or more of these basic contaminants.

Since chlorination is not mandatory for industrial organic wastes (they are generally free of pathogenic or coliform bacteria), this poses another economic question: "Is it less costly to collect and treat sanitary and process organic wastes separately, to combine them for treatment, or to separate sanitary wastes for discharge to a municipal treatment plant?" Costs of chlorination and operational costs of separate combined facilities are key considerations for the individual plant.

A basis of design should be developed for each section of the treatment plant. This should cover present plant conditions as well as expected future conditions. This should include a tabulation of the hydraulic load and the contaminant load, with expected variations. Particular attention should be given to the hydraulic loading on downstream portions of a treatment plant after any batch treatment operations.

Equalization is the single most important feature to be included in any planned industrial waste water treatment facility. Equalization involves smoothing out surges in hydraulic and/or concentration load. Treatment facilities generally are designed on the basis of two separate but interrelated parameters, namely, the amount of water to be treated in any given unit of time (usually gpm) and the amount of contaminant contained in this quantity of water. Some portions of a plant, such as clarifiers, are designed basically on the hydraulic load, while others, such as chemical feeders, are designed primarily on the basis of the contaminant load. This type of equipment has minimum as well as maximum effective operating rates.

Sizing of equalization facilities depends on the rate and the magnitude of the variation in influent characteristics. This basic information must be determined as part of the aforementioned fact finding survey. In the discussion of segregated collection facilities, mention was made of the possibility of needing separate collection facilities for concentrated solutions. In some cases, it has been found to be more economical, and just as effective, to size equalization capacity for the plant so that the concentrated solutions can be dumped into the same collection system as the usually occurring dilute solutions. As a general rule of thumb, equalization retention time should be at least 2 hr for a multiprocess manufacturing plant. Positive agitation should be incorporated into the equalization tank.

Hydraulic surges can result from a variety of causes. One of the most frequently encountered is periodic cleanup of equipment, which results in considerable quantities of water being discharged in a relatively short time. For this type of operation, it is usually desirable and more economical to install holding capacity so that portions of this high-flow water can be stored for a short period of time and then worked back into the system. For a gravity flow system, this can be accomplished with a high-water overflow to a holding tank, from which the water would then be pumped back into the system for treatment. For a pumped system, this can be accomplished by having surge capacity in the pump pit to hold this water temporarily until it can be worked back into the system by the pumps. If either system is used, it should include high-level alarms to indicate when the plant has received excessive quantities of water over extended periods of time and the hydraulic surge capacity is filled.

Contaminants in a plant effluent that may make it unsuitable for discharge can be grouped as follow:

1. Acids, alkalies (pH)
2. Suspended solids
3. Oil
4. Heavy metals
5. Organic
6. Nutrients
7. Settleable solids
8. Color
9. Taste and odor producers
10. Toxic compounds
11. Total dissolved solids
12. Heat
13. Radioactivity

Any given plant may have one or several of these present in its effluent. Some are handled in separate preliminary treatment operations, and sometimes two or more constituents can be handled simultaneously in one treatment operation. A treatment plant is thus the logical organization of a series of processes.

Table 3 lists contaminants and unit treatment processes. A brief discussion of each follows.

Neutralization This is the addition of an alkali to react with an acid or an acid to react with an alkali to adjust the pH of the solution to within the desired range so the water is suitable for discharge. Also, this basic treatment may be necessary to establish proper conditions for completion of an oxidation-reduction chemical reaction, for precipitation of heavy metals as hydroxides, for proper clarification, for better adsorption, etc. This is a basic reaction for a multiplicity of waste water treatment operations.

This operation is controlled on the basis of pH. By definition, pH is the logarithm of the reciprocal of the hydrogen ion concentration. A pH of 7.0 represents a neutral solution; values higher than this represent alkaline conditions, while values lower than this represent acidic conditions. The farther the pH value is from 7, the more alkaline or acidic is the solution. Automatic pH recorder-controllers are available for this usage.

Alkalies commonly used are lime, caustic, or soda ash. Lime is available as burnt lime, hydrated lime, or lime slurry. Lime is more difficult to handle than the other two alkalies; so the capital investment for efficient handling is generally higher. However, since the basic cost of lime is lower, higher capital costs may be warranted. Also, for some precipitation and clarification reactions, the presence of the calcium ion is desirable, thus making lime the preferred material. The reaction rate for lime is slower than for caustic; so more reaction time is required.

Caustic is available in solid form or as a 50 percent solution. The liquid form is less expensive. While liquid caustic use necessitates installation of bulk handling facilities, savings in chemical costs again may warrant this increased capital expendi-

TABLE 3 Pollution Control Processes

Processes	Contaminants																			
	pH	Acids	Alkalies	Settleable solids	Suspended solids	Heavy metals	Hexavalent chromium	Cyanide	Organic material	Oils	Phenol	Color bodies	Taste and odor	Herbicides and pesticides	Total dissolved solids	Thermal pollution	Nutrients	Phosphate	Nitrogen	Radioactive
Neutralization (pH adjustment)	X	X	X																	
Chemical oxidation or reduction							X	X			X	X	X							
Sedimentation				X	X					X										
Clarification				X	X	X	X		X	X								X		
Filtration				X	X	X														
Flotation					X					X										
Ion exchange						X	X				X				X			X	X	X
Lagooning				X					X	X	X	X	X			X		X	X	
Emulsion breaking										X										
Adsorption									X		X	X	X	X						X
Biological treatment									X		X	X	X					X	X	
Direct incineration								X	X	X	X									
Sludge dewatering				X	X	X			X	X	X	X	X							
Ultimate sludge disposal				X	X	X			X	X	X	X	X							
Deep-well injection	X	X							X						X					
Cooling towers																X			X	
Desalinization techniques															X			X	X	X

A variety of unit processes are applicable for treatment of different contaminants. Various combinations of these processes may be incorporated into any waste treatment system. For instance, an organic waste containing solids above 65 to 70 mg/1 would require clarification ahead of carbon adsorption.

ture. The reaction rate is practically instantaneous for neutralization of acid; so minimal retention times are required. In fact, in-line additions are suitable with proper control.

While *soda ash* also can be used for neutralization of acids, it cannot be used if precipitation of heavy metals is a part of the process.

Sulfuric acid is most commonly used for neutralization of alkalies and is available in liquid form. Concentrated 66° Be sulfuric acid is not corrosive; so it can be stored in steel tanks. Dilute sulfuric acid is very corrosive and requires special materials of construction for tanks, pumps, and piping.

Neutralization is accomplished as either a batch treatment or a continuous process, with properly designed controls and reaction tanks as shown in Fig. 5.

Chemical oxidation and reduction This process involves the addition of an oxidizing or reducing agent under proper chemical (pH) conditions. Classic examples are the destruction of cyanide by the alkaline chlorination method and the reduction of hexavalent chromium to trivalent chromium. It may also be used for the destruction of certain color producing bodies, taste and odor producing bodies, and phenols.

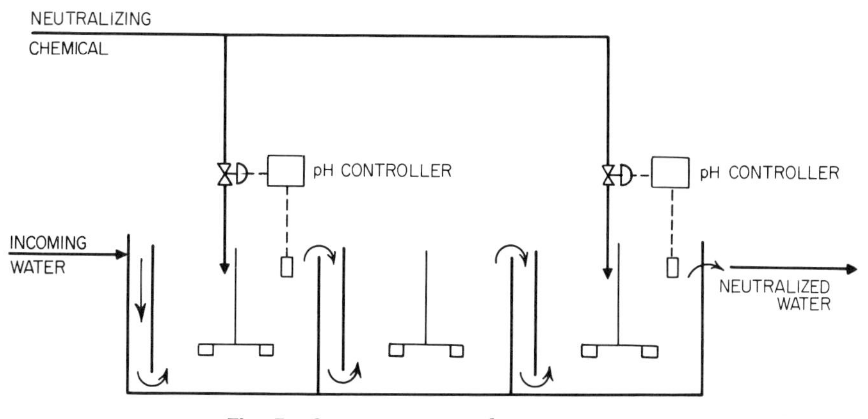

Fig. 5 Continuous neutralization system.

In the case of cyanide, the most frequently used process is alkaline chlorination. This involves adding chlorine under alkaline conditions (pH 8.5 or higher) to promote the chemical destruction of the cyanide. This is a three-stage reaction. One of the intermediate products is cyanate, which is not as toxic as cyanide and requires considerably less chemical. In a few rare instances, destruction to cyanate only has been permitted, but most often complete destruction is required.

For hexavalent chromium, a two-step process is required for removal. Hexavalent chromium is first reduced to the trivalent state and is then precipitated as a metallic hydroxide. The reduction step is a pretreatment step usually involving separate collection and treatment, as shown in Fig. 6. The pretreated water then can be mixed with other waters for precipitation and clarification.

Organic materials, such as phenol, color bodies, and taste and odor producing bodies may be destroyed or made innocuous by oxidation with chlorine, chlorine dioxide, potassium permanganate, or ozone. The effectiveness of any of these is best determined by simulating the treatment procedure in bench scale, batch type laboratory tests. Items to be considered are pH, oxidant chemical, dosage, and contact time. There is always a possibility that oxidation of an organic material, with chlorine, will produce a worse taste and odor problem than the original waste; so care should be taken in conducting these tests and evaluating these results.

Sedimentation This is a process which utilizes the natural separation tendencies of insoluble material in water. It is applicable to the removal of insoluble oils and/or particulate solids. Separation is effected by holding the water in a quiescent or a controlled low-velocity condition long enough for the oil to rise to the surface and the solids to settle to the bottom.

The major items that affect this separation rate are the size and specific gravity of the oil globules or the particulate matter. With constituents of a single specific gravity and a reasonable size, knowledge of the size will enable one to design required basins to accomplish a predetermined quality.

In most waste waters, however, there are many variables. Therefore, particle size distribution should be determined and expressed as a range of separation rates.

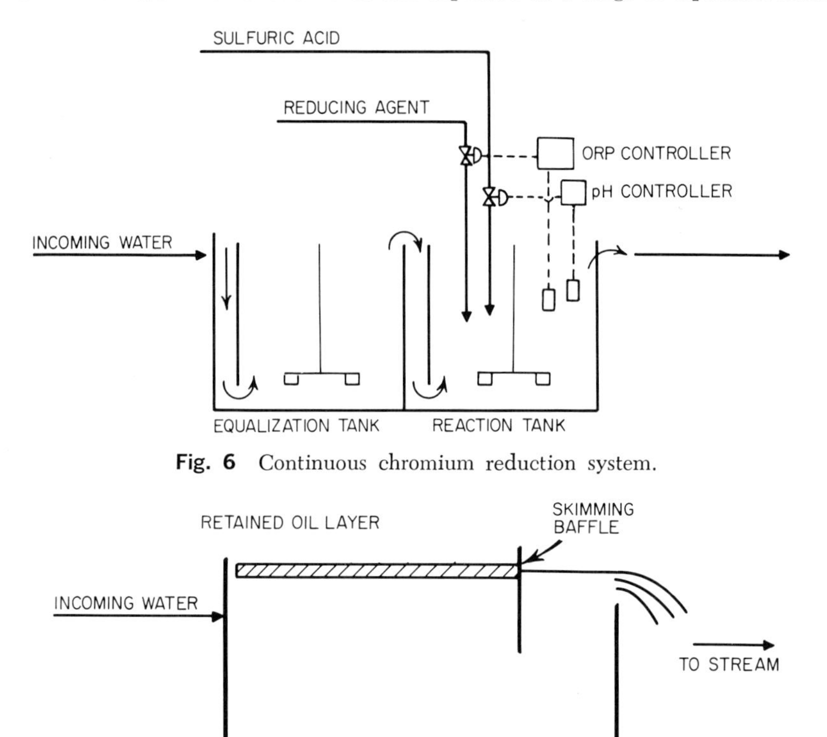

Fig. 6 Continuous chromium reduction system.

Fig. 7 Oil retention tank.

Procedures have been developed for determining this distribution for particulate matter. The same procedures should be applicable where oil removal is desired. (Oil globules have a tendency to grow because of contact with each other; so interpretation of these data may be more complicated than interpreting data from a settling rate on particulate matter.)

The separated oil must be removed from the surface of the separation channels. For irregular occurrences in very small installations, as shown in Fig. 7, the oil can be removed manually. The need for cleaning can be determined by regular visual observation. Devices for continuous oil removal are:

Slotted Pipes. This consists of a pipe with a lengthwise slot. The pipe is installed parallel to the water surface so it can be rotated manually around its axis. The level of the slot is set so the top layer of oil flows into the slot and onto a collection container.

Telescoping Pipes. This consists of one pipe inserted into another. The level is adjusted so that the top oil layer can flow by gravity into a collection tank for ultimate disposal.

Belt, Drum, or Rope Skimmers. These units operate on the principle that a belt, drum, or rope made of a material that has a high affinity for the oil continuously passes through the accumulated oil layer, thus picking up oil selectively. The conveyor is then passed through a set of squeezer rolls for removing the oil, which flows into a collection container.

Separated solids must be removed from the settling tank. For heavily loaded pits, such as a scale pit for a steel mill, the solids may be removed by bucketing with a crane. For other pits handling relatively lighter solids, it is usual practice to install flight scrapers or bottom scrapers with conveyor buckets or pumps for lifting the solids from the pit. Flight scrapers or bottom scrapers are slow-moving rakes that move the solids along to a collection and discharge point.

Clarification Clarification is preferred to sedimentation where suspended matter has slow settling rates and it is desired to reduce the size of required facilities. Chemical clarification also is applicable for producing effluents having an oil content of less than 15 mg/l. For metal processing operations such as pickling and plating, where metals are present in the water as soluble salts, clarification is necessary to promote more effective removal of the heavy metals as hydroxides by precipitation.

Clarification is an essential part of any process where a contaminant is removed by precipitation, such as phosphate removal. It also may be usable for removal of colloidal organic material including some color bodies by adsorption on formed chemical flocs.

Clarification involves the addition of chemical coagulants, coagulant aids, and pH adjustment to form a stable, rapid-settling floc which is then separated from the water by sedimentation. Chemical clarification improves settling rates and provides a more consistent effluent quality than is obtainable with straight sedimentation.

Coagulants used most often in waste water treatment are aluminum or iron salts such as aluminum sulfate, sodium aluminate, ferrous sulfate, ferric sulfate, and ferric chloride. These salts when mixed in water in the presence of an alkali under proper pH conditions precipitate as aluminum hydroxide or ferric hydroxide flocs. In the process small suspended solid particles are entrapped in the floc as they form and grow. Oils present in the water also will adsorb on the formed floc.

Clarification involves flash mixing, flocculation, and sedimentation. The treatment chemicals are added to the water and mixed so that they are completely dispersed. This solution then enters a gently mixed section where the floc forms and grows. The water then flows to a sedimentation section where the solids separate from the liquid by settling. In some available units, the functions are performed in different zones of the same unit, as shown in Fig. 8. Recirculation of settled sludge as seed for producing a better-formed floc has been found desirable for treating industrial waste waters.

Polymers are frequently used with coagulants to produce stronger, larger, better-settling flocs. Polymers are long-chain, high molecular weight, water soluble organic materials. Some of the materials have varying degrees of cationic or anionic charge, while others are basically nonionic. In some cases, the addition of one material will produce significant improvement, while in other cases two materials of different charge may have to be added.

Filtration This process in waste treatment is applicable for the removal of particulate matter from plant effluent, and the polishing of effluent from sedimentation basins and/or clarifiers.

Filtration involves passage of water through a packed bed for removal of suspended solids. The suspended solids fill interstices in the bed, and it gradually requires increasing pressure to pass the rated quantity of water through the same bed area. When the pressure drop across the bed reaches a partial limiting value, the bed is taken out of service and backwashed. Backwashing is the operation of passing water through the bed in a reverse or upward direction to remove the accumulated solids from the bed. The bed is then ready to be placed back on line.

The backwash water must be further treated before it is suitable for discharge. This can be accomplished in a batch holding tank, where the solids are allowed to settle out of the water, the water is decanted back through the filters, and the solids are removed for ultimate disposal. If the filters are preceded by a sedimentation tank or a clarifier, the backwash water may be transferred to a holding tank from which it is bled back into the system for solids removal and then through the filters for discharge or reuse. The backwash water is usually only 1 to 2 percent of the total throughput, but the rate at which it is used may be significant enough to affect the hydraulic loading on solids removal sedimentation and clarification units if it is passed directly back to these units.

Filters utilize media such as anthrafilt or sand. Current trends are to the use of mixed media, graded coarse to fine, in the direction of the water flow. The specific gravity of the media is so selected that backwashing and hydraulic grading

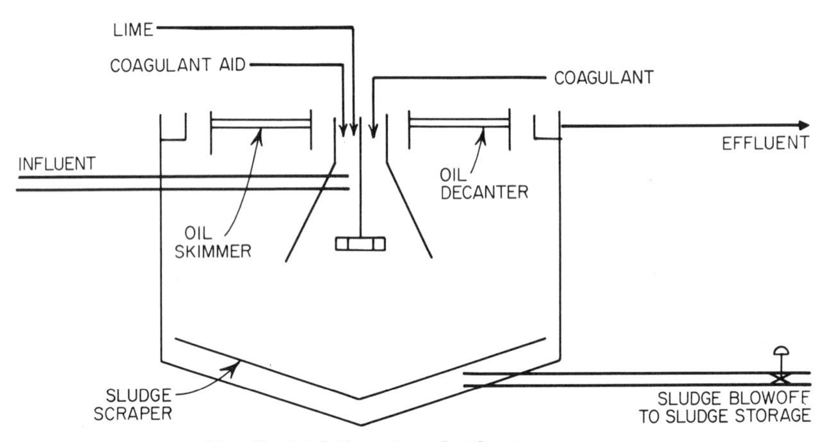

Fig. 8 Multifunction clarification system.

does not upset the layering of the graded media in the bed. Mixed media filtration systems have certain inherent advantages such as

1. Greater capacity per unit area of surface
2. Ability to handle a wider range of influent suspended solids concentrations
3. Longer filter runs

Another system that is gaining in use is the perforated drum filter. Water passes through the drum with the solids retained on the drum. Solids then are slurried from the drum and handled similarly to filter backwash water. Since slurry is produced at a more uniform rate, intermittent storage requirements may not be as critical as is the case with backwash water.

To promote filterability, it is often necessary to feed polymers to the waste ahead of this operation. It may be necessary also to include conditioning chemical feeds for the backwash water to attain solids removal when this water is being recycled.

Flotation The flotation process takes advantage of the natural tendency of oil globules to rise to the surface. It is also used for removal of some finely divided suspended solids.

Air is inducted into a pressurized water stream to achieve saturation, as shown in Fig. 9. The air-laden water is then passed through a pressure reducing valve, and the air released from the water as small bubbles. The bubbles attach themselves to oil globules or suspended particles and float them to the surface, from which they are removed for further handling. Chemicals, such as coagulants, polymer coagulant aids, acids, and/or alkalies, are often added ahead of the system to promote more complete removal.

Ion exchange A process used primarily for the concentration and recovery of valuable constituents, such as hexavalent chromium cyanide, copper, and other metals used in plating and other metals finishing operations. It is also used to purify and reclaim process solutions for reuse and to concentrate constituents for destruction.

Ion exchange is useful for a considerable number of special applications. Studies are currently being conducted on usage of ion exchange materials that have a high degree of selectivity for ammonia removal. These materials have been used also for concentration of radioactive salts from wastes and the contaminated exchange material is stored until natural decay makes it safe for handling.

This process is naturally applicable to the reduction of total dissolved solids when necessary. The concentrated regenerant, however, still requires further handling for ultimate disposal.

Ion exchange is particularly applicable where high-quality water is required as part of the processing operation, such as the final rinse on a chromium plating line.

Ion exchange operates as a fixed bed with the water passing through the bed, and the ion of concern replaced by the ion present on the exchange material.

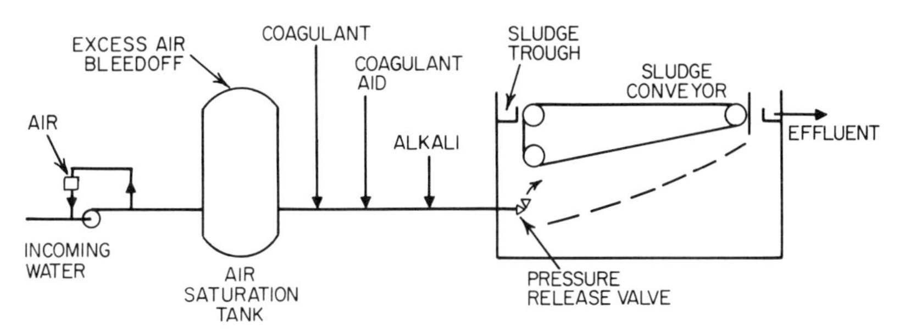

Fig. 9 Air flotation system.

For the purpose of describing the process, an illustrative example is presented here. Hexavalent chromium is present in rinse water from chromium plating lines, cooling tower blowdowns, metal finishing rinses, etc. When the water passes through an anion exchanger (assuming proper pH conditions), the hexavalent chromium is adsorbed on the resin. When exchange capacity is used up, the resin is regenerated with either caustic or salt and caustic, and the hexavalent chromium is carried off with the spent regenerant for recovery and reuse. The resin is then rinsed of excess regenerant and is ready to be placed back on line for removal of additional hexavalent chromium.

For waste water treatment, the ion exchange materials generally used are cationic or anionic synthetic polymers on which ion exchange sites are affixed. Because of the variation in cation and anion materials, specific information on material of choice should be developed for specific applications.

Lagooning This is the holding of waste water in manmade ponds or lakes for the removal of suspended solids and insoluble oils. Lagoons are used also as retention ponds after chemical clarification to polish the effluent and to safeguard against upsets in the clarifier; for stabilization of organic matter by biological oxidation; for storage of sludge, which is hauled out intermittently; and for cooling of water.

For solids removal, lagoons should have about a 24-hr retention plus long-term storage for the settled sludge. Where significant quantities of solids are present in the influent, it is not unusual to plan to use only the top foot of the lagoon for retention of the flowing liquid, with the remainder of the lagoon reserved for sludge storage. The lagoons should have sufficient sludge storage capacity to necessitate cleaning once or twice a year. Even with infrequent cleaning, it is advisable to have the lagoon so designed that half of it can be taken out of service, the water level dropped, and the sludge partially dewatered to simplify removal. Lagoons should be equipped with inlet and outlet distribution baffles to utilize the full settling area better. It is also desirable to have submerged bottom baffles at various points to retain the bulk of the settled solids at the inlet portion of the lagoon.

Lagoons for oil removal, assuming minimal suspended solids present, are also designed on the basis of a 24-hr liquid retention time, with depths of about 4 ft. Skimming baffles should be placed just beyond the inlet distribution box to retain the bulk of the oil in an area to facilitate removal. Other baffles should be installed near the outlet end of the lagoon to capture the oil that escapes the inlet baffles.

Lagoons have long been used for storage of clarifier sludge from metal processing plants because of lower initial capital cost and reduced operating costs. A disadvantage, however, is that this delays the problem of sludge handling. Heavy concentrations of metallic hydroxide sludges never seem to dry. A crust will form on the

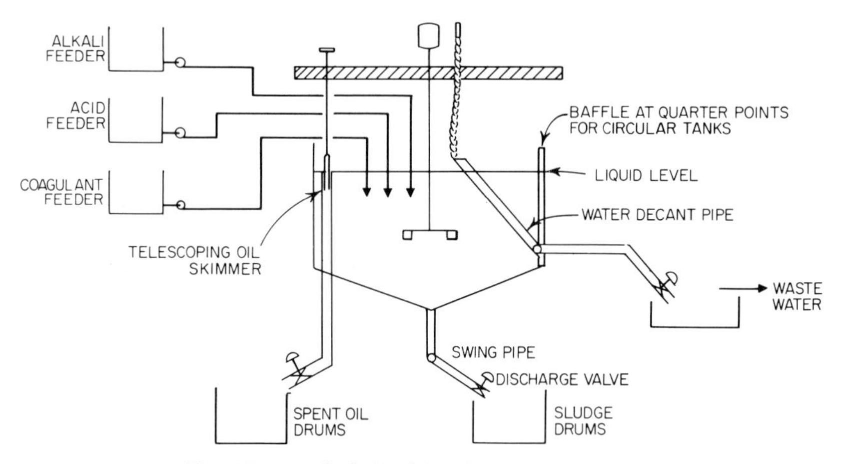

Fig. 10 Emulsified oil batch treatment system.

surface, but the lower portion still retains significant quantities of moisture. Thus, considerable acreage can be tied up indefinitely by this system.

Lagoons are used for biological stabilization of organic material. Biological activity occurs whether desired or not, because of the presence of soil bacteria and nutrients. Biological loading of a lagoon, therefore, must be considered to ensure that the natural aeration capacity is not overloaded. Overloading results in anaerobic conditions and development of unpleasant gases and odors such as hydrogen sulfide.

Emulsion breaking Emulsion breaking is a pretreatment step used for certain oil-water mixtures. Emulsified oils, i.e., soluble oils, are used in machine operations and as coolants. These are chemically dispersed in water at a ratio of 5 to 10 percent oil. These emulsions must be broken so the oil can be separated from the water.

Emulsion breaking usually is accomplished as a batch process as shown in Fig. 10. The spent emulsions are collected in a holding tank which is equipped with agitator(s) and skimmers. Tank contents are held undisturbed 2 to 8 hr to give insoluble oils time to rise to the surface for removal.

Tank contents then are agitated and emulsion breaking chemicals added. After the emulsion has been broken, the freed oil is allowed to rise to the surface for removal. The pH of the water then is adjusted and the waste clarified by flotation or clarification either alone or in combination with other wastes. The proper chemical and chemical dosage must be determined by experimentation since requirements can change from day to day. Usual steps tried are:

1. pH adjustment to 3 to 4.
2. Addition of iron or aluminum salts such as ferrous sulfate, ferric chloride, or aluminum sulfate (feed rates as high as 500 mg/1 of the coagulant are not unusual).
3. Salting out by addition of calcium chloride.

For some wastes only one of these steps, such as pH adjustment, may produce satisfactory results; for others, combinations of two or three may be required.

In exceptional instances, this treatment technique has failed to produce desired results. When this occurs, special studies are necessary to develop an effective treatment. (This emphasizes the point that after waste treatment plants are installed and consideration is being given to changing either the source of supply or the type of soluble oil used in a manufacturing operation, compatibility of the oil with waste treatment plant practices should be reviewed before a final decision is made.)

Adsorption Adsorption is a surface area phenomenon wherein soluble organic materials such as phenols, herbicides, pesticides, surfactants, and taste and odor producing bodies are removed from water and adhere to the adsorbent.

Adsorption is effective also with biologically treatable organic material as well as those organics resistant to biological treatment. Some low molecular weight organics, however, are less readily adsorbable by this process. Therefore, the adsorption characteristics of organic matter present in a waste water must be evaluated to determine the applicability of the process and the attainable effluent quality.

The most commonly used adsorbent is activated carbon in powdered or granular form.

Powdered carbon is used by adding the carbon to the flowing water stream, agitating it for intimate contact, and then separating the carbon from the water in clarification equipment as discussed previously. This method has been used for years to remove color and taste and odor producing bodies from water. However, it does not appear that powdered carbon will be as applicable to waste water treatment as granular carbon. Naturally, each specific case must be technically and economically evaluated.

Granular carbon is used in fixed or moving beds, as shown in Fig. 11. The water is percolated through the bed, where contact with the granules removes the organics. The carbon bed also acts as a filter for removal of suspended solids, but like a filter the influent suspended solids should not be greater than 50 to 65 mg/1. Higher solids loadings may cause excessive pressure drop and need for more frequent backwashing. The carbon becomes exhausted when the effluent contains predetermined concentrations of the organic. It is then necessary to either discard the carbon, chemically reactivate it in place, or remove it from the bed and thermally reactivate it for reuse.

1. *Throw Away.* This is feasible in small installations using limited quantities of carbon, where the material being removed is not amenable to chemical reactivation and where the cost of installing thermal regeneration facilities is not warranted.
2. *Chemical reactivation* is applicable to the removal from the carbon and recovery of a specific chemical, of which phenol is a good example. This is also applicable to the removal of organic material, which would be disposed of ultimately by liquid incineration, with the granular carbon acting as a concentrating mechanism only.
3. *Thermal reactivation* is feasible where adsorbed organics are to be disposed of and carbon reactivated for reuse. The activated carbon is dewatered, and then passed through a multiple hearth, controlled atmosphere furnace where the adsorbed organics are oxidized at temperatures of 1600 to 1800°F. The regenerated carbon is then water quenched and stored for return to the treatment system, as shown

in Fig. 12. The carbon loss during this regeneration cycle is about 5 percent by weight because of physical attrition.

Biological treatment This process is used to reduce, under controlled conditions, the oxygen demand of an organic containing waste water. When organic material is discharged into a receiving stream, a biological chain of events occurs, in which naturally present bacteria in the receiving stream metabolize and stabilize the organic material using oxygen in the process. The biological treatment process does essentially the same thing, but under conditions where the bacterial population can be controlled, and the bacteria produced can be separated from the water for ultimate disposal. It is basic to the process that sufficient oxygen be supplied so the proper

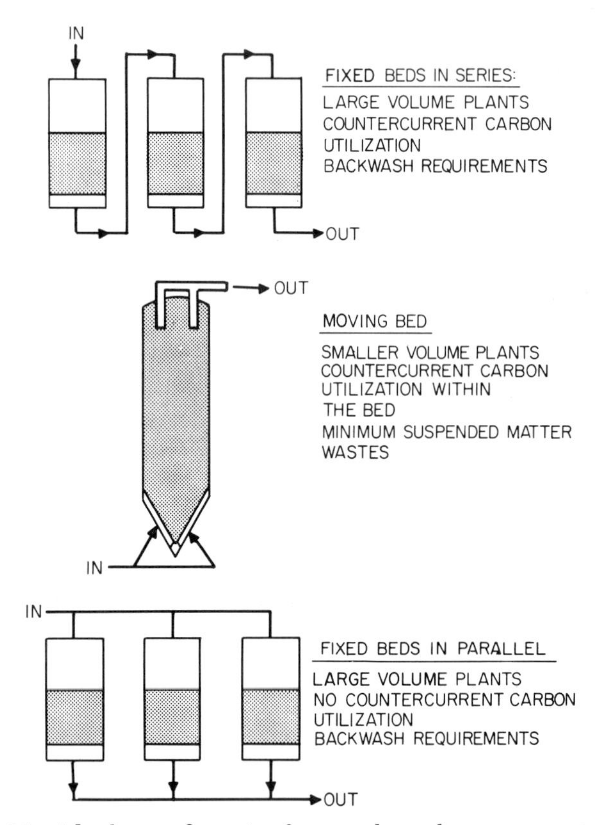

Fig. 11 Adsorber configuration for granular carbon waste treatment.

conditions for metabolism and stabilization of the organic material can be accomplished. This process is applicable in general to treatment of water requiring BOD reduction and for reduction and removal of specifically identifiable organic materials, such as phenol, which respond to treatment under proper conditions.

For most waste water, the destruction of organic material will take place under aerobic conditions in which a measurable oxygen residual is present. For certain concentrated organic wastes, anaerobic treatment (treatment occurring in the absence of oxygen) can be used, although its applicability is relatively limited, and it usually is followed by aerobic treatment for complete stabilization.

Instances have been noted where taste and odor producing bodies and color bodies have been effectively reduced by biological treatment. Small-scale experiments on representative samples of a waste can indicate whether the specific contaminant will be susceptible to biological treatment. In more than one instance, wastes have been treated effectively for BOD reduction with the color, for instance, passing through the treatment process unchanged.

Nitrogen and phosphorus are necessary constituents which enter into the bacterial metabolic process. Thus, for some wastes, nitrogen and phosphorus must be fed as nutrients to promote biological growth. For other waters, if the nitrogen and phosphorus are present in the proper concentration, biological activity will reduce

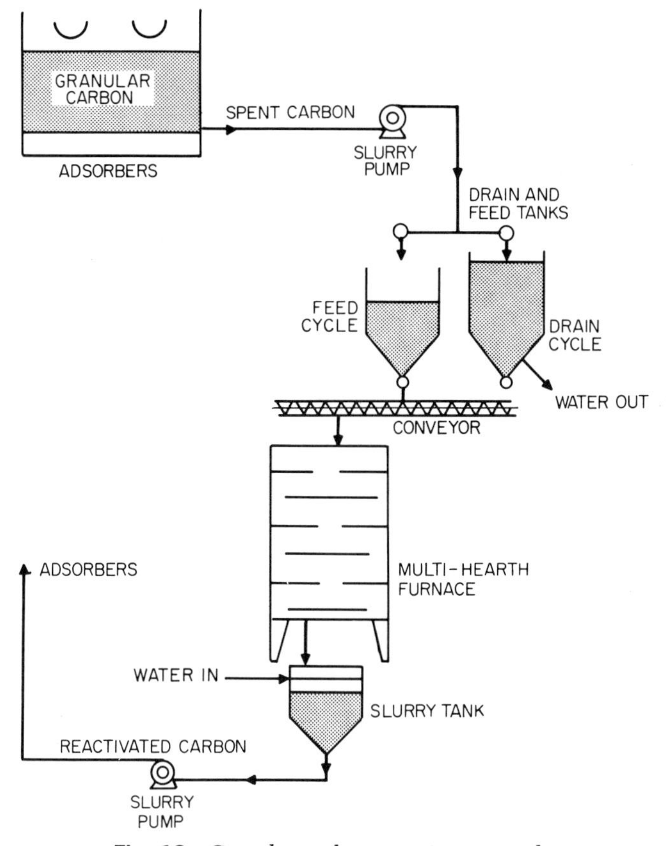

Fig. 12 Granular carbon reactivation cycle.

the concentration of these nutrients, and they will be removed with the biological sludge.

The most widely used biological process for industrial waste water is the activated sludge process. Incoming waste, with or without primary treatment for suspended and settleable solids removal, is mixed with return activated sludge and enters an aeration tank. This tank is aerated to maintain a dissolved oxygen residual, of about 1 to 2 mg/l, and biological growth and activity occur. If necessary, the feed of nutrient is made at this point. The wastes and the bacteria are held in contact long enough to stabilize the incoming organic material and accomplish the desired effluent quality. The mixed waste goes to a final settling tank where the solids (bacterial growth) settle from the water, the waste is discharged, and the solids are recycled to treat additional incoming water. Sludge will build up in the system in most cases to the point where some will have to be wasted for ultimate disposal.

Stabilization ponds (either with or without mechanical aeration) also have been used for biological stabilization of organic material. Without mechanical aeration, extremely low loadings in the range of 17 to 35 lb BOD/day/acre of surface should be permitted.

Trickling filters are another form of biological processing. Organic bearing water is sprayed over a packed bed of rocks with recirculation of trickling filter effluent and final sedimentation for removal of suspended solids. A biologically active film develops on the rocks which removes the organic material from the passing water, metabolizes and stabilizes it, and produces suspended solids, which slough off because of the hydraulic shear. Solids are removed in the final sedimentation tank.

Direct incineration A process applicable for disposal of materials that have a heat of combustion value sufficient to sustain combustion or require only minimal makeup fuel. Spent oils and solvents, and other highly concentrated organic solutions should be considered for disposal by incineration. For other organic wastes, incineration may be the only acceptable disposal method regardless of heat requirements.

Attention must be given to the air pollution potential of any material to be disposed of by incineration. For instance, chlorinated hydrocarbons produce hydrochloric acid when they are burned. Wet scrubbing with caustic, therefore, is necessary after incineration.

In the case of toxic waste disposal, temperature and combustion conditions must be determined in advance to eliminate production of toxic intermediate gases. In some instances, installation of an afterburner will be necessary to ensure that proper stack temperature is maintained.

Since liquid or slurry incineration systems may be involved, other plant wastes such as removed oils and combustible solids produced at a waste treatment plant also should be considered for disposal by incineration. Solids, of course, would be converted to slurry form by mixing with spent oils and solvents. In all cases, the incineration system should be designed to handle the predominant form of waste, i.e., liquid or solid.

For facilities with boiler plants or other incineration equipment, consideration should be given to their use for waste disposal. However, the organic waste should not be considered as an adjunct to the fuel supply, albeit in some situations, it may serve as an additional source of heat.

Sludge dewatering This is an intermediate process for concentration of sludge for disposal.

Sludge from a clarifier or a final biological sedimentation tank averages 1 to 3 percent solids by weight. Some thickeners handling particulate material yield solids running as high as 10 percent by weight.

The first step in dewatering is use of a thickener, which is a holding tank for settling the produced solids more compactly through gravity. These tanks generally have a 24-hr retention time based on the volume of sludge to be wasted from the clarifier or thickener. A thickener can increase solids concentration by from 3 to 10 percent for a metallic floc to as much as 15 percent for inert particulate solids.

There are other intrinsic values in use of thickeners. In addition, sludge usually is handled on less than a 24 hr/day basis. In general, clarifiers do not perform most efficiently when they are used for the dual function of clarification and sludge storage. Inclusion of thickeners in the system enables the operator to control the clarifier for optimum clarity of effluent, and still schedule further sludge handling.

With biologically active sludge, a digester functions also as a thickener and sludge reducer by further biological activity. Sludge digestion can occur aerobically or anaerobically. Retention times of 30 days are necessary, however, to get this double benefit from a digester.

The compacted sludge is then further dewatered by filtration or centrifugation.

The most commonly used filtration device is a rotary vacuum filter. This is a large rotating drum covered with a filter cloth or medium that is partially submerged in a trough containing the slurry. The vacuum on the inside of the drum causes a sludge layer to form on the filter media. The sludge is further dewatered as the drum rotates and is scraped from the drum onto a conveyor belt. The removed water is returned to the head of the treatment plant for processing prior to discharge. The solids content of the produced cake is usually around 20 to 25 percent by weight and has the consistency of wet clay.

Some sludges must be chemically conditioned before filtration. Representative sludge samples from the plant must be examined to determine chemical conditioning requirements and filter sizing.

For centrifugation, continuous operating units are used. Produced sludge has a solids content of 15 to 18 percent by weight. The water often contains fine solids and is returned to the head of the treatment plant for processing prior to discharge.

Sludge disposal is one of the major unresolved problems in water pollution control today. Although a variety of disposal practices are employed, none, except incineration, can be considered as a means of ultimate disposal. However, incineration has limitations since it cannot handle noncombustible wastes. And even with combustible wastes, supplementary fuel may be required, thus adding to operating costs.

Other disposal methods include:

1. Landfilling by layering sludge and earth in trenches. Organic type sludges will decompose and both organic and inorganic sludges intermix with the earth over an extended period of time.

2. Lagooning is used for disposal of inorganic sludges. It is only an intermediate solution, since further handling of the sludge will most likely be required at a future time. A chief disadvantage is the large land requirement.

Deep-well injection A process applicable to the disposal of low volumes of highly concentrated waste. However, proper geologic conditions must be assured. Deep-well injection is a system by choice, because of economic conditions, but seldom the only system applicable.

The waste is discharged into a selected stratum, sometimes under high pressures. The stratum selected must be below any potable water sources in the area and should be under an impervious stratum, so that the injected wastes do not find their way up into the usable potable water layer. The area also should be free of geological faults.

Pretreatment of the waste may be required prior to injection. This may involve filtration of the waste for removal of the suspended solids so they do not plug the receiving formation. There is also the potential problem of the compatibility of the waste being injected, the receiving stratum, and any waters present in this stratum. Chemical reactions may occur which will blind the receiving stratum. Thus, studies involving geological conditions and chemical compatibility are an integral part of any preliminary program before the final decision is made to proceed with deep-well injection. In most cases, test wells are required to confirm finally that injection can be utilized.

Although some toxic wastes now are being disposed of underground, the practice should be seriously questioned since breakthrough from the well could pose serious problems.

Cooling towers These are used for cooling heated waters to make them acceptable for discharges to a receiving stream and for air stripping ammonia from wastes.

Water is cascaded over a packed tower and intimately mixed with air. Heat is dissipated because of sensible heat loss and evaporation of the recirculating water. The water is generally cooled sufficiently to permit reuse, although in a few cases it is cooled only for discharge. If the water is recirculated, the system must be bled periodically to prevent buildup of dissolved solids.

Where ammonia is present as a gas, it escapes to the atmosphere as water is recirculated across the tower.

Desalinization Techniques

Desalinization is a general term for those processes used for the separation of dissolved solids from water. These systems have been developed to convert seawater or brackish water to potable water. They have been applied basically in water short areas. In the United States there are only a few operating installations, other than prototype and research installations, for the conversion of brackish water to potable water by reducing the dissolved solids content. The desired product in all of these cases is the water, which is then going to a distribution system for direct use.

Evaporation has been the most widely used system for producing potable water from seawater. In the United States this process has primarily been used for the production of a high purity water for a specific purpose, such as boiler feedwater. Studies on evaporation treatment of municipal waste waters indicate that steam distillable organics and inorganics (ammonia) are present in significant quantities in the finished water. Organic material present in the incoming water apparently causes some fouling of the heat-exchange surfaces. Activated carbon pretreatment for organic removal assists in reducing the amount of organic carryover and fouling of the heat-exchange surfaces. This is in addition to the normally expected problems of scaling caused by concentration of normally present dissolved solids. For the immediate future it appears that this process will continue to be used for the production of high-quality waters as contrasted to the general production of potable water.

Reverse osmosis is the newest process being investigated in this area. The stimulus for the development of this process was the government's desalinization program. Reverse osmosis involves the passage of molecules through a semipermeable membrane. The driving force is pressure. Selectivity and separation are accomplished by characteristics of the membrane. In general, smaller molecules and lower valence materials are the more difficult to separate from the water. This process is appealing because it is basically so simple.

At the present time the largest operating unit has a 100,000-gpd capacity. Thus, the throughput capacity of plants presently using this process is limited. The major questionable area for the process is the expected life of the membrane and desired membrane configurations. As the membranes age they lose throughput capacity. This appears to be accelerated by acidic or alkaline waters outside the pH range of 6.0 to 8.5. Suspended solids and biological growths also appear to cause fouling of the membranes.

Freezing involves the production of ice crystals by partial freezing from water. The ice crystals contain less dissolved solids than the brines from which they are produced. The problems involved in commercializing freezing involve efficient heat transfer, controlling the size of the produced crystals, and separation of the concentrated liquor from the produced ice crystals, with the resultant production of a suitable quality product water. It appears that inorganic salts can be removed more easily than organic material from the ice crystals; thus the resultant water still contains traces of organic material.

Ion exchange has been used extensively in water treatment for the softening of water and the removal of dissolved solids. Use of conventional ion exchange systems for dissolved solids removal has been justifiable in the past only where high-quality water is required. As a rule of thumb, demineralization has been the system of choice for water containing less than 600 mg/1 of total dissolved solids, whereas evaporation has been the system of choice for waters containing more than 600 mg/1 of total dissolved solids. A major item of expense is the chemical regenerant. This cost is increased by the regeneration levels required to produce the desired high-quality water. Currently, other regeneration systems are being investigated to extend the practical limit of applicability for this system and to facilitate disposal of the spent regenerant.

Electrodialysis involves forced migration of charged ions through cation-permeable or anion-permeable membranes by impressing an electric potential across a cell containing mineralized water. The electric power requirement is proportional to the number of ions removed from the water. The removed ions are concentrated in solution for removal from the system. This system looks promising.

The major problem encountered in using this system for treatment of waste waters is the fouling of the membrane, especially the anionic membrane. This appears to be caused by the presence of suspended solids, colloidal material, and possibly organic materials. This is in addition to the fouling problems that may occur due to the nature of the dissolved salts, such as calcium carbonate deposition, etc. Thus, extensive pretreatment, as well as internal conditioning of the water, may eventually be required. Studies are proceeding on these aspects.

SUMMARY

The various processes discussed have been considered individually as to where their use is applicable for removal of certain types of contaminants. For treatment of industrial wastes, it is not unusual to use two or more of these processes in series to be able to acheive consistently the desired effluent quality.

There is considerable concern regarding the changing objectives for water pollution control and the improvement in plant effluent quality that will be necessary to satisfy upgraded requirements. Part of this concern is based on the gnawing fear that equipment installed to meet today's requirements may be obsoleted by tomorrow's new requirements. While it is recognized that continuing upgrading of effluent quality in all likelihood will necessitate a parallel upgrading of treatment plants, it does not follow that entire plants or major portions of waste treatment plants will have to be scrapped. Instead, it is more likely that treatment processes such as carbon adsorption, ion exchange, reverse osmosis, and electrodialysis will be incorporated into existing plants. These treatment processes would produce a high-quality effluent which might blend with lower-quality effluent to produce an acceptable discharge.

In other words, since changes in effluent quality will occur on an incremental basis, treatment processes can be added incrementally to satisfy the new demands.

Obviously at some point in time, plant effluent may be of such quality that it will be too costly to throw away. There will be no recourse other than to return it to the manufacturing plant for reuse since, in many instances, in-plant water quality requirements are less stringent than those required for discharge. Also, the costs incurred in treating the waste water to a quality suitable for discharge really can be considered as a pretreatment cost which may be more favorable than treating raw water from a surface supply.

It becomes obvious that water pollution control is intimately intertwined with in-plant water use and must be considered a key part of a plant water management program. As such, it demands continuing attention to assure that objectives established for the program are achieved.

Chapter 7

Pollution Waste Control

R. D. ROSS

Vice President, Thermal Research & Engineering Corp.,
Conshohocken, Pennsylvania

INTRODUCTION

The total disposal problem of atmospheric pollutants, water contaminants, and solid wastes such as garbage and trash for a country of 200 million people is almost incomprehensible. For many years we have assumed that our atmospheric resources were unlimited, that the lakes, streams, and oceans of our land would accept whatever liquid waste we could pump into them, and that we could handle our solid waste problem forever by dumping it into the vast resources of our unpopulated land areas.

In the last 10 years a few people have begun to show some concern about these seemingly unlimited resources depleting at such a fantastic rate. This concern has passed on to others, and now the problem of pollution, whether it is air, water, land, thermal, or noise, has become one of national interest, even to the extent that both political parties made platform statements about their position on the war on pollution in 1968.

We now are beginning to see that our children and the generations that follow them must breathe the air, drink the water, and live on the land which we thought was unlimited in supply.

In order to understand the total problem better, we should perhaps first define what we mean by each type of pollution.

Air Pollution

Air pollution is the release of waste gases or odors from a biological or chemical process which contains substances which can be considered harmful to human life and confort, either because they are toxic, because they reduce the oxygen available for sustaining life, or because they are aesthetically undesirable.

Water Pollution

Water pollution is the release of materials into water sources or supplies which are damaging to life because of their toxicity, because of their reduction of the normal oxygen level of the water, or because they are aesthetically unpalatable.

Land Pollution

Land pollution is the misuse of land in a way which makes it unfit for man's future needs, such as the construction of buildings or the growth of food or other materials which he uses in his daily life and which could cause either dangerous toxic contamination of the air and water resources or give them a disagreeable appearance, taste, or odor.

In order to solve the total pollution problem, four basic steps are involved. The first is the recognition and definition of the problem in terms of the type of pollution, its constituents, and their effect on human life. The second is a determination of the practical limits of these contaminants which will permit freedom from concern for human, animal, or plant life. The third is the development of treatment methods for these effluents to a safe and acceptable level. The fourth is the development of disposal methods for materials that cannot be adequately or economically treated, or for the residue remaining after treatment, to levels that are safe and acceptable.

The purpose of this chapter is to discuss the fourth step, or the disposal of gaseous, liquid, and solid effluents in a manner which will not cause pollution of the atmosphere, water, or land resources.

It is difficult to draw the line of demarcation which indicates where treatment stops and disposal begins. The word disposal pertains to the final deposition of the material, either before or after a treatment method has been used, and in many cases where no treatment method has been used. For example: a waste gas may be treated by removing particulate matter through mechanical separation or wet scrubbing, and then vented to a stack; or it may be vented to the stack without any prior treatment and still provide acceptable ground level concentrations of the particulate in the atmosphere. In the same manner, liquid waste, which is highly aqueous in nature, may contain a few hundred parts per million of an organic or inorganic contaminant which could be removed by chemical or biological treatment, or which could be pumped into a large water body and diluted to a level well below acceptable limits. It is therefore readily apparent that the term "disposal" is relative concerning its use in any waste process.

Interpollution of Resources

Another problem which develops in pollution control is the differentiation between types of pollution. Many air pollution problems can be solved by transferring the pollutant from the air to a scrubbing medium, such as water, thus causing water pollution problems. Many solid disposal facilities, which in themselves are safe and satisfactory, can cause water pollution problems because of groundwater leaching of toxic materials from the solid wastes. Both solid and liquid wastes can emit strong odors because of biological degradation, thus causing air pollution. Improper incineration of solid and liquid wastes can cause smoke and particulate matter to pass into the air. Therefore, it is not just enough to solve the individual problem of either air pollution, water pollution, or ground pollution. We must also consider the transfer of the pollution problem to other areas. Many so-called safe treatment and disposal facilities have merely caused a problem elsewhere. In many cases the cure is worse than the disease.

While our history in the solution of pollution problems is brief, there has been an unsettling tendency for us to discuss each problem in its individual category (i.e., air, water, land) rather than as a total problem, and this has led to incomplete answers, which often result in worse situations cropping up in other forms.

For the purpose of this chapter we shall consider incineration as a disposal method, whereas it could well be considered a treatment method. In most cases, it is the last step in the train of events, and for this reason, we have considered it as disposal rather than treatment, but it should certainly be kept in its context, especially when discussing solid waste disposal, because a final residue will usually remain even after the incineration step.

Odor control is also covered in this chapter as a disposal method, when it may well be considered a treatment step. But in the case of odors we are usually concerned with aesthetic qualities and not toxicity, so that while the final effluent may be realeased to atmosphere without odor counteraction, it is usually desirable to apply this as a final control measure.

DISPOSAL OF GASEOUS WASTES

Disposal of gaseous wastes is usually more of a quantitative than a qualitative problem. Most gaseous wastes are easily treated from a chemical or biological viewpoint. They contain known contaminants which can be absorbed, adsorbed, oxidized, separated, or otherwise removed by existing conventional methods. Therefore, it is not the type of contaminant so much as the total quantity of gas which must be handled that makes the gaseous waste disposal problem a difficult one. For example, a single 100,000 lb/hr boiler, which is not large by today's standards, may emit 40,000 to 50,000 scfm of flue gas. If it is necessary for economic reasons to operate this boiler on either coal or a residual fuel oil which contains sulfur, and if it is necessary subsequently to remove this sulfur, then the physical size and cost of the treatment apparatus are dictated not by the amount of sulfur present in the flue gas rather by the total volume of flue gas that must be handled. This type of problem is present in most waste gas treatment systems. That is, the amount of contaminant is small but the amount of nontoxic carrier gas, usually air or inerts, is tremendously high in comparison. A similar example might be the scrubber required on a municipal incinerator. A good municipal incinerator should be able to produce a minimum of particulate matter, but the scrubber is not sized to the amount of particulate matter but rather to the total quantity of gas which must pass through it, and the design problems are based primarily on the total flow rather than on the quantity of pollutant material in the waste gas.

These two examples are noted for one reason only, to point out to the reader that a reduction in waste gas flow, however it may be effected, results in far greater savings than a reduction in the quantity of pollutant carried in the waste gas, assuming that both streams, regardless of their quantity, will be treated in the same manner.

Once a waste gas from an oven, a boiler, a furnace, an incinerator, a chemical process, a grinder, or any of the hundreds of possible systems has been collected in a duct, a flue, or some similar container, there are only several processes which can be applied as treatment before final disposal to the atmosphere is necessary. The first consideration should be mechanical removal of any particulate matter which is present in the waste gas, for any gas can be handled more easily as a pure gas rather than as a combination of gas and solids, and then final disposal methods can be applied. Particulate matter is usually removed through a filter or entrainment separator, a cyclone, an electrostatic precipitator, a fabric bag collector, or one of the many types of wet scrubbers. Solids are collected in dry form in mechanical systems or as a slurry in a wet scrubber. The resulting effluent is then ready for a disposal technique. At this point it is imperative that we know several things about the gas. These are (1) temperature, (2) volume, (3) chemical constituents, (4) dew point, and (5) permissible atmospheric tolerance levels for the constituents in the gas. With a knowledge of these factors we can then select the method which is best for the final disposal of the waste.

For many years the first consideration was not one of atmospheric pollution but merely getting rid of the waste gas through a stack into the atmosphere, with the hope that the effluent would not descend in the immediate area. Naturally this method is restricted in its acceptance today. However, it is not without value, and in a good many cases it is possible to vent toxic materials, such as sulfur dioxide, in small quantities through high stacks and maintain ground level concentrations which are below acceptable minimums (see Fig. 1). But high stacks are expensive, and since there is very little doubt that air pollution regulations will become more stringent in the future, careful consideration should be given to this choice of disposal. If the waste gas has organic materials which are combustible, then incineration should be considered as a final method of disposal. Direct flame, thermal, or catalytic oxidation of such wastes can produce an effluent of carbon dioxide, nitrogen, and water vapor which can be vented safely to the atmosphere. Economic considerations are paramount in the selection of incineration systems because of the high fuel costs when concentrations of organic constituents are low.

If the waste gas contains organic materials in relatively high concentrations, then vapor phase activated carbon adsorption may be a means of recovering valuable solvents, while at the same time removing the pollutant from the air at the lowest possible cost.

Fig. 1 High stack dispersion, 700-ft stack.

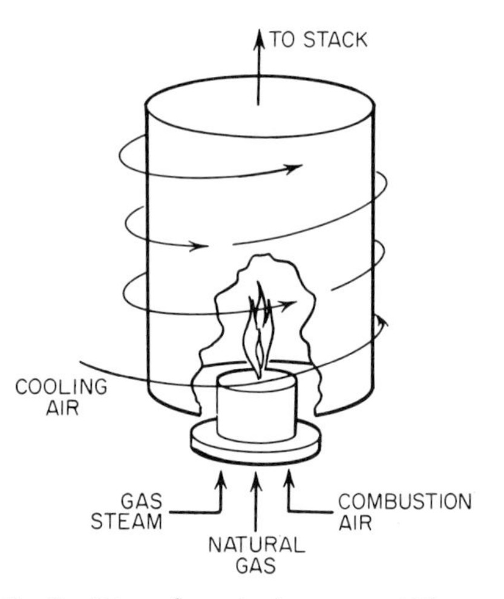

Fig. 2 Direct flame incinerator. (*Chemical Engineering, Oct. 14, 1968, p. 161.*)

Waste Gas Incineration

Assume that we have selected incineration as the best disposal method for a gaseous waste because it contains organic compounds which will rapidly oxidize at high temperatures. Then we must determine which type of incineration is most applicable to our problem. There are three basic types of waste gas incineration systems:

1. Direct flame
2. Thermal
3. Catalytic

Direct flame incineration Direct flame incineration is one method of incinerating waste gas. It is used normally with materials which are at or near their lower combustible limit when mixed with air. It is also used when the waste gas is a combustible mixture itself without the addition of air (Fig. 2).

In a well-designed commercial combustor or burner, gases having heating values as low as 100 Btu/cu ft can be burned without auxiliary fuel. Gases having even lower heating values, which are preheated to 600 or 700°F, often will sustain combustion without additional help from auxiliary fuel. Blast furnace gas is one example of a low heating value fuel which sustains combustion. Here we are talking about direct flame incineration, however, as differentiated from combustion, and while both operations are in fact combustion, in one case we are burning a fuel to utilize the heat, and in the other case, we are burning the waste gas to destroy certain toxic components of the gas. Hydrogen sulfide would be a good example of a waste gas with a heating value high enough to sustain combustion in the pure form when mixed with air, or at least with a significant heat release in lesser concentrations. While burning hydrogen sulfide does not entirely solve the air pollution problem, because it will create sulfur dioxide or sulfur trioxide, depending upon the final temperature, it is one step toward the solution of the problem.

Hydrogen cyanide, which is an extremely toxic gas, may be burned in air; carbon monoxide, which is also a deadly gas and a by-product of many partial combustion reactions, can be burned in this manner. Solvent vapors mixed in high quantities with air may produce an explosive or a combustible mixture which can be burned in a conventional forced draft combustion system.

When the amount of combustible material in the mixture is below the lower flammable limit, it may be necessary to add small quantities of natural gas or other auxiliary fuel to sustain combustion in the burner. But in either case, whether the material burns with or without the assist of auxiliary fuel, combustion occurs at high temperatures above 2500°F, good mixing is achieved with the oxygen in the air, and the resultant products of combustion are carbon dioxide, nitrogen, and water vapor. Here the contaminant, whether it is a solvent vapor or pure gas, is serving as a part of the fuel. It is contributing a significant portion of the total heat release to the system and can be burned with a minimum of auxiliary fuel and therefore a minimum of operating cost.

Direct flame combustion should be employed only where the amount of auxiliary fuel needed to sustain combustion is low and where the contaminant supplies at least 50 percent of the fuel value of the mixture.

Equipment for direct flame incineration may be a conventional industrial burner or combustor and combustion chamber, either forced or induced draft, or it may be a flare type burner as found in many petroleum refineries and petrochemical plants. Flare burners are of two basic types: the ground flare and the elevated or tower flare. The ground flare, as its name implies, is used at ground level where there is sufficient space around the flare for safety purposes to burn waste gas from an oil field operation or similar source. The tower flare, usually found in refineries, is elevated to keep the flame well above the level of surrounding process equipment, protecting the refinery against possible fires. Flares are basically open pipes which discharge a combustible gas directly to atmosphere, and the end of the pipe contains a flame holding device and a continuous pilot or pilots to ignite the waste gas (Fig. 3). Air for combustion is supplied by the surrounding atmosphere. Steam injection is often supplied to the flame of the flare to prevent smoking when burning waste hydrocarbon gases which have more than two carbon atoms. Flares are affected by atmospheric conditions, especially high winds. They cannot be considered an infallible method of waste gas disposal because unburned waste gases often escape from a flare system, but they are expedient and economical for high-volume discharges of combustible waste gases.

Thermal incineration Most waste gas incineration problems involve mixtures of organic material and air in which the amount of organic is very small. This means that if it were injected directly through a burner, along with auxiliary fuel such as natural gas, the amount of natural gas required to achieve complete combustion would be quite high. Most conventional industrial burners require temperatures of 2200°F or greater to sustain combustion, whereas thermal incineration can be carried out at much lower temperatures, sometimes as low as 900°F, but generally between 1000 and 1500°F. These weak mixtures of organic material and air will usually have very low heating values, on the order of 1 to 20 Btu/cu ft. Some of the most common applications may be found in drying ovens which drive off a solvent or plasticizer in low concentrations in air, or from lithographing ovens or other process drying operations. Here it is more economical to heat a combustion chamber, using a conventional fuel in an industrial burner, and inject the contaminated air into this chamber just downstream from the burner flame, or even into the burner flame (Fig. 4). Usually the waste gas is essentially air and therefore contains enough oxygen to complete combustion of the organic contaminant. But in some cases, where sufficient oxygen is not present in the fume, it can be added by means of a fan or blower, either by premixing with the fume or by injecting into the secondary combustion chamber along with the fume.

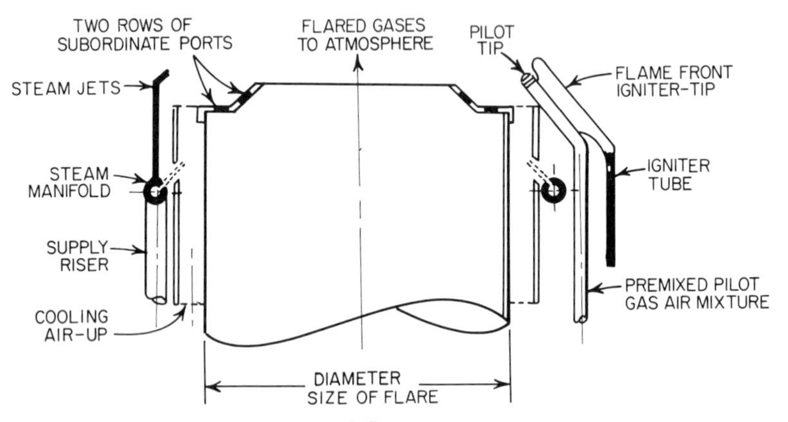

Fig. 3 Typical flare tip construction.

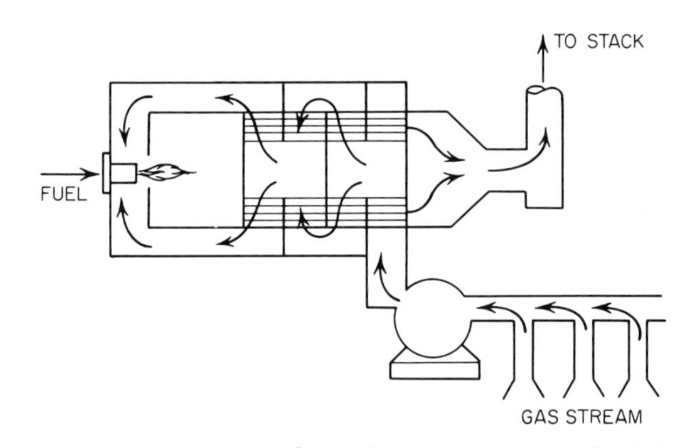

Fig. 4 Thermal incinerator. (*Chemical Engineering, Oct. 14, 1968, p. 161.*)

In such cases, the three "Ts" of good combustion are paramount: time, temperature, and turbulence. The optimum of each of these three should be considered for a good combustion system. The temperature must be high enough to oxidize the organic contaminant. Specifically, this is above the autoignition temperature of the pure contaminant and, for practical reasons in this type of system, usually several hundred degrees above the autoignition temperature (Table 1). The time is the period of residence of the waste gas in the combustion chamber. This will vary, depending upon the temperature and the turbulence, but should be between ¼ and 1 sec for most gaseous wastes. The turbulence is the degree of mixing achieved between the contaminant and air so that oxygen molecules reach each of the organic molecules within the combustion chamber at the required temperature. It can be readily seen that, if the temperature is high, less turbulence and less residence time will suffice. If the temperature is low, longer residence time and better turbulence will be required. If the turbulence or mixing is excellent, a reduction in temperature and residence time may result. While rates of reaction for most chemical compounds and oxygen can be calculated, empirical data for various materials derived from actual test results are the most reliable. Since good turbulence or mixing is the easiest and cheapest of the three factors to supply any combustion system, it should always be considered paramount. Long residence time makes the incinerator larger and more expensive, and higher temperatures utilize more auxiliary fuel, resulting in higher operating costs.

Incineration systems for thermal oxidation of gaseous wastes are of many different types and forms. Some utilize "line" burners when the fume contains sufficient oxygen for its own combustion. Here the waste gas passes over and through the flame of the "line" burner in a refractory lined duct. Other systems utilize an external burner, either natural, forced draft, or aspirating type. In this system

TABLE 1 Autoignition Temperature of Some Common Organic Compounds, °F

Compound	°F	Compound	°F
Acetone	1000	Hydrogen	1076
Ammonia	1200	Hydrogen cyanide	1000
Benzene	1075	Hydrogen sulfide	500
Butadiene	840	Kerosene	490
Butyl alcohol	693	Maleic anhydride	890
Carbon disulfide	257	Methane	999
Carbon monoxide	1205	Methyl alcohol	878
Chlorobenzene	1245	Dichloromethane	1185
Cresol	1038	Methyl ethyl ketone	960
Cyclohexane	514	Mineral spirits	475
Dibutyl phthalate	760	Petroleum naphtha	475
Ethyl ether	366	Nitrobenzene	924
Methyl ether	662	Oleic acid	685
Ethane	950	Phenol	1319
Ethyl acetate	907	Phthalic anhydride	1084
Ethyl alcohol	799	Propane	874
Ethyl benzene	870	Propylene	940
Ethyl chloride	965	Styrene	915
Ethylene dichloride	775	Sulfur	450
Ethylene glycol	775	Toluene	1026
Ethylene oxide	804	Turpentine	488
Furfural	739	Vinyl acetate	800
Furfural alcohol	915	Xylene	924
Glycerin	739		

the flame passes into the duct from the burner mounted in the duct wall, causing turbulence in the chamber, and the contaminated air passes through and around the flame and is heated to the reaction temperature. Such units may be vertical or horizontal and may be induced or forced draft, depending upon the physical arrangement most desirable for the system.

Catalytic incineration Catalytic incineration is applied to gaseous wastes containing low concentrations of combustible materials and air. Usually noble metals such as platinum and palladium are the catalytic agents. A catalyst is defined as a material which promotes a chemical reaction without taking a part in it. The catalyst does not change nor is it used up.

These catalysts must be supported in the hot waste gas stream in such a manner that they present the greatest surface area to the waste gas so that the combustion reaction can occur on the surface, producing nontoxic effluent gases of carbon dioxide, nitrogen, and water vapor. Commercially available catalysts come in many different forms, but those most widely used today are found in four basic configurations. The first is a Nichrome wire screen, where the wire has been coated with a precious metal. This screen, usually random in nature, is placed directly in the path of the effluent gas. A second form of catalyst consists of a number of airfoil-shaped alumina rods upon which the noble metal catalyst has been deposited, and over which the waste gas must pass in the system. A third form is a bed of spheres or pellets which have been coated with the catalytic agent; and the fourth form is an alumina honeycomb on which the catalyst has been deposited. In every case the free area through the catalyst is high, on the order of 90 percent, and the pressure drop across the catalyst is quite low, on the order of several tenths of an inch. Since most waste gases from ordinary industrial processes are at low temperatures up to 300°F, a preheat burner is required to bring these gases up to the reaction temperature.

The advantage of the catalyst is that the reaction temperature in catalytic systems is lower than it is in thermal systems because the catalyst promotes the reaction at a lower temperature. Most catalytic reactions can be carried out at preheat temperatures between 600 and 1000°F. This of course results in a fuel saving when compared with thermal systems but involves a much higher initial investment because of the catalyst cost. Catalytic incinerators usually operate at or below 25 percent of the LEL (lower explosive limit) of the organic material in the waste gas and below the normal oxidation temperature of the contaminant. Care should be taken, however, when analyzing the waste for catalytic combustion, that the waste gas contains a low enough concentration of the organic contaminant to prohibit burnout of the catalyst. Most catalysts are suitable for maximum operating temperatures of 1500 or 1600°F. A high concentration of organic contaminant in the waste gas, even with minimal preheat, may release enough heat on the surface of the catalyst to cause catalyst burnout. Therefore, catalytic systems are most applicable to low concentrations of contaminants where the temperature rise across the catalyst will be on the order of several hundred degrees.

Catalytic systems have been used widely in the oxidation of paint solvents, odors arising from chemical manufacture, food preparation, wire enameling ovens, lithographing ovens, and similar applications.

The physical configuration of the system may be very similar to direct thermal incineration, with the exception of the catalyst section. The catalyst is distributed across the incinerator downstream from the preheating area where the waste gas is preheated by a burner. This mixture then passes through the catalyst, which is generally only several feet deep, and out to atmosphere (Fig. 5). The cross section of the catalyst may be circular, rectangular, or square, depending upon the type of catalyst and the physical limitations of the system.

Catalyst poisons While catalysts do not directly enter into any oxidation reaction, they are not without maintenance. Periodic washing to remove atmospheric dust and dirt is necessary because these materials blind the active surface of the catalyst. In high-temperature ovens, especially paint baking ovens, certain paint pigments and other residues coming from the original coating materials may also have the same effect. The average life of a catalyst, under normal operating conditions, without any poisons in the waste gas is 3 to 5 years. They must then be reactivated. This is done by returning the catalyst and catalyst support to the manufacturer for reclamation. The noble metal is still on the catalyst support and has a definite value. It is much the same as a returnable soft drink bottle and will be credited to the user.

Materials such as phosphorus, silicon, and lead are known to shorten the active life of catalysts. Iron and vanadium are also damaging. Arsenic can also be considered a catalyst poison. Halogen compounds and sulfur compounds, and also halogens in the elemental form, will tend to suppress the functionality of the catalyst and lower its life.

Heat recovery Almost every waste gas incineration system is ideal for heat recovery. Generally, if incineration is selected as the disposal method, particulate matter, if present in the beginning, is removed by mechanical methods and corrosive or toxic gases have been eliminated because they cannot be discharged to the atmosphere. Therefore, after combustion we usually have inert gases containing carbon dioxide, nitrogen, and water vapor, and possibly some residual oxygen. As stated before, the temperature of these mixtures will be from 800°F to possibly 1600°F. While they may be discharged to atmosphere at these temperatures, it is foolish to throw away the heat, especially if there is a reasonable use for it. The most immediate use is to preheat the waste gases prior to incineration, especially if these gases are free of particulate matter and are noncorrosive.

By using the waste heat from the combustion reaction, we can preheat the incoming fumes to a temperature at least halfway between their exit temperature from the process equipment and the incineration temperature. Let us assume that we have fumes at 300°F coming from a process oven and that these fumes are well below the flammable limit. Experience indicates that the organic in the fumes will be rapidly oxidized at 1300°F, and this is the temperature which we should select for incineration. The fumes from the oven pass through one side of the heat exchanger or recuperator, where they are indirectly heated to 800°F. The

Fig. 5 Catalytic incinerator. (*Oxy-Catalyst, Inc.*)

heated fumes then pass into the incinerator, where they are heated the additional 500° to the incineration temperature of 1300°F by direct mixing with the products of combustion from the auxiliary fuel burner. The exhaust gases from the combustion reaction then pass over the other side of the heat exchanger supplying the heat to preheat the fumes, and the gas, finally vented to the stack, has a temperature of around 800°F instead of 1300°F, which would have been the temperature had we not used the recuperator.

It is quite obvious from the above example that the fuel savings in such a system are significant. We have preheated the fumes over a 500° range with the recuperator. If only direct thermal incineration had been employed, it would have been necessary to raise the temperature of these fumes 1000°; so in effect, we have cut the fuel consumption in half by adding the heat exchanger.

The heat exchanger, of course, is a considerable capital investment, and therefore the use of heat recuperation in such a system must be weighed against the cost of fuel and all the other factors which would be part of an economical evaluation.

As a general rule, the addition of heat recuperation to any waste fume incineration system will usually more than double the original cost.

Heat exchangers are not always used to preheat the incoming fumes. In many cases they may be used to preheat air. The preheated air can be used as makeup air for comfort heating in the building, or it can be recycled back to the oven to supply the heat normally required, through either fuel fired burners or electric heaters.

In many cases dual preheating roles are assigned to the waste gas. As mentioned above, it would be possible to preheat incoming waste fumes to 800°F from a 300°F oven discharge temperature, but this would only lower the stack temperature of the waste gas to about 800°F. There is still a considerable source of heat in this waste gas which could be used to heat air for comfort heating in the building, for heating hot water, or possibly for generating low-pressure steam. Steam generation is seldom considered because of the low temperature difference. European equipment manufacturers and process design companies have made far better use of waste heat than we have in the United States, primarily because of the higher cost of fuel. We are prone to disregard heat recovery because of its high initial cost, but often a recuperator can pay for itself within a reasonable amortization period despite our low fuel cost in the United States.

Recuperatuve heat-transfer devices take many forms. Perhaps the most commonly used is the waste heat boiler, which is used to generate low-pressure steam or pressurized hot water by the recuperation of heat from a process. The first type of incineration which we discussed, direct flame incineration, is carried out at a temperature which precludes the self-recuperative type of exchanger and lends itself best to the waste heat boiler approach. Thermal incineration and catalytic incineration are suited to the self-recuperative type system used for preheating fumes or makeup air. These heat exchangers take several forms. They may be plate type exchangers consisting of a series of corrugated metal plates having alternate flow patterns which are at right angles to each other. In other words, the fume to be preheated flows through a series of openings between the plates in one direction, while the high-temperature products of combustion from the incineration process flow through a series of openings which are 90° to the path of the cold waste gas. This exchanger has one very attractive feature. It can provide a high surface area in minimum space. Gas-to-gas heat-transfer coefficients are generally very low; consequently, recuperators for waste gas must contain large amounts of surface and the plate type exchanger provides this surface in a compact configuration. These heat exchangers are fabricated by brazing together a complete assembly of plates to prevent leakage, or by mechanically fastening plates together while providing that any leakage which occurs goes from the hot to the cold side. This prevents contamination of the stack gas. Such heat exchangers are limited in maximum-temperature operation and they are under severe thermal stress because of differential expansion as the incineration temperature increases. Generally, they should be limited to metal temperatures of 800 to 1000°F.

A second type of recuperator is the tubular recuperator. This takes many forms. It can look much like a conventional shell and tube heat exchanger with shell side or tube side expansion joints. It can take the form of a U tube arrangement, which is most acceptable because it has no expansion problems and internal stresses are minimum, or it can take the form of a coiled tube or series of coiled tubes which are designed to absorb their own stresses by the expansion or contraction of the coil. As in the plate type exchanger, heat-transfer coefficients are generally low, and large surface areas are required for relatively small heat-transfer rates. Tube type recuperators generally require higher pressure drops than plate type recuperators.

A third type of heat-transfer device, which finds wide use in fume incineration equipment, is the regenerative type heat exchanger. This can be of either the rotary plate type or the refractory checker wall type. The rotary type has a revolving set of metal plates which first pass through the hot stream absorbing heat and then into the cold stream where heat is desorbed to preheat the incoming waste gases. Such units have been used for years for the recuperation of heat in steel mills and power boilers. A shortcoming of this type is that there is always cross leakage of gas because the seals cannot be built to withstand such temperatures and pressures and still prohibit leakage. In actual practice, leakage from the hot gas to the cold gas side is usually between 8 and 15 percent.

The refractory regenerative heat exchanger is a cyclic system. Here a large refractory checker wall or similar device absorbs heat from the hot gases exiting from the incinerator; and once the refractory is up to temperature, the stream is reversed by means of a switching valve to another mass of refractory in the form of a checker wall or similar device, and the cold fumes are then run in reverse direction through the first or hot checker wall and then into the incinerator. Such systems are usually very large and expensive and involve a myriad of control problems; so they are seldom used as waste gas incineration equipment. Perhaps the most critical aspect of the system is the hot gas transfer valve.

In the field of fume incineration, there are a number of opportunities to recover heat, and we have mentioned some of the general types of exchanger designs. We should not forget, however, that fumes often must be collected from a variety of locations and that in some plants and operations, long ductwork must be constructed to bring the fumes from the process source to the incinerator. Often these fumes can be preheated in the duct which carries them to the incinerator by using a concentric duct arrangement where the products of combustion from the incinerator pass through an annular space between the cold duct and an outer duct wall. This type of arrangement may eliminate the need for a special heat exchanger and at the same time eliminate condensation of vapors in the supply duct. It is a very effective use of available heat-transfer surface at a minimum cost.

Materials of construction Incineration systems and heat-recovery systems may be fabricated from a wide variety of construction materials. The selection of construction materials depends upon several factors: (1) corrosion, (2) strength, and (3) temperature. Most fume incinerators are constructed of carbon steel material and are lined with appropriate alumina refractory to withstand the temperatures of the incineration process. Some catalytic units, however, as well as some thermal incinerators are fabricated without refractory, using only high-temperature stainless steel. The advantage of this approach is the elimination of the refractory, which will eventually need refurbishing or replacement. More costly materials such as Inconel, Incoloy, or Hastelloy are normally utilized only when the waste gas is corrosive to other materials.

The heat exchanger, on the other hand, must be carefully designed for both the operating pressure and the temperature. Most commercial applications today utilize carbon steel for temperatures up to approximately 600°F. Above this, aluminized steel is satisfactory to approximately 1300°F. Stainless steel is required for higher temperatures. Special materials are utilized only when the corrosive properties of the gas so dictate.

Refractories used in incineration systems are generally of the alumina type. Standard or superduty firebrick backed up by insulating brick is suitable in most cases. Castable refractories are also widely used. In short, the refractories which are used for most incineration applications are equivalent to those which would be used for high-temperature furnaces.

Economics Waste disposal of any type does not usually have an economic incentive associated with it. It is a problem which must be dealt with in order to satisfy local laws or to maintain decent relations with our neighbors. Therefore, it cannot be considered in terms of a normal payout period as most process equipment is evaluated. It can, however, be made more palatable by close examination of the economics involved so that we get the most for our money, not only in terms of initial capital investment but in terms of operating cost. In order to determine the least expensive system to be installed, we must know certain things. We must know how much fuel the device is going to use, and have a knowledge of the fuel cost. We must know how much electric power is going to be used and have a cost figure for this. We must know the amortization period which we wish to apply to the equipment and the interest rate for the money that we are going to borrow or spend, and we must have an estimate of the installed cost of the incineration system which we propose. With these figures, we can very easily calculate a total fixed cost and the total operating cost of a system. This is a rather simple task for a straight incineration system whether it be thermal, direct flame, or catalytic. However, when we add heat recuperation, we get into the problem of optimizing the size of the heat exchanger.

Earlier in this chapter, we pointed to a typical recuperator which would recover 50 percent of the waste heat of the system or, in effect, cut our fuel bill in half. It is also possible to design the recuperator to recover 60 or 65 percent of the heat in the system or, if we wish, only 25 or 30 percent. Actually, the more heat which is recovered, the more expensive the recuperator becomes because it requires greater surface area. While the overall heat-transfer coefficient remains approximately the same despite the size, the approach temperature varies and therefore the surface area must vary in order to recover more heat. The effect of surface area on the cost of the heat exchanger will depend greatly on the temperatures involved and the type of tube material which must be used. Two companies with essentially the same problem may find that optimum heat recovery for one is not optimum heat recovery for the other because of different amortization periods. Usually the chemical process industry cannot afford to put in large heat-recovery systems because its amortization period is quite short. Therefore, anything which increases the capital cost of the equipment greatly increases the payout time and the advisability of the investment. On the other hand, many of our basic metal industries can afford to get maximum heat recovery and install recuperative equipment which is designed to reduce fuel cost to an absolute minimum because their amortization period is long and their operating costs have a much greater effect on the total cost picture.

It is difficult to provide accurate figures on equipment costs in days of rising prices and especially in a competitive field such as fume incineration. Some companies fabricate and other companies purchase cheaply constructed marginal equipment, while other companies manufacture and their customers demand rugged, well-designed, expensive equipment. Some requirements dictate unattended automatic operation, while others will permit fully attended manual operation. Because of these factors, it is difficult to pin down absolute costs for this type of equipment, but we can take a typical case as shown in the following table, thermal and catalytic incineration both with and without heat recuperation. Costs for a similar direct flame system would depend on the combustible content of the gas.

Typical Cost Comparison, 10,000 scfm, Waste Gas Incineration System

	Thermal incineration	Catalytic incineration
Incinerator only	\$15,000–\$20,000	\$20,000–\$25,000
Fuel required (1,000 Btu natural gas)	10,000 scfh	6,700 scfh
Incinerator with 50% heat recuperation	\$30,000–\$45,000	\$35,000–\$45,000
Fuel required (1,000 Btu natural gas)	5,000 scfh	3,400 scfh

Comparison based on waste solvent air mixture at less than 25 percent of LEL, 300°F inlet.

Optimization of the heat-exchanger size is obviously a job for the computer, and many companies in the design and fabrication of incinerators and recuperators are developing computer programs for the optimization of the heat exchanger.

Typical applications Thermal and catalytic systems may be used for a variety of waste gas problems, but they have had excellent results in processing waste gases from the following processes:

- Wire enameling
- Resin manufacturing
- Potato chip drying
- Rendering plant effluents
- Coil and strip coating
- Phthalic anhydride manufacture
- Lithographing and printing
- Metal decorating
- Bonding operations
- Paint and varnish manufacturing
- Rubber manufacturing

Odor Control

Another aspect of the disposal of waste gases which might more logically be considered treatment is the control of their odors. We have already discussed incineration as a possibility for eliminating toxic compounds by oxidation, and the same is true of nontoxic compounds which are aesthetically undesirable because of their odor. If such materials are used in a process where they may volatilize in small quantities into air, the odor of these compounds is carried into the adjacent area. The odor may be present only in the process plant, or it may transcend the boundaries of the process plant and affect other industrial or residential neighbors.

If the waste material is toxic in nature, there is a definite requirement to incinerate or scrub out the toxic compound, but where the odor is nontoxic and often in low concentrations, a solution to the problem may be odor counteraction.

For many years before deodorants were invented, colognes and perfumes were in vogue. While their use today is to make the wearer have an attractive odor, in days past perfumes were used to overcome some of the more objectionable malodors of the human body. Incenses were burned in churches and other public places often to achieve the same results. In the same way today we can use an industrial odor counteractant to blanket, hide, or counteract odors which are aesthetically undesirable because they work on our olfactory nerves in such a way as to cause discomfort or possibly nausea and insomnia.

Odor intensity is a difficult thing to measure because there are so many different types of odors. Therefore, it should be sufficient to say that a disagreeable industrial odor is one which smells bad to one or more individuals, and if this odor crosses the boundary of the plant in which it is produced, it will undoubtedly cause complaints from the neighbors. It is difficult to say when an odor is pleasant or when it is unpleasant because it depends upon the decision of the recipient. Psychiatrists admit that the dislike or like for an odor often depends on the association of that scent with pleasant or disagreeable experiences. Several other interesting facts about odors are that weak odors are not perceived in the presence of strong ones, that odors of similar strength may blend together to produce an odor in which the original components are unrecognizable, and that a constant intensity of a certain odor may cause an individual to lose his sense of olfactory perception of that odor. For example, anyone who has smelled hydrogen sulfide would hardly have difficulty identifying it on a second occasion. Yet in strong concentrations it is a highly toxic and lethal gas, and deaths have been caused because of unawareness of a hydrogen sulfide leak because only seconds after the first noticeable odor, the olfactory nerve is paralyzed and the odor sensation disappears.

Odor counteraction is based on the premise that two or more odor substances may cancel each other, meaning that if they are properly mixed, the result will be an odorless gas.

In determining the proper use of odor counteraction methods, it is first necessary to classify the odors and identify the compound which is causing the odor. Six basic categories which have been suggested for the description of odors are spicy, fruity, resinous, flowery, decomposition, and burnt. The spicy odor might suggest cloves or allspice or lavender or possibly camphor or some similar cyclic organic compound. Materials such as acetaldehyde will be described as having a fruity odor. Resinous odors would be pine tar, asphalt, etc. Flowery or sweet odors would be material like ethyl ether or acetone. The odors of decomposition are readily available in composting operations and landfill, and burnt odors, of course, are self-evident.

Once the odor has been identified, we must determine where and why it is causing a problem. Two factors will be important in this determination. The first is the atmospheric condition present at the time and the point of discharge. High humidity has a tendency to make odors drop quickly to the ground, or high winds and low humidity may dissipate the odor without resulting in any complaints. Natural air dilution in many cases will change the characteristic of the odor and also the intensity, and some materials will be automatically oxidized in the normal air dilution which comes at the point of discharge. Reemphasizing the previous premise, the complaint on the basis of an odor will usually come when it is first released to atmosphere, since the olfactory sense becomes fatigued after long exposure. Thus residents near a plant which has been emitting an odor for a long period of time will tend to disregard or ignore the odor, whereas a casual visitor to the area may find the odor objectionable. Odor will always travel downwind; so complaints should never come from the upwind side of the source.

For every odorous material there is a concentration in air below which the odor is indiscernible. This is known as the olfactory threshold, and it is the level at which identification of the odor begins. As the concentration of the odor-causing compound increases, usually the intensity of the odor increases but not proportionately. A much higher increase in concentration is generally required to cause an increase in detection. There are a few people who have the ability to detect odors in very low concentration ranges of 1 ppm or less. Animals and many insects are much more sensitive to odors than the human. The relationship between odor intensity and concentration is expressed by the Webber-Fechner equation, which is

$$P = K \log S$$

where P = odor intensity
K = constant
S = odor concentration

Odor detection, assuming we are talking of nontoxic organic materials, is usually a function of the olfactory nerve of the human being most directly affected, and the level of odor concentration may therefore be rated from no odor to a very strong odor, with intermediate points on a scale of 0 to 100 or 1 to 5, but it usually depends on agreement on the intensity of the odor by more than one person. Once it has been established that such an odor exists, we must devise methods for counteraction.

We said that strong odors tend to mask weaker ones; then if we were to inject a chemical with a strong odor into a waste gas containing a chemical with a weak odor, we would probably smell only the stronger one. This would be known as odor masking. In the same way, certain odors in appropriate relative concentrations will tend to destroy the odorous properties of both, and the intensity of each is diminished. This is called counteraction or neutralization. Odor counteraction as it is used today is a scientific approach to the problem, and is based on the significant work done in the past by Zwaardemaker. Many industrial odors can be handled at the source by means of combustion, adsorption, filtration or aeration, or many of the other available methods previously discussed. However, there are times when complex industrial waste odors find their way to the atmosphere in concentrations high enough to become objectionable. These odors often come from bodies of water, landfill operations, ovens, storage tanks, and a variety of other places. They cannot be easily captured and ducted through a black box, which we call an odor counteraction unit, so that our real task in odor counteraction is generally to improve the odor of the surrounding area so that when the malodor crosses the fence into the neighbor's yard, it is undetectable. In many cases, the source of the odor and the complaining recipient of the odor may be several miles apart, depending upon atmospheric conditions. Since odors themselves are for all practical purposes always gases, the material which is used to counteract the odor must also be vaporized into the atmosphere, carrying the objectionable odor. This must be done by atmospheric dispersion. Commercial odor counteractants are vaporized into the air in much the same way as we would vaporize fuel in a burner, that is, through an atomizing nozzle. Drums of the chemical counteractant material are usually located at ground level in the area of the malodor; and through specially designed air atomizing nozzles, the liquid counteractant is finely atomized into the air. Such installations are not limited to ground level; they may be located on the roof of the manufacturing plant or at any point where the malodor is first released. Such systems are very easy to install and require a minimum of equipment. Only small amounts of the counteractant material are used, since usually the concentration of the material causing the odor is in very small quantities.

There is no universal odor counteractant. Each counteractant formula is designed for specific odors or groups of odors. The counteractant also is not designed to desensitize the human olfactory apparatus. Without the use of the sense of smell, an individual might be subjected to dangers from odors of toxic compounds which he would otherwise be able to identify. While the atomization of odor counteractants is a normal method of dispersion to the atmosphere, many liquid waste lagoons emit odors from the surface which can be better handled by a mixture of the counteractant and a light paraffin based oil applied to the surface in a thin film.

A similar approach is used on solid wastes, especially in landfill operations. The counteractant is actually sprayed onto the surface of the landfill to counteract the decomposition odors.

It can readily be seen that odor counteraction methods have their place in the final disposal of nontoxic malodorous gases but that the proper selection of counteractants is an art developed by several companies in the field rather than a scientific system.

There are few parameters to use for cost determination in odor counteraction methods because the type of counteractant and the quantity depend largely on the problem to be solved. Odor counteractants are expensive, but a little bit goes a long way. Counteractants used for diesel engine exhaust odor control will cost about $0.01 per 5 gal of oil, but the counteractant costs between $1.50 and $2 per pound.

Permanganate oxidation Another treatment method which bears mention here, even though it is not actually a disposal method, is the oxidation of odors by potassium permanganate solutions. $KMnO_4$ is a strong oxidizing agent, and it will attack oxidizable material under various conditions. When dilute solutions (about 1 or 2 percent) of permanganate are used in conventional low-energy scrubbing apparatus at a controlled pH level which is slightly basic, the oxidizing capability of the permanganate will greatly reduce the odor of various malodors.

Rendering plant odors, asphalt plant fumes, fish processing odors, sewage plant odors, and odors from many odorous chemicals can be substantially reduced or eliminated using this method.

The reaction is basically $2KMnO_4 + H_2O = 2KOH + 2MnO_2 + 3(O)$. The oxygen is not released in molecular form but immediately combines with the oxidizable contaminats. For example, the reaction with SO_2 would be $2KMnO_4 + 3SO_2 + 4KOH = 2MnO_2 + 3K_2SO_4 + 2H_2O$.

Permanganate treatment is not widely used. It is an expensive method compared with other oxidation methods if there is any substantial amount of contaminant in the waste gas; however, it can be economical for treating dilute odorous fumes. Permanganate solutions have also been used as a surface deodorant in barnyards or cattle feed lots to oxidize manure odors.

Stacks and Chimneys

There are very few things that can be done with a waste gas after it is treated. For purposes of this chapter, we stated that incineration was a final disposal method even though it may be considered a treatment method, and we further maintained that odor counteraction was a disposal method whereas it is truly a treatment. In the final analysis, after we remove any particulate material, and in some cases before we do, we must release the gas to the atmosphere. In order for gas to be released to the atmosphere, it should be relatively free of particulate matter, toxic compounds, and odorous compounds, although practice is usually far from this ideal. Air pollution laws in most states throughout the United States and many foreign countries are becoming so stringent that the use of the dispersion method for getting rid of toxic gases or those which contain high concentrations of particulate matter must soon be abandoned. The stack or tall chimney, however, has not completely lost its worth to industry if it is used with discretion.

Perhaps first we should differentiate between the casual term "stack" and a stack or chimney which is actually used for dispersion. A stack has come to mean any vertical vent pipe which discharges gases to atmosphere from any one of a thousand different processes. The gases may be warm or cold; they may have high exit velocities or low exit velocities, and the stack is merely a means of getting them out of the operating area into the atmosphere where they can be diluted and dispersed. These are certainly not stacks with any scientific method in their design. Some are designed as a forced draft system with a blower to pressurize the stack, while other stacks may be tall enough to provide a very small induced draft at their base. Such stacks are not designed for the dispersion of materials and should be considered only an extension of a duct or flue.

Tall stacks or chimneys have both a historical use and a present value. Tall stacks were designed to provide natural draft for combustion processes, but with no concern for the safe atmospheric dispersion of the various waste gases which pass through them. In fact, in most situations, the stack was designed to bring the smoke from the furnace out of the building instead of having it back up into the building. Today, however, we can see a different use for high stacks as an ultimate disposal method for certain wastes which would otherwise be difficult to treat. Stack heights throughout the world have been increasing in height for the purpose of providing better dispersion. One utility company stack measures 1,206 ft in height. Stacks in the Ruhr Valley of Germany are 500 to 600 ft in height. All these cost *big* money but obviously no more than a SO_2 cleanup system.

The purpose of a high stack is to convey waste gases containing either toxic or particulate material to a point high enough above the ground level so that, after normal dispersion at the top of the stack, the ground level concentration of any of the contaminants will be well below permissible levels for the contaminant. This applies to concentrations not only at the base of the stack but at any distance from the base of the stack. Tall stacks on the order of 300 to 400 ft are usually designed to reach above the inversion layer and to disperse the smoke or waste gas at a point where it will not be trapped by the inversion layer and can be dispersed into the atmosphere. Obviously, the taller we build the stack the greater will be the dispersion of the effluent and also the greater the cost.

Temperature inversion has a significant effect on stack operation. Normal atmosphere should decrease 1°C in temperature for every 100 m rise. This is known as the adiabatic lapse rate. On a clear day, solar radiation heats the land, causing a greater lapse rate per 100 m of height. This is known as a superadiabatic lapse rate. On a clear night the land is cooled by radiation and the lapse rate may become zero or even produce a higher temperature at a height increase. This is known as a temperature inversion.

The next morning, as the sun begins to heat the land once again, the air temperature also rises. The stable air in the inversion layer will cause any stack plume from a stack that does not extend through the inversion layer to descend to the ground rather than to rise, producing ground level concentrations ten to twenty times the predicted level (Fig. 6).

To determine the acceptability of a stack as a means of disposing of a waste gas, we must first determine the acceptable ground level concentration of the toxic constituents or the particulate matter in the waste gas. We should also have some idea of the topography of the area so that we can locate the stack properly with respect to buildings and hills which might introduce a factor of air turbulence into the operation of the stack. We should also be aware of the meteorological conditions prevalent in the area, such as prevailing winds, humidity, and rainfall. Finally, we should have an accurate knowledge of the constituents of the waste gas and its physical properties.

Determination of the permissible ground level concentration of the particular gaseous waste, or a particulate constituent in the waste gas, is generally available from literature and will vary from state to state and community to community, depending on air pollution regulations. Sulfur dioxide, for example, is usually limited by pollution regulations to concentrations below 0.5 ppm, while hydrogen chloride might be permitted in many areas up to concentrations of 5 ppm. While both these materials are certainly toxic in nature, recovery of such small concentrations is extremely expensive, and as long as safe dispersion methods can be used, they will probably be acceptable.

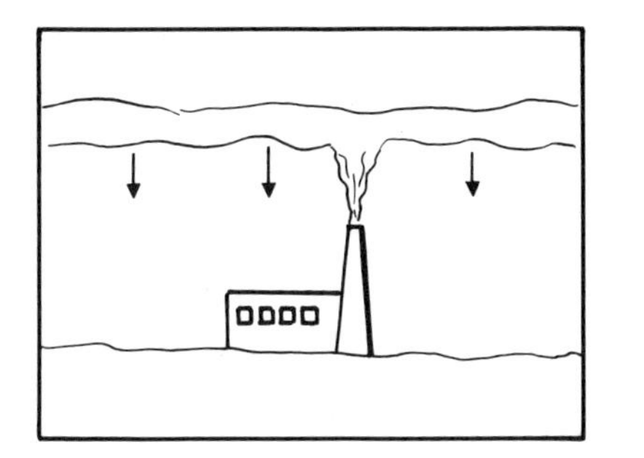

Fig. 6 Stack discharge below an inversion layer.

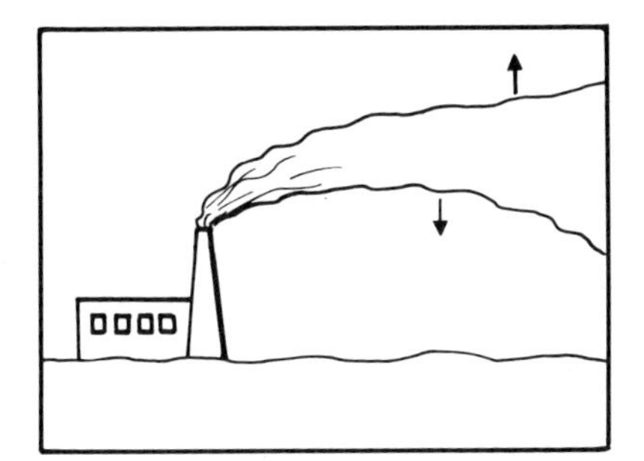

Fig. 7 The effect of thermal turbulence on stack operation.

A stack plume is usually conical in nature, having an angle of about 14° from the source. As the cone expands, it touches the ground at some distance from the base of the stack (Fig. 7). Since turbulence and diffusion affect the ground level concentration of the contaminant at any point, we need a convenient equation to estimate this concentration. It can be expressed by the Bosanquet-Pearson equation

$$C_0 \max = \frac{2.15\, q_A\, 10^5}{u H_e^2} \frac{p}{q}$$

The value of p/q, the diffusion parameter, is 0.63 and

H_e = effective stack height, which is the sum of the physical stack height plus the plume height resulting from the discharge velocity and buoyancy of the gas

$C_0 \max$ = maximum ground level concentration of the contaminant, ppm

q_A = emission rate of gas contaminant at ambient temperatures, fps

u = wind velocity, fps

The average distance from the stack to the point of C_0 max is about $10H_e$. To design a stack that will conservatively provide a stack to provide the desired C_0

max, we use the Davidson-Bryant equation

$$\Delta h - d\left(\frac{V_s}{u}\right)^{1.4}\left(1 + \frac{T}{T_g}\right)$$

where $\Delta h = h_V + h_t$
$H_e = h_s + h_V + h_t$ = effective stack height, ft
d = inside diameter of stack, ft
V_s = stack velocity, fps
T = temperature difference of effluent gas over ambient, °C
T_g = stack gas temperature, °K

Combining the Bosanquet-Pearson and the Davidson-Bryant equations, we get

$$h_s = 3.7 \times 10^2 \sqrt{\frac{q_A}{u\, C_0 \max}} - \left[d\left(\frac{V_s}{u}\right)^{1.4}\left(1 + \frac{T}{T_g}\right)\right]$$

where h_s is the actual stack height.

The effect of adjacent terrain and buildings on stack operation is one which is very difficult to predict. Many pollution authorities have run studies, using large models to help to determine the possible effects if a tall stack is constructed in a particular area. Inversion layers which occur in particular locations at certain times of the day should be carefully studied, and a stack to be located in an area where an inversion layer is a common occurrence should be so designed that it extends through the inversion layer. Otherwise the contaminants emanating from the stack will be dispersed along the bottom of the inversion layer and dropped to earth in higher than permissible concentrations. In all cases, the concentration at ground level should be based on the worst possible situation, especially where toxic elements or compounds are concerned, to avoid situations similar to the occurrence at Donora, Pa., in 1948, where smelter fumes caused a number of fatalities, or the deadly London smog which caused 4,000 deaths from SO_2 in 1952.

Brink and Crocker have suggested several rules of thumb for the design of any stack. These are:

1. The stack should be 2½ times the height of surrounding buildings or the surrounding countryside so that significant turbulence is not introduced by these factors.

2. The gas injection velocities from the stack should be greater than 60 fps so that the stack gases will escape the turbulent wake of the stack. In many cases it is desirable to have the gas exit velocity on the order of 90 or 100 fps if possible. There is a critical wind velocity for every stack exit velocity. Above this critical velocity the wind shears off the gas as it leaves the chimney and there is no corresponding rise of the waste gas due to the exit velocity. Then gas temperatures and flow rate no longer affect the ground level concentration, for which the exit velocity was initially designed. Figure 8 shows certain gas exit velocities vs. the critical wind speed.

3. Gases from stacks with diameters less than 5 ft and heights less than 200 ft will hit the ground part of the time, and ground concentrations may be excessive. In effect, we are saying that for these conditions, stack design is unpredictable.

4. The maximum ground concentration of stack gases subjected to atmospheric diffusion occurs about 5 to 10 stack heights from the stack in a horizontal direction. By this we mean that the maximum concentration will not be found at the base of the stack but at some distance equivalent to 5 or 10 stack heights away from the stack.

5. When stack gases are subjected to atmospheric diffusion and building turbulence is not a factor, ground level concentrations on the order of 0.001 to 1 percent of the stack concentration are possible for a properly designed stack. This is a wide range and gives a fair testimony to the degree of science involved in the design of stacks.

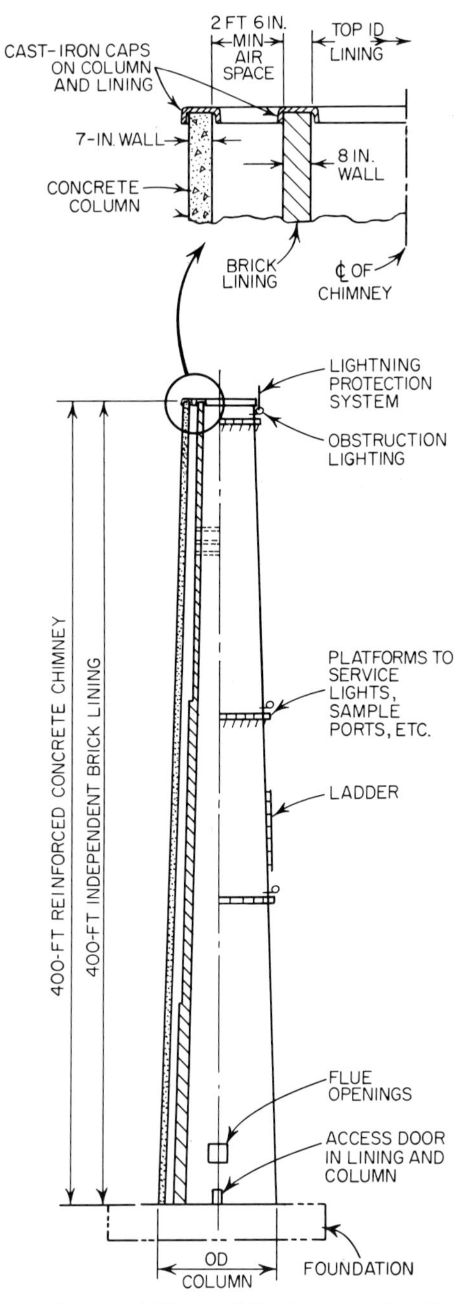

Fig. 8 Stack exit velocity. (*Chemical Engineering, Oct. 14, 1968, p. 168.*)

6. Ground concentrations can be reduced by the use of higher stacks. The ground concentration varies inversely as the square of the effective stack height.

Stack designs in the past have been based largely on empirical data, and along with the many trials there have been many errors. Using these rules of thumb, adequate stacks usually can be designed. However, the stack is never operated at precisely the design conditions because process fluctuations cause changes in the volume flow. In some cases it may be necessary to add air to the stack or chimney in order to keep the exit velocity at the optimum rate. In many cases stacks are designed for future conditions with no thought given to the low exit velocity, which will result in present operations producing ineffective results.

In crowded industrial areas, multiple stacks within a short distance from one another can cause a myriad of problems in ground concentration of various contaminants. For example, two stacks emitting sulfur dioxide, located 1 mile apart, will undoubtedly create ground concentrations of sulfur dioxide at some point which will be higher than either stack would have created by itself. Because much

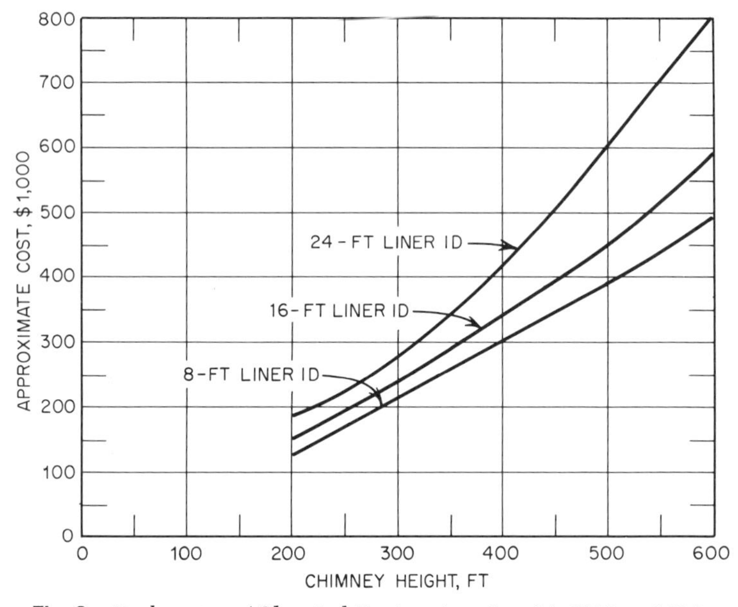

Fig. 9 Stack costs. (*Chemical Engineering, Oct. 14, 1968, p. 169.*)

of this information is empirical in nature, many companies are now devising computer programs for the development of safe stack design.

Materials of construction The design of a stack involves many different materials of construction. Many small-diameter stacks are constructed entirely of steel or stainless steel and are held in position by guy wires. In the chemical process industry, small-diameter stacks several feet in diameter have been installed up to 175 or 200 ft high. By proper guying and forced draft by means of a blower, such stacks can be made acceptable. But generally, when we are talking about a stack for dispersion, such as we would use for sulfur dioxide or similar waste gas, we are forced to consider a lined chamber, usually constructed of steel and lined with firebrick or suitable castable material which will withstand the temperature of the exit gases. Extremely tall chimneys are designed with two major sections, the liner and the column or outer shell. The liner is designed to withstand the temperature and the corrosive conditions of the gas, while the outside column, which may be either steel or masonry, protects and supports the liner. Large stacks are always free standing and consequently are much larger in diameter at the base than at the top.

Since tall chimneys or stacks do not really solve the pollution problem, but just make it livable, there is some question as to the future of this approach. They do not reduce the amount of pollutant released to the atmosphere by one gram. Undoubtedly they will be used for some years to come, but as pollution laws become more stringent, dispersion methods will be eliminated. There is, however, another factor which will gradually cause the disappearance of tall stacks, and this is the cost. Vent stacks which are entirely fabricated of steel and have no refractory lining will cost approximately \$1 to \$2 per lb. The addition of refractory lining to this type of stack increases the cost approximately by a factor of 2. These two forms of stacks or chimneys are relatively inexpensive and can be erected quickly with the help of a large crane. The self-supporting refractory lined stack, however, gets to be considerably more expensive. These usually consist of a carbon steel shell lined with either castable material or firebrick. The diameter of the stack at the base is often dictated by the load bearing requirements rather than the flow, but usually stacks of this type are built up to 200 ft in height and with an outlet diameter up to about 4 ft. Costs vary from \$200 to \$400 per ft, depending on the amount and type of refractory lining used.

Finally, the cost of high chimneys such as will be found in big smelting operations is shown in Fig. 9. Most stacks require some degree of field erection, even though they may be fabricated in modular sections and brought to the job site.

SO_2 and Nitrogen Oxides

There are several troublesome gases which contribute greatly to the difficulty in solving the air pollution problem today. Despite methods for absorption and incineration, which remove hydrocarbons from air, there are several "bad actors" which are especially difficult to treat or destroy.

The first is SO_2. This is created through the combustion of any sulfur bearing material from elemental sulfur and H_2S to fuel oils containing small amounts of sulfur. The threshold value for SO_2 is 5 ppm, and normal ground level concentrations should be kept at 0.5 ppm or less. High power plant stacks can keep ground concentrations in this range, but lead, zinc, and copper smelting operations contain too much sulfur to rely on stacks alone, and new treatment methods are being tried to eliminate this toxic gas. None have shown any real promise of economic feasibility, but they may be necessary if we intend to continue the use of fossil fuels and smelting operations.

A mixture of sodium and alumina oxides has proved capable of removing 90 percent of the SO_2 from coal or oil fired furnaces. Absorption of the SO_2 forms sodium sulfate. The spent adsorbent is then regenerated by contacting it with a reducing gas such as natural gas, hydrogen, or producer gas. This forms H_2S, which can then be processed to recover sulfur through the Claus process or similar sulfur recovery systems. The by-product sulfur can be sold, thus defraying the cost of the absorbing equipment.

Recovery of 90 to 95 percent SO_2 is possible by alkaline scrubbing with lime water or caustic. Lime is preferable because the calcium sulfate is less soluble and will precipitate.

A regenerative sodium sulfite process removes sulfur dioxide with aqueous sodium sulfite or bisulfite. The scrubbing liquor is then treated with ZnO, precipitating $ZnSO_3$ and NaOH. Calcination of the precipitate releases the SO_2, which can be used to manufacture H_2SO_4.

Ammonia can be used as an adsorbent, removing 90 percent of the SO_2, but the cost of ammonia or ammoniacal liquor is high.

Catalytic oxidation to SO_3 with subsequent conversion to H_2SO_4 will convert about 90 percent of the SO_2. It usually requires an electrostatic precipitator to remove the acid mist.

The Reinluft process from Germany adsorbs SO_2 on char formed by calcined coal at about 1100°F. The adsorbent can be reactivated by treating the char with H_2SO_4 and subsequent evaporation of the acid. The char is desorbed by heat, giving off a gas containing 40 to 50 percent SO_2, which can be liquefied or used to make H_2SO_4 or sulfur.

The other troublemakers are the oxides of nitrogen, primarily NO and some NO_2. Control of these can best be accomplished at the source. If the source is a fuel fired system, a combination of low excess air combustion and reduced flame temperatures will reduce the nitrogen oxide levels. There are no viable adsorption methods for nitrogen oxide removal at the present. Catalytic oxidation to promote the reaction between a hydrocarbon (natural gas) as the fuel and the nitrogen oxide as the oxidant has been successful when high nitrogen oxide levels are present. This method has been used to destroy tail gas fumes from nitric acid plants where the NO_x concentration is 5,000 ppm or more.

DISPOSAL OF LIQUID WASTES

When discussing the disposal of industrial gaseous wastes, we emphasized that the major problem was the large volume of waste which had to be handled in contrast to the amount of contaminant in the waste. Liquid waste, however, usually presents a different problem. While volumes of liquid industrial waste may be large and the percentage of contamination small, the major problem is the disposal of the variety of contaminants present. Industry has a historical tendency to combine all its liquid waste in a single area and subject it to a common treatment. This philosophy is found today in many large industrial complexes. It probably occurred because sanitary engineers familiar with municipal sewage systems were assigned the problem of waste disposal in the industrial plant. Since a municipal sewage system takes all sanitary sewage through a central treatment facility, a similar solution was naturally developed for the industrial facility. The point is that by handling industrial waste in this manner we often make a grave mistake. We combine a wide variety of wastes which treated individually might present simple disposal problems, but when combined they often create an insurmountable dilemma, even for the sanitary engineer.

As with gaseous wastes, we must first define what we consider to be disposal systems vs. treatment systems, and this line of demarcation is not too clear. There are many possibilities or options open to the engineer trying to solve a liquid waste disposal problem, some of which may be considered either treatment or disposal, but let us first consider the total problem. Assume that we have a liquid waste which contains solids, inorganic toxic materials, or organic materials and which for one or all of these reasons cannot be dumped into the nearest sanitary sewer or river. There are certain treatment methods which we can use to improve the conditions of this waste so that ultimately we can dispose of it in a satisfactory manner.

The first consideration should be the removal of solid material, and this can usually be handled in some type of filtration or centrifugation apparatus or system. The solids from the centrifuge or the filter can then be handled as a solid waste and be either buried, burned, or reclaimed. The resulting liquid can be concentrated to remove dissolved solids in a crystallizer or distilled to remove a solvent fraction. Liquid-liquid extraction or adsorption might also be employed to remove certain compounds. If the liquid is high in biodegradable materials, standard sewage treatment systems may be used. Neutralization with either caustic or acid may bring the pH to acceptable levels. Ion exchange can be employed to remove metallic contaminants and recover them for reuse, but when the liquid waste has finally been treated by one or several of these methods, either it must be suitable for returning to the water resources of the area or it must be disposed of in a way which will not cause harm to plant or animal life. It would therefore appear that there are several final disposal methods which can be considered suitable. All of them are in use today. These are

1. Incineration
2. Dilution

3. Land disposal
4. Deep-well disposal

Incineration

Incineration is one possibility for the destruction of liquid wastes, and with properly designed equipment, it can be made to function with a clean effluent in strict accordance with air pollution regulations. It may seem redundant to classify liquid waste into two types from a combustion standpoint: (1) combustible liquids, and (2) partially combustible liquids. Obviously noncombustible liquids cannot be treated or disposed of by incineration. The first category would contain all materials having sufficient calorific value to support combustion in a conventional combustor, burner, or other apparatus and would give products of combustion of carbon dioxide and water vapor when burned. The second category would include materials that would not support combustion without the addition of auxiliary fuel and would have a high percentage of noncombustible constituents such as water. A partially combustible waste may also contain material dissolved in the liquid phase which, if inorganic in nature, will form an inorganic oxide upon combustion and require further collection apparatus.

We must assume that either of these types of wastes is primarily organic in nature, even though the quantity of the organic material may be small. Incineration of such materials becomes essentially a straightforward combustion problem in which air must be mixed with the combustible at some temperature above its ignition temperature, causing rapid oxidation to occur and producing an effluent of carbon dioxide, nitrogen, and water vapor. As in waste gas combustion, the three Ts of good combustion are also of prime consideration here. Since we are starting with the waste not as a gas but rather as a liquid, we must supply the necessary heat for vaporization of the liquid in addition to raising it to its ignition temperature. Since liquids vaporize and react more rapidly when finely divided in the form of a spray, atomizing nozzles are usually employed to inject waste liquids into incineration equipment whenever the viscosity of the waste permits atomization. There are many wastes which might be classified liquid which are hardly liquid in nature. Slurries, sludges, and other materials of high viscosity can be handled in incineration systems but ones that are special in nature. These will be discussed later.

Combustible wastes In order that a liquid waste may be considered combustible, there are several rules of thumb which should be used. The waste should be pumpable at ambient temperatures or capable of being pumped after heating to some reasonable temperature level. By reasonable level we usually mean 400 or 500°F, since pumping a hot tar or similar material at higher temperatures is difficult. The liquid must be capable of being atomized under these conditions. If it cannot be pumped or atomized, it cannot be burned as a liquid but must be handled as a sludge or solid. Liquid waste incineration generally involves liquids having viscosities up to approximately 1,000 SSU, although lower viscosities are desirable.

In order to be considered a combustible waste, the material must sustain or support combustion in air without the assistance of an auxiliary fuel. This means that the waste will generally have a calorific value of 8,000 to 10,000 Btu/lb or higher. This does not mean that liquids with heating values below 8,000 Btu/lb will not sustain combustion by themselves, but as a general rule the borderline between combustibility and noncombustibility is a heating value of this approximate magnitude. Below this calorific value, the material would not exhibit properties which would enable it to maintain a stable flame in a commercial combustor or burner. Materials which fall into this category are light solvents such as toluene, benzene, acetone, ethyl alcohol, and heavy organic tars and still bottoms similar to residual fuel oil. The wastes may be combinations of both, which would give a mixture having an intermediate viscosity and heating value. These wastes come from cleaning operations in chemical plants and refineries or are the residues from distillation processes. The materials are usually not recovered for economic reasons, and therefore they may be burned.

The equipment which is used to handle this type of waste can also vary from manufacturer to manufacturer, but its basic form will be that of a combustor or burner designed to handle a liquid waste through a steam, air, or mechanical atomizing nozzle (Fig. 10). If the system is properly designed and sufficient open area is provided around the combustor, no secondary incinerator is required; in fact, even the stack may be omitted. Generally this involves forced draft types of burners or combustors having short flame characteristics and exhibiting stable burning conditions which will be unaffected by atmospheric conditions such as wind or rain. Such combustors must complete combustion within the device without the need for a secondary combustion chamber.

Where open space is not available, a similar combustor or a more conventional natural draft burner can be installed in a dutch oven or furnace containing sufficient combustion volume for the complete destruction of the waste. High heat release combustors require minimal secondary incineration chambers, but usually incineration is carried out in combustion chambers having volumes which provide for a heat

Fig. 10 Vortex burner. (*Thermal Research & Engineering Corp.*)

release of 25,000 Btu/(hr)(cu ft) of combustion volume. Residence times within an incinerator burning liquid waste will vary from 0.5 to 1 sec.

This secondary or primary combustion chamber, as the case may be, is usually cylindrical in shape and may be used in a vertical or horizontal arrangement. The vertical chamber has the advantage that the incinerator acts as its own stack, but obviously it is not well adapted to a tall stack arrangement (Fig. 11). Horizontal incinerators can be more easily connected to tall chimneys or stacks.

Many combustible liquid wastes can be utilized as fuel for a boiler, air preheater, or other heat recovery device which can turn the waste heat energy from the incineration system into profit. Heat recovery devices, however, are advisable only when the amount of heat recovered and the cost of the recovery equipment can be economically justified. If the waste liquid should contain noncombustibles such as inorganic salts or materials, which would be converted into corrosive compounds in the combustion reaction, such as chlorides or fluorides, then heat recovery is usually incompatible with this system and should not be considered.

While costs for this type of equipment will vary widely, depending on the manufacturer and the type of system employed, and also the waste to be handled, it should be possible to purchase adequate incineration equipment for a truly combusible waste at a figure of between $200 and $350 per gph capacity. In this case we are assuming that the wastes are completely combustible and that no auxiliary fuel, except for pilot operation, is required.

Partially Combustible Waste

Liquid wastes having a heating value of below 8,000 Btu/lb can be considered in the partially combustible category. We must again emphasize that this is a rule of thumb and that some materials as high as 10,000 or 11,000 Btu/lb will not sustain combustion by themselves. It is also important with this type of waste that the material handling method be compatible with the equipment selected. Viscosities should be reduced to the point where the material is pumpable and atomizable at either ambient or slightly elevated temperatures.

Fig. 11 Vertical liquid waste incinerator. (*Thermal Research and Engineering Corp.*)

Waste material in this classification is often aqueous in nature, consisting of organic compounds miscible with water. Such waste may also contain sulfur compounds, phosphorus compounds, or combinations of organic and noncombustible inorganics. These materials may have enough organic content to exhibit visible combustion in a high-temperature furnace, or they may be so low in combustible material that no visible combustion is apparent.

There are several basic considerations in the design of an incinerator for a partially combustible waste. First, the waste material must be atomized as finely as possible to present the greatest surface area for mixing with combustion air. Second, adequate combustion air to supply all the oxygen required for oxidation or incineration of the organics present should be provided in accordance with carefully calculated requirements. Third, the heat from the auxiliary fuel must be sufficient to raise the temperature of the waste and the combustion air to a point above the ignition temperature of the organic material in the waste. Unlike the combustible waste, which sustains combustion by itself, this waste may not always be injected through the combustor or burner but may rather be atomized into the secondary combustion chamber. If the waste material is marginal in combustibility, it may be fed directly through the burner or combustor along with the auxiliary fuel. Temperatures of 2200 to 2700° will result, complete combustion of the organic in the waste will occur, and the products of combustion can be vented to the atmosphere. Normally the auxiliary fuel is natural gas or some other gas fuel, since burners atomizing both a liquid fuel and a liquid waste would present problems in design. Dual nozzles do not lend themselves to concentricity. The second approach might be to use the larger combustion chamber discussed before, fired with a burner on a conventional fuel—either oil or gas—and into which the liquid waste is atomized and mixed with combustion air. The liquid waste may be injected through a single nozzle or multiple nozzles, depending upon the volume of waste and the best geometric arrangement for mixing. Obviously the system which provides the best mixing and the least excess air gives us the optimum arrangement and lowest fuel cost.

The equipment for handling this type of waste is usually a horizontal or vertical refractory lined cylindrical furnace with an auxiliary fuel burner firing at one end or tangential to the cylindrical shell. The size of the incinerator depends upon the heat release in the system, the amount of waste injected, and the amount of combustion air to be used. Mixing is accomplshed by baffles or a checker wall, and the temperature of the incinerator should vary, depending on the type and the amount of the waste. In most cases, it is possible to incinerate most organic aqueous mixtures below 1800°F and many in the range of 1200 to 1500°F. As with gaseous wastes, the autoignition temperature of the waste should first be determined, and the incinerator should be operated at a controlled temperature several hundred degrees above this point.

The products of combustion discharging from this type of incinerator are normally released directly to atmosphere either from the open end of the incinerator or through a stack. If the waste is highly aqueous, a visible steam plume will be evident under certain atmospheric conditions. However, the temperature of this plume is high enough that the evaporated water is usually dissipated long before its temperature is reduced to the dew point.

Heat recovery in this type of system is seldom used because of the low operating temperature. Stack temperatures normally at 1500°F or below require significant heat-transfer surface. There is minimal possibility for the self-recuperation that we find in gaseous waste incineration since we are dealing with a liquid waste feed. Makeup air heating, hot water heating, or other types of recuperation are possible but are not generally considered. The temperature of the heat recovery device, if used, should always be kept above the condensation temperature of the vapor. This type of system could utilize a waste heat boiler in this manner.

In certain situations it may be possible to concentrate a partially combustible aqueous waste by means of evaporation prior to incineration. Such concentration may enable the waste to be incinerated as a combustible liquid, which reduces the size of the incineration equipment. It may also effect savings in fuel cost. In the incinerator it is necessary to raise the temperature of all the water present in the waste to the temperature at which the combustibles will be destroyed, whereas preevaporation merely requires raising the temperature to the atmospheric boiling point of the water. This is usually a significant saving in terms of fuel, but evaporation is not always practical because of air pollution considerations.

Many liquid wastes which lend themselves to incineration cannot be incinerated without some secondary form of treatment, because incineration will produce products which might be toxic in nature and therefore cannot be released to the atmosphere. Normally, these wastes can be divided into three categories:

1. Waste which contains inorganic salts
2. Waste which contains halogen compounds
3. Waste which contains sulfur compounds

Those which contain inorganic salts dissolved in the liquid waste will produce the oxide of the metal ion of that salt upon combustion. The commonest inorganic metal ion is sodium, although potassium, or for that matter any other metal ion, may be found in the waste. The oxide which is formed in the combustion reaction usually will be in a finely divided form and will require subsequent removal by either mechanical methods or wet scrubbing. This type of product usually requires a high-energy scrubber of the venturi type.

The halogen ions most commonly found in organic liquid wastes are generally chlorine and fluorine, which are often part of the halogenated hydrocarbon before they are fed into the incinerator. Complete combustion of the organic portion of the waste may result in the production of chlorine or fluorine in the products of combustion. These are relatively insoluble in water and therefore cannot be removed by wet scrubbing as long as they are in this form. The type of waste should first be analyzed to determine the amount of hydrogen in the waste material, since the hydrogen will react with the halogen forming the halogen acid, if there is sufficient hydrogen present. In many cases, there is not sufficient hydrogen in the fuel to accomplish this conversion to the halogen acid, and therefore conversion must be accomplished by injection of additional fuel in the form of natural gas or other combustible fuel at an air rate less than stoichiometric. Once the halogen has been converted to the acid gas, it may be satisfactorily removed in a wet scrubber. Here the low-energy or packed tower type of scrubber is satisfactory. For example, if we assume that we have trichloroethylene as a major component in a waste effluent, it can be incinerated in accordance with the following reaction: $CHClCCl_2 + 2O_2 = 2CO_2 + HCl + Cl_2$.

While the hydrogen chloride which is formed in this reaction can be removed by scrubbing with water, the chlorine, which is relatively insoluble, will pass through the water and into the atmosphere. By the addition of natural gas or another hydrocarbon fuel, all the chlorine can be converted to hydrogen chloride as follows: $CHClCCl_2 + 3\frac{1}{2}O_2 + CH_4 = 3CO_2 + 3HCl + H_2O$.

Hydrogen chloride formed in this reaction can be removed in a wet scrubber. Because of the low calorific value of the trichloroethylene, natural gas or other auxiliary fuel would be required for combustion in any case. but the excess hydrocarbon fuel will cause complete conversion.

Sulfur compounds are often found in liquid wastes either as part of the sulfonated organic molecule or in the form of sulfates or sulfides. Complete combustion of these wastes with a minimum of excess air to prevent SO_3 formation will result in SO_2 in the products of combustion. Complete removal of SO_2 can be handled by caustic scrubbing or a number of more complicated processes designed ultimately to recover sulfur.

The incinerator design does not have to be limited to a single combustible or partially combustible waste. Often it is both economical and feasible to utilize a combustible waste, either liquid or gas, as the heat source for the incineration of a partially combustible waste which may be either liquid or gas. Multiple or dual fuel burners for combustible wastes can be utilized in a single incineration chamber. Combination systems can often reduce the operating cost in terms of auxiliary fuel and should be carefully evaluated in the overall waste treatment program of any process plant. Heat recovery in these systems is applicable on an individual basis.

Economics The economic considerations of liquid waste incineration as a disposal method are not very clear-cut. Unfortunately they depend greatly on the cost of alternative methods. One extreme would be a completely combustible solvent which could be fired in a conventional burner into a conventional boiler for the recovery of heat and the generation of hot water or steam. Here the disposal of such a waste would actually reduce the normal fuel bill required to produce the steam in the plant and would be a credit rather than a debit. On the other hand, a highly aqueous waste containing just a few percent of organic material might be disposed of by incineration methods which would require large quantities of natural gas or other fuel for the disposal. On a continuing 24 hour per day, 7 day per week basis, the fuel bill, even on a small incinerator of this type, gets to be astronomical. It is interesting to note, however, that waste liquids such as phenol-water mixtures have been incinerated, rather than treated, at concentrations as high as 97 percent water and 3 percent phenol, because the high BOD of the phenol made incineration more economically attractive than added sewage treatment facilities.

It is almost impossible to give meaningful equipment costs for the wide variety of noncombustible wastes which are found in industry. The percentage of combustible material in the waste greatly affects the initial cost and the operating cost, but the following calculations may provide some clue to cost calculations. In this example, we have used a nominal value for the cost of natural gas at $0.50 per 1,000 cu ft, but these costs will vary within the United States from approximately $0.25 per 1,000 cu ft to $2 per 1,000 cu ft, depending upon the location and the time of year. It can be seen that even the geographic location affects the advisability of incineration as a technique.

Problem. A waste solution of 7 percent phenol and 93 percent water is discharged from a plant operation at a rate of 800 gpd. It is desired to incinerate this waste daily over an 8-hr period, using natural gas at a cost of $0.50 per 1,000 cu ft. What is the fuel requirement and the cost of the system?

First we must decide on the incineration temperature. We select 1600°F because of experience with this material and also because it is above the autoignition temperature of phenol. The chemical oxidation reaction is $C_6H_5OH + 7O_2 = 6CO_2 + 3H_2O$.

The heat release from the combustion of the phenol is about 14,000 Btu/lb. Since we have roughly 100 gph × 8.5 lb/gal or 850 lb/hr of waste or 0.07 × 850 = 59.5 lb/hr of phenol, this is then 59.5 × 14,000 = 835,000 Btu/hr of heat released from the oxidation of the phenol. The heat required to incinerate the waste is

Sensible heat: 850 × 1 × (212 − 70)	=	121,000 Btu/hr
Latent heat: 850 × 1,000	=	850,000 Btu/hr
Superheat: 850 × 0.5 × (1,600 − 212)	=	590,000 Btu/hr
Total heat required	=	1,561,000 Btu/hr
Less the heat from phenol	=	835,000 Btu/hr
Net heat required	=	726,000 Btu/hr

We need 7 moles × 379 cu ft/mole or 2,650 cu ft of oxygen/mole of phenol. We have 59.5/94.0 = 0.635 mole of phenol. 0.635 × 2,650 × 100/21 = 8,000 cu ft of air/hr, or 133 scfm for the phenol. To be sure of proper combustion of the phenol, we should use about three times the theoretical amount of air, say 400 scfm, and we shall need additional heat to raise the temperature of this air to 1600°F. 400 scfm × 4.56 lb/(hr)(scfm) × 0.26 Btu/(hr)(°F) × 1530°F = 725,000 Btu/hr. 725,000 plus 726,000 is 1,451,000 Btu/hr heat required for the system. At 1600°F we have only 50 percent of our heat available at 10 percent excess air through the burner; so we need a supplementary fuel rate of 1,451,000 × 2 = 2,902,000 Btu/hr. The fuel cost will then be $1.45 per hr based on using 1,000 Btu/cu ft natural gas. The cost of the incinerator will vary, depending upon the manufacturer, but should be between $10,000 and $15,000. These costs can then be compared with other methods of disposal.

Other Oxidation Methods

Incineration of wastes which are not pure liquids but which might be considered sludges or slurries is also an important waste disposal problem. Because sewage is handled in sludge form, many of the processes and equipment previously applied to handling sewage sludge have found application in the industrial disposal field. The combustion principles are the same, but the manner of achieving the combustion is different.

Rotary kilns One of the oldest types of incinerator is the rotary kiln. It is applicable to both solid wastes and sludges. It consists of a long horizontal cylinder that rotates on steel "tires" which rest on trunnions. The waste material is fed at one end, tumbled by the kiln to provide attrition and exposure to combustion air and drying, and gradually burned as it moves toward the exit where the ash is removed. Afterburners are usually required for this type of system because the volatilized gases from the refuse cannot be adequately mixed with the air in the kiln. This type of brute force approach is used in central plant disposal facilities where there will be a wide variety of waste materials. It is expensive but flexible. Auxiliary fuel is normally required.

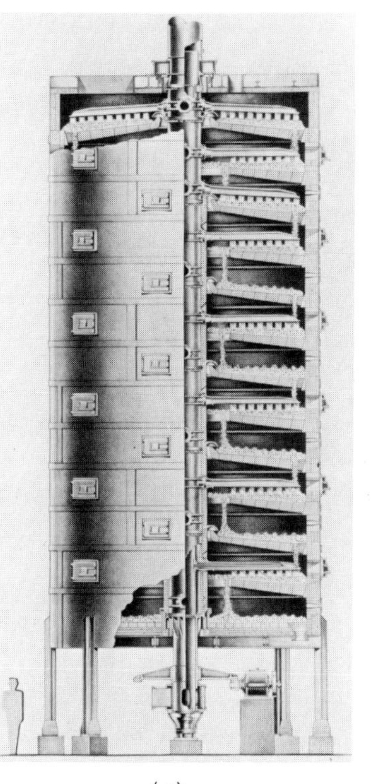

(*a*)

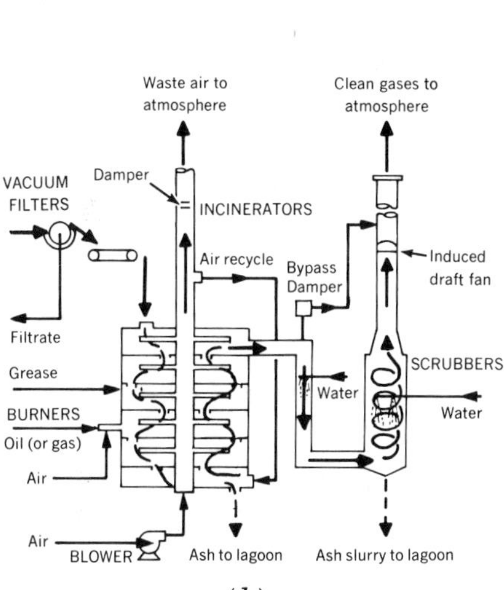

(*b*)

Fig. 12 Multiple hearth incinerator. [(*a*) *BSP Corporation;* (*b*) *Chemical Engineering, Oct. 14, 1968, p. 112.*]

Multiple hearth furnaces This is another outgrowth of the sewage field. First used in 1934, it is now a widely accepted method for sludge disposal and serves this purpose equally well in the industrial field. Liquid-solid waste sludges are fed by a conveyor to the top of the incinerator. The incinerator is a vertical cylindrical enclosure with a number of horizontal hearths, each having an opening to the hearth below (Fig. 12). The waste sludge is moved to lower hearths by the plowing action of air cooled metal rabble arms. The waste sludge is dried on the first two hearths and burned on the next three, and the ash is cooled on the lower hearth. The ash is removed for burial or reuse. Temperatures are generally 600 to 1000°F on the top hearth, 1400 to 1800°F on the combustion hearths, and 400 to 600°F on the cooling hearths. Depending on the calorific value of the waste, auxiliary fuel burners may be used only to ignite the waste or throughout the incinerator to burn the material. Waste gases may require afterburning to prevent air pollution, and these gases can often be used for preheating combustion air or as a source of heat for other processes.

The multiple hearth incinerator is a reliable system. It has moving parts at high temperature; so it is subject to regular maintenance. The hearths are refractory lined, as well as the shell. Initial equipment costs are generally higher than for straight liquid incinerators, but they are used in applications where other types of liquid waste incinerators would not work. Operating costs are generally nominal. Multiple hearth incinerators are used only when there is a high percent of solid content in the waste. Total costs per ton of dry solids removed range from \$8 to \$14, depending on the size of the system, and operating costs are from \$0.50 to \$5 on the same basis.

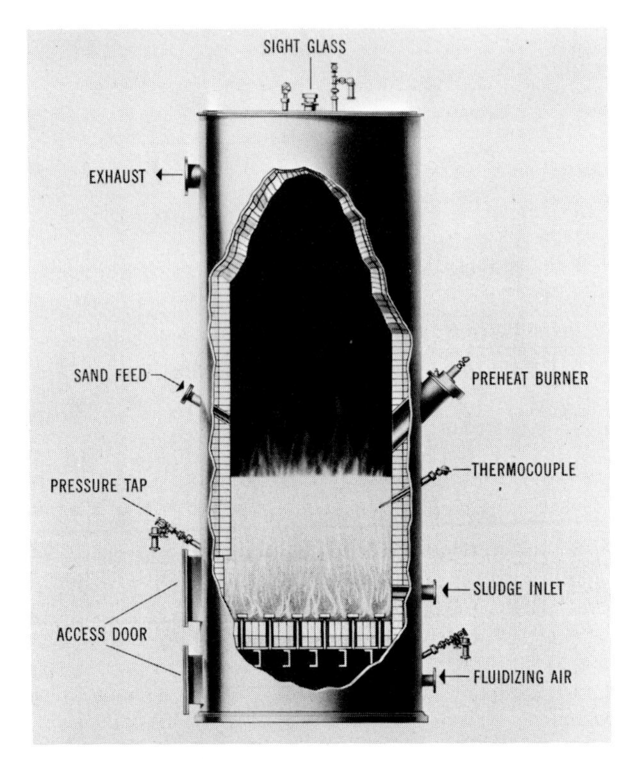

Fig. 13 Fluid bed incinerator. (*Dorr-Oliver, Inc.*)

Fluidized bed This type of incineration system is a relative newcomer to the field. It was first used commercially in 1962 and has had some difficulty competing with other methods because of cost and mechanical problems. It was first used on industrial waste situations, especially in the paper industry. The waste sludge is fed onto a bed of sand that is fluidized with air at 3 to 5 psig. The air has been preheated to 1000°F or more, and evaporation and combustion of the waste take place on the surface of the bed. Often auxiliary burners are located above the bed to provide supplemental heat and destroy noxious gases formed in the chamber. Excess air is held to about 25 percent, and combustion temperatures reach 1500 to 2000°F. The preheated air is generated by exchange with the stack gases. Ash passes out with the effluent gas and is removed in a cyclone. This incinerator has the advantage of a large heat sink in the sand bed, which promotes intermittent operation, and few moving parts to reduce maintenance (Fig.

13). Costs of operation vary widely. Sewage sludge units indicate total operating costs including the investment at about $20 per ton. Actual operating costs are $3 to $5 per ton.

Flash drying systems Recycling type flash driers have been used to incinerate wet sludges. The system is almost identical to the standard flash drier used for drying many chemicals. The solids in the system are dried, separated in the cyclone, and sent to a secondary incinerator for final destruction. The gases from the drier often require afterburning. Supplementary fuel is required. Sewage sludge may be dried in this manner, and the dry sludge can be used as a fertilizer.

Wet air oxidation This process oxidizes the sludge in the liquid phase without mechanical dewatering (Fig. 14). High-pressure high-temperature air is brought into contact with the waste material in a pressurized reactor. Oxidation occurs at temperatures from 300 to 500°F and pressures from several hundred to 3,000 psig. Once oxidation has started, the need for supplementary heat is minimal. The product is usually an easily drainable liquid. Wet air oxidation has been used

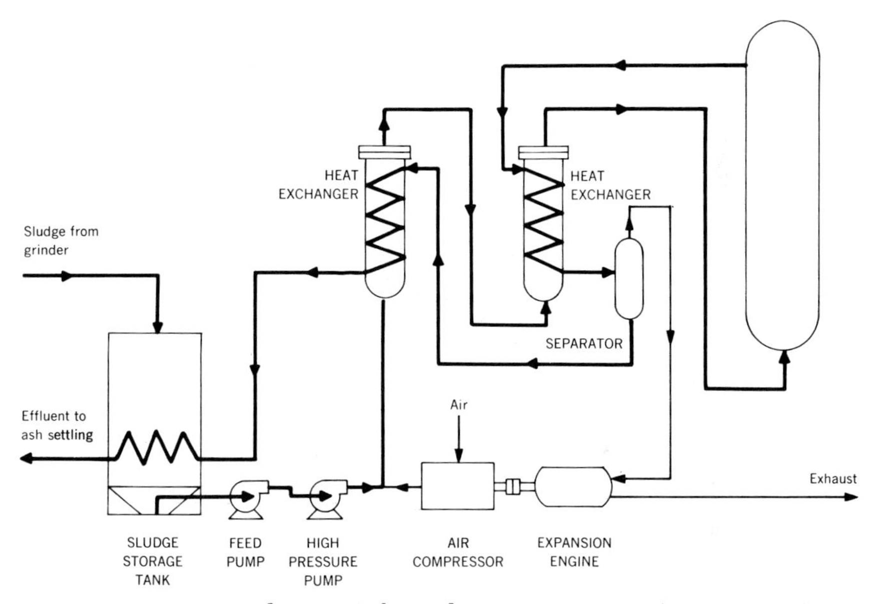

Fig. 14 Wet air oxidation. (*Chemical Engineering, Oct. 14, 1968, p. 116.*)

for pulp mill liquors, plating wastes, sewage, cyanide destruction, the treatment of sour waters containing sulfides and phenol, and the recovery of valuable products such as silver from x-ray film. It is an expensive process involving complex equipment and generally requires first-class operators.

Atomized suspension This is a pyrolysis process rather than a direct oxidation method. The waste material is atomized into a vertical stainless steel tube which is externally heated to 1500°F or more. The solids are dried in this manner, and the resulting volatiles are vented to afterburning if necessary. The process, developed by the Pulp and Paper Institute of Canada, was conceived to dispose of paper-mill waste but has never been successful in this regard because of high-temperature corrosion problems. One small unit was installed for processing sewage sludge and one for a process application. No valid cost comparisons are available.

Deep-well Disposal

Many liquid wastes are discharged from industrial process applications which do not lend themselves to normal methods of treatment, nor do they lend themselves to incineration because most of the contaminating material is inorganic or possibly all the contaminating material is inorganic. Let us consider, for example, a weak acid effluent from some process which is present in relatively high volumes. Acid effluents can always be neutralized with caustic or lime, but the mere neutralization does not prevent the possibility of water pollution. Neutralization of weak hydrochloric acid with sodium hydroxide or calcium hydroxide forms soluble sodium or calcium chloride, which is no more desirable when discharged into a body of fresh water than the original waste acid, except that the pH is more nearly neutral. Sulfuric acid wastes constitute similar problems, although in this case, calcium sulfate tends to precipitate because of its low solubility in water.

One method for the disposal of such waste was suggested by a practice common in the oil industry for some years, that is, deep-well injection. Large volumes of saline water, an unwanted by-product of oil production, were for many years returned to the subsurface through old dry deep wells. This brine, if allowed to accumulate on the surface, would kill all plant life in the area, but injection back into the earth provides a reasonable disposal method. There are estimated to be over 40,000 brine disposal wells in the United States today, but only about 100 wells used for the disposal of industrial wastes. This is not really a disposal method unless we consider that things which are out of sight are out of mind. It might more reasonably be called a storage method, using a convenient place that we hope we will not need for other purposes in years to come.

Industrial waste disposal experts only recently have begun to think in terms of deep-well injection, perhaps for several reasons. First, there are only certain areas in the United States where deep-well injection is practical. Second, it is not always an economical solution to waste disposal, and third, there are certain legal considerations concerning subsurface injection of any waste.

Approximately 35 states permit the construction of certain types of waste injection wells subject to specific laws and restrictions. In almost every case, a permit is required. Other states either do not permit deep-well disposal or have no specific laws on the books concerning this type of disposal.

Assuming that we have a waste which we would like to consider for deep-well disposal, we must first decide whether the local area is suitable for the installation of a deep well. A suitable deep-well site must have geological formations at the injection point with sufficient porosity, permeability, and area to act as a reasonable storage reservoir. The injection horizon, which is the level at which the waste liquid will be stored, should be well below the level of fresh water circulation, and the areas should be confined by an impermeable rock enclosure. Sandstones, limestones, or dolomitic rocks usually form a suitable horizon because they contain nothing more than saline water at deep-well levels. Many sedimentary rocks contain oil, gas, coal, and sulfur resources, and protection of these resources should be observed at all cost.

The impermeable confinement of shale, clay, or similar material should overlie and underlie the injection horizon to prevent vertical escape of the injected waste. In other words, we are trying to store the waste material below the surface of the earth in a storage basin having a definite depth.

We should also have some idea of the extent of the lateral movement of the injected waste into the well, since the waste could find itself in oil and gas wells which project through the same level at some distance from the waste injection well.

The depth of a deep well can vary from several hundred to over 12,000 ft. The disposal formation into which the waste is injected has been measured in some cases to be as high as 10,000 sq miles and the thickness of the formation as deep as 5,000 ft. Even in deep disposal formations, only a fraction of the depth may be suitable for disposal because of its permeability.

The Gulf Coast and the Atlantic coastal plain are particularly good locations for injection wells because they contain relatively thick sequences of salt-water-bearing sedimentary rock, and also because the subsurface geology of these areas is known because of previous extensive drilling for oil and gas. Areas where we find volcanic rock protruding through the surface of the earth are generally not suitable, because such rock has fissures and fractures which will allow the escape of the waste into fresh water sources.

Considering that waste water storage space in injection wells is limited, we should try to confine the use of such wells to only the difficult disposal problem. This means highly concentrated or highly polluted or untreatable wastes. The volume of any well is limited, and once the well is filled, other wells must be drilled at high cost.

Once the well is established, the waste material may be pumped into the well at a pressure high enough to cause it to be injected into the interstitial area of the formation. Injection rates may be high or low, depending upon the geology, but generally range from 100 to 300 gpm at pressure levels from 100 to 500 psi. The lower the injection pressure the more desirable the operation because of lower pumping costs and the pressure capacity of the well casing and tubes, and the damage resulting to the confining strata.

The nature of the waste to be injected into a deep well is important. Materials containing solids or suspended matter usually are not recommended for deep-well disposal because plugging can result in the pores of the injection horizon. Plugging can also be caused by entrained gas, bacteria, and mold. Any material which will form an insoluble solid should be removed before deep-well injection. There is often the problem of a reaction occurring between the injected waste and interstitial water which can form precipitates to plug the pores at the injection horizon. This can be determined only by sampling water at the injection horizon and mixing it with the waste prior to injection. Sometimes a buffer of nonreactive pure water can be injected ahead of the waste to prevent such reaction. Also the high pressure and the resulting high temperature used for injection of certain materials can cause polymerization or precipitation of solids, which can also plug the pores of the well.

Some reactions between the waste material and the layer at which the waste is injected can be beneficial. Carbonate materials, found in limestone and dolomite, are soluble in acid wastes. The reaction with these carbonates could be beneficial if no undesirable precipitates result, or if the generation of carbon dioxide does not cause a pressure buildup in the well at the injection horizon. If plugging occurs for any of these reasons, it usually can be seen by a buildup of the injection pressure.

Costs of deep-well disposal can be estimated most easily from experience in a particular area. The petroleum industry, of course, has a vast fund of cost knowledge on the drilling of wells, whereas industrial waste disposal applications have been relatively few. The fixed cost for drilling the well is usually a major portion of the total operating cost of the system and normally averages 75 percent of the cost. The remainder is the cost required for pumping and maintenance. For wells with a low injection rate, fixed costs may be as high as 90 percent of the total cost. Since there are many factors which influence the cost of drilling a deep well, such as the type of subsurface geology, which affects the drilling rate, it is only possible to give a range of costs, which increase with the depth of the well on a per foot basis. Deep wells generally cost between $10 and $30 per foot, with the lower figure applying to wells at the 1,000- to 5,000-ft level and the high figure applying to wells between 10,000 and 15,000 ft of depth.

Injection is accomplished by centrifugal, turbine, and piston type duplex and multiplex plunger pumps. Centrifugal pumps are most suitable for low-pressure injection, usually under 300 psig. Piston type pumps are utilized at pressures up to 500 psi for the duplex piston type and multiplex plunger pumps for higher pressures. Pretreatment of many wastes is an expensive operation, while other wastes may not require any treatment at all. Brine treatment costs on fuel oil operations will run at rates between $0.15 and $0.30 per gpd, on an installed plant basis. It is difficult to provide costs on deep-well installations because of the wide variation in conditions.

Deep wells are drilled generally with cable tool drills rather than rotary drills because drilling mud should not plug the injection horizon.

In the evaluation of a well for waste disposal we should examine such things as drilling cores and logs and make pumping or injection tests to determine permeability. We should determine the mineral content of the formation which will form our injection horizon, by taking drilling samples or cores, and we should measure the temperature of the formation and determine the rate at which injection can be made into the horizon.

Once the well is drilled, we must insert a well casing, cement the casing in place, and install our injection tubing and wellhead facilities. Often several casings are required, especially in deep wells. A large-diameter surface casing may extend

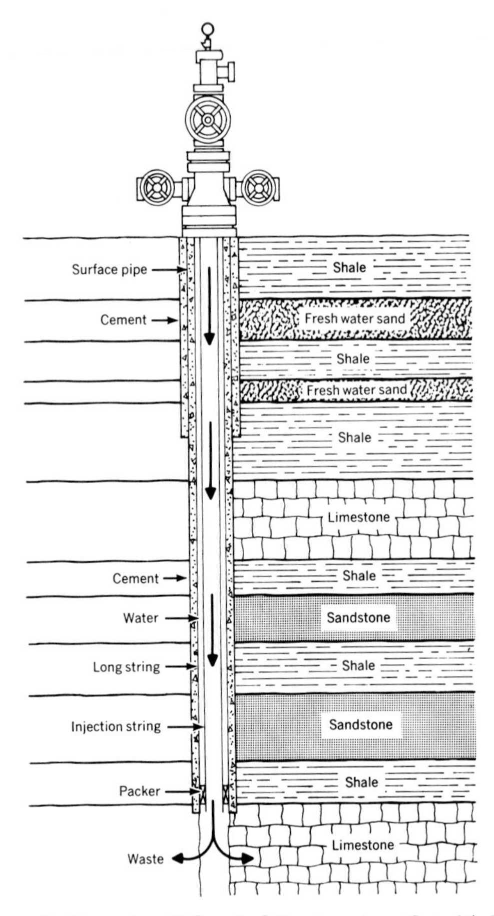

Fig. 15 Deepwell disposal. (*Chemical Engineering, Oct. 14, 1968, p. 109.*)

for the first several hundred or several thousand feet to protect fresh water at one level from saltwater contamination of the deeper level. The next casing will be smaller in diameter and is run through the surface casing to the top of the disposal horizon or to the bottom of the hole, depending on the method of completion of the well (Fig. 15). This closes off zones of soft rock and saltwater sources to protect gas and oil bearing strata.

Corrosion of the injection well casings can be a very serious problem. Casings can be coated with various plastics, corrosion resistant cements, etc., to inhibit corrosion, or even plastic-lined tubing can be used inside the casing. The annular area between the rock strata and the casing is cemented to prevent mixing of waters contained in aquifers other than the injection horizon to protect the pipe from external corrosion by subsurface water and to increase the casing strength. Cementing is carried out, using oil field techniques.

Wells can be treated to increase the rate of injection. Chemical treatment utilizing hydrochloric or hydrofluoric acid tends to dissolve calcium carbonate and other acid soluble materials. Acidizing is effective in limestones or dolomite rocks, but it is also used in sandstones to dissolve carbonate materials and to destroy clay minerals.

Mechanical means of increasing the well's capacity are accomplished by scratching or washing the surface, or reaming the bore of the well to open up plugged areas. Hydraulic fracturing, which is used in the oil well industry to increase the acceptance of brine injection, can also be used, but it may not be entirely safe for waste injection wells because it may open vertical fissures extended through the caprock. Once wells have been installed, they should be monitored constantly to see that the waste material is not escaping. Monitoring is accomplished by a continuous record of the injection rate, wellhead pressure, and volume of the injected waste. In a normal deep well used for waste disposal, the pressure needed to maintain the injection rate will gradually increase with time. A sudden increase in the intake rate of an injected well at a set pressure or at a decreased pressure indicates that some type of hydraulic fracture has occurred in the wall and that this may have occurred at the injection horizon or in the casing of the well. Sometimes an intermediate fluid is used between the column of waste and the outer casing. If a pressure increase is noted at this point, it also would indicate a leak was present in the discharge tubing. It is wise to actively monitor wells used for potable water in the general area of a deep-well disposal system to detect any possible leakage.

Deep-well disposal for industrial wastes is now being used for the disposal of such materials as pickle liquor from steel mill pickling applications, brine disposal in oil fields and potash mines, and acid wastes generated in petroleum refineries and chemical plants. Hydraulic fracturing, a technique used for years in the oil industry, is now being considered for the disposal of radioactive waste. The waste will be mixed with a special cement to form a grout, which when injected at high pressures into a shale formation causes a fracture to occur. It is estimated that the fracture may permit as much as an 8-in.-thick horizontal layer of grout extending 1,000 ft from the casing. This hardens, and the radiation is trapped safely below the earth.

Lagoons and Ponds

One of the earliest methods of handling large volumes of liquid waste was to dig a large earthen basin and collect the waste within its confines. If the waste contained a settleable solid, the solid matter would gradually sink to the bottom of the lagoon, and the clean or less polluted water could be drawn from the surface over a weir into a sewer or stream. Acid mine water which is pumped to the surface is neutralized with lime and then pumped into a lagoon to allow calcium sulfate and iron hydroxide to precipitate.

A lagoon may also be used to separate an oil or floating scum from water. The oil will rise to the surface, where it may be removed or ignited and burned, while the water can be drawn from the lagoon near the bottom. Lagoons are used for aeration to accomplish oxidation of organic materials and for evaporation to reduce the volume of the effluent. Solid materials which settle to the bottom of a lagoon must be removed periodically by mechanical methods and reprocessed if valuable, or disposed of in other ways.

Lagoons act as equalization and stabilization basins for a variety of liquid wastes. BOD is generally reduced by the period of retention, whether the waste is mechanically aerated or not.

The inlet of a lagoon must be baffled to prevent crossflow of the waste to the exit end of the lagoon. Baffles are also provided on outlet ends of the basin to collect oils and scums. Most lagoon designs are based on a given retention time, which may be as long as a week but usually exceeds 1 day. The basin's major function is sludge storage, and the amount of storage volume required is dependent upon the solids concentration of the feed and the rate of settling or biological degradation. Such basins are commonly used for industrial wastes which contain large quantities of solids.

In building a lagoon or earthen basin, the volume of sludge to be stored should be considered since it is not desirable to clean the basin at short intervals. The walls of the basin should be made of compacted soil which will withstand the hydrostatic pressure of the impounded material and will not let the waste leak to the outside. Sludge removal is usually accomplished by pumping, draglines, or clamshells, and the sludge is hauled away to a dump or some type of incinerator where it can be finally destroyed.

Maintenance of a lagoon system can be difficult, especially where animal life abounds. Lagooning is not a permanent disposal method but acts as a temporary holding basin to remove the bulk of solid materials or surface scum. Careful monitoring of the effluent is desirable. Lagoons have been used for separating rolling mill oil from water, separating calcium sulfate from neutralized mine water, and many similar applications.

Liquid Waste Dispersion

A method of disposing of liquid wastes which may be acceptable is merely dumping it back into the environment but with sufficient dilution that a nondangerous and nontoxic condition is the result.

Dilute waste waters are often discharged into rivers, lakes, or streams after appropriate treatment, but the proper dispersion of even this treated water is an important consideration. The location of the discharge point and the type of dispersion equipment used are significant in their protection of other water sources and in maintaining an overall desirable situation. Plants located near the ocean or a large lake or a large river may discharge their wastes through a pipe or ditch leading to the shore, but if the discharge of the pipe occurs above the main body of water, the formation of foam due to air entrainment, or incomplete dilution because of a low water condition may result, giving high concentrations near the point of discharge. A properly designed subsurface dispersion system, however, will permit the receiving body to assimilate the waste properly. This reduces treatment requirements. Some of these submerged devices include an open-end pipe with special nozzles or diffuser systems consisting of a series of smaller pipes with holes or slots. Waste should be discharged at the best angle to the flow of water in the main body to effect rapid dispersion.

Dispersion pipes should be located so that the discharge point is far enough from the shoreline that it will protect plant or other water intake systems. It should be directed so that existing currents and tides tend to disperse the waste into the main body rather than bring it back to the shoreline.

The same basic problem of liquid dispersion exists in this type of system that exists at the top of a tall stack, and the same basic considerations are important. The chemical characteristics of the waste, the discharge velocity, the turbulence of the receiving stream, tidal action, temperature variations, density differences, and flow patterns in the receiving water body are all important.

While this method of disposal is widely used, especially for sewage, it cannot be considered a reasonable approach for the future. Those methods which appear to be safe and economical now because of our seemingly inexhaustible resources must one day cause problems for other generations.

Not all dispersion methods utilize pipelines directly to the receiving source. Many wastes are deposited some miles at sea by barging the waste to a point beyond the sight of land.

While no serious results have been noticed to date from the piping or barging of wastes to sea, there is evidence of ecological changes which should act as a warning. This method of disposal is far more economical than other forms of treatment, and therefore it will remain a method of disposal until it is prevented by legislation. The economics are so attractive that plans are now being considered to pump industrial waste from Frankfurt am Main to the North Sea, a distance of nearly 300 miles.

Land Disposal

Many industrial waste waters contain constituents which have high BOD or COD requirements but are not toxic. Such waste waters may be considered for land disposal. Land disposal of sanitary sewage wastes is a process that has been used for thousands of years and is today a common practice in the rice paddies of the Orient. Such methods in the United States have been unpopular for aesthetic reasons and because the odor is often difficult to control, but they are nonetheless viable means of disposal. If land can be used as a means of dewatering and oxidizing sanitary sewage sludge, it should also be considered for industrial wastes that contain biodegradable materials or those inorganics which can contribute to the nutrients in the soil.

Wastes from food products and paper industries can often be disposed of by such methods. Such waste waters can be spread over farmland by tank trucks or by means of spray irrigation, or even by a network of pipes discharging the waste between ridges or furrows on the ground. Since a high solids content would not be desirable for this kind of disposal technique, prior filtration or sedimentation methods are often used. One Ohio paper company treats 3 million gpd by removing the solids (mostly clay and wood fibers) in a primary clarifier and spraying the remaining liquid over a 30-acre wooded hillside. The liquid waste is sprayed from the top of elevated spray nozzles onto the tops of the trees so that there is some absorption of the water before it reaches the ground. This prevents erosion and excess water runoff.

This method can be used for many crops and can be a beneficial method for the crop because it adds soil nutrients, and for waste water disposal because significant reduction in BOD occurs.

System costs will vary from one situation to another, depending on the quantity of waste water and the distance that it must be pumped or transported.

There are some shortcomings to this type of disposal, especially in colder climates. Freezing may occur in the winter, and the reduction of BOD is less in cold weather than in warm. In most cases, however, a reduction of 60 percent of the BOD can be accomplished before the water reaches groundwater level.

Loading rates for irrigation vary widely from 10,000 to 100,000 gpd/acre, but BOD loadings should be kept below 200 lb/day/acre.

SOLID WASTE DISPOSAL

In discussing gaseous wastes, we stated that the most difficult problem is the volume of waste to be handled. For liquid wastes the problem is the variety of constituents. In the discussion of solid wastes, we have a combination of both problems.

If we were to take all the refuse of our present society, which would include old newspaper, waste paper, beer bottles, tin cans, garbage, junk cars, old refrigerators, bicycles, automobile tires, plastic toys, and construction materials, and total their weights, we would find that it was accumulating at a rate of 165 million tons/year, which is approximately 4½ lb/day for every man, woman, and child in the entire nation. The U.S. Public Health Service predicts that 20 years from now the total national output of solid wastes will be up to 500 million tons, or three times the present rate. If we do not start thinking about what we are going to do with it soon, it is going to be too late.

The problem has been that most communities are handling garbage and wastes the same way they did 50 years ago. They take it out to the town dump and burn what will burn, and the rest just gradually must rot away. Fewer than half of all the communities of 25,000 or more persons maintain adequate facilities for the sanitary disposal of solid wastes, according to the Department of Health,

Education and Welfare; yet we are expending $3 billion a year toward the solution of this problem.

These figures sound astounding and they are, but they should also be frightening enough to make us begin to cope with the problem immediately from two directions. We do not have many alternatives. Solid waste disposal, because of its very nature, allows us to do basically three things. The first is to burn it, using some type of acceptable incineration method. The second is to bury it by digging a hole, throwing in the waste, and covering it over with dirt, and the third is to treat it, using biological methods to make a reasonable product which can be used elsewhere. Only incineration and landfill or burial methods are final disposal considerations, but the third, composting, is mentioned here because this method may offer a valuable answer to a desperate problem.

Pretreatment

There is a possibility of working on this problem from both ends. One, of course, is the disposal; the other is the reduction of certain types of wastes by changing our living habits. We must consciously go back to saving the string from the package or reusing the paper bag that the product was carried in. We must get more mileage out of our disposable containers or make them more easily disposable by presently acceptable methods. For example, 20 years ago returnable bottles were redeemed at 5 cents each. These bottles averaged 50 trips between the consumer and the plant. Now these bottles have given way to the no-return container, which obviously multiplies our waste glass problem by a factor of 50. The plastic container is being thrown away at a rate of 6 billion lb/year. Not only will the plastic container bounce when the shampoo is dropped on the bathroom floor, but it will not sink in water, it will not degrade in soil, and it causes fantastic maintenance problems in municipal incinerators, while increasing air pollution.

The steel industry's recent switch to basic oxygen furnaces has almost eliminated the need for scrap. This means an increase in the number of scrap cars, which must be disposed of in some other manner.

The price of old newspapers used to be enough to keep a tidy little business in operation, but today the paper industry finds it cheaper to grow trees than to reclaim the waste, and now instead of selling old newspapers and magazines, we have to pay to have them hauled away.

There are several things which we can do to pretreat the waste before we burn it or bury it, but these do not eliminate the next step; they only help ease the burden. The first is compaction. By compressing garbage and waste paper under high pressures, the volume can be reduced by 75 percent. If the resulting material is coated with an outside layer of concrete or similar material, it can be used in landfill, foundations, and sea walls, but it certainly does not improve its suitability for incineration since a more loosely packed material presents a more acceptable condition for incineration. Another thing that can be done which will promote better incineration is shredding or size reduction of combustible solids, which will allow an incinerator to be designed to handle the material at a given consistency and size. In this way grates and loading apparatus can be designed more efficiently, and combustion can be completed with minimal or no air pollution.

Solids Incineration

Solids incineration is not a true disposal method, because most solid materials contain noncombustibles and have residual ash. It is discussed here in brief because it is, in effect, considered disposal and in fact is perhaps the major disposal method for solid refuse used in the United States today.

Municipal and industrial incinerators probably account for 30 to 50 percent of the total trash disposal within the United States at the present time. The object of any incinerator is to provide complete combustion of the material fed into it. The complications are, however, the wide variety of materials which must be burned. We have everything from wet garbage with a heating value of approximately 2,000 Btu/lb to such plastics as polystyrene, which have a heating value of approximately

19,000 Btu/lb. Controlling the proper amount of air to give good combustion of both materials is difficult, to say the least, and with most currently available incinerator designs, it is impossible.

Certain basic principles for complete combustion of solid refuse with a subsequent low particulate emission from the combustion zone are as follows:

1. Excess air: Air quantities should usually be kept on the order of 50 to 150 percent above the stoichiometric requirements.
2. Minimum use of underfire air: This maintains low velocities and therefore reduces the particulate emission from the incinerator because it keeps small particles out of the gas stream.
3. Proper use of overfire air: This provides ample oxygen and turbulence in the combustion space above the fuel bed. The overfire air injected into the system may be as high as 50 percent of the total required.
4. Temperature: Temperatures in the furnace space should be between 1400 and 1800°F to reduce the rate of smoke formation and odor. Temperatures below 1400°F will produce smoke and allow odor to escape from the incinerator. Above 1800°F there may be sintering or fusing of the ash with the furnace refractories. Excess air is used to control the furnace temperature.
5. Sufficient combustion volume: The incinerator should have enough combustion volume to provide sufficient residence time for the burnout of all flying particulate matter. The average heat release per cubic foot of furnace volume should not exceed 25,000 Btu/(cu ft)(hr).
6. Secondary chamber: A secondary chamber zone should be provided in every incinerator and, in fact, is required in most municipal and state codes now being adopted.
7. Residence time: The residence time in the incinerator should be between 1 and 2 sec.
8. Reasonable loading rates: Low loading rates per square foot of grate surface should be adhered to, even in forced draft incinerators. They should be no more than 60 lb of refuse/sq ft/hr.

Early municipal incinerators were charged by a crane from a storage pit onto stationary grates. Some furnaces were hand stoked. Ash handling was arranged through undergrate hoppers in which the ash was water quenched and dumped into trucks and hauled away. Today, because of the consciousness of air pollution, more modern designs have been developed. A variety of modern stokers, which provide uniform and regular agitation of the feed, are used to feed the waste material onto the surface of the grate. The grate may be a fixed grate or a traveling grate system. Where the traveling grate is used, it is also considered part of the stoking apparatus.

Multiple chamber incinerators The modern two-stage incinerator should be employed in most municipalities and industrial plants today. The incinerator consists of the following basic components (see Fig. 16):

1. A primary chamber wherein preheating and combustion take place
2. A secondary chamber for combustion and expansion of gases
3. A chamber for settling fly ash
4. A stack which discharges the gases to the atmosphere, or in certain cases where state or municipal regulations require a scrubbing system

The combustion process in a two-stage or multiple chamber incinerator consists first of a primary chamber into which the solid waste material is charged. This is known as the ignition chamber. The charge is either manually or automatically fed into the ignition chamber through the charging door onto the grates, which are at the bottom of the ignition chamber. In this chamber, drying, ignition, and combustion of the solid refuse occur. As the burning proceeds, the moisture and volatile components of the fuel are vaporized and partially oxidized in passing from the ignition chamber through the flame port, which connects the ignition chamber with the mixing chamber. The flame port is an opening over the baffle wall which separates the ignition chamber and the mixing chamber. From the flame port the products of combustion and volatile components of the refuse flow through the mixing chamber, where secondary air is introduced. The combination of adequate temperature and additional air which can be supplied by secondary auxiliary fuel burners, if necessary, assists in initiating the second stage of the combustion process. Turbulent mixing resulting from restrictive flow areas, and abrupt changes in flow direction cause the combustion reaction to go to completion. In a third chamber known as the final combustion chamber, there is also a refractory wall or curtain wall between the mixing chamber and the final combustion chamber. Fly ash and other solid particulate matter are collected in the combustion chamber by impingement on these walls or by gravity settling. Finally the gases are discharged through a stack or combination of gas cooler, which is a water spray chamber, and an induced draft system. Whether forced or induced draft is used on the incinerator, careful air control is required to give complete combustion.

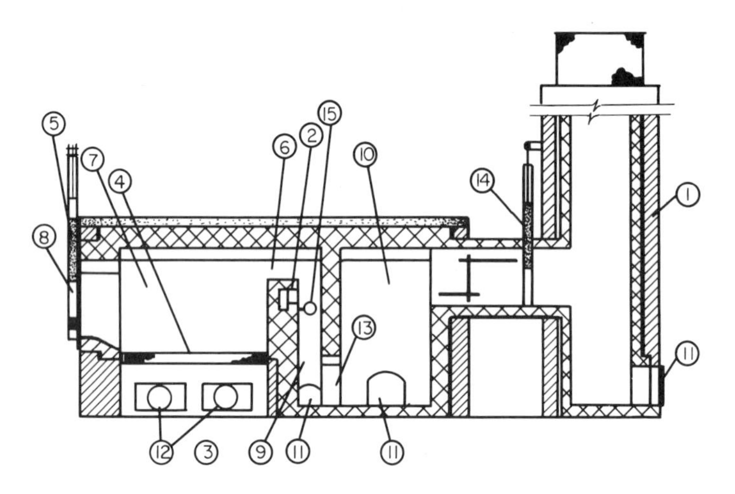

SIDE ELEVATION

1. STACK
2. SECONDARY AIR PORTS
3. ASH PIT CLEANOUT DOORS
4. GRATES
5. CHARGING DOOR
6. FLAME PORT
7. IGNITION CHAMBER
8. OVERFIRE AIR PORTS
9. MIXING CHAMBER
10. COMBUSTION CHAMBER
11. CLEANOUT DOORS
12. UNDERFIRE AIR PORTS
13. CURTAIN WALL PORT
14. DAMPER
15. GAS BURNERS

Fig. 16 Multiple chamber incinerator.

The importance of the arrangement of the various chambers and baffle walls in an incinerator is to provide for a number of changes of direction of the products of combustion, which will give good mixing, and to help trap any particulate matter before it is discharged to atmosphere.

There are two basic types of two-stage incinerators. The first is the retort type, where the arrangement of the chamber causes the combustion gases to flow through 90° turns in both lateral and vertical directions. This arrangement permits the use of a common wall between the primary and secondary combustion stages. The retort type is used for the smaller-capacity systems because of its simple boxlike construction.

The other type of two-stage incinerator is the in-line type where the flow of the combustion gases is straight through the incinerator with 90° turns only in the vertical direction. The in-line arrangement gives a rectangular plan to the incinerator and is readily adaptable to installations which require variation in sizes of either the mixing or the combustion or ignition chambers. All ports and chambers extend across the full width of the incinerator and are as wide as the ignition chamber. This type of incinerator is usually applicable for waste rates above 1,000 lb/hr.

Solid waste incineration equipment in small sizes is factory fabricated as a complete package up to several hundred pounds per hour. In large sizes, however, the unit is usually installed on the job site by the contractor. It may have a steel shell lined with refractory, or it may be constructed of several courses of brick, the innermost being high-temperature-resistant firebrick. There are many manufacturers of this type of equipment throughout the United States, some with little knowledge of the problem and others who are quite expert in the design and construction of solid waste incineration equipment. Incinerator designs and calculations are based on some of the parameters which have been stated previously, but are too lengthy for further treatment here.

Refractory One of the most important elements of a solid waste incinerator construction, other than the design, is the proper installation and selection of refractory material. It is imperative that manufacturers use suitable materials of construction and be experienced in high-temperature furnace fabrication and refractory installation, because faulty construction may well offset any benefits of good design. Incinerators for normal refuse service should be lined with either high-duty firebrick or 120 lb/cu ft castable refractory. These materials, when properly installed, have proved capability to resist the abrasion, spalling, slagging, and erosion resulting from high-temperature incineration.

From time to time, incinerators will be forced beyond their design capacity and given a severe increase in duty. Superior refractory material such as superduty firebrick or plastic firebrick should be considered in these cases. Stacks are normally lined with 2000°F refractory lining with a minimum thickness of 2 in.

Operation Perhaps the most important single aspect of a good pollution-free incinerator is the way it is operated. It must be charged properly at all times in order to reduce the formation of fly ash and to maintain adequate flame conditions within the unit. The ignition chamber is normally filled to a depth two-thirds of the distance between the grate and the top arch prior to light-off, and after approximately half the refuse has been burned, the remaining refuse should be carefully stoked and pushed as far as possible to the rear of the ignition chamber. The charge should be spread evenly over the grates so that the flame can propagate over the surface of the newly charged material. Variations in both underfire and overfire air will give the operator an opportunity to determine the best settings for various types of waste material, depending upon the stack emission.

Wherever mixing chamber burners are used to raise the temperature between the ignition chamber and the combustion chamber, they should be started prior to igniting the waste material so that the mixing chamber can be preheated to operating temperature.

Pit incinerator Several years ago a Du Pont engineer developed a simple and inexpensive incinerator for burning a wide variety of industrial trash. The development of this incinerator was instituted because of problems incurred in conventional single or multiple chamber incinerators when burning plastic materials such as nylon, polyethylene, polypropylene, and acrylics. Plastic materials, when charged onto the grate of a conventional incinerator, tend to melt before they volatilize and burn. This melted material then runs through the grate onto the hearth and burns from the surface of the hearth, causing high temperatures on the grate despite underfire air. The open pit incinerator eliminates the grate and uses only overfire air for combustion. Its construction consists of a refractory lined box with appropriate cleanout doors at either end and with an open top. An air manifold with a number of nozzles set at 30° to the top plane of the incinerator is fed by a forced draft blower which forces air down and across the refractory lined box. The waste material is dumped into the bottom of the pit to a height of approximately one-third of its total depth, ignited, and allowed to burn with the air circulating over its surface. The excellent mixing achieved by this forced air arrangement gives complete smoke-free combustion of the waste material, while at the same time the high heat release generated by the plastic is released from the enclosure by direct radiation to the sky, thus reducing the high temperatures found when burning simlar materials in an enclosed incinerator (Fig. 17). This type of incinerator has been highly successful on wood and plastic materials. Light materials tend to become airborne, and therefore particulate emissions for normal trash are somewhat higher than they would be in a two-stage incineration system. Costs, however, are generally one-fourth to one-third of the cost of an enclosed incinerator of comparable capacity.

Heat Recovery from Solid Waste Incinerators

Heat recovery from a solids incinerator is difficult but not impractical. The first approach was to recover waste energy by the introduction of the waste material into existing boilers. The attempt to burn trash material in boilers not only was

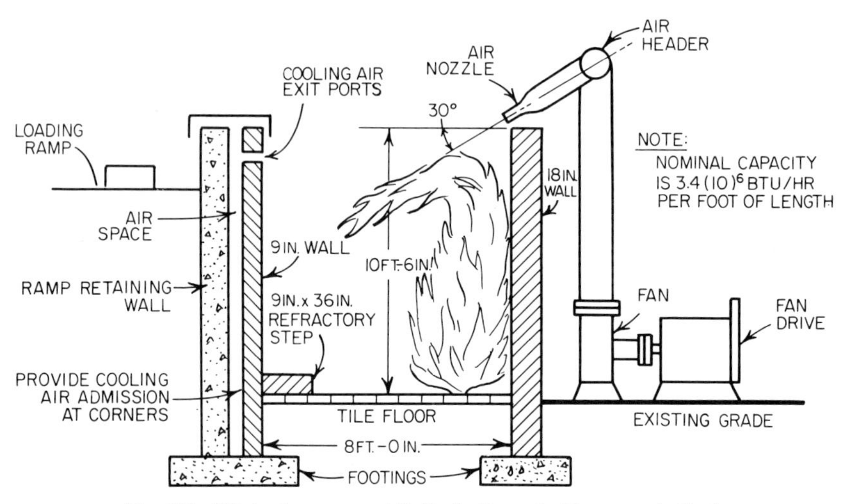

Fig. 17 Pit incinerator. (*E. I. du Pont de Nemours & Co.*)

unsuccessful, but now is illegal in most areas. A boiler burning chamber is not designed for the destruction of waste material but is designed to utilize the heat from a normal fuel in the most efficient manner possible. In addition, the interior of the boiler cannot withstand the effects of some partially burned waste materials. The fly ash from partially burned material can render a boiler inoperative or cause it to require abnormal maintenance, canceling out any savings derived from the energy recovery. Most incinerator designs in the past have been simply reengineered space heater concepts. The burner chambers of these incinerators utilize the same basic design as the boiler. The designs are dedicated to the extraction of heat and have lost sight of their main goal—destruction of the waste. A good incinerator should therefore be designed to generate, contain, and utilize this heat in a closed loop destructive process. A typical waste energy recovery system can provide meaningful dollar returns annually. A typical 5,000 lb/hr incineration system for general plant trash releases approximately 32.5 million Btu/hr. If this energy is compared with equivalent natural gas costing $0.05 per therm and the unit is operating 6 hr/day and 5 days per week, 50 weeks per year, the recovery is valued at approximately $21,000 per year. Waste heat boilers are designed to utilize the waste heat from the flue gases through a large heat exchanger, to heat buildings, regulate the temperature of waste sludges, and generate steam.

One type of system currently receiving a great deal of attention is the wet wall incinerator or boiler. This design appears to be slightly overengineered by the boiler designers and seems to be taking the wrong direction. These units are designed without refractory, and complete destruction of the waste material is difficult. Air pollution requirements are met by flooding the final chambers with water, removing the burned and unburned particulate from the flue gases. This does not seem to be efficient and has a doubtful future.

Composting

A method of waste disposal which has been used for the disposal of solid wastes overseas, but which has found very little application in America, is composting. Anyone who has planted a garden probably has started his own compost pile. It usually consists of a pile of all the organic refuse normally found around a garden. Plant cuttings, grass, fruits which have dropped to the ground and begun to rot, leaves, corn silage, and similar materials are accumulated in a heap and permitted to degrade biologically. Biological degradation is usually slow in cool weather and rapid in warm weather. The addition of biologically active material, such as manure or animal remains, will usually cause the compost pile to "work" at a much more rapid rate. Moisture is also an important ingredient of composting, since it helps in the degradation of the material. After a few months, if the surface of the pile is removed, a rich brown humus is usually found in place of the original material, and this can be spread over the surface of a garden or a planting area, providing nutrients to the soil and also conditioning the soil to make it more easily workable. Mature compost contains nitrogen, phosphorus, and potassium, and its main use is to feed the soil rather than to supply these mineral nutrients directly to crops.

In a society which often has been accused of wasting more than it uses, it is not surprising that we use very little of our wastes a second time. Our usual justification is that reuse is more expensive than the manufacture of new material, and in addition, it does not contribute to the growth of the economy. The synthetic fertilizer industry in America has been booming over the past 15 or 20 years. Ammonia, urea, phosphates, and potash are now sold in gigantic proportions to the agricultural industry, but it is time that we begin to think of our waste materials and the beneficial things we can do with them rather than just consider the economics of the alternatives.

Some attempts to build composting plants have been made in the United States. Several years ago in Florida there was a composting operation which appeared to be economically attractive; however, land costs were too high, the odor associated with the degradation of the material was bad, and the available market for the product was not sufficient to sustain the operation.

In the early 1960s, the city of Phoenix, Ariz., showed interest in the development of mature compost utilizing both sanitary waste and garbage. While the results were somewhat of a technical success, they were a financial disaster, and the plant was closed and has not been operated since.

In Europe, however, where the population density is much higher and where waste recovery, whether it is heat, water, or air, is more of a necessity, some significant work has been done on composting plants. A Danish corporation has been building composting plants with capacities from approximately 5 tons/day up to as high as 300 tons/day since 1939, and has installed these throughout most of the world with the exception of the Western Hemisphere. Their approach is a continuous and fully mechanized biological process within which the organic content of normal household refuse and sewage is transformed into a compost under hygienic conditions. Sewage sludge does not have to be processed along with the refuse, but it may be added if desired. More than 100 composting plants are now operating in more than 20 countries throughout the world.

Composting is a biological process where fresh organic wastes are transformed by decomposition into a stable humuslike substance. The processing of organic wastes, unlike the backyard compost pile, is accomplished in a mechanized process, resulting in an organic manure which can be used agriculturally. As with the backyard compost pile, the microorganisms which effect the decomposition are indigenous to the wastes themselves; but when there is a variety of domestic wastes, granulation of the dry waste is necessary along with thorough mixing to ensure homogeneity and definite control of moisture content and complete aeration of the material being treated. Under these conditions the wastes decompose rapidly and create decomposition temperatures which are lethal to pathogenic organisms.

A completely automatic system involves several steps. The crude refuse is dumped into a container onto a belt conveyor, and then iron or metallic particles are removed by a magnetic separator. There is also a means of extracting nonmagnetic metal materials which will not decompose. The degradation of the material actually occurs in a rotating cylinder similar to a rotary drier. This cylinder contains as much as 5 days' supply of refuse, depending upon the degree of treatment required, and these wastes are delivered to the cylinder in a moistened condition by the addition of water and/or sewage sludge. This cylinder rotates slowly on large "tires," much like a cement kiln, and the wastes move from one end toward the other. They are thoroughly mixed and granulated by abrasion. Air is added at low pressure and in controlled amounts throughout the length of the cylinder. In this manner an environment is created where the action of aerobic microorganisms ensures a rapid decomposition of the wastes under inoffensive conditions. Normal operating temperatures up to 140°F are spontaneously developed. The final process material is then delivered onto a conveyor belt and through a screen where the compost is separated from more durable matter and delivered into a storage container, where splinters of glass from bottles, etc., are extracted. From 100 tons of crude refuse it is possible to extract between 60 and 70 tons of compost.

The resulting compost is suitable as a fertilizer and soil conditioner for flowers, vegetables, shrubs, and fruit trees. It is usually applied during autumn and winter months to the ground at rates from 5 to 20 tons to the acre and then worked into the topsoil.

Cost information on composting operations is meager, but in no case should it be considered a moneymaking proposition. It is a means of getting rid of a troublesome waste and developing a usable product, but it must be subsidized by municipalities in the same manner that a sewage treatment plant would be subsidized. The city or town is the logical purchaser and operator of automatic composting equipment, and the cost of operation should be borne by the tax structure. A "guesstimate" of installed cost per unit of capacity based on American costs would range from $25,000 to $30,000 per ton per day of capacity for small plants down to $10,000 per ton per day of capacity for larger plants. Operating costs would appear to be minimal and involve electric power primarily. No direct information is available on maintenance.

Sanitary Landfill

One of the solid waste disposal methods which is receiving the most publicity today is the sanitary landfill method. For years open dumps were located in undesirable locations on land which would probably never be used for buildings or highways. As our population increases, we now begin to realize that almost any open piece of land eventually will be used for another purpose. At the same time we have found that we cannot burn all our solid waste materials in municipal incinerators and we are not permitted, in many cases, to burn materials on open dumps. To complicate this, open dumping is no longer aesthetically desirable nor is it sanitary. Landfill then becomes a matter of filling in depressions with trash and garbage, covering it over with 6 or 8 in. of soil, and allowing nature to do the rest.

Good landfill operations, properly conceived, can be beneficial. Low swampy areas, preferably near industrial complexes and not those used for wildlife reserves, can be filled in and leveled, using trash from domestic operations. Generally the earth is removed or scooped out by means of dredges or bulldozers and the trash trucks dump their refuse into the void. The bulldozer then distributes the waste over the area, compacts it, and covers it with 6 in. of soil. More trash can be added and covered in the same manner until the land has been brought up to a level suitable for building or construction.

There are several pitfalls in this operation, however, which should be carefully considered. First we should be sure that the landfill operation is not just an expediency for disposing of large quantities of trash, but that it will eventually serve a purpose of reclaiming lowland in a valuable area. Compacted material will usually provide much better landfill than noncompacted material and will prevent later problems of erosion, because the soil is covering a porous substratum in terms of the material which has been dumped into the void. It is therefore desirable that the waste be as uniform as possible and that it contain biodegradable material which will eventually become topsoil. If the material to be dumped into the landfill is not organic in nature, at least it should retain its original form and shape and not be compressed to any greater degree. Plastic bottles and similar containers constitute the real problem in this regard. Land is usually not usable until several years after a landfill operation is complete so that degradation of the material can take place. Compaction of the surface is necessary to ensure proper soil bearing qualities before construction can begin. Soil mechanics is an important consideration in landfill operations.

We should also consider the possibility of contamination of nearby water sources. Wastes which are now used for landfill and contain water soluble materials can be leached by rainwater to remove these materials, which will ultimately contaminate nearby springs and streams, possibly rendering water sources unusable. In some cases even toxic contaminating materials may be present in the waste and will be removed over a period of time. Our concern should be that the material placed in the landfill operation does not present this problem in the future.

Landfill operations always create the possibility of rodents and vermin because much of the material which we are putting into the landfill is domestic garbage upon which rodents feed. Odor problems can be kept under control by proper landfill covering operations.

Perhaps the most important consideration of a landfill operation is that of conservation of our wildlife and natural resources. Just as we can destroy natural resources by the indiscriminate use of a deep well, we may destroy some of the most valuable cover for birds and animals by filling in swamps and lowland meadows or even abandoned quarries which now act as sanctuaries. Certainly we need to reclaim land for the human race, but we should also consider the contributions made by this wildlife and do everything possible to protect its stake in the future.

The city of Philadelphia and state of Pennsylvania are presently considering a program which will permit Philadelphia to ship its trash out of town on rail cars after compaction and baling. This trash will be hauled to the anthracite strip mines of northeastern Pennsylvania and used to refill and reclaim abandoned mines. The cost of this operation will presumably be $2 per ton less than present incineration methods. Philadelphia is now burning more than 800,000 tons/year. This project is controversial, to say the least, and demands careful study before it is undertaken.

The overall economics of landfill or burial methods are affected by several factors. The first is the cost of transportation from the source to the burial site. The second is the cost of the land, and the third is the cost of the landfill operation itself. The latter is a minor part of the total, since it usually requires only one or two bulldozers and operators on an 8 hr/day basis. The waste, however, must be brought to the site and the resulting fill compacted. With the cost of transportation and land increasing at alarming rates, this method may soon be less attractive than when it was first proposed. While many quantitative estimates have been made for various landfill operations, there are few which have been accurate over a 5-year period. One small suburban municipality estimated that an abandoned quarry would provide sufficient landfill for 10 years of normal refuse. Only 2 years after that prediction, it appears that 4 years may be the maximum if the land is used wisely and selectively.

The costs of sanitary landfill have been estimated to range from $2 to $5 per ton for small operations and from $0.75 to $2.50 per ton for large operations.

Good sanitary landfill operations have made valuable contributions to the reclamation of otherwise unusable land. A specific project in Palos Verdes, Calif., which met with initial serious objections, has turned out to be a boon to the community by reclaiming a canyon which had previously been a scar on the landscape.

Chapter 8

The Coordinated Industrial-Municipal-Regional Approach to Air and Water Pollution Control

KENNETH S. WATSON*

Director of Environmental Control,
Research and Development Division,
Kraftco Corporation, Glenview, Illinois

INTRODUCTION

Unless proper attention is given to the operation of industrial manufacturing plants, there is a possibility of creating air and water pollution problems in the communities where the plants are located. Although there are similarities as well as differences between the two types of pollution, this chapter will consider the coordinated regional approach to both types of problems.

Water Pollution

The sources of water pollution and its effects on the welfare of people do not begin or end at political boundaries. Thus the approach of solving such problems by region has been practiced rather extensively in the United States.

* Formerly Manager, Industrial Environment Control, Research and Development Center, General Electric Company, Schenectady, N.Y.

The largest regions are usually regulated by interstate agencies developed to coordinate and correlate water pollution control efforts in an entire river basin. These agencies have been formed because their proponents reason that the control program for a basin or watershed should have a uniform approach in each of the states in the basin or at least that the programs of the various states should be coordinated. Table 1 is a summary of the interstate compacts now in effect.

TABLE 1 Interstate Pollution Control Compacts

	Ratified	Signatory states	Area, sq miles
Ohio River Valley Water Sanitation Commission	1948	Ind., W. Va., Pa., Ill., N.Y., Ohio, Ky., Va.	15,822,337
Interstate Commission on Potomac River Basin	1940	Md., W. Va., Va., Pa., D.C.	2,100,000
Interstate Sanitation Commission	1936	N.Y., N.J., Conn.	11,000,000
Interstate Commission on Delaware River Basin	1936,* 1961	Del., N.Y., Pa., N.J.	5,000,000
New England Interstate Water Pollution Control Commission	1947	Conn., Me., Mass., N.H., R.I., Vt.	6,739,309
Klamath River Compact Commission	1957	Calif., Oreg.	7,500
Bi-State† Development Agency	....	Mo., Ill.	3,567
Tennessee‡ River Basin Water Pollution Control Commission	1949	Tenn., Ky., Miss.	40,000

* Originally in 1936 but reenacted in 1961 to provide a broadened base of authority.
† No longer active in the field of water pollution control.
‡ States of Virginia, North Carolina, Georgia, and Alabama also eligible for membership. No direct action toward further ratification as a result of small area of involvement on part of remaining states and good progress through informal cooperation.

The smallest type of region is probably one consisting of a small municipal area in which the municipality has gotten suburban settlements in the area and the industries to work together toward solving their waste water problems.

In between the large and small regions mentioned are the many sanitary districts of the nation formed to handle the liquid wastes of an area.

When the coordinated regional approach is in use in the field of water pollution control, it indicates that the organizations with waste discharges in an area with geographic unity and common interests have agreed to cooperate in solving the waste problems of the total region. In addition, the approach is in many cases a joint one whereby the organizations in the area make use of common collection systems and treatment plants. In the case of the interstate agencies, the area encompassed is too large for the joint approach; so these agencies are an organizational mechanism by which a river basin program is carried out.

The joint approach is gaining in usage for a number of reasons. One of the major ones is that it represents the most economical means by which an area may provide waste water collection and treatment. In other words, treatment for the entire area can most economically be provided in a small number of area plants rather than each contributing organization providing its own treatment facilities. Such savings would apply to the provision of treatment facilities, their operation, and in many cases even the collection system.

Another justification for the combined approach is that it tends to put treatment facilities, which can have a direct impact on public health, in the hands of full-time professionals. If every small organization in an area has a treatment plant, the likelihood that all will be professionally operated is not outstanding. With a small number of larger treatment plants in operation, professonal operation is considerably more likely.

Air Pollution

Air pollution is a more recently recognized type of pollution affecting the common welfare; so the amount of time for the evolution of methods and techniques for the solution of such problems has been more limited than with water pollution.

One advantage of considering the two types of pollution in the same chapter is that in a considerable number of cases they are interrelated. Further, this approach permits one to review the water field for indications of how the rapidly evolving air pollution control field may develop.

There are fewer examples of the regional approach to air pollution control than water pollution control, because the former field is newer. In the final analysis, however, the regional approach to air pollution control may have greater ultimate significance than with water pollution. This appears true because the regional approach to air pollution control seems to be the soundest one available. The formation of localized pollution control programs seems to be dictated because most effective work can be done in a geographic area affected by the same or common sources of pollution.

In other words ambient air quality standards, if any, for a region are related to the stack emissions affecting it. These emissions will be regulated to protect the region. The proponents of the uniqueness of area doctrine strongly support the regional approach.

One of the differences between water and air pollution is that the former is restricted to channels into which it is discharged. Another difference is related to the regional approach. In air pollution control the regional approach covers the legislature, administrative, and supervisory mechanism under which a program is carried out. Each contributor to the area problem would seek to solve his problem by a sound means. In water pollution control, as already mentioned, the problem is often solved in common facilities.

THE JOINT APPROACH IN WATER POLLUTION CONTROL

In water pollution control the joint version of the regional approach has proved popular in the past and is destined to be more widely used in the future. The fact that this is a widely used approach is borne out by the fact that at least 42 states have enabling legislation authorizing the creation of districts for the handling of community waste water and other purposes. The states of North and South Dakota and Georgia, Alabama, and Mississippi appear to be among those where enabling legislation would have to be ratified before sanitary districts could be created. Since Hawaii and Alaska are relatively new states, they have no enabling legislation of this type to date.

Since the number of industries connected into district systems or making use of the joint approach is not available, the growth of this desirable trend can probably best be shown by summarizing statistics on district systems. Obviously where district systems are in use industry is making use of the joint approach.

The number of sewage disposal districts or joint approach areas has increased decidedly in recent years. Table 2 shows the increase between 1952 and 1962. It will be noted that the 937 districts in operation in 1962 are more than double the 429 such districts in operation in 1952. The decrease over this period in the number of such districts in operation in the Southwestern region probably resulted from combining smaller regions into larger ones.

To obtain a better understanding of the portion of the current treatment facilities which are for authorities and for which construction with federal funding was approved, some study was given to the project register of the Water Pollution Control Administration for the month of August, 1967. This review permitted the development of Table 3, which shows total approved projects of 66 compared with 17 of the authority type. Further, of the total estimated cost of almost $56 million, about

TABLE 2 Sanitary Districts or Authorities in the United States

Region	1962	1952
Northeast	337	36
Midwest	169	107
South	74	53
Southwest	6	19
West	351	214
Totals	937	429

TABLE 3 Proportion of Total Projects Approved during August, 1967, under Section 8 of the Water Pollution Control Act Which Were of the Sanitary District or Authority Type

States	Total projects		Sanitary district or authority	
	Number	Estimated cost	Number	Estimated cost
California	6	$10,025,692	3	$ 8,296,892
Colorado	5	633,000	2	148,800
Connecticut	3	5,453,000	2	1,586,000
Illinois	3	3,161,600	1	2,973,000
Kansas	3	1,276,100	1	84,000
Montana	3	262,700	1	91,300
New York	5	13,168,500	1	460,000
Texas	2	4,333,500	1	4,288,000
Virginia	1	748,100	1	748,100
Washington	6	2,936,100	4	1,187,800
Subtotal	37	41,998,292	17	19,863,892
Remaining states	29	13,584,700		
Grand total	66	$55,582,992	17	$19,863,892

$20 million, or some 36 percent, represents the cost of facilities for sanitary districts or areas in which the joint approach is being used.

The joint approach to water pollution control takes a variety of forms. In the interest of adequately covering the subject, three significant areas of the country which make use of the more conventional versions of the joint approach will be reviewed. Further, three less frequently used but interesting and significant versions of the joint approach will be considered in minor detail.

Conventional Cases

Chicago Metropolitan Sanitation District This is one of the oldest organizations of this type in the nation, having been created in 1889. The district covers some 858 sq mi and almost all of Cook County. This organization is a greater Chicago governmental agency responsible for all aspects of the collection and treatment of industrial, storm, and sanitary waste water delivered to it by the communities and industries of the area. It is therefore an example of metropolitan government handling one of the key services which proponents of greater than city government claim can best be handled in such manner.

This Chicago system serves three principal drainage basins and has three waste water treatment plants. Figure 1 is a flow diagram of a sewage treatment works. Figure 2 is a picture of the West-Southwest sewage treatment works. Northern

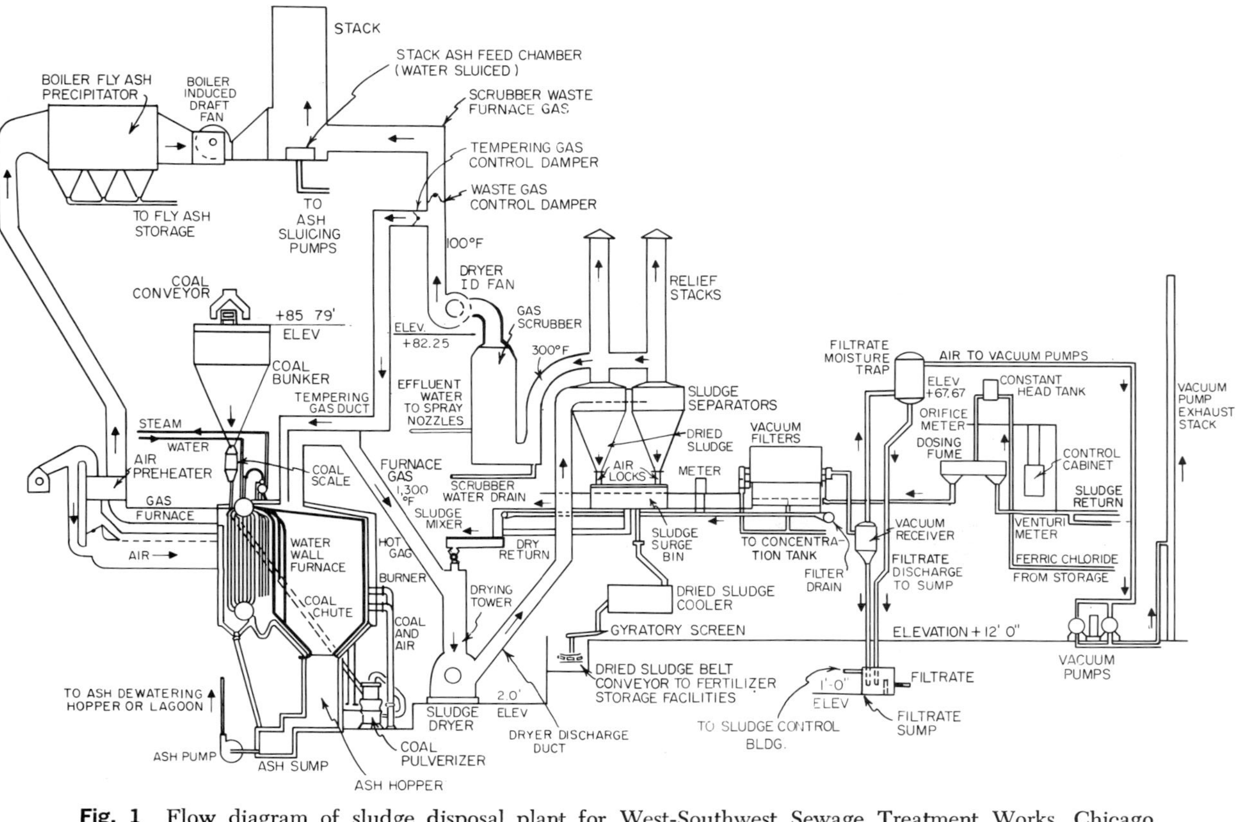

Fig. 1 Flow diagram of sludge disposal plant for West-Southwest Sewage Treatment Works, Chicago.

Chicago plus a total of 20 smaller local sanitary districts, an area populated by 5½ million people, is connected to this system.

The district constructs interceptor sewers only, and all local sewers and small trunks are constructed and operated by local governing bodies. The system is presently financed by ad valorem taxation, but study is presently underway to evaluate the desirability of adopting the surcharge ordinance approach.

Industries can connect directly into the district system or into the system through a local district or municipality. About 11,100 industries located in the district are at present connected to the metropolitan system.

Fig. 2 Chicago's West-Southwest Sewage Treatment Works services one drainage basin. The city's system includes three waste water treatment plants.

Cincinnati The city of Cincinnati, being the largest municipality in that area, elected to assume responsibility for providing sewer main and treatment plant capacity for industries and communities desiring to make use of such services. The city has for many years permitted wastes from industrial plants located both inside and outside the municipal limits to be connected into its sewer system. Preparation for the building of treatment facilities starting about 1949, however, led the city to give detailed consideration to all the factors necessary to make the joint approach a beneficial one for the entire area. An industrial wastes ordinance was passed in 1953.

The Cincinnati system serves some 215 sq mi of area. It covers three principal drainage areas and operates three treatment plants. Figure 3 is a photograph of the city's Little Miami treatment plant.

In addition to serving Cincinnati, the waste treatment system handles the waste waters from 29 surrounding cities and villages. A population of 850,000 people is presently served by the Cincinnati system.

About 2,000 industries are connected into the Cincinnati system. Industries can connect directly to the city system after appropriate negotiations or discharge into the system of a small community the waste water of which is in turn handled by the city.

Cincinnati used a sound and realistic approach for integrating industries of the area into the program. First, representatives of the Public Works Division met with all groups and organizations which represented a segment of industry to consider the joint approach. After the passage of the industrial waste ordinance, over 900 industries were asked to complete an industrial waste questionnaire. The original application of the surcharge was based on these questionnaire returns. The total industrial surcharge was determined by establishing the degree that the total waste

Fig. 3 The Little Miami treatment plant, Cincinnati, Ohio, is one of the three treatment plants. Cincinnati's system serves 215 sq mi with three principal drainage areas.

load exceeded that from sanitary origin. The individual industries were then charged based upon their portion of the total industrial load. As the industrial load changes, as established by analyses conducted by representatives of the city, individual plant charges are modified. After the charge basis was established, the industrial plants were given adequate time to determine whether they wished to provide pretreatment or reimburse the city for such treatment.

Allegheny County Sanitary Authority This authority built and is operating the sewer system and treatment plant under the Pennsylvania laws governing the establishment of municipal authorities. The authority is a public corporation with broad and independent powers which can borrow money and issue bonds and has many of the jurisdictions of municipalities. Such an authority can operate projects of many kinds in addition to waste water. The authority is operating as the metropolitan government for handling waste water in the greater Pittsburgh area. The city of Pittsburgh exercised leadership in launching this program by loaning funds for preliminary planning prior to the authority having any funds. As a result of the predominant population of the authority area living in the city, three members of the five-member governing board are named by the city.

The authority was formed in March, 1946. Thirteen years later, in April, 1959, the system and treatment plant were put into operation.

The authority is presently serving 71 of the county's 128 municipalities. Some 69 miles of intercepting sewers are involved as well as 30.5 miles of tunnel. The service area includes a population of 1,200,000 people, and all but 100,000 of these are presently connected to the system.

The area serviced encompasses 218 sq mi, or 75 percent of Allegheny County. The project had a cost of about $100 million, and the treatment is provided in a single plant. Figure 4 is a helicopter view of the plant in question.

The authority derives its entire operating, maintenance, and retirement of bonds budget from sewer service charges based on the quantity of water used by each customer. Municipalities connected to the system can elect to pay their charge as a lump sum or have the authority bill the individual premises within any particular city for services.

The authority uses the surcharge approach for industries with excessive concentrations of chlorine demand and 5-day BOD and/or suspended solids when compared

Fig. 4 Allegheny County, Pennsylvania, treatment plant serves 75 percent of the county, or 218 sq mi. This single plant installation cost approximately $100 million.

with typical sanitary waste water concentrations. Industries with effluent concentrations typical of the sanitary sewage of the area or lower pay on a volume basis. About 431 industries are connected into the system.

Unique Cases

South Charleston The joint approach in use at South Charleston, W. Va., is an interesting and different one. In this case, a small city and a major industry combined in the joint approach. By mutual agreement, the Union Carbide Chemical Company conducted the pilot study program, to establish technical feasibility of the approach, designed, built, and is now operating the treatment plant. To keep Carbide from acting as a public utility, the treatment plant is owned by the city.

The activated sludge treatment plant which was placed in operation in 1963 had a cost of $5.8 million. Figure 5 is an aerial photograph of this waste water treatment plant. The city's share of the cost was $1 million, and the industry's share the remaining $4.8 million. The city spent an additional $4.2 million for storm and interceptor sewers, and Union Carbide invested an additional $1.8 million for the same purpose.

This plant has the further unique features of providing neutralization for the industrial wastes prior to combining them with the city waste water. Presently the plant is being expanded by the addition of a new aeration basin and two final clarifiers. This $1.5 million expansion is scheduled for completion in early 1968 and will provide secondary treatment for the total discharge.

The plant serves a population of 21,000. In addition the industrial wastes population equivalent amounts to 700,000. The plant processes about 17.7 mgd with two-thirds of this quantity being of industrial origin.

Fig. 5 The joint approach of city and industry is demonstrated by South Charleston, W. Va., and Union Carbide Chemical Company. The company conducted pilot study, designed, built, and now operates the activated sludge treatment plant. City owns plant after investing less than 20 percent of cost.

There are about 60 miles of sewer in the city system. Carbide employs about 2 miles of flume to collect the wastes from several miles of sewer and convey them to the treatment plant.

Los Angeles County Sanitation Districts The concept being used in the districts approach is an imaginative and significant one which is being followed by other California counties. The approach of consolidating individual community sewage services into a county effort was authorized by the County Sanitation District Act passed in October, 1923. This act gives districts formed most of the classic powers extended by such legislation, for example, employment of necessary specialized personnel, authority to own and operate facilities, and bonding and taxation authority.

In addition to authorizing greater than city-area districts, the act granted authorization for districts to own and operate joint facilities. As a result of leadership on the part of personnel in the Los Angeles County sanitation districts, this joint approach authorization became the major reason for the success of the approach in this particular area. The act provides for (and the districts organization has in effect) two key agreements, namely, joint administrative agreement and joint outfall agreement. Under this program, both debt service and annual operating costs are paid for by a direct tax on the real property of the district.

Under the county districts approach, originating in 1923, outstanding accomplishments have been achieved. Twenty-four districts are presently encompassed in the county districts system. The system contains 795 miles of trunk sewers which serve 6,850 miles of laterals. This regional system handles 327 mgd of sewage originating from a 688 sq mi area which domiciles 3.7 million people served by the system. The area served contains 70 incorporated cities as well as large tracts of unincorporated area. About 2,121 industries are connected into the regional system (by virtue of being connected into the system of a local district).

The county districts organization operates the joint primary water pollution control plant, which treats about 310 mgd. Figure 6 is a picture of this treatment facility. Figure 7, on the other hand, is the flow diagram for this plant. In addition to the joint plant, this county organization operates eight smaller secondary treatment plants, which handle about 26 mgd of waste water.

As a point of interest, this regional organization is also operating four major ocean outfalls. The interesting investigations and developmental work going into the installation and operation of these ocean outfalls is beyond the scope of this chapter.

Since the districts organization has no police powers, it depends upon the ordinances of the various member cities and the County of Los Angeles for enforcement. Control of industrial wastes discharging ultimately into the regional system is handled

Fig. 6 Joint treatment plant of Los Angeles County District provides primary water pollution control for over two thousand industries in a regional system.

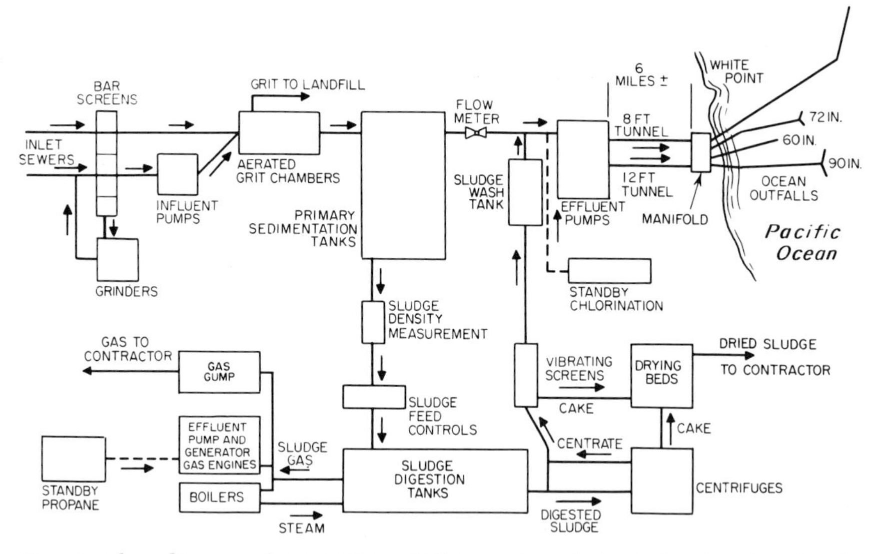

Fig. 7 Flow diagram of Joint Water Pollution Control plant, Los Angeles County Sanitation District. Plant treats 310 mgd.

in this manner. Industrial plants pay no surcharge for the waste water they contribute but are assessed on the basis of real property at the same tax rate as all other such property in the district.

Lyman, South Carolina This is another interesting example of municipal and industrial wastes being treated in a common system. The 9 mgd of textile finishing and dyeing wastes from the Lyman Printing and Finishing Company is combined with the 115,000 gal of discharge from the town of Lyman with a population of 1,600 people.

This treatment facility, which was placed in operation in late 1967, is included because it represents another example of the joint approach where the industrial contribution is considerably greater than that from the city. Figure 8 is a picture of the treatment plant consisting of aeration lagoons, settling basins, sludge aeration basin, sludge thickener, and sludge storage lagoons. As a result of earthen lagoon and basin construction, the treatment facilities had a cost of about $1.8 million.

Sewer Use Ordinance

The features of the sewer use ordinance are much the same when used by a community or by a joint sewer authority. If industrial wastes are involved, the ordinance should prescribe prohibitions and limits for such wastes. Such limits are specified to protect the sewer system as well as the treatment plant. In some cities and authorities, acceptance limits for industrial wastes are considered sufficiently significant that a separate ordinance, termed an industrial wastes ordinance, is in effect.

Fig. 8 Textile finishing and dyeing wastes are combined with domestic sewage in one treatment plant at small town of Lyman, S.C. (populaton 1,600).

Although sewer ordinances vary considerably from area to area, there are certain features which are common to most of them. Principal among these are:

1. Definitions of the terms used in the ordinance.
2. Regulations requiring the use of the sewer for properties inside and sometimes outside the city or district.
3. Regulations covering private sewage disposal systems.
4. Regulations covering specifications for building sewers and the connections into same.
5. Regulations controlling (or limiting) types of wastes which may be connected into a sewer system. This section normally prohibits the discharge of storm or cooling water into the sanitary systems.
6. Power and authority of inspection of those properties served.
7. Regulations covering protection from damage and penalties for violating usage specifications.

As a result of the special interest on the part of industrial representatives concerning discharge limits for connection into the joint sewerage systems, this subject will be given special attention in this chapter.

Protection of sewer system Each city sewer ordinance should have a section prohibiting the discharge of certain materials into the sewer system and setting concentration limits on others. The examples listed below are among those typical of this portion of the ordinance:

No person shall discharge or cause to be discharged to any public sewer any of the following substances, materials, waters, or wastes:

1. Any gasoline, benzine, naphtha, fuel oil, or mineral oil or other flammable or explosive liquid, solid, or gas.
2. Any liquid or vapor having a temperature higher than 150°F (65°C).
3. Any garbage that has not been properly comminuted or triturated.
4. Any waters or wastes which contain grease, oil, or other substance that will solidfy or become discernibly viscous at temperatures between 32 and 150°F.
5. Any waters or wastes containing emulsified oils and grease, exceeding an average of 100 ppm (833 lb/million gal) of ether-soluble matter.
6. Any cyanides, in excess of 2 ppm by weight as CN.
7. Any waters or wastes, acid or alkaline in reaction, having corrosive properties capable of causing damage or hazard to structures, equipment, and personnel of the sewage system. Free acids and alkalies must be neutralized, at all times, within a permissible pH range of 5.5 to 9.5.
8. Any ashes, cinders, sand, mud, straw, shavings, metal, glass, rags, feathers, tar, plastics, wood, paunch manure, hair and fleshings, entrails, lime slurry, lime residues, beer or distillery slops, chemical residues, paint residues, cannery waste bulk solids, or any other solid or viscous substance capable of causing obstruction to the flow in sewers, or other interference with the proper operation of the sewage system.
9. Any noxious or malodorous gas such as hydrogen sulfide, sulfur dioxide, or nitrous oxide or other substance, which either singly or by interaction with other wastes is capable of creating a public nuisance or hazard to life or of preventing entry into sewers for their maintenance and repair.
10. Radioactive wastes unless they comply with the Atomic Energy Commission Act of 1954 (68 Stat. Part D, Waste Disposal, Section 20.303 of the Regulation issued by the Atomic Energy Commission or amendments thereto).

Protection of treatment plant Each city sewer ordinance should have a section prohibiting or limiting the type of materials which might upset the operation of the treatment plant.

No person shall discharge the following described substances, materials, waters, or wastes into the sewer system if it appears likely in the opinion of the superintendent that such wastes can harm the sewage treatment plant:

1. Any waters or wastes containing toxic or poisonous solids, liquids, or gases in sufficient quantity, either singly or by interaction with other wastes, to injure or interfere with any sewage treatment process, constitute a hazard to humans or animals, create a public nuisance, or create any hazard in the receiving waters of the sewage treatment plant, including but not limited to cyanides in excess of 2 mg/l as CN in the wastes as discharged to the public sewer.
2. Any waters or wastes containing iron, chromium, copper, zinc, and similar objectionable or toxic substances; or wastes exerting an excessive chlorine requirement, to such degree that any such material received in the composite sewage at the sewage treatment works exceeds the limits established by the superintendent for such materials.
3. Any waters or wastes containing phenols or other taste or odor producing substances, in such concentrations as to exceed the limits which may be established by the superintendent for such materials.
4. Unusual concentrations of inert suspended solids (such as, but not limited to, fuller's earth, lime slurries, and lime residues) or of dissolved solids (such as, but not limited to, sodium chloride and sodium sulfate).
5. Excessive discoloration (such as, but not limited to, dye wastes and vegetable tanning solutions).
6. Unusual 5-day BOD, chemical oxygen demand, or chlorine requirements in such quantities as to constitute a significant load on the sewage treatment works.

Trend for future It would appear that most logical effluent standards or limits could be set on the treatment plant influent. Recent developments indicate some trend in this direction. In other words, the concentration of chromium or copper in the sewer lines themselves is not objectionable, if the concentration does not get high enough to create problems at the treatment plant.

The following is quoted from the Cincinnati Ordinance as representative of the newer trend toward permitting industry to make optimum use of community facilities in the practice of the joint approach:

> Materials such as copper, zinc, chromium, and similar toxic substances shall be limited to the following average quantities in the sewage as it *arrives at the Treatment Plant* and at no time shall the hourly concentration *at the Sewage Treatment Plant* exceed three (3) times the average concentration:
>
> | | |
> |---|---|
> | Iron as Fe | 15 parts per million |
> | Chromium as Cr (hexavalent) | 5 parts per million |
> | Copper as Cu | 3 parts per million |
> | Zinc as Zn | 2 parts per million |
> | Chlorine Demand | 30 parts per million |
>
> and with contributions from individual establishments subject to control in volume and concentration by the City Manager.

Figure 9 is a flow diagram of Cincinnati's Mill Creek treatment plant.

Plant Preparation for Joint Approach

There are a number of steps which may be taken to determine whether the joint approach is a sound one. Since modern industry should in most cases be taking such action anyway, these efforts will be considered in some detail.

These steps will be considered as applied to an existing plant, but they have equal application in building new plants. In considering the building of a new plant, industry should go through all the steps outlined and then plan and build waste treatment facilities as the plant is planned and built.

Characterize effluents By a program of sampling and analyses the nature and extent of plant wastes can be defined by plants which have not previously taken such action. The definition can be carried out by plant or company personnel or a qualified consulting organization. Since waste control is one of the costs of doing business today and will remain so in the future, waste definition by company personnel appears to have some advantages. By this approach personnel intimately familiar with the manufacturing operation gain a detailed knowledge of the nature and variation in plant wastes; thus the stage is set for a realistic control program.

Discharge approaches available Concurrent with the definition or immediately thereafter, an evaluation should be made of the advantages and disadvantages of the methods of discharge available to the plant. In general the plant will be able to discharge waste water for final disposal, depending upon degree of contamination and treatment, into:

1. Surface waters
2. Municipal sanitary systems
3. District sanitary systems
4. Storm sewers
5. Combination of methods listed

Information developed in the definition survey is essential in establishing the optimum mode of discharge. In completing this step, a detailed knowledge of the sewers and treatment plants of the area is necessary. A conference with the appropriate city or district sewer system representative will yield accurate information. Further, consulting firms are qualified to make a contribution in this area if called upon.

Segregation Most of the facts have now, through steps already outlined, been collected to permit plant personnel to consider an optimum waste collection and treatment approach. Uncontaminated water should be segregated from contaminated water to reduce the total industrial discharge and hence the size of treatment

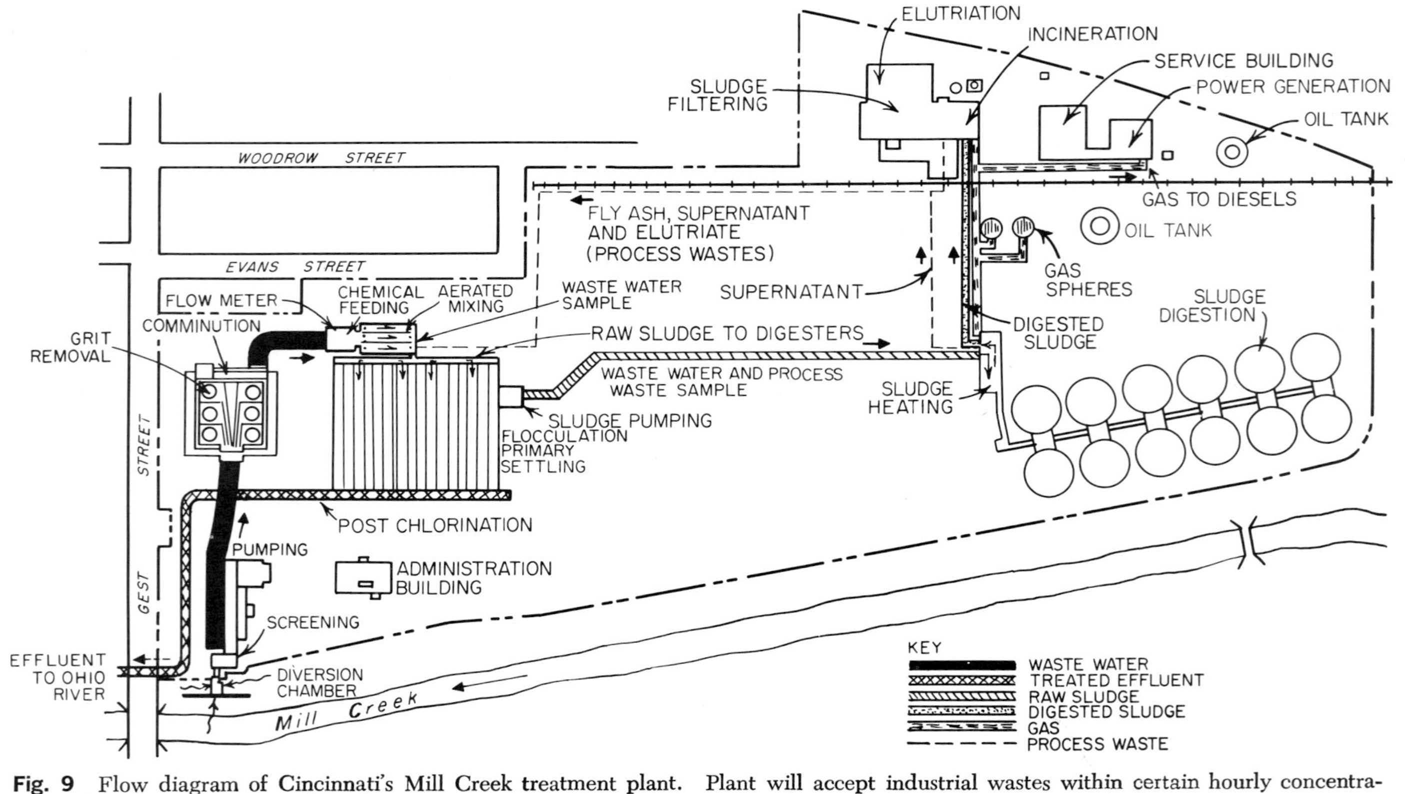

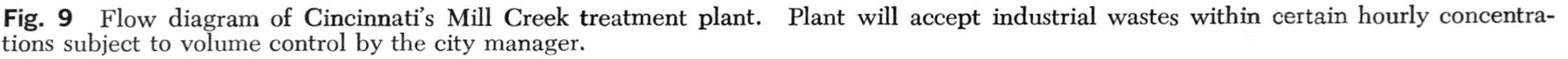
Fig. 9 Flow diagram of Cincinnati's Mill Creek treatment plant. Plant will accept industrial wastes within certain hourly concentrations subject to volume control by the city manager.

facilities. Some types of effluents will need to be segregated from others to permit optimum treatment or pretreatment. In other cases some separated effluents, such as acid rinse waters from one area of the plant and alkali rinses from another, will need to be combined.

Recirculation Since cooling water should be kept separate, the cost of the separate system as well as water costs of the one-pass usage approach should be thoroughly evaluated. In most cases it will be more economical in cost of water and equipment to provide for recirculation and reuse of water than to use a one-pass cooling system. Today and in the future the proper use of a vital natural resource is going to dictate the more extensive use of recirculation systems.

Treatment approach As plant personnel, in many cases aided by a consulting firm, become familiar with all the details of the waste problem, treatment and pretreatment will come up for study. The discharge approaches available, among other things, have a significant bearing on the extent to which treatment is necessary.

In cases where the waste problem is complex, some bench top and pilot scale

Fig. 10 Pretreatment plant at General Electric's Appliance Park processes waste water prior to discharge into Louisville, Ky., and Jefferson County Metropolitan Sanitary District's system.

work may be necessary. Efforts and funds spent in this area usually yield outstanding dividends because they develop all the necessary information to permit the final design to tailor the treatment facilities to the problem.

There are some types of wastes, primarily of an organic nature, which can be discharged without pretreatment into the community sanitary system. Such a course can particularly be followed if the industry has evaluated the alternatives and elected to pay a surcharge to the community system for waste handling service. Figure 10 is a picture of General Electric's Appliance Park Treatment Plant, which pretreats process waste waters for discharge into the Louisville and Jefferson County Metropolitan Sanitation District's system.

In other situations perhaps the plant may decide to provide complete treatment facilities and discharge into the nearest stream. Such facilities, depending upon the type of wastes involved, could handle both the process and sanitary wastes. In some cases, dual facilities must be provided to handle the wastes separately. Figure 11 is a picture of General Electric's Erie Treatment Plant, which treats the process wastes from the Erie operation for discharge into the lake. At this location the sanitary wastes are connected into the city sewer system.

The final choice After evaluating all the pertinent factors, of which most of the major ones have been outlined, it will be possible to decide upon an optimum course of action for the existing or proposed plant. This decision would next be implemented by designing and building proper treatment facilities. Industry which does not have a central engineering organization would probably retain a consulting engineering firm to design the waste treatment facilities.

After the facilities have been built, they must be properly operated. This is particularly important in order to justify the expenditure and further to protect a vital natural resource, the nation's surface waters. Such sound operation should be expedited by assigning definite responsibility for this area just as responsibility is assigned for safety and quality control.

Fig. 11 Dual systems, one for process waste (photograph of General Electric's Erie, Pa., treatment plant), and plant sanitary waste connected to Erie's city sewer system handle wastes separately.

Financing the Operation

Everyone is concerned when it comes to financing a sewerage system for his area, and this is perhaps even truer for a regional system. Greater interest is perhaps indicated in a regional system because the area to be served is larger in size, more organizations are involved, and thus it becomes a more complex undertaking to achieve completely sound costs for everyone concerned.

Persons concerned with equitable rates are continually challenged by the fact that the sewer systems and treatment plants must handle the volume of waste water generated and reduce the objectionable characteristics to the point where they do not unduly degrade the receiving stream and the fact that such facilities are quite expensive. One of the key features of this challenge is that the waste load from each contributor can be defined, as can the total load reaching the treatment plant. Thus it should be possible to develop a formula requiring a contributor to pay his equitable part of the cost of the sewer system, the plant, and its operation.

With the degree of sophistication in monitoring the quantity and quality of waste water available today, it seems that the matter of equitable rates is becoming more and more a function of the willingness to give the matter adequate attention.

Industry is no exception when it comes to being concerned with the financing formula, because the quantity of waste water discharged by industry is often large and at times the concentration may be of considerable magnitude.

In the roughly 10,000 communities throughout the nation which have sewerage facilities, many different methods are used for raising the revenues required to construct and operate them. Perhaps in an introductory vein, it would be desirable to enumerate the common methods.

The ad valorem tax on property is the oldest method being used. Probably the next oldest approach is the special front footage assessment method, which is still widely used as a method of financing collecting sewers which abut individual properties.

There have been many cases of special negotiated contracts between municipalities and industry so that the community could accept and handle the industrial discharges.

The flat rate method of financing has been in use for many years, and since it has long been the accepted approach of many municipalities without an extensive program of water metering, it will remain in slowly declining use for years to come. The special assessment method has also been used in the past and will no doubt receive some limited use in the future.

The method of charging for sewer service on the basis of water used has grown in acceptance in recent years as a sound basis for charge. As a point of interest, it would appear that the "sewer service charge" designation usually used in referring to this method of financing is not in optimum form from a semantics standpoint. In more accurate terminology it should be called "sewerage service charge" because the service encompasses the use of both the sewers and the treatment plant.

In the last few years the availability of federal and state grants has reduced the amount of funds which must be spent by local jurisdictions for sewerage facilities.

Perhaps more important than the individual techniques of financing is the overall method by which these techniques are used singly or in combination to permit the political subdivision to provide this important service. Thus the most significant approaches now in use will be briefly discussed and the apparent future potential of each delineated.

Flat rate formula This method of financing is based on the premise that financing of a sewerage system can be equitably worked out without being concerned about the quality of wastes generated by each contributor. Originally, for example, the rate which the individual contributor paid was based upon the number of plumbing fixtures in the home. In the case of industry, however, the rates could be based upon the number of fixtures, the square feet of factory area, the number of employees, or some other method of arriving at an estimate. Prior to extensive information being available on discharge quantities or analytical characteristics of the waste water, this represented a method probably in general no better or worse than the ad valorem tax method.

Today and undoubtedly in the future, since more extensive knowledge is being gained on the characteristics of various waste waters, it is obvious that more sophisticated formulas will be used in financing waste treatment.

Volume modification Since more detailed information has become available over the years on the volume of waste water generated by various industries, the flat rate in many cases has been based on volume. This represents a step up in accuracy because many features in a sewer system and treatment plant are related to hydraulic loading. Further, based upon sound principles of setting rates in the water industry, the sliding scale concept of charging larger industrial users for sewage service at a lower unit rate has rather wide acceptance.

In spite of this volume approach, some thought concerning the subject and its compexities indicates that a completely equitable service charge basis cannot be derived based upon quantity of discharge alone. Further, it is apparent that, as an effort is made to factor the strength of the wastes into the picture, the formulas must become more complex in the interest of accuracy.

Quality-quantity formula In the quality-quantity approach, commonly called the Q-Q approach, the concentration of the waste water in terms of 5-day BOD, suspended solids, and sometimes chlorine demand is factored into the basis for charging the contributor.

An appreciation for the use of this type of formula can be gained by assuming a particular case and working out the particular charges. In the interest of simplicity, we will take the approach used by the British and assume that treatment costs can be divided equally three ways between the volume of the waste water, 5-day BOD, and the suspended solids.

Our hypothetical sewerage system was designed for 20 mgd and is up to capacity. Further, the municipal waste water has an average concentration of 300 mg/l of 5-day BOD and 250 mg/l of suspended solids. The combined debt service and operating costs amount to $600,000 annually. The industry under consideration is connected into this system and discharges 2 mgd of effluent with an average concentration as follows: 5-day BOD 600 mg/l and suspended solids 400 mg/l.

Therefore, the volume cost of $200,000 must be divided by the annual discharge to obtain a unit volume cost as follows:

$$\frac{\$200{,}000 \times 10^3 \text{ gal}}{2 \times 10^7 \times 365 \text{ days}} = \$0.0273 \text{ per } 1{,}000 \text{ gal}$$

The cost of treating the 5-day BOD can be calculated similarly:

$$\frac{\$200{,}000 \times 10^6}{2 \times 10^7 \times 300 \times 8.33 \times 365} = \$0.0109 \text{ per lb 5-day BOD}$$

When computations are completed on the suspended solids in the same manner, a cost of $0.0132 per lb is obtained.

Now, since the unit costs for handling each characteristic have been determined, the annual cost per year can be computed for the industry:

Volume $$\frac{2 \times 10^6 \text{ gal per day} \times 365}{10^3} \times \$0.0273 \text{ per } 1{,}000 \text{ gal} = \$19{,}929$$

5-day BOD $$\frac{2 \times 10^6 \text{ gal per day} \times 8.33 \times 600 \times 365}{10^6}$$

$$\times \$0.0109 \text{ per lb 5-day BOD} = \$39{,}769$$

The cost of handling the suspended solids using the same approach is $32,107. Thus the total annual charge for sewer service for the industry would be $19,929 + $39,769 + $32,107 = $91,805.

Although the use of the Q-Q formula alone is not fully equitable for industrial contributors to a system, it introduces the basis on which large industrial users can be surcharged based on the degree to which their effluent concentration in 5-day BOD and suspended solids exceeds that for sanitary sewage. This is a significant step toward an equitable approach because the handling of such excessive loads requiring additional plant capacity will be partly at least financed by the contributors. The adoption of a formula of this type makes it possible for the industries in a region to determine whether it is more economical to pay the regional authority for treating their total waste waters or pretreat in their own facilities prior to connecting into the regional system.

Joint group approach Although the Q-Q approach was a decided improvement over past practices in establishing equitable rates, experienced people still realized that it did not make provision for charging the underdeveloped properties to be served. Such properties should pay their proportional portion of the cost of the system which improves the entire area as a place in which to live and operate. As a result of such conclusions, a joint group or committee was created in about 1948 to study the matter of rate structures. This committee issued a report in 1951 titled "Fundamental Considerations in Rate and Rate Structures for Water and Sewage Works." This group consisted of committees of the American Society of Civil Engineers, the Section on Municipal Law of the American Bar Association, as well as representatives of six other organizations including the Water Pollution Control Federation participating.

After giving the rate matter considerable thought, the joint group outlined in its report what were construed to be the fundamental principles involved. Some of these are:

> The needed total annual revenue of a water or sewage works shall be contributed to by users and non-users (or) by users and properties, for whose use, need, and benefit the facilities of the works are provided, approximately in proportion to the cost of providing the use and the benefits of the works.
>
> It is considered that rates and rate structures for water and sewage works should achieve fairness through payment by each user or beneficiary of his fair share of the total annual cost of the works required and no more.

Thus the joint group performed a real service by extending the quantity-quality approach so an equitable method of charging future users and undeveloped property became available. Certain cities and consulting engineering firms had already started to move in this direction prior to the issuance of the joint group's report. This report nevertheless served to consolidate the soundness of such an approach.

In 1951, Dr. George J. Schroepfer, Professor of Sanitary Engineering at the University of Minnesota, published a classic article in the *Sewage and Industrial Wastes Journal* (vol. 23, no. 1493, December) which completely spelled out how the joint group concept could be used to establish an equitable basis for charge for sewer service. J. F. Byrd of the Procter and Gamble Company has also written rather extensively on the merits or benefits of this approach.

Interest in the joint group philosophy and the most equitable approach to charging for sewer service continues. This is borne out by the fact that under the leadership of the American Society of Civil Engineers a new committee has been appointed to summarize again the best thinking on the basis for charging for newer service.

This committee, which consists of three representatives of each of the following organizations, has had two meetings in a series of meetings leading to a final report:

American Society of Civil Engineers

Water Pollution Control Federation

American Public Works Association

Dr. George J. Schroepfer of the Universtiy of Minnesota is chairman of this committee. Since the subject is complex, some time will be required before a report can be issued. Since the area under consideration is significant and the approach proper, the new committee deserves brief mention here.

Some detailed consideration needs to be given in an industrial wastes handbook, because industry should be informed when the joint approach, which is growing in acceptance, comes up for consideration. With proper background on the subject, industrial plants can take a positive position on such an approach if plant wastes are compatible with such a course of action.

The annual revenue for the operation of a sewerage system is required for the maintenance of the collection system, the maintenance of the intercepting sewers, and the operation and maintenance of the treatment plant. The final major item is the fixed charges or debt service, which includes interest and amortization of the bonds required to finance the construction of the project.

Since the joint committee makes the point that properties should pick up part of the cost of providing a sewerage system, perhaps this point demands brief consideration. In exploring this thought, surface and subsurface drainage is related to property or area and has no relation to use in terms of the quantities of waste water generated by the present user. Further, the provision of additional capacity in the sewer system and treatment plant to provide for development of property and future growth of the community should not equitably be charged only to the present users. Thus these two items are ones which should equitably be at least partly charged against the area or property.

In contrast to these types of items, a portion of the collection and interceptor sewers is provided to take care of the present user, and he should finance this.

Much of the cost of operating the plant and to a lesser degree its fixed charges can be soundly charged to the user.

The consulting engineer who has been retained to design the sewerage system and in many cases to develop a satisfactory finance formula for the city system becomes the key man in selecting an equitable approach. In many cases, experienced city waste water treatment personnel working with the consulting engineer will make a major contribution in developing the finance formula. This type of combined approach is indicated because a detailed analysis of at least the following types of information must be completed:

1. A characterization of the typical sanitary sewage and the contribution of each major industry
2. Waste water discharge broken down into volume of flow from each major source
3. Assessed valuation of total community as well as that of major industries
4. Detailed breakdown of the cost of the interceptor sewers and the treatment plant
5. Operation and maintenance costs of the sewer system and treatment plant
6. Detailed breakdown of fixed charges

Since this approach is complex, it will only be possible, as a result of length limitations, to cover major considerations here. Perhaps, therefore, the matter can most succinctly be treated by assuming that a rate schedule is to be developed for a community with a present population of 50,000 and with three major industries connected into the city system. The treatment plant and intercepting sewers have recently been constructed. Collecting sewers were also included but will be omitted from these assumptions in the interest of simplicity.

The present volumes from the various sources as well as the basis of design for the new plant and intercepting sewers are summarized in Table 4.

TABLE 4 Flow Information on the Sewer System and Treatment Plant

Source	Present avg discharge, mgd	Basis of design, mgd	
		Plant	Sewers
Sanitary waste water	7.2	10	20
Major industries:			
1	0.9	1.8	3
2	0.6	1.2	2
3	0.5	1.0	2
Infiltration	0.5	0.7	2
Storm water	....	....	5
Public use	0.3	0.5	1
Total	10.0	15.2	35

With the operation, maintenance, and fixed charges figures available, the next assignment is to distribute these costs equitably against the users and properties benefiting from the system. This is not a simple procedure, and extensive technical judgment is involved in some areas.

Table 5 shows how the fixed charges on the sewer system can be broken down against the properties and users. Further, Table 6 shows how the fixed charges on the treatment plant can be broken down in accordance with the types of wastes handled. For example, the proportion of the fixed charges chargeable to the future and thus the property insofar as BOD and suspended solids are concerned can be estimated based upon the percent that the design load is greater than the present load.

into play, industrial plants located in the geographic regions in question will be participating in this type of program.

1967 Air Quality Act The new air quality act passed in November, 1967, by Congress recognizes the significance of the regional approach by specifying that the federal program will encourage cooperative activities by the states and local governments as well as the formation of interstate compacts. As a further encouragement of the regional approach, the Secretary of Health, Education and Welfare can pay two-thirds of the cost of establishing such agencies and up to one-half the operating cost. A region is described in the act as including the area of two or more municipalities.

The act further authorizes the Secretary, in the formation of interstate air quality control regions, to pay the total cost of the same for a period of 2 years, after which up to three-fourths of the operating costs can thus be financed.

Since none of the interstate compacts have thus far been approved by Congress, this body is apparently waiting for the Public Health Service to take action directed prior to granting such approvals. The Air Quality Act specifies that the Secretary will, within 18 months after date of enactment and after consultation with all pertinent authorities, designate air quality control regions. Apparently the areas falling under the jurisdiction of the compacts proposed will need to be negotiated by interested parties to be sure that the present geographic limits agree with those arrived at by the Public Health Service, prior to being approved by Congress. Conceivably some compact areas will need to be adjusted in order to obtain congressional approval.

Federal grants In order to develop a further feel for the growth of the regional approach in air pollution control, federal grants for 1967 were reviewed. During this year, five grants were made to regions involving more than one county. Thirteen grants were made to city-county combines and 24 grants to counties. Grants to cities and states totaled 80; so over one-third of the total number of grants went to political subdivisions representing some type of region.

National sampling efforts The Public Health Service set up the National Air Sampling Network (NASN) to measure solid pollutants in the air in 1953. This approach, in which the Service provided equipment and local agencies provided manpower, grew to 66 stations by 1957.

By 1960, it was apparent that gaseous pollutants in the air also represented a threat to health and welfare; so in 1962 the Service initiated the Continuous Air Monitoring Program (CAMP). The very sophisticated CAMP stations were set up in six cities. At present the CAMP effort is being supplemented to a certain extent by state and local sampling networks. In the interim the NASN net has grown to about 200 urban and 30 nonurban stations.

This sampling effort is necessary and significant because it has permitted health officials to begin to develop an accurate feel for the magnitude of a problem which has only begun to receive significant attention in the past 15 years. The significance of this effort becomes more apparent as a result of federal, state, and local officials working together in the direction of refining this approach. These sampling efforts are mentioned here because the data being generated by these networks and associated state and local programs represent the basis for regional programs covering both municipalities and industries of a region.

Interstate Programs

As in the case of water pollution control, regional air pollution control efforts for the most complex and oftentimes largest geographic regions have been projected under the jurisdiction of interstate compacts. Interstate compact efforts obviously do pertain not exclusively to industry but to all sources of contamination located in the interstate area in question. Since the compact concept appears to be popular, however, it deserves brief coverage here, since industry will feel its impact in the future. Compact efforts are to date for the most part projected, because in only one case has an actual program been carried out by such an interstate agency. In the interest of developing a better understanding of activity in the interstate area, the principal efforts of this type to date will be briefly outlined.

Interstate Sanitation Commission This commission was created in 1936 for the abatement of existing and control of future water pollution in the tidal waters of the New York metropolitan area. As a result of authorizing legislation by New York and New Jersey with the approval of Connecticut, this commission made a study in the summer and fall of 1957 and determined that an interstate air pollution problem did exist.

Following up on this study, Interstate Air Pollution Control legislation was drafted and ratified by New York and New Jersey in 1961. Although this Interstate Act contained no enforcement powers, the following activities were among the most significant which it authorized:

1. To conduct studies and undertake research, testing, and development
2. To trace sources of air pollution by appropriate sampling and analyses
3. To refer complaints and other pertinent matters to the appropriate state agency
4. To make recommendations and reports to the governors and legislatures of the participating states

Funds first became available for the regional effort on Jan. 1, 1962. As a result of continuing funding from the states and lack of ratification of the Mid-Atlantic Compact, this regional program continues in operation. During its period of activity, studies have been conducted covering sulfur dioxide and other odor producing materials, with findings being transmitted to the participating states. Further, the commission developed a Regional Air Pollution Warning System for the New York metropolitan area which is still in effect.

As a result of a conference called by the Secretary of Health, Education and Welfare for the New York–northern New Jersey metropolitan area on Jan. 3, 1967, definite conclusions were drawn. The two major findings were:

1. An appropriate interstate agency must be vested with adequate legal authority in order to deal with this bistate air pollution problem.

2. The conferees (other than participants from Health, Education and Welfare, who abstained) agreed that the Interstate Sanitation Commission should be expanded by the states for dealing with interstate air pollution problems.

When the matter of enabling legislation for an interstate approach was again considered by the states, however, the decision was reached to establish a different type of compact.

Mid-Atlantic States Air Pollution Control Compact This compact, which was ratified by New Jersey and New York in 1967 but has not been ratified by the federal government, has some unique features:

1. It covers primarily New Jersey and New York.

2. The states of Delaware and Connecticut and the Commonwealth of Pennsylvania are authorized to join if desired.

3. The governors of the states are exofficio commissioners and the federal commissioner will be appointed by the President.

4. Each member of the commission shall appoint an alternate.

5. The compact grants rather extensive legislative authority which in no way abrogates or impairs application of state authority. Further, the compact instructs the commission created to consult with state agencies in positions of authority.

6. The commission is granted authority to establish standards of air quality and emission control and conduct the necessary surveys required in such an effort.

Interstate regional control thinking was going on elsewhere in the nation during the period while the greater New York City approach was evolving. In two areas these efforts resulted in the drafting of compacts. It thus appears appropriate to consider briefly the Illinois-Indiana and the West Virginia–Ohio compacts.

Illinois-Indiana This compact was ratified by the states in 1965. The United States Congress has not thus far approved the compact. Some of the main features of the compact can be outlined as follows:

1. Seven commissioners from specified interested spheres of activity will be appointed by the governor of each state.

2. The commission may make interstate air pollution control studies and prepare reports where considered appropriate documenting findings.

3. Reports covering objectionable conditions shall be transmitted to state and local air pollution control agencies. No less than 6 months after the report is furnished, if local air pollution control agencies have not taken sufficient action, the commission can issue an order after an appropriate hearing.

4. No mention is made in the compact concerning representation on the part of the federal government or approval by the United States Congress.

Ohio–West Virginia This interstate air pollution control compact was passed by the two states during 1967. To date, the United States Congress has taken no action on the compact. Since this compact is in many respects similar to Indiana-Illinois, only specific points of difference will be listed:

1. Provision is made for five commissioners from each state to be appointed by the governor.

2. Reports on objectionable interstate air pollution will be transmitted to pertinent state and local control agencies; then after a reasonable time enforcement action will be taken by the commission if the local agencies have not taken appropriate action.

3. The compact will go into effect when ratified by the two states and approved by the United States Congress.

Other compacts It is understood that interstate compacts were drafted during 1967 between the states of Illinois and Missouri and Missouri and Kansas. The Missouri state legislature passed both compacts, but no action was taken by the other states or the federal government.

State Programs

Since it appears that the states will continue to exercise an important role in air pollution control, the problem will be reviewed from the viewpoint of what is underway at this level of government.

Much legislative activity is underway at state level, as is the case at federal and interstate levels. This is borne out by the fact that 12 states adopted major air pollution control legislation during the first 5 months of 1967. Today the total number of states having legal authority to control air pollution is 37 compared with 16 in 1963.

At present, 14 states have implemented their authority to control air pollution by adopting air quality or emission standards. This compares with 7 as of 1963.

The increase in activity can to some considerable degree be gaged by increases in expenditures. In 1963, less than $13 million was spent in total state and local programs. As of 1967, this figure had grown to approximately $26 million. At the local level alone, less than $10 million was being spent in 1963, and this compares with more than $15 million being spent annually at present.

Regional Developments

The Los Angeles County program of air pollution control is an outstanding one for regions of county size. Under this program industrial hydrocarbon emission in the county has dropped from a figure of 2,400 tons/day in 1940 to 800 tons/day in 1966. This reduction has taken place in spite of the fact that the industrial establishments increased from 5,900 to 18,500 over this 26-year period. Over the same period, however, the contribution from automobiles has increased 3.1 times. The 1967 budget for the Los Angeles County Air Pollution Control District was $3,758,227.

Another approach which the county district has originated and refined will no doubt see use in many regions in the future. This is a toxic air pollutants alert system. This system went into effect on June 20, 1955; so through Sept. 1, 1967, 62 first alerts, all based on ozone readings, had been pronounced in the district. Notices go out to all mass media, and 450 industries are directed to prepare to close down should a second alert be declared. Should a second alert stage be reached, the Emergency Action Committee and members of the Air Pollution Control

Board are notified. If a third alert should be declared, the governor will be requested to declare a state of emergency, which permits appropriate action to be taken under the California Disaster Act.

The San Francisco Bay Area Air Pollution Control District is an interesting example of the regional approach in operation in greater than a county region. This region went into operation in 1955, and at present six of the nine eligible counties are active. This district had a budget of $1,126,591 in 1967.

Two regional approaches of greater than county magnitude are in operation in Massachusetts. The Boston metropolitan region went into operation in 1960 and the Springfield metropolitan area in 1966.

Two regional programs involving more than one county are underway in North Carolina, and a single program of this type is active in Florida. There are probably other regional efforts which have just been launched, encouraged by federal funding, but they are so new that no information is available on them to date.

After this point is reached, the next step is that of allocating the fixed charges against the interceptor sewers and components of the treatment plant. This action is summarized in Table 7.

Since it is generally not possible completely and accurately to apportion certain items such as the cost of providing water and electrical service for the control building, these items have in the table been broken out from the cost of the treatment plant. As a general rule, these items are apportioned by the average percentages for all other units.

TABLE 5 Items by Percentage against Which Fixed Charges on the Sewer System Can Properly Be Charged*

	Percent
Property:	
Storm water	16.8
Groundwater infiltration	6.7
Public use	2.9
Growth capacity and use for property yet to be developed	26.6
Subtotal property	53.0
Users:	
Sanitary waste water	25.6
Commercial waste water	10.8
Industrial waste water	10.6
Subtotal users	47.0
Grand total	100.0

* Collection sewers omitted in interest of simplicity.

The allocation of operation and maintenance costs against the structures and functions of the system is also necessary. Such effort is summarized in Table 8. After the details of this approach are evaluated to this point, the allocation of the fixed and operation charges should be summarized. Such action is outlined in Table 9. The charges against the property of the area can be distributed using a number of approaches, but the ad valorem tax method would probably be used in the interest of equity. Through the use of this method, one part of the sewage rate structure has been completed.

TABLE 6 Items by Percentage against Which Fixed Charges on the Physical Treatment Plant Can Be Charged

	Items affected by		
	5-day BOD, %	Avg vol, %	Suspended solids, %
Users:			
Sanitary waste water	28	32.8	30
Commercial waste water	15	14.5	17
Industrial waste water	30	13.2	20
Subtotal	73	60.5	67
Property:			
Growth capacity	24	34.2	28
Infiltration	...	3.3	
Public use	3	2.0	5
Subtotal	27	39.5	33
Total	100	100.0	100

An allocation and distribution of total annual revenue requirements is next indicated, and such a summary of information is outlined in Table 10. Since the total annual charges to users have been broken down into charges based on volume,

TABLE 7 Allocation of Fixed Charges

Components	Total fixed charges, $	Chargeable to property		Chargeable to users, $	User's share chargeable to					
					Volume		Suspended solids		5-day BOD	
		%	$		%	$	%	$	%	$
Intercepting sewers.	200,000	53	106,000	94,000	100	94,000				
Treatment plant:										
Broken down into components such as:										
Pumping station.	10,000	53	5,300	4,700	100	4,700				
Screen, grit chamber.	30,000	53	15,900	14,100	65	9,165	35	4,935		
Preliminary settling tanks.	17,000	40	6,800	10,200	80	8,160	20	2,040		
Activated sludge.	65,000	23	14,950	50,050	15	7,508	. . .	. . . . . .	85	42,542
Secondary settling tanks. .	20,000	30	6,000	14,000	50	7,000	. . .	. . . . . .	50	7,000
Chlorination.	15,000	30	4,500	10,500	40	4,200	. . .	. . . . . .	60	6,300
Digesters.	20,000	27	5,400	14,600	. . .	. . . .	100	14,600		
Vacuum filters.	25,000	27	6,750	18,250	. . .	. . . .	100	18,250		
Subtotal.	202,000	32.5	65,600	136,400	30	40,733	28	39,825	42	55,842
Other items:										
Water supply.	6,000	32.5	1,950	4,050	30	1,215	28	1,134	42	1,701
Electrical service.	8,000	32.5	2,600	5,400	30	1,620	28	1,512	42	2,268
Roads and grounds.	7,000	32.5	2,275	4,725	30	1,418	28	1,323	42	1,984
Main control building.	15,000	32.5	4,875	10,125	30	3,038	28	2,835	42	4,252
Subtotal.	36,000	32.5	11,700	24,300	30	7,291	28	6,804	42	10,205
Total plant	238,000	32.5	77,300	160,700	30	48,024	28	46,629	42	66,047
Total fixed charges.	438,000	42.0	183,300	254,700	55	142,024	19	46,629	26	66,047

TABLE 8 Allocation of Operation and Maintenance Costs

Structure or function	Total operation and mainte-nance costs	Chargeable to property		Chargeable to users, $	User's share chargeable to					
					Volume		Suspended solids		BOD	
		%	$		%	$	%	$	%	$
Breakdown somewhat as follows:										
Intercepting sewers.......	13,100	58	7,598	5,502	60	3,301	40	2,201		
Pumping stations.........	21,300	23	4,899	16,401	100	16,401				
Primary treatment........	19,400	48	8,312	10,088	50	5,044	35	3,531	15	1,513
Secondary treatment......	46,100	17	7,837	38,263	10	3,826	10	3,826	80	30,611
Effluent chlorination......	18,400	16	2,944	15,456	10	1,546	....	......	90	13,910
Sludge disposal...........	63,200	8	5,056	58,144	100	58,144				
Supervisory..............	20,300	16	3,248	17,052	61.4	10,475	6.4	925	32.2	5,652
Collection and billing.....	19,700	17	3,349	16,351	61.4	10,012	6.4	899	32.2	5,440
Total..................	221,500	20	43,243	177,257	61.4	108,749	6.4	11,382	32.2	57,126

5-day BOD, and suspended solids, unit rates can now be computed. The rates should cover cost of handling the waste load per 1,000 gal of volume and 100 lb of 5-day BOD and suspended solids. This provides the basis for establishing an equitable rate for any user depending upon his contribution.

TABLE 9 Summary of Fixed Operation and Maintenance Costs from Tables 7 and 8

	Chargeable to			
	Users		Property	
	$	%	$	%
Fixed charges:				
Intercepting sewers	94,000	47	106,000	53
Treatment plant	160,700	67.5	77,300	32.5
Operation and maintenance	177,257	80	44,243	20
Totals and averages	431,957	65.5	227,543	34.5

TABLE 10 Allocation and Distribution of Total Revenue Requirements from Tables 7 and 8

	Fixed charges		Operation and maintenance		Total annual charges	
	$	%	$	%	$	%
Chargeable to property	183,300	42	44,243	20	227,543	34.5
Chargeable to users:						
Volume	142,024	32.4	108,749	49	250,773	38.0
Suspended solids	46,629	10.6	11,382	5.2	58,011	8.8
BOD	66,047	15	57,126	25.8	123,173	18.7
Subtotal	254,700	58	177,257	80	431,957	65.5
Grand total	438,000	100	221,500	100	659,500	100

In accordance with the joint group approach, then the total industrial charge as well as the charge for each plant can be determined. This is done by adding the property charge, volume charge, suspended solids charge, and BOD charges together. The property charge is simply the industrial evaluation times the community tax rate. The volume charge is the industrial volume times the cost of handling waste water per 1,000 gal. The industrial BOD and suspended solids costs are determined by multiplying the industrial plant load by what it costs the city to handle such loads. More details on the use of this approach are outlined in Dr. George Schroepfer's article already mentioned.

REGIONAL AIR POLLUTION CONTROL

Impact of Federal Program

The federal program and the legislation covering it do not specifically mention industrial plants, but a review of these areas does give some indication of how the regional approach is evolving. Obviously, when such regional efforts came

Chapter 9

Research Programs for Air and Water Pollution Control

JOSEPH B. ZUZIK

Supervisor, Industrial Hygiene and Safety,
American Cyanamid Company,
Stamford, Connecticut

The need for a dynamic program in air and water pollution research for the strict purpose of finding answers to control industrial pollution is now. Too much time has already elapsed and has been wasted on pieces and bits of scattered works, many of which have been duplications of past efforts.

Uncoordinated probes have seemingly little finite direction to be of value. There is a definite need for a coordinated effort by federal, state, and local governments in conjunction with and with the support of industry to combine the many disciplines that play a vital role in seeking out the right answers to the so many perplexing questions concerned with industrial pollution. Development of a strong centralized program is necessary.

The information in this chapter should be of value to the plant manager, the professional engineer, and the general public.

Air pollution affects the health and well-being of all citizens of the globe. It is not confined by any political line of demarcation or by any subdivision of a geological boundary. It in fact exists as soon as anything foreign is introduced into the atmosphere other than the basic constituents of air itself.

CAUSES OF AIR POLLUTION

The chemical, toxicological, and physical complexities of air pollution are as yet poorly defined. Problems causing air pollution involve gaseous emissions and mixtures of gases and/or particulate material of many types from single or multiple sources. Practically every activity on the surface of the earth, human or otherwise, in today's world does in some way pollute the earth's environment. Frequently a long finger is pointed toward industry as the chief culprit in causing environmental pollution, with little or no regard to the true sources of pollution. Man, by his very existence, is the chief polluter.

NATURE OF AIR POLLUTION

Before we can cope with pollution problems, we must first be able to realize the nature of pollution. Air is the gaseous mantle that surrounds the planet earth. This mantle is comprised of a number of basic gaseous constituents, the most abundant of which are nitrogen (78 percent) and oxygen (about 21 percent). The remaining 1 percent of the earth's atmosphere is made up of the other gases such as argon, neon, helium, krypton, xenon, hydrogen, water vapor, carbon dioxide, and ozone. The elements of sulfur dioxide, nitrogen dioxide, ammonia, and carbon dioxide should be regarded as contaminants rather than as constituents of normal air, since they generally arise from industrial sources. There are many other elements incumbent in air such as methane, formaldehyde, and suspended particulates.

The air we breathe is invisible, odorless, and tasteless. It is essential to human life and is consumed at the rate of about 30 to 40 lb/day.

The air that man breathes has always been contaminated to some degree, by either natural or manmade pollution such as dust, smoke, gases, mists, vapors, and odors. Early sources of natural contamination were due to earthquakes, erupting volcanoes, plant pollen, and forest fires, which ravaged many acres of land as far back as the history of man. Man produces airborne wastes by practically everything he does to sustain living, such as cooking his food, heating his home, producing his goods, cultivating the soil, building construction, garbage disposal, and transportation.

Air pollution may be defined as a mixture of one or more contaminants of solids, liquids, or gases discharged into the air by nature and/or by man in such quantities and of such duration which may be, or may tend to be, injurious to human, animal, or plant life or property, or which may interfere with the comforts of life itself. Air contaminants may be dispersed rapidly in large quantities into the air or may accumulate in various concentrations, depending on the topography of the geographical area and the existing meteorological conditions at the time of dispersal. Air contaminants in themselves may be defined as a mixture of a variety of contaminants changing in form and concentration from time to time and place to place.

The problems confronting the scientist and the general public in air pollution concern the economic and social relationships which appear to be the most difficult to control. Many areas of academic discipline are being utilized in the physical and biological sciences to help understand the technical complexities of air pollution. With the application of ingenuity and hard work, and with adequate time, the perplexing questions will be answered.

LEGISLATION

Pollution of one of our most important natural resources, which is the air we breathe, is a matter of national and international concern. Barring natural causes, air pollution is a product of our industrial and urban societies. It will steadily increase as the affluence of our society increases. Over the past 20 years, it has become increasingly evident that the trends of urbanization, economic growth, and technological change have caused a serious adverse effect on the quality of the air we breathe. In 1955, Congress passed legislation authorizing a federal program of research in air pollution and technical help to state and local governments. This legislation established the policy which continues in effect that state and local governments have a basic responsibility for dealing with community air pollution problems and, further, that the federal government has an obligation to provide leadership and support. During the period 1954 to 1964, great strides were made toward improved scientific knowledge of the nature and geographical extent of the air pollution problem, its impact on public health and welfare, the existence of techniques for controlling many sources of air pollution, and the need for new and better techniques.

By 1963, it was increasingly evident that progress toward a better understanding of the problem of air pollution was not being matched by real progress toward better control, primarily because most state and local governments were still not equipped to cope effectively with regional air pollution problems.

In December, 1963, Congress passed the Clean Air Act, which enabled state and local governments to join the federal government in an expanded and highly vigorous attack on air pollution. The Clean Air Act authorized two major new federal activities, the awarding of grants directly to state and local agencies to assist them in developing, establishing, or improving control programs and federal action to abate interstate pollution problems, which are essentially beyond the reach of individual states and cities. The Clean Air Act also called for an expanded federal research and development program, and it placed special emphasis on investigating two of the most important aspects of the national air pollution problem, motor vehicle pollution and sulfur oxides pollution arising from the combustion of coal and fuel oil.

In the 4 years that followed enactment of the Clean Air Act, expanded research efforts helped to demonstrate the need for and have hastened the development of new and improved control technology. In 1967, the Air Quality Act was adopted, which calls for a coordinated attack on air pollution on a regional basis. It provides a blueprint for action by all levels of government and among all segments of industry.

The federal responsibility for assuming leadership in the prevention and control of air pollution rests with the Department of Health, Education and Welfare. In order to carry out the operating responsibility, the Department established the National Center for Air Pollution Control in January, 1967, formerly known as the Division of Air Pollution of the Public Health Service. In its present form, it is comprised of three major units—criteria and standards, control technology research, and development and abatement and control.

PROBLEM AREAS

The problem of air pollution is basically regional in nature for several reasons: First, air pollution knows no boundaries because prevailing air currents can carry pollutants for great distances from their source. Second, the continued urbanization of America means that pollution sources will continue to spread over larger geographical areas. Because air pollution is a regional problem, the most effective way, then, is to attack it on a regional basis.

The Air Quality Act of 1967 sets up a system which hinges on the designation of regions where two or more communities, in either the same or different states, have a common air pollution problem. One such region is presently being set up which will include areas of New Jersey, greater New York City, and most of Fairfield County in Connecticut. A public meeting concerning the establishment of this region was held at the United States Mission to the United Nations on Sept. 30, 1968.

Meteorology

Meteorological factors play a primary role in determining air pollution levels. For a given pollutant emission, the resulting ground level concentration of the pollutant can vary greatly depending on the meteorological conditions that prevail at the time. For long time periods, e.g., for several years, there are large areas of the country over which the average meteorological conditions affecting the transport and diffusion of air pollution are quite similar. The National Center has defined these areas. There are eight in number covering the entire United States. These were delineated on the basis of long-term values for the important meteorological factors, frequency of occurrence, persistence, height variations of the stable (inversion) layers of air, and the frequency of various wind speeds. Since these factors affect the transport and diffusion of pollutants released into the atmosphere and thereby influence the geographical extent of urban air pollution problems, atmospheric areas are one of the factors considered in the designation of air quality control regions.

Problems in defining specific numbers to delineate harmful quantities or concentrations of air pollutants must be solved. Air quality criteria represent a scientific evaluation of the extent to which individual pollutants or combinations of pollutants are harmful to health and damaging to property. By providing an indication of the predictable effects of various levels of pollutant concentrations on public health and welfare, the criteria provide guidelines for the establishment of air quality standards.

RESEARCH AND DEVELOPMENT ACTIVITIES

The state of the art is such that research on gaseous and particulate control devices for air pollutants is primarily concerned with abatement improvement development coupled closely with cost reduction and development of new and different methods for doing the job.

New and better ways must be found for collecting sulfur dioxide, which is one of the major contributors in polluting the air. Considerable attention has been given to using low sulfur content fuel. Because of the scarcity of such fuels and the cost in reducing the sulfur content of standard grade fuel, a more practical approach must be made. Research into the removal of SO_2 from stack gas effluents is the best approach. Some experience has been gained in this direction, whereby sulfur oxides are removed from the stack through sophisticated scrubbing techniques and reclaimed by formation into sulfuric acid, which in turn has a marketing value and can help underwrite the cost of processing. The present cost of such a reclamation procedure bears further study to increase capacity yields and cost reduction. Areas of investigation into the removal of SO_2 from flue gases include absorption using alkalized alumina or manganese oxide, adsorption using activated carbons, and further studies using oxidation catalysts. An absorption study using alkalized alumina at an elevated temperature above 600°F and then converting the oxides to elemental sulfur by reduction of the spent absorbent has taken place. Results of such investigation bear promise.

The following is a list of flue gas desulfurization techniques that are under study and in some cases in use:

1. Acoustic coagulation of sulfuric acid mist in a sound field with optimal frequencies at 16 and 22 kHz
2. Wet scrubbing with ammonia and then adding sulfuric acid to convert the ammonium sulfite bisulfite to ammonium sulfate, an end product which can be used as a fertilizer
3. Liquid purification processing using limestone, ammonia, or sodium sulfite
4. Electrostatic precipitation
5. Hydrogenation

No method is known at the present time which assures the prevention of emissions of sulfur oxides and at the same time is truly economical in operation. Investigations continue into the desulfurization of coal and fuel oil, the addition of additives to the combustion process, and the gasification of fuels. An economically feasible method of sulfur removal is hoped for.

Research into the following problem areas is needed:

1. Removal of carbon monoxide from industrial flue stacks
2. Removal of nitrogen oxides and hydrocarbons from plant stack gases and engine exhaust
3. Practical ways of removing sulfur oxides and gaseous end products from combustion processes with the intent of complete removal or conversion into by-products of commercial value

COMPOSITION AND CLASSIFICATION OF AIR POLLUTANTS

Classification schemes have been devised covering a variety of pollutants that may be present in the atmosphere. Pollutants may be classified as to origin, state of matter, and chemical composition and may exist in either particulate or gaseous form.

On the basis of origin, air pollutants are classified as primary and secondary in quality. Primary pollutants are those which are emitted to the atmosphere as a consequence of a process. This type of pollutant exists in air in the same form as it was discharged. Secondary pollutants are formed as products of some reaction; generally, an existing pollutant reacts with some other substance in the atmosphere. This existing substance may be some other pollutant, or it may be a natural substance of the atmosphere.

On the basis of state of matter, air pollutants are classified as gaseous or particulate types. Gaseous pollutants are present in the atmosphere as contaminants which behave similarly to the air itself. Once diffused, such gases do not settle out. Examples of gaseous pollutants are nitrogen dioxide, nitric oxide, sulfur dioxide, sulfur trioxide, carbon monoxide, carbon dioxide, hydrogen sulfide, hydrogen fluoride, hydrogen chloride, ammonia, and hydrocarbon gases and odors. This list is by no means complete and would contain any other gaseous constituents not considered as normal components of natural ambient air.

Particulate pollutants include dust of any type, smoke, fumes, mist, and sprays and are classified as solids or liquids. These pollutants are generally finely divided and readily dispersed in the atmosphere. The larger-sized particulates tend to settle out quickly. The smaller size, however, may tend to act as a gas and stay in suspension.

Generally speaking, when the words air pollution are mentioned, it is immediately felt that such pollutants are all harmful. In most cases this is true; however, other pollutants exist which are nontoxic in nature. Examples of such are the pleasant odors of perfumes and the wonderful aroma of freshly baked bread or the smell of freshly cut hay. In some cases, odors may be nontoxic but also very obnoxious, as evidenced by a decayed animal along the highway.

PROBLEM EVALUATIONS

Problems associated with gaseous pollutants are many. Procedures for the identification of specific compounds contributing to air pollution are largely tedious and expensive, and few laboratories are equipped to undertake such evaluations. The chemical composition and physical state in which air pollutants occur are of increasing importance, especially as they relate effectively to the physiology of life's breathing apparatus. A need exists for better analytical methods to determine the magnitude of air contamination, especially for less sophisticated methods that would be less expensive than those in existence. A larger spectrum of analytical measurement is necessary so that the more complex compounds and the like can be measured more simply and more frequently. Our modern society subjects us to an ever-widening group of air contaminants, increasing in quality and quantity. Technological history indicates that the composition of emissions and sources is rapidly changing and that the quantity will increase at an increasing rate in the future; unfortunately, the direction in which such changes occur or will take place cannot be precisely predicted.

A significant source of pollution in urban areas is caused by motor vehicle emissions. The exact interrelationship between hydrocarbons and oxides of nitrogen when photochemical smog is produced in the presence of sunlight needs further clarification.

Identification of participating primary pollutants and determination of the relative importance of each in the photochemical process are not only necessary but required. Our changing technology necessitates a fundamental study in photochemistry. An identification of primary reactants associated with secondary toxicants will enhance the development of more effective and less costly controls. A thorough and adequate understanding of the intermediate and secondary products must be first acquired before their biologic effects can be fully analyzed and understood.

More information must be obtained on the synergistic effects and relationships between ambient existing air constituents and those extraneous additives emitted into the atmosphere.

Present mortality statistics by themselves are not sensitive enough to reflect the subtle influence of minute concentrations of toxins in the atmosphere. An assessment of illness due to air pollution requires a far greater sophistication in information retrieval covering more than the present routine physical examination. Objective measures of physiological processes and their changes have been made; however, more are needed.

A causal relationship between air pollution and chronic obstructive ventilatory diseases, which appear to be increasing in incidence in this country, has not yet been fully established. Certain isolated findings indicate that asthmatic attacks occur more frequently on days with heavy smog and that patients suffering from emphysema improve on breathing filtered air after having had exposure to smog.

A definite need exists for an integrated research effort utilizing analytical and laboratory techniques, statistical methods, and an epidemiologic approach, along with meteorological, physical, and engineering studies which would provide the hope for a solution to the vast complexity of air pollution problems. It is the combination of many disciplines that will accomplish this aim, and no one method will attain this end.

CONTROL OF AIR POLLUTION

Regardless of the air pollution problem to be attacked, whether it be a community-wide problem or merely a single-source problem, there should be one basic approach to control, that is, to control the pollutant at the source so that it does not enter the atmosphere. Another method of control is by natural dilution of the pollutant after its admission into the atmosphere to such a concentration that it will not harm man, animals, vegetation, or building materials. It should be borne in mind, however, that this method does not eliminate pollution; it merely dilutes it to a level that is more acceptable.

Control of the pollutant at the source can be accomplished by several methods: raw material changes, operational changes, modification or replacement of process equipment, adopting alternate methods, and more efficient operation of existing equipment.

There are several basic methods of reducing air pollutants entirely or to tolerable levels. Some of these are by destroying the pollutant and by masking, counteracting, or collecting the pollutant.

Destroying the pollutant can be accomplished by the use of fire and/or by using catalytic burners. These methods are generally limited in their application in that they are limited to the destruction of only those wastes which are combustible. This method would not be recommended for the destruction of effluent gases containing heavy elements of metals or oxides of sulfur.

Pollutants may be masked by superimposing a more pleasant odor into the air. This method certainly does not eliminate the pollutant but only creates a new odor less offensive to the senses. It would not be acceptable where toxic odors are involved.

Counteracting the pollutant is not a good approach, since it does not eliminate the pollutant but simply cancels out the odor by the synergistic effects of two odors when in combination with each other and when such combinations would be a specific to neutralization by their intermediate reactions.

Collecting the pollutant is by far the method of choice and is the most widely used method of abating air pollution. This method is to trap the pollutant by use of collection equipment, thus preventing its escape into the atmosphere. To illustrate, the following example is given and is concerned with the removal of a gaseous pollutant.

Removal of Chlorine from a Gas Stream

A development program is in progress in which a chemical reaction, liberating about 12.5 lb/hr of chlorine, is being studied. If this were a full-scale commercial installation, the chlorine would be recovered and recycled to extinction, but such recycling is not desirable at such a small level of operation.

The total gas throughput is approximately 137 scfh, consisting mainly of nitrogen and oxygen; the balance is chlorine. Chlorine is removed by means of a two-stage scrubber. The first stage is a stirred tank containing a caustic (sodium hydroxide) solution, and the second stage is a small packed tower. The caustic scrubber is similar in principle to reactors used in the preparation of sodium hypochlorite or more commonly called household laundry bleach. The removal of chlorine differs from the hypochlorite process in that (1) the caustic solution used is more concentrated and (2) the chlorine is mixed with other gases.

The first-stage arrangement consists of a 125-gal tank fitted with an agitator and an exhaust duct. The gas to be scrubbed is admitted via a dip leg; it bubbles through the vigorously agitated solution, is picked up in the exhaust stream, and then is fed to a packed tower, where any residual chlorine is removed. The caustic is prepared by diluting with water 50 gal of liquid caustic (approximately 50 percent NaOH) to 25 percent. This theoretically provides sufficient caustic to remove the chlorine for approximately 5½ hr of operation, but actual operating experience has shown that 3½ to 4 hr is a practical upper limit. Overworking the caustic solution may cause it to boil out of the scrubber, with a resultant loss of solution and possible liberation of free chlorine.

The use of full-strength (50 percent) caustic would increase the capacity of the scrubber, but it has been found that this concentration of caustic causes a serious reduction in the chlorine removal efficiency accompanied by a higher concentration of chlorine in the scrubber off gas. Whether this is due to instability of the hypochlorite solution at higher caustic concentrations, to a reduction in the reaction rate between the chlorine and the caustic, or to some other cause has never been established, but the problem has been eliminated by diluting the caustic as above described.

A cooling loop has been installed in the scrubber to remove the heat of dilution of the caustic and also to dissipate the heat of reaction between the chlorine and the caustic. This heat of reaction has been estimated at about 625 Btu/lb of chlorine; thus the coil must remove about 31,500 Btu/hr—certainly not a very large quantity of heat—but if it were not removed the resulting temperature rise would reduce the efficiency of chlorine removal. This could become serious if the caustic boils.

The gases which come from the caustic scrubber have had the major portion of the chlorine removed. These gases are the feed to the second-stage scrubber—the packed tower. This consists of a section 8 in. in diameter and 4 ft long, packed with ¾-in. polypropylene pall rings. The tower is continuously flushed with about 3 gal of water/min, at ambient temperature. There is no detectable odor of chlorine in the off gas, and the removal efficiency is very high.

A second, less widely accepted choice of air pollution abatement is by dilution of the pollutant in the atmosphere before it can reach the population in harmful concentrations. Such a control method would include the use of tall stacks designed so that the discharge is high enough to allow the natural movement air currents to dilute and disperse the pollutants in such a way that they would not be harmful when they reached the ground. This method does not remove the contaminant; it only reduces the concentration per unit volume and is applicable to both particulate and gaseous emissions.

Zoning the use of air is another method of attempting to accomplish natural dilution of air contaminants. This method involves community planning to prevent harmful ground concentrates from occurring within designated areas.

Meteorological control of process operations is control involving curtailment or cessation of manufacturing operations during periods of adverse meteorological conditions, especially in times involving stagnant air conditions and temperature inversions. Such controls should be imposed as soon as adverse meteorological conditions appear.

Removal of Particulates

Particulate pollution abatement can be accomplished by several methods such as gravity settling, filtration, electrostatic attraction, and particle conditioning. It must be noted here, however, that as good and efficient as filtration seems, there is no means of absolutely removing 100 percent of all the particles; thus unfortunately, a small portion of the total amount of the particles, especially the smaller particulate, does get into the ambient air. The problem of complete removal exists. Improvement of design characteristics will tend to increase further the existing capacities, especially in systems employing filters and water scrubbers.

MONITORING

Sampling and monitoring the ambient atmosphere for particulate or gaseous pollutants can be accomplished by a variety of techniques. Selection of the technique will depend upon the needs of the investigator, the type of pollutant to be sampled, and the type of analysis to be performed. Monitoring programs are established to ensure that certain regulatory requirements established by federal, state, or local governing agencies are maintained and to evaluate potential hazardous conditions which may harm or endanger life or property. Monitoring may be done on a continual or part-time basis, depending upon the type of information desired.

Long-term effects of air pollution on man, on other biological systems, and on property must be quantitated accurately in order to establish the necessity, feasibility, and economic practicability of control measures designed to abate these effects. Research is still sorely handicapped by the lack of techniques sensitive enough to detect minimal changes. To determine the levels of air pollutants above which biologic effects can be expected, extensive toxicologic, pharmacologic, and physiologic investigations are necessary.

Although some progress has been made in identifying and measuring general classes of pollutants, there is a real need for intensive research into methods for identifying and measuring these classes of pollutants. A need for intensive research into methods for identifying and measuring the individual substances that make up these classes is also sorely needed. Simpler and less expensive procedures should be a prime requisite for this purpose. Our knowledge is far from complete regarding air pollutants; their identity, specific sources from which they are derived, factors governing their dispersion, and chemical and physical changes in the atmosphere; their effects singly, in combination, and also synergistically.

Many types of data have been collected and much more data needs to be collected before definitive action can be instituted. Some very serious situations arise in monitoring programs out of the lack of adequate and trained personnel to do the monitoring. In many cases state and local programs are nonexistent or are grossly inadequate. Technical work in the field directed toward appraisal of problems and development of solutions is at a level far below that required for maintenance of an acceptable level of air quality.

The federal government has and continues to foster cooperative joint federal, state, and local programs and has provided tools and techniques in a form useful in attaining a solution to questions of concern. The federal government is making available financial help to state and local agencies for the purpose of surveillance and in operating control programs, as widely evidenced by the grants made to many municipalities. State and local health agencies have primarily been the recipients of this aid.

Monitoring programs are of paramount importance in determining what levels of contamination exist in the atmosphere; however, of similar importance is the ability to monitor accurately. Too many different types of analytical procedures exist which tend to be based on assumption and not on fact. Several types of monitoring are herewith listed.

Particulate Pollutant Monitoring

These include pollen grains, fungus, spores, metal oxides, dusts, fly ash, smoke, mists, and vapors.

Routine investigation and analysis of atmospheric particulates occurring in the ambient air generally are done with the aid of fallout buckets, high-volume samplers, and/or tape samplers. In certain cases particulate samples may be collected in a liquid medium or on a greased surface by particle impingement. By far the most accurate type of particulate sampling is by the use of high-volume collection on a filter paper or by tape sampling wherein a known volume of air is passed through a filter medium upon which the particulate collects and is then weighed. In the case of tape sampling, a light source is passed through the deposited material and the reduction in the transmission of light through the spot by means of an illuminator determines the quantity of solids deposited on the filter.

Dustfall or, more appropriately, fallout samples are general indicators of particulate concentration. This type of sample may give erroneous answers in that there is no way to tell whether some particulates were lost or blown away during the sampling period. This type of sampling is usually set up on a month-to-month basis. A sample of very small area is considered to be representative of a large segment of the geographical community. The short sampling period does not lend the sample to be useful in the detection of intermediate peak deposition periods.

Other types of particulate samplers are inertial devices and electrostatic and thermal precipitators.

Gaseous Pollutants Monitoring

There are numerous materials which have been used for absorbing gaseous air pollutants. Some of these are carbon particles, activated carbon, silica gel, activated alumina, and synthetic high-porosity materials such as sodium or calcium aluminosilicate of high porosity. Other methods of collection are by freeze-out or condensation of collected material. In this method an air sample is drawn through an intensely cooled chamber and condensable contaminants are collected. The chamber is cooled by immersion in a bath of low-temperature liquid, such as liquid nitrogen or liquid air. This method of collection is not very desirable, since the efficiency of this method must be accurately determined.

Grab samples are methods of choice in certain situations where electricity and laboratory facilities are not available. This type of sampling is useful when concentrations of pollutants vary considerably over a period of time and it is necessary to obtain a sample at a specific time. Grab sampling has a serious limitation; the sample obtained is generally not large enough to detect very small quantities of materials except by the most sensitive techniques.

In the application of any method for monitoring air pollutants, despite the exercise of care and skill, it is found that successive measurements will differ among themselves. The extent of agreement between duplicate, triplicate, or other multiple measurements is expressed in terms of the size of the deviations in comparison with the mean value reported. A complete expression of precision must be statistical in nature and must include the percentage of individual measurements which can be expected within a stated limit of confidence.

Several basic problems in setting up adequate monitoring programs for the express desire of obtaining a valuable picture of air pollutants in a given local area or in a larger area such as a state or region are the ability to cover a representative area and to gather all the facts concerned with the sources of pollutants for the area concerned. It is indeed difficult to obtain full and complete cooperation from all local inhabitants. Help from local industries is a vital link in establishing a source inventory list of contaminants. Additional help from other source polluters is equally helpful.

Psychology plays a vital role in establishing good rapport between investigating teams and private concerns and other polluters.

VEHICLE EMISSION CONTROL BY A CATALYTIC METHOD

Laboratory scale experimentations are presently being carried on for investigating the use of catalysts to control emissions from motor vehicles, including automobile exhaust control and diesel exhaust odor control.

Experimentation of this nature takes into consideration the compatibility of catalysts with associated hardware systems and the necessary requisites of appropriate equipment.

Many variables exist in the development of catalysts, some of which include metal content, type, composition, and physical structure. Catalysts must have the ability to withstand hostile environments. Capabilities of these agents must be ascertained under the most stringent conditions simulating as exactly as possible the environment to which they will be subjected.

Equipment and Source of Exhaust Gas

Special type equipment is needed to determine and measure the quality and quantity of engine exhaust effluents. Motor vehicle engine exhaust effluents are used which are produced by operating standard engine components under stresses encountered in normal vehicle operations.

Analytical Equipment

Exhaust components of concern for air pollution control are hydrocarbons, carbon monoxide, and oxides of nitrogen, which are identified and quantitated by flame ion detection and gas chromatographic analysis. Other analysis specified for measuring carbon monoxide and oxides of nitrogen is measured by oxidation reactions and ultraviolet detection.

The only really good piece of equipment available for studying the effect of catalysts on diesel odors is the human nose. This may be accomplished by selecting an elite panel of people who are asked to grade the odors on a scale of 0 to 5, zero being the state at which no odor is detectable and five being the state to which the highest odor level is attributable.

There are numerous problems in analysis, especially when trying to qualitate or quantitate the components of a sample. The method of choice should be selected on the basis of accuracy, measurability, consistent reproducibility, and ease of performance. Wet analyses are used to measure components such as aldehydes and sulfur dioxide. Gas samples containing these components are collected in suitable flasks, which are then processed using standard techniques.

Specific measurements are made on catalyst materials before and after treatment and use. Of particular interest are pore volumes, surface areas, and lead accumulation.

Screening Tests

A catalyst sample must be screened through certain steps before consideration is given to motor fleet testing. Certain conditions should be met to determine the activity of the material for converting hydrocarbons and carbon monoxide. A reduction in these components and the ability of the catalyst to maintain its integrity would signify its usefulness as a pollutant reductor. Catalysts showing promise must be road tested under varieties of operating conditions, with the concentration of exhaust emissions being measured as a function of time, and then comparing the final results with a known standard.

SULFUR DIOXIDE

Combustion of coal and oil for power generation is the major source of sulfur dioxide emissions to the atmosphere. Fuel substitution such as nuclear fuel and fuel desulfurization has been advanced as a possible solution to the sulfur dioxide pollution problem. Another alternative is the removal of sulfur dioxide from flue gas itself. Some of the more promising approaches to fuel desulfurization follow.

Reinluft process A thermally regenerable charsorption process developed in Germany, presently on a plant scale prototype operation.

Catalytic oxidation Sulfur dioxide is oxidized to sulfur trioxide by use of solid catalysts such as vanadium oxide. A feature of this process is the possibility of recovering sulfuric acid or ammonium sulfate.

Dolomite injection Powdered dolomite is directly injected into power plant boilers. Sulfates thus formed appear to have no market value and are therefore discarded.

Alkalized alumina Metal oxides are reacted with sulfur dioxide in the flue gas. Products recovered upon regeneration of the metal oxide may be sulfur, sulfur dioxide, or sulfuric acid.

Miscellaneous processes Other approaches actively being considered for flue gas desulfurization are electrolytic, sulfite solution scrub, molten carbonate, electrochemical, and the modified Claus method.

Table 1 shows that much of the air pollution research in the United States conducted by universities, the federal government, and industry represents an expenditure of about $25 million for 1966.

Table 2 shows the type of organization providing financial support of air pollution research activities. The federal government supplies a large portion of the total financial support. Industries generally do not permit the release of information concerning funding of their complete air pollution research programs.

TABLE 1 1966 Classification by Organizations Conducting Research in Air Pollution*

Organizations conducting air pollution research	Projects reported		Projects reporting 1966 funding		
	No. of projects	% of total	No. of projects	Funding, thousands of dollars	% of total funding
Universities	242	48	210	8,462	33
Federal agencies	151	30	148	10,414	41
State agencies	5	1	5	93	<1.0
Local agencies	13	3	6	653	3
Industrial organizations	57	11	44	4,570	18
Private (nonindustrial)	33	7	27	1,158	5
Totals	501	100	440	25,350	100

* "Guide to Research in Air Pollution," 6th ed., U.S. Department of Health, Education and Welfare, 1966.

Investigations relative to a better understanding of the chemistry and physics of the air surrounding us, and source and source emission studies, together represented an expenditure of more than $5 million in 1966. Research projects directed toward studying the effects of air pollution upon the health of man and animals including epidemiologic, physiologic, pathologic, and toxicologic studies represented an expenditure of $4.3 million in the same year.

TABLE 2 Classification by Organizations Supporting Research, 1966*

Organizations supporting air pollution research	Projects reported		Funding reported, 1966		
	No. of projects	% of total	No. of projects	Funding, thousands of dollars	% of total funding
Universities	29	6	22	499	2
Federal agencies	369	74	349	20,283	80
State agencies	5	1	3	55	<1.0
Local agencies	11	2	4	72	1
Industrial organizations	39	8	28	1,479	6
Private (nonindustrial)	27	5	18	2,264	9
Federal and university	14	3	12	535	2
Federal and state	1	<1.0	1	5	<1.0
Federal, industrial, and private	1	<1.0	1	100	<1.0
Not classified	5	1	2	58	<1.0
Total	501	100	440	25,350	100

* "Guide to Research in Air Pollution," 6th ed., U.S. Department of Health, Education and Welfare, 1966.

The states of California, New York, Ohio, and Pennsylvania accounted for a large segment of the total dollars funded for air pollution research in the United States, which accounted for approximately $12,707,000 in 1966.

An exact dollar figure expended on air pollution research is difficult to ascertain; however, it can be said with certainty that expenditures of this type will increase as long as the problem exists.

PROBLEM AREAS IN WATER POLLUTION

Water pollution touches upon every living human being and every animal and plant organism. Life in whatever form is a polluter and, by the same token, suffers from pollution. Pollution affects the quality of our environment and by so doing creates one of man's biggest problems, which is the ability to control water pollution economically and effectively. Additional problems lie in the area of information gaps. There is a serious and dire need for knowledge, for precise, accurate, and reliable data which can be utilized for specifying system design, performance, and management.

It is often stated that the necessary technology for successful abatement of water pollution is at hand, that such technology has not always been applied because of a lack of financial resources or because of institutional barriers of one type or another. A close scrutiny of the subject shows that this simply is not true in many instances. For example, there is no present technology available for economically gross treatment of mine drainage wastes and only partial alleviation of municipal sewage wastes can be accomplished. In fact, in the future more efficient reductions in pollution abatement will depend more and more on new and improved technology as the artificial barriers to application are removed. New technology not necessarily developed for pollution will have a definite and major impact in the forthcoming years.

More basic and applied research is needed upon which to build both understanding and techniques. Exploratory development is necessary to bring new techniques to actual trial, and a definite need exists for final development to produce techniques in a form suitable for general use.

The greatest problem concerning water pollution is the problem of maintaining a supply of clean, pure, fresh water. The world population can no longer afford the prevalent, widespread illusion that its water supplies are drawn from a limitless source. Comprehensive steps must be taken to conserve and protect water resources in order to ensure the country's continued growth and prosperity. The world is not running out of water. There is as much water today as there ever was, but there is a limit to a clean, developable, dependable water supply, and water needs are rapidly approaching this limit. If the existing clean water is to serve all the purposes for which it is needed, then indeed these same waters must be used over and over. To do this, an effective water resources management program in both quantity and quality will be needed. An effective program of pollution prevention and control is necessary to maintain a high level of water quality in the ground, lakes, streams, and coastal waters. This program must be instituted where nonexistent. It must be maintained, and above all, it must be constantly improved to satisfy the needs.

Most of our present water pollution problems have resulted from too little attention in the past. A lack of interest by industry and by the federal government has in the past enhanced the problem. It has been only recently that the federal government has started to deal with the problem.

Some municipalities and industries have resisted constructing waste treatment facilities on the basis of being an unnecessary or unbearable financial burden. Too much reliance has been placed on the philosophy that water purifies itself.

In the past, water pollution control dealt for the main part with problems caused by sewage, industrial wastes of known toxicity and behavior, and natural organic pollution. Pollution control was primarily aimed toward protecting downstream public water supplies, abatement of local nuisance conditions, and protection of aquatic life. Generally, streams provided dilution capability to prevent serious pollution problems, and waste treatment plants, where provided, were designed to take full advantage of the self-purifying capacity. Water purification plants provided the safety barrier for the water consuming public. Pollution problems in the past were principally local in extent and their control a local matter. Today, national growth and change have altered this picture. Urbanization, increased living standards, and encirclement of industry by the municipality have increased the volumes and strengths of municipal wastes.

Increased production of goods and the institution of newer and more exotic industries have greatly increased the amounts of common and rare type industrial wastes. New technologies are producing complex new wastes and products that defy our present ability to treat or control them or even detect their presence in water. The increased application of commercial fertilizers and the development and widespread use of a vast array of new pesticides are resulting in a host of new pollution problems from land drainage. The growth of the nuclear energy field and use of radioactive materials may further enhance a serious water pollution problem.

Conventional biological waste treatment processes are hard pressed to hold the pollution line, and for a growing number of our big city areas these processes are not adequate. There is a growing concern over the ability of water purification plants to protect the public adequately against the mass of biological and chemical pollutants entering plant intakes.

Our expanding urban growth and massive population densities in sharply defined locales has placed a tremendous burden on existing sewage disposal facilities. Sewage construction has not matched either this growth rate or its movements. As a result of the lack of central sewage treatment facilities, a large portion of our population must rely on industrial septic tanks for its waste disposal. Such large usage of this type of disposal results and enhances a serious pollution of groundwaters which often must serve the same population. Metropolitan areas find it extremely difficult, if not impossible, to escape this serious problem.

Groundwater pollution is a serious problem enhanced by an ever-increasing intrusion of seawater, especially along many of our coastal plain areas. Oil field brine disposal practices are also adding to the pollution of groundwaters, especially seepage from lagoons and evaporation pits located in the mid- and southwestern parts of the United States. Industrial waste oxidation ponds, sewage ponds, and waste storage pits enhance groundwater pollution especially when not properly lined to avoid seepage or percolation into permeable substrata. The use of substrata as a disposal site for industrial waste, including highly toxic and radioactive materials, has gained some acceptance from industry. Such a practice, however, should be discouraged unless it is absolutely ascertained that no deleterious effects will be induced to the existing quality of the available groundwater supply in the area of concern.

Many coastal and inland metropolitan areas have discharged their waste effluents into nearby streams, rivers, and shoreline areas with little or no treatment. Such practices have led to increasing pollution, which at present will be almost overwhelming to correct.

SOURCES OF INDUSTRIAL WATER POLLUTION

Mining Wastes

Large quantities of effluent wastes result from mining. Today it is estimated that mine drainage contributes over 4 million tons of acid pollution to about 11,000 miles of major and minor streams. Acid drainage enters our streams and lakes, destroying fish and fish food organisms and wildlife. Acid mine water is associated with surface and underground mining operations. About half the acid production comes from abandoned mines, the rest from operating mines.

Sulfur bearing minerals mix with air and water to form sulfuric acid. The greater portion of this pollution is caused by coal mining operations.

There are two ways to control pollution from acid mine water. One is to prevent the water from ever becoming polluted in the first place. The other is to remove the acid and other impurities from the water once they get into it. The preventive technique is the most effective and more reasonable. It certainly is cheaper to prevent pollution from occurring than it is to remove the pollutants from every gallon of water in acid ridden streams.

Waterflow control appears to offer the greatest promise for long-term correction of the problem and involves preventing the entry of water into the acid producing strata or removing rapidly the water which does gain entry, so as to minimize its contact with acid forming materials. Removing acid from the water after it has become polluted may be accomplished by neutralization, reverse osmosis, distillation, and electrodialysis. Successful use of these methods to date has been limited by high cost of chemicals and energy necessary to treat the large quantities of waste water involved. Economic problems have yet to be solved for the disposal of effluents from treated mine water, which contain concentrated acids and salts.

Solution of the acid mine water problems requires development of new procedures and technology. Generally four stages should be involved and would include

1. Research
2. Single process demonstration
3. Area cleanup demonstration
4. Full-sized action

Research into problems relating to acid mine drainage in the U.S. Department of the Interior is as shown in Table 3.

Chemical Pollutants

The tremendous growth in the production of synthetic chemicals for all kinds of uses is producing an entirely new type of pollution problem. Wastes from the manufacture and use of these chemicals are reaching natural waters in significant amounts. Their effects on plants and animals are poorly understood, and many do not undergo decomposition as readily as do most biologically produced materials.

Synthetic chemical pollutants are known to create taste and odor problems that are difficult and expensive to solve. Many of these chemicals are suspected of interfering with aquatic food chains. Others have questionable carcinogenic proper-

TABLE 3 Research Funding for Acid Mine Drainage Pollution Problems

Agency and projects	Fiscal year 1966	Fiscal year 1967	Fiscal year 1968	Fiscal year 1969
Bureau of Mines:				
Chemistry, sealing, and draining studies	$ 100,000	$ 154,000	$ 118,000	Continue
Survey of surface mining drainage	250,000	250,000		
Bureau of Sport Fisheries and Wildlife field studies (Indiana)	....	....	25,000	Continue
Geological Survey:				
Mine hydrology	23,000	23,000	23,000	Continue
Mine drainage problem areas	....	51,000		
Broad research programs	....	....	....	Begin
Office of Saline Water desalination methods tests	17,500	60,900	140,000	Continue
Federal Water Polution Control Agency	127,000	191,000	164,000	Continue
Enforcement program:				
Monongahela	117,000	141,000	144,000	Pending recommendation of enforcement conference
Colorado	10,000	50,000	20,000	
Comprehensive program:				
Ohio Basin	160,000	170,000	230,000	Pending recommendation of enforcement conference
Chesapeake	70,000	70,000	160,000	Continue
Susquehanna	90,000	100,000	70,000	Continue
Research and development	2,721,463	1,572,255	2,701,000	
Treatment of mine drainage	....	48,950	85,000	Continue
Methods of control at source	2,668,020	1,489,000	1,861,000	Continue
Drainage control	....	....	100,000	
Effects of mine drainage	31,999	32,745	380,000	Continue
Identification	1,444	1,560	275,000	Continue
Grand total	$3,398,963	$2,472,155	$3,401,000	

SOURCE: Water Pollution 1967 Hearing, 90th Congress on S. 1591 and S. 1604.

ties. Although the kinds of chemicals and their effects remain largely unknown, it is known that on a gross basis, many if not most of them pass through the water treatment processes and reach the consumer in his drinking water.

Far more research is required on the identity, fate, and biological and nonbiological transformation of these chemical pollutants. Means must be developed to minimize such pollution by new or improved manufacturing techniques, by new methods of waste treatment, and by the substitution of more readily degradable or less objectionable chemicals.

To assess pollution properly, research must disclose the species composition and requirements of healthy aquatic communities in all parts of the United States.

We cannot consider pollution in any absolute sense. It must be considered in the light of what we want in amenity and usability of the water. Individual sources of pollution may in themselves be quite innocuous, but the sum total of all such innocuous sources may be so great that much of the life of the stream is killed and the amenity of the water is lost. Because there are so many different kinds of pollutants, any assessment of pollution must vary with time and place, and no single set of criteria or standards can be applied to all situations.

Broadly speaking, pollutants may be grouped into various different principal categories with some overlap between these categories: poisons, organic materials, suspended solids, heat, nontoxic salts, pesticides, and radioactive isotopes.

Poisons, such as metallic ions, may be present in industrial effluents and may be poisonous to aquatic life at low levels of concentration (lead 0.3 ppm, copper 1 to 3 ppm, zinc 0.2 to 0.7 ppm, mercury, and many others). Complexes of these ions may be affected by secondary alterations through chemical or physical action such as changes in temperature and oxygen content. Frequently temperature has a direct influence on toxicity. A rise of 10°C may halve the survival time of aquatic life. Organic material, usually sewage, but also from many industrial processes such as the preparation of foodstuffs and the manufacture of papers and synthetic fibers, may be toxic; chemicals may cover the bottom of the stream so as to smother the fauna or may enrich the water with phosphates and nitrates to a point at which algae growth becomes objectionable and difficult to control. Also, the biochemical oxygen demand may become so great that the concentration of dissolved oxygen is so depleted that most animals cannot inhabit the area. Suspended solids usually accompany organic materials from domestic and industrial sources but may also arise from mining and quarrying operations, from poor control of erosion on agricultural lands, and from milling processes and construction. Increases in temperature usually result from the use of large quantities of water for cooling industrial plants such as generating plants powered by steam or by nuclear energy. Nontoxic salts as waste products from certain kinds of mining and industry, from oil drilling, or from irrigated agriculture may increase the salt content to quite high levels of 100 to 500 or more parts per thousand. Pesticides used in the control of plant or animal parasites may be washed into waterways following heavy rains, where they may be highly toxic and frequently are cumulative. Radionuclides from industrial reactors are a potential source of pollution, especially in the event of accidental discharge of cooling waters from nuclear facilities. Concentrations of these isotopes should not exceed certain limits discharged to receiving streams above natural background levels. Allowable discharge levels are listed in Title 10, Code of Federal Regulations, Part 20.

The amounts of heavy metals and the levels of dissolved gases toxic to aquatic organisms are generally known and can easily be measured by rather simple chemical procedures, and simple detection devices can reveal the amounts of radioactivity known to be harmful to man. However, in many instances, the effects of pollution may be so subtle that they cannot be detected until some parts of the biota have been eliminated. A substance which may be toxic to one organism may aid in the growth of another organism in the same community. Similarly, different salts of a particular metal may have quite different effects on the same kind of organism.

Compounds such as organic phosphates and chlorinated organic mixtures used as insecticides have been shown to be lethal to fish and many other aquatic organisms in doses as small as 0.1 to 0.3 ppm for DDT and 0.01 or less for endrin.

Each pollutant has a primary effect on various organisms either by killing them directly or by so changing their physiological processes that the composition and dynamics of the aquatic community are drastically changed. It is therefore evident that a thorough knowledge of the biota of the aquatic environment must be at hand to assess the effects of pollution. Adequate and accurate assessment of pollution and interpretation of the data are necessary. Special sampling programs must be designed for each situation, and each situation must be appraised individually.

"Water Quality Criteria" by McKee and Wolff, State Water Quality Control Board, Sacramento, Calif., Publication No. 3-A, contains a wealth of data on water quality and the effects of contaminants on aquatic life. This document has been of significant help in serving as a manual and handy reference for water quality criteria.

Treatment and Disposal of Industrial Waste Waters

Available methods for treating and disposing of industrial waste waters by themselves are generally quite like those for domestic sewage. They include screening, lagooning, sedimentation with or without neutralization, coagulation or precipitation, biological treatment, and ultimate disposal of the treated liquids into receiving waters or onto land. The exact mechanics in performing the above procedures are well documented in the literature. There are, however, certain procedures which are difficult to perform under ordinary circumstances, and as yet many deleterious and unwanted constituents in waste waters still go unabated.

Waste waters contain mineral impurities, organic impurities, both organic and mineral impurities, and radioactive waste and are generally classified into three categories: waste from manufacturing processes, waters used as cooling agents in industrial processes, and finally waste from sanitary uses.

One big problem in minimizing the effects of industrial wastes in receiving streams and treatment plants is to reduce the volume of such wastes. This may be accomplished by conservation of water, by changing production to decrease waste, and by revising both industrial and municipal effluents for raw water supplies. The reduction or elimination of materials from liquid effluents of industrial plants by process changes, equipment modifications, careful housekeeping, and other in-plant measures can significantly reduce the waste load to be treated and which eventually reaches the stream.

RESEARCH DEVELOPMENT

Industries produce a wide variety of types of industrial wastes, the predominant sources of which are from metal production, fabrication and finishing operations, and coal and oil production. The cost of controlling and treating wastes from these manufacturing processes has always been of deep concern to industry and has become a very important factor in the overall manufacturing operations, not only to comply with existing legislation but to economize in waste collection, treatment, and disposal methods.

Many industrial water pollutants do not respond to biological treatment and may even be harmful to such processes; therefore, sanitary waste treatment facilities do not have the capability of removing industrial pollutants from plant effluents.

Industrial waste may contain any number of components which, as an example, might contain the following: emulsified and free oil, metals such as copper, zinc, chromium (hexavalent or trivalent), nickel, iron, cyanide as CN, nickel, other metals, organic solvents, acids, and alkalies. These wastes must be handled separately and cannot be discharged indiscriminately into a storm or sanitary drain.

To date many industries have their wastes trucked away by an outside vendor without regard to the final disposition of the waste. Areas for discharge of such waste are rapidly decreasing and are coming under close scrutiny of local health authorities.

Ordinances regulating the use of public sewerage and drainage systems are being passed by municipalities governing the use of such systems. Such regulations may prohibit the user from discharging any substance that may be deleterious in a mechanical or biological way to the system.

Industrial wastes should be treated at the plant site with a resultant quality acceptable to the local municipality.

Problem areas requiring research should be directed to find:

1. Economical means of removal of solvents from spent liquors in chemical production facilities.

2. Economical methods of reclaiming acid wastes used in metal pickling operations. Some research was done in this area by the author in using certain strains of iron bacteria. Particular emphasis was placed on sulfuric acid solutions used in steel mill operations.

3. Practical methods in removing sulfuric acid and other sulfur compounds from acid mine drainages.

4. Methods to concentrate industrial liquid wastes by removing all noxious constituents through flocculation and/or precipitation.

5. Methods to use all end products so that no waste is generated.

WASTE EVALUATION

Toxicological and epidemiological studies are an essential part in the assessment of the public health importance of new waste, and a protocol of effective research must be established. Data on which to base the environmental requirements of aquatic life are needed and must be developed through research. These requirements, which include items such as temperature, dissolved oxygen, carbon dioxide, and pH, should be based on scientific fact so that reliable water quality criteria can be established to help restore and maintain a suitable aquatic environment. A waste may impair water treatment processes by retarding flocculation or increasing chlorine demand, or it may have some other effect. Methodology which is reliable to predict such effects should be developed to be used as a guide in water purification practices. Objectionable taste and odor in drinking water are the commonest manifestations of industrial pollution, which is difficult to identify and trace. Methods for predicting a waste's taste and potential are needed.

Knowledge is needed on the effects of mineral salts, organic compounds, and domestic and industrial wastes as well as the etiological agents of infectious diseases. Chronic low levels of exposure to trace elements of molybdenum, selenium, vanadium, nickel, zinc, copper, and other elements should be ascertained especially as to their toxic effects on animal and aquatic life. Many of the waste substances now entering water supplies are known to be toxic in sufficient concentrations, but unfortunately there are many others whose toxicity is unknown. Increased attention should be directed to the development of rapid screening tests for waste materials which carry toxicity hazards in waters used for human consumption.

TREATMENT OF WASTES

Conventional treatment methods remove about 40 to 60 percent of incumbent pollution in wastes, but a real desire to develop entirely new processes that will approach actual purification is needed. This is a particular requirement especially in heavily populated areas and where the same water must be used over and over again to meet the needs of consumption.

Biological systems of waste treatment must be improved to select and adapt microorganisms to metabolize new organic compounds that will increase the effectiveness of existing biological treatment systems. It is important for municipalities and industry alike to know whether organisms can metabolize a given compound or not so that determinations can be made as to whether or not a waste would be acceptable in a treatment facility and the receiving water.

Use of stabilization ponds is a recent low-cost treatment procedure for water. Unfortunately, not enough is understood of the natural forces such as solar energy, respiration, and wind properly to evaluate these effects on design criteria. Knowledge is needed on the effects of hydraulic loading and the character of wastes to determine pond efficiencies in removing pathogenic and other biologic organisms.

Present Interest in Waste Treatment

Areas of present research interest in waste treatment encompass such procedures as adsorption, solids extraction, foam fractionation, freezing, ion exchange, oxidation, filtration, and membrane processing. Major concerns of such treatment approaches revolve around eventual economic feasibility of these operations. Active participation in programs concerned with the utilization of these techniques by universities, private research institutions, and industry is not only desirable but essential.

Disposal of Waste Effluents

Dilution of waste into receiving streams is not good practice and raises pertinent questions regarding justification in such action. Discharging liquid industrial and radioactive wastes into deep porous strata emphasizes the need for more information on this method as a means of final disposal. Burial or discharge into the sea only propagates a worsening situation. The ultimate in the handling of wastes is to have no waste at all. By-products, whether they be from industry or other sources, must be made useful.

WATER QUALITY SURVEILLANCE

A minimum criterion for effective water quality surveillance is necessary. An acceptable program must be based on stream data that accurately reflect the stream's condition. Sampling and analytical procedures should be dependable and inexpensive. Applications of our most advanced laboratory methods presently provide only a partial picture of water quality standards. Because of the importance of such data in water quality management programs, emphasis should be given to research satisfying this need. Simpler, more rapid, and more specific procedures for identifying contamination by human wastes must be developed. Improved techniques for detecting and measuring virus contamination must be developed, especially in light of numerous outbreaks of infectious hepatitis which have been traced to contaminated drinking water. Widespread use of organic chemicals has introduced these new and highly complex agents into our water resources for which no methodology exists in determining accurately their quality or quantity. Development of more effective methods for capturing concentrations and identifying and measuring organic contaminants represents an important need in water quality surveillance and in development of controls. Improved, highly refined instrumentation for microchemical analysis needs to be developed for this purpose.

WATER TREATMENT

A well-operated, modern water treatment plant removes suspended particles and tends to destroy microorganisms which manage to survive flocculation and filtration. It does not, however, remove dissolved impurities efficiently and is unable to handle satisfactorily problems involving soluble organics. Carbon is used to remove odor and taste but is selective in its action. To increase the efficiency and economy of water treatment, it is necessary to develop a fundamental understanding of physical-chemical principles applicable to the removal of foreign substances in water.

INSTRUMENTATION

The complexity of water pollutants and their characteristics requires that the old methods of analysis be replaced by new, modern, highly advanced electronic analysis. Rapidity and accuracy are a must to obtain quick information, especially in detecting and analyzing chemical constituents and biological criteria.

Objectives for the Federal Water Supply and Pollution Control Program are the conservation of water quality and quantity and to assure continuously an adequate supply of water suitable in quality for public and industrial use, propagation of fish and aquatic life and wildlife, recreation, agriculture, and other legitimate uses. Government, industry, education, private research, and the population at large must strive to abate water pollution.

Section 2

Pollution Control by Industry Problem

Chapter 10

Pollution Control in the Steel Industry

FRANK N. KEMMER

Manager, Pollution Control Department, Nalco Chemical Company, Chicago, Illinois

INTRODUCTION

Few people outside the steel industry are aware of the enormous volumes of water and air required to produce iron and steel from raw materials to finished shapes. Figure 1 is a graphic representation of the materials required in the primary iron producing facility, the blast furnace, in producing 1 ton of finished iron. About 3.5 tons of air must be supplied to the blast furnace to reduce the ore to metal, and this amount of air produces about 5 tons of blast furnace gas. So it is apparent that the tonnage of air and gas far exceeds the solid charge to the furnace and the products tapped from the furnace.

WATER REQUIREMENTS AND TYPICAL CONTAMINANTS

If the raw ore is taken to finished steel products, the water required is even more impressive than the air requirements, generally ranging from 20,000 to 40,000 gal/ton of steel produced. In the older mills of the East and Middle West, most of this water is used on a once-through basis; in the West, where water is less plentiful, a high percentage of the water is collected and reused.

For the year 1964, the U.S. Census Bureau tabulated water intake for the steel industry as 3,815 billion gal with gross water usage of 5,510 billion gal, corresponding to a water reuse of about 42 percent.

By far the largest portion of the water requirement for steel manufacturing processes is for cooling services. The various areas requiring cooling water are tabulated in Table 1.

Because of the large consumption of water for cooling in an integrated steel mill, the heat absorbed by the water and discharged to receiving streams can be considerable, usually in the range of 15,000 to 20,000 Btu/hr per daily ton of production. Evaporative cooling is sometimes required to control the temperature if the receiving body of water is classified as a stream required to support fish life and the stream temperature below the mill may exceed specified limits during drought conditions.

The temperature rise across various units using cooling water usually ranges from about 10°F for blast furnaces and surface heat-exchange units, such as turbine condensers, oil coolers, and vapor condensers in the coke plant, to as much as 50 to 60°F for open hearth furnaces and annealing and reheat furnaces. The temperature reached by water used for roll cooling varies considerably depending on the material being rolled and the degree of recycle. Since workmen are operating in these areas, there is a tendency to limit the temperature rise in this service, because as the water becomes heated, the humidity in the area builds up and working conditions become unpleasant.

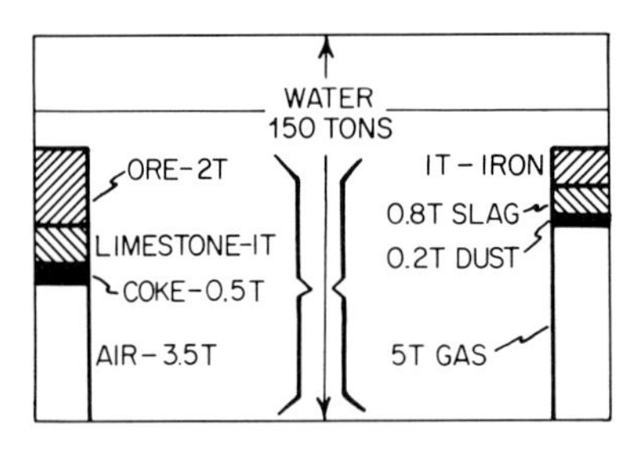

Fig. 1 In order to produce 1 ton of finished iron in the normal iron producing facility, the blast furnace requires 150 tons of water. A troublesome by-product is the 0.2 ton of dust that contributes to air pollution.

Another large use of water in the steel mill is that required for washing blast furnace gas, which requires about 2 to 3 gpm per daily ton of production. Since the blast furnace gas is distributed throughout the plant for use as a fuel in soaking pits, annealing and reheat furnaces, and the boiler houses, it is important that this gas be cleaned of debris—mostly fines from the blast furnace charge—so that valves and lines can be kept open and combustion equipment maintained in good condition. Of course, the amount of dust collected is very sizable (Fig. 1); so it has value for reprocessing in the sinter plant, for recharging to the blast furnaces. The gas is usually cleaned first in a simple drop-out drum, followed by scrubbing with water, cooled with additional water, and finally treated through electrostatic precipitators before being distributed to the system. Most of the washers in this kind of service are of the venturi design; electrostatic precipitators are usually the wet type, using some water for cleaning and cooling.

TABLE 1 Cooling Water Uses in a Steel Mill

Utilities (both at boiler house and at local use areas)	Turbine condensers, bearings, compressor jackets, glands
Iron production	Blast furnaces, valves on stoves, gas cooling, pig machine
Steel manufacture	Furnaces, degassing condensers, gas quenching
Rolling and shaping	Rolls, bearings, guides, saws, straighteners, forges
Heat treating	Furnaces, quenching vats
Coke plant	Gas coolers, vapor condensers, coke quenching

In some plants, open hearth furnace dust may be scrubbed from the open hearth gas in a system similar to that used on the blast furnace. However, the operating conditions are quite different, the blast furnace dust being basic and the open hearth dust acidic. The open hearth scrubber systems must therefore be of corrosion resistant construction.

A final large-scale consumption of water takes place in the rolling mills, where water is used for bearing cooling and for sluicing scale from the rolling stands

TABLE 2 Water Pollutants in a Steel Mill Complex

Description	Source	Disposition
Oil:		
Rolling oils......	Rolling mills, cold and hot rolling	Part adheres to scale; free oil collected for incineration
Lubricants.......	Motor power: steam engines, forge and hammer pistons, gear drives, electric motors. Fabrication machinery: machining, forging, drawing, etc.	Floor spills soaked up on adsorption compounds; cutting oils and other emulsions segregated for incineration; oily water skimmed at source, and free oil collected for incineration
Hydraulic oils...	Motive power pumps for positioning devices or for press operation	Spills segregated for pickup and incineration
Quench oil.......	Heat treatment	Seldom replaced; can be incinerated
Fuel oil..........	Boiler plant, furnaces, soaking pits	Spills segregated by dikes around storage tanks; incinerated
Solvents.........	Paint shops, degreasing operation	Collected for incineration
Tars and pitch...	Coke plant and by-products recovery	Collected for incineration
Suspended solids:		
Scale............	Rolling mills	Sinter plant for recovery as sinter
Sand............	Foundry	Slag pile, landfill
Burden fines.....	Air washers at sinter plant and skip hoist charging area; blast furnace gas washers	Sinter plant for recovery as sinter
	Open hearth and BOF fines	Sinter plant if Zn is low; otherwise buried in slag pile
Fly ash..........	Coal fired furnaces	Cement or cement block additive
Coal and coke...	Coke plant	Collected and burned
Chemicals:		
Pickle liquor.....	Acid pickling	Regenerated or neutralized
Acid sludge......	By-products plant	Regenerated or burned
Caustic wash....	By-products plant	Incinerated
Lime............	Mold or stool coating; water softener sludge	Recovered for pickle liquor treatment; controlled release to waste water
Brine...........	Zeolite regeneration	Reclaimed or controlled release to waste water
Cleaners.........	Surface treatment, degreasing	Segregated for oil breakout or incineration
Toxic chemicals	Coke plant, gas line drip legs, metal treatment	Chemical or biological destruction; incineration
Heat:		
Cooling water....	Furnaces, heat treatment, roll cooling, air conditioning, heat exchangers	Cooling towers
Boiler blowdown	Steam plant	Recover to heat feedwater
Sanitary wastes:		
Domestic water	Change rooms, toilets, cafeterias, etc.	Segregate and treat by standard methods

into flumes leading to scale pits. Water is also used at the rolling mills for scale breaking and for washing and quenching the gases evolved during hot scarfing.

A lower-volume but higher-quality water requirement is makeup to the steam generating boilers in the various boiler houses usually located throughout the steel mill complex. If the plant supplies all its own steam and power requirements, the boiler feedwater usually amounts to approximately 500 gal per daily ton of production, with a large percentage of this being recovered condensate. This includes waste heat boilers at the open hearth or BOF (basic oxygen furnace) shops as well as the main boiler houses.

Miscellaneous water uses in the mill include the water used on barometric condensers in vacuum degassing operations, rinse water and process water for surface preparation and finishing, potable water for the change rooms and for the drinking fountains and cafeterias, and makeup to the various air scrubbing units which are located in such production areas as the foundry and the burden handling areas at the charging level of the skip hoists. Miscellaneous uses also include makeup to the high-pressure water jets used for cleaning castings and equipment, supply to the various maintenance shops—such as garages and locomotive maintenance yards, sprays for dust control in such areas as the slag aggregate handling and grading plant, and sprays for cooling or granulating hot or molten slag.

The contaminants typically found in steel mill waste water and their sources are shown on Table 2.

AIR REQUIREMENTS AND CONTAMINANTS

As indicated earlier, the largest use for air is in the conversion of iron ore to pig iron, with additional large volumes being required by all the fuel burning operations as combustion air. The blast air for the blast furnace operation is usually compressed by turboblowers, although older mills continue to use reciprocating compressors. The pressure varies depending on the design of the furnace and the desired top pressure, but usually amounts to approximately 30 psi, with the air being heated to approximately 900°F in stoves for introduction into the tuyeres of the blast furnace.

Gases other than air, particularly high-tonnage oxygen, are used in steelmaking, and become a part of the gas discharge from manufacturing operations.

Air handling equipment in the areas where dust control is required, such as the ingot mold foundry, the sinter plant, and the skip hoist loading areas, is sized on the basis of the dust loading usually encountered in these areas.

The varieties of contaminants encountered in the steel mill are shown in Table 3.

TABLE 3 Typical Air Contaminants

Contaminants	*Sources*
Particulates	Stacks on boiler plants, steel making furnaces, sinter plant, coke ovens, foundries, slag plant, ore preparation, skip hoist area
Gases:	
Sulfur compounds	Combustion of sulfur-bearing fuel
Carbon monoxide	Blast furnace gas
Cyanides	Blast furnace and coke oven gas
Fluorides	Steelmaking furnaces
Benzene compounds	By-products plant

SOLIDS HANDLING

An integrated steel mill must stockpile and handle tremendous tonnages of solid material, to supply the burden to the blast furnace, to transport semifinished products within the mill, and finally to dispose of almost 0.8 ton of slag for every ton of iron produced. To accomplish this, there is an intricate system of roadways and rail trackage within the complex, with special cars built to handle the ore, limestone and coke required for the burden and to transport hot metal, ingots, billets, slabs, finished products, scale, and dust returning to the sinter plant, and slag discarded at the slag pile or in the slag processing area. Solid waste, including refuse and trash, timbers, and similar debris, is usually carted to the slag dump and put in an area segregated for this particular disposal purpose.

WASTE WATER

Blast Furnaces

All the iron produced in an integrated steel mill originates at the blast furnaces, which are vertical, water cooled furnaces ranging from about 16 to 28 ft in diameter and up to 100 ft tall. A blast furnace may require from 1,000 to 5,000 gpm of cooling water and process water for washing.

The individual water streams leaving the blast furnace area comprise the furnace cooling water, gas wash water (which includes water used on the electrostatic precipitators), wash water from air scrubbers, and cooling water used to cool the blast air valve leaving the stoves.

The cooling water on the blast furnace itself is collected from the various furnace zones and discharged through an air gap—so that the operators can verify the flow of water through all the circuits—into a collection trough which carries it to the main sewer. This water leaves the furnace essentially as received except for the heat added. If the heat is not objectionable in the receiving stream, this water can pass directly to the stream without treatment.

If the heat is objectionable, however, it may be necessary to collect the blast furnace cooling water and pass this over a cooling tower, with the discharge going directly to the receiving stream or being reclaimed for recycle to the blast furnace. One of the difficulties in handling the heat load is the relatively low temperature rise and the very small driving force to produce evaporative cooling in a high-humidity environment. In order for an evaporative cooling system to operate effectively, the temperature level must be built up by recirculation of the cooling water directly from outlet to inlet at each blast furnace. An alternative would be to recirculate a portion of the water from the cooling tower back to the blast furnace supply line, which would cause a gradual elevation in the cooling water temperature to a point where evaporative cooling can occur at the cooling tower.

Gas wash water is always clarified to remove the dust carried over from the blast furnace. The solids consist of burden fines having a specific gravity of 3 to 4 and being relatively fine with about 50 percent smaller than 10 microns. The concentration of suspended solids varies considerably, not only from plant to plant but even from hour to hour on the same furnace. A typical range is 5,000 to 20,000 ppm. A well-designed clarifier can reduce this to less than 50 ppm. The dust is thickened as a sludge and usually dewatered on vacuum filters to produce a relatively dry cake (usually over 80 percent solids), which can be easily handled in rail cars and returned to the sinter plant. In the sinter plant, the sludge is blended with other fines and fired on grates in a furnace to produce a clinkerlike material which can be recharged to the blast furnace.

The gas wash water also dissolves contaminants in the vapor phase, which include ammonia, phenol, cyanide, and carbon monoxide. The wash water dissolves alkali from the dust as well; so there is a net increase in sodium and potassium bicarbonates along with the solution of these other contaminants.

In some steel mills, the overflow from the wash water clarifiers discharges into a sewer, where it combines with the total plant outfall. The dilution effect may reduce the contaminants in the wash water to an acceptable level. If not, then either the total flow must be processed to remove these contaminants or the gas washer system must be closed up by recycle. Since recycle will build up the temperature of the system, such systems may require the inclusion of a cooling tower in the circuit. To avoid deposits in the system brought about by the buildup of alkalinity, acid may be fed to these closed systems to maintain an acceptable stability index. This may result in a pH in the range of 6.5 to 7.5, at which level the pickup of cyanide is minimized and the stripping of cyanide at the cooling tower is promoted. If the pickup of alkali salts is relatively moderate, stabilization

(*a*)

(*b*)

Fig. 2 Two views of clarification equipment used to remove fine solids, such as blast furnace dust and fine mill scale. Recovered solids must be scraped to the center and pumped to vacuum filters or centrifuges for dewatering. (*Inland Steel Co., East Chicago, Ind.*)

of the recirculated water with polyphosphate to control deposits is quite practical and eliminates the need for acid feeding.

With a closed system of this type, a wash water flow of about 20,000 gpm, for example, may be reduced to a discharge of only 100 to 200 gpm, depending on rate of concentration buildup and the ability to control deposits through acid treatment. With this method of operation, then, if removal of contaminants is required to meet water quality criteria, the stream to be processed is reduced to on the order of 1 percent of the gas wash water flow.

The clarification of the wash water is extremely important to maintaining a clean system. Various chemicals, such as coagulants (pickle liquor may sometimes be used) and polyelectrolytes, are helpful not only in reducing overflow turbidity but also in compacting the sludge and assisting in production of dry filter cake on the clarifier—generally requiring weir spill-over rates below about 15 gpm/lineal ft and rise rates of less than 2.5 gpm/sq ft. This discharge flow will contain less than 50 ppm suspended solids and the underflow will reach about 12 to 20 percent solids content. These results can be improved by chemical treatment.

Casting Operations

When the blast furnace is tapped, molten iron is directed through sand-lined sluices, controlled by gates in the casting floor, to the ladles. After the molten metal ceases to flow, slag continues to discharge from the furnace, and this is directed, by controlling the gates in the casting floor, to slag pots. The molten iron is taken to various production areas. It may be poured into pig iron or converted to steel in the open hearths, the basic oxygen furnaces, or electric furnaces.

In the pouring of cast iron into pigs of about 100 lb each, water involved in the operation is used for the spraying of heavy lime solution onto the molds of the pigging machine to prevent sticking of the pig iron in the molds and to ensure release from the molds after the metal has solidified and the molds are inverted. Additional water is sprayed onto the molds and the pigs for cooling.

In the teeming of molten steel into ingot molds, no water is involved except for the preliminary coating of the stools on which the ingot molds rest during pouring.

Some steel mills are installing various types of continuous casting machinery. Details on the performance of all these machines are not being fully disclosed. However, water is used for cooling and quenching, and this water contains fine scale and surface debris removed from the slab as it is being cast. It may also contain oil or synthetic mold release agents. The water is recirculated over a cooling tower to minimize loss and reduce the volume of the bleed stream for later treatment.

Foundry

Most integrated steel mills operate foundries, which may cast ingot molds for use within the mill or may supply castings for machinery to be built by heavy equipment manufacturers. Even in the most modern foundries, dust is a common problem, and it is necessary to provide adequate ventilation to hold the dust concentration to acceptable levels. The air may be scrubbed before discharge to the outside in water spray type units or filtered through bag filters, in which case the collected dust must be wetted before it can be removed and hauled to a disposal area. The waste water from air scrubbers is very high in suspended solids and must be processed for removal of these solids for final disposal.

Where the foundry pours castings in addition to ingots and pigs, it is usually necessary to clean the metal surfaces after shake-out. This is done with a high-pressure water jet. The water is collected, and the large particles are separated so the water can be reused. However, a small volume of bleed-off is necessary to remove mud, and this is usually very high in suspended solids in the range of 1,500 to 25,000 ppm (for details on foundry controls, see Chap. 11).

Open Hearth Furnaces

The open hearth furnace produces steel (below 1 percent carbon) from a charge of steel scrap, iron (about 4 percent carbon), and slagging materials such as lime and fluorspar. These furnaces produce about 100 to 300 tons per heat of 8 to 12 hr duration.

As in the blast furnace area, open hearth furnaces use large quantities of water for cooling, generally ranging from about 750 gpm for small furnaces to as much as 1,500 gpm for larger units.

Also, as in the blast furnace area, the gases from the open hearth furnaces contain a high concentration of dust, which may be removed by washing, precipitation equipment, or both. This dust is removed from the wash water by clarifiers and may be combined with the recovered dust from the blast furnaces for sintering if the zinc content is sufficiently low. The dust loading varies appreciably during the heat from a range of about 0.2 to 0.6 grain/scf during charging to about 2 to 3 grains/scf several hours after start of oxygen lancing.

The dust particles from open hearth furnaces are much finer than from the blast furnace, and chemical coagulation is needed to cause them to separate from the waste water. Oxygen lanced furnaces discharge a higher dust load than the older, conventionally fired furnaces.

Whereas the wash water from the blast furnace is alkaline, that from open hearth scrubbers is quite acidic, so that special materials of construction are required in closing up open hearth washer systems. However, because the pH of the circulated water is in the range of 2.5 to 3.0, the system can be completely closed without fear of deposits, the only water loss being that associated with the collected sludge or filter cake.

It may be necessary to select the scrap charge for the open hearth furnace carefully if the dust is recovered for sintering, since a high zinc content can be damaging to the refractory lining of the blast furnace when the sinter is recharged to the furnace. The zinc, which comes from galvanized sheet or white metal in the scrap, may pass through the refractory and sublime on the metal shell, gradually expanding and loosening the refractory.

A third source of waste water from the open hearth is the blowdown from the waste heat boilers, which recover heat from the exit gas and produce by-product steam. Water from the continuous blowdown is generally of such small volume that it does not significantly affect the composition of the collected waste water from the open hearth area, where this includes the furnace cooling water. However, mass blowdown from the boilers may show up as periodic high pH or high conductivity in the combined waste water. If there is a separate water treatment plant to supply makeup for the waste heat boilers, wastes from regeneration of the water treatment system may also constitute one of the pollution loads from the blast furnace area.

Basic Oxygen Furnaces

The BOF shop is a facility for conversion of iron to steel at much higher rates than attainable in the open hearth. In the conversion of iron to steel in the basic oxygen furnace, the molten metal charge is reacted with oxygen, introduced through a water cooled lance, which burns off the impurities in a period of about 20 to 25 min. Total heat requires about 50 min compared with the 8 to 12 hr required for steelmaking in the open hearth furnace. Most BOF units produce in the range of 100 to 300 tons of steel per heat. Because of the high heat release, gases leaving the furnace hood during lancing are very hot. The hood may be cooled with circulating water, or heat may be recovered from this gas through a boiler mounted directly above the hood. Because the heat is released at an uneven rate, and because little heat is released during the charging and tapping period, the flow of steam from the generating unit is irregular and must be modulated by discharge through an accumulator, which then releases steam at a relatively steady rate into the plant steam system at a pressure below that of the waste heat boiler.

Because of the very high temperature service, the lance cooling water is specially treated to prevent corrosion or scale formation at high temperatures; this circuit may be completely closed and heat extracted from the lance cooling water through a secondary cooling water circuit using a surface type heat exchanger.

The gas is cleaned in equipment similar to that used on the open hearth, although there are variations in methods of washing and removing accumulated dust from the washers and precipitators. The dust loading is usually appreciably higher in the BOF discharge (about 50 to 60 lb/ton of steel and in the range of 0.5 to 5.0 grains/scf in the gas entering the spark box) than in either the blast furnace gas (about 30 lb/ton) or the open hearth (about 20 lb/ton). There is appreciable variation in temperature and pH of the wash water during the heat as different ingredients are added to the charge during the oxygen blow.

Electric Furnaces

A third process for manufacturing steel, the electric furnace process, can either produce the common grades of low-carbon steel from scrap or can be charged with alloying materials to produce special steels such as stainless steel or tool steel. Electric furnaces are also used for production of the ferroalloys which are the alloying elements for these special steels, such as ferromanganese, ferrovanadium, and ferrochrome.

In the operation of the electric furnace, fluxing materials are added to the charge along with the metallic components, so that a slag is produced as in the other furnace operations. The charge is reduced by an electric current, and the violent activity in the area of the electrodes causes a rather high discharge of dust to the atmosphere, amounting to approximately 25 to 30 lb/ton of production. The duration of each heat is 2 to 4 hr.

Water is used to cool the furnaces and certain components of the electrical gear. The dust load in the shop is generally controlled by air washers, usually with some of the wash water being recirculated through the scrubber system. The suspended solids content of this water is usually extremely high, on the order of 5,000 ppm.

Rolling Mills

Although there is a variety of rolling operations in the steel industry, each having its own characteristic waste problem, the types of wastes can be classified into categories as to the size of the particles of scale flexed from the steel surface during rolling and by the type of oil used for lubrication of the sheet passing through the rolls. Both the scale and oil find their way into the circulating water.

The largest pieces of scale and debris are produced on the slower-speed scale breakers and primary rolling mills which reduce ingots to blooms, billets, and slabs. The scale or metal chips dropping into the flume below the stands of these mills may be chunks of heavy metal or fine flakes of 10- to 20-mesh particle size. Although there will be some particles finer than this, the percentage will usually be small. Oil in the water from these mills is generally lubricating oil from the bearings and is usually nonemulsified, readily breaking from the water surface in a short period of time. Some oil, however, adheres to the scale and metal chips. Because the particles of scale from primary rolls are large, the scale pit overflow is usually quite low in suspended solids—about 100 to 250 ppm, with about 10 to 25 ppm oil—and the overflow rate may be as high as 15 to 20 gpm/sq ft of separation area.

As the metal is further reduced in thickness through roughing stands and finishing stands, the scale is reduced in particle size to the micron range, producing a particle which settles very slowly in water.

On hot strip mills and structural rolling mills there is generally less oil found in the waste water than in the blooming mills, and it is also a nonemulsified oil which separates quite readily from water.

In an integrated mill, as much as 20 percent of the steel tonnage may be cold rolled sheet. When steel is rolled cold, the scale is quite similar to what is found with hot rolling, but the oil problem is considerably more difficult to handle, since emulsified rolling oils are added directly to the water sprayed on the metal as it is being rolled. Because this is a very difficult problem, the trend in modern mills is to recirculate this water, separating the free oil from the emulsified oil in the recirculation stream by centrifuging a bleed stream, and removing fine scale particles by filtration. In the older mills, however, the rolling oil may be added directly to a once-through flow of water being sprayed onto the metal so that this emulsified oil remains in the water collected from the mills and creates a difficult disposal problem.

Accessory equipment related to the rolling mills includes reheat furnaces, shears,

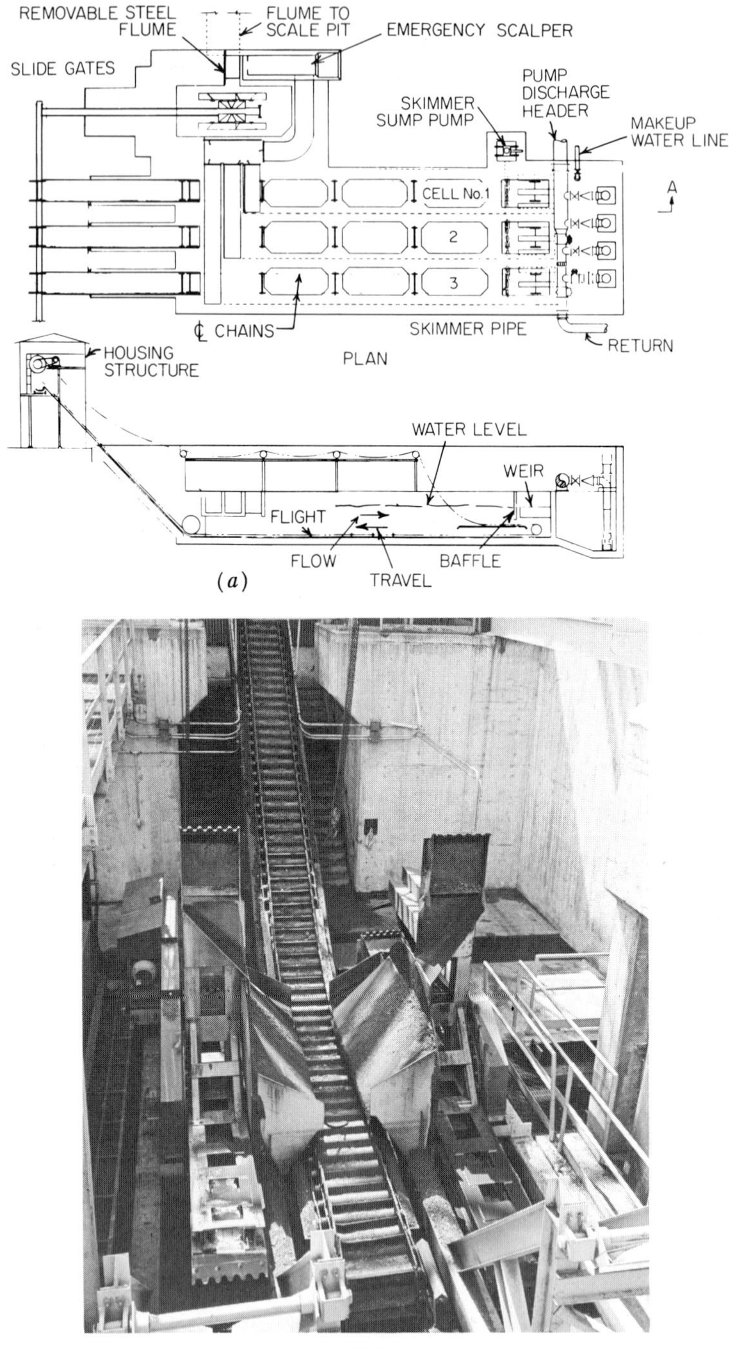

Fig. 3 (*a*) A three-cell scale pit serving a hot strip mill. The scalper in this diagram is shown in Fig. 3*b*. (*Inland Steel Co., East Chicago, Ind.*) (*b*) Scalpers remove coarse scale to reduce load on scale pit in Fig. 3*a*. (*Inland Steel Co., East Chicago, Ind.*)

hot scarfing machines, cold saws, and hot saws. All these require water, that used for the cooling of the reheat furnaces being by far the largest in volume. This cooling water generally leaves without picking up contamination, but the water from each of the other auxiliary operations contains fine metal and mill scale and may also contain some oil which leaks into the water sumps from hydraulic oil lines and from lubricated bearings.

Often the water from these auxiliaries is directed into a scale pit at one of the rolling mill stands, overloading the scale pit and complicating the task of separating oil and scale from water.

Diversion of a large volume of water from reheat furnaces from the scale pit is a first step in improving the separation efficiency of the scale pit or in reducing the volume of contaminated water which must be treated as it leaves the scale pit.

Sinter Plant

To reclaim mill scale from the scale pits and blast furnace fines from the gas washer system, and to use some of the fine ore which cannot be charged directly

Fig. 4 Modern scale pits are more than basins in the ground. They are equipped not only to produce an acceptable effluent but also to deliver the collected scale and oil in recoverable or disposable condition. (*Bethlehem Steel Corp., Bethlehem, Pa.*)

to the blast furnace, these materials are mixed with fine coal and fired on a traveling grate. This produces a sinter, or fused mass of material, which can then be charged directly to the blast furnace without undue loss of fine material. This sinter plant area is quite dusty because of the nature of the materials handled, and the air is cleaned either through washers or in a baghouse. Very little water is used in the area except for that required for preparing the mix before it is fed to the furnace grate and for miscellaneous cooling purposes.

Heat Treatment

To produce special physical properties in certain grades of steel, the metal may be put through a series of heat treatment operations. These would include heating in a furnace, annealing at a carefully controlled temperature for a specified period of time, quenching in water or oil, and final cooling in air on hot beds.

Generally, the temperatures in the annealing furnaces are not so high as to require water cooling of the furnace elements, but some water cooling may be

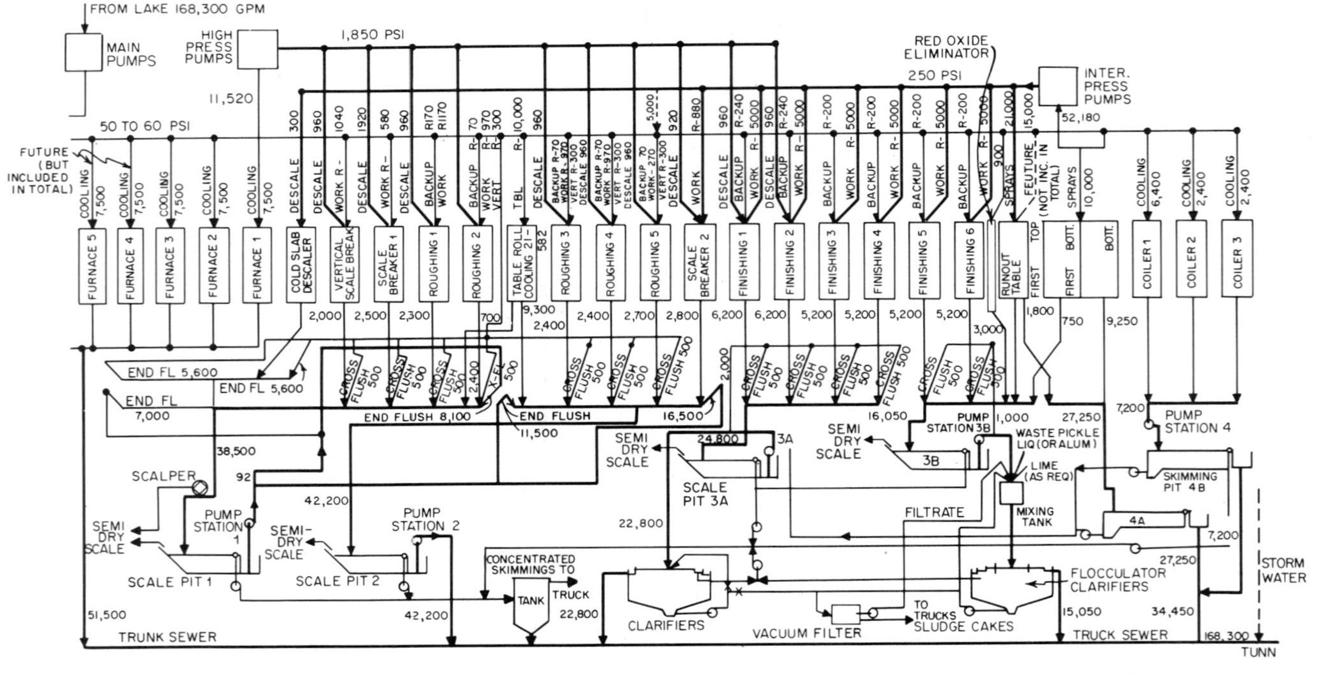

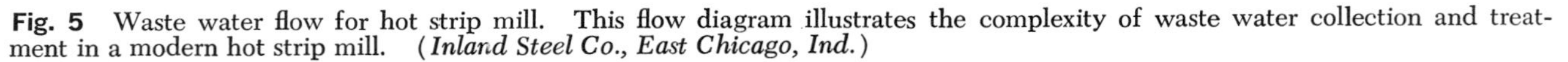

Fig. 5 Waste water flow for hot strip mill. This flow diagram illustrates the complexity of waste water collection and treatment in a modern hot strip mill. (*Inland Steel Co., East Chicago, Ind.*)

needed in high-temperature furnaces. The temperature of the quench oil or water quenching tank must be carefully controlled; so the contents of the quench tank are usually recirculated through a heat exchanger to remove the heat brought into the system from hot metal and to pass this on to cooling water flowing across the other side of the heat-exchange surface. The possibility of oil contamination must be guarded against, such as could arise by overflowing the quench oil system or by rupture of a heat-exchange tube in the oil-water heat exchanger. Oil is generally the only likely contaminant to arise in the heat treatment area.

Acid Pickling

The treatment of steel in an acid bath, known as pickling, removes oxide from the metal surface and produces a bright sheet stripped down to bare metal and suitable for finishing operations, such as plating, galvanizing, or other surface coating. Both sulfuric acid and hydrochloric acid are used in the industry, with the latter growing in popularity as more hydrochloric acid becomes available as a

Fig. 6 Dewatering of sludge is most commonly done on vacuum filters. The sludge cake may be returned to the sinter plant. Chemical treatments are usually used to reduce moisture and increase cake yield. (*Inland Steel Co., East Chicago, Ind.*)

by-product, reducing its cost. Regardless of which acid is used, the disposal of spent pickle liquor is a serious problem which has still not been satisfactorily solved for the steel industry at reasonable cost.

Pickling may be done either batchwise or as a continuous operation. Usually the acid is prepared at about 5 to 15 percent strength, depending on the work to be processed in the pickle tank and the type of acid used for pickling. As the acid works on the oxide surface, there is a gradual buildup of iron compounds in the pickle bath and a depletion of the free acid strength. When the iron content reaches a level which begins to slow the pickling operation, the bath is either dumped or reprocessed. In some pickling operations, the acid is continually withdrawn for disposal or reprocessing, in order to maintain a fairly constant ratio of iron to free acid in the pickle vat, thereby maintaining uniform pickling conditions for all the steel passing through the operation.

The metal leaving the pickle vat will drag out some of the liquor with it and carry this liquor into the subsequent rinsing and neutralizing operation. The loss of acid by dragout varies depending on the type of work being done, the shape of the steel products being pickled, and the speed of the operation, but it can run as high as 20 percent of the acid used.

Appreciable rinse water must be put into the rinse tank and withdrawn continuously for discharge through a treatment facility.

In some cases the dragout of acid from the pickling vat may be so great that the iron contamination level in the pickle vat never reaches an objectionable level, so that the strength of the vat is controlled entirely by makeup of fresh acid and water to bring the acid to its proper dilution. Some plants maintain elevated temperatures during pickling by injection of steam into the pickle vat, which complicates the control of the chemical conditions in the vat considerably. This also produces a corrosive atmosphere in the environment of the pickling tanks and may also evolve harmful vapors.

The discharge from the pickling area generally includes, then, spent strong pickle liquor and acidic rinse water which must be neutralized before it can be discharged. In some cases, there is a final treatment with lime slurry, with a final rinse so that this rinse water may contain excess lime and iron precipitate.

Since about 50 percent of the products of an integrated mill may be acid pickled, the steel industry is one of the major consumers of acid in the United States. The consumption of acid varies over a wide range from a low of about 10 lb/ton to a high of about 50 lb/ton. The volume of water used for rinsing is in the range of about 50 to 100 gal/ton of steel pickled.

Vacuum Degassing

Molten steel is often degassed under vacuum to improve its physical properties. In many plants the vacuum is obtained through barometric condensers. Since the vapor will contain the volatile components from the metal and the flux, the discharge from the vacuum degassing operation usually contains iron, manganese, and fluorides.

Coke Plant

Coke for reduction of iron ore in the blast furnace is produced in ovens from a carefully blended mixture of high- and low-volatile coals. Each coke oven is relatively narrow—about 15 to 24 in. in width—but may be approximately 30 to 40 ft in length and about 10 to 15 ft high. The mixed coal is charged into the hot oven and is then heated by the external combustion of coke oven gas until all the volatile materials have been driven off and the remaining coke has been fused into an incandescent mass. At the end of the coking period, the incandescent coke is pushed out of the furnace into a coke car and taken directly to a quenching area where water is sprayed on the hot mass and is evaporated into steam, which is carried up a stack directly above the coke car.

The gas produced during the coking of the coal contains valuable chemicals which are recovered in the by-product plant. These include ammonia, benzene, xylene, and toluene. In some plants, phenol and naphthalene are also recovered. As part of the recovery process, the gas is first washed with water, producing a weak ammonia liquor. This represents a very difficult pollution control problem, since the liquor is very high in ammonium chloride, is strongly colored, and contains such other contaminants as phenol, cyanide, and thiocyanates.

The bulk of the water used in a coke plant is for cooling purposes, but the aqueous wastes from process operations are very strong and difficult to handle. The strong wastes produced in the coke plant include spent acid from washing the light oil, spent caustic from the neutralization step, and spent adsorber solution, such as sodium carbonate, used for removal of sulfur from some of the coke oven gas which may be required to have a low sulfur content.

In some plants the coke is quenched with the weak ammonia liquor as a means of disposing of this waste, but this has a disadvantage in that the contaminants are accumulated on the coke and then transported to the blast furnace, where they then add to the contamination load in the blast furnace gas wash water.

Slag Plant

Various useful products can be recovered from slag, some being produced from the slag in a molten form and others after the slag has solidified.

Molten blast furnace slag can be quenched with water in a mill to produce lightweight expanded aggregate for the manufacture of cinder block. It can also be spun into mineral wool insulation. The solid slag is crushed to various sizes for use as track ballast, highway foundation, and similar structural purposes.

In plants handling slag, there is a high dust loading in the surrounding atmosphere which tends to cake on conveyor belts or interfere with proper operation of mechanical equipment. Water washing of the air is sometimes practiced, or water may be used for washing conveyor belting, creating a high suspended solids waste water.

Open hearth, BOF, and electric furnace slags are very high in iron content. Therefore, they are usually broken up and reclaimed for recharging to the blast furnace.

Almost all steel mills have a large slag pile which has accumulated over the years, since very few mills can convert all their slag into usable by-products. Storm

Fig. 7 Biological treatment may be suitable for destruction of soluble contaminants such as phenol. This unit treats weak ammonia liquor at a coke plant. (*Bethlehem Steel Corp., Bethlehem, Pa.*)

water percolating through the slag pile dissolves minerals from the slag. There may be a continuous seepage of artesian water from the base of the slag pile which is high in such contaminants as dissolved solids, sulfides, phenol, and cyanide. Water which has contacted slag being reclaimed from open hearth, electric furnace, or BOF furnaces may be quite high in fluorides.

Utilities

Because of the useful, combustible gases produced at the coke ovens and the blast furnaces, steel mills are able to produce a large percentage of their total power requirements by burning this fuel in boiler houses. In addition to boilers which are directly fired with these by-product fuels, other boilers reclaim heat from gases discharged from the open hearth furnaces, the BOF shop, and gas engines.

The steam generated from these facilities is used throughout the complex for driving turbines, powering presses and forges, and providing heat wherever it may be required.

Some of this steam may be treated with steam cylinder oil ahead of steam engines and presses, producing an oily condensate which requires treatment for disposal.

In addition to production of power from steam, some by-product gases are used directly through gas engines to produce electrical power or mechanical energy directly for such uses as compressing air for the blast furnaces. Other utilities include air compressors, vacuum pumps, and pump stations to supply the enormous quantities of water needed for the operation.

The water requirements for these utilities in the steel mill are similar to those of other industries. High-quality water must be produced for makeup to the steam generators, and these in turn concentrate the water, which is then removed by

Fig. 8 This unit recovers phenol from coke oven waste at Jones & Laughlin's Pittsburgh Works. (*Jones & Laughlin Steel Co.*)

blowdown at a relatively high salinity level. The water treatment facilities required for producing this high-quality water have their own waste, such as lime sludge from lime softening operations, brine from zeolite softener regeneration, or spent acid and caustic from demineralizer regeneration.

Cooling water is used by these utilities for such purposes as condensing turbine exhaust, cooling compressor jackets, cooling bearings on various types of powerhouse auxiliaries, and conveying ashes from furnaces fired with coal.

Maintenance Shops

The integrated steel mill has its own rolling stock and a variety of materials handling devices such as dump trucks, fork lift trucks, and devices for handling tote boxes. Regular maintenance of locomotives and trucks is conducted in repair shops, creating a source of oily waste common to such maintenance operations. Oil will range from 500 to 1,000 ppm. These wastes are usually generated by such operations as steam jet cleaning, degreasing of machine parts, and washing of air filters used on the diesel locomotives.

AIR POLLUTION

The picture of a steel mill in the public mind is of a giant complex characterized by tall stacks spewing smoke and dust into the air to be dispersed across the countryside for miles around. This scene is changing rapidly, though the recollection may linger because of the public's exposure to polluted air from the steel mills for many years. Today it is unusual to see emissions from the blast furnace flapper valves; the red dust from the open-hearth furnaces is diminishing as electric furnaces and the basic oxygen furnace are forcing retirement of these older units. Those open hearths remaining in service are being provided with dust collection devices. The coke plant alone has resisted attempts to control dust emissions, though improved housekeeping and closer attention to good charging procedures have brought about some improvement.

Gas Cleaning

Gases produced in the steel mill can be classed as those having useful value as fuel and those which are simply waste products of a manufacturing operation.

The useful gases are generated by the blast furnaces and coke ovens. Since these are distributed great distances throughout the plant complex and are carefully regulated for optimum combustion efficiency at the point of use, cleaning for removal of troublesome dust is a necessity for reliable operation. (Although BOF gas also has heating value as it leaves the ladle and enters the hood, for practical reasons it can be used as a fuel only in the hood itself, and therefore does not require cleaning for this purpose alone.)

The waste gases are produced in the open hearth furnaces, at the sinter plant, in boiler houses, and as the final discharge from combustion of the useful fuel gases in various types of furnaces. These must be cleaned to meet requirements for air quality in the environment of the plant.

The basic cleaning devices include simple gravity separators (such as cyclones), wet scrubbers, electrostatic precipitators, and baghouse filters. The selection is not a simple choice; it is based on such factors as dust loading, allowable emission levels, characteristics of the gas (such as temperature, pressure, chemical composition), final disposition of the collected dust, available space, and accessibility for maintenance. Most systems are designed for a final dust loading of less than 0.05 grain/scf.

The treatment of blast furnace gas uses a train of cleaning devices including gravity separation drums, venturi type wet scrubbers, and electrostatic precipitators. The chemical composition of the gas is relatively uniform, but the dust loading and the nature of the particulates vary somewhat as the furnace is charged and tapped, occasionally showing a burst of dust if the charge bridges and then suddenly breaks through.

Coke oven gas is washed with water, acid, and oil. The water removes tar and dust and cools the gas; the acid removes and recovers ammonia; and the wash oil absorbs organic chemicals (benzene, toluene, and xylene) which are more valuable as by-products than as fuel. There may be a final treatment to remove sulfur compounds. The gas is quite uniform because of the large number of individual ovens discharging into the single common gas line, but some variations occur because of changes in coal composition and in the ratio between high- and low-volatile stocks.

Open-hearth furnace gas is treated in a wet system similar to that used for blast furnace gas.

There are various combinations of wet and dry systems to control dust emissions from BOF shops, ranging from completely wet scrubber systems to those which use water only to the extent necessary to cool and humidify the gas prior to electrostatic precipitators.

Air Cleaning

Dust laden air in foundry areas, the sinter plant, and burden charging areas may be processed wet or dry.

Wet scrubbers may do an acceptable job of emission control and yet be troublesome because of the high solids loading in the waste water. Therefore, these simple units are not useful unless they can be tied into a waste water clarification or disposal system. It may be found practical, for example, to return wash water from air scrubbers in the skip hoist area to the gas washer system, or to use wash water excess in the sinter plant to prepare the sinter mix.

Where wet scrubbers cannot be used, baghouse filters may be applicable. These are much more costly, but they avoid a water pollution problem. They have been successful in cleanup of air pollution from electric furnace shops and sinter plants. Care must be taken in the design of the system to wet down solids as they are discharged for transportation to a disposal area to prevent a dust nuisance during hauling or dumping in the landfill site.

Chapter **11**

Pollution Control in Foundry Operations

R. P. BARZLER
Senior Mechanical Engineer

D. J. GIFFELS
Project Director

EDWARD WILLOUGHBY
Chief Civil Engineer,
Giffels Associates, Inc., Architects, Engineers, Planners,
Detroit, Michigan

INTRODUCTION—AIR POLLUTION PROBLEMS

Among industrial plants producing air pollutants, foundries must be placed with the foremost. They contaminate the air both around and inside them. It is possible, however, to build, ventilate, and operate foundries, both new and old, so that they can be regarded as "clean" in the practical sense of the word.

Control of In-plant Air Pollution

Quite obviously, if all the pollution could be restrained inside the foundry, generating it, pollution of the surrounding atmosphere would not occur. Likewise, if every place inside the foundry where an air contaminant is generated could be so controlled that dirt could not escape to the inside atmosphere, pollution of the inside air would not occur. Process exhaust systems, in the large sense, are designed to effect this control.

A "clean" foundry can be maintained so only by the provision and diligent operation of the following three interdependent factors.

Process exhaust ventilation This includes the containment of all points where some form of contaminant is generated. This is normally done by hooding the points as completely as possible and drawing room air into the hood through the remaining openings with sufficient velocity to contain the contaminant. In nearly all instances this air is ducted to an air cleaning device, cleaned, and discharged to the outside atmosphere.

Adequate makeup air It is impossible to exhaust air from a building without drawing in replacement air. If air is not provided by design, the building will

Fig. 1 Chrysler Corporation's Huber Avenue foundry, Detroit, Mich.

assume a pressure somewhat lower than the outside pressure, and air will leak in through every crack and opening. The lower pressure is produced by exhaust fans that are designed to operate against relatively high pressure differentials. The lower building pressure will significantly reduce the capacity of other exhaust systems in the building which have been designed for very low fan differential pressures. Moreover, outward air flow through gravity roof ventilators and stacks will be reduced or in severe instances reversed, resulting in the backflow of heat, noxious gases, and other pollutants. Makeup air supply systems provide a positive means to combat this situation. Outside air is heated under controlled conditions, usually filtered, sometimes even cooled in summer, and delivered to the building where it will do the most good. Working stations and operating levels thus are provided with clean, comfortable atmospheres.

The air volume so provided is about the same that would otherwise leak in. Inasmuch as this air will be heated to building temperature, no matter which way it comes into the building, the fuel costs will be approximately the same. Makeup supply air systems have proved to be well worth the investment.

Good housekeeping Although good process exhaust and supply air ventilation is extremely important, it cannot by itself keep a foundry clean. Constant house cleaning effort must be applied if a reasonably clean plant is to result. The breakdown of either aspect will shortly result in filthy conditions which can quickly get out of hand.

Housekeeping includes maintenance of the ventilation equipment such as a lubrication of moving parts, changing of filters, removal of collected wastes, and keeping access and cleanout doors closed. It also includes the maintenance of the many potentially dirty processes, such as sand handling conveyors, chutes, bins, and hoppers, and involves the day to day dusting, vacuuming, sweeping, and scrubbing that the terms ordinarily bring to mind.

The foundry has a variety of sources of pollution that can be controlled by process exhaust ventilation. Some, because of their very nature, can be contained quite simply and inexpensively. Others almost defy containment and require large capital expenditures to control.

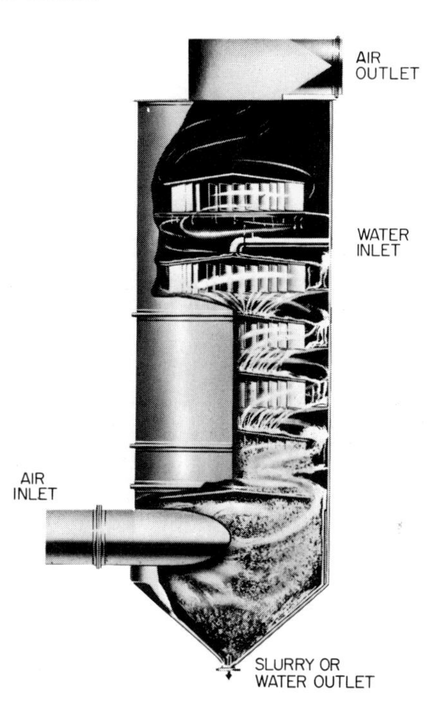

Fig. 2 Mutiple-stage wet centrifugal collector which uses centrifugal force to throw heavier particles against wetted surfaces. (*Claude B. Schneible Co.*)

Before one sets out to design an exhaust system—even the simplest one—one should possess a knowledge of the basic principles involved. There are many publications that set forth the principles of hooding, duct design, construction specifications, fan selection, economical system layout, resistance pressure computations, etc. One such publication is the "Foundry Environmental Control Manual" of the American Foundrymen's Society. Another is "Industrial Ventilation" published by the American Conference of Governmental Industrial Hygienists.

Although in-plant air pollution control is necessary and important, and results in the very exhaust discharges that could pollute the atmosphere outside the plant, it is not in the scope of this chapter to delve deeply into the designs of those systems. The sources of the contaminants do, however, have a direct bearing on the selection of the equipment for prevention of outside air pollution and must be considered for that purpose.

Moreover, it is most probable that the one who designs the exhaust system will at the same time also select the pollution prevention equipment. In fact, this equipment is such an integral part of the whole system that one cannot be considered without the other.

SELECTION CRITERIA PECULIAR TO FOUNDRIES

In foundry applications, the air pollutants for the most part are particulate in nature—that is, heavy and light dusts, smoke, and fumes. For this reason, dust collectors make up the bulk of the air cleaning equipment.

Types of Collectors

Electrical precipitators are very seldom used. Except for some special metallic fume exhaust systems, they have proved to be too expensive for the size of application normally encountered. Inasmuch as precipitators can attain extremely high efficiencies, they should not be ruled out completely in view of the trend toward cleaner ambient air quality.

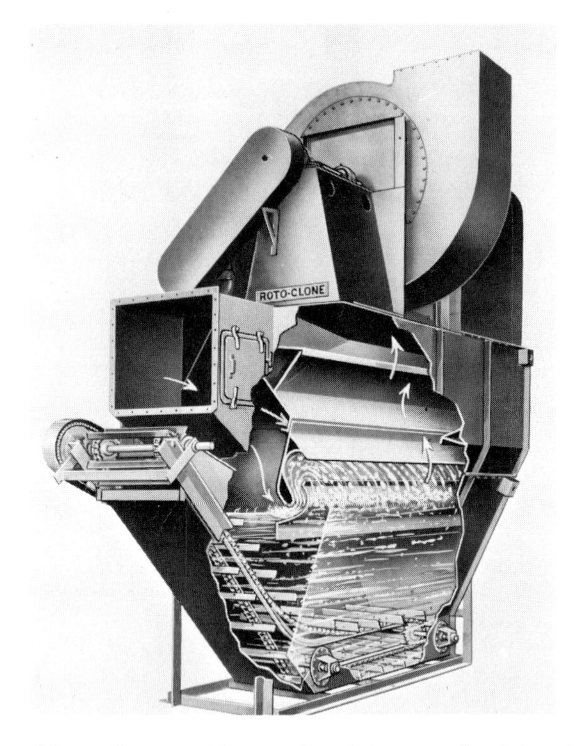

Fig. 3 Wet scrubber. Dust particles enveloped in water droplets. (*American Air Filter Co.*)

The other three types of dust collectors, the mechanical separators, the wet scrubbers, and the fabric arresters, are used extensively for foundry systems. Each of these three general types is manufactured in many types and styles. In fact the principle of mechanical separation is used in the design of most scrubbers and fabric collectors.

Mechanical inertial separators such as cyclones are used where the smallest particle size of the dust is not less than 5 microns (1 micron equals 1/25,400 in.). Inasmuch as most dusts generated in a foundry do not qualify for this limitation, the cyclone and less efficient gravity traps (useful for particles of more than 10 microns) are used primarily as precleaners, either to reclaim usable sand, or shot, or to reduce the dust load on the final collector.

Dust Collector Size

As final collectors, wet scrubbers and fabric arresters have their advantages and disadvantages. Because the scrubbers use high velocities to accomplish their cleaning action, they are relatively compact. For the same air volume the fabric arresters, which by comparison use very low through velocities, will be much larger. Offsetting this size advantage, the scrubbers cannot remove the very fine particles,

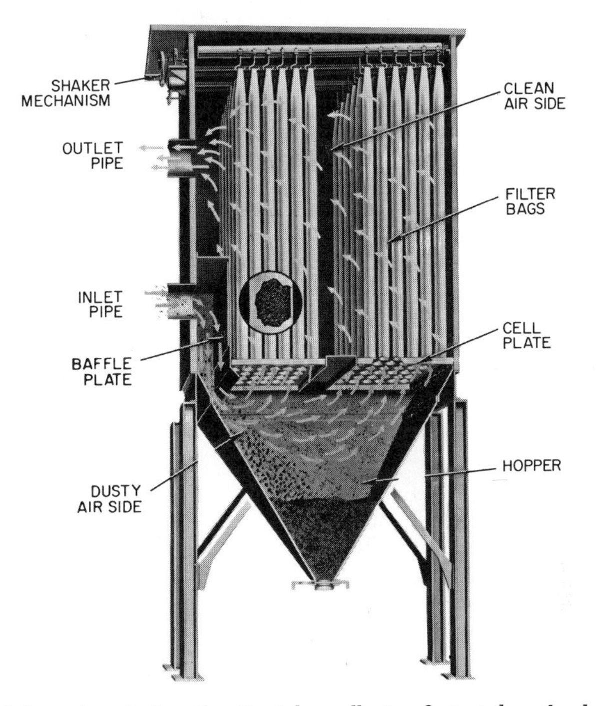

Fig. 4 Dust-laden air entering the Dustube collector first strikes the baffle plates, where the sudden change of direction and reduction in air velocity cause heavier particles to drop immediately into the hopper. All materials large enough to abrade or scuff the filter fabric drop out of the airstream so that only the very finest float dust reaches the tubes. (*Wheelabrator Corp.*)

such as metallic fumes, from the airstream with the same degree of effectiveness as the fabric arresters, except by the application of very high energy.

Working Principles

The wet scrubber brings the dust particles into contact with atomized water droplets with such force and direction as to envelop each dust particle in a water droplet. The water droplet is then easily removed from the airstream by mechanical means. Greater force is required to remove the very fine particles so that they will overcome the surface tension of the water droplets. This force may be applied to the water, to the airstream, or to both. There are almost as many ways to accomplish this with simple, unique, and economical designs as there are manufacturers serving the market (see Figs. 2 and 3).

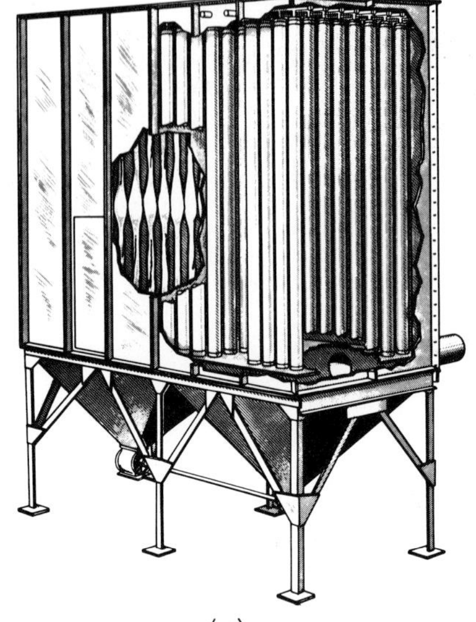

(*a*)

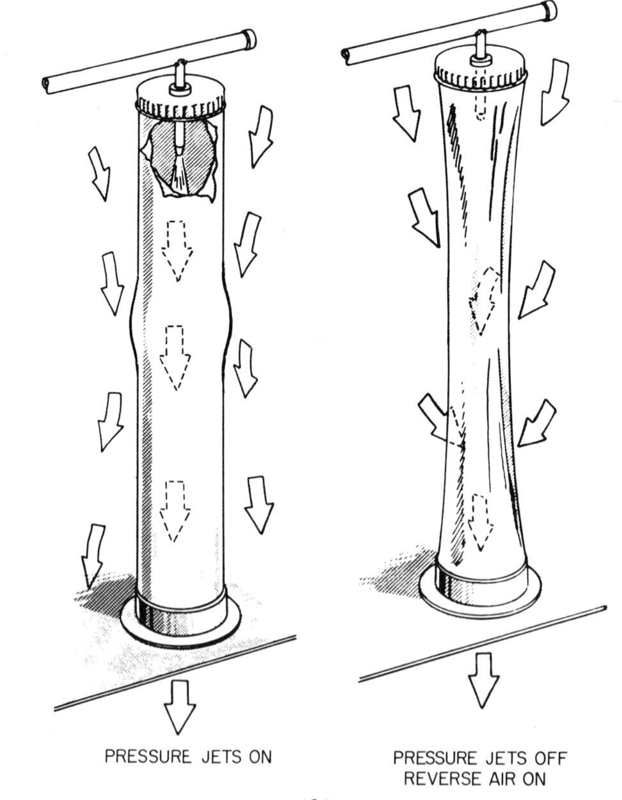

(*b*)

Fig. 5 (*a*) Bag-type collector with (*b*) pressure jet cleaning.

Experience dictates how much energy must be applied to accomplish the required degree of cleaning. It is therefore necessary to know the character of the dust—especially the percent of fine particles—if the designer is to be able to anticipate the degree of cleanliness of the discharged air.

Fabric arresters also come in many designs, shapes, and sizes, differing in the fabric and the method of removing the dust from the fabric. The first and by far the most common is the *woven* fabric, which acts as a screen to catch and hold dust particles. The dust then builds up and forms an effective filter that will remove all but the very finest dust from the airstream. The second is the *felted* fabric, which is more truly a filter itself. Felted fabrics will clean the air at higher through velocities than the woven type and therefore will require a smaller baghouse for a given volume of air.

Collected Solids Handling

The dust is shaken (Fig. 4) or blown off (Fig. 5*a* and *b*) the fabric and dropped into hoppers. These are used as storage for intermittent dust removal, or simply as chutes when the installation includes facilities for continuous dust removal. The airflow must be stopped when the dust is removed from the fabric, a fact that has resulted in a variety of designs and arrangements to accomplish a fairly constant system airflow.

Careful consideration must be given to the handling of the dust after it is removed from the fabric arrester hoppers. It is, of course, in a dry state. Unloading and transportation facilities, such as closed conveyors or covered carts to ensure dustless conditions, should be provided.

Wetting down the dust before moving it would be desirable, but the cost and difficulty in doing it have generally forestalled foundry applications.

On the other hand, the dust discharged from the scrubbers is in a wet condition either as a sludge or, where central dewatering equipment is available, as a slurry. This obviously makes it transportable without recontaminating the air. This fact alone has been paramount in the minds of some designers when they have been required to choose a type of dust collector for general application in a new foundry. Recirculated water systems can be included in the original designs to supply water to the dust collectors and return the slurry to a pond or settling tanks. The water is cleaned enough for recirculation, and the dirt is trapped and removed at one convenient and accessible place. Some of the areas of consideration of such a system will be discussed later.

Application Limitations

Temperatures above 550°F cannot be tolerated by fabric, and that only by fiber-glass cloth bags. Other synthetics that can be used up to 275°F are available. Except for applications on melting equipment, however, there is practically no system in the ordinary foundry where the temperature would exceed that which can be tolerated by cotton sateen or wool felt fabric.

Where the fabric arrester is applied to melting equipment, such as to a cupola (Fig. 6), it is necessary to reduce the incoming gas temperature to a level that can be tolerated by the fabric. Where it is not possible to provide enough interconnecting ductwork to act as a heat exchanger with the outside atmosphere, spray chambers can be used to reduce the temperature of the gas. Care must be exercised, however, to ensure that condensation does not form in the baghouse.

Wet scrubbers used in the same type of installation are also normally provided with spray chambers, not because of temperature but to reduce the volume of air, and consequently the size and cost of the scrubber. The effect of corrosion is reduced as well when the gas temperatures are kept down.

High relative humidity in the airstream may limit the use of fabric arresters. Certain sources of contamination in foundries expel considerable water vapor as well as dust. The mold line shake-out is an example. When castings are removed from sand molds shortly after they have solidified, such as on production mold lines, the moisture released from the mold sand does not have time to dissipate and is therefore removed by the dust exhaust system. The exhaust air attains a relatively high dew point. When such a condition exists, water will condense out the moment the airstream is exposed to a colder surface. Not a few baghouse owners have had to face the problem of a clogged dust collector. In fact, fabric arrester vendors, in view of past experience, will not propose their equipment if they suspect high airstream humidity.

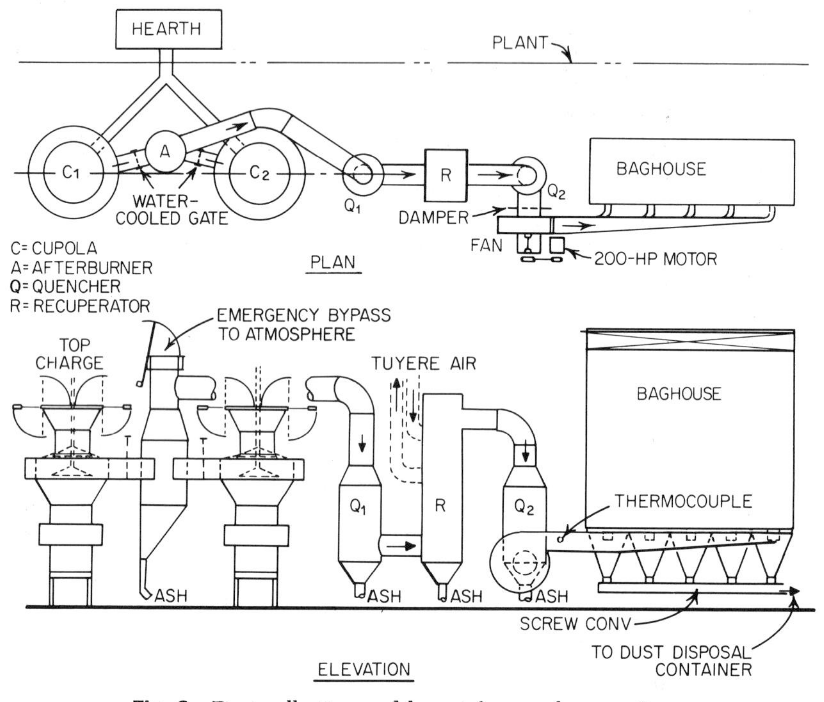

Fig. 6 Dust collection and layout for cupola operation.

On the other hand, wet scrubbers are not troubled by high humidity. Rather, they are subject to freezing conditions, and care must be exercised in that area. Scrubbers with water reservoirs in particular must have special consideration when located in a subfreezing environment. They will operate well as long as the airstream temperature is well above freezing, even though the unit is located outside. However, during shutdown periods, the reservoirs must be either drained or supplied with heat to keep the water temperature up. Where available, steam is often used for this purpose. Of course, if the airstream is itself below freezing, a wet scrubber is certainly the wrong choice, as ice will build up on the internals until it cannot perform its function.

A special problem arises in cleaning gases from melting operations where fluorspar is added for slag control. There may be enough fluorine released to attack fiberglass fabrics and shorten the useful bag life from as much as 1½ years to a few weeks. Fluorine corrosion can be avoided by not using the high temperatures that demand the use of glass fabric, limiting the temperature to a maximum of 275°F, and using other filter materials. The other alternative, the wet scrubber, is not affected by temperature limitation, but where there is the possibility of chemical reaction causing corrosion, the internal surfaces of the scrubber should be made of or coated with a suitable material to combat deterioration.

Certain types of wet scrubbers should be avoided where the airstream, in whole or in part, exhausts from shot blast equipment or other sources from which the dust is likely to contain iron.

The iron oxide formed in the presence of the water has a tendency to build up at impingement points or in small orifices, making it necessary to clean the collector periodically. Scrubbers using tower packing, sprayed marble beds, or other types of small orifices to attain the scrubbing action could prove to require much maintenance.

Initial Cost

For a given gas source the initial investment will tend to be lower if a wet scrubber is selected instead of the fabric arrester. The latter is usually a much larger machine, requiring more material and generally more intricate internals. The cloth cleaning apparatus takes many different forms. Whether it is a shaker device, a traveling reverse jet, a compressed air pulse, or some other arrangement, there is no counterpart in the wet scrubber. In fact, other than possibly a pump, most wet scrubbers have no internal moving parts but attain gas cleaning by means of air interaction with water.

It must be realized, however, that the dust collector itself does not constitute the complete package. For instance, the fan motor, by comparison, probably will be larger when a scrubber is used. This is because the scrubber requires more pressure to attain its collection efficiency. Water, of course, is required to operate a scrubber and, in most cases, is not required with the baghouse. The water cost and availability must be considered.

Where water is not abundant or must be conserved for economic reasons, the cost of water conservation equipment becomes part of the initial investment. Where one or very few systems are being considered, wet scrubbers can usually be provided with sludge ejecting conveyors which continually remove the collected material from a water reservoir inside the scrubber. However, where many systems are in view and the dust collectors are scattered about the foundry, a recirculating water system, the cost of which may be prorated against each of the units, may very well prove economical as it consolidates and centralizes the material disposal equipment.

If, on the other hand, water is abundant and can be circulated on a once-through basis, consideration must be given to cleaning the water before it is discharged. The problem of air pollution must not be merely changed to one of water pollution. Settling ponds or tanks, where sufficient time is available to settle out the fine particulate, are necessary. In fact, clarifying equipment may very well be necessary for final water treatment before it is discharged to public waterways. This type of waste treatment is covered in more detail later in this chapter.

Dust disposal equipment from dry fabric arresters also must be included in the total initial investment appraisal. If the dust collector can be located where removal trucks can be loaded directly from the hoppers, this item is minimal; where not, some means such as chutes, screw conveyors, or pneumatic conveying systems may become necessary. The initial cost of dust disposal equipment, when analyzed, may quite easily be the deciding factor in determining which system is least expensive to buy.

Operating Costs

Before making a decision based on cost to buy and install dust collection equipment, it is well to consider how much it will cost to operate it. For instance, the collected material, which in most foundry applications is not reusable, must be removed from the plant. The man-hours required to remove the waste material from the various points where it is deposited by the collectors is an operating cost variable that may influence the choice.

The cost of maintaining the equipment is another factor. The multiplicity of moving parts in the baghouse required to remove the dust from the fabric must be kept in good repair, or the whole system will fail. Bags must be replaced, eventually, and the cost of replacement bags, and their installation, should become a budget item. This cost is very difficult to anticipate, and only experience can determine it precisely. Anticipating complete replacement of the bags in 2 years will provide a basis upon which to arrive at a budget figure.

Power costs are also an operating expense. The fabric arrester, in addition to the fan motor, will have relatively small motors to operate the shaking devices or reverse jet blowers. The motors to run any dust removal conveying equipment may be significant, however, and should be considered.

The wet scrubber, as previously mentioned, will normally require greater power to move and clean a given quantity of gas than the fabric arrester. This difference is usually small in most foundry applications. However, where metallic fumes or other fine particulates must be removed from the airstream, the wet scrubber will not do the job unless much greater power is applied. Here again, motors are required to operate the disposal equipment, whether it is a drag conveyor, pumping and water clarifying equipment, or both.

Character and Sources of Contaminants

The selection of air pollution control equipment will depend on the character of the contaminant. It may be an odor, a gas, or an aerosol. The latter category includes both liquid and solid particulates. The solid particulates may be further classified as dust, smoke, or fume.

The fabric arrester, assuming it is kept in good repair, has a high collection efficiency for solid particulates. Liquid aerosols, as mentioned before, may cause clogging trouble. Gases and odors cannot be filtered out.

Collection efficiency of the wet scrubber will vary with the size of particle to be captured. Moderate-energy scrubbers will remove 99.8 percent of foundry dusts 2 microns in size and larger (see Fig. 7).

Dust from foundry shake-out and return sand handling systems is composed mostly of sand which has been fractured by heat, binders such as bentonite, cereal, seacoal, clay, etc. Dust from casting cleaning equipment, such as shot blast machines, tumblers, and grinders, will include the same but will have a larger percentage of iron as well. For these sources of contaminants, either the fabric arrester or the moderate-energy scrubber will have sufficient collection efficiency to meet existing air pollution control codes as well as those in the foreseeable future.

To maintain good cleaning efficiency on smaller solid particles such as smokes and fumes, where wet scrubbers are to be used, it is necessary to apply very high energy either to the gas stream or to the scrubbing liquid. In the foundry, arc furnaces, cupolas, etc., emit high percentages of these very fine particulates.

Liquid aerosols, in general, will be trapped by the scrubbing action of the wet units, and some odors will be reduced to an unobjectionable level. It is very seldom, however, that a scrubber unit is applied to an airstream for the sole purpose of reducing an objectionable smell. Rather, it is generally considered to be an additional benefit derived from the application of the wet equipment.

Airstream Conditions

It has already been mentioned how extreme temperatures and high relative humidities in the airstream may have a bearing on the selection of air cleaning equipment. Another airsteam variable is also possible and may also affect the final choice; that is, volume fluctuations. For instance, where sections of the duct system at times may be shut off by dampers and then returned to service, the dust collector must be able to maintain its ability to remove the dust through the cycle.

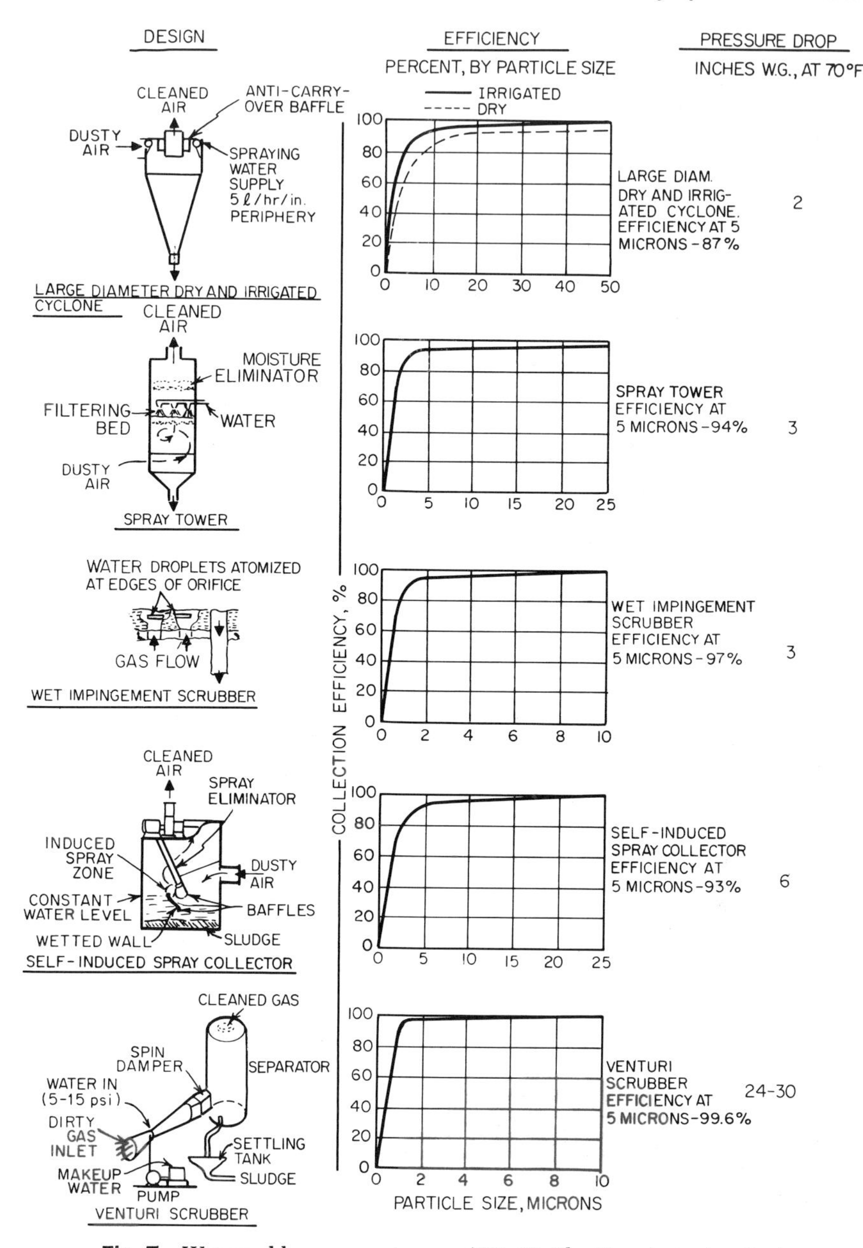

Fig. 7 Wet scrubber comparisons. (*W. W. Sly Manufacturing Co.*)

Fabric arresters generally are not influenced if they are sized to handle the largest anticipated volume. The mechanical cyclone collectors and many of the types of wet scrubbers that utilize the mechanical principles to effect the intimate air-water contact will be drastically influenced by a sharp reduction in air volume. The final moisture eliminator section on the scrubbers normally employs mechanical separation and will also lose effectiveness. Where such equipment is used, it is better to design two exhaust systems, each of which will always operate at full capacity.

It is normal practice to operate the dust collection systems continuously throughout the entire working period during which the foundry machinery served by the systems is running. However, where fabric arresters are to be used, it is necessary to stop the airflow periodically to dislodge the accumulated dust from the fabric and allow it to drop into the hoppers. Where a system must be kept in continuous operation, the baghouse must be divided into sections, each capable of being removed from the airstream for a while. This is done by dampering in sequence, usually with automatic timers.

Initial operating and maintenance costs savings can be realized if the system can be shut down long enough to shake the fabric. A single section baghouse can then be used. Lunch hours and shift change times will often provide all the time necessary to dislodge the dust which has accumulated in the preceding period. Delayed action interlocks between the production equipment (blast cabinets, screens, grinders, etc.) and the systems fan that will allow some time for the dust system to purge itself before the fan stops are recommended. The fabric cleaning mechanism is activated immediately upon fan shutdown, with time allowed for the dust to settle in the hoppers. The equipment is then ready for further service.

It should be noted that as the dust builds up on the fabric with this type of service, the pressure drop through the dust collector increases. This results in a gradual decrease in the volume of air handled. It is important, therefore, to design the system with a maximum allowance for dust collector resistance so that the air volume handled will always be adequate.

Available Space

As the air from all the pickup points included in a projected dust collection system must be collected and delivered to the dust collector, one of the first things to be considered is the location of the collector. All too often, especially in existing foundries, the available locations will not be ideal. The system designer will be forced to choose a location that best suits the requirements. It is possible that the type of dust collector chosen will be considerably influenced by the available location.

In order to evaluate a proposed dust collector location, one should consider the following.

The size of the space Obviously there must be room to accommodate all the equipment with enough room for access platforms, ladders, and the like. It is advantageous to locate the fan unit in the same area in order to localize maintenance and keep the connecting ductwork as short as possible. The possibility of freezing a wet scrubber will be greatly reduced if an inside heated space can be utilized.

Accessibility for maintenance All too often the frequency with which the equipment is inspected and maintained has a direct relationship to the difficulties encountered in getting to it.

Accessibility for the removal of collected material This material, whether wet sludge or dry dust, must by some means be removed from the premises. A location next to an aisle or roadway or position above such a location from which chutes may be used should be sought for this purpose. In some foundries, existing dust handling systems such as slurry sluicing systems can be used if the chosen space affords access to it.

Nearness to pickup points A short system will ordinarily be the least costly to install and to operate.

Availability to services The adjacent space must be such as to allow the passage of the entering and leaving ductwork, power lines, water, compressed air piping, etc.

Structural adequacy The space must afford a way to support the equipment properly. Existing buildings may not have been designed to carry the loads that this type of equipment imposes. Where the units can be placed on the ground, there is no problem, but if, in order to conserve floor space, the equipment must be placed in the upper part of a building, the reinforcement that becomes necessary may add to the erection costs.

Surroundings Nearby structures, air intakes, windows, etc., which may require that the discharge from the exhaust system be carried to an excessively high level or distant point should be considered.

Appearance There have been not a few instances when a dust collector location that satisfied the above requirements had to be rejected because it was on that side of the plant which was exposed to the view of the public and would be detrimental to the foundry's public relations policy.

Available power The selection of air pollution control equipment may well be influenced by the power that is available. The amount of electrical energy demanded by the exhaust equipment, though not necessarily a large percentage of the whole plant demand, can overload existing facilities and require additional transformer equipment. The power to move the air through the system will vary with the pressure drop through the pollution control equipment. Usually this pressure drop is lower with fabric arrester equipment. Where the particulate to be removed is very fine, such as fume from the melting processes, wet scrubber equipment may very well be designed to use a high pressure drop in order to get adequate cleansing action. In cases such as this, the power requirements will be usually high.

Applicable Codes

There is a danger that the air pollution control equipment required to clean an exhaust gas stream will be selected by checking the local laws to find out what will just suffice. It is regrettable that economic and political pressures all too often have caused such "watering down" of those codes that it has been legal for air pollution to grow in intensity throughout the world. The inevitable result has been the public clamor for more stringent rules.

Most urban communities in the past have enacted regulations designed to give the public power to demand, through their government, that public nuisances such as air pollution be curbed. The purpose of any such codes or regulations, be they local or national, is to define a minimum acceptable condition of the air and to attain that condition.

The danger in just "getting by" existing codes can be seen in the demonstrated fact that what will be good enough today may be judged inadequate in the foreseeable future. It is better to select equipment that will do as good a job as practically possible in the knowledge that not only will the law, both existing and future, be satisfied, but that the public relations image of the foundry will be enhanced.

POLLUTION CONTROL ON FOUNDRY OPERATIONS

Foundries differ from one another in many ways. They range in size from small operations to the large production type that turns out thousands of tons of castings each day. The majority fall in the small to medium size range, many being classed as jobbing foundries. They differ in the size of casting produced, ranging from a few pounds to multiton. They differ in the material melted, such as gray iron, steel, brass, or aluminum, and where continuous casting is not employed, in the type of molds and cores used. They may be highly automated, completely manual, or somewhere in between. In fact each foundry has an individual character of its own that will make its requirements for exhaust ventilation differ from those of other foundries.

There are, however, foundry departments some or all of which may be found in all foundries in one form or another. Each usually presents its own peculiar exhaust ventilation problem. They are the melting, pouring, molding, casting cleaning, core making, patternmaking, casting repair, and maintenance departments. An investigation of the pollutant sources in each department is necessary in order that the appropriate pollution control equipment may be selected.

Melting Department

Cupolas If one item of foundry equipment were to be singled out as the largest cause of air pollution, it undoubtedly would be the cupola. This is borne out by the fact that air polution control officers, throughout the country, have chosen the cupola as their primary target. This is not surprising in view of the fact that the cupola is usually the tallest object to be seen in the plant and seems to be emitting the most smoke.

The emission from the cupola consists of a variety of contaminants, the most bothersome from a control standpoint being the metallic oxides, which may be

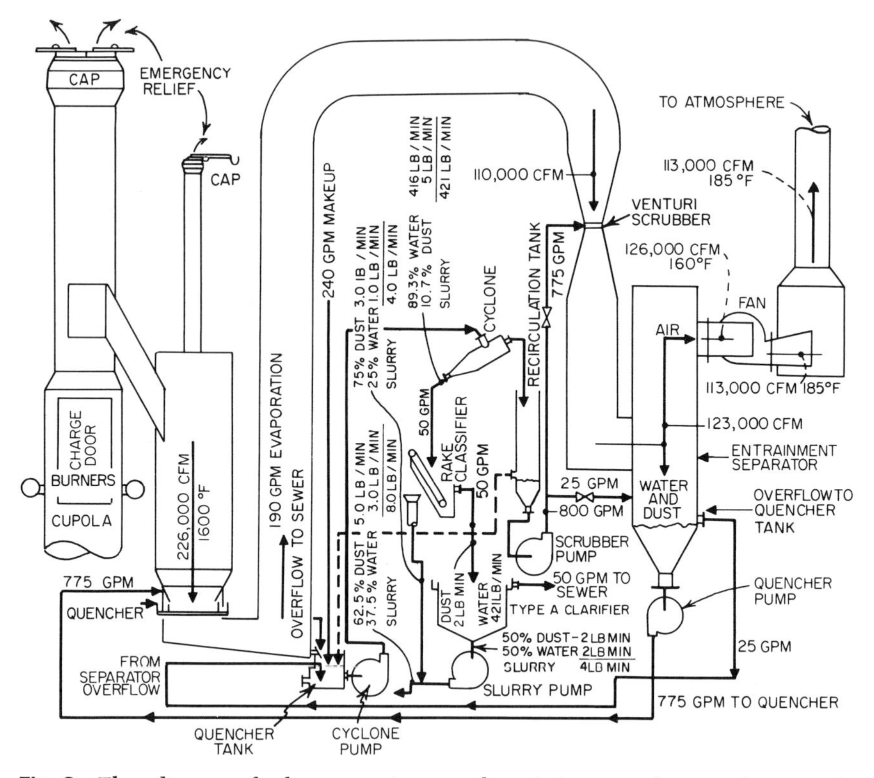

Fig. 8 Flow diagram of a large gray iron cupola emission control system (wet type).

submicronic in size, unburned hydrocarbons, and carbon monoxide. The traditional cupola control contrivance is the familiar "wet cap" set at the top of the open top cupola. The gases are directed through a water curtain by thermal impetus. Only the larger particles are removed. The resultant discharge is quite dense, even though it may be of a lighter color.

Because the industry did not seem to be inclined to do a better job of cleaning the cupola effluent, air pollution control authorities have been inclined to write the regulations stringent enough to outlaw the "wet cap" and demand a more efficient control system.

Various types of systems have been developed for the gray iron cupola (Fig. 8) by pollution control equipment manufacturers, who have gone to considerable research and obtained valuable practical experience in the process. Mechanical separators, scrubbers, and fabric arresters have been applied, the latter two attaining control to meet the most stringent codes. The mechanical separators have satisfied the less severe regulations allowed for cupolas that are used in the "jobbing foundry" and are in use only a few hours each day. One typical code limits the particulate matter in the exhaust to 0.4 lb/1,000 lb of gas. At this rate, such a cupola will not contribute as much pollution to the air as a "production" cupola that is operated on a continuous basis and emits the maximum allowed by the same code of 0.1 lb/1,000 lb of gas.

To satisfy the more stringent code, however, requires a much more elaborate system. Where a fabric arrester is used, the most succesful systems have employed an afterburner to complete the combustion of the carbon monoxide and other combustibles, a cooling section where water sprays are used to lower the temperature of the gases to around 500°F, fiber-glass cloth tubes in the baghouse, and an elaborate control system capable of maintaining the baghouse temperature. Equipment safeguards, such as a baghouse bypass which would open to protect the fabric if the temperature exceeded the fabric temperature limit, are normally included. This might be expected to happen at "burn-down" time. The air pollution control is lost during these periods, but the equipment is protected. Where the fiber-glass bags are used, it is important to limit the use of fluorspar in the cupola charge to periods when the baghouse is not in use lest the fluorine ruin the bags. Fortunately, the need for using fluorspar can usually be limited to "burn-down" periods. Because this fabric cannot endure stress without danger of rupturing, the dirt removal mechanism must be such as to shake the bags as gently as possible. Even so, if the high efficiencies of air pollution control are to be met on a continuing basis, the operator must be continually vigilant to detect and repair leaks in the fabric.

One of the dangers in the use of fabric arresters on a cupola discharge system is the possibility, particularly in colder climates, of condensing out water in the baghouse. The entering airstream has generally been cooled by water sprays and consequently has a high relative humidity and dew point. The controls must hold the gas temperature at a point above the condensation level and below the maximum allowed by the fabric. Where temperature reduction can be done without using water, the danger of plugging up the baghouse is greatly reduced.

Because of the small particulate size in the cupola discharge, it is necessary to apply high energy where the wet scrubber is used. The venturi scrubber is uniquely suited to this application because it can be readily designed for the required pressure drop necessary to attain the cleaning efficiency desired and can be automatically adjusted to maintain, as the gas volume varies. Other designs, such as the flooded disk and units that apply the energy to the water rather than the gas stream, can be used. Normally the gases are cooled to reduce the volume that must be handled. Again, this is done by water sprays, but with the scrubber downstream, the water carryover causes no problem.

It is automatic that the initial cost of the pollution control system will vary with its volumetric capacity. The volume of the gases handled depends on the temperature of the gases, the position on the cupola from which the gases are withdrawn, and whether afterburning is employed. Where the take-off position is above the charging door, there seldom is any dependable restriction to airflow into the door. Consequently, a volume of air approximately equal to the blast air volume will be added to the cupola blast air and to the products of combustion. Under these conditions afterburning, to complete carbon monoxide combustion in order to minimize the possibility of explosions, should be provided. Water vapor is added to the gas by the cooling sprays. The resultant volume, corrected for temperature, must then be handled through the emission control equipment, fan, and stack.

As large volumes require large pollution control equipment, and result in high initial and operating costs, the trend, at least for new cupolas, is to locate the take-off position for wet scrubbing control systems below the charging door.

Where the take-off position is below the charging door, the unburned charge built up above the melting level acts as a restriction to the air entering into the door. The volume of gases is thus reduced to the sum of the blast air, the products of combustion, and the volume from the charge door, which can be maintained at about 10 percent of the blast air. If adequate controls are employed, afterburning, which increases the air volume, can be omitted and the resultant volume, corrected for temperature, can be passed through the scrubber, fan, and stack. In locations where the carbon monoxide must be controlled, afterburning as a final stage is employed (see Fig. 9).

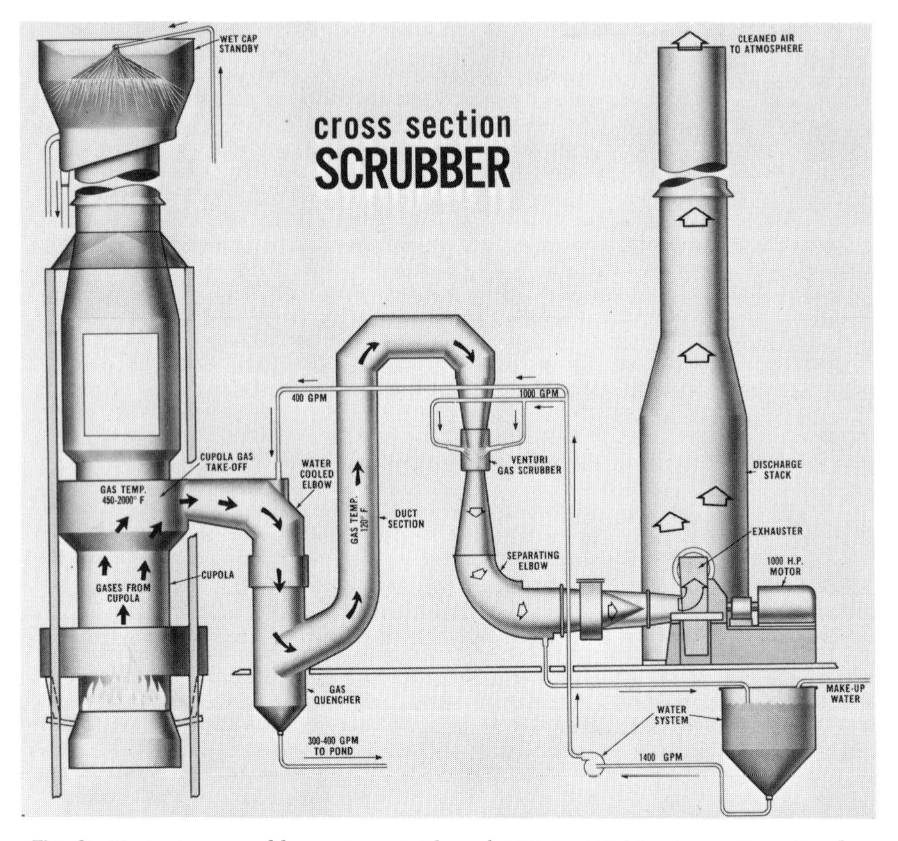

Fig. 9 Venturi gas scrubber system at Chevrolet Motor Divisions's gray iron foundry at Saginaw, Mich. Gases from the melting furnace at lower left pass through a water cooled elbow and gas quencher en route to the venturi scrubber, where dust particles are water soaked and separated before the cleaned gases are exhausted through the stack at right. Chevrolet Motor Div., Detroit, Mich. (*Courtesy of Michigan Challenge.*)

There are many facets in the cupola pollution control system that must be considered which vary with each individual cupola. The composition of the charge, the condition during burn-down time, the height of the stack, the available space, the climate, and many other variables tend to make each emission control problem unique.

Electric arc furnaces In climates where makeup air need not be heated, arc furnace buildings have been exhausted from the roof level. However, abnormally large exhaust air volumes must be moved to attain satisfactory in-plant conditions. And, as the air should then be cleaned before it is discharged to the atmosphere, the initial and operating costs of control equipment are high. In most cases the exhausted air volume is kept to a minimum for economic reasons. Local hooding will attain effective control of the furnace emissions with much lower exhaust volumes than general exhaust.

The arc furnace emits smoke, particularly when the power is applied, from the openings in the furnace. Local hooding should control the smoke as it comes from the three electrode openings, the pouring spout, and slagging openings.

Various types of hoods are commercially available to control the smoke emissions. They usually employ a combination of the canopy and side draft principles, and are designed to exhaust enough air to keep the exhaust temperature in the 250°F range. The hood may be made in two parts with mating breakaway flanges, the one part mounted on the furnace roof ring and the other on the furnace shell. This type is used on furnaces where the roof is moved aside for charging. An alternate to using a hood on the arc furnace is to draw the fumes directly from the furnace interior. A duct take-off, made of material capable of withstanding the high fume temperatures, is inserted through the furnace roof. This, in effect, makes the furnace become its own hood, as an in-draft through all the openings is induced. It is important to maintain a stable negative static pressure in the furnace in the range of 0.02 to 0.05 in. (water gage) to keep the inflow adequate but as low as possible. Controls that sense this pressure and automatically operate a damper in the exhaust duct to correct the ever-changing condition are required. These controls must be extremely fast-acting and capable of withstanding the high temperatures involved.

It is best to have the exhaust duct connected to the furnace hood at all times. However, as the furnace must tip for pouring and slagging, revolving duct connections and possibly even telescoping duct sections may be required. A simpler method is to use mating breakaway flanges between the furnace hooding and the exhaust duct. As this disconnects the exhaust system from the hood, it results in loss of control during the pouring operation. This is not serious on furnaces where the heating is cut off before the metal is poured.

The exhaust is most often cleaned with fabric arrestor equipment, provided that it is of the continuously operating type and that the temperature of the air is below the maximum temperature that the fabric can withstand. Fabrics such as Dacron, which can withstand 275°F, or silicone coated glass fiber cloth if higher temperatures are expected, should be used.

Safeguards to protect the fabric from overly hot gases are normally employed. They may be a simple damper arrangement to bypass the baghouse until the danger is over, or a dampered opening in the duct to bleed in atmospheric air for cooling when the need arises. The furnace operator prefers the former because it does not reduce the volume of air removed from the hood; the air pollution control officer prefers the latter. Where the dust collector can be located a good distance from the furnace, the long connecting duct will act as a heat exchanger. This diminishes the danger of overheated gases and reduces the volume that must be handled by the equipment.

Electric induction furnaces Electric induction furnaces have been used for years, mostly for melting in the nonferrous industries. Recently they have been installed in iron foundries as well, finding use as holders, melters, and duplexers.

Holding furnaces normally require no exhaust ventilation because the method of melting is clean and the metal itself has had most of the impurities removed during the already performed melting process.

Melters are subject to considerable fume emission, particularly when they are charged. The amount depends on the condition of the charge material. Either canopy or side draft hooding to control the emission during and following the charging operation can be used. The type of hooding is determined by local conditions and the required exhaust volume by the type of hooding.

Inoculation of the metal in the duplexing furnace may require some attention. In the nodular iron inoculation process, for instance, where, among other ingredients, magnesium is placed in the inoculation ladle and the metal is poured thereon, there is a definite possibility of excessive fume emission, especially if the operation is not performed properly. Inasmuch as the ladle is usually on a monorail or bridge crane, the side draft type hood, with a relatively high exhaust volume, is required.

As with the arc furnaces, the fume emission may be cleaned by fabric arresters, and the same design criteria would, in general, apply.

Other than cupolas and electric furnaces, there are the direct flame and indirect flame fuel fired furnaces that emit smoke. Hooding of some type designed for the particular condition of the furnace installation is necessary. Usually an adaptation of the canopy type hood can be used. As in all hooding designs, the equipment should be enclosed as much as possible, consistent with the operations involved. In many instances, the furnace hoods have been served by a gravity stack, discharging the smoke to the atmosphere. In those instances, however, where cleaning the effluent is deemed necessary, exhaust fans capable of overcoming the resistance of the dust collector must be provided.

Pouring and Mold Cooling Department

The molten metal is generally transported to the pouring area in a transfer ladle, where it is poured into the pouring ladle, if not directly into the molds. The poured molds are then allowed to cool for a period of time, and then the solidified castings are removed from the mold, rough cleaned of mold material, and allowed to cool until the cast metal is cold enough to handle. During this time the metal emits a light smoke, and except on production type mold lines where the work is done at definite stations, the smoke is best removed by exhausting from above the pouring monorails. Foundries differ widely in the method by which these operations are carried out, and the hooding must be designed for the particular condition.

On production type mold reels, the pouring station is provided with a hood and the fumes are discharged from the area. The mold reel, depending on the type of conveyor used, is also covered or provided with a suitable exhaust hood for the length of travel when the poured mold is on it. These hoods are usually discharged directly to the atmosphere without cleaning, inasmuch as so little pollutant is present that dust collection equipment that could remove the fine particulate involved could not be economically justified.

After the casting is removed from the mold, very little, if any, pollutant is forthcoming. Many times, a cooling conveyor carrying hot castings is provided to expose them to outside conditions, possibly above the roof until they are ready for the cleaning room.

Molding Department

The sand handling system that returns, conditions, and supplies the basic ingredient in the sand mold has many points where dust is generated. Usually systems recirculate used sand. Hoppers below the shake-outs, etc., discharge to a system of conveyors, elevators, chutes, screens, coolers, bins, mixers, and the like.

From the first place the poured mold is broken, and at every subsequent point where work is done to remove the solidified casting from the mold, exhaust ventilation should be provided. The route that the hot, heat-fractured sand follows on its return trip to the mixer (where it is again conditioned for molding) must be ventilated at each point where the sand is disturbed. Some of the points, such as the shake-out, plows, and mixer, need special hooding to accommodate workmen required on the process. Other points, such as transfer points where the sand falls from one conveyor to another, can be enclosed as much as possible.

Throughout the design, care should be exercised to locate the duct take-off point of the hood so that the conveyed material is not thrown into the take-off. The hood take-off should also be tapered to keep the point where the exhaust air reaches the duct velocity away from the disturbed material. Too much of the fine sand, so important to the molder, may otherwise be removed.

On the other hand, the exhaust system may be required to serve a drum screen cooler, an aerator, or some other equipment that naturally will tend to introduce excessive sand in the airstream. A trap should be placed in the exhaust duct as close as possible to the hood take-off to remove all but the fine dust. A means of removing the collected material or, better, returning it to the sand return system is required. This should be considered when the trap is located.

Either wet scrubbing or dry mechanical collection should be used for this type of system. Where good collection efficiency is desired, the latter type may prove to be inadequate. Fabric arrester equipment may tend to clog because of moisture condensing from the airstream.

Casting Cleaning Department

When the castings have cooled enough to be handled, the sprues and cores must be removed. This is often done by simply knocking them off (or out of the casting) by workmen using handtools. A vibrating table is sometimes provided to aid the operation for higher production. In any event, the station where this is done must be exhausted to protect the workmen and remove the dust. The refuse generated at this operation is often removed by conveyor, either to a sprue mill where the metal is prepared for remelting or to a point where the waste material is deposited. The sprue mill and the conveyors carrying the dusty waste material require dust exhaust. The exhaust air cleaning device for this service may be either a medium-efficiency wet scrubber or a fabric arrester.

Following the knockout, the casting usually requires further cleaning. Depending on the type and the size of casting involved, this may be done by grinding, abrasive blasting, tumble mills, chipping, sawing, cutting, powder washing, etc. All these methods tend to create a dusty condition and require exhaust ventilation.

Grinding

Grinding wheels, whether they are mounted on a stand, supported from a swinging frame, or portable, need hooding that is designed with care, because the rotating wheel, in a sense, acts like a fan and tends to throw dust in all directions. Most of the dust is thrown, however, in the first quadrant from the point where the grinding wheel contacts the casting. It is good, in this case, to locate the duct take-off point so that the dust is thrown, as much as possible, directly into the take-off. A trap is usually included to drop out the heavy particles and reduce the abrasive erosion on the exhaust ductwork. Swing frame grinders should be arranged so that the dust is thrown into an exhausted booth, or where practical, a local hood may be mounted on the frame and exhausted using a flexible duct. Hand grinders are possibly the most difficult to hood. Where possible, grinding bench hoods with down- and back-draft exhaust will afford environmental dust control.

Abrasive Blasting

Relatively large exhaust air volumes are usually associated with abrasive blasting equipment. The equipment will vary in size, method of operation, type of abrasive used, how the abrasive is impelled, how the castings are handled, whether continuous or batch, and in other ways to suit the particular work to be done. The manufacturers of this equipment normally recommend the air volume that should be exhausted from each of the connections provided. Inasmuch as no moisture is involved in the process, the exhaust airstream will not have excessively high relative humidity, and the fabric arrester type of dust collector is usually used. Mechanical precleaners, usually simple traps, are recommended to reclaim the usable abrasive and at the same time reduce the dust load on the dust collectors.

Where steel shot is used for such abrasive cleaning, certain types of wet scrubbers should not be used to clean the exhaust air. The type that depends on the impingement principle, where the airsteam must suddenly change direction in order to wet the particulate, will tend to build up ferrous oxide deposits at the points where the change of direction occurs. Cleaning the scrubber then becomes a periodic maintenance chore.

The dust collection equipment that is deemed suitable for the cleaning room exhaust systems mentioned above will be equally suitable for systems serving tumble mills, chipping, and other cleaning room operations.

Core Room

Sand is usually the basic ingredient used to make cores. It is therefore important to investigate the route the new sand takes after it is brought to the plant until it is prepared for the mold. The sand may be dumped, conveyed, stored, dried, and mixed in much the same way as in the molding department. Wherever dry sand is disturbed while being conveyed, transferred, plowed, mixed, etc., it is recommended that exhaust ventilation be provided. The method of collecting and cleaning the exhaust airstream for this type of system will be similar to the sand handling systems found in the molding department. However, one would not expect excessively high humidity in the core room exhaust. Consequently, the fabric arrester equipment would be more acceptable for the core room than it would be for the molding department.

The air exhausted from core grinders and waste core removal and crushing equipment should also be cleaned before it is discharged to the atmosphere. The basic ingredients used to make core wash quite often are dusty when dumped into the mixer. The resulting condition may very well require that exhaust ventilation be applied to the mixer. Either wet or dry dust collection equipment will generally be suitable to clean the airstream on these applications.

The core blowers, delivery conveyors, and ovens do not normally emit particulate matter to the extent that dust collection equipment might be justified. However, certain quick-hardening types of core blowing equipment emit odor and irritants in such quantity that it becomes necessary to provide exhaust systems for the benefit of the operators. This odor is normally not objectionable when diluted, so that it is usually not necessary to treat the airstream before it is discharged. If conditions do dictate treatment, wet scrubbing can be applied.

Pattern Shop

Where woodworking equipment is installed and used to any great extent, a local exhaust system to capture the sawdust, chips, etc., should be provided. This sort of system can be fitted with floor sweeps to facilitate housekeeping as well. Normally a mechanical type of dust collector such as a cyclone is sufficient to remove the entrained particulate because the dust particles are much larger than encountered in other foundry exhausts. The saws, planers, shapers, sanders, drills, etc., may have built-in hoods which require only a simple duct connection. If no hood is provided, local hooding, designed according to accepted practice and provided with adequate suction, should be furnished.

Maintenance Shop

Much of the machinery used in foundry maintenance shops, such as grinders, mills, drills, and the like, perform their operation under wet conditions and therefore do not require exhaust. Where such work is done under dry conditions, the possible need for local exhaust should be investigated. Dry grinders in particular should be exhausted. Where such equipment is located so that a general dust collector system is too remote, a small packaged dust collector unit of the fabric arrester type that cleans the air and then discharges it into the room may well be the best solution.

Welding and soldering operations should be done under controlled conditions. An exhausted booth to gather and remove the smoke and fumes generated should be provided and used whenever such operations are performed. Normally no air cleaning equipment is needed, inasmuch as the air volumes used dilute the exhaust to an acceptable level.

WATER POLLUTION PROBLEMS

The greatest water pollution problem in foundries occurs from efforts to clean the air emissions. While sand sluicing systems can often introduce phenols used in bonding agents into the waste waters, the tons of airborne solids collected by scrubbers on cupola stacks and dust collection systems constitute the principal contaminants.

General Considerations

As in any industrial waste water treatment problem and in particular with foundry waste water, the desired results constitute the principal criteria of the system design. Whether the water is to be reused in the foundry processes or wasted to a stream or sewer system, certain guidelines must be first established. For this reason, it is essential that, when required, regulatory agencies be included in the initial water pollution control planning.

The experience of such agencies in dealing with the principal pollutants contained in foundry waste waters can often be beneficial in establishing a treatment system concept. Their codes governing removal of suspended and dissolved solids, pH control, and toxic constituents such as phenols often establish the system criteria very readily.

If the foundry is located in an area serviced by municipal sewers, it is quite possible that the waste waters may be treated to a lesser extent than would be required to meet various stream regulations. An immediate investigation should reveal what can be accepted by the city sewer systems.

Most cities have strict regulations governing the properties of waste waters discarded through their sewage system. In addition to the matter of quality of waste waters which they will accept, there may be limits on the amount of waste waters that can be accepted. Foundries use considerable amounts of water, and their waste waters can often be sufficient to overload hydraulically any existing city sewage treatment facility. Therefore, it is wise to take the local authorities into confidence when planning waste water control. Many older cities have sewers constructed of materials such as brick which are poor with respect to carrying the abrasive materials often contained in foundry waste waters.

This may be the controlling factor on the amount and kinds of waste waters city sewers will accept, rather than the possibility of problems occurring with waste water treatment facilities farther down the line. Nearby industries may also be discharging waste waters which will have a detrimental effect on the capacity of the sewers.

The kind of waste treatment processes that the city uses must also be considered. Some of the more sophisticated biological secondary treatment processes are particularly susceptible to upset by pH changes. Also, sedimentation facilities preceding such secondary treatment processes may be incapable of handling the amount of solids discharging from a foundry. Very often, however, a city will permit a limited amount of solids, dissolved and suspended, to be discharged into the sewer because the city sewage treatment system will be able to handle that amount of waste. If this is the case, such concessions should be obtained in writing prior to the design of an industrial waste treatment system for the foundry. This may, in effect, enable the foundry waste treatment system to get by with a somewhat lesser degree of treatment than would be required should its discharge be to a stream. In any event, the availability of sewers or streams that dispose of waste waters should be made a part of the original site-selection criteria when a new foundry is proposed.

Waste Water Treatment Criteria

As a rule of thumb, many states have provided industry with the criterion that the water must be returned in at least as good or better condition than it was in the source from which it was withdrawn. In addition, no states will permit the oxygen residual in a stream to be depleted, and state codes may provide other restrictions on industry. Other industrial wastes in the streams may complicate matters, caused by the fact that certain dissolved solids in foundry wastes may react with other wastes to cause rapid oxygen depletion or solids deposition in the receiving water. Also, the use of water by downstream users must be carefully considered. The discharge of waste waters containing high degrees of turbidity, acids, oils, etc., may cause downstream users to suffer; for instance, dissolved solids could be taken in by the water plants of downstream industries, causing scaling, corrosion in boilers, and so forth. It is not unreasonable to suspect that damages could be sought by such other water users, even in the event that stream codes were followed. Most states now have stream classification standards by areas, in which each stream is classified as to degree of water quality to be maintained. Generally, these classifications are determined by the ultimate use for which the stream is intended, such as water supply or recreation.

It cannot be overemphasized that it is vitally important for initial planning to be done with the regulatory agencies involved if it is intended to discharge waste waters to a stream or sewer. This will prevent disappointments resulting from the selection of unacceptable water pollution control methods and resulting design work being carried too far before the state regulatory agency is consulted.

Funding

Generally, few funds are available for industrial waste treatment, be it a foundry or any other industry. Managements have not been prone to appropriate monies for such nonproductive purposes; however, it is quite possible that current tax regulations, which permit limited tax write-offs on industrial waste treatment equipment, will be improved to permit further tax write-offs. Full advantage should be taken of these potentials if foundry waste treatment facilities are to be built at reasonable costs.

It is also possible to enter into lease-back arrangements for industrial waste treatment facilities through several organizations. This can be done for foundries as for many other types of industrial wastes. Arrangements can be made to provide for the on-site construction of the necessary sedimentation basins, sludge gathering devices, sludge dewatering devices, and so forth, which facilities can be leased by the industry with option to purchase at the end of the agreed lease period. In the funding considerations, the operating costs of the foundry waste treatment facilities are very important. Although skilled operators are generally not necessary, the persons who are in charge of running facilities should be made fully aware of the function that the facilities play in what constitutes good or poor waste water treatment. The need for operating personnel can be limited by automatically controlled sensing elements to regulate the foundry waste water treatment facilities; however, all such facilities must have optional manual control to prevent gross pollution in the event of failure of the automatically controlled system.

The problem of funding may be alleviated if it is determined that a possible by-product can be reclaimed. The reuse of treated waste waters in the foundry process can offer substantial savings or at least defray part of the waste water treatment costs. The sands and grits separated from foundry waste waters may often be used as fills under inconsequential structures quite successfully; however, heavy foundations and structures of value cannot usually be placed on such fills because of the possible lubricating value between the particles when deposited as fills. All opportunities for the disposal of waste materials removed from the water through sale or reuse rather than dumping should be investigated.

Space requirements for any proposed foundry waste treatment system are a matter of concern, for they are substantial. In addition to the space required by the actual equipment, space must often be found for the disposal of solids by burial, dumping, or filling.

Building costs, land costs, chemical costs, and power costs as well as operating costs are proper matters for consideration. When considering funding, amortization periods as well as initial costs are important. Short-term amortization of waste water treatment costs can run up the cost per pound of finished product to unacceptable limits; whereas, if the long amortization period can be justified, it may be easier to obtain the monies required for the proper waste water treatment equipment. Future process changes must be kept in mind when establishing an amortization period. Very often, changes in castings and types of casting materials will force changes in industrial waste water treatment equipment before it is fully amortized. The cost of money must also be taken into consideration, since money spent on foundry waste water works can be used in other perhaps more productive fashions. When all factors have been considered, it must be possible to relate the cost of waste water treatment to the cost per pound of finished castings.

This may be difficult to determine, because it is not only based on the tons of castings manufactured but will also be affected by the configuration and size of the various castings involved. It is very unlikely that two foundries would end up having exactly the same waste water treatment cost per pound of finished product because of all the variables previously mentioned.

Foundry Site Selection

When the site for a new foundry or the site for waste water treatment facilities for an existing foundry is considered, the following checklist should be considered:

1. Is it possible to dump treated waste water into a sewer rather than into a creek?
2. Is it possible to store waste waters originating in varying quantities and treat them on a round-the-clock basis, permitting the actual waste water treatment facilities to be constructed to a minimum size?
3. Is the area under consideration large enough? Space required for waste water treatment facilities is substantial.
4. Consider the slope of the terrain. Is it possible to eliminate the pumping considerations, or can the land provide readily filled areas in which lagoons can be constructed and the waste of years of foundry work allowed to accumulate?
5. Does the climate make heated buildings mandatory? Some of the waste water treatment equipment used in foundry work requires protection from the winter. Vacuum filters used for dewatering sludges invariably have to be housed in a building. Sedimentation ponds and clarifiers, though, usually do not require buildings; however, the heat loss from such equipment must be investigated to be sure there is no danger of detrimental freezing.
6. Proprietary processes exist which require less space; however, these must be carefully investigated to be sure they will work, since there are cases in which there have been failures when waste waters varied from a norm with respect to quantity and quality.

Design of Waste Water Treatment Systems

Waste waters from a new foundry present a problem to the waste treatment designer in that he must allow a considerable margin for error in estimating his waste water quantities. However, the quantities of waste water to be treated are readily determined for an existing foundry. Measurement of flows by means of weirs or other water consumption records are usually obtainable; however, care must be taken to be sure that no storm water or sanitary sewage inadvertently is included, since this will be sufficient to give false readings and will complicate the waste treatment processes. The properties of the waste waters such as pH, temperature, and turbidity can also be readily determined for an existing foundry by measurement; however, a proposed new foundry waste must be estimated by means of materials balance. In other words, nearly all sand, seacoal, limestone, and other materials will eventually appear in atmospheric or water wastes. It may often be presumed that the waste water treatment facility will eventually collect all dust from the foundry atmosphere, and it is therefore not unreasonable to design on the basis of all materials received by the plant showing up either in the waste waters or in the finished products, or in other dry waste materials which may be removed from the plant.

The principal pollutants in waste waters emitted by foundries are generally in the nature of very fine solids. At times, other polluting materials, such as phenols arising from bonding agents used in molding sand, some acids, and cleaning wastes, may also be found. Phenols in concentrations of only 1 part per billion have been known to affect water supplies. If present in foundry waste waters, biological or chemical oxidation may be required.

Lagoons

The least expensive way in general to remove the major contaminants prior to discharge of the waste waters is by using large lagoons; however, land costs or land availability around existing foundries may make this way not usable. Lagoons properly designed with sufficient sedimentation time can do a very satisfactory job of cleaning up turbid waters or waters full of solids, dust particles, fine slagging wastes, or sand. The success of lagoons in this use is dependent on a combination of several factors: the shape of the lagoon, the design of the dike, particular attention with respect to inlet and outlet devices, and the ability to skim the outlet to prevent discharge of oils or any floating wastes. Detention times will run into days. The actual efficiency of removal of fines by lagoons can be predicted only by testing. Many of the colloidal-sized particles settle out only with very long detention times. As a rule it is desirable to have at least 30 days' detention of waste water in a lagoon; however, reasonable results can be expected with as little as 10 days' detention time. These times should be computed realizing that lagoons will gradually fill with solids, thereby decreasing detention time. In addition, the required efficiency of the lagoons would also be influenced by the manner in which the water is to be discharged. If it is to be reused in the foundry, even though it is colored, slightly turbid, or offensive looking, it could still be quite acceptable. If the water must be discharged through a sewer, however, usually a higher degree of solids removal is required. If the water is to be discharged to a stream, clear water, well within regulatory agency restrictions, must be obtained.

The usual practice is to provide a set of two or more such lagoons, alternating their use, permitting one to dry while the others are used. After the material in the lagoon has dried (usually many weeks are required), sand and solids that have accumulated can usually be removed with a dragline and trucked away.

If sufficient area is available, the time between lagoon cleanings might well amount to years. The dikes surrounding such devices must be stable against any hydraulic loads imposed upon them and should be sodded or otherwise finished to prevent erosion. Care should be taken that burrowing animals do not go through the dikes and into the lagoons.

As a rule, however, when the wastes are collected in the lagoons, neutralization is not required. The detention time alone will often provide some pH adjustment because of the action of winds and sunlight. Generally, this is sufficient since waste waters discharged from foundries are only slightly acid or alkaline in nature.

Clarification and Filtration

Lacking land to provide sedimentation lagoons, foundry wastes generally have to be cleaned up by means of mechanical devices. These devices fall into four basic classifications:

1. Short-term detention chambers retaining the waste water for 10 to 15 min where the heavy solids drop out, followed by
2. Mixing tanks, into which coagulation aids may be added if necessary, followed by
3. Mechanical clarifiers with sludge collection equipment, followed by
4. Vacuum filters to permit the sludge collected by the mechanical clarifier to be dewatered. All waste water treatment systems should be reviewed against regulatory agency criteria regarding such equipment. Such criteria with respect to rise rates, overflow rates, filtration rates, etc., are generally conservative and may be used with safety.

Regulatory agencies desire adherence to conservative detention and settling criteria for normal sanitary sewage. However, in the case of foundry wastes, this is almost prohibitive in cost. The particles to be settled are generally at least 2½ times the specific gravity of normal materials settled out of domestic sewage. So, it is quite reasonable that settling rates could be increased to at least twice the suggested rates of most state codes. At this point good efficiencies can be expected in the removal of particulate matter, particularly if coagulant aids have been added.

Furthermore, the rates established by the previously mentioned codes such as state standards are based on a flow which is two to three times the average flow. This is proper for domestic sewage in normal municipal loads. In a foundry, however, the flow rates are practically fixed at a constant rate, since this is often a 24-hr around-the-clock operation or at least a uniform operation through a normal 8-hr working day. As a result, since no surges are anticipated, high loading rates for precipitation, sedimentation, clarification, etc., are often practical. It cannot be emphasized too much that the basis of success of any industrial waste treatment system and particularly foundries is immediate coordination with state regulatory agencies. If they will accept such criteria and reasoning, it may be possible to keep the size of components and the cost to a relatively low level.

If coagulation aids are used in such systems, the pH of the waste waters must be carefully controlled to permit maximum effectiveness of ofttimes costly coagulant aids. Very often, laboratory work is necessary to determine the pH adjustment which may be required and the amounts of acids or alkalies needed to correct the pH. This can also require a high degree of instrumentation or sophistication in the system design if the system is to operate with little attention.

One of the principal sources of contaminants in foundry waste waters is the dirt collected by cupola emission control systems such as shown in Fig. 8. The quencher tank overflow indicated would probably have to be given treatment, as previously described prior to discharge.

Many established waste water treatment suppliers are beginning to cater to the waste water needs of the foundry industry. Many newcomers, however, are endeavoring to supply proprietary devices which claim to simplify the treatment of waste waters from foundries. Often these devices combine the coagulation, settling, and sludge removal functions outlined previously in a single device rather than in separate components. Such devices should be viewed with reservations, since they often attempt to short-step well-established principles of waste water treatment. It cannot be overemphasized that any foundry wishing to treat its waste water properly should rely on two basic sources of aid. The first is engineering talent, well equipped and knowledgeable of sedimentation and coagulation processes; and the second is equipment furnished by established waste water treatment equipment suppliers, with a long record of the design and successful performance of equipment involved in the sedimentation, coagulation, and precipitation of solid wastes from turbid waters.

Package Treatment Systems

Foundry waste water treatment systems can be provided on a turnkey basis. This is usually done by specifying system performance requirements. Since generally the foundry waste waters will vary in quantity and quality in the course of the year, a system which is designed tightly will prove inadequate in many instances. Hence systems bought on the basis of price alone may prove inadequate at many times. Only by prolonged, active sampling of wastes of the existing foundry can any such performance specification be intelligently written. When such procedures are used with respect to new foundries, the design of a system based on anticipated wastes incurs a great risk with respect to ultimate satisfaction, since the wastes cannot be predicted satisfactorily. All waste water treatment system designs must be conservative in nature, since the foundry industry is dealing with problems of such a widely varying nature that it must provide waste water treatment systems that are not only conservative in design but are readily flexible and may be changed to meet the variety of conditions that will be encountered.

Waste Water Piping Systems

The waste waters from a foundry cannot be properly treated unless they are collected by a positive method. The collection piping is the beginning of the waste water treatment. The handling of waste water systems in a foundry prior to treatment, particularly those used in recirculating systems in which the waste waters are reused for various purposes, presents problems with respect to piping erosion. The waters flowing in a dust collector water recirculating system generally contain the products taken from the air, such as dust, grit and sand. Many times, even slagging wastes are included. These abrasive particles have great cutting power when moving in the stream of water. As a result, all waste water piping within foundries should be designed for the minimum velocity which will suspend the particles in the flow stream. Velocities should not be any higher than required. These velocities will be dependent on the size and nature of the particles involved. In general, the particles are inorganic and will have a specific gravity of about 2½ times the specific gravity of organic materials so that the velocity in the piping system must be considerably faster than those normally used in sewers. At the same time, however, these relatively high velocities will encourage pipe erosion. Any solution to this problem must, of course, be a compromise. As a rule these velocities should be in the range of 6 to 7 fps. Also, many of the particles will be of such size that they cannot be kept in suspension and will merely be dragged about along the bottom of the conveying piping or sluiceway.

It is wise to avoid unnecessary bends or changes in the direction of piping or sluiceways, for it has been found that the bends and fittings are points of severe erosion. Initial failure takes place at these points. Also, since the larger particles are dragged along the bottom, it has been found that the inverts of such piping or sluiceways are particularly susceptible to erosion.

Since pumping of these abrasive wastes presents such a problem in the design of a water collection system for a foundry, it is strongly recommended that the pumping be done on the clean side of the system. In other words, if there is a recirculating system reusing the foundry waste water, the pump should be placed on the clean side of any waste treatment system and the relatively clean water should be pumped to a point where it can flow through the entire contaminant collection process and back to the sedimentation point by gravity. This will greatly prolong the life of pumps and prevent many problems in the waste water treatment process.

If dirt filled waste waters must be pumped, the selection of the pumps is of critical importance. Centrifugal pumps will be subject to particular distress in the volutes and in portions of the impellers that have high velocities in them. The use of hardened steel impellers or pump protection by use of some of the newer synthetic materials or even rubber coatings may prove to be beneficial in preventing erosion. Furthermore, the pipes should often be lined with materials that will prolong their life. Rubber lining has been suggested as a possible solution, and other materials are available. The expense, however, prohibits their use in many cases.

As an alternate in some foundry designs, it has been found wise to provide for actual abandonment of the pipe once it wears out. This is most easily achieved by utilizing quick-coupling devices. This will permit the piping to be taken down and changed in a few hours if necessary should leaks develop. As a sideline benefit to such quick-coupling types of pipe connectors, the life of the piping can often be prolonged. As mentioned previously, the pipe erodes principally at bends and inverts; and by turning the pipe over 180°, bottom side up, the wear can be equalized throughout the pipe, and often the life of the steel pipe carrying these abrasive particles can be greatly extended.

Erosion is not the only problem which is a threat to foundry waste water piping. Corrosion cannot be neglected. To compound the matter, the waste waters are often warm and the rate of corrosion is thereby accelerated.

The use of open conduits to carry waste waters wherever possible in a foundry should be considered. Two benefits accrue. As long as gravity flow can be used, there is no need for a top on such systems and the systems are wide open and easily cleaned. Since it is almost impossible to achieve sufficient velocities in foundry piping to buoy or convey large particles, an open system permits manual cleaning of these sluiceways should the particles settle to an extent where they impede the flow. In the design of such sluiceways they must be given generally excessive slopes to promote high velocity and keep the flow of the particulate matter moving along to its point of ultimate collection or disposal. In designing flumes or sluices for carrying foundry waste waters, a careful hydraulic analysis relating to flow quantity, velocity, hydraulic radius, easy bends, etc., is in order. Not only will good velocities be obtained, but by promoting smooth laminar flows, erosion will be minimized.

It is not impossible that certain flumes or sluiceways can also be lined with steel plates or clay liners such as used in large storm sewers. These can have the benefit of prolonging flume or sluiceway life.

As a result of all these conditions the importance that must be given to waste water conveyance in a foundry cannot be overemphasized. Careful alignment of piping, careful blocking of piping to take care of thrusts resulting from hydraulic pressure, and minimizing of fittings and pumping all can prolong the life and use of waste water piping within a foundry.

Summary

No two foundries have exactly the same industrial waste water treatment problem. An existing foundry with known flow rates and known quantities can have its waste water treatment system designed specifically for those wastes. However, other limitations exist, such as minimum space remaining for such a system. On the other hand, a proposed new foundry in which all quantities and qualities must be estimated in advance can create a conservatism in design which is well founded but often very expensive.

The conservative approach is usually well justified by the fact that when waste waters are reused in dust collection systems, the success and long life of scrubbers, fans, etc., will often be directly related to the quality of treated waste water returned for reuse.

The foundry industry has faced up to the fact that it is under the gun with respect to air pollution. It is an odd fact that the air pollution problem will often be solved by creating a water pollution problem. Water pollution considerations will play an increasingly important part not only in the operating cost of existing foundries but in the site selections for new foundries built in the future.

Chapter 12

Pollution Control in Plating Operations

LESLIE E. LANCY

President, Lancy Laboratories, Inc.,
Metal Finishing and Waste Treatment Engineers,
Zelienople, Pennsylvania

WATER POLLUTION PROBLEMS

The common denominator of the electroplating industry, when viewed from a water pollution angle, in general terms covers the water pollution effects encountered in:

1. Cleaning, which is the removal of surface oils, greases, buffing compounds, etc.
2. Removal of oxides, rust, scale, etc.
3. Electrochemical or chemical processing to provide the basis metal with a surface coating consisting of a plated metal or a chemically deposited, so-called conversion coating, such as phosphate, oxide film as in blackening, etc.

In general the aim is to change the surface of a product in such a manner that the corrosion resistance is improved, or the appearance is changed to a more pleasing appearance; improved hardness is imparted; wear resistance is increased; surface conductivity changes finishes to either improve the appearance or suit specific engineering applications. It is evident that the first steps in the process—that is, the cleaning and oxide removal—are mainly preparatory steps for good adhesion and receptivity for the subsequent finishes to be employed, and while these preparatory steps may be similar to the procedures used in the primary metal manufacturing industries, or the manufacturers of certain finished products, such as automobiles and appliances, these preparatory steps are a major part of the activity in an electroplating plant, while they are of relatively minor importance for the manufacturer.

An electroplating plant may also be engaged in mechanical finishing activities, such as polishing and buffing, sandblasting, or wire brushing, or it may also have cleaning or painting processes employing solvents, the main activity being with various chemical solutions using water as a solvent for the chemicals and water as a rinsing medium between the various process solutions through which the work progresses.

With an activity centered around the use of various processes employing water as a solvent material, it is evident that water pollution problems will be encountered whenever an effluent is discharged. The severity of the pollution naturally will depend on the source of waste, the type of process employed, the size of the installation, the relative concentration of the effluent, etc.

Sources of Waste

Dumping of waste process solutions The cleaning and descaling process solutions are so formulated that they will have the ability to remove soil or scale, sur-

Fig. 1 Typical plating plant.

face metal film, and will hold the removed material without depositing it back on the work that is being processed. Naturally, slowly the capacity for additional soil or metal removal will be reduced in view of the soil or metal content acquired by the cleaning solution, and the time will be reached when the particular cleaning solution is considered spent. At this time the process solution is dumped. The dumping can occur as a batch waste discharge or perhaps a continuous, slow wastage to maintain a certain uniform concentration of active cleaning compounds or acids and maintain a uniform contaminant loading or metal concentration to avoid the necessity of batch dumping.

These batch discharges occur periodically; the relative volume of waste is usually not large; but as the chemical concentration is relatively high, the pollution effect may be considered also relatively serious.

The cleaners employed in metal finishing are usually compounded with various alkali phosphates and relatively high concentrations of wetting agents to provide fast and complete oil, grease, and soil removal. The various acid solvents for metal removal may contain higher acid concentrations than normal in the primary manufacturing industry in view of the greater demands regarding utmost cleanliness of the metal film for subsequent processing and also because the usual demands for bright finishes require high concentrations of acid solutions and frequent dumping to maintain a low metal contaminant level in the acid cleaners in use.

Most of the electrochemical process solutions in use, such as electroplating and anodic treatment processes, can be maintained in working order by periodic or continuous filtration, purification, or additions of various chemicals for replenishment, but many process solutions employed in finishing either cannot be completely purified or the purification is uneconomical, in which case the process solution itself will reach a point where dumping is necessary and a new process solution has to be made up. Under this category would come, as an example, chromium plating solutions contaminated by iron, copper, nickel, etc., anodizing solutions for aluminum processing, some of the cyanide type plating solutions, and chromating or phosphating type conversion coating processes.

From a pollution hazard standpoint, these wasted process solutions may be considered primary subjects for waste treatment. In view of the periodic or infrequent discharges, the relatively small volume, and ample time available for proper treatment, the waste treatment effort is relatively small. As the pollution effect, in view of the high concentrations of chemicals to be discharged, can be most severe, it will be evident that the chemical consumption for treatment may also be significant. The considerable time available between batch discharges and the usually small total volume to be treated, on the other hand, may allow small-sized equipment to provide proper treatment.

Accidental discharges of process solutions The second most severe pollution hazard in connection with plating operations is accidental discharges of key process solutions. The concern shown for the treatment of the periodically dumped process waste should be multiplied with regard to the accidental loss of process solutions, because this hazard not only is on hand for the few process solutions that are assumed to have finite life but may affect the contents of processing vats which under normal conditions could be maintained by the usual purification maintenance practiced in the particular plant.

Nearly every process solution in the plating plant is prone to be discharged through an accident, mainly because in the past an engineering effort was never directed toward the avoidance of these accidents. The usual plating plant is so laid out that the entire plating area drains on the floor and the floor is only an extension of the sewer system leaving the plating area.

While it is not common that a plating tank would spring a leak of such magnitude that the entire plating solution could leak away undetected, many plants are operated on a haphazard basis so that a slow leak amounting to a solution loss of 1 to 2 in./day could go undetected for months. Also it is common practice to make up evaporation losses by adding water with a hose to a process solution or turning on a spigot and returning to the process tank only after it is overflowing the rim of the tank for a while.

Filter hoses, heat-exchanger connections, and pumped process lines are all prone to leak. Deterioration of the hose and normal engineering would anticipate a certain frequency of accidental spillage, depending on the general maintenance in the particular plant.

Steam coils or heat exchangers undergo slow corrosion reactions, and it may be anticipated that pinpoint corrosion or a corrosion cracking perforated the barrier between the process solution and heat-exchange medium, that is, steam condensate or water. As the steam condenses, vacuum forms in the heating coil or in the jacket of the heat exchanger, drawing in the process solution through the voids created by the corrosion action. Proper waste treatment engineering would therefore concern itself with the accidental contamination of either the steam condensate or the cooling water utilized in the particular process.

Some state codes require containment of the most toxic process solutions, such as cyanide and chromic acid plating vats, by requiring a surrounding outer container capable of holding the entire volume of the process solution in case of accident. No doubt these state codes reflect on experience accumulated with various plating operations. The fallacy of these regulations is that, as enumerated above, there are many additional ways for serious accidents to develop, causing severe pollution conditions for which the containment for the process vat would have been no insurance. The fact that there is protection for the least common occurrence, a serious leak in the process tank itself, has only added to the cost of the installation of a plating plant but did not provide the insurance required. On the other hand, the

TABLE 1 Common Waste Discharges Due to Accidents in Metal Finishing Plants

Source	Method of detection	Correction or prevention
1. Process tank overflow *a.* Unattended water additions *b.* Leak of cooling water into solution from heat exchanger or cooling coil	High-level alarms in floor collection system to signal unusual discharges	1. Provide proper floor construction for floor spill segregation and containment (curbs, trenches, pits) 2. Provide treatment facilities for collected floor spill 3. Use spring-loaded valves for water additions 4. Provide automatic level controls for water additions
2. Process solution leakage *a.* Tank rupture or leakage *b.* Pump, hose, pipe rupture or leakage, filtration, heat exchanger, etc. *c.* Accidental opening of wrong valve	Same as above	Same as 1 and 2 above
3. Normal drippage from workpieces during transfer between process tanks	Inspection	1. Provide drainage pans between process tanks so that drippage returns to the tanks 2. Collect floor spillage
4. Process solution entering cooling water (heat exchanger leak)	1. Conductivity cell and bridge to actuate an alarm 2. Use of the cooling water as rinse water in a process line where the contamination will be immediately evident	
5. Process solution entering steam condensate (heat exchanger or heating coil leak)	Conductivity cell and bridge to actuate an alarm	Use conductivity controller to switch contaminated condensate to a waste collection and treatment system
6. Spillage of chemicals when making additions to process tanks or spillage in the chemical storage area	Make the solution maintenance man responsible for chemical additions	1. Careful handling and segregation of chemical stores 2. Segregation and collection of all floor spillage

awareness of the regulatory agencies of some of the potential hazards with the so-called "accidents" in a plating installation should help in a waste treatment engineering effort aiming for the utmost safety considering the particular plant and processes under scrutiny.

Table 1 lists some of the more common accidental discharges together with methods for detection and prevention or correction.

Contaminated rinse water effluent When generally discussing waste treatment in connection with electroplating processes, one normally assumes that the topic will be the elimination of the toxic constituents from the rinse water effluent. As discussed above, the most severe hazards are not with the discharge of an untreated rinse water effluent.

Electroplating in particular and metal finishing in general require copious quantities of water to wash away the remaining chemical film on the work surface dragged out from one process solution before it enters the next process. The reason for this is that water is the common solvent for the remaining chemical films carried by the workpieces. Secondly, the removal of this tenaciously adhering chemical film can be more easily accomplished with fast-flowing water, providing agitation around the work surface. A chemical film dragged out from one process and remaining on the surface may react with the next process solution, with which it may precipitate on the metal surface insoluble salts as barriers for good adhesion of the subsequent electrodeposit, all affecting the brightness, appearance of the finish as cloudy spots, roughness, etc. The chemicals that may be dragged from one process into the other could cause contamination of the subsequent process solution due to the slow accumulation of dragged in impurities, chemical constituents of the previous process. The need for good rinsing finally may be due to the fact that any chemicals remaining in the pores of finished work may lead to later discoloration, tarnishing, or corrosion, destroying the desired finish.

In view of the relatively large volume of rinse water discharged by the average plating plant, it is then understandable that water pollution concern is mainly due to this visible volume of water flow, even though the total contaminating chemical discharge is less than in the relatively smaller volume of the periodic discharges due to batch dumping and accidents.

The rinse water effluent by an electroplating plant will carry the various dissolved solids originally contained in the film or droplets carried out on the work surface leaving the manyfold chemical processes employed in the finishing sequence. The dragout from alkali cleaners by now is mixed with the dragout from acids, pickling solutions, and the various plating processes. While the total dissolved salt concentration in the water may not have increased appreciably, the effluent carries, as either dissolved or precipitated suspended solids, the various metal salts, cleaning compounds, and anions of the acids utilized and possibly a small quantity of the oils and greases originally removed by the cleaners from the work surface.

The relatively large volume of the effluent discharged by an electroplating plant makes the treatment of the rinse water effluent the major problem. An additional complication is the fact that after all the various process rinses are mixed, the proper chemical treatment becomes much more complicated or maybe even impossible. In these cases, it may be necessary to segregate the rinse waters leaving the plant into various chemical groupings to be able to provide proper treatment, taking the various individual groups and mixing the total effluent only after the segregated effluent waste streams have previously received specific chemical treatment.

Toxic, Potentially Harmful, and Corrosive Chemical Compounds Commonly Used in the Plating Industry

The cleaning, pickling, and processing solutions may contain a variety of chemical compounds, most of which may be toxic to aquatic life in very low concentrations. In higher concentrations, they may be also toxic to humans. As the discharge usually is directed to either surface waters or to a municipal sewage treatment plant, the toxicity considerations and the permissible limits of these toxic compounds in the effluent discharged will be dependent on their effect on aquatic life, that is, fish or lower organisms in the water or the bacteria in the sewage treatment plant.

Anions Very low concentrations of cyanides, chromic acid, and soluble chromates were found to be toxic to aquatic life. The bacterial colonies in the sewage treatment process can develop a tolerance for both cyanide and chromate compounds. Cyanides may be actually destroyed by bacteria. As usually the cyanide is complexed with a heavy metal cation such as copper, the bacterial tolerance is limited. The removal of chromic acid and chromates in the sewage treatment process is only partial. Therefore, even if bacterial tolerance is developed, the fact that these compounds may appear in the effluent from the sewage treatment plant limits the permissible concentration in an effluent discharged into the municipal sewer system. Similar limitation may be set also for cyanide concentrations in view of the hazard of incomplete removal in the sewage treatment plant and the hazard for workmen in the sewer system and sewage treatment plant due to the possible exposure to hydrogen cyanide gas. Fluoride concentration in the discharge is limited because of the toxic concentrations which may be reached in potable water supplies. The assumed toxicity to human beings is between 1 and 2 ppm.

Cations Most heavy metal compounds, such as arsenic, silver, beryllium, cadmium, lead, chromium, copper, zinc, and nickel, are toxic to aquatic life. Ammonia would be another cation toxic to fish in low concentrations. Iron and tin impart only color to the discharged effluent, but in addition, iron would also tend to remove dissolved oxygen from the stream. For these reasons, the permissible limit for all these cations is near the range of 1 ppm maximum but may be as low as 0.01 ppm for cadmium, lead, silver, and 0.1 ppm for chromium, etc.

Nitrates and phosphates Nitrates and phosphates are anions coming under the heading of nuisance effects rather than toxicity. Limitation for nitrates is mainly because the maximum limit in potable water supplies is 45 ppm. The limitations to phosphates are due to their being an important nutrient for algal growth; they may be limited to 1 ppm in an effluent discharge.

Solvents Solvents create an explosion hazard in the sewer systems. The chlorinated solvents create toxicity hazards in the sewer lines; oils and greases coat the sewer lines and therefore may lead to restrictions in the sewer canals; also none of these materials is amenable to bacteriological sewage treatment, and they are therefore completely excluded from acceptance into the municipal sewage discharge. It is evident that concentrations of more than 10 to 30 ppm of oily materials will not be acceptable as a discharge to surface waters.

Acidity and alkalinity As the materials of construction in the municipal sewer system are corroded by acids and high concentrations of alkalies, the pH limits are usually in the range of 6.5 to 9. A somewhat higher low limit for the pH may be stipulated for discharge to surface waters.

Suspended solids Suspended solids, such as precipitated metal hydroxides, tumbling and burnishing media, metallic chips, and paint solids, may settle and clog the municipal sewer system. Limits of 100 to 600 ppm of suspended solids are usual. The suspended solids level in the discharge to surface water usually has a lower limit of 30 to 40 ppm maximum because of the bad appearance of an effluent with high suspended solids content.

Treatment Quality Requirements

Discharge to the municipal sanitary sewer system From the foregoing, it will be evident that the requirements are more lenient, first because the dilution effect of the total flow reaching the municipal sewage treatment plant will reduce the concentration of the toxic elements to below the nontoxic limits. Secondly, there is a certain amount of treatment effect available in the sewage treatment plant, such as a sufficient alkalinity to precipitate metal hydroxides and incorporate them with the primary treatment solids, removing these metals from the effluent discharged by the sewage treatment plant. The aerobic treatment in the secondary treatment system of the sewage treatment plant will provide opportunities for the complete elimination of trace quantities of cyanides. Small quantities of chromic acid may be reduced by the organic waste or sulfides present in the total waste undergoing treatment. As a rule, one may say that the rinse water effluent from a plating plant, considering a small- to medium-sized installation, may not require treatment in a municipality of 100,000 or more inhabitants, provided that the treatment of batch waste and accidental process solution discharge is properly attended to.

Discharge into storm sewers and relatively large rivers Such discharge allows the consideration of the available dilution effect of the receiving stream, and therefore the quality requirements for the effluent may be more lenient.

Effluents discharged to small streams, storm sewer ditches, and lakes This discharge usually has to meet the minimum standards insofar as the toxic chemical concentrations are concerned as stipulated for drinking water supplies. If aquatic life in streams and lakes has to be protected, the concentrations allowable in the U.S. Public Health Service drinking water standards will be reduced in view of the fact that the toxicity of most of the anions and cations is far greater for aquatic organisms than for humans. Similar consideration may also affect discharges to tidal bays, especially if they are used for oyster fishing or other marine harvesting. Table 2 gives typical requirements to be expected for this type of discharge.

TABLE 2 Typical Effluent Requirements*

pH	6.5–9.0
Cations:	*ppm*
Ammonia (as NH_3)	< 2.5
Arsenic	< 1.0
Barium	< 1.0
Cadmium	< 0.05
Chromium:	
Hexavalent	< 0.5
Trivalent	< 1.0
Copper	< 0.5
Iron	< 2.0
Lead	< 0.1
Manganese	< 1.0
Nickel	< 2.0
Selenium	< 0.01
Silver	< 0.05
Tin	< 2.0
Zinc	< 1.0
Anions:	
Chloride	< 150
Cyanide	< 0.05
Fluoride	< 1.5
Nitrate	< 45.0
Sulfate	< 250
Suspended solids	< 20.0
Oil	< 10

* Maximum allowable concentrations in an effluent being discharged to a storm sewer or stream.

Neutralization

Most of the electroplating waste effluents are on the acid side and require the addition of an alkali to neutralize the waste. The chemistry of the neutralization is rather simple, and therefore we restrict ourselves to discussing the most often used chemicals.

For alkali addition, lime or liquid caustic is used. Lime can be added in the dry form or as a lime slurry. If added in the dry form, care has to be exercised that the lime utilization is efficient. With insufficient agitation for rapid addition of lime, the lime particles may be coated with gypsum, the precipitate of calcium sulfate from neutralization of sulfuric acid with lime, in which case 50 percent of the lime may not be utilized. The efficiency of lime usage is better with lime slurry, but the slurry should be continuously recirculated to avoid clogging of the pipelines. Lime will create additional sludge, since most of the sulfates and all the carbonates will precipitate. On the other hand, lime is a cheaper chemical, and it will be the preferential chemical to be used in the event that the effluent is high in phosphates or fluorides. Neutralization, using marble chips or a limestone bed, can be considered only if a plating system uses only hydrochloric and nitric acids. Since the effluent volume is usually considerable, the size of the neutralization system would have to be greatly increased, if efficient neutralization is anticipated by flowing through a limestone contact chamber.

The control point of the neutralization system should aim for a high pH, 8.5 to 9, for efficient precipitation of the metal hydroxides. The unprecipitated, heavy metal ions, such as zinc, nickel, and copper, may appear in concentrations around 3 to 5 ppm if the pH of the neutralization is held below 8.

Treatment Methods Commonly Used

Batch treatment Figure 2 shows the schematic presentation of batch treatment for cyanide elimination. The batch treatment tanks have to be so sized that approximately 8 hr of flow will be collected while the previous 8-hr discharge is being treated. Chemical additions and testing will require nearly 8 hr, including draining of the batch treatment tank.

When treating concentrated wastes, it has to be recognized that most chemical reactions, such as chlorination of cyanides, reduction of the chromic acid, and neutralization of acids, generate heat which may necessitate slow treatment, allowing natural cooling effects to dissipate the heat buildup or dilution of the waste so that the water volume has suitable heat uptake capacity to avoid chemical treatment at elevated temperatures.

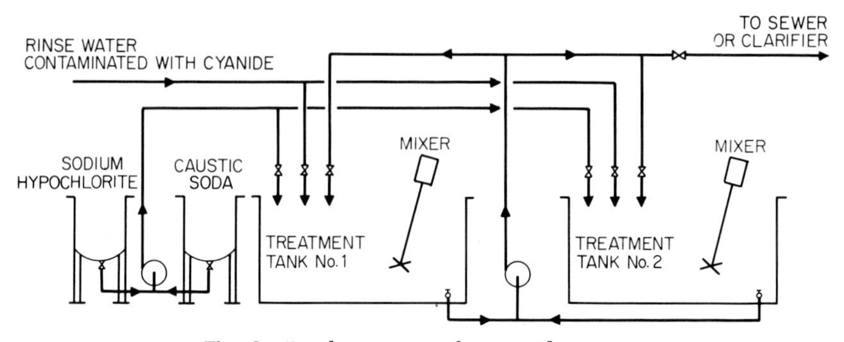

Fig. 2 Batch treatment for cyanide waste.

Continuous treatment Figure 3 is a schematic presentation of a continuous flow-through treatment system for cyanide and chromium contaminated rinse waters. Electronic control instrumentation and suitable sensing electrodes are available to control and monitor a waste stream, adding the necessary quantities of treatment chemicals and thereby rendering the total effluent acceptable for discharge.

The usual scheme is to split the effluent streams into three components:

1. Cyanide containing rinse water effluent
2. Chromic acid and hexavalent chromium containing rinse water
3. Rinses after all other chemical processes, such as acid plating baths, cleaners, and acid pickles

In this scheme, the cyanide stream is treated by raising the pH and adding the necessary quantities of chlorine chemicals, measuring the oxidation reduction potential of the resulting waste. The hexavalent chromium containing stream undergoes an acidification controlled by a pH meter, and an ORP controller controls the quantity of sodium metabisulfite solution or SO_2 gas entering the reaction tank.

The treated cyanide waste and treated chromium containing waste are afterward mixed with the total effluent where the entire volume undergoes pH adjustment to precipitate the various metal cations and chromates dissolved in the effluent.

Integrated treatment Figure 4 is a schematic presentation of typical integrated treatment systems following cyanide and chromic acid process solutions. The design idea in this process is that the dragout film and droplets containing the harmful chemical compounds are first rinsed in a chemical process solution to eliminate these compounds before the workpieces are water rinsed. The chemical treatment solution is specific for the type of chemicals which have to be treated. There is a process for cyanide treatment, for the treatment of chromium chemicals, for the elimination

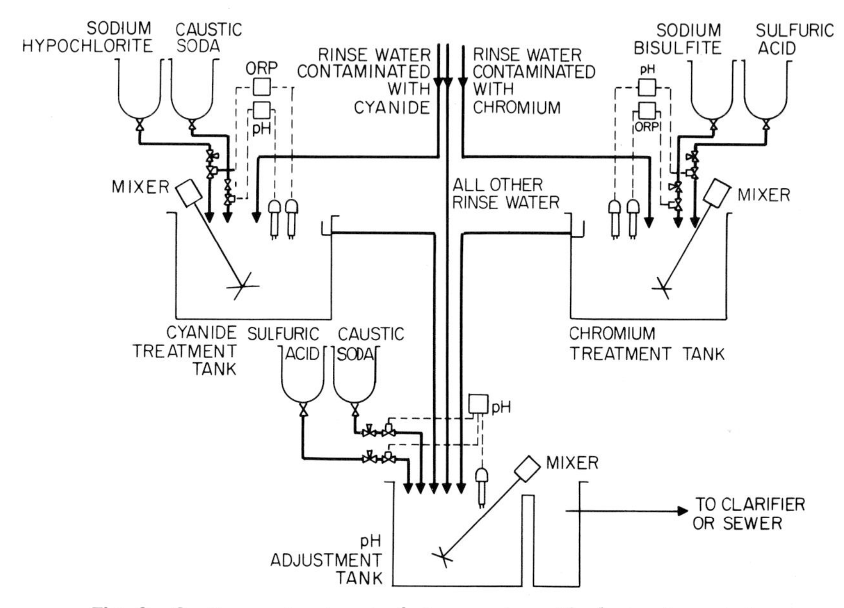

Fig. 3 Continuous treatment of rinse wastes with electronic controls.

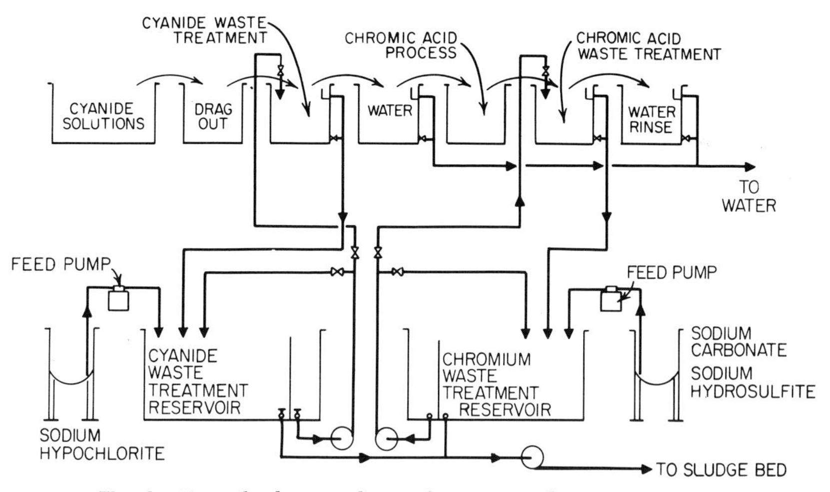

Fig. 4 Typical schematic layout for integrated waste treatment.

of copper from an acid process, etc. The treatment solution is recirculated between the various treatment wash tank stations and a reservoir tank where the precipitated metal hydroxides are settled. Chemicals are added as required to make up for consumption. The advantages of the integrated treatment system are minimal chem-

TABLE 3 Waste Treatment Costs

Waste solution	Treatment method	Treatment cost per gal of waste solution	
Typical cyanide type plating solution at 6 oz/gal CN^-	Integrated system	30 oz Cl_2	$0.125
		34 oz NaOH	0.105
			$0.23
	Batch or continuous system. 1 gal waste in 800 gal water; treatment to cyanate only, pH 11.5, 5 ppm excess Cl	20 oz Cl_2	$0.081
		4.94 lb NaOH	0.247
			$0.328
	Batch or continuous system. 1 gal waste in 800 gal water. Complete destruction to CO_2 and N_2; first stage pH 11.5; second stage pH 8.0; 5 ppm excess Cl_2	3.5 lb H_2SO_4	$0.084
		48.5 oz Cl_2	0.197
		6.96 lb NaOH	0.348
			$0.629
Typical chromium plating solution at 2 lb/gal CrO_3	Integrated system	1.92 lb SO_2	$0.125
		(or 2.86 lb $Na_2S_2O_5$)	(0.17)
		2.3 lb Na_2CO_3	0.128
			$0.253
	Batch or continuous system. 1 gal waste in 800 gal water, pH 3.0, 5 ppm excess SO_2	5.6 oz. H_2SO_4	$0.009
		1.95 lb SO_2	0.127
		(or 2.9 lb $Na_2S_2O_5$)	(0.174)
		14.7 lb Na_2CO_3	0.588
			$0.724
		Using 10.3 lb lime instead of Na_2CO_3—total	$0.362
Typical 10% H_2SO_4 acid dip at 6.5 oz/gal Fe^{++}	Batch system; waste treated in concentrated form	3.58 lb NaOH	$0.179
		or 1.65 lb Ca(OH)	$0.0363

TABLE 4 Effluent Quality

Contaminant	Batch treatment system or continuous flow-through treatment system, ppm*	Integrated treatment system, ppm
Cyanide	0.5	0.02
Copper	1.5–2	0.15
Chromium	0.1	0.01
Fluoride	18–20	1.0
Lead	0.5	0.01
Nickel	3–4	0.5
Zinc	3–4	0.5

These figures are the concentrations of some specific contaminants which may be expected in a plating shop rinse water effluent when using the treatment method indicated.

* These values are based on the assumption that the treatment is performed on the total effluent just prior to discharge. If dilution water is available following these treatments, the values will be correspondingly lower.

ical consumption costs; the precipitation and containment of the metal hydroxides and chromates so that the effluent discharged does not have to be clarified, filtered, etc.; the effluent usually exceeds the stipulated metal concentrations for drinking water; and in view of the utmost purity of the effluent, water reuse is available but for the pumping costs. Table 3 gives some comparative cost figures for the various treatment methods discussed, and Table 4 gives the usual concentrations of various contaminants which will be found in the effluents produced by these methods.

Waste Abatement Practices

Recovery of dragout Countercurrent rinsing with multiple rinse tanks provides an opportunity to reduce the total quantity of rinse water needed for efficient rinsing and may make it possible to balance evaporation losses with the rinse water effluent flow and thereby return the rinse water after each process back to the process from where the dragout originated.

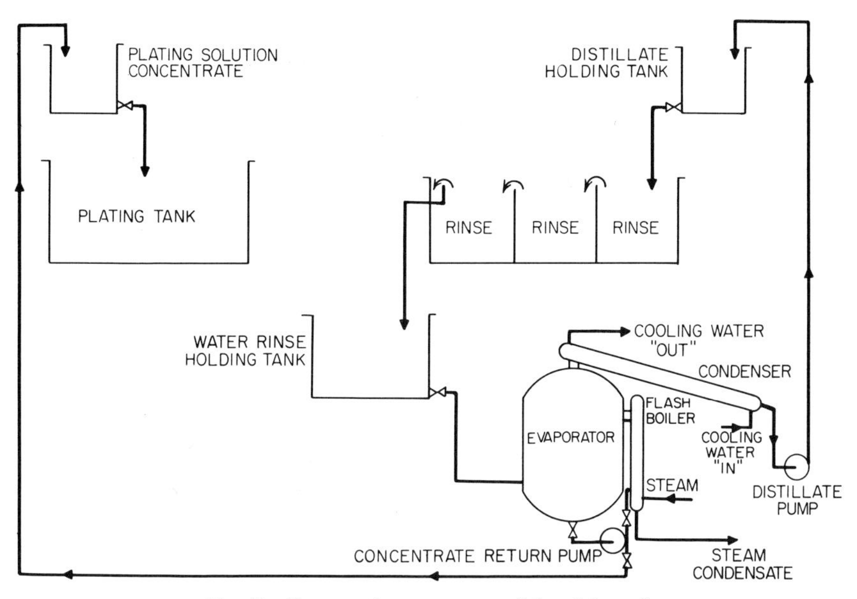

Fig. 5 Evaporative recovery—"closed loop."

Reconcentration of rinse waters When the process solution is operated at room temperature, or the quantity of rinse water exceeds evaporation losses, the reconcentration of the rinse water by evaporating the excess water is many times feasible. Vacuum evaporation reduces the heat input requirements otherwise needed to evaporate the volume of water to allow the return of the rinse water to the process solution. Usually the distillate is used in these systems as the rinse water in the process, and therefore provides improved purity of water for rinsing as an addition gain. Figure 5 is a schematic presentation of such a system.

Recovery of chemicals or purification of process solutions by the use of ion exchangers The use of ion exchange equipment to remove the dissolved solids, including the toxic or harmful chemicals returning to the plating plant pure water and the solids removed from the waste stream in a relatively high concentration, thereby limiting the size of the waste treatment installation, is also possible. The chemical cost and maintenance cost of such a closed system are relatively high, and it is not widely practiced. Specific uses of ion exchangers to provide purification of process solutions which otherwise would have to be discarded, or to provide continuous or periodic metal removal to maintain optimum metal concentrations, will many times have economical justification. The use of ion exchange systems in connection with evaporative recovery is also highly beneficial, eliminating the chances for impurity accumulation which would otherwise accompany the elimination of dragout. Moving bed ion exchangers may have the greatest future in these types of applications, because the large quantities of solids to be removed require either large ion exchange capacities or a high frequency of regenerations. A moving bed ion exchanger also has the advantage of allowing continuous regeneration and elimination of the parallel systems otherwise needed. Figure 6 shows schematically the operation of such an ion exchanger.

Solution regeneration Process solution regeneration applying dialysis, electrodialysis, or reverse osmosis techniques is yet in its infancy. Based on claims in the technical iterature and issued patents, a number of the process solutions which at this time have to be discarded periodically may be regenerated, and thereby it can be assumed that in the future many of the spent wastes will yield to one kind of

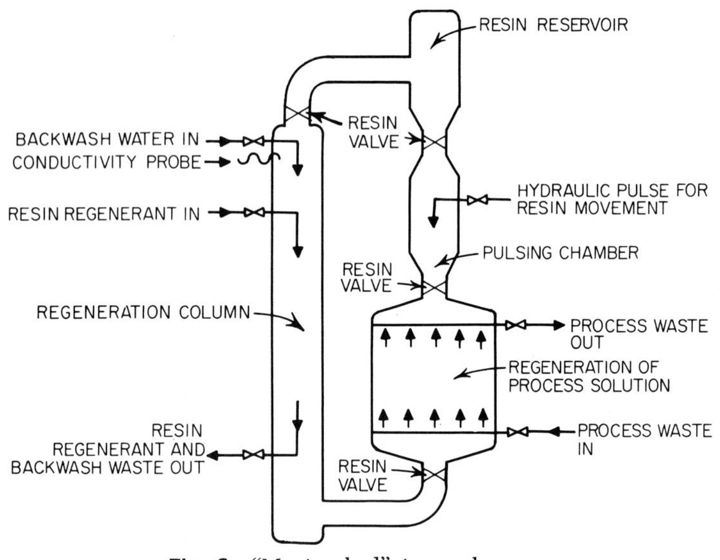

Fig. 6 "Moving bed" ion exchanger.

recovery method or another, avoiding waste treatment cost and recovering valuable process chemicals and metals.

Organic Wastes in the Plating Room Effluent

Most process solutions, and especially plating electrolytes, contain minor amounts of organic addition agents such as brighteners and wetting agents. The quantity of the organic substances from rinse water effluent is negligible. The only time that treatment of organic wastes may become a requirement is with the dumped process solutions that are relatively high in cleaning or emulsifying compounds or oils.

Soaps and detergents Dumped cleaners constitute a relatively minor volume of waste, and if these are treated as a batch waste with dumped acids, emulsions may be broken; oil content can be freed; and the organic content, if slowly discharged with the rinse water effluent, would have negligible effect on the quality of the effluent. Tumbling, burnishing, and peen plating wastes, on the other hand, have a very high concentration of similar compounds; however, since the volume of these discharges is usually relatively minor, discharging this waste to the sanitary sewer system can be practiced without any pretreatment except settling of the settleable, suspended solids.

Oils, solvents, paint solids These should be kept out of the effluent and can be most easily removed manually for haulage to a community incinerator or other suitable means of inoffensive discharge.

Separation of Suspended Solids from the Treated Effluent

This is perhaps the most critical problem with metal finishing waste treatment. An effluent that is discharged to the surface waters or storm sewers will usually have to meet the requirements of low suspended solids content and should have no color to be aesthetically acceptable. Discharge to the sanitary sewer system, on the other hand, may not require the separation of the precipitated solids, as usually an effluent containing 100 to 600 ppm solids is accepted.

When the clarification of the treated rinse water containing precipitated metal hydroxides is required, this normally is accomplished by one of the following:

Settling tanks or clarifiers, which are normally large tanks with a depth of 5 to 8 ft to allow settling and quiet flow of the effluent to avoid turbulence. Retention of 2 to 4 hr may be necessary to provide the clarification and settling of the suspended solids. Freshly precipitated metal hydroxides have a specific gravity very close to 1, and the settling and densifying of the sludge is a problem. Polyelectrolytes added to the effluent stream greatly improve the settleability of these solids, but the cost of the polyelectrolyte additions, in view of the relatively large volume of flow of discharged rinse waters, can significantly add to the chemical cost of waste treatment.

Filtration using pressure filters is proposed often. The metal hydroxides are usually not easily filtered and tend to blind the filter media fast. A copious use of filter aid such as diatomaceous earth or cellulose fibers is recommended, but the cost of filtration, depending on the particular effluent, may be costly from both the maintenance labor and filter aid consumption standpoint.

Sludge Handling and Drying

The precipitated metal hydroxides after they are settled in a clarifier, or with the integrated treatment system in the individual reservoir tanks, require further handling to allow removal of the solids without having to handle great quantities of water with the solids waste. For the drying of these sludges, various means are available.

Sludge beds can be earthen lagoons, usually rather shallow structures, and their performance will depend on the seepage rate through the ground, as in most areas the evaporation loss equals the precipitation gain in water during the seasons.

Sludge filters, depending on the settling and slow filtration of the supernatant water into the adjacent ground or sewer system.

Vacuum filters, which will perform reasonably well, depending on the preconditioning of the metal hydroxide sludges. The sludges should be sufficiently thick, 2 to 5 percent dry weight, and either they should be of such nature that they do not tend to blind the filter cloth, or a precoat filter has to be used.

Centrifuge is theoretically feasible but may be far too costly for this type of separation, at least from the general metal finishing waste standpoint.

Incineration, that is, the evaporation of the water and drying of the sludge through the use of an incinerator, which would be suitable practice if the particular plant has large quantities of waste oil or other waste products providing sufficient quantities of heat generated in burning of these wastes to support the energy requirements to evaporate the water contained in the metal finishing sludge.

Solid Waste Discharge

The dry sludge resulting as a waste product from metal finish waste treatment, and containing mainly metal oxides and metal hydroxides, can be in the form of a completely dry sludge which is haulable to a dump site, or a wet sludge which is to be disposed of by tank truck means of haulage to a dump.

No deleterious aftereffects regarding ground contamination can result from the discharge of sludges from metal finishing waste treatment provided that the treatment process was complete and aimed for a suitable and nontoxic final product. Under no circumstances should untreated sludges be dumped, that is, waste that can be shoveled from a process tank without any treatment. Many of the sludges, such as calcium fluoride, chromates, or various metal salts, may contaminate groundwater if the treatment was not complete or the sludge was not in the form that ensures extremely low solubility in rainwater. Care also has to be exercised that the dumped sludges will not be washed by rainwater into the nearby streams.

Water Conservation and Water Reuse

In view of rising water costs, water scarcity in certain areas, and the cost of sewer usage charges, water conservation and water reuse are now practiced more widely. There are many simple and inexpensive methods with rinsing programs in electroplating processes that may allow significant reduction of the volume of water needed for the process.

Countercurrent Rinses. Each additional rinse tank through which the water flows in a countercurrent manner allows a theoretical reduction of water usage by one-tenth. The practical value may be somewhat less. Figure 7 shows a typical three-tank countercurrent rinsing system. Theoretically it should be possible to reduce the

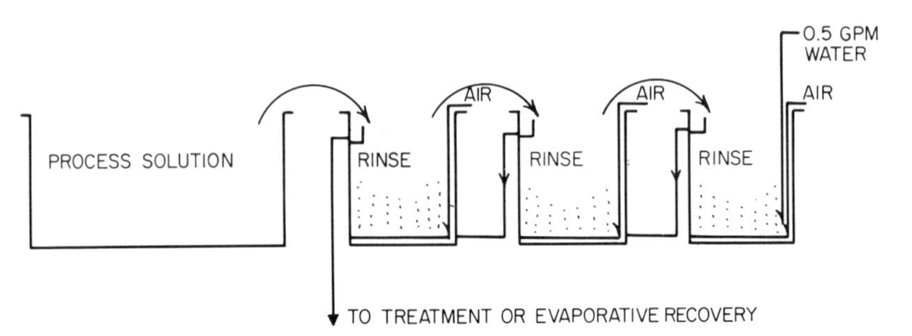

Fig. 7 Countercurrent rinsing. If a single rinse tank with 20 gpm water flow were substituted for this countercurrent system, the concentration of contaminant in the single rinse would be the same as the concentration in the last tank of this system.

rinse water flow by using a sufficient number of rinsing tanks in countercurrent sequence to match the evaporation losses in the process. The only question will be whether the dragout losses which have maintained low concentrations of impurity buildup will be missed if all water, and thereby the impurities, are also returned to the process. In a practical way, double countercurrent rinse tanks are used quite extensively in industry.

Conductivity Bridges. Sensing the dissolved solids concentration in the water, the conductivity bridges operate a solenoid valve adding water to the rinse tank at a predetermined flow rate. This system overcomes the usual open valve practice in electroplating operations, disregarding the question of whether the water is needed or not.

Ion exchange waste treatment systems and *evaporative recovery* provide clean rinse water as a by-product.

Water reuse is widely practiced in connection with the use of the integrated waste treatment system. This process allows an effluent purity suitable for return to the plant for reuse, both because it can meet the requirements from a chemical purity standpoint and because the rinse water is not contaminated with the usual metal hydroxide precipitates. Water recirculation in the range of 75 to 80 percent is usually practiced.

AIR POLLUTION PROBLEMS

Toxic Fumes in the Exhausted Air Discharge

Fumes from nitric acid Nitric acid is a strong oxidizing agent, and as the reaction proceeds on the surface of oxidizable metals, nitrous and nitric oxide fumes are generated which are highly toxic. The most violent reaction, and therefore the most copious fuming, will occur when using nitric acid for descaling, pickling, or bright dipping on copper and cuprous alloys. This may be the most severe atmospheric pollution problem encountered in electroplating during the preparatory cleaning stages or in metal finishing operations. Considering copper and cuprous alloy basis metals, a similar reaction occurs when using nitric acid on iron or steel, but there are only a few cases where this process would be used. On aluminum and aluminum alloys, the attack is not nearly as severe; therefore, the quantity of fumes generated is also greatly reduced, but yet some nitrous and nitric acid fumes are released to the atmosphere in these aluminum cleaning operations.

The use of nitric acid in stainless steel, titanium, and zirconium pickling operations creates atmospheric contamination hazards, but not because of the release of the gaseous breakdown products of nitric acid. The contamination is rather due to only the entrained nitric acid in the exhaust vapor over these cleaning solutions.

Hydrochloric acid in the exhaust discharge Hydrochloric acid is often used in the metal cleaning stages prior to electroplating. The vapor pressure of the hydrochloric acid gas is sufficiently high to allow considerable quantities of acid to be released from the solution even at room temperature when the concentrations are higher than 50 percent by volume of the commercial hydrochloric acid. At elevated temperatures, even low concentrations of hydrochloric acid solutions will release the acid in the gas form.

Entrained acids and alkalies such as chromic acid, caustic soda, in the exhaust air discharge In electrolytic applications such as chromium plating, anodizing, and electrolytic applications for sulfuric acid or caustic soda. such as descaling and cleaning, hydrogen gas is generated at the cathode and oxygen may be generated on the anode surfaces. The freshly generated gases escape from the solutions with sufficient velocity to carry along substantial quantities of the solution itself in the form of droplets to cause atmospheric contamination. These conditions may be considered entrained chemicals in the exhaust airstream, and pollution could be avoided if the solutions would be reprecipitated and recollected.

Dust from grinding, polishing, sandblasting Considerable air pollution is caused by solid particles which are usually carried by the exhaust airstream.

Solvent vapors from cleaners and paint solvents Solvent vapors from cleaners and paint solvents, also chlorinated solvent vapors, are usually encountered in connection with electroplating and metal finishing operations.

Equipment Usually Utilized

Fume washers Fume washers depend on the exhaust air passing through a water curtain which acts both as a temperature condenser of the vapors and for washing down the entrained liquids precipitated in the contact chamber. Recirculation of the water is feasible, and the use of an alkaline solution washing fumes from, as an example, sulfuric acid exhaust systems may be preferable.

Packed tower fume scrubbers Gases and fumes such as nitric and nitrous oxide and hydrochloric acid can be best removed from the exhaust airstream by passing them through a packed tower fume scrubber. The absorbing liquid is usually a caustic soda solution which is continually recirculated and periodically discharged. Careful design is required for the removal of the nitrous oxide, for which no high-efficiency absorbing liquid is known. Contact time and tower design will be the main criteria for the efficiency of the removal.

Dust collectors Either of the Rotoclone precipitation type or various filter media are used to remove the entrained dust from exhaust systems. Water spray is sometimes also employed.

Solvent recovery and catalytic oxidation systems The discharge of solvent vapors is under continuous scrutiny, and most often complete combustion is required for the solvent vapors removed from paint systems for large cleaning operations utilizing solvent cleaners. Chlorinated solvents cannot be easily removed from the airstream, and for these, catalytic combustion systems are not useful. Careful design of the chlorinated solvent vapor degreasers and vapor recovery still is perhaps the only easy way of solving these atmospheric pollution problems by avoiding their occurrence. If chlorinated solvents are in the exhaust stream in sufficient quantities to require treatment, carbon filtration may be the only suitable means of rectifying the conditions.

Chapter 13

Pollution Control in Metal Fabricating Plants

B. BASKIN
Senior Mechanical Engineer

D. J. GIFFELS
Project Director

E. WILLOUGHBY
Chief Civil Engineer,
Giffels Associates, Inc., Architects, Engineers, Planners,
Detroit, Michigan

THE METAL FABRICATING PLANT

Metal fabricating plants may generate any single contaminant or combination of contaminants because metal fabrication, considered as a division of product manufacturing, involves a very broad spectrum of processes, products, and materials. The scope of metal fabrication encompasses ferrous and nonferrous metals undergoing a wide variety of operations and treatments affecting the basic size, shape, and configuration of the material as well as surface preparation and finish to best suit the product or assembly. Metal fabrication embraces any metalwork which concerns itself with the following functions:

Metal removal, such as in a machining plant
Metal forming, such as in a stamping plant
Metal joining, such as welding, brazing, and soldering
Metal finishing, such as solvent degreasing, priming, and painting

Fig. 1 High-speed galvanizing line. (*United States Steel.*)

GENERAL CONSIDERATIONS AND OBJECTIVES

It is not within the scope of this chapter to develop the principles of metal fabricating plant layout design. However, the design of the plant can have a profound effect upon the water and air pollutants which are generated and upon the effectiveness of the systems which are incorporated to control these pollutants. Therefore, it is obvious that pollution control and the plant design are mutually interdependent.

Since the generation of pollutants is the result of a process within a plant, the study of the process with a view to alteration to mitigate the source of pollution is a necessary part of any pollution control study.

Pollution control in its broadest sense includes the maintenance of hygienic conditions within the plant itself, for the benefit of the employer as well as the employees. In many cases, the solution of this problem will result in considerable reduction or complete elimination of a pollution problem outside the environs of the plant. For this reason. considerable attention must be given to the solution of the in-plant problem. In general, this is primarily a problem of air pollution.

Air contaminants in a metalworking plant may be classified on the basis of origin or method of formation into the following principal groups:

1. Solid particulate matter—dust, fumes, and smoke
2. Liquid particulate matter—mists and fogs
3. Nonparticulate matter–vapors and gases

In addition, these air contaminants may be classified by chemical nature–organic or inorganic; by optic properties—visible or invisible; by particle size—submicroscopic, microscopic, and macroscopic; by state—particulate or gaseous; and by effect on humans—toxic or harmless.

In metal fabricating plants, the range of the particles responsible for air contamination is considered to vary in mean size from ultramicroscopic (0.7 micron) for metallic fumes to microscopic (100 microns) for metal grinding particulate. Charts show the range of particulate sizes, concentration, and collector performance of a wide range of contaminant particles associated with this industry as well as the primary metal industries (American Air Filter Co.).

THRESHOLD LIMIT VALUES

Unless control of the contaminants of the air is exercised, dispersion throughout the plant will occur, with the possibility of toxic effects. Although absolute atmospheric purity in the plant is seldom attained, the individual is able to assimilate small quantities of contaminants without permanent injury because the physiological flushing, excretion, and repair processes operate at a higher rate than the deleterious effects of the contaminants of the air when present at a tolerable level.

Thus the human body, living in a hostile industrial environment during the working day, develops adaptive mechanisms to maintain its vitality. Considerable effort has been expended to determine the limits of tolerance and the levels of exposure to airborne contaminants without harmful human effect. These are known as threshold limit values and represent a set of values established by industrial toxicologists as a result of experience in metal fabricating plants and other industrial operations.

Threshold limit values are time-weighted average concentrations of air contaminants for a normal workday rather than maximum allowable concentrations. These values are a result of experience, research, and medical data and are subject to constant revision. They constitute an approximation of airborne concentration of contaminants which represents threshold conditions under which daily exposure will not result in discomfort. Responsibility for the promulgation and revision of the information rests with the Committee on Threshold Limits of the American Conference of Governmental Industrial Hygienists. Threshold limit values apply not only to metal fabricating plants but to any plant or atmosphere into which airborne pollutants may be received.

VENTILATION AND POLLUTION CONTROL

The objective of the plant ventilation system is to control the airborne contaminants or to correct unsuitable thermal conditions or to do both. The purpose may be mainly to remove a disagreeable or toxic atmospheric contaminant and thus eliminate a hazard to health. General ventilation consists of a flushing of the interior of the plant with clean outdoor air to dilute contaminants to predetermined, tolerable concentrations. The air supplied sweeps the undesirable contaminants away from the worker at the work station, and the ventilation fans exhaust the diluted waste products to the exterior. To counterbalance this air, makeup air is supplied so that a condition of air balance is obtained wherein the air exhausted is equal to the air supplied. Many ventilation designers prefer to create a slight imbalance in supply exhaust rates so that the plant is slightly "oversupplied" and pressurized. Pressurization tends to make exhaust fans operate more efficiently, helps eliminate cold drafts around building openings, and improves the operation of combustion equipment.

As an alternate to general dilution ventilation, local exhaust may be employed to remove the contaminant at its point of origin before it escapes into the general plant atmosphere. In local exhausting, the contaminants are captured close to the point of emission and are conducted by a ductwork system to an exhaust fan or to a pollution control device. The practice of locally exhausting a contaminant for pollution control rather than utilizing general dilution ventilation results in the conservation of heating in cold weather since much smaller quantities of air are required. In addition, local exhaust systems permit the capture of pollutants before discharge to the atmosphere. In local exhaust ventilation, the operation is enclosed as completely as possible, or a close-fitting hood near the contaminant source is used.

Design of a successful local exhaust system depends on the correct estimate of the rate of airflow into the exhaust hood. The required flow rate depends on the character of the contaminating process, on the type and dimensions of the hood and its placement relative to the process, and on the amount of contaminant generated and its degree of toxicity.

The final design of the ventilation system will be governed by considerations regarding the nature, distribution, and intensity of the air contaminants. In the design of a ventilation system in which control of the contamination will be effected by dilution, the overall ventilation estimate is made as simply as in this example: A process generates 1 cfm of vapor the concentration of which in the air must not exceed 1 part in 10,000 parts of air. The required overall ventilation rate is, then, 10,000 cfm. Should this process be scattered through the plant, then local exhaust for contamination control is indicated as a more effective technique. However, general ventilation may provide a more economical system than local exhaust if the total quantity of contaminant is small enough to be diluted effectively by a general ventilation system.

MACHINING

Machining of metals invariably involves the use of coolants, cutting oils, and lubricants, each of which is a potential pollutant of air or water or both.

Lubricating oils can be separated within the plant by skimming off the top layer in a separator. Metal chips and fines can be pulled off from the bottom of the settling chamber.

Coolants such as soluble oils should be retained without excessive wastage, in order to minimize the pollution control requirements for the plant waste. However, in order to continue to reuse the coolant, it is necessary to prevent the breakdown due to the action of anaerobic bacteria. Aeration by use of air injection into a holding tank can prevent putrefaction so that the useful life of the coolant is extended. Coolant which clings to the cutting chips can be separated by centrifuging and can be returned to the coolant system.

Cutting oils, lubricants, and coolants tend to break down during the machining operations, with a resultant fume emission. With most of these, the fumes pose no particular problem and are easily picked up by hoods positioned close to the machining operation. When sulfochlorinated mineral oils are used as lubricants or coolants, toxic chlorine bearing or sulfur bearing vapors may be released. These must be captured by use of mist eliminators and filters. It is essential that any stream of air carrying oil must be subjected to a separation process to prevent the creation of a fire hazard due to the accumulation of oil in ductwork and to preclude the possibility of damage to the built-up roofing in the plant.

Cast iron machining produces dust particles ranging in size from chips to fine particles smaller than 5 microns. Graphitic carbon is also a product, which deposits on the entire exhaust system. The machining operations are typically hooded very closely to increase the effectiveness. Exhaust ductwork is generally of flanged construction, to permit cleaning and replacement. Normally the type of collector used is a high-efficiency dry centrifugal type, although wet collectors and bag type filters are also used occasionally. Cyclone collectors are rarely, if ever, used, as the efficiencies attainable are not adequate for the removal of the fine dust.

SURFACE FINISHING

Production grinding, scratch brushing, and abrasive cutoff operations result in light concentrations of particles which in the main are larger than 15-micron size. A wide variety of designs of hoods to capture the dusts generated have been developed to suit every conceivable arrangement of equipment. Transport velocities in the ductwork conveying the coarse materials captured must be in the range of 4,500 fpm to ensure that the particles will not settle out in the exhaust ductwork. Because such a large percentage of the particles are fairly large size, a fairly good overall efficiency of collection can be attained with the use of a cyclone. The cyclone collector is much less efficient on the finer particles and requires the use of a final filter to prevent the discharge of metal dust outside the plant. Metal dust deposition, with attendant rust spotting, is so offensive to the plant's employees and neighbors that it is well worth the cost to install higher-efficiency collection equipment such as wet collectors or bag collectors.

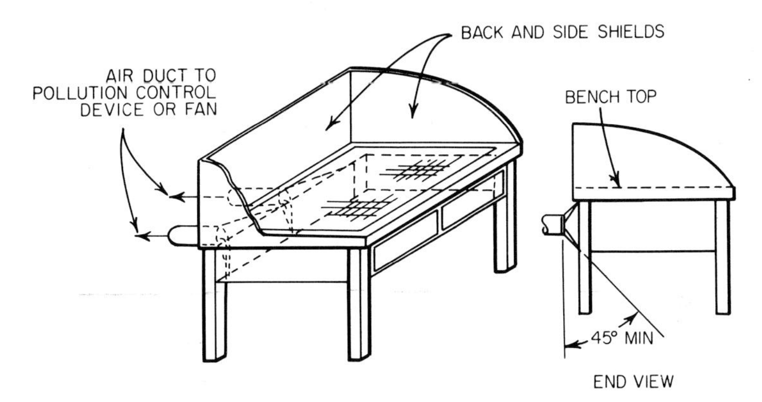

Fig. 2 Portable hand grinding. Workbench with built-in air duct. (*From Industrial Ventilation.*)

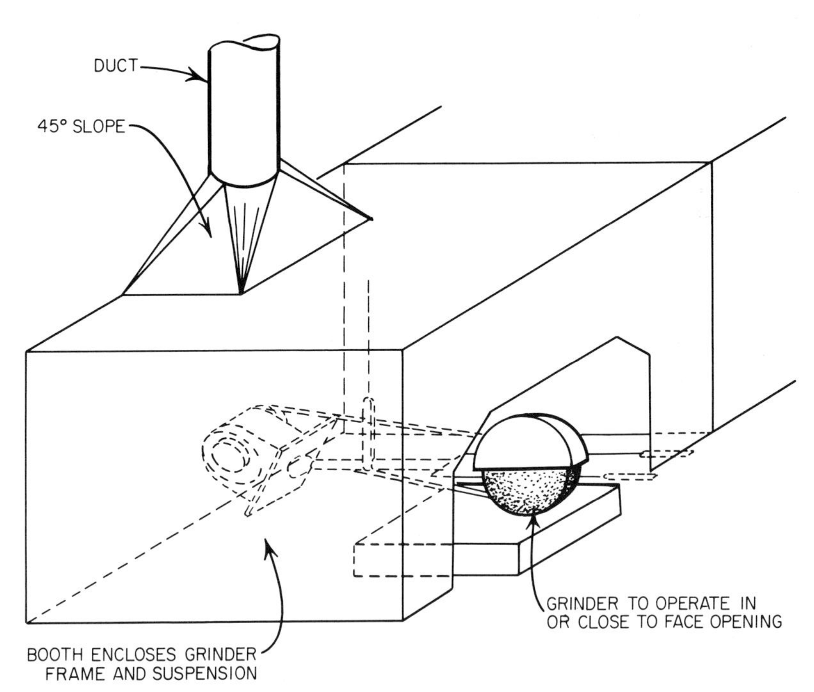

Fig. 3 Swing grinder dust hood design. (*From Industrial Ventilation.*)

Portable hand grinding (Fig. 2) is accomplished within a hood wherever possible, with the air drawn off from the underside of the workbench into ductwork at the rear. Swing frame grinders (Fig. 3) are also normally enclosed completely within a hood. The particle sizes resulting from this type of grinding operation are usually somewhat smaller, in the 5- to 15-micron range. It is therefore even more important that the higher-efficiency filters be used. For this reason, the cyclone collector is rarely used for this purpose. Instead, wet collectors or bag collectors are commonly utilized.

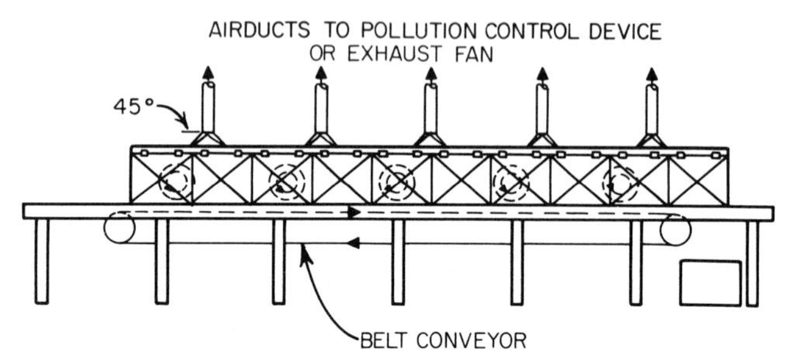

Fig. 4 Straight line automatic buffing dust hood design. (*From Industrial Ventilation.*)

Buffing operations introduce new pollutants in addition to the metallic dusts already considered, and these pollutants introduce special problems. Buffing wheels are made up of a collection of cloth disks, which are rotated at high speed. Such abrasive as is needed to remove fine scratches from the workpiece is applied intermittently to the buffing wheel by rubbing a cutting compound in bar or stick form against it as it rotates. The buffing operation results in the formation of lint particles from the buffing wheels, sticky buffing compounds and metal dust being thrown into

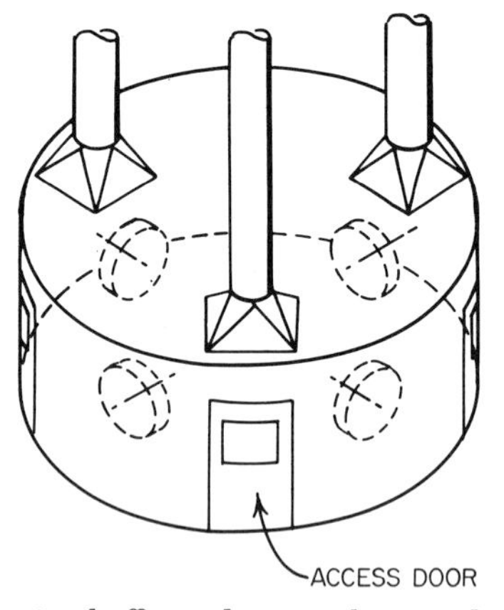

Fig. 5 Circular automatic buffing dust enclosure design. (*From Industrial Ventilation.*)

the air. The entire operation must be completely hooded, but completely accessible for cleaning, as the buffing compounds and lint in combination leave a deposition which can be a serious fire hazard. Hoods and exhaust systems must be provided with automatic fire protection as required by codes or insurance regulations. Examples of hoods for buffing operations are shown in Figs. 4 and 5. High-efficiency centrifugal collectors are not normally used, because of the possibility of plugging by the linty mass, nor are bag type filters utilized, because of the fire hazards. Wet collectors are frequently used for this purpose, as are cyclone collectors of a type which will not plug.

HEAT TREATING

Heat treatment is an operation or combination of operations that involves the heating and cooling of a solid metal or an alloy for the purpose of obtaining certain desired physical properties. These physical properties are attained through changes in the nature, form, size, or distribution of the structural constituents. These changes occur because of the effect of temperature on phase equilibrium and include grain growth, recrystallization, and diffusion or migration of atoms, or from changes in the composition of the material, such as carburizing or nitriding. In the heat treatment process, the metal is subjected to a definite time-temperature cycle, which may be conveniently divided up into three parts: (1) heating, (2) holding, and (3) cooling. The chemical effect of the environment is of great importance during heating, as in carburizing or nitriding, or the controlled environment (oxidizing, reducing, or neutral) inside the heat treating chamber of a furnace. Cooling may be done by immersion in a fluid (air, water, oil, molten salt, or other media) after an elapsed period of time at the holding temperature in order to attain the desired properties. Heat treating methods include hardening, annealing, stress relieving, tempering, normalizing ferrous metals, and refining grain of ferrous metals. Surface changes in steel by casehardening include such methods as carburizing, nitriding, carbonitriding, cyaniding, flame hardening, induction hardening, and siliconizing.

Sources of air pollution in heat treatment areas may be attributed to:

1. Malfunctioning of the furnace combustion system
2. Issuance of vapors and fumes because of unclean surfaces on the work
3. Oil mist or vapor generation during cooling depending on the medium being used
4. Molten salt pot fumes
5. Toxic gases emitted from furnaces under special atmospheres

During oil quenching, conditions are created which are similar to machining operations in that oil mist and vapors are released. However, these mists and vapors are released in far greater quantities than during machining, adding to the difficulty of abating this pollution. The primary source of toxic and hazardous emissions is the carbonitriding process known as "gas cyaniding." The cyanide salt decomposes into carbonates, and if these carbonates are carried into the atmosphere uncontrolled they will combine with atmospheric moisture to form toxic compounds.

There is a widely held belief that heat treat furnaces are major contributors to the emission of many toxic pollutants—cyanides, carbon monoxide, nitrogen, methane, and metallic oxides.

Effective prevention and control of emission resulting from heat treating operations may be accomplished by the following techniques:

1. Correction of operating procedures combined with selection of the proper fuel and burner, resulting in elimination of smoke.
2. Removal of organic material adhering to the work by degreasing prior to heat treatment.
3. Proper selection of the oils and control of oil bath temperature, which will materially reduce the mists and fumes generated during oil quenching.
4. Removal of fumes from molten salt pots by cloth arresters in baghouses. Baghouses may require constant duty operation because of the hygroscopic and corrosive properties of the salt fumes.
5. Use of flame curtains covering the ends of the heat treat furnace, preventing escape of the special atmosphere into the heat treating department.
6. Use of properly designed canopy hoods above the heat treat furnaces to capture effectively the convection updraft of hot air.

METAL JOINING PROCESSES

In metal joining processes, respiratory protection by means of adequate ventilation must be provided. Pollutants are created as a result of the use of the various base and coating materials in welding, cutting, and soldering operations. Actual hazard or damage usually does not occur unless an individual inhales these materials in substantial amounts over an extended period of time. The generation of toxic materials will depend upon the type of welding, the filler, the base metals, and the presence of contamination on the base metal or of volatile and reactive solvents in the air. The following list shows the contaminants for the various welding processes:

1. Shielded metal arc welding.
   a. Use of plain carbon or low-alloy steel electrodes results in emission of iron oxide and flux material fumes.
   b. Use of low-hydrogen electrodes results in emission of fluoride compound flux fumes.
   c. Use of stainless steel and high-alloy electrodes causes emission of chromate and fluoride flux fumes.
2. Submerged arc welding.
   a. Electrode coverings vary in fluoride content between 2 and 5 percent, resulting in toxic fumes emission.
3. Gas metal arc welding.
   a. Causes the generation of ozone.
   b. Causes the decomposition of trichlorethylene or perchlorethylene fumes upon contact, thus producing phosgene and hydrochloric acid.
   c. Causes carbon dioxide decomposition when it is used as a shielding gas and the resultant generation of carbon monoxide.
   d. Creates high concentrations of nitrogen dioxide when nitrogen gas is used for shielding.
   e. Metallic fumes produced by the higher temperatures.
4. Resistance welding does not usually involve significant air pollutants, depending on the amount of metal release and its toxicity.
5. Base metals may be extremely hazardous if they have been coated with toxic materials such as cadmium, lead, or mercury, or paints containing toxic materials. Beryllium welding may produce an extremely hazardous atmosphere.
6. Some silver brazing alloys contain cadmium which may be volatilized if heated appreciably above its melting point. The yellow-brown fumes are extremely dangerous, producing severe and even fatal lung damage on short exposure. The amount of cadmium fumes produced depends largely on the temperature reached. Low concentrations have been reported when using a gas-air torch, higher with oxyacetylene.
7. Fluxes containing fluorides are frequently used in silver brazing and for brazing and welding aluminum, magnesium, and their alloys.

Fume Control Measures

General ventilation Natural ventilation is usually considered adequate to prevent the accumulation of excessive amounts of fumes when welding on mild steel with electrodes which do not contain fluorides in the coating, provided that the room volume is not less than 10,000 cu ft per welder, the ceiling height is at least 16 ft, and cross ventilation is not blocked by partitions or equipment. In spaces which do not meet these requirements, general mechanical ventilation at a rate of 2,000 cfm per welder is considered adequate to prevent the excessive accumulation of fumes. The presence of fluoride compounds in the coating, or of other toxic materials, introduces more stringent requirements, depending on individual circumstances.

Portable circulating fans are sometimes desirable to dilute fumes which might otherwise tend to accumulate in the breathing zone of the welder. These may be useful in semiconfined areas or when the welder is positioned directly over the work.

Local exhaust ventilation This should be used whenever fume control as opposed to fume dilution, is required. Its use is not always practicable, but where it can be used, it provides the most efficient and economical means of fume control. It may be applied by:

1. A fixed hood with a top and at least two sides which surround the welding or cutting operation, having sufficient airflow to maintain a velocity of not less than 50 fpm away from the welder.

2. A freely movable hood with a flexible duct, intended to be positioned by the welder as close as practicable to the weld, and provided with sufficient airflow to produce a velocity of 100 fpm in the zone of welding. Airflow requirements range from 150 cfm when the hood is positioned 4 to 6 in from the weld, to 600 cfm at 10 to 12 in. This is particularly applicable for bench welding but may be used for

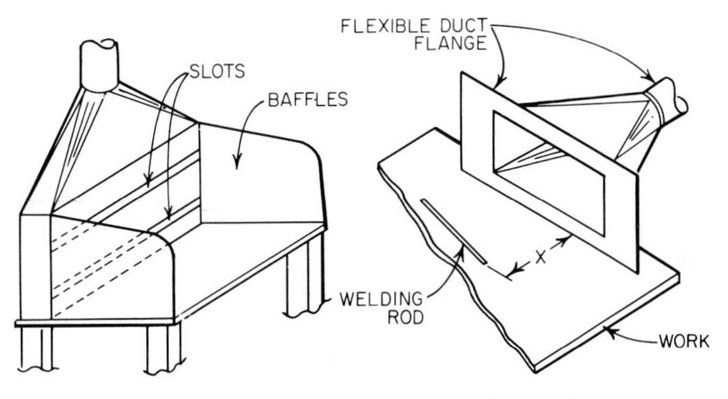

Fig. 6 Soldering and arc welding workbench design with fume collectors. (*From Industrial Ventilation.*)

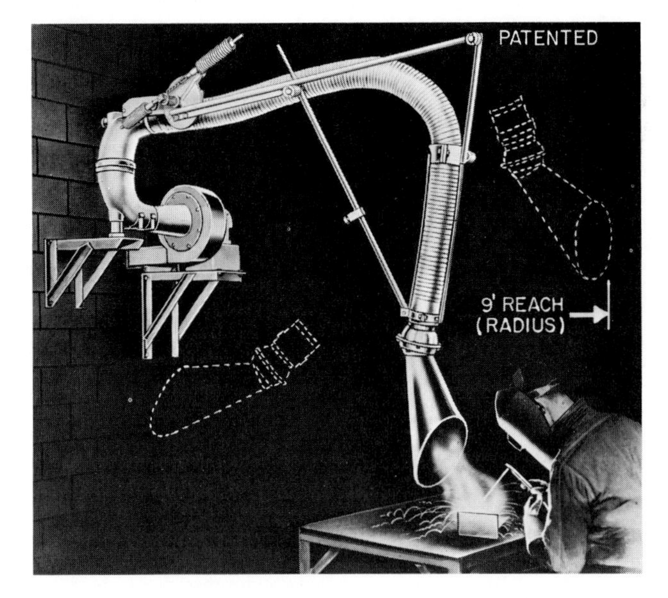

Fig. 7 Fume collector for welding station. (*From Ruemelin Manufacturing Co.*)

any welding provided the hood is moved as required. A velocity of 100 fpm will not disturb the gas shielding in gas shielded arc welding when normal gas flow rates are used, but care must be taken to see that the velocity does not exceed 100 fpm. Figures 6 and 7 illustrate local hoods which may be used for this purpose.

Respiratory protective equipment Respiratory protective equipment may be used where the use of local exhaust ventilation is not practicable or, in the case of very toxic materials, to supplement local exhaust ventilation. Air-line respirators or hose masks will give adequate respiratory protection for all types of contaminants and are generally the preferred equipment.

Air-supplied welding helmets are available commercially but have found little acceptance among welders. Filter type respirators, approved by the U.S. Bureau of Mines for metal fumes, are effective if properly used and maintained. Their general use is not recommended because of the difficulty in determining that these provisions have been carried out. They are not used for protection against vapors more toxic than lead. No filter or cartridge respirator will protect against mercury vapor, carbon monoxide, or nitrogen dioxide.

METAL FINISHING OPERATIONS

Metal finishing operations include cleaning and surface preparation, surface coating, and painting operations. Tabular data shown in Table 1 survey the airborne contaminants released by metallic surface treatment and metal cleaning operations.

Solvent degreasing is used in the metal fabricating industries as a means of cleaning metal to remove soil, usually in the form of oil or grease, prior to electroplating, painting, or other surface finishing. Nonflammable chlorinated hydrocarbon compounds most commonly used for this purpose are perchlorethylene and trichlorethylene. Choice of the solvent is dictated by the temperature required for the work. Trichlorethylene boils at 189°F and perchlorethylene boils at 249°F.

Air pollution results from emission of slightly toxic solvent vapors at the degreaser or from the carry-out. The vapors are of fairly high specific gravity, and hence hooding of the tank is usually not practiced. However, laterally slotted hoods are sometimes used to capture the small amount of solvent escaping to the atmosphere. The quantity of air used in slotted hoods is low, which precludes overloading the control device used to capture the vapors in the airstream. The control device currently used is an activated carbon adsorber which has been known to reduce solvent usage by as much as 90 percent, thus effecting an economy in usage as well as a reduction in air pollution. Recent experience in the installation of a vapor adsorption system to control solvent vapors from a degreasing operation proved to be desirable from an economic as well as employee morale viewpoint. Simply covering an open degreasing tank will significantly reduce loss of solvent and air contamination. Figure 8 illustrates concepts of solvent recovery for degreasing tanks.

Ventilating air from the degreasing operation should be kept away from welding operations to prevent thermal decomposition of the solvent vapor which results in the formation of toxic phosgene and corrosive hydrochloric acid. Strong cross drafts in the area will be responsible for a high loss of solvent due to greater evaporation under these conditions. Complete physical isolation of this operation is accordingly recommended from an economy and safety viewpoint.

SURFACE COATING

Surface coating is a continuation of the metal finishing process, starting with work which has been subjected to a metal cleaning operation and includes priming and painting.

Painting operations include dipping, spraying, flow coating, and electrostatic finishing. The coatings applied in these operations vary widely as to composition and physical properties. In spraying operations, compressed air operated spray guns

TABLE 1 Airborne Contaminants Released by Metallic Surface Treatment and Metal Cleaning Operations[a]

Process	Type	Component of bath which may be released to atmosphere	Physical and chemical nature of major atmospheric contaminant	Rate of gassing	Usual temp range, °F
Surface treatment	Anodizing aluminum	Chromic-sulfuric acids	Acid mist	Medium	95 and up
	Black magic	Conc. sol. alkaline oxidizing agents	Alkaline mist, steam	High	260–350
	Bonderizing[b]	Boiling water	Steam	Medium-high	140–212
	Chemical coloring	None	None	Nil	70–90
	Descaling[c]	Nitric-sulfuric, hydrofluoric acids	Acid mist, hydrogen fluoride gas, steam	Medium-high	70–150
	Dulite	Conc. sol. alkaline oxidizing agents	Alkaline mist, steam	High	260–350
	Ebonol	Conc. sol. alkaline oxidizing agents	Alkaline mist, steam	High	260–350
	Galvanic-anodize[d]	Ammonium hydroxide	Ammonia gas, steam	Low	140
	Hard coating aluminum	Chromic-sulfuric acids	Acid mist	High	120–180
	Jetel	Conc. sol. alkaline oxidizing agents	Alkaline mist, steam	High	260–350
	Magcote[e]	Sodium hydroxide	Alkaline mist, steam	Low-medium	105–212
	Magnesium predye dip	Ammonium hydroxide-ammonium acetate	Ammonia gas, steam	Low	70–180
	Parkerizing[b]	Boiling water	Steam	Medium-high	140–212
	Zincate immersion[f]	None	None	Nil	70–90
Metal cleaning	Alkaline cleaning[g]	Alkaline sodium salts	Alkaline mist, steam	Medium-high	160–210
	Degreasing	Trichlorethylene-perchlorethylene	Trichlorethylene-perchlorethylene vapors	High	188–250
	Emulsion cleaning	Petroleum-coal tar solvents	Petroleum-coal tar vapors	Low-medium	70–140
	Emulsion cleaning	Chlorinated hydrocarbons	Chlorinated hydrocarbon vapors	Medium	70–140

[a] New York State Department of Labor, Division of Industrial Hygiene and Safety Standards.

[b] Also aluminum seal, magnesium seal, magnesium dye set, dyeing anodized magnesium, magnesium alkaline dichromate soak, coloring anodized aluminum.

[c] Stainless steel before electropolishing.

[d] On magnesium.

[e] Also Manodyz, Dow-12.

[f] On aluminum.

[g] Soak and electrocleaning.

are used. A spray booth enclosure, ventilated by a fan, provides a means of preventing explosive concentrations of solvent vapor and creates an atmosphere acceptable from a health viewpoint.

Flow coating machines automatically subject the conveyor supported work to a flood of paint. The excess paint drains to a basin from which it is recirculated to the flow nozzle. This process is used principally for prime coats or protective finishes and minimizes the amount of organic solvent vapors discharged to the atmosphere.

In dipping, excess paint is drained back into the paint reservoir by use of drain boards sloped to return the drippage or by suspension above paint reservoirs after immersion.

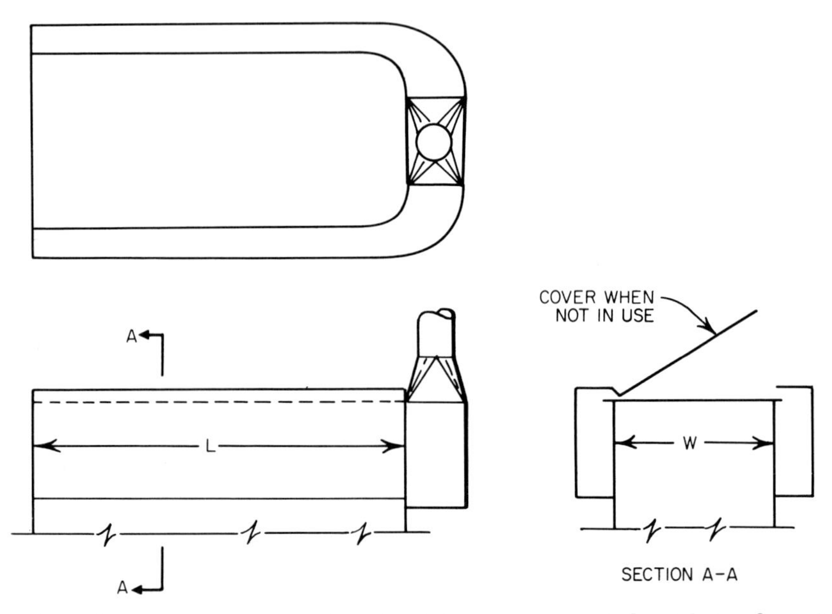

Fig. 8 Solvent degreasing tanks fume hood design. (*From Industrial Ventilation.*)

Paint roller machines operate in a manner similar to printing presses. One roll is immersed in paint and transfers paint by contact with a second parallel roller. Work passes between the second and a third roller.

Air contaminants from paint dipping, flow coating, and roller coating exist only in the form of organic solvent vapors since no particulate is formed. These paint systems frequently operate without ventilation hoods. A canopy hood may be installed when local ventilation is desirable.

Electrostatic painting uses atomized coatings which are impressed with an electrostatic charge similar to the atomizer but opposite to the work on the conveyor to create the conditions for electrostatic deposition resulting in the formation of a homogeneous integrated coating. There is practically no overspray in this method.

Contaminants from paint spraying operations consist of the organic solvent vapors as well as the particulate matter. In water wash spray booths, considerable overspray is a result of the inherent nature of the spray operation plus the necessity of adequate coverage of the work. Amount of overspray varies from 30 to 90 percent. Particulates in paint spray booths are controllable by baffle plates, filter pads, or water spray curtains. Baffle plates are effective in removing enamel spray particulates but are very inefficient in collecting lacquer spray particulates because of rapid drying and consequent slight adhesion to the baffles. Water spray curtains trap the overspray and separate the paint and water into three layers. The light floating layer consists of the vehicle and solvent. The heavy lower layer is pigment, and the middle layer is mainly water which is recirculated. Particulates trapped in water curtains must be treated as a waterborne industrial waste material.

In the dry type booths, the airborne particulate must be removed by filtration to prevent committing nuisance due to paint spotting in the neighborhood. Spray booth equipment manufacturers have enlarged their scope of work to provide packaged self-contained equipment which includes makeup of air supply unit, air filtering and humidification equipment, and all necessary fans as an adjunct to the paint spray booth. Fans used for spray booth ventilation are built with explosionproof motors and sparkproof or coated wheels.

The organic solvents used in paint formulation are eventually evaporated to the atmosphere. The solvent concentration in the spray booth varies from 100 to 200 ppm. These solvents are not controllable by filters, baffles, or water curtains. They are subject to photochemical reactions which create smog and cause eye irritation. Adsorption by use of activated carbon is the only feasible method for solvent control provided the particulates are removed from the contaminated airstream prior to adsorption. The activated carbon adsorption media may be either fixed or movable. The fixed bed may be housed in either a vertical or a horizontal cylindrical vessel, or the media beds may be placed in a multilayer arrangement. Adsorption units may be arranged to suit service conditions so that regeneration may be accomplished as required.

The average velocity across the face of the spray booth is not less than 100 fpm, which is adequate for positive movement of the spray particles toward the exhaust fan. Water curtains and sprays and filters remove paint particulates with efficiencies up to 95 percent.

Paint mix rooms are used for paint formulation, and usually have laboratory facilities for routine operations required for quality control of the paint to be applied. These areas receive special treatment with respect to ventilation and fire protection because of the organic solvents used. Ventilation of the space recognizes the heavier than air characteristic of the solvent by supplying air at a high level and exhausting air through a system of ductwork which begins near the floor in order to pick up these heavy vapors. The hazardous nature of the solvents used makes it necessary to use a fire protection system, such as carbon dioxide, to supplement automatic sprinklers. Generally the concentration of solvent vapors in the exhaust air is low enough, because of the need for large-scale general ventilation, to make scrubbing unnecessary.

PAINT BAKING

After the surface coating has been applied by some method of painting, the work is subjected to a drying and hardening process in an oven, resulting in the evaporation of organic solvents. The air pollution and safety aspects of this process must be viewed in the light of the following:

1. The lower explosive limit (LEL) of the organic solvents vaporized must not be exceeded. The lower explosive limits relate to the allowable concentration of vapors within an enclosure and are usually much higher than the threshold limit values. A factor of safety, varying from 4 to 12 and higher, is applied to the LEL in the calculations made to determine the amount of dilution ventilation required.

2. During loading and unloading of the oven, the atmosphere within the oven must be below the level of toxicity.

3. The oven must be free of smoke and products of incomplete combustion.

4. Aerosols produced from the partial oxidation and polymerization of the organic solvents and resins are obnoxious from the standpoint of odor nuisance and eye irritation.

The standards of the fire underwriters for paint baking oven installations must be observed with respect to the lower explosive limits and dilution ventilation requirements to attain these limits.

Direct fired vapor combustion devices operated at temperatures of 1400°F or higher can be used for control of the organic solvent vapors, odors, and aerosols emitted from paint baking ovens.

Frequently at some substantial cost savings catalytic combustion systems are used to oxidize the paint fumes completely. Figure 9 illustrates a typical system. The paint solvent vapors are drawn off into the catalytic combustion equipment, where they are oxidized. The gases are mixed with recirculating air from the bake oven and makeup air for temperature control. The advantage of this system is not only that the pollutants are eliminated but that their combustion provides a supplementary heat source for the oven. Generally, this heat recovery can be economically justified if solvent vaporization exceeds 10 gal/hr. However, in special applications

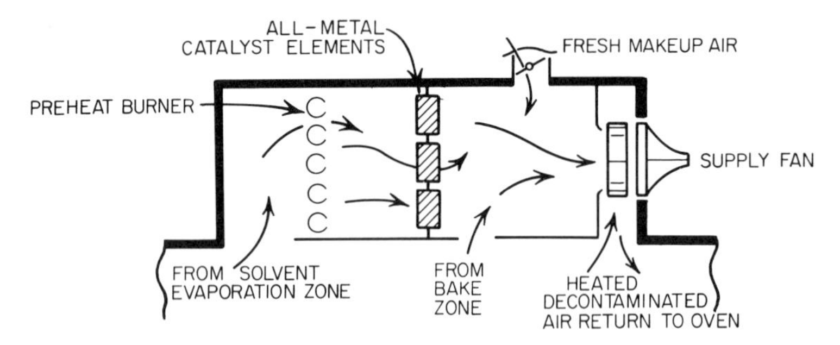

Fig. 9 Typical catalytic combustion system for paint bake oven. (*From UOP Air Correction Div.*)

such as in wire enameling ovens, even smaller rates of solvent release can justify this method. In some cases, the oven size is so small that the heat release from the solvent combustion is sufficient to be self-sustaining for the oven.

PAINT STRIPPING

Paint stripping is used to remove accumulations of paint solids on hooks, hangers, masks, and other supports for work to be spray painted. Stripping separates these accumulated solids so that the support may continue to be used. There are several methods in use today in industry: mechanical, chemical, or molten salts. Mechanically, stripping is accomplished by sanding, sandblasting, chipping, or scraping. Chemical or solvent stripping uses either a hot or a cold method. Cold strippers may be further classified by material used into:

1. Organic solvents
2. Emulsion type
3. Acid type
4. Combination of the other types

The molten salt stripper, operated at 850°F, is an example of hot stripping. Gas fired immersion burners equipped with an automatic flame protection system and automatic ignition are used for tank heating. Immersion in the salt bath is followed by a hot water rinse. The cleaning and rinsing may be a batch or conveyorized operation.

The mist and fumes generated in this operation are removed in an air scrubber. The operation is usually isolated in a separate building or separate room in the plant service area. Water passing through the scrubbing medium is treated in the process waste treatment plant.

AIR POLLUTION CONTROL EQUIPMENT COST

Air pollution control equipment cost, especially on an installed basis, is difficult to estimate. In general, costs will be greater if the dust collection installation is constructed as an alteration in the existing plant as opposed to initial construction in a plant expansion program or as a service in a newly constructed plant. Cost determination and cost comparisons require a careful analysis and evaluation of the components and operational features offered.

Some emission control devices include exhaust fan, motor drive, and starter. In other designs, these items and their supporting structure must be secured from other sources by the purchaser. Likewise, while storage hoppers are integral parts of some devices, they are not provided in other types. Duct connections between parts of the equipment may be included or omitted. Recirculating water pumps or settling tanks may be required and may not be included in the price.

Efficiency of operation must be so stated that fair comparison can be made

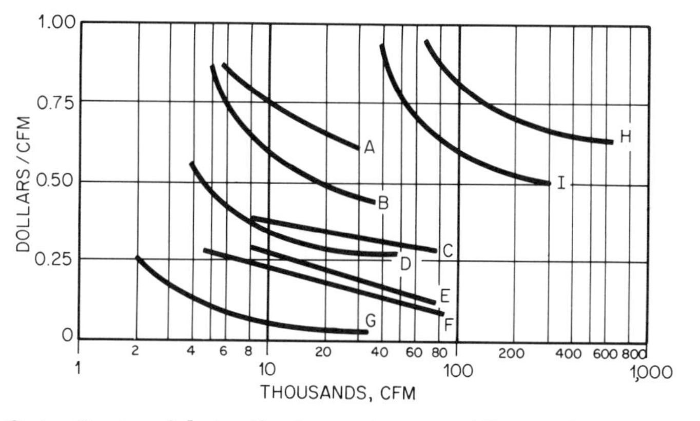

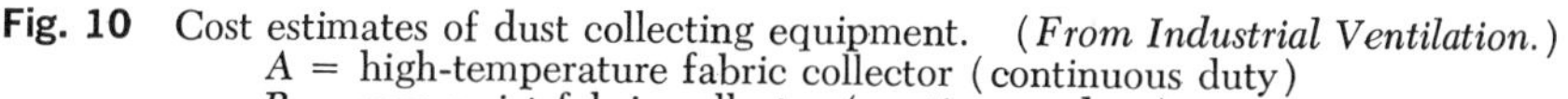

Fig. 10 Cost estimates of dust collecting equipment. (*From Industrial Ventilation.*)

A = high-temperature fabric collector (continuous duty)
B = reverse jet fabric collector (continuous duty)
C = wet collector (maximum cost range)
D = intermittent duty fabric collector
E = high-efficiency centrifugal collector
F = wet collector (minimum cost range)
G = low-pressure drop cyclone (maximum cost range)
H = high-voltage precipitators
I = high-voltage precipitators (minimum cost range)

Cost based on collector section only. Cost does not include ducting, water requirement, power requirement, or exhausters (unless exhaust is integral part of secondary air circuit).

Cost of continuous duty sectional fabric collector approaches cost of reverse jet continuous duty collector.

Price of electrostatic precipitators will vary with the contact time and the electrical equipment required. Prices shown are for fly ash installations when high velocities of 300 to 600 fpm are usual. Precipitators for metallurgical fumes, etc., will be considerably higher in cost per cfm.

between competing equipment. An actual or an assumed basis of inlet loading imposed on the control device must be stated as well as the tolerable outlet loading. The services of a consulting engineer with experience in preparing emission control equipment specifications are recommended so that the purchaser's interest in installing properly functioning equipment is protected. Information available from public regulatory bodies concerned with pollution abatement should be used so that equipment design features from other sources are incorporated in the equipment.

A higher than usual cost may be due to a decision to install a pollution abatement facility which will be adequate to accommodate future plant growth, thus introducing an additional cost charge above that which would have been required for the original installation only.

The cost of installation of equipment can equal or exceed the cost of the manufacture of the equipment, depending on the degree of shop assembly, such as completely assembled, partly assembled, or completely knocked down, depending on the location, which may require expensive rigging to put in place, and depending on building alterations required such as supporting steel, access platforms, and roof or wall openings. Cost can be measurably influenced by the need for water and drain supply, special or extensive electrical work, and expensive material handling equipment for collected material disposal. The per unit cost of collected material disposal will vary inversely as the volume of material handled.

Costs are usually based on an assumption of standard or basic construction. Increase in cost for corrosion resisting material, special high-temperature fabrics, and insulation and weather protection for outdoor installation can introduce a multiplier from one to four times the standard construction cost.

An illustration of the costs, expressed in dollars per cubic foot per minute plotted against cubic feet per minute of air, is shown in Fig. 10. These data are for new installation and include only those items shown in the notes.

A recent survey reveals cost data showing the average total annual equipment operating costs for air pollution control equipment are made up of the following cost items: power, fuel, water, replacement parts and materials, maintenance labor, and collected waste disposal. Of these, collected waste disposal is the largest single item of cost. Obviously, there is a need for salvage of collected emission products to reduce this inordinately high cost.

WASTE WATER TREATMENT IN METAL FABRICATING PLANTS

Generally the waste waters originating in metal fabricating plants are similar regardless of whether they originate from small shops or large production facilities. For the most part the waste waters from such plants contain small amounts of metal particles, free and soluble oils, and various cleaning compounds used in cleaning either the product or the shop itself. Some plants have waste waters from air pollution control devices for painting operations. These residues are treated in a manner similar to free oils. Normally, the wastes are slightly acid or alkaline, and they are usually opaque, milk-colored, and contain some free (nonsoluble) oil.

The simplest form of treating such wastes is by means of gravity skimming tanks. In many cases such skimming will be sufficient to permit discharge of the industrial waste to a sewer. Optimum gravity skimming tanks are generally designed along standards set forth by the American Petroleum Institute which relate tank dimensions to particle size, oil rise rates, and detention time as well as to the nature of the type of oil to be skimmed. Such tanks operate best when supplied with a limited, uniform influent flow rate. As a result it is often wise to optimize the skimming tank design by the API Standards and then utilize large holding or equalizing tanks upstream of the actual skimming tanks to permit pumps to deliver the waste water to the gravity skimmers at a fixed design rate. The gravity skimming will remove free oils. The metal particles will settle in the bottom of the tank and can be removed manually at infrequent intervals, or automatically by means of drag conveyors if such metal particles are deposited in significant quantities.

If the metal fabricating plant waste water contains considerable soluble oil, generally used as a machining coolant, further treatment will be required in addition to gravity skimming. Usually waste waters from a metal fabricating plant will be contaminated with soluble oils, since even if soluble oils are not used in plant processes, they may well develop because of the mixing of free oils and metal cleaners or emulsifying agents when brought together in the waste water collection system.

Batch Treatment

Under these conditions a basic decision must be made when considering such waste treatment: whether or not the qualities and quantities of the waste water are such that a batch system is in order or whether a constant flow system utilizing continuous skimming and other oil removal methods would be preferred. Generally speaking, small quantities of less than 2,000 gpd can best be handled by batch type systems, with larger quantities being treated by a continuous system. However, in the event there is a possibility of toxic contaminants such as plating waste waters, a batch treatment system becomes essential. A batch type system more readily permits testing and additional treatment to be applied to the waste waters if this

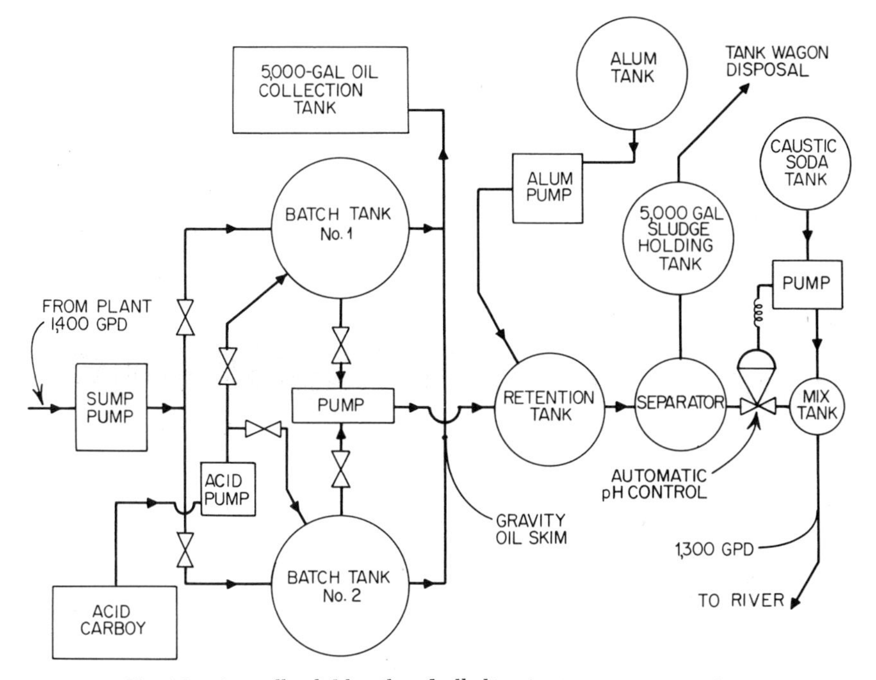

Fig. 11 A small soluble oil and alkaline waste treatment system.

becomes necessary. Certain cleaning compounds can mix with the soluble oil waste and convert it into a jellylike material which will defy treatment by usual methods. Such complications can be accommodated only by a batch system. These treatment systems consist of parallel batch collection and skimming tanks such as those indicated in Fig. 11. When sufficient waste waters have been accumulated in one batch tank, the free oil is skimmed and conveyed to a waste oil collection tank. Miscellaneous grit and metal particles can be allowed to accumulate in the batch tanks until manual cleaning is required. The remaining waste waters containing soluble oils are then cracked in either the batch tank or a separate retention tank to break the oil emulsion. This is usually done by means of adding acid or alum until the pH has been lowered sufficiently to cause the emulsion to break down. The free oil is then skimmed or decanted in a separator and the remaining waters are neutralized by means of caustic soda or lime. If sodium hydroxide (caustic soda) is utilized the clarified waste water may be high in dissolved solids consisting mainly of the soluble sodium salts resulting from neutralization. Frequently such waste cannot be discharged directly to a stream, but it is usually acceptable in municipal sewer systems.

If lime is used for neutralization the waste waters will contain few dissolved solids because of the relative insolubility of the resulting calcium salts. However, neutralization with lime will precipitate a sludge consisting of the calcium salts of the acid. This presents the problem of disposal of the resulting sludge. As shown in Fig. 11 this sludge can be collected and stored in a holding tank for ultimate disposal. Not infrequently it is wise to dry this sludge on vacuum filters, since with its high water content it can present a difficult problem with respect to storage and/or disposal of the precipitate. Removal of most of the water will reduce the volume of the precipitate or sludge to a more easily handled quantity. While lime neutralization of acid waste waters is less expensive than caustic soda neutralization, the cost of sludge handling or drying may far outweigh the savings achieved by the use of lime. The volume and concentrations of waste waters must be known and the resulting sludge volumes calculated before a comparative economic study can be made.

Several new proprietary compounds have recently been marketed which will crack oil emulsions. Such chemicals usually will not require subsequent neutralization with possible resulting sludge complications.

Continuous Treatment

Perhaps the most successful of all metal fabricating plant continuous waste treatment systems are those involving air flotation. Such a system is indicated by Fig. 12. In this process the soluble oil contaminated waste waters are collected in large hold or equalizing tanks. This permits one-shift operation of the air flotation unit at fixed input rates. Operation for three shifts would permit smaller hold tanks to be used but might not provide the quantitative or qualitative equalizing of the waste waters. After collection the waste waters are conveyed by transfer pumps to the

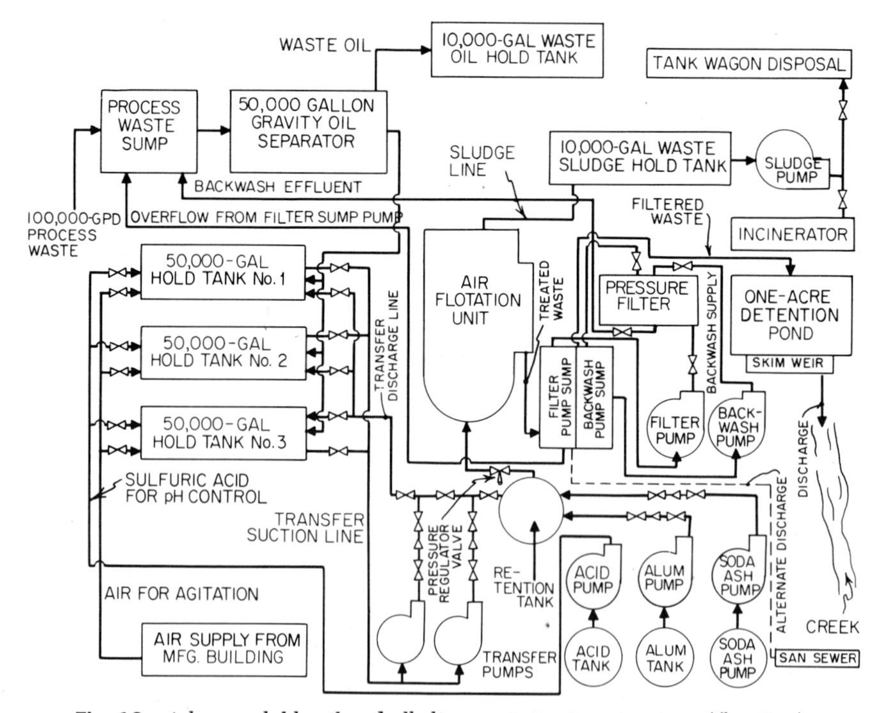

Fig. 12 A large soluble oil and alkaline waste treatment system (flotation).

retention tank. Acid, alkali, and soda ash are added as indicated, and the waste is detained in the retention tank under air pressure long enough to absorb quantities of air. Upon release through a pressure regulator valve into the air flotation unit the cracked soluble oil is floated to the top surface along with the minute floc quantities and swept into a sludge hold tank by scrapers or flight conveyors. The relatively clear effluent can then be disposed of as indicated and the heavy oil filled floc disposed of by incineration or tank wagon disposal methods. As shown in Fig. 12, the effluent may require further filtration or sedimentation in detention ponds to meet stream requirements. The less stringent requirements for discharges to sanitary sewers will usually obviate needs for filters or detention ponds.

Incineration of resulting precipitates or sludge is an excellent method of disposing of this waste product. Incinerators are available which develop temperatures high enough to burn the residues completely and which will present few air pollution hazards. In the event toxic material such as the metal salts of plating baths is in the sludge, this method is most desirable.

The oil content of the sludge fed into such a device is generally high enough to be self-sustaining in combustion: however, it is always necessary to provide supplementary fuel such as gas or oil to permit the incinerator to be heated initially to a high enough temperature to destroy the waste product when it is first introduced into the incinerator. If the water content of the sludge is high or the oil content is low, continuous use of supplementary fuel may be required.

The probable fuel costs for disposal of sludges and similar wastes by means of incineration range from $0.02 to $0.04 per gal. However, a saving in total operating cost can be achieved if the treated waste water which was separated from the sludge is reused in the plant for nonpotable uses such as in toilet rooms and process water. It must be kept in mind, though, when considering such uses, that trace amounts of contaminants may still remain in the effluent.

The other means of continuously removing contaminants from metal fabricating plant waste water is by means of automating a batch treatment system, e.g., settling out the metal particles, skimming the free oils, chemically cracking the soluble oils and again skimming, neutralizing the now oil free waters using caustic soda or lime as discussed previously, and dumping the clarified waters into a sewer or stream. This amounts to a continuous flowing, instrumented version of the batch system discussed before and shown in Fig. 10. This would require pH sensing devices in the batch tanks to regulate the acid pump. Level switches in the batch tanks would be needed to activate the pump to the retention tank and the alum pump. Also, automatic monitoring and recording of the final effluent properties would be required. There are many points in this type of system which must be instrumented to achieve the required automatic control using pH and level sensing devices, and it may therefore be difficult to achieve reliability. The clinging of residual oils and/or sludge to the electrodes of pH sensing devices and the possibilities of unexplained materials finding their way into the system and fouling the chemical treatment and sensing devices almost preclude a trouble free automatic system.

As a result, for most plants the most practical approach to an automatic system is the previously mentioned air flotation system. Several waste water treatment equipment manufacturers construct such devices, and in general most of them are satisfactory. However, care must be taken not to overload an air flotation machine. A slight overload of the machine will cause substantial reductions in the pollutant removal efficiencies. A 10 percent hydraulic overload may cause as much as 50 percent reduction in oil removal efficiencies and result in carryover of oil bearing floc with the clarified effluent. All automatic or batch systems must be examined to be sure that valves, piping materials, tanks, coatings, and linings are suitable for the oils, pH variations, and chemicals used in the waste treatment process. Such considerations will be important to the cost of the facilities.

System Sizing

As in all industrial waste treatment problems, the question of anticipating waste quantities and strengths in conjunction with the design of a new metal fabricating waste water treatment facility is much more difficult than measuring the known quantities or properties of a waste that occurs in existing facilities. As a result, the waste water treatment facility for a new proposed metal fabricating plant must be designed much more conservatively than one for an existing plant in which the effluent quantities and properties can be measured.

A possible solution, if it can be achieved, is to go into production on a new facility and, once waste quantities are determined, design the waste water treatment equipment. This takes cooperation on the part of state and regulatory agencies and may require tank wagon disposal or scavenger service for the first few months of operation.

Since waste water treatment is usually an overhead cost, all possible means to defray expenses must be examined. In the case of metal fabricating plant waste waters, several possibilities exist. The recovered waste oil may be reused or at least sold for road oil, and the treated waste water may be reused as previously discussed to satisfy plant process, cooling, or other nonpotable water demands.

Carried to the ultimate end, such reuse may virtually eliminate plant waste water discharges, thereby negating the involvement with waste water regulatory agencies. As a result, no waste treatment operational records for agency review are required, no permits must be obtained, and in some states, no licensed waste treatment plant operator need be hired.

SOURCE CONTROL

Control of pollution sources strives for the elimination of potential air contaminants contained in the raw material. Source control means the revision of a process or the substitution of another ingredient for a deleterious material. An example of this would be the substitution of chlorinated solvents of less toxicity for carbon tetrachloride, or the removal of most of the sulfur from crude petroleum to prevent the emission of contaminants to the atmosphere.

Source control entails the search for substitution in process components, chemicals, or procedures to solve a problem. In metal grinding operations, grinding wheels of natural sandstone were used prior to the development of synthetic abrasive materials such as silicon carbide (carborundum) and fused alumina (alundum). These materials are not only superior for grinding, but more importantly, they do not contribute to the incidence of the lung disease silicosis. The substitution of synthetic abrasives is an example of source control. This substitution, plus adoption of standards for grinding wheel hoods and exhaust systems to collect grindings. brought this problem under control. Lowering oven temperature markedly diminished the incidence of eye irritation observed and also improved the quality of the product. The search for optimum temperature for oven operation is an example of controlling pollutants at the source.

In spray painting operations in the metal fabricating plant, organic solvents are introduced first as a paint ingredient. During spraying this same solvent is introduced into the atmosphere. The solvent is finally captured in an emission control device such as an afterburner to prevent atmospheric contamination. This is an example of three different steps involving the use of a material requiring special consideration. The development of painting methods based on electrodeposition principles eliminates the use of organic solvents and promotes emission control through source control. This underlying principle should serve as a guide to indicate a possible solution to a contamination problem by applying technology to seek substitutions which offer a solution to the problem.

PLANT LAYOUT AND THE CONTAMINATION CONTROL PROBLEM

The plant layout and arrangement of processes and manufacturing sequences in metal fabricating plants must include consideration during the planning stage of the problem of atmospheric and waste water contamination emitted as a result of the plant operation. This applies to the construction of a new plant as well as the alteration or expansion of an existing plant. Machinery or operations having a common waste product are amenable to a centralized system of contaminant collection and disposal. There are advantages to a machinery layout in which smaller operations are grouped, such as the convenience in supplying the utilities and services.

When a centralized contamination control layout is adopted, allowances for future requirements, such as capped studs for future connections and sizing of dust collection headers to accommodate the ultimate requirements, can readily be made. In the absence of a centralized contamination control system, unit emission control devices will be required, resulting in greater equipment and installation costs and inferior pollution control results, at higher costs.

The practice of isolating equipment within an enclosure on a waste emission basis will minimize the size of the area affected and will reduce the amount of diluent ventilation required as well as the size of pollution control device. Thus, if a lubricant in a metal forming or machining operation must be applied as a spray to be effective, and hooding of the machine to capture the oil spray locally is impractical, then the area enclosure is the only method which will prevent widespread dissemination of the emitted contaminant.

The case history of a small parts supplier illustrates how the centralization of all dust producing machinery to facilitate dust collection and the control of plant air quality can become a production asset. Nine dust generating machines, originally spotted in three different locations, were repositioned around an abatement system. This was an improvement over the ineffective individual blower attachment on each machine provided to collect particulate matter and deposit it in holding boxes. The dust, created by the machining of commutators and collector rings, and consisting of particles of mica, copper, and plastic, was responsible for undue wear of motors, tools, and machinery. The installation of centralized collection around the dust collector avoided the expense of ductwork and the expense of isolation of the operation by partitions to prevent the spread of the dust, and resulted in a reduction of manpower required when tool adjustments were necessary. The abatement equipment consisted of a cyclone separator to remove the large particles supplemented by a small afterfilter to remove submicroscopic particles. The dust laden air was returned to the plant after filtration.

The chemical reaction relationship between emissions from adjacent operations is a matter of concern to the plant layout. The interaction of gas metal arc welding and vapor degreasing using trichlorethylene and perchlorethylene has already been described. Vapor degreasers using these solvents should not be located in the same room near the welding operations unless there is a positive air movement from the welding area toward the degreasers. Here again, isolation of the degreasing operation within its own enclosure subject to an atmospheric pressure lower than the remainder of the plant is advisable.

In the selection of manufacturing equipment for the plant, special attention must be directed toward the equipment specifications by staff technicians or consultants so that any pollution abatement accessory acquired at that time will be the most effective available. Without this emphasis and orientation, the pollution abatement equipment may fail to attain the objectives desired.

A summary should be made showing atmospheric contamination data describing existing conditions or conditions to be imposed on the pollution abatement equipment to be used. These data should set forth the following criteria:

1. Processes creating contaminants
2. Qualitative properties of contaminants
3. Quantitative properties of contaminants
   a. Gradation of loadings (dust)
   b. Concentration of airborne wastes
4. Toxic properties
5. Possibility of process alteration as a means of source control
6. Possibility of marketable by-products due to material recovery
7. Regulatory agencies concerned with this problem; their standards and their potential for assistance in solving problems

8. Possibility of incorporation of heat reclamation equipment to economize on cost of equipment
9. Education of purchasing and operating personnel toward cognizance of a pollution-abatement orientation

SUMMARY

Fundamental to any plant layout and planning considerations is the attitude of management, which must be predicated on the concept that the resolution of the plant pollution problem is an inescapable concomitant of the plant operation. The solution of these problems is equally as important as the supplying of the utilities and services required for manufacturing. The significance and the priority assigned to problems of atmospheric and industrial pollution control must be elevated to the status of the more directly related production factors such as good housekeeping, enforcement of proper safety practices, efficient material flow, production processes, and machine operations. Management's thinking and planning must acquire a "pollution-oriented" attitude so that the short-term goal of control and abatement of presently contaminated air and water and the long-range objective of abatement of atmospheric pollution will be realized for the good of the plant, the industry, and the society it services.

Chapter 14

Pollution Control in the Chemical Industry

EMMET F. SPENCER, JR.

Pollution Control Engineer, FMC Corporation, Chemical Divisions, New York, New York

INTRODUCTION

The chemical industry is characterized by its great diversity in chemical products, processes, and wastes. The number of chemicals commercially produced reaches into thousands, each having its particular variants in the manufacturing process.

Since World War II the chemical industry has become so diversified that it is difficult even to classify it accurately. Today many chemicals are produced by petroleum producers and paper companies, and even manufacturers of electrical machinery and farm equipment are chemical producers.

To further complicate the pollution control aspects of the chemical industry, the industry is constantly changing, using different raw materials and new processes, and producing new products.

In many cases the process and its wastes are unique and pollution abatement systems must be individually engineered.

However, many processes are free of wastes. A survey by the Manufacturing Chemists' Association in 1962 revealed that 33 percent of the nearly 8,300 processes in use at 879 reporting plants had no routine discharges of gases, mists, or dusts to the atmosphere. Similarly, many plants reported no waterborne wastes.

AIR POLLUTION: GENERAL CONTROL METHODS

The air pollution control measures applied in chemical manufacturing are for the most part the everyday application of chemical engineering unit operations using mechanical collectors, bag filters, electrostatic precipitators, wet collectors, catalytic oxidation or reduction units, and direct flame incinerators or adsorbers, alone or in combination.

The selection of the control device for the particular process is a function of the degree of control necessary, the nature of the effluent, the capital and operating costs of the control equipment, the effect of the control system on the process itself, reliability, and the ultimate disposal of the collected waste material.

In addition, process or equipment modifications may be specifically incorporated to eliminate or minimize the emission of air pollutants. This is the preferred procedure and is particularly applicable to new plants or processes. Thus the control device alone may not fully indicate the attention given to air pollution control.

There is a small but growing body of technical literature about industrial emissions to the atmosphere and their control. Since 1962, the Manufacturing Chemists' Asso-

Fig. 1 Overall view of a chemical plant.

ciation has been engaged in a cooperative study with the U.S. Public Health Service on atmospheric emissions from selected chemical manufacturing processes and the performance of devices used to control these emissions. Two reports have been published from this work, one concerning sulfuric acid and the other nitric acid. Similar studies are currently in progress on both wet and furnace process phosphoric acid, hydrochloric acid, and chlorine–caustic soda production.

MINERAL ACIDS

Sulfuric Acid

Of all the chemicals produced in the United States sulfuric acid is probably produced in the greatest quantity. Estimated 1967 production in the United States is 30 million tons.

All sulfuric acid is made by either the chamber or the contact process. Over a long period of years the number of contact plants has increased, while the number of chamber plants has decreased. A 1963 survey showed 163 establishments using the contact process and 60 using the chamber process.

Chamber plants The primary source of emissions in this process is the final Gay-Lussac tower. Emissions include oxides of nitrogen, sulfur dioxide, and sulfuric acid mist.

Sulfur dioxide concentrations in these exit gases range from about 0.1 to 0.2 volume percent. Oxides of nitrogen concentrations are in the same range. About 50 to 60 percent of the total nitrogen oxides is nitrogen dioxide, which is responsible for the characteristic reddish-brown color of the exit gas.

Sulfuric acid mist in the exit gas varies from 5 to 30 mg/cu ft. The sulfuric acid mist contains about 10 percent dissolved nitrogen oxides. Greater than 90 percent of the acid mist particles are larger than 3 microns in diameter. Loss of acid mist in the exit gas varies with the design of the Gay-Lussac tower. These losses are usually much less than 0.1 percent of the acid produced.

Recovery equipment is seldom used for pollutants in the exit gases from the final Gay-Lussac tower. In one known instance, however, water scrubbing the exit gases reduced the sulfur dioxide by 40 percent and the oxides of nitrogen by 25 percent.

Contact plants The major source of emissions from contact sulfuric acid plants is the absorber exit gases. This gas stream is predominantly nitrogen and oxygen but also contains unreacted sulfur dioxide, unabsorbed sulfur trioxide, and sulfuric acid mist and spray. When the waste gas reaches the atmosphere, sulfur trioxide is hydrated with atmospheric moisture to sulfuric acid mist and forms a visible white plume with a bluish cast.

The major emission is sulfur dioxide from the absorber exit stack. The catalytic conversion of sulfur dioxide to sulfur trioxide is never complete, and some sulfur dioxide is discharged. Most contact plants are designed with a conversion efficiency of 96 to 98 percent. The mean conversion efficiency for 31 tests at typical contact plants is 97.3 percent.

The concentration of sulfur dioxide in the exit gas is a function of the initial sulfur dioxide concentration of the gases entering the converter and the efficiency of the catalyst. The sulfur dioxide concentrations in the absorber discharge gases range from 0.13 to 0.54 percent. The mean for 33 tests is 0.26 percent sulfur dioxide (Fig. 2).

Equipment to remove these relatively low concentrations of sulfur dioxide released is generally not economically feasible, and it is generally adequate to vent the exit gases through a tall stack so that diffusion into the atmosphere can dilute them to an acceptable level.

In Germany a contact plant process has recently been developed which reportedly reduces the exit concentration of sulfur dioxide to 0.01 to 0.03 percent. The process consists of the addition of an intermediate absorption tower and heat exchanger just ahead of the final stage of conversion. Removal of sulfur trioxide at this point results in a reported overall conversion efficiency of 99.7 percent. Twenty-six such double contact plants are now in operation in Europe.

Sulfuric acid mists result from the presence of water vapor in the process gases fed to the converter. The drying towers of most contact plants are able to dry the air or sulfur dioxide gas to a moisture content of about 3 mg/scf. The remaining moisture combines with sulfur trioxide after the converter, when the temperature falls below the dew point of sulfur trioxide. The acid mist so formed is very difficult to remove in the absorber, and most of it passes to the atmosphere.

Mists may also be formed from water produced by the combustion of organic impurities in the sulfur. Dark sulfur may contain up to 0.5 percent organic matter and may generate more mist from the combustion of the organic matter than from moisture in the dried air.

Even with no moisture present in the inlet gases, it is possible to form mists in the absorbing tower because of excessive gas velocity or shock cooling.

Acid mist content of the absorber discharge gas in 33 tests varied from 1.1 to 48.8 mg/scf of stack gas with an average value of 12.9 mg.

The concentration of sulfuric acid mist can be reduced by the use of electrostatic precipitators, glass fiber filters, wire mesh filters, and Teflon packed gas cleaners. When designed for high performance, glass fiber eliminators and electrostatic precipitators show a high degree of recovery, with efficiencies ranging from 92 to 99.9 percent. Two-stage wire mesh eliminators show good results, 92 percent recovery, when the acid mist particles are predominantly larger than 3 microns diameter but substantially lower results when the proportion of small particles is greater.

The reported efficiency of the Teflon packed gas cleaner is 98.9 to 99.6 percent with 60 percent of the mist less than 3 microns in diameter. Pressure drop through the collector was reported to be 8 to 10 in. W.G.

Stack impingement may effectively reduce the acid mist content. Condensation and agglomeration within a 250-ft stack reduced the mist concentration by 91 percent, resulting in an exit concentration of 2.5 mg/scf.

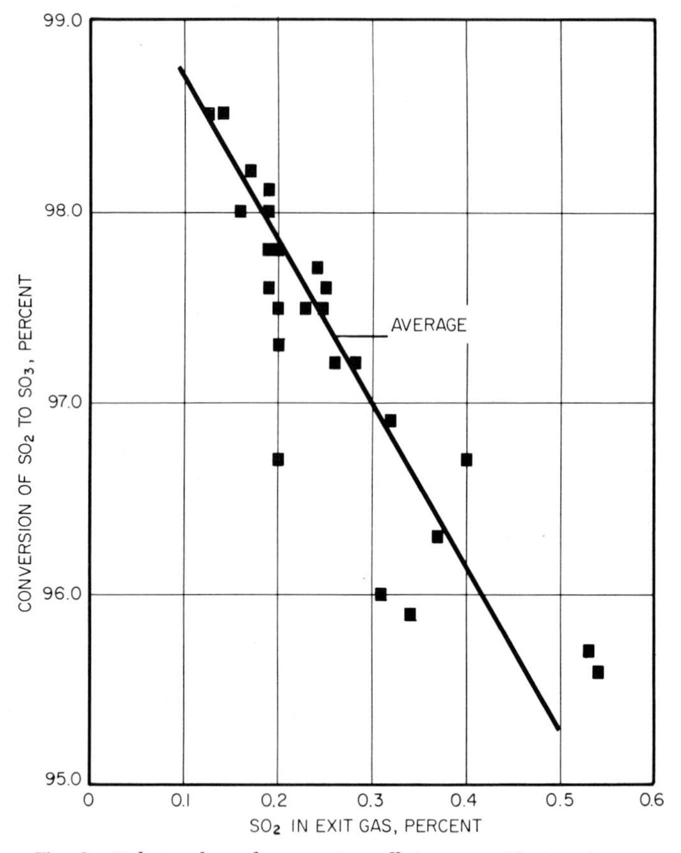

Fig. 2 Relationship of conversion efficiency to SO_2 in exit gas.

Nitric Acid

Most of the nitric acid made commercially in the United States is produced by the high-temperature oxidation of ammonia with air over a platinum catalyst to form nitric oxide. The nitric oxide is then oxidized to nitrogen dioxide and absorbed in water to produce an aqueous solution of nitric acid. Three types of plants are used: atmospheric, 0 to 10 psig; intermediate, 20 to 60 psig; and pressure, 80 to 120 psig. Ninety-one percent of the nitric acid manufactured in the United States is produced by the pressure process.

The tail gas from the absorption tower containing unabsorbed oxides of nitrogen is the main source of atmospheric emissions. These oxides are largely in the form of nitric oxide and nitrogen dioxide. Trace amounts of nitric acid mist are also present.

The tail gas is reddish-brown, because of the nitrogen dioxide. Nitric oxide is colorless. Effluent gases containing less than 0.03 percent nitrogen dioxide are essentially colorless, while concentrations of 0.13 to 0.19 percent by volume produce a definite color in the exit plume.

Recent tests at 12 plants showed an emission range of 0.1 to 0.69 volume percent nitrogen oxides with an average of 0.37. Limited data indicate that nitric oxide accounts for approximately one-half to two-thirds of these values, the balance being nitrogen dioxide. The plants tested were not equipped with waste gas treatment equipment.

The typical composition of the tail gas from the usual pressure process is:

	Percent
Total nitrogen oxides ($NO + NO_2$)	0.3
Oxygen	3.0
H_2O	0.7
N_2, etc	Balance

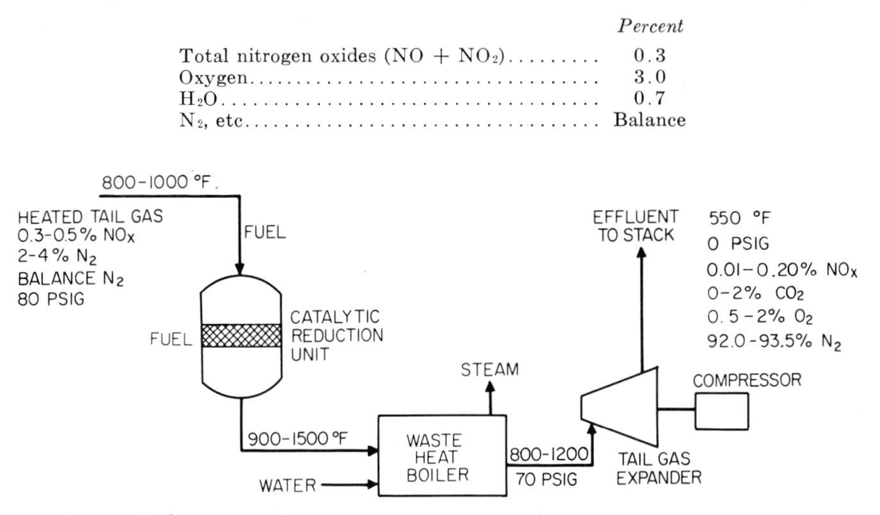

Fig. 3 Typical nitric acid plant tail gas catalytic reduction unit utilizing natural gas.

The emission of oxides of nitrogen may be reduced most commonly by catalytic reduction with certain fuels and occasionally by absorption.

Reduction Catalytic reduction is particularly suitable for the pressure ammonia oxidation process because the tail gas is of uniform composition and flow, is under pressure, and can be reheated by heat exchange to the necessary ignition temperature. Efficiencies above 90 percent are possible.

The tail gases from the absorber are heated to the necessary catalyst ignition temperature, mixed with fuel such as hydrogen or methane or a mixture of the two, and passed over the platinum or palladium catalyst bed. The reactions are:

$$CH_4 + 2O_2 = CO_2 + 2H_2O \quad (1)$$
$$CH_4 + 4NO_2 = 4NO + CO_2 + 2H_2O \quad (2)$$
$$CH_4 + 4NO = 2N_2 + CO_2 + 2H_2O \quad (3)$$

Reactions (1) and (2) proceed rapidly and generate considerable heat, which can be recovered economically. The tail gas is now colorless as the nitrogen dioxide is converted to nitrogen oxide. The reaction of additional fuel with nitric oxide, reaction (3), takes place more slowly. When this reaction is complete, total abatement is achieved. A typical catalytic reduction unit is shown in Fig. 3.

The cost of a catalytic reduction unit, including catalyst, varies from $1 to $2 per scfm of tail gas. Depending on the amount of steam produced and fuel costs, the systems will recover a substantial portion of their capital cost. The steam can be used for plant process steam or can furnish some of the power to drive the nitric acid plant air compressor.

Large nitric acid plants so equipped may require no outside power after startup. The heat generated by the oxidation of ammonia and in the catalytic reduction unit provides all the energy required to drive the large air compressor.

In one known commercial installation, ammonia is used as the fuel in the catalytic reduction of nitrogen oxides. Temperature rise across the catalytic reactor is very small because the ammonia selectively reduces the oxides of nitrogen without simultaneously reacting with the oxygen in the tail gas. A platinum catalyst supported on ceramic pellets is used. The reactions are:

$$4NH_3 + 3NO_2 = 3\tfrac{1}{2}N_2 + 6H_2O \quad (4)$$

$$2NH_3 + 3NO = 2\tfrac{1}{2}N_2 + 3H_2O \quad (5)$$

These reactions are carried out within a temperature range of 410 to 520°F. Below 410°F it is possible to form ammonium nitrate; above 520°F oxides of nitrogen are formed by the oxidation of the ammonia. The potential presence of ammonium nitrate is a deterrent to the use of this process where the tail gas is subsequently used to drive a turbine expander.

Removal of 94 percent of the nitrogen oxides with ammonia in only slight excess over the oxides of nitrogen has been reported.

Absorption Water absorbers are not very effective control devices, since the tail gas streams represent the effluent from aqueous absorption systems and the oxides of nitrogen concentration are too low for efficient absorption.

Absorption in alkaline solutions is more effective than in water. Under proper feed stream compositions, nitrate and nitrite salts can be formed as follows:

$$2NaOH + 3NO_2 = 2NaNO_3 + NO + H_2O \quad (6)$$

$$2NaOH + NO + NO_2 = 2NaNO_3 + H_2O \quad (7)$$

Disposal of the by-product salt solutions may be a problem.

An alternate scrubbing system utilizes two stages, with water in the first stage and sodium hydroxide in the second. A 91 percent reduction in nitrogen oxides is reported with an outlet concentration of 0.03 percent nitrogen oxides. This same reference reports a scrubber using a sodium carbonate solution to be 94 percent efficient with a discharge concentration of 0.15 percent nitrogen oxides.

Phosphoric Acid

Phosphoric acid is produced commercially by two methods, acidulation of phosphate bearing rock and the electric furnace process.

Electric furnace process In the electric furnace process, elemental phosphorus is first produced from phosphate ore, coke, and silica in an electric furnace. Phosphoric acid is then produced by burning phosphorus with air to form P_2O_5, which is cooled and reacted with water to form orthophosphoric acid.

The recovery of phosphoric acid cannot be accomplished in a practical way by burning phosphorus and bubbling the resultant P_2O_5 through either water or dilute phosphoric acid. When water vapor comes in contact with a gas stream containing a volatile anhydride, such as P_2O_5, an acid mist consisting of liquid particles of various sizes will be formed almost instantly. A recent investigation indicates that the particle size of the phosphoric acid mist is small, about 2 microns or less. This investigation showed a mass media diameter of 1.6 microns, with a range of 0.4 to 2.6 microns. Media particle size has been reported as 1.6 microns, with 13 percent of the particles less than 1 micron in diameter.

Extremely high acid mist loadings from the acid plant are common. Acid mist loadings to the mist collection equipment as high as 72 grains/scf have been reported.

Five types of mist collection equipment are generally being used in furnace grade phosphoric acid plants: packed towers, electrostatic precipitators, venturi scrubbers, fiber mist eliminators, and wire mesh contractors. The choice of the type of control equipment is dictated by the desired (or required) collection efficiency, materials of construction required, pressure drop across the control device, and capital and operating costs. The trend is toward control equipment with very high efficiencies.

Packed towers using Raschig rings or coke have been used for many years. The mist bearing gases are contacted countercurrently in the packed tower, which is irrigated with either water or dilute acid. A collection efficiency of 90 percent at a superficial gas velocity of 6.5 fps is reported.

Higher efficiencies of 98 percent or better have been obtained by following the packed tower with one or two additional packed towers or filter units. The overall pressure drop through this system is about 30 in. W.G. The packed scrubber must be thoroughly and uniformly wetted and must have uniform gas distribution to achieve this high collection efficiency. Good gas distribution is also mandatory for the filter units, and a superficial gas velocity of less than 100 fpm is recommended.

The Tennessee Valley Authority used electrostatic precipitators to reduce the emission of acid mist. Severe corrosion was always a problem with these precipitators. It is reported that the problem has been partially solved by reducing the tail gas temperature to 135 to 185°F. The relatively low gas temperatures and resultant infrequent failure of the high-tension wires are given as the reason for the high mist recovery from the gas stream. The acid lost amounts to about 0.15 percent of the P_2O_5 charged to the combustion chamber as phosphorus.

A commercial installation reports a collection efficiency of 99.15 percent using two electrostatic precipitators in series. Superficial gas velocity through the precipitators is 6.8 fps.

The venturi scrubber is widely used for mist collection. An overall plant efficiency of 99.8 percent of the phosphorus burned has been reported. Pressure drop across the venturi scrubber was 25.2 in. W.G. and 1.9 in. W.G. across the entrainment separator. Others have reported a venturi scrubber efficiency of 98.4 percent with a total pressure drop of about 40 in. W.G.

The venturi scrubber is particularly useful for acid plants burning sludge. Sludge is an emulsion of phosphorus, water, and solids carried out in the gas stream from the phosphorus electric furnace as dust or volatilized materials. The impurities vary from 15 to 20 percent. The venturi scrubber is able effectively to collect the acid mist and the fine dust discharged in the exhaust from the hydrator.

Two very high efficiency control devices have recently been developed. Fiber mist eliminators type control devices have been applied successfully to phosphoric acid plants. The mist eliminators delivered 99.98 percent efficiency on particles less than 3 microns in diameter and 100 percent on larger particles for an overall efficiency of 99.998 percent. Pressure drop across these high-efficiency collectors varies from 5 to 15 in. W.G.

A collector using two wire mesh contactors in series was introduced commercially in 1964, and several additional units have been installed. The first contactor agglomerates the fine acid mist into larger particles which are collected by the second stage and separated from the gas stream. Superficial gas velocity through control device varies from 15 to 29 fps, resulting in a pressure drop varying from 21 to 42 in. W.G. Reported collection effciency is as follows:

Pressure drop, in. W.G.	Acid concentration, grains 100% H_3PO_4 per scf		Removal efficiency, %
	Inlet	Outlet	
39.8	48.7	0.0167	99.965
39.8	53.8	0.0133	99.975
41.9	71.7	0.0108	99.985

Capital cost was reported to be $1.48 per cu ft of gas cleaned.

Wet process Wet process acid is made by reacting pulverized, beneficiated phosphate ore with sulfuric acid to form calcium sulfate and dilute phosphoric acid. The insoluble calcium sulfate and other solids are removed by filtration. The weak phosphoric (32 percent P_2O_5) may then be concentrated in evaporators to acid containing about 55 percent P_2O_5.

The reaction of the phosphate ore, essentially a fluorapatite with the formula $Ca_{10}(PO_4)_6F_2$, with sulfuric acid can be more conveniently written as

$$3Ca_3(PO_4)_2 \cdot CaF_2 + 10H_2SO_4 + 20H_2O = 6H_3PO_4 + 10CaSO_4 \cdot 2H_2O + 2HF \quad (8)$$

The HF then reacts with silicon dioxide in the phosphate rock, producing silicon tetrafluoride gas.

Gaseous fluoride compounds are also released from the gypsum filter and the phosphoric acid concentrator. The following typical fluoride effluent concentrations have been reported:

Acidulation off gas: SiF_4, 200 to 500 mg/cu ft
Filter: SiF_4, 10 to 30 mg/cu ft
Concentrator: SiF_4 and HF, 2 to 5 percent F

Control methods for gaseous fluorides are discussed under Gaseous Fluorides.

The concentration of wet process phosphoric acid is carried out in vacuum evaporators, submerged combustion equipment, or other direct fired evaporation equipment. Emissions from vacuum evaporators are generally controlled with a barometric condenser or scrubber.

High-temperature direct contact methods are attractive because serious evaporator tube scabbing problems are avoided and steam generating facilities are not required. However, direct fired equipment generates a dense plume of phosphoric acid mist contaminated with HF and SiF_4. Development work has been reported on a suitable mist eliminator fiber, compatible with both phosphoric acid and gaseous fluorides, for eliminating this problem area.

The preparation of the phosphate ore generates dust from drying and grinding operations. The dust from these operations is generally controlled by a combination of dry cyclones and wet scrubbers. The material collected by the cyclones is recycled, and the scrubber water is discharged to the waste gypsum ponds. Simple spray towers and wet cyclonic scrubbers are the type most frequently used. At one location the dry cyclone is followed by an electrostatic precipitator.

Hydrochloric Acid

Hydrochloric acid, commonly referred to as muriatic acid, is a solution of hydrogen chloride in water. In addition some anhydrous hydrogen chloride gas is produced, usually for immediate consumption in nearby plant processes.

The manufacture of hydrochloric acid involves the generation of hydrogen chloride gas and its absorption in water for the aqueous solutions. There are three principal processes used to produce hydrogen chloride: reaction of salt with sulfuric acid producing hydrogen chloride and sodium sulfate (Mannheim process); burning chlorine in a slight excess of hydrogen; and as a by-product from the chlorination of organic compounds. In 1961, 10 percent was produced from salt, 13 percent from chlorine-hydrogen, and 77 percent as by-product acid.

Since Mannheim plants are now few in number and are being steadily retired from service, the air pollution control aspects will not be discussed in detail. Emissions include hydrogen chloride gas, sulfuric acid mist, and entrained dust.

The chlorine-hydrogen process is used where chlorine and hydrogen are available as by-products from other nearby plant operations, such as electrolytic caustic cells. Chlorine is burned in a slight excess of hydrogen, producing hydrogen chloride. The facilities consist of a combustion chamber, chlorine burner, necessary control and safety devices, and hydrogen chloride processing equipment. If hydrogen chloride gas is the desired product from these facilities, gas cooling equipment to reduce the temperature from 2200°F is included. If aqueous hydrochloric acid is to be the end product, facilities are added.

The production of chlorinated organic chemicals by the reaction of chlorine and an organic compound often results in the generation of hydrogen chloride as a by-product. If hydrogen chloride gas is the desired product, the reaction products are fed to a still or condenser to separate the organic products from the hydrogen chloride. Absorption equipment is added if aqueous hydrogen chloride is the desired product. By-product acid usually is contaminated with organic impurities and may require further purification.

Hydrogen chloride gas combines readily with water as it has a relatively high solubility. Its solution in water generates considerable heat, so that adequate cooling facilities must be provided for the design of the absorption facilities.

Absorption facilities may comprise a packed tower system or a cooled absorption tower system followed by a packed tail gas tower. The cooled absorption tower system has proved to be a more efficient, economical, and compact unit than the packed tower system and is rapidly supplanting this method for the absorption of hydrogen chloride. Commercial designs include water jacketed packed towers with countercurrent flow and impervious carbon "falling film" cocurrent flow towers.

Emissions from absorption facilities during normal operations are exceedingly small because water is the scrubbing medium in the tail tower at the end of the system, resulting in practically complete removal of hydrogen chloride from the exit gas stream. Unabsorbed gases are vented through tall stacks so that unacceptable ground level concentrations are avoided. Upsets in operating conditions, however, may result in temporary localized problems.

INORGANIC GASES

Sulfur Dioxide

The principal sources of sulfur dioxide emissions by the chemical industry are the fossil fuel fired power or steam generators, fossil fuel fired furnaces and kilns, sulfuric acid plants, and small miscellaneous sulfur based processes. The concentration of sulfur dioxide in sulfuric acid plant waste gases and in the stack gases from the combustion of fuel is low. In such instances the economic recovery and removal of sulfur dioxide are difficult, and only a few installations for its removal at these low concentrations exist.

Where the dilute sulfur dioxide waste gas stream also contains particulate matter (fly ash, dust, etc.) even less useful technology has been fully developed.

In general, when the exit sulfur dioxide concentration is less than 1 percent, tall stacks have been used to disperse the sulfur dioxide and provide satisfactory ground level concentrations.

Low-concentration gas streams can be effectively scrubbed using an alkaline solution of potassium permanganate, buffered with sodium carbonate, by the following chemical reaction:

$$2KMnO_4 + 3SO_2 + 4KOH = 2MnO_2\downarrow + 3K_2SO_4 + 2H_2O \qquad (9)$$

Hydrogen Sulfide

Hydrogen sulfide is principally produced in petroleum refining operations, the coking of coal, the purification of natural gas, and the evaporation of black liquor in the kraft pulping process. For many of these operations hydrogen sulfide removal and recovery processes have been installed for years. The collected hydrogen sulfide is converted to sulfur dioxide for subsequent sulfuric acid manufacture or to elemental sulfur.

For small plants or processes where recovery systems cannot be justified, a number of air pollution control techniques are available.

Small dilute streams of hydrogen sulfide can be burned in boiler fireboxes or flared. It must be borne in mind that hydrogen sulfide has a lower explosive limit of 4.3 percent and that the end product is sulfur dioxide, another air contaminant.

A catalytic oxidation system for the removal of hydrogen sulfide and other odorous sulfides has been reported recently; 300 cfm of the odorous gas is heated from

160 to 950°F, the level required for maximum efficiency. The gases are then oxidized as they pass through the platinum-alumina catalyst bed. After heat exhange, the exhaust gas is put through a water scrubber. The scrubber effluent is reported to be odor free.

The reported cost of the system is $7,500. The catalyst bed is replaced every 9 months, at a cost of $650.

Hydrogen sulfide is a reducing acid gas under normal conditions and can be neutralized or oxidized. Oxidants react with hydrogen sulfide forming sulfate, if the pH is 7 or less, and thionates, sulfites, sulfur, or sulfates at pH 7 or greater depending on the oxidant used and concerntration of reactants.

If the oxidant solutions contain chlorine, the following stoichiometric relationships are found:

$$H_2S + 4HOCl = H_2SO_4 + 4HCl \tag{10}$$
$$Na_2S + 4NaOCl = Na_2SO_4 + 4NaCl \tag{11}$$
$$Na_2S + H_2O + NaOCl = NaCl + 2NaOH \tag{12}$$

The effect of pH on hypochlorite requirements is shown in Fig. 4. It is reported

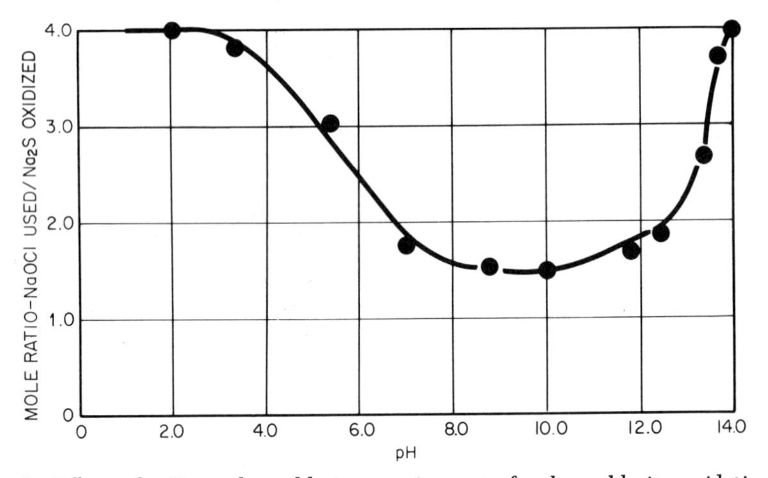

Fig. 4 Effect of pH on hypochlorite requirements for hypochlorite oxidation of sulfide.

that hypochlorite solutions at pH 12 or greater are effective for removal of hydrogen sulfide from gas streams, while acid hypochlorite solutions at pH 12 or greater are effective for removal of hydrogen sulfide from gas streams, while acid hypochlorite solutions have poor absorption efficiency relative to basic hypochlorite solutions. This reference reports basic hypochlorite solutions have on the order of ten times the absorption coefficients of acid hypochlorite solutions and that gaseous chlorine will always be lost to the atmosphere when using acid chlorine solutions.

Hydrogen sulfide is quantitatively and virtually instantaneously oxidized to sulfate by alkaline permanganate solutions. This scrubbing process is also useful for removing other sulfurous compounds, which may accompany the hydrogen sulfide.

Sodium hydroxide reacts with hydrogen sulfide in water solutions by the following neutralization reactions:

$$H_2S + NaOH = NaHS + H_2O \tag{13}$$
$$NaHS + NaOH = Na_2S + H_2O \tag{14}$$

Hydrogen sulfide remains in the liquid phase as sodium hydrosulfide or sodium sulfide if sufficient hydroxide is present. Care and judgment must be exercised in the disposal of the spent caustic solution containing sulfides. Subsequent acidification of the spent solution releases the hydrogen sulfide with potential nuisance, health, or explosion problems.

For batch operations with relatively low concentrations of hydrogen sulfide, the iron oxide–dry box process is useful. The gas to be treated passes through a bed of iron oxide sponge, which usually consists of wood shavings impregnated with iron oxide. The hydrogen sulfide reacts with the iron oxide to form an iron sulfide. After the bed is exhausted, the sponge is replaced or treated with air to regenerate the iron oxide and form elemental sulfur.

Oxides of Nitrogen

Nitric oxide, NO, and nitrogen dioxide, NO_2, are the common forms of nitrogen oxides emitted from industrial processes. In the chemical industry exhaust gases from nitric acid plants, nitration processes, chamber sulfuric acid plants, nylon intermediate plants, oxidation reactions using nitrates, and combustion processes contain small amounts of nitrogen oxides.

Control systems used in the United States on nitric acid plant tail gases have been reported in detail under Nitric Acid. A Canadian nitric acid plant is reported to obtain a collection efficiency of 50 to 85 percent using a two-stage collection system. The first stage is a raschig ring packed tower circulating a sodium hydroxide and waste ammonia solution. Ammonium nitrate and nitrate particles are formed, producing a dense white plume. These particles are removed by a second-stage venturi scrubber operating at a pressure drop of 22 in. W.G.

Little technology is available for scrubbing gas streams with low concentrations of oxides of nitrogen. Relatively enormous amounts of water would be required to absorb appreciable amounts of nitrogen dioxide in concentrations below 2,000 ppm. For temperatures below 100°F, reports give no appreciable difference in absorption between 20 percent sodium hydroxide solutions and pure water.

The adsorption of oxides of nitrogen by silica gel or activated carbon is satisfactory for nitrogen dioxide concentrations higher than 0.1 percent and for nitric oxide concentrations higher than 1 to 1.5 percent. Silica gel containing adsorbed nitrogen dioxide catalyzes the oxidation of nitric oxide, and the resultant nitrogen dioxide is removed by adsorption on the silica gel. The silica gel is periodically regenerated by heat, producing a higher concentration of nitrogen dioxide, which can be used in a nitric acid plant.

This process requires a large investment and is not suitable for gas streams containing particulate matter, which fouls the silica gel beds. This method has not been used commercially for reducing the oxides of nitrogen.

Gaseous Fluorides

Fluorine contaminants may be emitted to the atmosphere by a wide variety of industrial processes in which fluorine compounds are manufactured, utilized as catalysts or fluxes, or are present as impurities in the process materials.

The most common fluoride emitted is hydrogen fluoride. Gaseous silicon tetrafluoride, evolved when materials containing calcium fluoride or calcium fluorapatite are calcined or treated with acid in the presence of sand, is readily hydrolyzed by the moisture in the air to give hydrogen fluoride.

Water has a high affinity for hydrogen fluoride, and absorption of the gas with water is the generally used control method. However, if elemental fluorine is present in the gases, pure water should not be used as the danger of explosions exists under some conditions. Collection efficiencies of 95 to 99.9 percent can be achieved with a variety of wet scrubbers. The higher efficiencies are obtained by using two- or three-stage scrubbers.

The types of scrubbers most commonly used are spray, venturi, wet cyclone, impingement, and packed beds as a part of multistage units. Problems develop with certain processes and in certain types of equipment as a result of the precipitation of silica or calcium fluoride.

In some processes the effluent scrubber liquor is treated with lime slurry to remove the fluoride ion as insoluble calcium fluoride.

Lump limestone beds have been used for removing gaseous hydrogen fluoride. Portions of the bed are withdrawn from the tower periodically to remove the insoluble calcium fluoride coating.

One of the more difficult fluoride problems is the reduction of emissions from triple superphosphate curing sheds, because of the low concentration of fluorides and very large exhaust gas rates.

An overall fluoride collection efficiency of 99 percent has been reported for a system installed on a curing shed. The installation consists of four cyclonic scrubbers in parallel, handling a total airflow of 350,000 cfm. Gypsum pond water is used for scrubbing at a rate of 3,000 gpm. Reported capital cost for the system is $950,000.

Another installation using four parallel scrubbers reports a collection efficiency of 98+ percent. The spray tower scrubbers are packed with staggered pads of loosely knitted polypropylene netting. The system handles 100,000 cfm, and gypsum pond water is recycled through the scrubbers at a combined rate of 1,800 gpm. Gas pressure drop through the scrubbers is about 3 in. W.G.

Pressure drop through these scrubbing systems is an important economic consideration because of the large volumes of gas treated.

Gas streams containing elemental fluorine and fluoride gas should not be scrubbed with pure water, because explosions may occur under some circumstances. Therefore, more expensive processes such as the sodium hydroxide process must be used when elemental fluorine is involved. In this process, fluorine and hydrogen fluoride are continuously absorbed in a 5 to 10 percent solution of sodium hydroxide. The effluent liquid is treated with a lime slurry. The precipitated calcium fluoride is removed continuously in a settling tank.

Chlorine

In the manufacture of chlorine a mixture of inert gases and chlorine, termed "sniff gas," must be vented. The principal sources of sniff gas are the vents from the main liquefication unit and the vents from the tanks of air used in the transfer of chlorine. The chlorine concentration in this mixture will vary from 20 to 50 percent. Similar losses occur in many chlorine consuming operations. For example, many chlorinations require an excess of chlorine to go to completion, and the excess chlorine must be vented.

Chlorine producers and consumers use various methods for the recovery or disposal of sniff gas. One method is to reduce the concentration of chlorine in the mixture before disposal. This is most commonly done by condensing at high pressure or low temperature or a combination of both. This condenses additional chlorine, and the leaner sniff gas is then vented.

Often the sniff gas is vented from a tall stack for dispersion to acceptable concentrations.

The residual chlorine can be absorbed in cold water in a packed tower. However, the disposal of the chlorine saturated water presents another problem, and this method is not generally suitable.

Another process reacts the chlorine with a solution of sodium hydroxide or lime. The reaction goes substantially to completion with the use of excess caustic. Again disposal of the resultant hypochlorite solution is a problem, unless there is some in-plant use for it.

Other methods less commonly used are the reaction with ferrous chloride to form ferric chloride, followed by the reaction of ferric chloride with scrap to convert back to ferrous chloride. This process is economical if one has use for the resultant ferrous or ferric chloride. Sulfur has also been used in a similar manner to form sulfur chloride. Again, the process is used only where there is a use for the sulfur chloride, as its disposal is more difficult than waste chlorine.

A process which economically recovers the chlorine from the sniff gas has been reported. Three of these recovery plants were in operation in 1956, and it is reported that chlorine was recovered as a liquid at an average total cost of $19 per ton.

The process is based on the differential absorption of chlorine in carbon tetrachloride at a pressure of 100 psi, followed by the recovery of the chlorine from the carbon tetrachloride by stripping at a pressure of approximately 35 psi.

In another type of unit, the chlorine is adsorbed on silica gel, from which it is subsequently recovered by heating.

PARTICULATE MATTER

Particulate matter includes both solids and liquids, such as dust, smoke, condensed fume, and mist. The chemical industry emits an infinite variety of particulate matter; so a general discussion of the design considerations for the control of particulates will be made. In all but the simplest cases, the control system and device must be specifically engineered for the particular process.

Particulate matter may be evolved from the chemical reaction or from such processes as drying, calcining, screening, grinding, conveying, mixing, or packaging.

With the exception of nearly complete control of very fine particles, technology is adequate, and a wide variety of control devices are available. These include cyclone collectors, baghouses, a wide variety of scrubbers, electrostatic precipitators, demisters, agglomerators, tall stacks, and gas fiber filters. Generally at least two types of control equipment are suitable for any given process, and an evaluation should be made to determine the optimum system.

The first consideration is the degree of abatement required to meet established or anticipated regulatory requirements. Then an intimate knowledge of the gas stream and the nature of the particulate matter must be developed.

Important properties of the gas stream are its temperature, density, dew point, and volume. The temperature, for example, may be too high for certain types of control equipment, and cooling may be required. Waste heat boilers may be economical if the gas temperature exceeds 1200°F. Other cooling methods include radiation columns, evaporative coolers, and dilution with ambient air. The latter types are generally a poor choice, as the gas volume to be treated is increased, requiring a large control system.

The dew point of the gas must be known to assure that condensation will not occur within the system. Water dew points of 190°F are common; sulfur trioxide dew points in excess of 300°F have been encountered. Insulation of the equipment or heating the gas stream may be indicated.

The volume of the gas stream to be treated is an important consideration. Is it constant or does it vary widely? The system should be sized for the worst conditions, that is, the greatest gas volume.

Many properties of the particulate matter must be determined including particle size distribution, particulate concentration, density, corrosiveness, abrasiveness, whether the material has noxious properties, the angle of repose of the material if it is a dust to be collected dry, and whether the material is sticky or hygroscopic. Additional information may be required for certain applications. For example, the wettability of the particle is an important consideration for scrubbing, and the electrical resistivity of the particulates must be known when considering electrostatic precipitation.

Corrosion problems must be considered, even if the particulate matter itself is inert. For example, trace halogens, such as chlorides and fluorides, materially shorten the life of glass fiber filter beds or bags. Sulfur trioxide condensation must be considered and avoided.

The appearance of the plume discharged from the process must be considered in jurisdictions where equivalent opacity regulations are enforced. Conspicuous steam plumes contaminated with trace amounts of residual particulates may be a problem. Condensed sulfur trioxide mist may result in a detached plume.

A very important consideration is the proper disposal of the collected material. Can it be sold or recycled within the process, or does it require ultimate disposal? Should the material be collected wet or dry?

Using the above design parameters, alternate control systems should be evaluated. A well-designed control system must meet code requirements; must be properly sized for the gas volume; must fit in the available factory space; must be reliable, durable, and constructed of suitable materials; must be compatible with the basic process; and may require instrumentation to ensure smooth operation and long life.

If these criteria are met, an economic evaluation of the capital and operating costs of the alternate systems studied should indicate the type of control device most suitable for the particular process.

Sometimes it is advantageous to consider two different types of collectors in series, for example, a cyclone to collect valuable coarse material, followed by a high-efficiency scrubber, baghouse, or electrostatic precipitator to collect the fine material discharged from the cyclone. In the carbon black industry electrostatic precipitators are used to agglomerate the very fine particles, which are subsequently collected by a baghouse. It may be advantageous to collect the coarse dust in a cyclone collector and discharge the effluent containing smoke and/or sulfur dioxide from a tall stack for atmospheric diffusion.

ORGANIC COMPOUNDS

The air pollution potential of the great variety of organic compounds produced in the United States differs greatly from that of the inorganic compounds, which are frequently inert. The organic compounds are generally combustible, may be toxic in varying degrees, and are often odorous, and certain types of organic compounds react with oxides of nitrogen in the atmosphere to produce the oxidizing Los Angeles type smog.

Solids and Tars

Side reactions in many organic chemical processes produce waste solids and viscous heavy tars which must be disposed of. Such material is frequently incinerated and therefore has an air pollution potential.

A variety of incinerators are available for this type of waste. Auxiliary fuel is used to start the combustion, which is frequently self-sustaining once combustion temperature has been reached. For most wastes an exit gas temperature of 1600°F or greater will provide complete combustion and a clean effluent.

The heavy wastes may be blended with other residues to adjust the viscosity before firing, or to dissolve the solid wastes. Nonsoluble solids, after size reduction in a hammer mill or other device, can be slurried with other liquid residues and be fed to the incinerator.

An open top incinerator recently developed has been effective with a number of solid chemical wastes, such as nitrocellulose, synthetic fibers, plastic scraps, and various polymers. It consists of a rectangular container, which may be a pit in the ground or an above-ground vessel. The principal feature of the unit is the admission of all air for combustion through closely spaced topside nozzles. Adequate mixing with combustion air is achieved. The system is simple with a relatively high capacity.

Liquids

Volatile organic liquids may be released to the atmosphere because of storage tank breathing losses, the opening of pressure relief valves and leaks from valves, pump packing, and pipe flanges.

Several techniques are available to minimize evaporation losses from storage tanks. Both floating roof tanks and vapor recompression systems of fixed roof tanks materially reduce such losses. Emissions during the loading of tank trucks and tank cars can be greatly reduced by submerged filling techniques. Evaporation loss calculation methods for loading operations have been developed.

Depending on the nature of the process and the organic chemical, vapors released through pressure relief valves may be vented to the atmosphere for dispersion or burned in a flare stack.

Pump gland losses can be reduced by the use of mechanical seals. A good preventive maintenance program with frequent inspection and prompt repair of piping system leaks helps minimize such losses.

Refractory organic wastes which are toxic or nonbiodegradable are frequently burned, to prevent possible water pollution problems. Both horizontal and vertical combustion chambers are used, and aqueous and organic streams may be fired simultaneously, through separate burners. Mechanical, air, or steam atomization is recommended depending on the nature of the waste and utility costs.

Exit temperatures range from 1000 to 2000°F depending on the type of combustible. If the waste is completely organic, it can often be burned completely to carbon dioxide and water, without the use of auxiliary fuel. Heat recovery is often practiced with this type of incineration because in most cases the liquid waste is an acceptable fuel. Waste heat boilers or combustion air preheaters are often practical.

Partly combustible wastes are usually aqueous solutions with miscible organic compounds present. They may also contain noncombustible inorganic matter. Auxiliary fuel is generally required for burning aqueous wastes. In many cases a partially combustible aqueous waste may be concentrated by evaporation prior to combustion. Such concentration may enable the waste to be burned directly and in many cases will reduce the size and cost of the incineration equipment.

Liquid wastes containing sodium salts present several problems. Complete combustion results in the emission of very fine sodium carbonate fume. Depending on the concentration of the sodium salts, it may be necessary to disperse the plume from a tall stack or install additional control equipment, usually a scrubber.

Special consideration must be given to the refractory lining of the incinerator when burning sodium salts. High-alumina brick fails rapidly because of sodium attack. Chrome refractory brick is recommended because of its superior resistance. The high thermal conductivity of chrome brick requires the use of a low-conductivity brick lining to prevent excessive incinerator metal temperatures.

Reported capital cost for an incinerator handling a waste stream consisting of 1,800 lb/hr of mixed organics dissolved in 6,000 lb/hr of water was $70,000. Natural gas auxiliary fuel of 3,000 cu ft/hr is required to maintain an exit gas temperature of 1600°F.

The incineration of halogenated wastes also requires special consideration. Usually the halogen is chlorine and occasionally fluorine. Since chlorine and fluorine are highly toxic and only slightly soluble in water, the products of combustion from the halogenated organic are reacted with hydrogen to form hydrogen chloride or hydrogen fluoride, which can be removed from the gas stream by scrubbing with water or caustic. Hydrogen is provided by injecting excess auxiliary fuel into the combustion chamber. A conventional scrubbing tower can then be used to remove the hydrogen chloride or hydrogen fluoride.

An incinerator-scrubber system for disposal of heavily chlorinated wastes has been reported. No auxiliary fuel is required once the unit reaches operating temperature of 1300°C. The exit gases are cooled and scrubbed and the effluent scrubber liquid is neutralized before disposal. Capital cost is reported to be $250,000 for a unit destroying an average of 3,000 lb/hr of residues.

Gases and Vapors

A wide variety of gases and vapors are released in the production and use of organic compounds. A number of control techniques are available and usually must be specifically engineered for a specific process. These include dispersion, flare stacks, direct and catalytic combustion, absorption, adsorption, compression, and refrigeration. The most suitable techniques depend on the physical and chemical nature of the vapor, its concentration, exhaust volume, flammability, explosive characteristics, toxicity, and economic value as a raw material or product.

Tall stacks are often suitable for dilute, low-concentration sources. Tall flare stacks are commonly used in petroleum refineries and chemical plants to destroy intermittent releases of concentrated gas and vapor streams, caused by the venting of reaction vessels.

Complete design of flares involves a number of important considerations such as height, tip configuration, pilot burners, ignition systems, knockout drums, water seals, and smoke reduction. Steam injection into the waste vapors before ignition produces smokeless flames. The amount of steam required varies from 0.1 lb steam/lb of saturated combustibles to 0.8 lb steam/lb of combustible for a gas containing approximately 20 percent unsaturates.

Frequently the use of an elevated flare is necessary. Even with an elevated flare, potential hazards exist in case of flame failure. Prevention of flame failure cannot be guaranteed. Design techniques have been developed to size flare height to prevent dangerous concentrations of contaminants over level terrain if flame failure occurs. This work was extended to predict the behavior of dense stack gases, i.e., with gas densities greater than that of air.

Incineration, using either direct flame incineration or catalytic oxidation, has been widely applied to the oxidation of waste organic vapors to carbon dioxide and water vapor. The direct flame incinerator usually has lower capital costs but higher operating (auxiliary fuel) costs, while the catalytic unit has a higher capital cost but lower fuel requirements. Either will provide a clean odorless effluent, if the exit gas temperature is sufficiently high. Heat-exchange equipment is often justified for either type of incinerator to reduce operating costs.

The catalytic units are not adaptable to gas streams containing appreciable amounts of heavy metals, liquids, or solids. Chlorinated hydrocarbons, arsenic, lead, and zinc are particularly poisonous to most catalysts.

Absorption of organic vapors and gas in a wide variety of liquid media is commonplace.

Adsorption of organic vapors on porous solids such as activated carbon, activated alumina, and silica gel is frequently used where the vapor can be economically collected and regenerated for further use. Fixed bed adsorbers have been widely used for years, and continuous fluidized recovery plants are finding more uses.

With either fixed bed or fluid bed solvent vapor recovery systems, the capital cost is primarily a function of the airflow to be handled. The air volume in turn is usually determined by the need to avoid explosive vapor compositions. Usual design concentration is 25 to 50 percent of the lower explosive limit (LEL).

An automatic fluid bed recovery unit using activated carbon for the adsorption of carbon bisulfide vapors has been described. The unit treats 250,000 cfm with a reported adsorption efficiency of over 98 percent.

Compression and/or cooling of concentrated vapor to condense and recover more of the organic is practiced in numerous production processes. Vapor recompression is also used to minimize storage tank evaporative losses.

Odors

The techniques and equipment generally used to reduce the emission of gases and vapors are also applicable to the elimination of odor problems. However, the concentration of the odorous compound may be several orders of magnitude lower than encountered in controlling gaseous or vapor emissions; so modifications of these techniques may be necessary.

No one has yet developed an instrument better than the human nose when it comes to detecting odors. In the past a number of studies have been made on odor thresholds, but they have been conducted in a variety of ways and as a result have produced a number of values. Different studies of the same chemical compound have produced odor threshold values varying as much as a thousandfold in concentration where the odor becomes noticeable.

To provide for a more uniform body of odor threshold data, the Manufacturing Chemists Association, Inc., has sponsored a project by Arthur D. Little, Inc., for the determination of the odor threshold of industrially important chemicals. The odorants were presented to a trained odor panel in a static air system using a low-odor background air as the dilution medium. The odor threshold is defined as the first concentration at which all panel members can recognize the odor.

The recognition odor thresholds reported in this study were developed under ideal laboratory conditions using single commercially pure compounds. No attempt was made to define the degree of objectionability, if any, of the odorant chemical, or the intensity at which it becomes objectionable.

Because of the low background odor in the test chamber, the reported thresholds may be several orders of magnitude lower than recognition levels in the ambient atmosphere, where such factors as the level and type of background odor, olfactory fatigue, synergism, and neutralization interact. The data determined to date are shown in Table 1.

TABLE 1 Recognition Odor Thresholds in Air (ppm Volume)

Compound	ppm	Compound	ppm
Acetaldehyde	0.21	Ethyl acrylate	0.00047
Acetic acid	1.0	Ethyl mercaptan	0.001
Acetone	100.0	Formaldehyde	1.0
Acrolein	0.21	Hydrochloric acid gas	10.0
Acrylonitrile	21.4	Methanol	100.0
Allyl chloride	0.47	Methyl chloride	Above 10 ppm
Amine, dimethyl	0.047	Methylene chloride	214.0
Amine, monomethyl	0.021	Methyl ethyl ketone	10.0
Amine, trimethyl	0.00021	Methyl isobutyl ketone	0.47
Ammonia	46.8	Methyl mercaptan	0.0021
Aniline	1.0	Methyl methacrylate	0.21
Benzene	4.68	Monochlorobenzene	0.21
Benzyl chloride	0.047	Nitrobenzene	0.0047
Benzyl sulfide	0.0021	Paracresol	0.001
Bromine	0.047	Paraxylene	0.47
Butyric acid	0.001	Perchloroethylene	4.68
Carbon disulfide	0.21	Phenol	0.047
Carbon tetrachloride (chlorination of CS_2)	21.4	Phosgene	1.0
Carbon tetrachloride (chlorination of CH_4)	100.0	Phosphine	0.021
Chloral	0.047	Pyridine	0.021
Chlorine	0.314	Styrene (inhibited)	0.1
Dimethylacetamide	46.8	Styrene (uninhibited)	0.047
Dimethylformamide	100.0	Sulfur dichloride	0.001
Dimethyl sulfide	0.001	Sulfur dioxide	0.47
Diphenyl ether (perfume grade)	0.1	Toluene (from coke)	4.68
Diphenyl sulfide	0.0047	Toluene (from petroleum)	2.14
Ethanol (synthetic)	10.0	Tolylene diisocyanate	2.14
		Trichloroethylene	21.4

SOURCE: Manufacturing Chemists' Association, Inc.

Odors may be eliminated by oxidation, either chemically or by combustion. In many instances vapor incinerators are the most practical means of eliminating odor problems.

Equipment falls essentially into two categories, direct flame incineration and catalytic incineration. Direct flame burner unit consists essentially of fuel feed system and burners, a combustion zone, and a means for exhausting the products of combustion. Natural gas is the fuel most commonly employed, although propane, butane, and some fuel oils are also adaptable. Critical design factors are flame contact and temperature, turbulence, and residence time (the so-called three Ts—time, temperature, and turbulence).

For odor problems most direct flame units operate with an exit gas temperature of 1200 to 1500°F. Good mixing in the combustion zone can be promoted by various means. Residence times of the gases in the incinerator unit of 0.25 to 0.50 sec have been found to give satisfactory cleanup of most odors when other factors are as specified. Where a large volume of dilution or cooling air is admitted after the burner section, it is particularly important that residence time be adequate. Otherwise, quenching of the flame can occur with resultant incomplete combustion.

With any of the combustion type control devices it is important that combustion be as complete as possible; otherwise partially oxidized compounds such as aldehydes and organic acid may be released which are more noxious than the original compound.

A special case of the direct flame burner is that in which an exhaust gas stream is fed directly to the firebox of a steam generator or similar heat-utilizing fired equipment. In this case, the burner flame serves both as a heat source and as a fume burner. This arrangement is most successful when the volume of boiler flue gas is considerably larger than that of the contaminated gas stream.

The successful use of a flare stack to eliminate a malodorous amine discharge has been reported.

The essential advantage of a catalytic incinerator is found in its lower fuel requirements. Adsorption type catalysts generally are operated in the range of 600 to 900°F, and vapors and gases can be burned in them without the need of a spark or flame. Catalytic units are applicable to gaseous air contaminants and to those that will vaporize at the operating temperature of the catalyst surface. They are not adaptable to gas streams containing appreciable amounts of heavy metals, liquids, or solids. Chlorinated hydrocarbons, arsenic, lead, and zinc are particularly poisonous to most catalysts.

Heat-recovery systems should be employed wherever economically feasible with either type of incinerator. Effluent gases from these units can be used to preheat the vapors to be incinerated, or recovered waste heat can be used in various other ways.

Summarizing, vapor incinerators can do an effective job in abating odor problems for most organic vapors. Operating costs for fuel are high, particularly if waste heat recovery equipment cannot be justified.

Odors can also be oxidized chemically. Ozone effectively abates certain odor problems. Other oxidizing agents, such as potassium permanganate, chlorine, and chlorine dioxide, added to the circulating waters of a gas scrubber, are effective on a wide variety of odors. The oxidation reaction either destroys the odor or modifies it to something more pleasant and agreeable.

Other applications of scrubbing or absorption are quite common. Water or aqueous solutions are suitable absorbing liquids for many gases, including sulfur dioxide, hydrogen chloride, and chlorine. Many other gases or vapors, particularly petroleum hydrocarbons, are absorbed in petroleum absorption oils, while other organic or inorganic solvents are effective for other gases.

Another effective wet device is the common spray chamber or contact condenser. It is particularly useful for steam laden streams containing malodors. Such condensers are quite acceptable where cooling water and sewage facilities are abundant. They require 15 to 20 gal of cooling water for each gallon of steam condensed. This odor laden water cannot be run through a cooling tower and therefore cannot be reused. Effluent condensate volumes can be greater than sewage facilities will allow. Nevertheless, contact condensers dissolve and condense much of the odorous fraction of the exhaust stream, such that the remaining small volume of uncondensed gases can be effectively incinerated in a small afterburner or boiler firebox. The effluent waste water should be discharged to a sanitary sewer for subsequent treatment to prevent a water pollution problem.

Surface condensers are more acceptable than contact condensers in most areas. Water cooled and air cooled surface condensers are used with success to remove 95 percent of the exhaust volume. Condensate is less voluminous and richer in odorous materials. As would be expected, the remaining uncondensed gases are generally more odorous than those from contact condensers. Again, the remaining small volume of noncondensed gases should be incinerated.

Activated carbon is highly effective in deodorizing gaseous effluents. However, carbon will not adsorb malodorous gases adequately at temperatures in excess of 120°F, and the stream must be relatively free of particulates which will foul the adsorbent. Whenever activated carbon is used for odor control, care must be taken to assure that the carbon is regenerated before it becomes saturated and ineffective. Odorous materials displaced on regeneration should be incinerated or otherwise controlled to prevent discharge to the atmosphere.

The tall stack for atmospheric dispersion to below odor threshold levels is useful for eliminating low-concentration odors and as a secondary device following a primary control device, when dealing with a compound with a very low threshold concentration.

In addition to proper control devices, it should be emphasized that good housekeeping and prompt maintenance are an important part of any good odor control program.

WATER POLLUTION

The large number of commercial chemicals and the diversity of their effect on water make it impractical to generalize for the entire chemical industry. The wastes from a chemical plant may be inorganic, organic, insoluble, soluble, inert, toxic, or any combination thereof.

Both organic and inorganic wastes have water quality effects essentially different from sanitary sewage, and cannot be described properly in sanitary sewage terms such as the fictitious "population equivalents." Simple yardsticks such as oxygen demand, acidity, or dissolved solids are not always applicable.

The abatement technique applied by the chemical industry for its pollution problems bears the stamp of the industry's own technology. Most waste treatment facilities are unique and individually conceived and constructed. Minimizing or eliminating wastes is an important consideration in any new process or plant. Water pollution control should be considered in every stage of chemical plant development from research through plant design or modification and operation.

To protect the beneficial uses of the receiving waters adequately, it is necessary to minimize the waste, characterize the effects of the waste on the receiving waters, and have knowledge of the assimilative capacity of the receiving waters.

Waste Characterization

An intimate knowledge of the characteristics of a waste stream is necessary to protect the receiving waters and to treat the wastes adequately. Needed information includes, but is not limited to, biological oxygen demand (BOD), toxicity, suspended and settleable matter, insoluble oil films, taste, odor, pH, temperature, and volume of the waste stream. Some of these characteristics may vary with concentration, temperature, or acclimatization of bacteria.

Ideally each waste stream should be studied individually. Trace constituents may have synergistic effects which adversely affect biota either in the treatment plant or in receiving waters. Some ions in the waste may react and precipitate in the receiving waters.

The analytical and experimental methods used to determine the various parameters have been standardized, and the use of these standard methods is recommended.

Waste effluents from organic chemical plants are usually a heterogeneous mixture whose composition will not be reliably known until the plant has been in operation for some time. In considering a new process, evaluation of the projected waste streams is necessary. Sources of data may be an existing similar plant, estimates by analysis of pilot plant effluents, testing of synthetic mixtures, and estimates based on anticipated composition of wastes.

BOD values for a number of pure compounds and other information that may be useful in preliminary estimates are shown in Table 2. A reasonably exhaustive literature search is clearly indicated. A recently published critique reviews the existing literature on the aquatic behavior of organic chemicals in the aquatic environment. Some 600 pertinent articles were critically reviewed. Among the conclusions drawn in this report is that much research is needed in this area, not only to fill existing

TABLE 2 Waste Characteristics of Some Dissolved Organic Chemicals

Chemical	Biochemical oxygen demand[a]		Chemical oxygen demand,[b] lb/lb	Theoretical oxygen demand,[c] lb/lb	Taste and odor threshold (tent.), mg/l	Toxic threshold (approx),[d] mg/l
	Test conc., mg/l	lb/lb				
Acetaldehyde	....	1.3	....	1.82	.....	< 60
Acetic acid	....	0.34–0.88	1.01	1.07	25	>100
Acetone	<400	0.5–1.0	1.12	2.20	40	>500
Acetophenone	....	....	....	....	0.17	
Allyl alcohol	....	0.2	....	2.20		
n-Amyl acetate	440	0.9	....	2.34	0.08	
n-Amyl alcohol	....	1.6	....	2.72	.....	>350
Isoamyl alcohol	....	1.6	....	2.72		
Aniline	1.5–3	1.5	....	3.09	70	>200
Benzaldehyde	....	....	....	....	0.002	> 10
Benzene	....	0	0.25	3.07	0.5	> 10
Benzoic acid	....	1.37	1.95	1.97	.....	> 50
Benzylalcohol	....	1.6	....	2.52		
n-Butyl alcohol	1–1,500	1.5–2.0	1.90	2.59	2.5	<250
Butyraldehyde	....	1.6	....	2.44		
Isobutyraldehyde	....	1.6	....	2.44		
Butyric acid	....	0.89	1.75	1.82		
Carbon tetrachloride	....	0	....	0.21		
Catechol	....	....	1.89	1.89		
Chloroform	....	0.008	....	0.33	.....	<10
o-Cresol	....	1.6	2.4	2.52	0.65	>0.5
m-Cresol	....	1.7	2.3	2.52	0.7	>0.5
p-Cresol	<440	1.45	....	2.52	.....	>0.5
p-Cumylphenol	....	....	2.6	2.8		
Cyclohexanol	....	0.08	2.5	2.72		
Cyclohexane	....	....	....	....	.....	>5
Dibutylphthalate	....	0.43	....	2.24		
Diethanolamine	2.5	0.10	....	2.13		
Ethyl acetate	1–1,000	0.6–0.10	....	1.82		
Ethyl alcohol	1–1,500	1.0–1.5	2.0	2.1	.....	High
Ethyl ether	....	0.03	....	2.59		
Ethylene dichloride	....	0.002	....	0.97		
Ethylene glycol	10	0.6	....	1.29		
Formaldehyde	<260	0.6–1.07	1.06	1.07	50	>10
Formic acid	....	0.15–0.27	....	0.35		
Fumaric acid	4–8	0.65	....	0.83	.....	<440
Furfural	2–20	0.77	....	1.66		
Furoic acid	....	....	1.25	1.29		
Gasoline	5	0.08	....	....		
Glycerol	2–500	0.7	....	1.56	.....	Very high
Kerosene	10	0.53				
Maleic acid	....	0.38	0.83	0.83		
Malic acid	....	....	0.68	0.72		
Melamine	....	Nil	....	3.04		
Methacrylic acid	....	0.89	....	1.67		
Methyl alcohol	1–1,000	0.6–1.1	1.5	1.5	.....	>8,000
Methylethylketone	....	2.1	....	2.4		
Methylphenylketone	....	0.5	....	2.5		
Monochlorbenzene	....	0.03	0.41	2.06		
Monoethanolamine	2.5	0.8–1.1	....	2.4		
Morpholine	....	Nil	....	2.6		
Naphthalene	....	0	....	3.0	....	<10

TABLE 2 Waste Characteristics of Some Dissolved Organic Chemicals (Continued)

Chemical	Biochemical oxygen demand[a]		Chemical oxygen demand,[b] lb/lb	Theoretical oxygen demand,[c] lb/lb	Taste and odor threshold (tent.), mg/l	Toxic threshold (approx),[d] mg/l
	Test conc., mg/l	lb/lb				
1, 4-Naphthaquinone	....	0.8	....	2.1	....	<0.3
α-Naphthol	....	1.7	....	2.55		
β-Naphthol	<3	1.8	2.5	2.55	1.3	
Nitrobenzene	<440	0	....	1.95		
n-Octyl alcohol	....	1.1	....	2.95	0.13	<65
Oleic acid	....	....	2.25	2.89		
Oxalic acid	....	0.12	....	0.13	....	>90
Pentaerythritol	....	....	1.23	1.3		
Phenol	1–10	1.6	....	2.4	0.15	0.1–15
Phthalic acid	....	0.87	1.45	1.45		
Polyethylene glycol	....	<0.08	....	1.29		
Propionic acid	....	0.84	....	1.51		
Isopropyl alcohol	....	1.45	1.61	2.40		
n-Propyl alcohol	....	0.47	....	2.40	.....	>350
Pyridine	....	....	0.02	3.03	0.8	<1,000
Quinoline	....	....	....	....	0.7	
Resorcinol	....	1.15	....	1.89		
Sodium acetate	2–900	0.35	....	0.78	.....	>5,000
Sodium benzoate	....	....	....	....	.....	>585
Sodium formate	....	0.04–0.10	0.23	0.24	.....	>2,000
Sodium oxalate	2–900	0.1	....	0.12		
Sodium propionate	....	0.5	....	1.17		
Sodium stearate	2–900	1.2–1.7	2.49	2.72		
Stearic acid	....	0.8	....	2.93		
Thymol	....	....	2.2	2.77		
Toluene	....	0	0.7	3.13	.....	<22
o-Toluidine	1–2.5	1.4	....	3.14		
p-Toluidine	2.5	1.6	....	3.14		
Triethanolamine	....	Nil	....	2.04		
2, 4, 6-Trinitrophenol	....	....	0.92	0.98		
Urea	....	....	....	....	.....	>17
Xylene	....	0	....	3.16	2.2	>22
1, 3, 4-Xylenol	....	1.5	....	2.62		
1, 3, 5-Xylenol	....	0.82	.....	2.62		

SOURCE: Manufacturing Chemists Association, Inc.

[a] Data are 5-day BODs in all cases. Standard dilution method.

[b] APHA method including use of silver sulfate catalyst.

[c] Stoichiometric oxygen demand to CO_2 and H_2O. When nitrogen is present, values are computed to HNO_3.

[d] Toxic threshold to sensitive aquatic organisms.

gaps but to develop and refine techniques which will provide reliable and useful results. Much of the existing information cannot be reconciled because of variations in purpose and method.

It should be emphasized that literature searches are no substitute for experimental determinations for a specific waste stream.

Process Changes

Wastes and their treatment are nonprofit operating cost items, and prevention of wastes has the potential of reducing costs, reducing or eliminating the capital and operating costs of abatement facilities, and possibly eliminating refractory or relatively toxic wastes. A number of techniques are available to minimize or eliminate wastes from a given process.

Segregation of process streams is often indicated. For example, cooling water may be discharged separately, rather than be contaminated by a small volume of waste. The concentrated waste may have raw material values or may have heating values and can be burned. A concentrated waste stream can generally be treated more economically than a dilute stream.

Direct contact condensers may produce dilute waste streams. Indirect condensation or cooling can often be justified on the basis of eliminating possible water pollution.

A change in raw materials is often beneficial. The use of a higher-purity or different raw material may eliminate a troublesome waste stream. If in-process neutralization is required, different waste characteristics result from the use of lime, caustic soda, or ammonia, or the use of sulfuric, hydrochloric, or nitric acid. An economic evaluation of total costs, including waste control, may indicate a change in reagents is warranted.

Retention basins or lagoons are desirable to provide the controlled release of large quantities of chemicals to the sewers or receiving waters. A large lagoon capable of handling several days' supply of waste water and which can be checked before being released is sometimes used. Lagoons are also used for impounding wastes until they can be released and satisfactorily assimilated during periods of high stream flow. Adequate holding and blending capacity is desirable for startups and shutdowns and for cyclic batch operations to prevent shock loads to subsequent waste treatment operations.

Analysis may reveal that one product, process, or subprocess is responsible for a disproportionate share of the plants' total waste treatment costs. Product discontinuance or process retirement may be justified if improvements cannot be made.

Housekeeping

Good housekeeping practices, if overlooked or neglected, can seriously affect the pollution potential of a plant. Poor or improper maintenance can materially increase the quantity and strength of plant effluent. Accidental spills or upsets may upset waste treatment facilities, kill fish, impair the quality of public water supplies, or create other damages and hazards.

The Manufacturing Chemists' Association states:

> A program to prevent accidental spills involves first, a careful field survey to determine possible sources; second, the development of preventative measures and controls; third, equipment additions or alterations to minimize spills; fourth, operator education; and fifth, development of emergency procedures and a warning system.
>
> Common sources of spills are boilovers, overpumping, discharge valves in unprotected areas, operator error, tank car and truck loading and unloading, improper or inadequate maintenance, piping or storage tanks exposed to impact by vehicular traffic, etc.
>
> Even though correctives are evident, hazards may go unrecognized until an incident occurs. Consideration should be given to providing diked areas equipped with sump pumps, high level alarms in process vessels, and holding lagoons to prevent unwanted materials being inadvertently released.
>
> A plant employing good housekeeping practices will keep waste discharges to a minimum. Instructions should be standard in a plant when equipment must be cleaned out for any purpose. Suitable containers should be provided for waste solvents and oils, and floor washdowns. Whenever organic matter can be segregated in a manner to permit incineration, the waste characteristics of the plant effluent will be improved.

Receiving Waters

In order to prevent deleterious effects on the water receiving wastes, the nature of the receiving body must be known. Important considerations are beneficial uses of the water resource, its quality, its flow rate, and the variability of its flow.

Recognized beneficial uses of the nation's water resources include aesthetics; water supply for agriculture, industrial, and domestic use; waste assimilation, sanitary and industrial; propagation of fish and other aquatic wildlife; recreation; water power; and navigation. Generally drinking water supply and other domestic uses are of primary importance. The relative importance of the other legitimate uses varies with the economy of the area and the desires of the general public in the area.

In accord with the provisions of the Federal Water Quality Act of 1965, all states have classified and established water quality criteria for interstate waters, and many have issued criteria for intrastate waters. Many states require permits to discharge into the waters of the state.

Whether permits are required or not, early contact with state and local control agency officials is recommended. Potential water pollution problems are of mutual concern, and early and frank discussions are desirable. With a knowledge of volume and characteristics of the waste and the degree of treatment required, costly errors can often be avoided. For existing plants, it is wise to keep in touch with regulatory agencies and stay up to date with their requirements.

The treatment method selected for a given plant will depend on the properties of the waste(s), its volume, and the flow rate, character, and important downstream uses of the receiving waters.

TREATMENT METHODS

Many chemical wastes are compatible with sanitary sewage and can be discharged into sewage systems for subsequent treatment in the municipal sewage treatment plant. Generally such wastes are governed by local sewer ordinances which regulate the discharge of industrial wastes and usually specify the manner by which the waste may be admitted. Pretreatment may be required to remove toxic substances, flammable compounds, heavy metals, or to adjust pH prior to sewering.

It available, municipal sewage treatment is often the most practical and economic solution for dissolved organic wastes provided that secondary treatment is provided by the facility, that there is sufficient excess oxidative capacity above that required for domestic sewage, and that the organic wastes are readily biodegradable.

The biggest advantage in joint treatment for both the community and industry is lower costs. Increased waste flows will generally reduce the basic per unit cost of treatment. For industry, a joint treatment plant owned and operated by the community can eliminate capital costs and operating manpower, substituting instead a service charge as a normal operating expense.

Another advantage to industry is that municipal sewage often aids in the treatment of industrial wastes, through dilution and the addition of nutrients that speed the biological processes which break down the wastes to harmless substances.

There are also possible disadvantages to combined treatment. In some cases, industrial wastes are considerably more complex than municipal wastes and may be toxic or nonbiodegradable.

The combined treatment approach is not a cure-all for every waste treatment problem. Obviously there are times when it is not the right approach or when it is abused. Many factors must be considered to develop an effective, mutually satisfactory municipal-industrial cooperative program.

If the waste is not compatible with municipal wastes because of its nature, volume, or other factors, in-plant treatment is usually required before the waste is discharged to the receiving waters. A number of treatment methods are available which may be broadly classified as physical, chemical, biological, and ultimate disposal. Frequently a combination of two or more methods is required, and the waste disposal system may be tailor-made for a given waste. The various types of treatment equipment and methods are described in detail in Chap. 5.

Despite a growing body of literature on the treatment of chemical wastes, one should not expect to find a treatment method specifically applicable to one's particular problem. Extensive pilot plant or laboratory work is frequently needed to determine properly the treatment method(s) for a given waste prior to discharge to a given receiving water.

Physical Treatment

The first treatment procedure is frequently a physical separation of insoluble or immiscible materials from the waste water. Clarification, the process of removing turbidity, sediment, and floating material, is usually the first step in treatment since these impurities are objectionable and interfere with any subsequent treatment.

Pretreatment Pretreatment ahead of sedimentation may include screening and comminuting, degritting, as well as grease and scum removal. Pretreatment usually starts with screening to eliminate easily removed solids and any oversize material. Comminuting devices reduce particle size by cutting or shredding and leave the particles in the waste water. Grit removal units take out small coarse particles of sand, gravel, or dirt. Grease, scum, and floatable solids are skimmed from the liquid surface.

Screening is the simplest way to remove suspended and oversized material which might damage equipment or adversely affect subsequent treatment processes. Coarse screens, commonly called bar screens or racks, are those with openings of ⅛ in. and larger. Fine screens intercept smaller solids using woven wire cloth in the size range of 6 to 60 mesh. A wide variety of coarse and fine screening equipment is available.

Grit removal units are often installed after screening equipment to trap solids with a particle size in the range of 6 to 150 mesh and a specific gravity of 2 or greater, to protect subsequent equipment from abrasion, to avoid pipeline clogging, and to reduce the sedimentation load on the primary clarifier. Gravity basins and centrifugal separators are the two basic types of equipment normally used.

The gravity basins depend on gravity for solids separation, and size is based on waste water flow rate, overflow rate, and velocity.

Centrifugal units are liquid cyclones similar to those used for air pollution control. The heavier particles are discharged as underflow, while the finer particles and the greater volume of the waste water are discharged as overflow.

Clarification The remaining settleable suspended solids are removed by sedimentation in clarifiers. The degree of solids removal depends on the characteristics of the raw waste and whether or not flocculation and chemical coagulants are used. Clarifiers are sized on the basis of settling rate and detention time. Clarifier shape may be either circular, square, or rectangular, and weirs, baffles, and feed wells are used to ensure good inlet distribution and effluent collection. The settled solids are normally carried across the floor of the basin by bottom scraping mechanisms which continuously carry the solids to the discharge point.

Flocculation Flocculation is the agglomeration of finely divided suspended matter and floc resulting from the gentle agitation or stirring of the waste water. The resulting increase in particle size increases the settling rate and improves removal by providing better contact between suspended solids, dissolved impurities, and chemical coagulants.

Many clarifier designs combine mixing, flocculation, and coagulation in one tank. Designs of this type are more economical and can produce better-quality water than the conventional approach of using separate treatment units.

Flotation Flotation is basically sedimentation in reverse to remove floatable materials and solids with a specific gravity so close to water that they normally settle very slowly or not at all. The waste water or a portion of the clarified effluent is pressurized to 40 to 60 psi in the presence of sufficient air to approach saturation. When this pressurized mixture is released to atmospheric pressure in the flotation unit, minute air bubbles are released from solution. The flocs and suspended solids are floated by these minute air bubbles. The air-solids mixture rises to the surface, where it is skimmed off.

Gravity separators Gravity separation is used to remove liquid materials which are insoluble in water, such as petroleum oils or organic liquids. Most such materials have a specific gravity lower than water and will rise rather than settle. Free oils will separate from water by gravity alone, unstable emulsions separate very slowly under the influence of gravity, and stable emulsions will not separate even over an extended period.

Gravity separators are usually rectangular, although circular units are also used. Basic elements include an inlet distribution, internal baffles, and a skimming device to collect the immiscible phase. When both free and emulsified oils are present, it is most economical to remove as much free oil as possible, and then use chemical coagulation to break up the remaining emulsion. Alum is effective on many emulsions. The emulsion is absorbed by the alum floc and then removed in a conventional settling basin.

Filters Filters are used to remove suspended impurities and organics, if activated carbon is used. The degree of filtration depends on the fineness of the filter media and the types of filter aids used. A wide variety of filter designs are commercially available.

Adsorption Granular activated carbon has had long usage in filtering equipment to remove color and turbidity. It is also effective in adsorbing organic contaminants from waste water measured in terms of BOD, COD, or odor. It also has the ability to remove soluble organic compounds which are not biodegradable.

Therefore, activated carbon filters are often added to existing clarifiers to improve primary treatment or used for tertiary treatment following biological treatment systems. Spent carbon from filters can be regenerated thermally for reuse. About 5 percent of the carbon is lost during each cycle of use and reactivation.

Miscellaneous Other physical treatment methods may be applicable to some organic chemical waste streams. Methods which should be evaluated for some somewhat unique chemical waste streams include solvent extraction, steam stripping, concentration and evaporation, distillation, and molecular sieving.

Chemical Treatment

Chemicals and chemical processes are playing an increasingly important role in waste water treatment. Major chemical treatment systems in use include neutralization, oxidation and reduction, coagulation, precipitation, ion exchange, and membrane technology.

Neutralization Probably the oldest and most frequent of chemical methods is the neutralization of acidic or alkaline waste streams to keep the pH of the effluent in the range of 6 to 8 required by many regulatory agencies. Many chemical waste streams exceed these limits and fluctuate sharply with time.

Acidic wastes are commonly neutralized with waste alkaline streams, lime, dolomite, ammonia, caustic soda, or soda ash. The choice of the alkaline reagent used depends on the volume of the waste stream, the variability of pH, and the price of the neutralizing alkali. Lime is most often used, despite the frequent formation of precipitates or suspended solids which must be separated by sedimentation or filtration before the waste is discharged to the receiving waters because of its low cost.

Highly alkaline wastes usually require treatment with a waste acidic stream, sulfuric acid, hydrochloric acid, or flue gas containing carbon dioxide.

Neutralization is usually accomplished in two steps since pH varies in concentration increments of 10: rough neutralization with waste streams or with cheap chemicals, followed by final polishing, often with instrumented controls, using caustic soda solutions and/or sulfuric acid.

Where large volumes of waste are involved, recovery may be more economical than neutralization. A recent process involving recovery of a waste hydrochloric acid stream has been reported. The by-product acid gas is dissolved in an adiabatic absorption column, and the solution's heat is used to separate impurities. The acid is then cooled and passed through activated carbon and an ion exchange unit to remove further contaminants. An acid of commercial purity is produced, eliminating a waste disposal problem.

Oxidation-reduction Chemical oxidation-reduction reactions are useful for treating several types of wastes. Chlorine, hypochlorites, and chlorine dioxide are all strong oxidizing agents that have found application in waste treatment. Aside from small or specific requirements, liquid chlorine is generally preferred to the other forms because it is less expensive per unit of available chlorine and is easy to use.

Chlorine provides the equivalent of ½ mole of oxygen per mole of chlorine, and also reacts with complex organics to form addition products. It has wide application in bleaching colored wastes, prevention of slime growths, disinfection, prevention of septicity, and to a lesser extent as an aid to settling, clarification, and coagulation.

Ozone has much the same potential application as chlorine. Although more expensive, it has the advantage over chlorine of not forming intermediate addition products such as chlorophenols. It is not an effective oxidant for saturated hydrocarbons and aliphatic acids, but it readily attaches such groups as $-SH$, $-NH_2$, $-CN$, $=S$, $-OH$ (phenolic), $=NH$, and unsaturated organics. Ozonation may be useful as a tertiary treatment technique or for small volumes of waste water.

Oxidation-reduction reactions are widely used in the metal plating industry for the treatment of plating wastes. The same techniques are useful in the chemical industry for the treatment of cooling tower blowdowns. Many plants use chromates as a corrosion inhibitor for cooling towers and associated heat exchangers. The blowdown is often a treatment problem, as the usual hexavalent chromium concentration of 15 to 20 ppm may require reduction to less than 0.05 ppm to meet receiving water standards.

The most widely practiced technique is reduction to trivalent chromium followed by precipitation as chromium hydroxide. The reducing agents most commonly used are ferrous sulfate, sodium metabisulfite, and sulfur dioxide.

With ferrous sulfate, an excess dosage of 2½ times the theoretical must be made and the pH must be kept below 3.0 to obtain a rapid, complete reduction of the chromium to a trivalent state. With sulfur dioxide or sodium metabisulfite the reaction is nearly instantaneous at pH levels below 2.0.

The pH is then raised to 8.0 to 9.0 to precipitate chromium hydroxide. Caustic soda, lime slurry, and soda ash are most frequently used for precipitation. Finally the waste flows into a settling tank or lagoon where the precipitated solids settle and clarified effluent can be discharged.

Batch treatment is recommended for small systems, whereas a continuous system is preferable if the blowdown is continuous and exceeds 25 to 30 gpm.

Cyanide wastes, which under acidic conditions will hydrolyze to form toxic HCN, can be treated by the alkaline chlorination method. Cyanide waste treatment can be accomplished in either batch or continuous treatment systems.

Batch treatment is accomplished by first feeding caustic soda to the waste to raise the pH to 10.5. Then chlorine is added at a constant rate and caustic soda is simultaneously added to maintain the pH above 10.5 while chlorine is being added. This completely oxidizes the cyanides to harmless nitrogen and carbon dioxide,

$$2NaCN + 5Cl_2 + 8NaOH \rightarrow 2CO_2 + N_2 + 6H_2O + 10NaCl \qquad (15)$$

Excess chlorine is used to complete the reaction. Instrumented systems are frequently used with automatic chlorine and caustic feeders, a pH controller, and an oxidation-reduction potential (ORP) recorder-controller. The ORP recorder-controller indicates when the end point of the reaction is reached, shuts down the chlorinator, and signals the operator. The batch continues to circulate another 30 min, and if the ORP recorder still shows the reaction complete, the treated waste can be dumped.

In continuous treatment systems ORP can be used to control the addition of chlorine. In the presence of excess chlorine, complete oxidation is assured if the pH and retention time are maintained at their proper levels.

Continuous treatment systems are best when large amounts of waste containing relatively low concentrations of cyanide contaminants are being treated. Nevertheless, in some states, continuous treatment is not allowed, particularly when the effluent discharges into a stream.

Another reported use of chemicals for oxidation is the control of septic odors resulting from anaerobic conditions. Sodium nitrate and sodium hypochlorite are added to supply oxygen and oxidize the odorous hydrogen sulfide.

Coagulation and chemical precipitation Coagulation is employed to improve or make possible the removal of materials in suspended or colloidal form. Colloids are represented by particles over a size range of 10^{-5} to 10^{-7} cm. These particles do not settle out on standing and cannot be removed by conventional physical treatment systems.

As the colloidal particles add to the turbidity of receiving waters, their reduction is indicated. Since the stability of a colloid is primarily due to electrostatic forces, neutralization of this charge is necessary to induce flocculation and precipitation.

Coagulants such as aluminum sulfate, sodium aluminate, ferrous sulfate, ferric chloride, and organic polyelectrolytes dissolve and form ionic charges. These neutralize the repelling charges on the colloids, which agglomerate, producing a jellylike floc. The enormous surface area of the floc mass adsorbs particles of turbidity, organic matter, and bacteria.

Coagulant aids, such as silica, clay, and organic polymers, are added to stimulate floc formation and improve its settling characteristics.

The most popular coagulant is alum. However, the newest and most versatile coagulants are the polyelectrolytes. These are water soluble, high molecular weight polymers which dissociate in water to give large highly charged ions. Three types are available: anionic, cationic, and nonionic.

Choice of coagulant, its dosage, coagulant aid, and optimum pH depend on the waste water analysis, temperature, type of clarification equipment, and the end use of the treated effluent. A number of simple jar tests conducted under actual operating conditions will normally indicate a coagulation program giving optimum results at lowest cost.

Precipitation The development of an insoluble salt or an immiscible oil phase subsequently removed by physical means has found some application for reducing soluble organics from aqueous wastes. The methods are many and the following may be of value:

- Development of an insoluble organic salt such as by the use of calcium
- Liberation of an organic phase or precipitate by pH changes or "salting out" methods
- Development of an insoluble reaction product

Ion exchange Ion exchange is a versatile process with constantly expanding uses. In waste water treatment it is used to remove or recover anions or cations depending on whether they are valuable, undesirable, or both. Cations are exchanged for hydrogen or sodium and anions for hydroxyl ions.

Most cation exchange resins in present use are synthetic polymers containing an active ion group such as SO_3H^-. The common anion exchange resins are synthetic resin amines. The reactions which occur depend on chemical equilibria situations in which one ion will selectively replace another on the ionized exchange site. When all the exchange sites have been substantially replaced, the resin can be regenerated by passing a concentrated solution of ions through the bed, which reverses the equilibrium. Regenerants commonly used include brines, sulfuric acid, and sodium hydroxide.

The process is such that the pollutant ion may be recovered for use. In the chemical industry ion exchange has found application in the recovery of chromic acid from cooling tower effluents, recovery of phenol, and the removal or recovery of small concentrations of dissolved electrolytes from solution.

Membrane processes Membrane processes may be widely applicable to waste treatment. Three important processes are dialysis, electrodialysis, and reverse osmosis or ultrafiltration.

Dialysis involves the separation of solutes by means of unequal diffusion through membranes. The rate of diffusion is based on the concentration driving forces and makes separations on the basis of molecular sizes, or diffusion rates of the solvents. The membrane primarily separates feed and product streams and in some cases acts as a molecular filter.

Dialysis has been used to recover sodium hydroxide from textile wastes containing hemicelluloses. With a waste containing 38 to 40 percent NaOH, the discharge concentration was reduced to 2 to 3 percent NaOH.

It has also been used for the separation of acids and metal salts. Acid recoveries as high as 75 percent have been reported.

Electrodialysis is similar but uses electromotive driving force rather than a concentration gradient to make separations. The ability of membranes to pass either cations or anions selectively is of importance.

Electrodialysis is finding increased use for tertiary treatment of biologically treated waste water to remove dissolved solids content.

Reverse osmosis, or ultrafiltration, uses a pressure gradient to make a separation based on the relative solubilities and diffusivities in the membrane of the various liquid phase components. Water with an undesirable content of dissolved solids is placed in contact with a suitable membrane at a pressure in excess of the osmotic pressure of the solution containing the dissolved solids. Under these conditions, fresh water or water with a small amount of dissolved solids flows through the membrane and is collected for use.

Reverse osmosis is useful for tertiary treatment or reclamation and for the recovery of a valuable constituent present in a waste stream. Projected capital and operating costs for flow rates of 20,000 gpd have been reported.

Since reverse osmosis can remove organic materials, viruses, and bacteria and because it can lower total dissolved solids, application of the process to tertiary sewage treatment for total water reuse is an attractive possibility.

Biological Treatment

Biological oxidation is widely practiced for treating soluble organic chemicals dissolved in water. A biological sludge consisting of bacteria, fungi, algae, and other microorganisms can convert many soluble organic compounds into bacterial cells and inorganic compounds as follows:

$$\text{Organic matter} + O_2 + NH_3 \xrightarrow{\text{cells}} \text{new cells} + CO_2 + H_2O \qquad (16)$$

Basic tests should be performed by specially qualified chemists and biologists to determine if the waste is amenable to biological oxidation. If the waste can be biologically decomposed, it often is acceptable to the municipal sewage treatment plant if it can handle the load. Usually treatment at the sewage plant is less expensive than at a separate treatment plant.

If separate biological treatment is to be employed, several methods are adaptable:

Trickling filters
Activated sludge
Oxidation ponds
Other biological methods

In all biological treatment systems microbes must have all the elements they need to grow. Domestic sewage normally contains all that the bacteria require. But nitrogen and phosphorus, primary nutritional elements, may be missing from chemical wastes; and a supplemental feed of these is needed. Minimum quantities of nitrogen and phosphorus required are 4.0 lb N/100 lb BOD removed and 0.6 lb P/100 lb BOD removed. Ammonia and phosphoric acid are commonly used to provide the required nutrients in wastes which are deficient in these elements.

A relatively narrow effective pH range will exist for most bioxidation systems. For most processes this covers a range of pH 5 to 9, with optimum rates occurring in the range of pH 6 to 8.

Variations in temperature affect all biological processes. The rate of biological reaction will increase with temperature to an optimum value, approximately 30°C for most aerobic waste systems. This means BOD removal efficiency falls off in the winter and improves in the summer.

Toxicity in biological oxidation systems may be caused by:

1. An organic substance which is toxic in high concentrations but biodegradable in low concentrations

2. Substances, such as heavy metals, which have a toxic threshold depending on the operating conditions

3. Inorganic salts and ammonia which exhibit a retardation at high concentrations

There are specific hydraulic and biological parameters for the proper sizing and design of biological oxidation systems.

Trickling filters A trickling filter consists of a bed of stone, gravel, or slag from 4 to 8 ft deep covered with slime growth over which the waste is distributed evenly. As the waste passes through the filter, organic matter present in the waste is removed by the biological film. Maximum hydraulic loadings to the filter of 0.5 gpm/sq ft will yield BOD removal efficiencies of about 85 percent.

Recently, molded plastic packings have been employed in depths up to 40 ft with hydraulic loadings as high as 6.0 gpm/sq ft. Depending on the hydraulic loading and depth of the filter, BOD removal efficiencies as high as 90 percent have been obtained on some wastes.

The maximum BOD concentration which can be effectively handled is approximately 500 mg/l applied to the top of the filter. High organic loadings are usually diluted by recirculating a portion of the filter effluent. The recirculation technique is also used during periods of low flow so that the filter is always wet.

The main advantages of a trickling filter are its basic simplicity, low power consumption, and ease of operation and maintenance. Disadvantages include the large surface area required and sensitivity to operating temperature.

Activated sludge Activated sludge is the most versatile biological treatment since it can be tailored to handle a wide variety of wastes and effluent requirements. The activated sludge process utilizes the same type of bacteria as the trickling filter process. In activated sludge, however, the bacteria are suspended in water and aerated in the presence of the organic wastes. With 4- to 8-hr retention time, activated sludge may provide up to 90 percent BOD removal.

Oxygen may be supplied by mechanical or diffused aeration systems, such as diffused aeration units, swing diffusers, surface aerators, turbine aerators, and spargers.

Since activated sludge is a flexible process, a number of modifications have been devised. These include the conventional system, tapered aeration, step aeration, contact stabilization, high-rate, extended aeration, and complete mixing. The various activated sludge systems and oxygen transfer devices are discussed in detail in Chap. 5.

Advantages of activated sludge, compared with the trickling filter, include its smaller area, lower capital cost, higher degree of treatment, greater flexibility, and more precise control.

Main disadvantages are higher power costs and the greater complexity of properly operating the system.

Both trickling filters and activated sludge produce a biological sludge which must be disposed of. This sludge contains the excess bacteria developed during the oxidation process. The sludge can be partially digested anaerobically and filtered followed by burning or can be filtered directly and burned or buried.

Oxidation ponds The oxidation pond is widely used where land is available and cheap and the climate is temperate. An oxidation pond is a large shallow lagoon or basin with waste water added at one point and stabilized effluent removed from another. Depth is usually 2 to 4 ft. Oxygen is supplied by respiration of algae and from wind action on the surface of the pond.

Deeper ponds, 3 to 6 ft deep, are called facultative ponds and are divided by loading and thermal stratification into an aerobic surface and an anaerobic bottom. Sludge deposited on the bottom will undergo anaerobic decomposition, producing methane and other gases. Odors will be produced if the aerobic layer is not maintained. Eventually the pond must be dredged to remove settled inert matter and microorganisms.

BOD removal is at a rate of about 50 lb/acre/day, without forced aeration. Biochemical activity is related to sunlight and temperature, and BOD removal efficiency therefore varies seasonally and is negligible with ice cover.

Aerated lagoons using mechanical surface aerators or diffusers are able to treat much greater organic loads. Floating suface aerators, turbine aerators, and brush aerators are commonly used. The method is a form of the activated sludge process without sludge recirculation with a relatively dilute concentration of microorganisms.

The organic loading can be increased by a factor of 5 compared with an ordinary oxidation pond. Raising the organic load, however, also raises the microbe population. In order to produce a low BOD effluent, these microbes must be removed. These suspended biological solids are commonly removed in a second-stage lagoon or pond.

Anaerobic digestion Anaerobic digestion is widely used to stabilize concentrated organic solids removed from settling tanks and from aerobic biological treatment systems. The waste is mixed with large quantities of microbes, and oxygen is excluded. Under these conditions, anaerobic bacteria thrive, converting up to 90 percent of the degradable organics into methane and carbon dioxide. High process efficiency requires elevated temperature and heated reaction tanks, with 95°F considered optimum.

The high degree of destruction minimizes the problem of excess sludge disposal. Power costs are reduced because oxygen is not required. In addition the methane gas is a source of energy for heating or for the generation of electricity. Disadvantages include the fact that a relatively high temperature is needed for efficient operation. Dilute wastes may not produce enough methane to take care of the heating. Also, methane producing bacteria grow at a slow rate, making it difficult to start the process and adjust to changing operating conditions. The system is well suited to treating concentrated wastes with BODs in excess of 10,000 mg/1.

Costs The manufacturing Chemists' Association reports that

> Biological oxidation capital costs range from $50 to $150 per pound of BOD treated for the basic elements. The cost of collection systems, any chemical pretreatment and sludge disposal will be extra depending on methods used. Operating costs approximate 2 to 10 cents per pound BOD per day. Oxidation ponds appear to be economical at land costs up to $5,000 per acre; however ponds can be considered only for small treatment systems.

Ultimate Disposal Methods

Other methods have been developed where the waste effluents, after proper treatment, are not discharged into adjacent surface receiving waters. In certain circumstances these alternate methods may be the most logical and economical.

Barging to sea Certain wastes, such as sewage sludges, acidic wastes, miscible organics, and atomic wastes, have been effectively disposed of at sea. With the exception of atomic wastes, these wastes are assimilated by the saline waters without detrimental effect.

Discharge at sea should not be started without sufficient test work to assure the discharger and the appropriate authorities of the anticipated effects. Despite the fact that no actual authority exists to prevent dumping beyond the continental limits of the country, good practice dictates that appropriate agencies be notified and asked for advice.

If the wastes contain material toxic to marine life, a biologically safe concentration should be determined in water similar to that of the dumping site. Then the dispersion rate can be determined so that this safe concentration can be reached quickly. Mixing to the desired concentration is accomplished by control of discharge rate and barge speed. The actual rate of dilution can be determined by adding tracer dyes to the waste.

Controlled rate subsurface dischage of spent sulfuric acid has been practiced for many years in outer New York harbor without detriment. In fact, the so-called "acid grounds" is a favorite spot for party fishing boats because of the variety and quantity of fish in the area.

A similar acid disposal program has recently been announced. Some 2,000 tons/day of sulfuric acid will be barged 110 miles from the plant for controlled rate subsurface discharge. The announced cost of the barge is over $1 million, and anticipated operating costs for storage tanks, pumps, and pipelines are roughly $1,000 per day.

The new program has the blessings of the state governor, the U.S. Department of Interior, the U.S. Fish and Wildlife Service, and all appropriate local, state, and federal authorities.

Burial Burial of chemical wastes in nearby sanitary landfills is appropriate for certain wastes. The producers of chemical wastes have, as a rule, excellent knowledge of the safety or hazard inherent to each waste product, together with the knowledge of their proper, safe handling.

Wastes should be carefully segregated and labeled using such classifications as flammable liquid, flammable solid, spontaneously combustible, dangerous when wet, oxidizing agent, organic peroxide, poison, acid, caustic, nonhazardous, and emitting a noxious odor.

The best system is one where the producer has complete control and responsibility for the identification, segregation, transportation, and disposal of the wastes. Where this is not feasible, it is mandatory to provide adequate communication between the waste producer, the hauler, and the receiving landfill operator.

Deep-well injection The disposal of chemical wastes into underground strata which are not satisfactory for other purposes is gaining in importance. A summary of data on 110 industrial waste injection wells has recently been published.

Each deep well for waste disposal must be considered as an experiment until test drilling shows that underground strata are suitable. An expert opinion by a competent geologist is essential.

A suitable stratum for waste disposal should be porous, of little or no economic value, be located between nonpermeable strata so that contamination of other groundwaters cannot occur, and be located in a seismically stable region. Areas meeting these criteria are limited but are fairly well defined.

Typically in regions where deep-well disposal is feasible, a permit is obtained to drill a test well, with core samples being submitted to appropriate state agencies.

Then if, and only if, the core samples indicate that the geology and the nature of the waste are compatible, an injection permit is given.

Capital costs vary from $30,000 for an 1,800-ft-deep well, where no surface equipment was required for treatment of the waste, to $1,400,000 for a complete system, including a well 12,000 ft deep, a treating plant with a clarifier, dual filters, and four piston type displacement pumps.

Operating costs are quite variable and depend on the required pumping pressure and the degree of pretreatment required. The latter may be quite expensive, since the formations into which the waste is pumped are usually quite dense and cannot accept any appreciable amount of suspended matter. Also the waste must be compatible with the connate waters so that no precipitate will plug the formation.

Obviously subsurface disposal must be approached with caution. However, quite a number of highly successfully deep wells are in operation where the economics and geology are favorable.

Incineration Incineration techniques have been discussed. It should be stressed that adequate attention should be given to proper operation to ensure that air pollution problems do not result, which may be more serious than the water pollution problem being corrected.

Chapter 15

Pollution Control in Textile Mills

HERBERT A. SCHLESINGER
Manager, Pollution Control Division

EMIL F. DUL
Engineering Specialist, Pollution Control Division

THOMAS A. FRIDY, JR.
Project Director, Water and Waste Treatment,
Civil and Marine Engineering Department,
Lockwood Greene Engineers, Inc.,
New York, New York and Spartanburg, South Carolina

INTRODUCTION

Textiles, the finished, purchasable product, have their origin in wool, cotton, synthetic fibers, or combinations thereof. The processing which the raw fibers undergo aims to:

1. Remove natural impurities (in the case of cotton and wool)
2. Impart particular qualities of sight, touch, and durability

As the degree to which these aims are carried, so varies the overall effluent quality

(*a*)

(*b*)

Fig. 1 Overall view of new type textile mill.

from a textile mill. The following sections indicate processing methods, waste sources and qualities, and treatment methods associated with textile manufacture.

PROCESSING

Cotton

Cotton processing consists of two basic steps—weaving and finishing (see Fig. 2). *Weaving* entails mostly dry processing wherein:

1. Trash and foreign matter are manually or mechanically removed from the raw fibers (opening, cleaning, picking, carding, combing).

2. The fibers are joined, straightened, drawn into thread, and wound on spools (drawing, roving, spinning, winding).

3. The thread is run through a starch solution (sizing or slashing) and then dried so that it has the strength and stiffness required to withstand the abrasion and friction generated in the weaving operation.

4. The strengthened thread is woven into cloth (greige goods).

Pollution problems resulting from this processing are predominantly dust generation in the high-speed mechanical operations and a minimum amount of liquid waste generated from the spills and washout of the slashing boxes and the preparation of the size solution.

Dry dust collectors of the vacuum type are well suited to keeping a dust free working area.

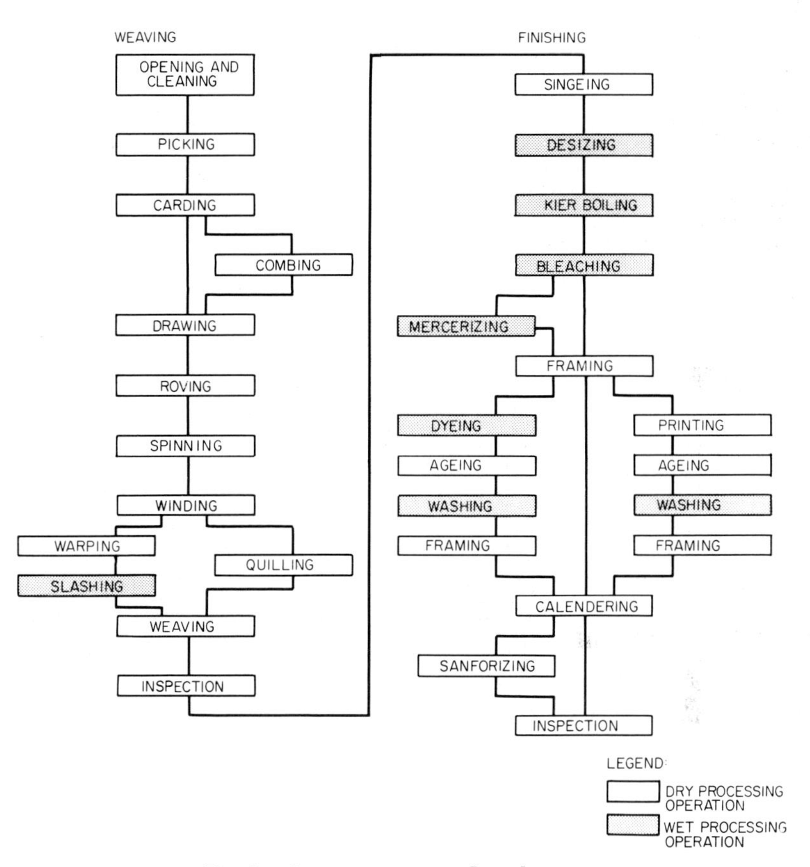

Fig. 2 Cotton processing flow diagram.

Finely divided lint is present in quantity in the air in the vicinity of various operations associated with spinning the yarn and weaving the cloth. Most modern mills have air tunnels designed and built under the floor to effect a downward flow of air around the machines which produce the largest volumes of waste lint (carding, roving, spinning, twisting, weaving, etc.). In opening and picking rooms in a cotton plant, stock is usually handled by overhead ductwork. Lint and dust are collected by the vacuum system as close as possible to their point of origin. Automatic traveling vacuum cleaning systems are frequently used in conjunction with the underfloor system. The suspended lint and dust are conveyed by air movement through ducts to a centrally located filter system for removal.

Automatic dry type traveling disposable air filters are extensively used for this purpose. Since any filtering media tends to hold the lint fibers tenaciously, a low-cost disposable paper medium has been extensively used. A typical unit operates in a fashion similar to a window shade (see Figs. 3 and 4). The paper medium, supported by a screen, moves downward at a rate depending on air contamination. A pressure differential switch can be set for the desired operating resistance. When the pressure drop exceeds the set point, the drive motor will move and expose enough new filter surface to reduce the pressure drop by about 0.1 in. W.G. A media runout switch, indicating that a new roll is required, can activate a visible and/or audible alarm.

A typical filter section is about 4 ft wide and 5 to 15 ft high, constructed of galvanized steel and provided with a flat flexible stainless steel screen to support the paper roll filter medium. One or more sections can be used, depending on the air volume to be handled. A 4-ft-wide paper roll about 800 ft long (single-ply) costs about $10 and may last 1 to 3 months depending on air contamination.

Fig. 3 Front view of automatic traveling lint filter. (*American Air Filter Co., Louisville, Ky.*)

Operating air resistance is usually about 1 in. W.G. (single-ply medium). Recommended air velocity is 300 to 600 fpm. Single-ply paper media have been found satisfactory for textile mills.

Some 5 to 8 lb of waste lint per 100 lb of cotton processed can be recovered by the vacuum system. This collected lint is salable at a low price.

The pollution load from the slasher boxes is relatively small when compared with a finishing plant, but it can become a very troublesome problem. Presently, many plants use circulating systems of starch and water plus other additives in small quantities to feed cookers which, in turn, supply the starch size boxes. In other instances starch is cooked in kettles of several hundred gallons capacity and is circulated hot through a supply system to the slasher boxes. In either instance it is possible to have a bad mix or some other emergency which will necessitate dumping some 500 to 1,000 gal or more of the concentrated size formula. Depending on the mix being used, this can produce (for the 1,000-gal dump) from 1,000 to 2,000 lb of 5-day BOD. This can easily incapacitate a relatively small treatment system. Many of the plants are located in isolated areas, and the usual treatment system provided is not capable of handling this shock load. One solution has been to dump these batches to a storage tank, add enzymes to reduce the viscosity of the mix when it becomes cold, and convert the starch form. From this storage tank the waste is metered uniformly over a week or more to the treatment plant. The increased 5-day BOD loading contributed to the treatment plant under these conditions is tolerable and can be provided for.

The *finishing* mill receives the greige goods and must process them to satisfy variable client demands. The processing is essentially wholly of the wet type, wherein:

1. Natural (wax, pectins, alcohols, etc.) and acquired (unremoved size agent, dirt, oil, grease, etc.) impurities must be initially removed (to variable end points) to make the cloth suitable for chemical processing.
2. Wet chemical processing is performed to impart to the cloth the desired properties and appearance (again, to variable end points).

This dichotomy also represents the major sources of liquid wastes—natural impurities and process chemicals.

Finishing operations incorporate the following (see Fig. 2):

Singeing The opened, brushed cloth passes between heated plates or rollers or across an open gas flame at a high rate of speed (300 yd/min) to burn off the loose fibers that are attached to the surface of the cloth. Sparks are quenched by then passing the singed cloth through the water box of the singer.

Fig. 4 Rear view of automatic traveling lint filter. (*American Air Filter Co., Louisville, Ky.*)

Desizing To prepare the greige goods for dyeing and further processing, the size (starch) added to the cloth in the sizing (slashing) step in the weaving mill must be removed. Solubilizing the starch by enzymatic (1 percent commercial enzyme by volume, salt, penetrant) or acid (0.5 percent sulfuric acid solution) methods is the typical approach employed. The desize solutions are usually contained in the water box of the singer (enzyme desize is destroyed by heat and therefore should be preceded by a water box for spark quenching), and contact of cloth and desize takes place. Contact is followed by a detention time of 3 to 12 hr (detention chamber is similar to a wooden crate), during which time the starch is hydrolyzed and expressed liquor escapes. Fresh water rinsing of the greige goods completes the desize operation. The goods are then run through a caustic and penetrant bath, and are ready for kiering.

It should be noted that desizing must *always* be performed on the greige goods before any further processing.

Kiering This operation consists of cooking the greige goods with a hot alkaline detergent or soap solution to remove cotton wax, dirt, and grease in order to develop a white, absorbent fiber that is essentially pure cellulose. This operation may be performed by either of two major methods—pressure kier boiling or continuous scour. The former is the traditional method, whereas the latter is coming into more prominence because it requires less time and space and reduces water and chemical use, while producing the same results.

Pressure Kier Boil. The caustic and penetrant-laden greige goods received from the postdesizing wash operation are "plaited down" (sinusoidally packed) in steel or stainless steel pressure vessels (kiers) which range in diameter from 8 to 9 ft and in height from 10 to 12 ft. The kier has a capacity of 2 to 5 tons of cloth and is equipped with closed steam piping (for heating) and a grated support structure located a few feet above the tank bottom. The grate supports the cloth while additionally forming a reservoir at the tank bottom for kier liquor (alkaline scour agent) storage for recirculation.

The chemicals used to prepare the kier liquor for the pressure kier boil are:

Sodium hydroxide—1 to 8 percent based on weight of cloth or fiber
Sodium carbonate—1 to 3 percent of dry cloth weight
Sodium silicate—¼ to 1 percent of dry cloth weight
Pine oil soap—removes natural waxes
Fatty alcohol sulfates—provides a melting action

These chemicals are dissolved and mixed in preparation tanks preheated to 120 to 160°F and are added to the kiers while heat (via steam piping) is applied. Operating conditions entail pressurized (5 to 15 psi) cooking of the cloth and caustic liquor at up to 250°F for a period of 2 to 12 hr with variations in the parameters reflecting variable end points of quality of impurities removal for variable end uses of the cloth. For example, a cloth which is to be finally marketed as white requires more time and care than one which is to be marketed as black. During the cooking, the kier liquor is constantly recycled from the reservoir to the top of the kier. Completion of pressure kier boil involves reducing the tank pressure, drawing off kier liquor from the tank bottom, and starting the hot water rinse. The kier liquor drawoff has an opaque brown color, with progressive rinse waters containing less of the kier liquor. Rinsing is usually continued until no brown color remains in the rinse water, although this is subject to some variation.

Continuous Scour. The goods received from the desize wash are passed through a caustic solution (3 to 6 percent caustic soda plus wetting agents) until saturated and are then passed into a heated (live steam) J box for a period of 1 hr at a temperature of approximately 210°F. When kiering is complete, the kiered cloth is rinsed to remove kier liquor.

Bleaching Following kiering, bleaching of the cloth with chlorine, hypochlorite, or peroxide may be performed to whiten (remove natural coloring) the fabric to a high degree. The cloth is first water rinsed and then rinsed in an antichlor (sodium bisulfite and weak solutions of sulfuric or hydrochloric acid) bath. This is followed by another water rinse, and finally the cloth is passed through a hypochlorite solution, squeezed lightly, and stored in "white" bins for up to 24 hr at temperatures between 55 and 65°F for contact and drying.

Mercerizing This process may be performed on bleached cotton to swell the cotton fiber, add luster, and increase dye affinity. It consists of carrying the cloth through a cold 10 to 30 percent solution of caustic soda, while the cloth is kept in tension. The caustic action takes between 1 and 3 min and is followed by a water rinse, water wash, acid dip, and final water wash.

Dyeing This process can be performed in a variety of ways. Because of its variability, details of this operation will not be covered. This information can be obtained from the "American Cotton Handbook." The various types of dyes used are classified according to the method of application. A listing of the more prominent dyes follows.

Vat. The dye is put on the goods in its reduced state and is then oxidized. These dyes have excellent light and wash fastness.

Developed. The dye is applied to the cloth and diazotized, and the color is developed with a secondary chemical called the developer. These dyes have good wash fastness.

Naphthol. The naphthol is applied to the fabric and passed through the developer for coupling. This produces bright colors and good fastness to light, wash, and bleach.

Sulfur. The dye is put on the cloth in a reduced state and is then oxidized, producing good fastness to light and washing.

Aniline Black. The aniline is oxidized on the goods by air or steam aging, producing excellent fastness to light.

Direct. Applied directly to the cloth. These are usually low-cost dyes, easy to apply but not very fast.

Printing This process is used to impart a colored pattern or design to the cloth. The operation is performed on a roller print machine or a screen print machine wherein printing pastes, made in the textile plant color shop, are transferred to the fabric. After printing, the goods are steamed, aged. or otherwise treated to fix the color. Again, detailed descriptions of printing paste compositions are available in the "American Cotton Handbook."

Final finishing In order to impart a smooth appearance and desired stiffness to the cloth, a final size (starch base) or resin is applied to the cloth. The size mixes are made up in a 10 to 30 percent concentration. This operation is followed by the final operations of calendering and rolling. Specialty treatment may be provided to yield a cloth which is preshrunk, waterproofed, fireproofed, etc. These specialty finishes are of little waste generation significance.

Wool

Wool is a natural fiber of animal origin (sheep) and is classed as a protein, keratin. The wool fiber consists of five basic elements:

Carbon—50 percent
Hydrogen—7 percent
Oxygen—22 to 25 percent
Nitrogen—16 to 17 percent
Sulfur—3 to 4 percent

In the raw state wool contains impurities from natural, acquired, and applied sources.

Natural impurities Yolk is the term applied to designate all glandular secretions of animal origin adhering to the fleece. It consists of two parts—*suint* and *wool grease* (the water soluble and water insoluble portions, respectively).

Acquired impurities These consist of dust, dirt, straw, vegetable matter, and earth.

Applied impurities

1. Substances used for treatment against disease and insect pests
2. Substances used for identification (tar or paint)

Wool is insoluble in water under ordinary conditions, but at temperatures over 250°F, water under pressure will dissolve wool. The wool fiber expands when wetted but will contract to its original size upon drying. Wool is amphoteric and is easily damaged by caustic or acid solutions. Thus care must be taken when subjecting the wool fiber to the various alkaline and acid treatments it undergoes in finishing operations.

It should be noted that of the wool removed from the sheep only 40 percent is usable fiber. Stated in another way, for every pound of fiber (scoured wool) produced, 1½ lb of impurities are removed from it, much of which is discharged to waste.

As in the case of cotton, wool must undergo both weaving and finishing operations—the former in order to develop a continuous cloth from the bits of raw fiber, the latter in order to impart the desired characteristics to the cloth.

Weaving entails numerous mechanical and manual dry operations (see Fig. 5), while only two operations are of the wet type (scouring and slashing). A brief review of each operation will be given, and a detailed discussion will be afforded only the scouring operation.

Weaving operations include sorting and blending of raw fibers, scouring (to remove natural grease and foreign matter), drying, carding (opening and paralleling of fibers), rewashing, oiling (for reducing static electricity), gilling (paralleling of fibers), combing (to remove trash), roving (or drawing of fibers into rope form, spinning (or further drawing of fibers into threads), slashing (starch sizing of wool thread to protect fiber against abrasion and friction generated in the weaving operation), weaving (threads are woven into cloth), and mending.

Scouring is performed by detergents or solvents.

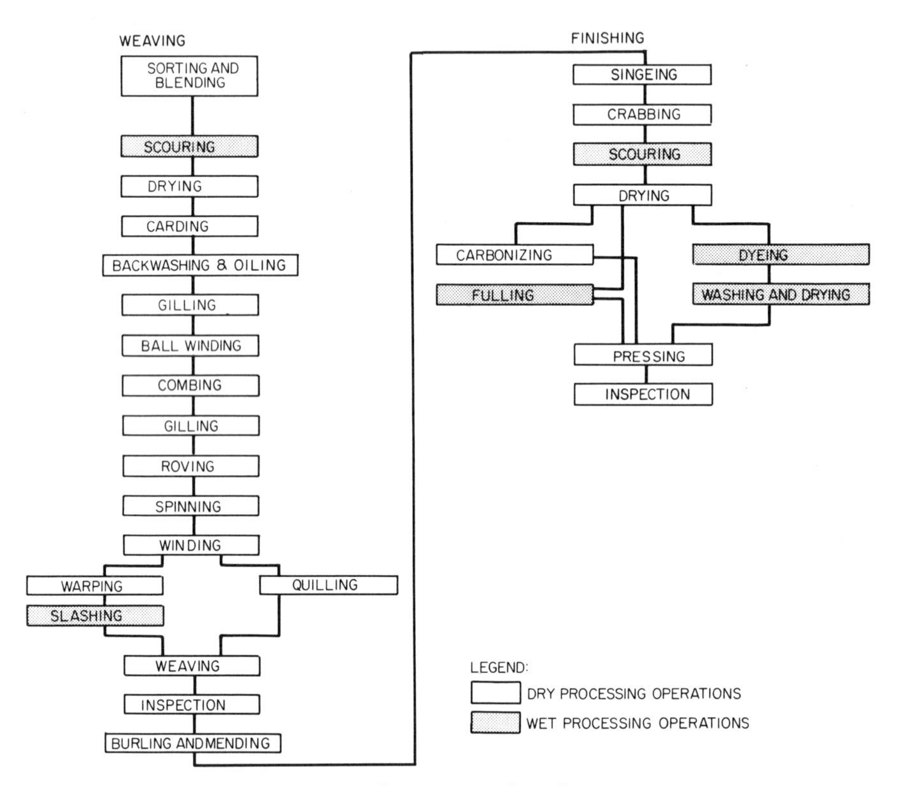

Fig. 5 Wool processing flow diagram.

Detergent scouring This is the predominant method. However, recently some plants have gone to solvent scouring. The process consists of using three to five bowls, each with a capacity of 1,000 to 3,000 gal. Bowl 1 is used for steeping (desuinting), bowls 2 and 3 are the soap-alkali and grease removing units, and bowls 4 and 5 are rinse units.

Steeping (Desuinting) Bowl. Suint is the water soluble portion of the yolk, or glandular secretions. Desuinting incorporates detention of the wool fiber in water at temperatures ranging from 90 to 110°F, wherein the suint is dissolved and heavy dirt particles are washed out of the cloth. A continuous cleanout is provided in this bowl to prevent heavy buildup of salts and dirt.

Scour Bowls. These bowls contain soap and alkali (usually sodium carbonate) solutions of varying concentrations, usually less than 7,500 ppm. The alkali achieves saponification of the fatty acids of the wool grease to form natural soaps. The soap emulsifies wool fat and forms suspensions of light dirt particles. Temperatures are kept between 125 and 130°F. The greatest source of pollution in the wool textile mill is generated in this scouring process.

With the advent of synthetic detergents, which are used instead of soap, and the more recent substitution of neutral agents, such as sodium sulfate or chloride, for sodium carbonate, there has been an increase in operating temperatures to 150°F. Although much less damage to wool results from the use of these newer agents, little overall change results in the pollution potential of this process (see Table 1).

Rinse Bowls. Removal of any residual soap, grease, alkali, or dirt is effected by using water at temperatures of 115 to 120°F.

Typical production rates through this operation are 1,000 to 3,000 lb of fiber/hr when a four-bowl train (48 in. wide) is used.

TABLE 1 Analyses of Wastes from Woolen Mill*

Item	pH range	Total solids range, ppm	BOD range, ppm
Grease scouring:			
Soap-alkali method:			
Bowl 1	9.5–9.8	42,116–76,950	11,900–27,000
Bowl 2	10.1–10.5	16,650–32,532	2,340–7,350
Bowl 3	9.7–9.8	834–1,424	150–400
Detergent—Na_2SO_4 method:			
Bowl 1†	7.6–8.0	47,108–91,456	11,000–25,000
Bowl 2†	7.4–9.1	5,024–7,856	775–1,560
Bowl 3†	6.4–7.8	1,052–2,406	115–260
Stock dyeing:			
Acetic acid used	4.8–8.4	2,418–5,880	1,440–3,450
Ammonium sulfate used	5.0–8.3	7,344–9,160	140–1,020
Wash after fulling, first soap:			
Soap used for fulling	9.0–10.7	11,270–23,120	3,900–24,000
Detergent used for fulling	9.7	4,516–5,144	4,000–4,000
Neutralization following carbonizing:			
First running rinse	1.9–2.4	494–1,988	20–35
First soda ash bath	7.9–9.0	8,678–10,884	21–36

* From Masselli, Masselli, and Burford, "A Simplification of Textile Waste Survey and Treatment," New England Interstate Water Pollution Control Commission.

† These samples were taken early in the day and give a false impression of low values over the soap-alkali method. Actually, at end of day results between processes should be similar.

Solvent scouring Because of the high pollution loads generated from detergent scouring (see Table 1), attention was given to solvent extraction of the wool grease. Since dirt is not removed by solvents, solvent scouring must be followed by a detergent wash. Solvent scouring is performed on a batch or continuous basis using organic solvents (benzene, CCl_4, ethyl alcohol, methyl alcohol, isopropyl alcohol) to dissolve and remove the wool grease from the fiber. The process is performed in a kier, and the grease laden spent solvent is recovered, redistilled, and reused. The cloth is then washed to remove suint and traces of solvent.

Finishing operations are shown in Fig. 5. The major wet processes are scouring, dyeing, fulling, washing, carbonizing, and bleaching.

Dyeing Following scouring, dyeing is performed in either open or pressure type machines. Pollution potential is a function of the dye used. A dramatic comparison of pollution load can be made (see Table 1) when acetic acid dye is compared with ammonium sulfate dye. Approximately 2,400 ppm BOD is reported in the former vs. approximately 580 ppm BOD in the latter.

Fulling This wet process imparts a felt quality to the wool. Soap is used as the felting agent, and little waste is generated since the fulling solution remains in the cloth.

Washing At this point in the processing, the wool cloth contains considerable process chemicals. The cloth is washed in a "first soap," squeezed between rollers, and then washed in a "second soap." Rinse of the cloth in a water bath completes the removal of process chemicals.

Carbonizing A final attempt is made to remove residual vegetable matter or impurities in the wool fabric by sulfuric acid treatment, heating, and mechanical dusting. The wool fabric is impregnated with 4 to 6 percent sulfuric acid and then oven-dried at 212 to 220°F. During drying, water is evaporated and the acid becomes more concentrated, charring any remaining organic matter. The cloth is then passed through rollers where the charred matter is crushed and then conveyed into a mechanical dusting machine to "knock out" the char dust. The fabric is then rinsed in water, passed through a bath of 2½ percent sodium carbonate for residual acid neutralization, washed again, and then dried.

This process generates very fine carbon particles which appear as smoke, as well as some fumes and odors. These fumes probably include residual sulfur oxides (if sulfuric acid has been used for carbonizing) as well as organic decomposition products, and are generally very corrosive.

Corrosion resistant stainless steels and/or plastics have been used in the collecting systems, which usually exhaust the fumes to the atmosphere, sometimes through a dry cyclone collector. However, since a significant quantity of the particulate matter is in the submicron range, a visible emission may be apparent at the cyclone discharge. In the future, more efficient removal of the carbon (smoke) may be required, as well as reduction of the sulfur oxides or other carbonizing chemicals. The severity of the problem will depend to a great extent on location, topography, and local meteorological conditions. One possible solution to the problem would be to vent the carbonizing machine exhaust to the main boiler stack, or to use a high vent stack. Other solutions could be wet scrubbing and fabric filtration, perhaps after a dry cyclone to remove larger particles. Wet scrubbing would reduce particulate carbon as well as residual gases such as sulfur oxides.

Bleaching Bleaching with sulfur dioxide or hydrogen peroxide may be performed to whiten the natural yellow tint of the wool.

Synthetics

Synthetic fibers are manmade. The most prominent (by generic classification) are acetate, acrylic, nylon, polyester, and rayon. Others, produced in lesser quantity, are glass, metallic, Olithon, rubber, Saran, Spandex, triacetate, and Vinyon. When received at the finishing plant, these fabrics require no processing for the removal of natural impurities because of their manmade origin, but they do require treatment for removal of sizing, antistats, and lubricating oils used in the weaving operations. There is considerable difference in the sizing requirements of spun vs. continuous filament yarn. Spun yarn is handled like natural fibers and requires a quantity of size to impart strength and a protective coating for the yarn in weaving. The continuous filament yarn requires considerably less size of different types. Because of the low moisture uptake of synthetic fibers as compared with wool and cotton, static electricity problems are encountered in yarn preparation and weaving operations. Thus, antistatic oils and lubricants are applied to the fiber before weaving. Typical sizes, antistats, and lubricants used are:

1. Polyvinyl alcohol
2. Styrene-base resins
3. Polyalkylene glycols
4. Gelatin
5. Polyvinyl acetate

The finishing processes are generally similar for all these fabrics and include:

1. Scouring—for removal of process chemicals used in weaving operations
2. Initial rinsing

3. Dyeing or bleaching
4. Second rinsing
5. Final finishing (waterproofing, shrinkproofing, etc.)

Detailed descriptions of finishing methods are included under Overall Waste Character and Residual Waste Treatment. Waste analyses from the various operations are given in Table 2.

TABLE 2 Analyses of Wastes from Synthetic Fiber Finish Mill*

	pH range	Total solids range, ppm	BOD avg, ppm	BOD % OWF avg†
Rayon processing:				
Scour and dye	8.2–9.0	1,042–5,572	2,832	5.7
Salt take-off	6.8–6.9	3,388–7,256	58	0.1
Waterproof	....	....	960	1.9
Acetate processing:				
Scour and dye	8.3–8.5	1,534–2,022	2,000	5.0
Scour and bleach	8.9–9.6	766–946	750 (Estimated)	1.8 (Estimated)
First rinse	7.0–9.1	108–188	Peroxide	
Second rinse	6.8–7.3	80–88	Contained peroxide	0.0
Nylon processing:				
Scour	9.3–12.6	1,492–2,278	1,360	3.4
First rinse	8.2–10.7	150–954	90	0.2
Second rinse	6.5–8.2	106–932	25	0.1
Dye	7.8–9.0	318–1,016	368	0.9
Last rinse	7.3–7.6	106–134	11	0.0
Waterproof	....	....	450	1.1
Orlon processing:				
First scour	9.5–10.0	1,350–2,470	2,190	6.6
First rinse	6.4–8.7	102–294	109	0.4
First dye	2.2–6.5	170–1.950	175	0.5
Second rinse	4.1–6.5	116–300	42	0.1
Second dye	1.3–1.7	130–3,002	995	3.0
Second scour	5.9–7.7	612–1,824	688	2.0
Third rinse	6.3–7.4	82–152	50	0.2
Waterproof	3.7–4.3	896–2,318	2,110	6.3
Dacron processing:	(Estimated from OWF concentrations as listed)			
Scour	....	....	650	
Dyes:				
o-Phenylphenol (10% OWF)	....	....	6,000	18.0
Benzoic acid (40% OWF)	....	....	27,000	81.0
Salicylic acid (40% OWF)	....	....	24,000	72.0
Phenylmethylcarbinol (30% OWF)	....	....	19,000	57.0
Monochlorobenzene (40% OWF)	....	....	480	1.4

* From Masselli, Masselli, and Burford, "A Simplification of Textile Waste Survey and Treatment," New England Interstate Water Pollution Control Commission.

† % on weight of fiber, a weight percentage based on dried cloth weight.

WASTE SOURCES, QUANTITY, AND QUALITY

Cotton

Each of the unit operations or processes performed on the greige goods during finishing generates a liquid waste with its own peculiar character, at each particular installation. This can be seen by inspection of Tables 3 to 8, wherein the average values of BOD, total solids, and flow, as reported by various authorities for the same general operations, vary widely. This is mainly because processing, although easily

TABLE 3 Analysis and Pollution Potential of Various Textile Mill Solutions*

Item	pH range	Total solids range, ppm	BOD range, ppm
Experimental desize solutions, first rinse water following detention time for enzyme action	5.0–9.5	4,968–9,474	2,600–5,450
Caustic kier:			
First boil	11.0–13.1	33,804–58,126	6,100–14,100
Second boil	11.5–12.9	5,790–48,458	1,525–2,300
Continuous scour:			
Drippings from J box	11.6–12.8	66,428–97,468	13,800–15,700
First rinse following caustic	11.0–12.3	1,858–2,370	425–630
Second rinse following caustic	9.9–11.0	174–264	24–46
Bleaching:			
First rinse following hypochlorite bleaching	7.3–81	140–548	6–100
First rinse following peroxide bleaching	7.5–9.5	214–368	26–78
Rinse following mercerizing (after fourth box)	12.1–12.5	2,032–4,952	7–54

* Adapted from Masselli, Masselli, and Burford, "A Simplification of Textile Waste Survey and Treatment," New England Interstate Water Pollution Control Commission.

TABLE 4 Pollution Effect of Cotton Processing Wastes*

Process	Wastes,† ppm		Gal. wastes per 1,000 lb goods
	pH	BOD	
Slashing, sizing yarn‡	7.0–9.5	620–2,500	60–940
Desizing	....	1,700–5,200	300–1,100
Kiering	10–13	680–2,900	310–1,700
Scouring	....	50–110	2,300–5,100
Bleaching (range)	8.5–9.6	90–1,700	300–14,900
Mercerizing	5.5–9.5	45–65	27,900–36,950
Dyeing:			
Aniline black	....	40–55	15,000–23,000
Basic	6–7.5	100–200	18,000–36,000
Developed colors	5–10	75–200	8,900–25,000
Direct	6.5–7.6	220–600	1,700–6,400
Indigo	5–10	90–1,700	600–6,000
Naphthol	5–10	15–675	2,300–16,800
Sulfur	8–10	11–1,800	2,900–25,600
Vats	5–10	125–1,500	1,000–20,000

* Adapted from "An Industrial Waste Guide to the Cotton Textile Industry," U.S. Public Health Service.

† Composite of all wastes connected with each process.

‡ Cloth weaving mill waste.

TABLE 5 Cotton Desize Waste Characteristics*

Waste	Flow, gal†	BOD, ppm	Pop. equiv.‡
Desizing bath	140	8,000	54
Wash water	200–1,800	840	42
Total	1,100	1,750	96

* From N. L. Nemerow, Oxidation of Enzyme Desize and Starch Rinse Textile Wastes, *Sewage and Industrial Wastes.*
† Per 1,000 lb goods.
‡ Per 1,000 lb goods per day.

TABLE 6 Quantity and Sanitary Characters of Enzyme Desize and Starch Rinse Wastes*

Item	Enzyme desize	Starch rinse
Volume (gal/100 lb goods)	2.5	720
pH	7.35	7.1
5-day BOD, ppm	4,375–5,063	997
Total solids, ppm	8,946–10,548	1,482–1,830
Suspended solids, ppm	197–272	244–268
Total dissolved solids, ppm	8,682–10,276	1,238–1,562
Color,† ppm	400	80
Turbidity, Jackson, ppm	....	250

* From N. L. Nemerow, Oxidation of Enzyme Desize and Starch Rinse Textile Wastes, *Sewage and Industrial Wastes.*
† Standard color wheel.

TABLE 7 Average Raw and Treated Kier Waste Characteristics*

Item	Raw waste	Treated waste†	
		Unseeded	Seeded‡
pH	12	8.2	8.3
Alkali, ppm as $CaCO_3$	2,174	869	906
Color	Brown	Tan	Tan
Turbidity, ppm	205	813	712
Total solids, ppm	6,323	5,275	4,908
Total suspended solids, ppm	55		
5-day BOD, ppm	2,140	695	507
Ratio, vol/BOD	1.3:1	2.93:1	3.37:1

* From N. L. Nemerow, Oxidation of Cotton Kier Wastes, *Sewage and Industrial Wastes.*
† Aeration for 24 hr at 20 to 25°C, following neutralization, and ammonium salt addition.
‡ With seed adaptation in domestic sewage.

TABLE 8 Kier Liquor Waste Composition*

Component	*Volume,† gal*
Caustic soda waste	25.0
Bleach waste	16.3
Wash water waste	4.4
Total	45.7

* From N. L. Nemerow, Oxidation of Cotton Kier Wastes, *Sewage and Industrial Wastes.*
† As found in typical North Carolina mill, in gallons per 100 lb of cloth processed.

described in general, is actually quite complicated, since each process may be performed in several steps. Further, any one operation may be performed:

1. Batchwise or continuously
2. In kiers, jigs, or other machines
3. To variable end points of client demand

Thus, only broad generalizations can be made regarding the character of a plant's waste. These generalizations should not be construed as precluding the need for an intensive in-plant survey of waste quantities and qualities.

Data in Table 3 are reduced to those in Table 9, wherein the BOD loads from various operations were computed as percent OWF (that is, a weight percentage based upon dried cloth weight) in order to develop a general relationship among the pollution potentials of these operations. Inspection of Table 9 reveals the following:

Total BOD from mill (lb) = [0.067 × lb (dry) of cloth desized] + (0.045 lb of cloth scoured) + (0.005 × lb of cloth bleached) + (0.006 × lb of cloth mercerized) + (0.025 × lb of cloth dyed and printed)

More interesting, however, is the fact that when ratios are taken of the percent OWF BOD potential of all operations, it is seen that desizing, which must *always* be performed on greige goods received at the finishing mill from the weaving mill, contributes a minimum of 45 percent of the BOD load from a textile finishing mill.

TABLE 9 BOD Contribution of Cotton Processes*

Process		BOD potential % OWF†	BOD contribution,‡ %	BOD sources
Desizing	Avg	6.7	45	Glucose from starch hydrolysis
Scouring:				
Pressure kier, first boil		5.3		Natural waxes, pectins, alcohols, etc. (80%), penetrants and assistants (20%)
Pressure kier, second boil		(0.8)		
Open kiering		(0.4)		
Continuous scour		4.2		
	Avg	4.7	31	
Bleaching:				
Continuous, hypochlorite		0.8		Penetrants
Continuous, hydrogen peroxide		0.3		
Open kier, hydrogen peroxide		0.4		
	Avg	0.5	3	
Mercerizing	Avg	0.6	4	Penetrants
Coloring:				
Dyeing		0.5–3.2		Sodium sulfide, sulfite, acetic acid, etc.
Printing (including color shop wastes and wash after printing)		1.9–4.2		Starch mainly, plus glycerol, reducing agents, detergents, soap
	Avg	2.5	17	
Grand total (based on averages)		15.0	100	

* From Masselli, Masselli, and Burford, "A Simplification of Textile Waste Survey and Treatment," New England Interstate Water Pollution Control Commission.

† % on weight of fiber or lb BOD/100 lb fiber or cloth.

‡ Based on grand total of averages (15.0 percent OWF).

Also note that, in the event that all the operations listed in Table 9 are not performed, the BOD contribution from desizing to the overall plant waste will increase. This fact has sparked intensive research in the area of finding substitutes for starch which will produce less pollution but with the same cloth results. Table 10 shows a 50 percent reduction in the BOD of one textile finishing mill's waste when carboxymethyl cellulose (CMC) was used as a substitute for cornstarch in sizing and soap was replaced by detergent in the yarn dyeing process. The same end effects on the textile resulted. This figure is conservative and has been confirmed by Lockwood Greene in two large finishing operations recently studied. One plant finished mostly purchased goods with starch sizing, while the other had a controlled source of greige goods using CMC. The 5-day BOD average was approximately 60 percent lower for the CMC plant. Some adjustment must be made for the absence of a comparable printing operation in the latter plant. Earlier, it was shown that for the same solution strengths under similar solution preparation conditions, CMC exerted a much lower BOD than starch (see Table 11). It was also shown that if the BODs exerted by starch and CMC are compared (see Table 12), CMC exerts a much lower initial oxygen demand (lag) and therefore less of an initial demand on the

TABLE 10 Pollution Reduction Due to Process Chemical Changes*

Item	Before changes	After changes†	Reduction
5-day BOD, ppm	400	210	190
pH	11.5–12.0	10.0–10.5	2.0
Total alkalinity, ppm	1,600	560	1,040
Hydroxyl alkalinity, ppm	....	180	
Carbonate alkalinity, ppm	....	380	

* From Hutto and Williams, Pilot Plant Studies of Processing Wastes of Cotton Textiles, *Proceedings of 9th Southern Municipal and Industrial Waste Conference.*

† Changes were: (1) starch replaced by CMC (carboxymethyl cellulose) and (2) soaps replaced by detergents.

treatment plant and receiving stream. In addition, because CMC costs more than starch, studies were undertaken to determine the optimum ratio of starch to CMC to be used in sizing (see Table 13). The findings were:

1. The BOD developed by mixtures of CMC and starch is in direct proportion to the substitution percentage. Thus a 50 percent starch replacement indicates nearly 50 percent reduction of 5-day BOD. (See Table 11.)

2. Actual practice and the economics of the operation have shown about 65 percent starch, 35 percent CMC to be normally applicable.

3. Since CMC itself is water soluble, when used with starch in the proper proportions, it does not require enzymes for removal (desizing). Thus, only water washing is required and only about *half* of the total water is necessary for removal of the starch-CMC mixture.

4. Actual mill practice has shown the desized cloth from the starch-CMC mixture to be equal to or better than the use of starch alone.

To direct attention back to Table 9, it is very interesting to note that natural impurities can account for 31 percent (all operations performed listed in Table 9) to 41 percent (only desizing and scouring performed) of the BOD load. Thus, 59 to 69 percent of the total waste load finds its source in process chemicals. Table 14 is a tabulation of cotton finishing process chemicals consumption and BOD.

TABLE 11 BOD, ppm, of 0.1 Percent Solutions of Starch and Carboxymethyl Cellulose*

Days	Starch†	Carboxymethyl cellulose†
1	245	1.58
2	549	4.08
3	672	7.38
4	760	7.98
5	812	10.7

* From Dickerson, A Solution to Cotton Desizing Waste Problems, *Proceedings of 4th Southern Municipal and Industrial Waste Conference.*

† Starch and CMC samples were boiled for 1 hr and then samples taken for BOD determinations, in order to simulate mill conditions.

TABLE 12 BOD, ppm, of 0.1 Percent Solutions of Starch and Carboxymethyl Cellulose*

Days	Starch	Carboxymethyl cellulose†
1	245	1.58
2	549	4.08
3	672	7.38
4	760	7.98
5	812	10.7
6	851	13.5
7	895	42.4
8	911	49.2
9	928	55.0
10	945	57.5

* From Dickerson, A Solution to Cotton Desizing Waste Problems, *Proceedings of 4th Southern Municipal and Industrial Waste Conference.*

† CMC acclimated with sanitary sewage. Nutrients added at the ratio of BOD:N:P = 100:5:2.

TABLE 13 BOD, ppm, for Various Starch-CMC Ratios*

Days	80% starch, 20% CMC	65% starch, 35% CMC	50% starch, 50% CMC	35% starch, 65% CMC
1	34	31	25	14
2	309	341	200	145
3	536	396	272	218
4	592	507	342	246
5	672	539	364	272
7	...	606	415	306
10	...	636	460	361

* From Dickerson, A Solution to Cotton Desizing Waste Problems, *Proceedings of 4th Southern Municipal and Industrial Waste Conference.*

TABLE 14 Cotton Finishing Process Chemicals, Consumption and BOD[a]

Chemical	Amount used, lb/1,000 lb goods	BOD, %[b]	BOD lb/1,000 lb goods
B-2 gum	22	61	13.4
Wheat starch	16	55	8.8
Pearl cornstarch	14	50	7.0
Brytex gum No. 745	4	61	2.4
KD gum	4	57	2.3
Slashing starch	96[c]	...	53.0
Total	150	...	82.9
Carboxymethyl cellulose	......	3	
Hydroxyethyl cellulose	......	3	
Tallow soap	20–100	55[d]	11–55
Nacconal NR	1	4	0.04
Ultrawet 35 KX	......	0	0
Acetic acid, 80%	27	52	14.0
Mixture of 18 dyes	37	7	2.6
Cream softener, 25%	20	39	7.8
Formaldehyde-bisulfite condensate	14	27	3.8
Glycerin	3	64	1.9
Sodium hydrosulfite	11	22	2.4
Urea	13	9	1.2
Finish T.S	8	39	3.1
Kierpine extra	5	61	3.1
Merpol B	4	44	1.8
Glucose	......	71	
Gelatin	......	91	
Caustic, 76%	118	[e]	[e]
Soda ash	42	[e]	[e]
Ammonia	7	[e]	[e]
Potassium carbonate	3	[e]	[e]
Trisodium phosphate	2	[e]	[e]
Sodium perborate	3	[e]	[e]
Sodium silicate	6	[e]	[e]
Liquid soda bleach	4	[e]	[e]
Hydrogen peroxide	5	[e]	[e]
Sodium chloride	7	[e]	[e]
Sodium dichromate	6	[e]	[e]
Sulfuric acid	10	[e]	[e]
Hydrochloric acid	6	[e]	[e]

[a] From Masselli, Masselli, and Burford, "A Simplification of Textile Waste Survey and Treatment," New England Interstate Water Pollution Control Commission.

[b] Based on weight of chemical; for example, 1 lb of B-2 gum (61 percent BOD) would require 0.61 lb of oxygen for stabilization.

[c] Calculated from analytical survey.

[d] Apparently contained high water content; dry soaps averaged 130 to 150 percent BOD.

[e] Negligible BOD assumed.

Some interesting observations from this table are:

1. 750 lb of chemicals are used to finish each 1,000 lb of cloth.
2. 150 lb of starch or modified starch compounds are used per 1,000 lb of cloth (substitutes previously considered).
3. Tallow soap has a high pollution potential because of its high consumption and high BOD. Consideration should be given to substitution of low BOD material for tallow soap.

4. Plant acid consumption is high. The BOD effect of acetic acid can be reduced by using sulfuric or some other low BOD acid as a substitute.

If substitutes are used for the major process chemicals causing high BOD loads (starch, soap, acetic acid), BOD reductions as high as 50 to 75 percent may be obtained.

Wool

Table 1 presents results of waste analyses from woolen mill unit operations and processes. At this point, it should be recognized that grease scouring removes all the natural and acquired impurities from the wool fiber. Some of these impurities have economic value. Wool grease, when purified, is the source of *lanolin,* which has the unique ability (among waxes) to form water-in-oil type emulsions. At least one major combing company is producing lanolin from this source. Suint is a source of potassium salts, but recovery is usually uneconomical.

Table 15 presents the pollution potential (percent OWF) of various individual wool processes and compares the overall pollution potentials of various combinations of processes. The greatest pollution load is developed in the scouring operation with 55 percent OWF and 75 percent OWF, respectively, developed in methods 1 and 2. The next greatest source of pollution is the wash after fulling, which generates 33 percent OWF and 22 percent OWF, respectively, in methods 1 and 2. Thus, in the case of wool waste water discharges, it can be generalized that scouring will generate greater than 50 percent of the overall BOD load, while the remainder will be generated by process chemicals.

Further inspection of Table 15 shows how process change can effect savings in treatment plant costs. Method 4, utilizing grease and suint recovery, dyeing with a low BOD agent, and substitution of low BOD detergent for soap, indicates that approximately 90 percent BOD reduction can be effected via process change. Of course, when such changes are contemplated, a detailed study should be made with plant personnel to determine fabric quality change, if any.

OVERALL CHARACTER AND RESIDUAL WASTE TREATMENT

Cotton

Table 16 indicates the average character of cotton finishing plant wastes as reported by various sources. Thus, the cotton finishing mill using conventional process chemicals (that is, no substitution made for starch, soap, acetic acid), discharges a waste which is decidedly alkaline, colored by the dye which predominates, with a BOD of approximately 300 ppm (mostly soluble) and a volume of approximately 70,000 gal/1,000 lb of finished cloth.

Recent test data from some Southeastern finishing operations vary considerably from the above. Water usage is lower (30,000 to 40,000 gal/1,000 lb of cloth) and BOD is understandably higher, being 400 to 600 ppm for starch sized goods and 150 to 300 ppm for plants processing goods sized with CMC. For new operations, the target, though seldom realized, is 20,000 gal of water/1,000 lb of goods. Types of processing, materials, styles, and finish make this a broad variable.

Some finishing mills have been able to reduce water consumption from 1 to 5 percent of those previously noted by performing desizing and kiering in the same bath. The BOD of such a waste water would be from 50,000 to 100,000 ppm.

Residual wastes from a cotton finishing mill are generally treated by biological methods. Chemical coagulation (using a coagulant and lime) is generally unfit for cotton wastes because of the high content of soluble BOD. It has been shown that chemical coagulation was not economically feasible for color removal because of the high alum requirement (600 ppm, or 5,000 lb/M.G.) for nearly 100 percent color removal. However, in the event that only small quantities of strictly chemical wastes are encountered, attention should be given by the treatment process designer to coagulation.

Cotton finishing wastes, although amenable to biological treatment methods, have

TABLE 15 Pollution Potential of Woolen Processes[a]

	Scour and finish mill, BOD			Finish mill BOD		
	% OWF[b]	% of total[c]	% reduction[d]	% OWF[b]	% of total[c]	% reduction[e]
Method 1:						
Scour with soap	25.0	55.4				
Stock dye with acetic acid	4.9	10.9	...	4.9	24.4	
Card with 100% BOD oil						
Full with soap	15.0	33.3	...	15.0	74.6	
Wash with soap						
Neutralize after carbonizing	0.2	0.4	...	0.2	1.0	
Total	45.1	....	...	20.1		
Method 2:						
Scour with 12% BOD detergent	22.1	74.6				
Stock dye with ammonium sulfate	0.9	3.0	...	0.9	12.0	
Card with 20% BOD oil						
Full with 12% BOD detergent	6.4	21.6	...	6.4	85.3	
Wash with 12% BOD detergent						
Neutralize after carbonizing	0.2	0.7	...	0.2	2.7	
Total	29.6	....	34	7.5	....	63
Method 3:						
Solvent scour, recover grease only	0	0				
Wash out suint salts and dirt with 12% BOD detergent	10.0	58.9				
Stock dye with ammonium sulfate	0.9	5.3	...	0.9	12.9	
Card with 3% BOD oil						
Full with 12% BOD detergent	5.9	34.7	...	5.9	84.3	
Wash with 12% BOD detergent						
Neutralize after carbonizing	0.2	1.2	...	0.2	2.9	
Total	17.0	....	62	7.0	....	65
Method 4:						
Scour with methyl and isopropyl alcohols, recover grease and suint	0	0				
Wash out dirt with detergent	1.0	19.6				
Stock dye with ammonium sulfate	0.9	17.6	...	0.9	22.0	
Card with 3% BOD oil						
Full with 12% BOD detergent	3.0	58.8	...	3.0	73.2	
Wash with 12% BOD detergent						
Neutralize after carbonizing	0.2	3.9	...	0.2	4.9	
Total	5.1	....	89	4.1	....	80

[a] After Masselli, Masselli, and Burford, "A Simplification of Textile Waste Survey Treatment," New England Interstate Water Pollution Control Commission.
[b] Based on ovendry wool.
[c] Based on the total of that particular method.
[d] Based on the total of method 1 (45.1 percent OWF).
[e] Based on the total of method 1 (20.1 percent OWF).

peculiarities which may cause difficulties in biological processes, and therefore these will be presently explored.

Some difficulties which may be encountered are:

Toxicity Toxic substances may enter the waste from the following sources:

1. Mildew depressants. These are fungicides applied to the cloth where storage of cloth is practiced.
2. Chromates—from dyes.

3. Chlorine, hydrogen peroxide—from bleaching operations.

Toxic waste waters may require segregation and separate treatment.

Nutrient deficiency The normal BOD:nitrogen:phosphorus ratio for optimum efficiency of aerobic biological processes is approximately 100:5:2. However, considerable latitude from this ratio has been tolerated in a number of sizable operations. When required, nutrient addition via seeding or dilution (with domestic sewage) or chemical addition (ammonia, phosphoric acid, ammonium phosphates) can be practiced.

pH Proper operating pH range for biological processes is between 6.5 and 9.5. Thus, pH correction may be required.

TABLE 16 Average Character of Overall Cotton Finishing Plant Wastes

Item	1[a]	2[b]	3[c]	4[d]
Turbidity, ppm	. . . .	125	Grayish colloidal	Gray colloidal
Color	Variable and dark	(Dyes)	(Dyes)	(Dyes)
Oxygen consumption, ppm	. . . .	300		
Alkali, total, ppm	. . . .	500	300–900	600
Alkali, hydroxide, ppm	. . . .	100		
pH	10.5–11.9	9.0	8–11	10–11.5
Suspended solids, ppm	. . . .	100[e]	30–50[e]	40[e]
Settleable solids, %	. . . .	0.25		
BOD, ppm	500–800	175	200–600	300
Total solids, ppm	. . . .	. . . . . .	1,000–1,600	1,300
Chromium, ppm	. . . .	. . . . . .	Up to 3.0	2.0
Volume, gal/1,000 lb finished	30,000–40,000	. . . . . .	30,000–93,000	70,000

[a] Lockwood Greene Engineers.

[b] From Bogren, Treatment of Cotton Finishing Waste at the Sayles Finishing Plant, *Sewage and Industrial Wastes.*

[c] From "An Industrial Waste Guide to the Cotton Textile Industry," U.S. Public Health Service.

[d] From Masselli, Masselli, and Burford, "A Simplification of Textile Waste Survey Treatment," New England Interstate Water Pollution Control Commission.

[e] Note should be given to the low suspended solids. Primary settling gives little removal of BOD.

Shock loads Because of the great variation in waste volume, strength, and character over the day, equalization and/or intimate mixing and dilution are necessary treatment procedures. Equalization is especially important in the activated sludge process, which is much more susceptible to variations in BOD and pH than trickling filters or aerated lagoons.

Synthetic detergents With up to 200 ppm of synthetic detergent, there is little effect on settling of suspended solids; with up to 500 ppm, there is little or no effect on bactericidal properties, except in the case of cationic detergents, which are bactericidal in concentrations as low as 1 ppm.

Waterborne waste fibers All cotton finishing waste contains some fine fibers. Using normal precautions (bar screens and, in some instances, mechanically cleaned fine screens), small fibers still remain in the waste. Some are settleable and give problems in dewatering sludge. Some fibers are so small that they stay in suspension in sufficient quantities to coat filter beds and seal sand beds designed to dewater sludge or polish effluent. These fibers have a tendency to clog equipment and have the ability to absorb certain chemicals, which will result in delayed changes in pH. Some textile finishing wastes contain sufficient quantitites to render trickling filters an impractical treatment process even after clarification. They can create unsightly downstream conditions and, as is the case with other natural fibers, can increase the long-term BOD.

To date, water washed mechanical fine screens have proved extremely difficult to clean and maintain and have a short life. The best solution found thus far to the fiber problem has been to select treatment processes which will tolerate the fiber, settle as much as possible in a final clarifier (by using a low overflow rate), and dilute the balance in a volume of stream water.

As a result of the difficulties that may be encountered in the treatment of cotton finishing wastes, the process design should incorporate some or all of the following:

Equalization Operations such as kiering and desizing yield intermittent slug discharges of waste. Biological treatment processes are sensitive to sudden variations of quality or quantity. The trickling filter, although much more tolerant of such variations than the activated sludge process, can react in the form of "ponding," wherein the active slime "sloughs" from the filter media surfaces, clogs the void spaces, and plugs the filter. Because of the serious effects that may result, fluctuations in quality are dampened via holding tanks which minimize waste quality variations. Coincident with this operation, however, the holding tank is often used as a surge tank to provide a constant flow rate of waste through the treatment plant. This feature is desirable when a plant operates only part of the 24-hr day and less than 7 days a week.

Segregation When the results of a waste sampling program are compared with effluent required, it may be found that some wastes will require substantially less treatment or no treatment. Herein, then, lies the possibility of reducing waste treatment costs by reducing waste volume. Additionally, toxic substances, for example, may require segregation from wastes that will be biologically treated.

pH adjustment This may be accomplished via segregation and/or equalization (to give a uniform pH treatment plant influent) and, if necessary, by pH monitoring and automatic acid feed (cotton finishing wastes are strongly alkaline). Some authors have reported the potential use of waste boiler flue gases, which are high in carbon dioxide, as a neutralizing agent for the alkaline waste.

Nutrient addition In order to maintain the required BOD:nitrogen:phosphorus ratio (100:5:2) in the waste stream, addition of domestic sewage or chemicals may be practiced. The engineer should always consider the economic possibility of discharging the plant waste to a municipal sewage system. Typical nutrient chemicals are ammonium salts and calcium nitrate as sources of nitrogen, and phosphate salts or phosphoric acid as sources of phosphorus.

Table 17 lists treatment removal efficiencies for various treatment methods used on cotton finishing wastes.

TABLE 17 Treatment Process Removal Efficiencies*

	Removal efficiency, %		
Removal method	BOD	Suspended solids	Total dissolved solids
Screening	0–5	5–20	0
Plain sedimentation	5–15	15–60	0
Chemical precipitation	25–60	30–90	0–50
Trickling filter	40–85	80–90	0–30
Activated sludge	70–95	85–95	0–40
Lagoon	30–80	30–80	0–40
Aerated lagoon	50–95	50–95	0–40

* From FWPCA, "The Cost of Clean Water," vol. III, Industrial Waste Profile No. 4, *Textile Mill Products*, September, 1967.

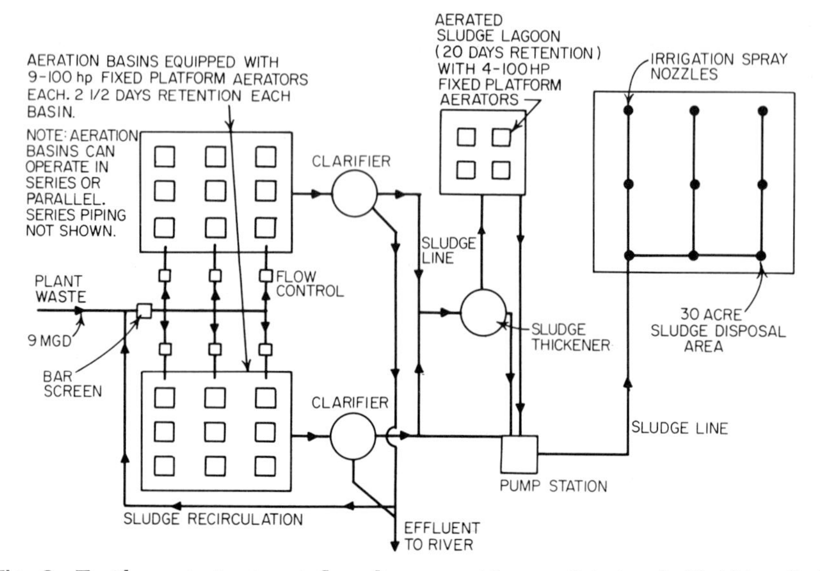

Fig. 6 Textile waste treatment flow diagram. (*Lyman Printing & Finishing Co.*)

Figure 6 shows the treatment plant flow diagram developed for a large cotton finishing mill, and Fig. 7 shows the installed and successfully operating final plant.

Wool

The equalized residual waste from the *wool scouring and finishing mill,* under the worst conditions of processing (that is, method 1, Table 15, in older plants), will

Fig. 7 Aerial view of Lyman Printing & Finishing Co. waste treatment plant. (*Lyman Printing & Finishing Co.*) (1) Lyman Printing and Finishing Company; (2) Town of Lyman; (3) Water filtration plant; (4) Pilot plant; (5) Middle Tyger River; (6) Plant's waste discharge flume; (7) Splitter box in flume; (8) Aeration basin 1 (primary treatment); (9) Aeration basin 2 (primary treatment); (10) Common wet well; (11) Clarifiers 1 and 2; (12) Telescoping valves; (13) Sludge aeration basin; (14) Sludge thickener; (15) Central pumphouse; (16) Sludge spray disposal field; (17) Sludge storage basin 1; (18) Sludge storage basin 2; (19) Sludge storage basin 3.

be characterized by a flow rate of 60,000 to 70,000 gal/1,000 lb cloth, a BOD of approximately 1,000 ppm, 3,000 ppm total solids, pH of 9 to 11, approximately 100 ppm suspended solids, and a brown color. The *wool finishing mill* will have an effluent free of the natural impurities contained in the scouring waste and will have concentrations on the order of one-third to one-half of the above quantities. The pH will be approximately the same.

Chemical coagulation, without pretreatment for grease removal, has been widely practiced on waste waters from wool scouring and finishing mills. Such treatment has yielded BOD removals in the range of 20 to 85 percent. The trend, today, however, is toward biological treatment.

Modified activated sludge treatment, coupled with influent pH adjustment, can achieve BOD removals consistently on the order of 90 percent.

When detergent scouring is practiced, biological treatment would be preceded by a chemical coagulation step (see Fig. 8) to reduce BOD loadings. The need for this

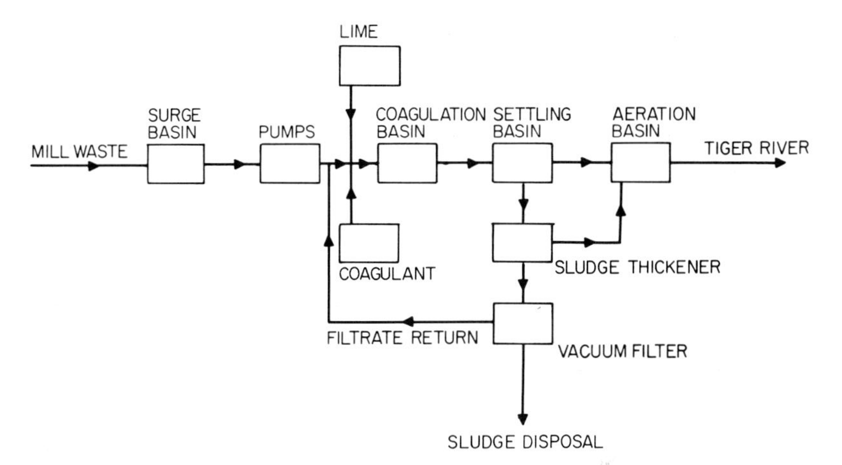

Fig. 8 Combination chemical coagulation and aerobic biological waste treatment plant flow diagram.

pretreatment step may, however, be eliminated by scouring with solvents which would result in little or no grease (a source of BOD) going into the plant effluent. Table 18 indicates various treatment process efficiencies.

If applicable, chemical treatment of wastes is advantageous in that it can be operated on a batch as well as continuous basis and on any strength of waste without requiring pretreatment.

The flow diagram for a wool waste treatment plant would incorporate some or all of the following:

1. Grease recovery.
2. Screening—for large solids removal.
3. Flotation.
4. Coagulation, consisting of (*a*) Flash mixing—wherein the waste flow is intimately mixed with coagulants, coagulant aids, and chemicals for pH adjustment in usually less than 2 min. (*b*) Flocculation—wherein paddles gently stir the waste and coagulant mixture for typically 15 to 45 min and cause agglomeration of the impurities to form a quickly settleable floc. (*c*) Settling—used to provide a quiescent state, typically 1 to 6 hr, wherein settleable floc separates from the waste by gravity. Steps *a*, *b*, and *c* may be combined into one unit by the solids contact process.
5. Biological treatment.

Coagulants used successfully to treat wool wastes are alum (aluminum sulfate), copperas (ferrous sulfate), ferric chloride, and calcium chloride. Each of these substances when mixed with a particular waste has a pH at which optimum coagulation and waste removal efficiency will occur.

Materials such as bentonite and commercially available polyelectrolytes are also used as coagulants to "weight" the floc and enhance its formation.

Experimentation with anaerobic digestion of wool scouring wastes has been reported. A pilot plant, consisting of 2,500-gal anaerobic digestion tanks has been described by Singleton (Experiments in Anaerobic Digestion of Wool Scouring

TABLE 18 Treatment Process Removal Efficiencies*

Treatment method	Normal reduction %				
	BOD	Grease	Color	Alkalinity	Suspended solids
Grease recovery:					
Acid cracking	20–30	40–50	0	0	0–50
Centrifuge	20–30	24–45	0	0	40–50
Evaporation	95	95	0	0	
Screening	0–10	0	0	0	20
Sedimentation	30–50	80–90	10–50	10–20	50–65
Flotation	30–50	95–98	10–20	10–20	50–65
Chemical coagulation:					
$CaCl_2$	40–70	.....	.....	.....	80–95
Lime + $CaCl_2$	60	97	.....	.....	80–95
CO_2 − $CaCl_2$	15–25	.....	.....	.....	80–95
Alum	20–56	.....	75		
Copperas	20				
H_2SO_4 + alum	21–83				
Urea + alum	32–65				
H_2SO_4 + $FeCl_2$	59–84				
$FeSO_4$	50–80				
Activated sludge	85–90	0–15	10–30	10–30	90–95
Trickling					
Filtration	80–85	0–10	10–30	10–30	90–95
Lagoons	0–85	0–10	10–30	10–20	30–70

* From FWPCA, "The Cost of Clean Water," vol. III, Industrial Waste Profile, No. 4, *Textile Mill Products*, September, 1967.

Wastes, *Sewage Works*). At loadings of 0.06 lb of volatile organic matter/(cu ft)(day), gas yields of 8.6 to 14 cu ft/lb of volatile matter were obtained while volatile matter reduction was 65 percent; 90 percent was grease reduction and 82 percent was BOD removal. This system operated for 2½ months on wool scouring wastes with an average BOD of 28,000 ppm.

Synthetics

Rayon finishing Rayon is not a true synthetic since it is composed of regenerated cellulose. It can be manufactured by the viscose, cuprammonium, or nitrocellulose processes.

Finishing operations entail:

1. First bath—wherein scouring and dyeing are performed coincidently
2. Salt take-off bath—wherein rinsing of the cloth with a salt solution is performed

Specialty finishing may follow these operations. The composited waste from operations (1) and (2) yields an average flow of 5,000 gal/1,000 lb of cloth, an average BOD of 1,445 ppm (5.7 percent OWF), 2,000 to 6,000 ppm salt, and a pH of 7 to 9.5. Waterproofing, if performed, would yield an additional 1.9 percent OWF BOD from the wax, solvent, gelatin, and hydrolyzed acetic acid used in the bath.

Acetate This synthetic is manufactured from cellulose acetate fiber. *Finishing* entails:

1. Scour and dye—performed in the same bath, yielding a waste discharge with 2,000 ppm BOD (5.0 percent OWF). The sources of BOD are similar to that in the case of rayon finishing and are antistats, sulfonated oil (for lubrication of fiber), detergent, and softener.
2. Rinse(s)—follows and yields a low BOD waste which finds its source in carryover from the scour and dye operation.
3. Specialty finishing—the equalized waste from operations (1) and (2) yields an average BOD of approximately 700 ppm with waste water generated at the rate of 9,000 gal/1,000 lb of cloth processed. The waste is alkaline with a pH in the range of 7 to 10.

If a white acetate is desired, the dye operation is replaced by a bleach operation.

Nylon Fiber 66, which is a polyamide, is the most widely known synthetic fiber. It is developed from processes using hexamethylene diamine and adipic acid. *Finishing* operations include:

1. Scour—again, for process chemical (antistats, soaps, etc.) removal. Waste strength is approximately 1,350 ppm BOD (3.4 percent OWF).
2. Rinse—low BOD generation resulting from carryover from scouring.
3. Dyeing—generates a waste with an average BOD of 375 ppm (0.9 percent OWF).
4. Rinse.
5. Final specialty finishing.

The equalized waste water will have an average BOD of approximately 350 ppm BOD (4.3 percent OWF) and an average flow rate of approximately 15,000 gal/1,000 lb of cloth. The waste will be alkaline, with a pH of approximately 7 to 9.5. Waterproofing, if performed, will generate an additional 0 to 12 percent OWF BOD.

Orlon An acrylic fiber differing from other synthetics in that it will respond only to light shade dyeing unless high-temperature (pressure) dyeing is performed or "carriers" are added to the dye solutions. "Carriers" used are copper, phenols, etc. Thus, as differentiated from other synthetic fiber finishing waste water, those from orlon finishing may include toxic and taste-producing substances.

Finishing operations incorporate first scour, rinsing, first dye, rinsing, second dye, second scour, rinsing, and specialty finishing. The equalized waste water will have the following average characteristics:

BOD—575 ppm (12.1 percent OWF)
pH—1.5 to 3.0 (due to acid dyeing)
Copper—25 to 50 ppm
Phenol—100 to 150 ppm

Dacron A polyester fiber usually manufactured by the esterification of ethylene glycol with terephthalic acid and similar compounds. It may be dyed with carriers (6 to 40 percent OWF phenylphenol, phenylmethyl carbinol, salicylic acid, benzoic acid, etc.) or using pressure dyeing without carriers.

One procedure for finishing polyesters would be scour, rinse, dye, scour, rinse, and special finishing. In the absence of direct analyses, it is estimated that BOD loads are as shown in Table 2.

Synthetics generate a waste water amenable to biological treatment. Table 19 indicates treatment removal efficiencies which can be expected. Attention must, however, be given to the fact that toxic materials (see orlon finishing) can be present in the waste water.

TABLE 19 Treatment Process Removal Efficiencies*

Treatment method	Removal efficiency, %		
	BOD	Suspended solids	Total dissolved solids
Screening	0–5	5–20	0
Sedimentation	5–15	15–60	0
Chemical precipitation	25–60	30–90	0–50
Trickling filter	40–85	80–90	0–30
Activated sludge	70–95	85–95	0–40
Lagoon	30–80	30–80	0–40
Aerated lagoon	50–95	50–95	0–40

* From FWPCA, "The Cost of Clean Water," vol. III, Industrial Waste Profile, No. 4, *Textile Mill Products*, September, 1967.

WATER AND CHEMICAL RECOVERY AND REUSE

Biologically or chemically treated water may be suitable for reuse in processing or as cooling tower makeup. In some cases, use of a recirculating cooling water system as a biological treatment facility for untreated or partially treated wastes may be feasible, as has been done in several cases in the petroleum industry.

Some European work has been done on reuse of treated textile mill effluent for soil irrigation. A large plant in the Southeast has been irrigating grass plots for about 2 years with the thickened sludge from aerated lagoons. Results have been satisfactory.

Reuse would result in reduced water consumption and waste water volume. The plant water and waste treatment systems would be smaller and cost less to build and operate. (See page 15-29 for waste water treatment with activated carbon.)

Cotton

In cotton processing, a government sponsored industry survey has estimated that in 1964, 16 percent of process water was reused. This will tend to increase as newer machinery, employing continuous countercurrent processing, is installed in new plants or replaces older methods. In mercerizing, reuse of caustic bearing rinse water is practiced in some plants. Dialysis and evaporation can concentrate the dilute caustic for subsequent reuse in mercerizing or kiering.

Laboratory work on cotton bleachery waste revealed that after 85 to 90 percent BOD removal by activated sludge, the effluent water could be reused for rinsing after bleaching.

Desizing waste is a source of glucose, which might be recovered by multiple effect evaporation as in the sugar industry. The evaporated water could be reused as process or heating steam. Plant BOD reduction could be greater than 50 percent. Recovery of other compounds from the natural impurities in the cotton, such as pectins and waxes, might be technically and economically feasible.

Wool

Use of continuous countercurrent processing can significantly reduce water consumption. Countercurrent scouring can reduce the quantity of water by up to 6,000 gal/1,000 lb of wool. Rinse water reuse can reduce consumption by about 4,000 gal/1,000 lb of wool.

Lanolin has been recovered via solvent extraction, and potassium salts could be recovered from suint. The main drawback to recovery of these materials has been the lack of an economically favorable market. If both lanolin and suint were recovered, a 50 percent BOD reduction could be effected in scouring and finishing plants. Recovery of soap for reuse and BOD load reduction may be feasible.

Synthetics

It has been estimated that the synthetic textile finishing industry reused about 10 percent of its process water in 1964.

Recovery of zinc from viscose rayon processing by ion exchange, precipitation, and flotation methods has received attention and is practiced at some plants.

WASTE SURVEYS AND PILOT INVESTIGATIONS

Initial portions of this chapter have chiefly indicated historical fiber processing techniques, waste quantities and qualities, and treatment techniques together with some indications of recent innovations in this field. Since more emphasis has been placed in recent years on the development of new synthetic fibers, on pollution reduction, on improving residual waste processing techniques, on reducing water consumption, etc., the engineer should remain broadminded and aware that each pollution problem is a separate entity and requires a custom-made solution. The solution should be flexible enough to incorporate possible new techniques that will be developed in the future, and changes in waste quality and quantity due to technological changes in the industry. Even historically this was true, since no two plants used the same processing chemicals and quantities, production rate, production equipment, or operators. Thus, this part of the chapter has been developed to outline a basic logical route that will help develop a scope of activities and investigations for any particular textile waste problem.

Initial investigations in any waste treatment study should incorporate developing waste water quantity and quality data. The information and data required should include an integrated sewer map for the plant and measurements of waste water volumes and qualities at all sources and at appropriate system junctions. The information developed may indicate locations where possibilities exist for segregation, volume reduction, and strength reduction. Although a prerequisite for further investigation, the initial survey, especially in older existing plants, can become time-consuming and costly. Older plants have undergone expansions, equipment changes and relocations, changes in processing technology, etc., which in many cases have not been properly recorded. The accuracy and precision to which the survey is to be performed should be considered by the enigneer before an inordinate amount of money is spent in just generating data.

When investigations regarding waste segregation, volume and/or quality reduction, process revisions, etc., are completed, the engineer can develop a relationship between waste quantity and quality and units of production. This relationship together with a production projection for the future (usually 5 to 10 years, because of possible changes in the treatment variables) will yield the basic parameters required for sizing a treatment plant for the various stages of plant growth.

The process to be used for upgrading the waste water quality to the applicable "standards" should then be determined.

After laboratory investigation of the waste characteristics and treatability, alternative treatment methods should be considered. Doubtful design criteria can be investigated in a pilot plant. If properly constructed and operated, a pilot plant offers an unparalleled opportunity to establish operating parameters of a treatment process or processes. In addition, ideas or innovations can be tried to prove their worth or limitations. Process refinements made during pilot operation may result in substantial initial and operating cost reduction.

In the pilot plant, the waste treatability can be determined. An indication of this can be obtained from laboratory tests; however, variations in quality and potential operating difficulties limit the usefulness of laboratory investigations.

Consideration should be given to the pilot plant's size to give extendable treatment results. Experience to date indicates good results with flows ranging from 10 to 50 gpm. Reasonably good scale-up of treatment results has been shown in plants with flow rates of 10 million gpd.

Extreme caution should be taken in scaling up certain pilot results. For instance, in the case of mechanical aerators, the efficiency and even design may appreciably change from the low-horsepower units up to and through 100-hp units (see Fig. 9). Some manufacturers have facilities for full-scale testing of this type of equipment. Although costly, the investment may be well worth while, particularly where thousands of horsepower are to be used in the full-scale plant.

Good results are not necessarily dependent on sophisticated instrumentation or on the use of more expensive construction materials. Before deciding on the exact form of the pilot plant, one needs to make a basic decision as to its life expectancy. If, as is often the case, the pilot plant is to be used only for several months to a year prior to design and construction of the treatment plant, less expensive and less durable construction should be considered. If the pilot plant can remain operable after

Fig. 9 Aerator (5 hp capacity) installed in 150,000-gal textile waste pilot aeration plant.

the treatment plant is in operation, it can continue to be the "testing ground" for new operating procedures, the determination of new treatment parameters for a planned change in waste quality, or the testing of treatment processes that may be developed. At the present stage of development, it is highly unlikely that treatment processes will remain unchanged. Therefore, it is desirable that plants designed today include flexibility for the adaptation of additional treatment processes and/or new methods at a later date.

The recent history of textile waste is one of continuous quality and quantity fluctuation. For most plants, growth has exceeded predictions, and for many, waste volumes and new cloth treatment technology have changed several times. With such dynamic changes, a permanent pilot plant for testing the results of process modifications and increases in waste output would be an asset.

Some useful points in setting up pilot plant operations are:

1. For recording flow rates, use weirs with simple float actuated clock operated recording cylinders. Permanent Parshall flumes, venturi tubes, or other metering devices can be installed if full-scale treatment plant location and layout can be determined.

2. For simultaneously metering waste water to different pilot units, a constant head tank with multiple discharges has proved useful. The tank is fed by a pump delivering a waste volume in excess of that needed to feed all units. The excess discharges to waste over a weir providing control of water level and thus accurate flow metering to the various pilot units.

3. Automatic samplers—the bucket or pump type for obtaining composite samples is a necessity if accurate samples are to be obtained. An approach to this can be made with sufficient numbers of grab samples during a 24-hr period, but it may not

be practical to obtain the same number of grab samples that one would get with automatic devices. In the case of a pipe discharge to the pilot plant, where the waste encounters a gravity fall, we have rigged a coat hanger and allowed the waste to drip down the coat hanger, and have successfully obtained a continuous small feed for making a composite sample. There is plenty of room for ingenuity both in the laboratory and in the pilot plant, and some very simple ideas can accomplish the same end point as that achieved by more sophisticated and more costly equipment.

4. Tests on the influent and effluent and the discharges from the various treatment processes in the pilot plant should be made daily, as a minimum, and sometimes more frequently if conditions require. Unusual variations at various shifts or times of the day can dictate separate testing. Minimum tests for the pilot plant would correspond to those which one would normally make for operational and record control of the full-scale operation. These should include but not be limited to pH, 5-day BOD, COD, temperature, alkalinity, settleable and suspended solids, notations of special conditions, and tests for any objectionable chemicals which might be present (i.e., such as chrome, phenol, or mercury).

Upon completion of satisfactory pilot studies, the full-scale plant can then be designed, constructed, and put into operation.

The preceding has offered a logical approach for attacking the pollution problem. Often, however, the time element may preclude pilot testing. In such cases, the client must rely on the experience of the consultant or his own in-house capability to design and construct a waste treatment plant directly from laboratory treatability studies. In this case, maximum flexibility should be incorporated in the design.

Process efficiency and economics are the two key elements of process design. The former can be maximized via laboratory and pilot plant studies. Investment can be minimized via laboratory and pilot studies and, possibly, via joint cooperation with other waste water generators.

The following will offer some thoughts which have on occasion resulted in economic savings and/or other advantages in waste treatment projects. At the point in the waste treatment study where the individual plants' waste quantity and quality has been determined, the engineer has three basic alternatives which should be investigated—individual treatment of the plants' waste, joint treatment of the plants' waste water with that of a nearby municipality, and joint treatment of the plants' waste water with that of one or more nearby industries. The first alternative is that most commonly employed. The latter two methods can possibly realize economic savings and ease of plant operations in that:

1. Combinations of wastes can produce a more equalized waste water, lessening shock loads to treatment facilities.
2. Advantageous changes in pollutant concentrations may result in improved treatability.
3. Compensating effects regarding nutrient content and pH of the mixed waste water can result.

The final decision to be made by the engineer from the alternatives available should be based upon the client's needs and desires, and the optimum blend of economics, and efficiency and flexibility.

WASTEWATER TREATMENT WITH ACTIVATED CARBON

The use of activated carbon for treatment of textile finishing or dye house waste should be considered where one or more of the following conditions exist: raw water and/or effluent treatment costs are high; existing pollution control and/or makeup water treating facilities are overloaded or marginal in operating efficiency; there are restrictions on water withdrawal or water discharge or a lack of water.

One carpet mill in Pennsylvania is using carbon adsorption to treat and then recycle 80 percent of about 350 gpm of waste water back to the mill. The carbon is thermally regenerated to reactivate it and burn off the adsorbed organics to CO_2 and water. Since no biological sludge is formed in the process, solid waste disposal is minimized and space is saved. Costs are stated to be competitive with conventional biological treatment.

AIR POLLUTION CONTROL

Lint filtration from spinning, drawing, carding, and twisting operations has been described under cotton processing. The problem of fumes from wool carbonizing has been dealt with. Another potential textile dyeing or finishing mill air pollution problem is odor.

The odors emanating from dyeing and finishing operations are usually confined to the immediate vicinity of the plant and do not ordinarily give cause for complaints. However, odors of acetic acid, formaldehyde, acids, dyes, and other organics may exist in higher than threshold concentrations and may require control. Odor thresholds in air for acetic acid and formaldehyde, as two examples, are reported to be 1 volume ppm.

The field of odor control is complicated by these factors:

Odor like and dislike may vary widely.
Nature of odors can change with dilution.
Odor thresholds of individuals vary.
Odor perception changes with temperature and humidity.
There is at the present time no instrument other than the human nose capable of measuring odor.
The sense of smell is fatigued with prolonged exposure to an odor.

Thus, an odor coming from a plant may not be perceived by many people; those who perceive it may not find it objectionable.

If a plant has received complaints, a field odor survey may be desirable to establish that the odor is from the plant in question. This consists of an observer (or observers) noting the following:

Date, time, and location of observation
Odor intensity

0	No odor	3	Moderate
1	Very faint	4	Strong
2	Faint	5	Very strong

Odor character such as flowery, fruity, burnt, decomposition, chemical, and other descriptive terms (rotten egg, vinegar, onion, etc.)
Temperature and local wind direction

The survey could be performed by plant personnel who do not work in an odorous environment, or by a consulting engineering company specializing in air pollution control and/or odor abatement.

As an aid to establishing the odor source, wind direction and speed might be ascertained by a local meteorological station or nearby airport. The use of 6-in.-diameter helium balloons in locating an odor source has been described. This procedure involves releasing a balloon at the point of odor complaint during the nighttime, since an odor "plume" can travel many miles under conditions of an evening inversion. The wind direction is ascertained by tracking the balloon with a strong flashlight and using a compass. Triangulation can be used to locate the source.

If, because of local meteorological and topographical conditions, an odor problem does exist, the control methods include masking with another chemical, combustion (catalytic or thermal), scrubbing with water or other liquids (with or without chemical additives), and adsorption of activated carbon or other materials. Details on these methods can be found in the air pollution control literature.

In lieu of the above control methods, it may be possible to reduce, change the nature of, or eliminate an odor by switching to a different chemical, or by equipment design modifications (enclosures and hoods).

A control procedure that may be used is a tall vent stack. If the chemical causing the odor is known, together with its discharge rate and odor threshold, one of the diffusion equations (such as Sutton's) could be used. This procedure would determine vent stack height required to reduce ground level concentration at a particular location to below the odor threshold. Another possibility would be to duct the fumes to the main boiler stack or a nearby vent stack.

Chapter 16

Pollution Control in Food Industries

L. C. GILDE

Director—Environmental Engineering,
Campbell Soup Company,
Camden, New Jersey

EFFECT OF THE ENVIRONMENT

The leaders in the food industry have stood behind a realistic approach to pollution abatement. The industry needs a water supply that can be utilized for processing purposes with satisfactory quality that can be reasonably treated, and at the same time, the industry has an obligation to be sure that the effluents from its operations are creating no problems for downstream users. Because the industry is involved with the total environment from the farm to the consumer, it recognizes the tremendous complexity of the complete problem. Unfortunately, emotion has entered the issue in recent years, clouding the total picture and at times resulting in uneconomic or technologically unreasonable programs.

There is a need to study, evaluate, and know the effect of the entire environment. We too frequently fail to understand or comprehend that the original land—the entire country as originally settled by our forefathers—was covered by forest, woods, marshland, and grassland. We have altered this basic environment in order to develop and utilize our natural resources and become a more prosperous nation. Forests were removed, marshlands were drained, and the entire landscape was altered in all the rich valleys surrounding our major streams. This, then, altered the original natural state of our streams and rivers. The effect of this alteration of the environment is only now beginning to be studied and evaluated. The key point here is that more needs to be known about our total environment in order to make realistic decisions and establish practical programs for the control of our environment for the benefit of all people. This is not an effort to place the blame or responsibility on others but rather is a voice of caution that there is a need for more research in order that well-informed people, given the total picture, will prevent improper legislation through superficial knowledge, lack of study, and irrational thinking. There is a real danger that, in seeking reason and facts and an orderly control of pollution where real benefits are obtainable, one can be branded wrongly. This inhibition is a detriment to economically sound programs in controlling our environment.

WATER

Supply Requirements

The food industry not only is involved with the total environment from the standpoint of obtaining its raw supplies, but also it is subject to the conditions of the environment from the standpoint of one of its major requirements—water. The food industry as a whole uses prodigious quantities of water, but this is not a consumptive use. The water is required for washing of all forms of food, blanching, pasteurizing, cleaning of processing equipment, and cooling of finished product. The quality of the water that touches the product is of extreme importance. As an example the author's company does not even use water supplied directly by a city without further treatment, in order to assure complete freedom from taste and odor, and to obtain uniform characteristics. This latter point is very important in order to be assured that the product made in any state in the country and, in fact, in any country in the world has the same quality and taste.

It is apparent, now, that the food industry is affected by both sides of the water cycle. It is vitally interested in large quantities of high-quality water, and then is faced with the problem of treating the water to return it to the stream and the environment. Another point that is sometimes misunderstood is that it is not practical to bring the stream to the food processing plant in a condition satisfactory for use without extensive treatment. The food industry is interested in high-quality water, but there should be no distortion of the fact that incoming water can be purified to a point that it would be saisfactory for all uses within a food plant.

Use, Reuse, and Segregation

Historically, little consideration was given to the reuse of water since the original land was blessed in most parts of the country with an abundance of water; therefore, it was practical many years ago to be wasteful. In our present society this is no longer possible. In a business sense, we are extremely aware of our water needs. Because water is so generally important and vital to food production, most companies work hard to avoid wasting water. Water is reused as much as possible. For example, water used as a cooling medium to cool cans and for air conditioning is later used for the primary washing of vegetables and other products. The same water later is used for fluming waste material, and finally, a portion of it is used to cool ashes in the powerhouse. Figure 1*a* shows the typical reuse of water within a tomato processing and pea processing plant. Table 1 lists typical use and reuse ratios of water in the food industry generally.

As previously indicated, the drinking water supply of most towns and cities is not of sufficiently high quality for immediate use in food plants, and the water is further treated to guarantee the highest quality. In many plants it is desirable to segregate water into a number of classifications, depending upon their eventual uses. The highest quality of water is naturally used wherever it may touch product or enter into product. This segregation or classification of waters is especially desirable when dealing with some surface water supplies. In many places throughout the country there has been a progressively worsening condition in streams concerning

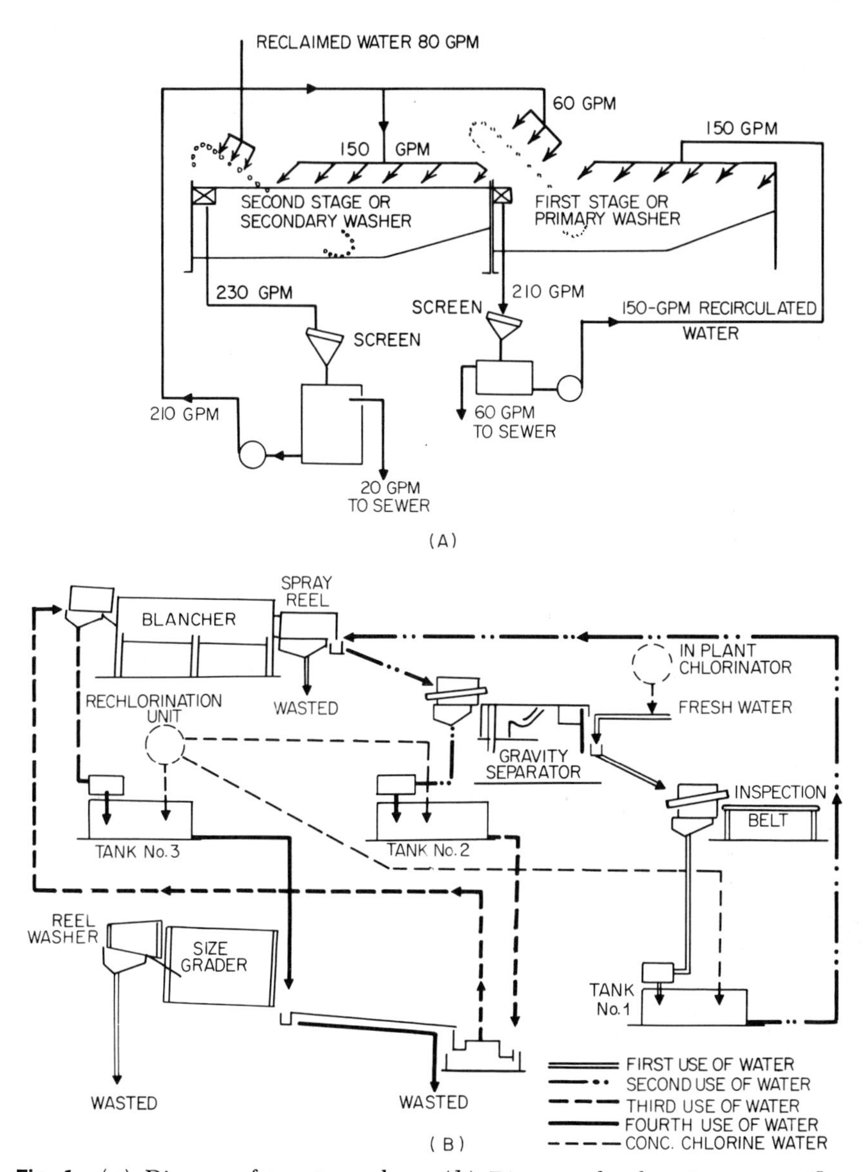

Fig. 1 (*a*) Diagram of tomato washer. (*b*) Diagram of a four-stage counterflow system for reuse of water in a pea cannery. Direction of water flow is counter to movement of peas.

tastes and odors. Some of the facts that have contributed to taste and odor situations in surface supplies have been nutrients, periodic algae blooms, low river flow, and trace organics. The occurrence and intensity of these problems vary constantly. Therefore, there has been a need for many safeguards to provide maximum protection of the finished water supply. In some cases, the additional treatment required has necessitated installation of chlorine dioxide generators, potassium permanganate feeders, multistage pressure granular carbon filters, and replacement of sand filters with reverse gradation MicroFloc filters, which in turn has permitted use of powdered carbon directly on the filters. Figure 2 indicates a comprehensive, modern water treatment facility of Campbell Soup Company which ensures the highest possible quality of water at all times from a surface supply.

TABLE 1 Reuse Ratios for Food Industries*,†

Industry	Ratio
Beet sugar	1.48
Corn and wheat milling	1.22
Distilling	1.51
Food processing	1.19
Meat	4.03
Poultry processing	7.56
Sugar, cane	1.26

* From "Water in Industry," National Association of Manufacturers, January, 1965.
† For each gallon of water withdrawn as an example, in the meat industry the water was used 4.03 times before ultimate disposal.

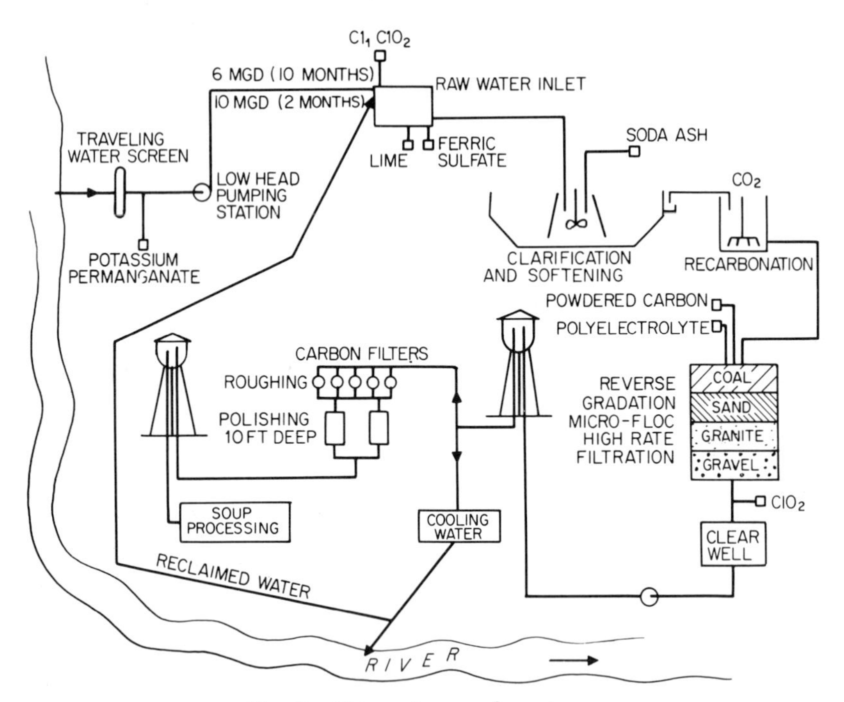

Fig. 2 Ohio water supply system.

SOURCE AND CHARACTERISTICS OF WASTE

In general, the food processing industry has a raw waste effluent before treatment that is extremely high in soluble organic matter. Even the smallest seasonal plant is likely to have waste loads with a population equivalent of 15,000 to 25,000 people. Some of the larger food processing plants around the country have raw waste loads before treatment with a population equivalent of approximately a quarter million people. Fortunately, the waste effluent from most food processing plants is not a health problem, from the standpoint of either disease or toxicity in the receiving stream. In most cases, the waste from food processing plants is amenable to biological treatment. Frequently, food wastes provide nutrients essential for good biological treatment in municipal plants.

A food processing waste discharged to a large enough stream would provide nutrients and food for the multitude of biological life, microscopic organisms, and fish. There are very few streams, however, that can assimilate large organic loads such as this, and it is necessary to treat and stabilize the waste so that it does not degrade the stream. If the stream is too small and organic waste too large in volume, the organic waste will utilize the dissolved oxygen in the process of being stabilized and will pollute or degrade the stream by reducing the dissolved oxygen value below that required by normal aquatic organisms. As an organic waste is discharged to a stream, a definite pattern is obtained in the receiving stream. This pattern is determined by the relationship between the rate at which oxygen is consumed to stabilize the organic matter and the rate at which oxygen is dissolved from the atmosphere and added from land drainage and aquatic forms.

The amounts of waste and the quantity of organics and solids discharged from processing operations depend a great deal upon the type of individual processing steps and water use and reuse in each plant. There is a great variation in waste load from plant to plant depending upon the layout of the plant and the manner in which foods are handled. In general, waste loads from specific products fall into more or less general categories of loads.

Cannery Wastes

The canning industry soon will pack approximately a billion cases of foods in a year. The average amount of clean water required to pack this food is approximately 50 gal per case, which means that the total requirement for the industry will reach 50 billion gal of water per year. Although the industry at large has made great strides in conservation and water reuse, there is not a tremendous potential at the present time for a diminution of water requirements because of food quality improvements and more rigid definitions of cleanliness. It must be expected that in the future water consumption will increase. One reason for increased water use is that raw foods must be thoroughly washed to remove toxic chemical residues as well as the natural contaminants present on field grown crops. Under today's economic conditions there is a tendency toward development of large bulk handling methods which require hydraulic transportation of the produce from one piece of equipment to the next, in addition to the large amounts of water required for washing and cleansing of the produce. Care has to be taken to limit the amount of recirculation that can be done in the washing and fluming, since it is known that microbial counts for food in the can are influenced by the sanitary conditions as the produce is flumed or pumped.

The canning industry is faced with disposal of two types of waste—solid waste and liquid waste. Solid waste can be quite considerable. As an example, tomato waste solids may reach 15 to 30 percent of the total quantity of product processed. Peaches and pears may have a solids content representing 20 percent to peaks of 40 percent. Peas and corn can be in excess of 75 percent. The disposal of solid waste and liquid waste is entirely different, and wherever possible the solid waste should be kept out of the water. The addition of waste solids to water for fluming and conveying from one point to another appreciably increases at times the concentration of soluble organics in the waste water. Also, by keeping the solid wastes separate, they are handled in a drier condition and therefore are more readily usable for by-products or feeding purposes. The waste water from a cannery will have an appreciable amount of solids, primarily because of the washing of food products, the cleaning of equipment, the hydraulic conveying of the product itself, spillage of foods and ingredients, and washing of floors and all areas. Wherever possible, the rejects, trimmings, and inedible parts should be handled in a dry state.

The liquid portion of the waste is normally screened as a first step in any treatment process. Solids from these screenings can be collected with the other solids from the plant in order to be trucked away as garbage or to join other solids in a by-products recovery program. Unfortunately, too much material is carted away as garbage. A great deal of research should be conducted in order to develop new markets and new by-products where these materials can be utilized. Considerable work has been done by the National Canners Association on composting methods in order to convert this organic material into an innocuous humus. The biggest problem to date has been to produce some by-product economically that can serve a useful purpose in the world market. The low financial return on by-product recovery has certainly been a key factor in preventing more canners from entering this field. Additional problems are the seasonal nature of vegetable crops and the reluctance of the food processors to attempt to take on the major capital expenditures and operating difficulties of by-product systems during the peak of the operating seasons.

An example of by-products from fruit and vegetable wastes are ensilage from pea and corn waste; dehydrated animal feed and molasses from citrus waste; vinegar, powdered pectin, pectin concentrate, jelly stock, and apple cider from apple waste; and tomato pomace, which is used in animal and dog foods, from tomato waste. Apricot pits are used as a raw material for the manufacture of sweet and bitter oils for use in the food, pharmaceutical, and cosmetic industries. Charcoal briquettes are also made from pits.

Although solid waste can be a major problem for the canner, the liquid waste generally represents the most critical disposal problem.

There are a number of sources of liquid wastes in a cannery operation. The greatest source of liquid waste is normally the fruit and vegetable washing facilities. This, except for final sprays, is normally reclaimed water, as shown in Fig. 1. Such washings are counterflow, and therefore the final use of the water on the initial washing is normally high in soluble organics and contains suspended matter, much of it inorganic, from the soil. The National Canners Association Research Laboratories have contributed important research on conservation techniques which have shown that more effective washing can be done with less water. For example, a study of tomato washing operations proved that the extent of removal of natural contaminants and chemical residues was dependent not upon the volume of water used but on the method of applying the water to the raw product. In this case, the location of sprays, water temperature, volume of the water, and pressure were key factors in reducing water consumption.

Other sources of waste come from the peeling operations and contain large volumes of suspended matter—primarily organic in nature. The amount of suspended matter varies with the type of peeling. Also, it is dependent upon whether the vegetables have been blanched or lye-treated prior to peeling. In the latter cases, the liquid waste contains considerable amounts of dissolved or colloidal organic matter.

Blanching of raw foods is commonly practiced in order to expel air and gases from vegetables, to whiten beans and rice as well as to soften and precook them, to inactivate enzymes that cause undesirable flavor or color changes in food, and to prepare products so that they are easily filled into cans. Little fresh water is added to the blanching operation over a normal 8-hr shift, and therefore the concentration of organic materials becomes high because of leaching out of sugars, starches, and other soluble materials from the product that is blanched. Although small in volume, the blanch water becomes a highly concentrated solution that frequently represents the largest portion of the soluble components in the liquid wastes of an entire food processing operation. The actual amount of dissolved and colloidal organic matter varies depending upon the type of equipment used.

Caustic peeling solutions also represent a very strong waste. The use of hot caustic is confined to a recirculated system where the strength of the caustic is kept up to normal by the continual addition of concentrated sodium hydroxide. The food is thoroughly washed after the caustic bath, thus imparting a high alkaline waste load to the sewer. In addition, periodic discharge of the entire caustic bath is a severe problem.

The final major sources of liquid wastes are from washing equipment, utensils, cookers, etc., as well as washing of floors and general food preparation areas. Cleanup periods after production will normally alter the characteristics of the waste in that a great deal of caustic is used, thus increasing the pH considerably above the normal character obtained when the plant is in regular operation.

After the product has been placed in cans and is cooked, it is cooled—an operation requiring a large volume of water. The water used for cooling is relatively high in temperature but has essentially no pollution load. As shown in Fig. 1, normally this cooling water is reused in washing vegetables. Where the cooling water cannot be reused, it can be sent directly to a stream without further treatment. In the latter case, it is important to keep this water segregated from the other liquid waste in the plant, as diluting this liquid waste with the cooling water will not normally help in providing better treatment but simply compounds the problems.

TABLE 2 Volume, BOD, and Suspended Solids of Cannery Wastes*

Product	Volume per case, gal	5-day BOD, ppm	Suspended solids, ppm
Apples	....	1,680–5,530	300–600
Apricots	57–80	200–1,020	260
Asparagus	70	16–100	30–182
Beans, baked	35	925–1,440	225
Beans, green or wax	26–44	160–600	60–150
Beans, kidney	18–20	1,030–2,500	140
Beans, lima, dried	17–29	1,740–2,880	160–600
Beans, lima, fresh	50–257	190–450	420
Beets	27–70	1,580–7,600	740–2,220
Carrots	23	520–3,030	1,830
Corn, cream style	24–29	620–2,900	300–675
Corn, whole kernel	25–70	1,120–6,300	300–4,000
Cherries, sour	12–40	700–2,100	20–605
Cranberries	4	2,250	105
Grapefruit	5–56	310–2,000	170–285
Mushrooms	6,600†	76–850	50–240
Peaches	1,300–2,600†	1,350–2,440	600
Peas	14–75	380–4,700	270–400
Peppers	....	600–1,220	
Potatoes, sweet	82	1,500–5,600	400–2,500
Potatoes, white	....	200–2,900	990–1,180
Pumpkin	20–50	1,500–6,880	785–1,960
Sauerkraut	3–18	1,400–6,300	60–630
Spinach	160	280–730	90–580
Squash	20	4,000–11,000	3,000
Tomatoes	3–100	180–4,000	140–2,000

* N. H. Sanborn, "Disposal of Food Plant Wastes," NCA Research Laboratories, Washington, D.C.

† Per ton.

The strength of canning wastes varies considerably from plant to plant, even on the same type of materials—raw product—because of the type of operations within the plant. Table 2 gives a list of typical waste loads and water use figures for various cannery wastes.

As noted in Table 2, there is a wide variation in the concentration of the strength of the waste as concerns BOD and suspended solids. Some of this strength variation is attributable to the volume of water used. In general, the larger the volume of water used, the weaker the waste. However, there are a number of things which greatly increase the concentration of the waste, such as fluming of the waste, screening solids, trims, and rejects, dewatering of waste solids in presses or cyclones without separate disposal of the liquor so created, and comminution of solids in grinders. The research laboratories of the National Canners Association have been conducting research on the segregation of strong wastes from comparatively clean waters with the purpose of possibly separate treatment or disposal of a smaller volume of liquid containing most of the organic load, otherwise diluted in the composite waste flow. An example of this is shown in Fig. 3.

From an economic standpoint, it is normally less costly to treat a high-strength low-volume waste rather than a large-volume diluted waste. The food processor discharging to a municipal system, faced with the additional cost of paying for treat-

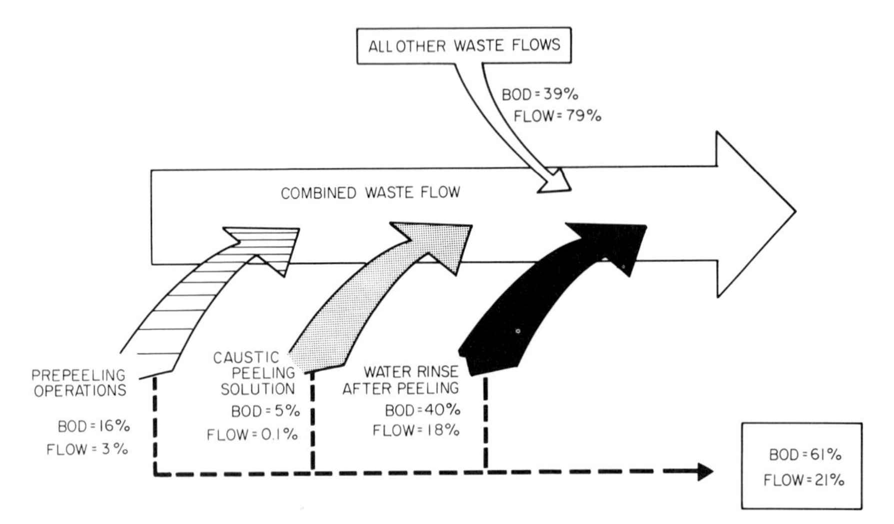

Fig. 3 Source and relative strength of wastes from peach canning.

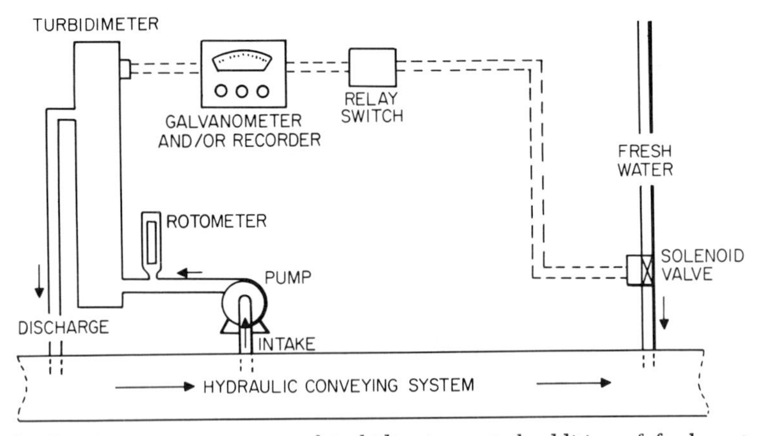

Fig. 4 Continuous measurement of turbidity to control addition of fresh water to hydraulic conveying system. (*NCA Research Bulletin.*)

ment on a BOD basis, should also consider the advantages suggested in Fig. 3 whereby a low-volume high-strength waste may be treated in a relatively small system at the plant, in order to reduce charges by the municipality. Some processors have done an excellent job of employee education, requiring that spilled materials and other wastes be shoveled up rather than flushed down the sewer.

Figure 4 shows a recent development by the National Canners Association Laboratories at Berkeley to control properly the quality of water in a hydraulic system or washing system for fruits and vegetables. This represents a new method of instrumentation in order to assure water conservation on an automatic basis through the evaluation of the quality of water by measurement of changes in transmitted light. Such a system requires establishment of guidelines in order to ensure good sanitary conditions at all times.

The treatment of cannery wastes will be discussed in more detail in a later section of this book. Figure 5, which has been prepared by the National Canners Association Research Laboratories, lists an outline of cannery waste treatment disposal methods.

Since World War II, there has been a tremendous mushrooming of the freezing industry with the advent of the home freezer. The freezing of products originated in canning plants, and in some cases both canning and freezing are still performed in the same plant. Except for the final step of preservation, the preparation of foods for freezing, starting with the planting of seeds, harvesting of crops, washing,

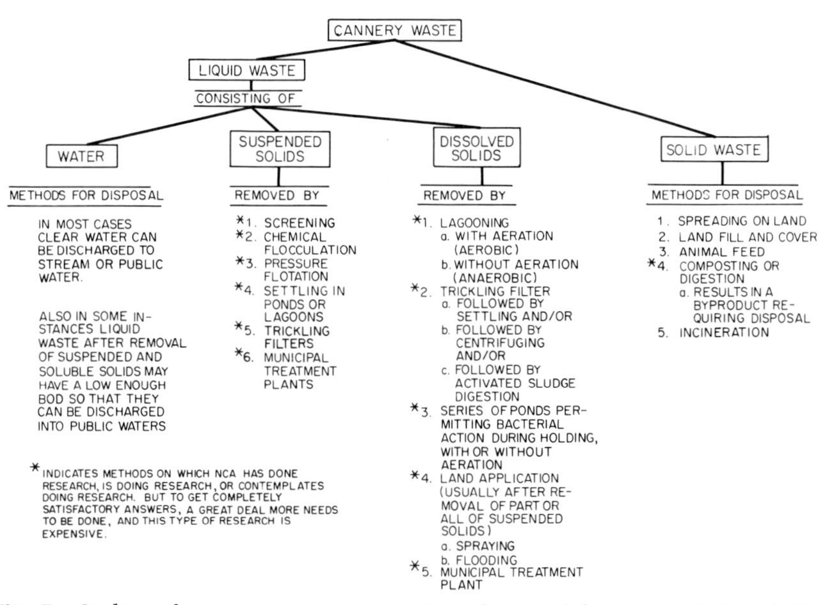

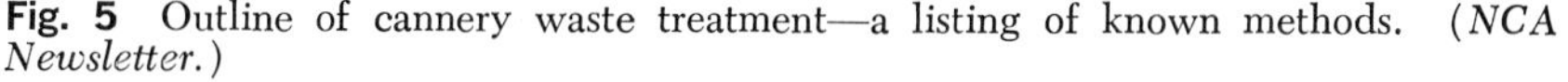

Fig. 5 Outline of cannery waste treatment—a listing of known methods. (*NCA Newsletter.*)

blanching, and all other preparatory procedures, is the same as for canning. In the latter case, the final preservation step is accomplished by heat sterilization, whereas in the former, freezing techniques are utilized.

Poultry Wastes

The poultry industry today handles approximately 10 billion lb of chickens and 2 billion lb of turkeys. This industry does over $1½ billion worth of business annually. The processing plants vary in size from approximately 50,000 birds per day to in excess of a quarter million birds per day. The basic poultry operations consist of receiving and storing, killing, defeathering, evisceration, packing, and freezing. The types of poultry handled vary somewhat. The great majority of chickens used today are called broilers, which range in live weight from 2.5 to 4.5 lb. "Fowl" refers to laying hens, which can weigh from 4.0 to 8.0 lb. Over 95 percent of all poultry is processed as eviscerated pack. When not eviscerated, the operation is known as "New York dress."

A great deal of waste is created in the first step of the operation, which consists of the receiving and storing of the poultry in coops, batteries, or pullmans. This waste consists primarily of manure and unconsumed feed along with water used to wash the cages and the entire storage floor area. Figure 6 outlines and diagrams a typical poultry operation. The first step in the operation is attaching the birds to shackles and suspending them from a moving overhead conveyor line. The birds

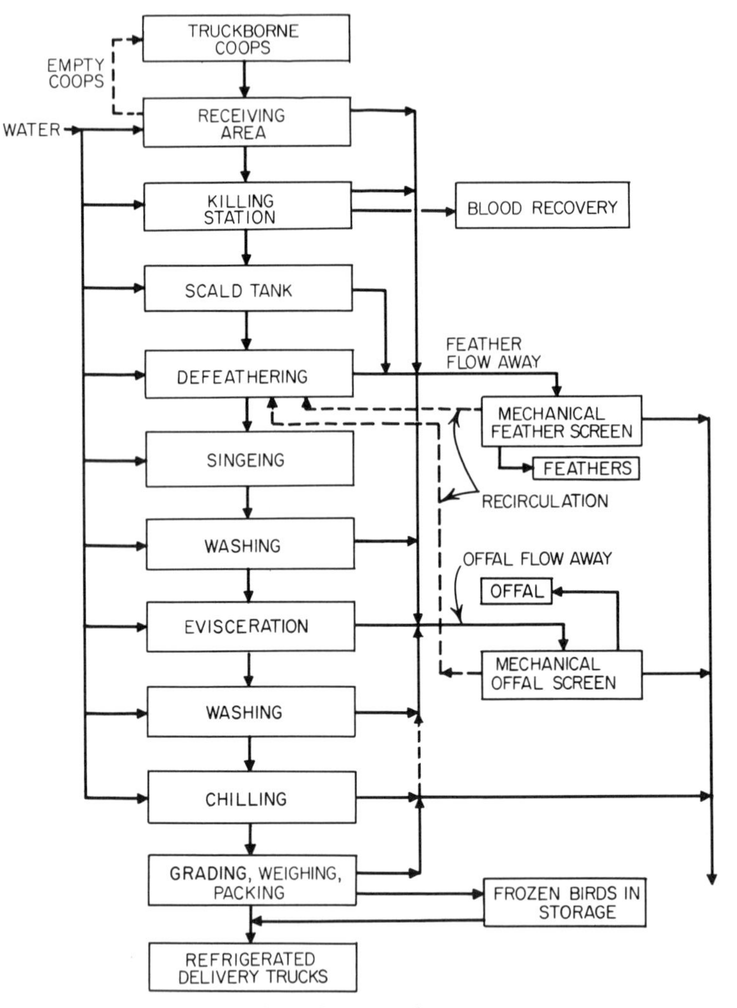

Fig. 6 Flow chart—poultry operation.

remain on this shackle from the time they leave their cages until they reach the eviscerating operation where, after the feet are cut off, the birds are rehung upside down on another line.

A major reduction in pollution load can be achieved if the manure and spilled feed and feathers from the receiving area can be handled in a dry fashion, that is, swept up and trucked away in solid form rather than being hosed down into the sewers. This material can be disposed of as fertilizer either directly to the field or to fertilizer manufacturers. The cleaning of cages before they are put on trucks and returned to the farms is another major source of pollution. The volume of wash water being utilized here can be reduced by using high-pressure sprays under carefully controlled conditions.

The first major step is the killing operation, which permits the bleeding of the poultry. The blood of chickens is reported to contain more than 90,000 ppm BOD. Table 3 indicates the strength of waste from a typical killing and eviscerating operation. The tremendous concentration and strength of blood dictates that this material should be kept out of the plant sewers to reduce the pollutional load from a poultry operation. In order to do this, it is important that the blood drainage time in a blood tunnel be adequate and that the blood be collected in containers for separate disposal to a by-products or rendering plant. The only blood that should be permitted to get to the sewers is the final cleanup of the blood tunnel at the end of the day.

TABLE 3 Composition of Combined Poultry Plant Wastes*

	Range
5-day BOD, ppm	150–2,400
COD, ppm	200–3,200
Suspended solids, ppm	100–1,500
Dissolved solids, ppm	200–2,000
Volatile solids, ppm	250–2,700
Total solids, ppm	350–3,200
Suspended solids, % of total solids	20–50
Volatile solids, % of total solids	65–85
Settleable solids, ppm	1–20
Total alkalinity, ppm	40–350
Total nitrogen, ppm	15–300
pH	6.5–9.0

Processing Plant Waste Loads per 1,000 Chickens

Type of plant	No.	Waste, gal	BOD, lb	Suspended solids, lb
Flow-away	10	7,000†		
With blood recovery	5	......	25	13
All blood wasted	5	......	41	23
Nonflow-away	3	4,500		
With blood recovery	1	......	23	12
All blood wasted	2	......	35	21‡

* U.S. Department of Health, Education and Welfare, *Technical Report* W62-3.

† Present trend in sanitation practices is toward greater water use. Current operating data indicate average water use of about 8,000 gal per 1,000 birds.

‡ Estimated.

After the chickens emerge from the blood tunnel, they are lowered or conveyed into a scalding tank. The scalder water conditions the bird for easy feather removal. This water is a source of pollution since it contains feathers, blood, manure, and general dirt adhering to the bird. Water temperature of 130 to 140°F is normally used.

After passing through the scalder tank, the birds are defeathered by machines which are normally continuous in operation and permit almost complete stripping of the feathers from the birds. Sometimes this is done in a series of automatic pieces of equipment consisting of roughers and pickers, mounted with rubber fingers and counterrotating steel drums. Continuous water sprays are utilized in the equipment in order to flush away the feathers. Water conservation enters at this point, in that after the feathers are removed from flume water by fine mesh screening, the water is returned to the mechanical defeathering operation for fluming away the feathers. After the automatic defeathering machines, the remaining pinfeathers are removed by hand picking and singeing. The singeing also removes the remaining fine hair in addition to the pin feathers. At this point, the birds receive a heavy spray wash, sometimes chlorinated for better sanitary control.

In poultry plants it is necessary to keep various operations isolated in order to avoid cross contamination from live birds or wastes from previous operations. In the eviscerating room, the lower leg portion is removed and the birds are trussed on the conveyor line and suspended in order to assist in the removal of the viscera. The bird then goes through a series of operations in which the oil gland is removed as a first step. A major incision is made, and the entrails and other inner organs are pulled out and left dangling so that they can be inspected along with the interior of the carcass by the federal inspector from the U.S. Department of Agriculture. The heart, liver, and gizzard are removed and are processed separately, requiring special cleaning and handling. For instance, the gizzard requires splitting, washing out the contents, peeling of the inner lining, and a final wash. As the bird continues along on the conveyor line, the inedible viscera are cut loose and discarded, normally in a flow-away flume system. The lungs and other material remaining in the carcass are removed by a vacuum suction system.

The material handled by the suction system provides a waste product which is stored in a large vacuum tank to be released to a tank truck for processing in a rendering plant. Sometimes this same vacuum system is utilized to remove blood from the blood tunnel in the initial killing step. The flow-away system for the viscera and offal creates an increased potential load, and where possible it is desirable to use a dry conveying system. Most plants, however, use the flow-away system as a more convenient and nuisance-free operation in handling the waste. After the offal flow-away system leaves the area, it has to be screened in order to remove the solids. These solids are then sent to a rendering plant with the other wastes from the operation previously discussed. All the solids are utilized in making chicken feed.

The final step in the eviscerating area consists of an inside washer utilizing high-pressure sprays to flush out the bird. This washer, as well as the previous outside washer, and hand-washing facilities for the operators, along with the evisceration line, are normally chlorinated in order to promote the most sanitary conditions.

Birds are then removed from the shackles and are dropped into chill tanks where they are cooled to approximately 40°F. These chill tanks may be batch type arrangements packed in ice or may consist of large mechanical chillers operating under very exact controlled conditions. In one type, a rotating drum is normally perforated to permit free movement of water. The birds are cascaded forward in a counterflow to the ice water. This operation then creates an additional pollutional load, but it is not as large as that from the earlier operations. The government regulation is ½ gal per bird overflow rate. From this point, the birds are processed for their intended use, which may mean they are cut up and frozen, cooked and packed, or shipped in a chilled fashion to other users. The subsequent operations require sizing the birds into proper weight categories and grading and packing them.

Meat Packing

As in the previous cases described, the meat industry is concerned with the total environment, from the farm to the processor to the consumer. There is a major concern about waste from the feed lot, the stockyard, the slaughterhouse, and the packinghouse. All these wastes are largely organic in character and normally are highly putrescent and malodorous. None of these wastes can be discharged directly to a stream without treatment, as they will cause a rapid depletion of the dissolved oxygen, thus damaging the aquatic life in the stream and causing aesthetically objectionable sights such as floating scum and sludge deposits.

The feed lot and stockyard wastes consist primarily of animal manure with urine, straw, and unconsumed feed, along with general dirt, drain water, and water used for cleaning and flushing purposes. Preferably these types of waste should be handled in a dry form. If they have to be flumed or flushed away, they should be screened before being admitted into any sewer system. However, the flushing of these materials introduces a very high BOD load into the sewage that can be avoided if the materials are handled in a dry form.

The meat packing industry utilizes approximately ½ billion gal of water per day in the federally inspected meat packing plants. Most water reuse is accomplished in the by-products area. Water conservation methods are under constant vigilance, but the industry does have a limitation on reducing water use as in the case of other food processing industries because of the requirements and limitations set up by good sanitary practices.

TABLE 4 Approximate Range of Flows and Analyses for Slaughterhouses, Packinghouses, and Processing Plants*

Operation	Waste flow, gal per 1,000 lb live weight slaughtered	Typical analysis, ppm		
		BOD	Suspended solids	Grease
Slaughterhouse	500–2,000	2,200–650	3,000–930	1,000–200
Packinghouse	750–3,500	3,000–400	2,000–230	1,000–200
Processing plant	1,000–4,000	800–200	800–200	300–100

Approximate waste loadings

Operation	lb per 1,000 lb live weight slaughtered		
	BOD	Suspended solids	Grease
Slaughterhouse	9.2–10.8	12.5–15.4	4.2–3.3
Packinghouse	18.7–11.7	12.5– 6.7	6.3–5.8
Processing plant	6.7	6.7	2.5–3.3

* From "An Industrial Waste Guide to the Meat Industry," U.S. Public Health Service Publication No. 386, revised 1965, p. 6.

The volume and organic content of meat wastes vary appreciably according to the type of operation and the degree of by-product recovery practice. Some plants are involved only in the slaughter and therefore fall into the classification of slaughterhouse operations, whereas other plants are completely integrated, including slaughterhouse operations with food packing and processing (see Table 4).

In the slaughterhouse, the animals are killed and the meat is dressed for distribution. As in the case of the poultry operation, the blood from the killing operation is excessively high, approximately 100,000 ppm BOD, and must be handled separately in order to avoid excessive pollution in the sewer. The blood does have a commercial value and therefore should be recovered. Whole blood has a number of by-product uses, such as fertilizer, livestock feed, and adhesives.

Another major problem in the slaughterhouse is the disposal of the paunch manure. This should be handled in a dry fashion if at all possible but is frequently flumed away and removed on screens. This operation again adds considerable BOD and suspended solids to the liquid waste from the plant.

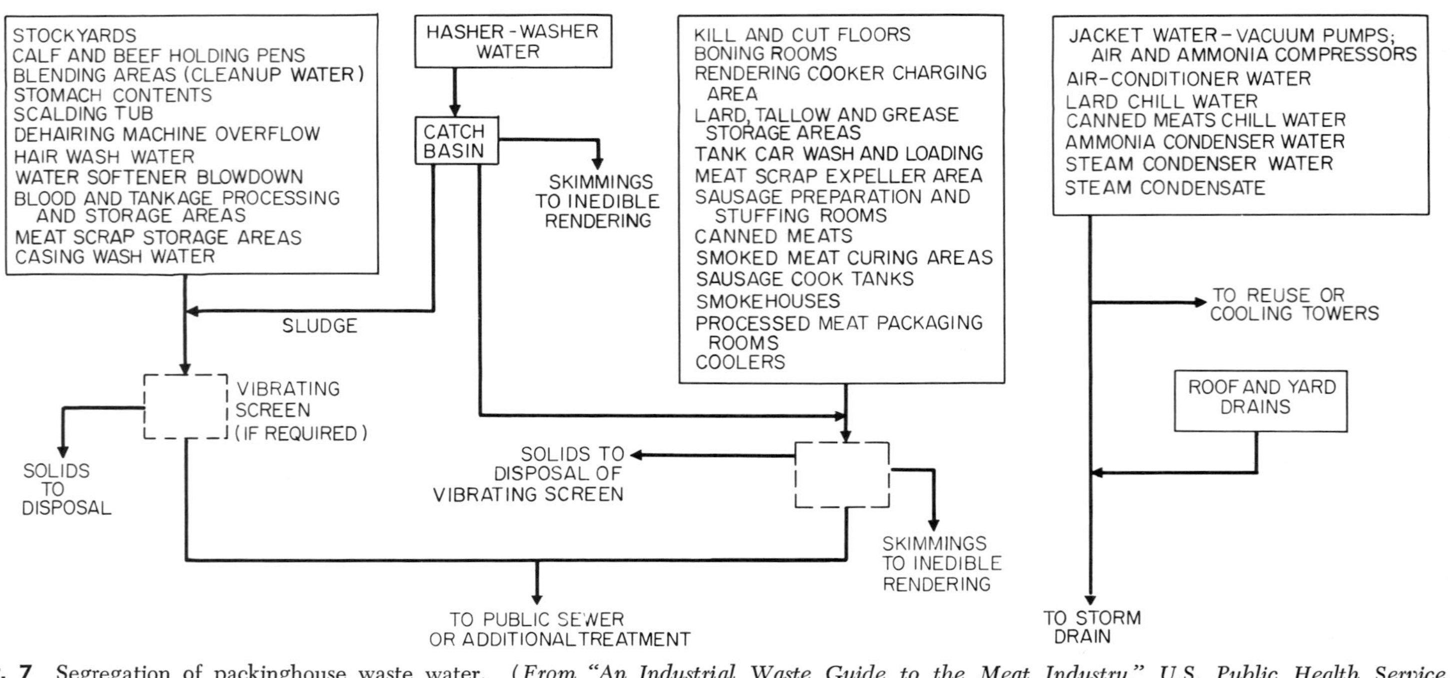

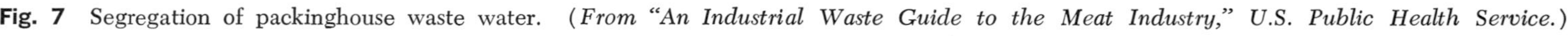

Fig. 7 Segregation of packinghouse waste water. (*From "An Industrial Waste Guide to the Meat Industry," U.S. Public Health Service.*)

Packinghouses frequently kill and dress, but in addition the meat is cooked, cured, canned, and frozen. These operations result in a rather strong waste.

Where possible, solids are recovered for by-product use. The main recoverable item in packinghouse and slaughterhouse wastes, in addition to those already mentioned, is grease. The removal of free-floating grease is important to avoid clogging of sewer lines and fouling of biological treatment plants. Table 5 indicates the types of treatment employed by meat packing processors and indicates the number of plants that incorporate grease recovery by either gravity flotation or air flotation methods, including the use of chemicals.

Slaughterhouse and packinghouse wastes are amenable to treatment. Frequently these wastes are treated in municipal sewage treatment plants; however, prior to release into city sewers, pretreatment practices of screening, sedimentation, and gravity flotation are normally practiced. Figure 7 shows a typical segregation of packinghouse waste water. The methods of treatment for these wastes are discussed in further detail in a later section.

TABLE 5 Types of Waste Treatment Employed by 108 Meat Packing and Processing Plants*

Method	*No. of Plants*
Screening	59
Sedimentation	71
Filtration	2
Flotation (air)	11
Flotation (gravity)	87
Flocculation	2
Evaporation	2
Chemical coagulation	1
Trickling filter	1
Activated sludge	12
Anaerobic digestion	5
Septic tank	13
Irrigation	2
Stabilization pond	7

* From "Water in Industry," National Association of Manufacturers, January, 1965.

Other Food Wastes

It is not possible in a handbook of this nature to discuss all the different types of food waste to be treated, such as corn products, pickling, potato wastes, rice, coffee, and citrus. A brief discussion will be given, however, on dairy and beet sugar wastes.

In the *dairy industry* most plants consist of several operations, and the types of waste vary accordingly. Among these operations there may be receiving stations, bottling plants, creameries, ice cream plants, cheese plants, and condensed and dried milk product plants. As in the rest of the food industry, controlled product losses serve the double function of improved yield and efficiency and at the same time reduce potential waste pollution problems. Table 6 gives the approximate quantities of BOD in the various processes. Because of the method of processing and the products which are produced, there are at times, with various operations, surpluses of separated milk, buttermilk, or whey, as well as occasional batches of sour milk. Unfortunately, there is no simple, economical method to reclaim and utilize these materials as by-products, and therefore the disposal of this material becomes a very serious problem. Indiscriminate dumping of this material into the sewers should be avoided, and where possible, these extremely strong wastes should be treated separately or disposed of by hauling away.

The treatment of milk wastes is normally handled through municipal plants. These wastes are amenable to treatment. Various methods of treatment will be discussed later. Pretreatment by screening is a good practice. In some cases, grit removal also should be utilized.

Another major food processing waste is created at *beet sugar* refineries. The beet sugar industry in the United States is approximately 100 years old and has most of its production facilities located in the irrigated areas of the Western United States. Sugar beets are harvested in the fall and are processed continuously for 3 to 4 months during the fall and early winter.

TABLE 6 Approximate Quantities of BOD in Various Process Wastes*

Process	*lb of 5-day BOD per 10,000 lb Milk or Milk Equivalent*
Receiving and cooling milk	6
Tank truck delivery to and from plant, including washing tank truck	1
Storage of fluid product in tanks	1
Evaporating whole milk, floor waste	2
Evaporating whole milk, entrainment loss	1
Canning and sterilizing evaporated milk	2
Spray drying whole milk	1
Cream separating	2
Cream pasteurizing, cooling, and can filling	2
Separated milk pasteurization, cooling, and can filling	2
Cottage cheese or casein manufacture	16
American cheese making, unwashed curd	10
American cheese making, washed curd	16
Separated milk condensing, plain, floor waste	3
Separated milk condensing, plain, entrainment loss	0.5
Separated milk condensing, sweetened, including barreling floor waste	6
Separated milk condensing, superheated	12
Separated milk condensing, sweetened, entrainment loss	0.5
Separated milk drying, spray process	1
Separated milk drying, roll process	16
Whey condensing, sweet, floor waste	8
Whey condensing, sweet entrainment loss and volatile	2
Whey condensing, acid, floor waste	8
Whey condensing, acid, entrainment loss and volatile	4
Whey drying, spray	5
Buttermilk condensing, floor waste	8
Buttermilk condensing, entrainment loss and volatile	4
Milk pasteurization, cooling and bottling (glass)	8
Milk pasteurization, cooling and bottling (paper)	6
Ice cream mix making, vat	4
Ice cream mix making, pan	4
Ice cream freezing	0.5
Cultured buttermilk making	5
Butter churning and washing	2

* From "An Industrial Waste Guide to the Milk Processing Industry," U.S. Public Health Service.

Waste waters from such an operation are extremely high in dissolved organic matter, and unfortunately the processing season occurs during the normal dry season when streams are at relatively low flows. Therefore, the wastes must be treated to avoid exerting a tremendous oxygen demand on the receiving streams, to protect aquatic life, and to avoid odor problems.

A typical flow pattern in the manufacture of beet sugar begins with the fluming of beets from storage piles to the plant. After the beets are thoroughly washed and all extraneous foreign material is removed, they are sliced into small, noodle-shaped strips. The strips are then passed into a diffusion apparatus where the sugar is extracted by means of a countercurrent system utilizing a 15 percent solution of hot water and sugar. Lime and carbon dioxide are utilized on the heated juice in a double filtration system to purify the sugar. The juice then goes to a multiple evaporation station, where it is filtered and crystallized. The crystals are separated from the syrup in a centrifuge and then dried in a granulator system and stored. The syrup is recycled for further evaporation and crystallization. The Steffen's process is one used by many plants to extract additional sugar using lime and carbon dioxide.

The largest use of water in the above flow pattern is for fluming the beets from storage into the plant and for washing. These two steps account for about 50 percent of the water demand at a beet sugar plant. Condensing the vapor produced in the multiple effect evaporators and the vacuum pans in which crystallization is effected demands almost as much water as does fluming and washing. A much smaller volume of water is required for dissolving sugar from sliced beets, and an even smaller quantity is needed to transport the lime solids filtered from the sugar juice.

The largest amount of waste water comes from the fluming and washing operation and contains large quantities of soil, suspended beet fragments, stems, roots, leaves, and dissolved organic materials from the beets. This waste stream has a significant BOD and a high but variable concentration of suspended solids. If the beets are in

TABLE 7 Representative Values of Unit Process Wastes from Beet Sugar Manufacture*

Source of waste	Flow/ton beets, gal	BOD, ppm	Suspended solids, ppm	BOD/ton sliced beets, lb
Flume water	2,600	210	800–4,300	4.5
Pulp screen water:				
Bottom dump cell	240	980	530	2.0
Side dump cell	1,420	500	620	5.9
Continuous†	400	910	1,020	3.0
Pulp press water	180	1,710	420	2.6
Pump silo drainage	210	7,000	270	12.3
Lime cake slurry	90	8,600	120,000	6.5
Lime cake lagoon effluent	75	1,420	450	0.89
Barometric condenser waste	2,000	40	....	0.67
Steffen's waste	2,640‡	10,500	100–700	231(3)

* Adapted from Tables 2 and 3 in "Beet Sugar—An Industrial Waste Guide to the Beet Sugar Industry," Federal Security Agency, U.S. Public Health Service, 1950.

† Pulp transported by water.

‡ Per ton molasses processed on a 50 percent sucrose basis.

good condtion, the waste will have a BOD of approximately 200 ppm. The strength will vary according to the condition of the beets and becomes appreciably higher if the beets are decomposed because of freezing and other factors. Table 7 gives the unit loadings of beet wastes from various operations.

Some of the water used for extracting sugar from the beets goes to the spent pulp and is known as pulp screen water and pulp press water. This waste contains sugar, other dissolved organics, and finely divided suspended pulp. The volume of this waste is relatively small, but the BOD is comparable with that of the fluming and washing wastes.

Lime cake slurry is the waste produced from mixing and conveying the lime cake. This waste is also of low volume, but it has a high BOD and a high suspended solids concentration. Almost in direct contrast, however, is the condenser waste water stream, which has a high volume but a low BOD.

Plants which use the Steffen's process have an additional waste to contend with, and this is known as Steffen's waste or Steffen's filtrate. Although of low volume, this waste stream has a high organic content.

The most conventional means of treating beet sugar wastes is lagooning. The wastes usually discharged to the lagoon are the flume and wash water and the lime cake slurry, although in some plants the lime cake is reburned and reused in the process. Most plants dry the spent pulp, thus eliminating the need for treatment of this liquid waste, and condenser waste water has a low enough BOD to allow it to be discharged, without treatment, to a receiving stream. A lagoon allows enough retention time for solids settling and a partial BOD reduction. Lagoons are discussed in a later section. Further work is needed to learn how to control lagoon odors so that lagooning practices may become more acceptable. Conventional secondary treatment such as activated sludge and trickling filter systems are, at present, impractical because of the seasonal nature of beet sugar manufacture and the large variations in quantity and quality of the wastes produced. Fortunately, most beet

Fig. 8 Water reuse equipment.

sugar plants are located in rural areas where it is practical from an odor and land cost standpoint to utilize lagooning.

Water reuse is an important concept in the beet sugar industry as well as in the food industry as a whole. Not only does water reuse reduce the demand for fresh water, but it also reduces the amount of waste water generated for ultimate disposal. As indicated by Fig. 8, the strength of the waste is exceedingly high. Every effort should be made to reduce the volume of waste by as much reuse as possible and to avoid the fluming of waste solids. Even if there is mechanical conveying of pulp and elimination of pulp solids drainage, there is still a tremendous BOD load, which normally is treated in lagoons. Further research work is required in the area of by-products recovery. In recent years, plants utilizing the Steffen's method have developed solids disposal similar to the citrus industry, where citrus molasses is used to fortify dry citrus pulp feed when other outlets for the molasses are not profitable,

Some typical water reuse steps in the beet sugar industry are as follows: (1) condenser water can be used as makeup water for both beet fluming and washing; (2) flume and wash waters can be partially recirculated after screening is used to remove solids; (3) pulp screen and pulp press waters can be recirculated to the diffuser using condenser water as makeup; (4) lime cake can be reburned and reused; (5) in plants utilizing the Steffen's process, the Steffen's waste can be evaporated and used for the production of livestock feed.

Since condenser water can usually be discharged without treatment, the major advantage in its reuse is a reduction in fresh water demand. Reuse of flume and wash waters, on the other hand, offers a reduction in the quantity of waste water that requires treatment. Elimination of the pulp waters and the lime cake eliminates a large portion of the organic load that would otherwise require treatment.

WASTE TREATMENT METHODS AND CONTROL

In-plant Control

In previous discussions on individual food processing operations we have noted the importance of water conservation and reuse. There is a need to keep solids out of water by screening and grease recovery.

Wherever possible, solid waste should be handled in a dry state in order to reduce the contamination of water. Such considerations follow good water conservation practices. The most important factor in any food plant is sanitation and the reduction of potential spreading of organisms that can contaminate the product. Automated mechanical systems such as that outlined in Fig. 4 are a tremendous assistance in providing proper sanitation within the plant. Where a food processing plant has its own treatment system, it is to its best interest to know the source and strength of various waste streams within the plant and, wherever possible, to keep the excessively strong waste isolated for separate treatment.

The National Canners Association Laboratories have suggested as a program of assistance in solving problems the following steps in in-plant control:

1. Reuse of clean or relatively clean water in appropriate operations
2. A reduction in the volume of water used for product transport
3. Removal of solid wastes by hand or mechanical means rather than flushing them to the gutter
4. Segregation of highly concentrated waste streams for separate treatment or disposal
5. Separation of can cooling or other clean waters for disposal without treatment to reduce the volume of waste
6. Recombining, under appropriate conditions, clean waters with treated waters to give dilution at the point of final discharge

Screening

Perhaps one of the most common methods of pretreatment of food processing wastes is the use of screens to remove solids. The technique will vary in type of equipment utilized based primarily on the character of the waste to be screened and the degree of solids removal desired. In some municipalities the waste need only be screened to remove identifiable objects. In other areas, screening through 40 mesh or greater is a necessity by ordinance. In the former case, an automated mechanical bar rack or barminutor may be all that is required. In the latter case, a fine screen capable of being continuously cleaned is a necessity.

Actually, to some degree, screening may be considered a lost art. The initial installations of sewage treatment plants utilized screening to a relatively high degree. Unfortunately, except for industrial waste this method of removing BOD loads has until recently become almost defunct in the municipal field.

James Paterson,[1] of Sacramento, has shown that fine mesh screening can provide

[1] Paper, California Water Pollution Control Conference, Apr. 29, 1966.

BOD reductions for both domestic and industrial waste or combinations thereof greater than can be accomplished in primary settling tanks. More development work is required in this area, but it points to the capability of having relatively high BOD reductions with equipment that is small in size and requires a much lower capital investment than do primary settling tanks. This is of considerable value to food processors, who are required to reduce BOD prior to discharging to a municipal system, as it affords a facility that does not take up a lot of valuable space nor does it create the secondary problem of sludge handling, since screened solids are normally much easier to handle than sludge from a primary settling tank. Frequently, the sludges from primary settling tanks are rather putrescible and become a major odor and disposal problem.

Where coarse screening is required, it is possible to utilize bar racks. These normally have an open spacing of 1 to 2 in., although if it is necessary to remove smaller solids the opening can be reduced to less than ½ in. For maximum efficiency and dependability, units should be equipped with mechanical racks to remove the solids continuously. One of the advantages of a mechanically operated bar rack is that it takes up very little space and can be installed "on stream" in the pipeline. The equipment can be installed in a small pit designed to maintain a minimum velocity of 1 fps so that solids do not settle out and also designed to avoid velocities in excess of 3 fps in order to prevent wedging of solids between the bars (see Fig. 8).

Fine mesh screening is utilized where the local municipality dictates such requirements through local ordinances or where the industry is desirous of removing the maximum BOD load possible for its own treatment system or a municipal treatment system. There are a number of arrangements for fine mesh screening consisting of disk screen, rotating drum screen, and vibratory type screens. The disk screen is a large circular disk submerged approximately one-third to one-half in the waste. As the screen rotates, the solids cling to the screen mesh or the collector arms. The screen is backwashed before it returns to the waste water stream. The backwash water carries away the solids to be handled in a separate system. These screens vary in size from approximately 4 ft in diameter to in excess of 12 ft in diameter.

The rotating drum screen consists of a cylinder frequently made of perforated metal or coarse wire mesh as a structural frame with the fine mesh screen being supported on the structure. One type of drum screen has the waste enter into the center chamber and pass through and drop out the bottom. These solids are carried up and flushed away with a back spray arrangement. Another arrangement for solids removal with this type of screen is an internal helix to screw the solids forward for removal in a relatively dry state. Another form of rotating drum screen has the waste water coming from the outside and passing the screened water to the inside, and the waste solids are removed in a waterborne arrangement. Most drum screens are horizontal, although a vertical arrangement drum screen is available. They vary in size from 2 to 8 ft in diameter and are 4 to 12 ft long.

The vibrating screen has several arrangements. One is rectangular or square, set at a slight incline so that the waste is applied at the top and the solids are vibrated down to the end, where they fall off in a relatively dry manner, with the water passing through the screen. Another type of screen available is circular in nature and utilizes eccentric weights to give a three-way gyrating motion.

Both types of vibrating screens normally have a capability of discharging the solids in drier fashion than either the disk or rotating drum screen. The type of screen to be utilized is clearly dependent upon the character of the waste to be screened and the degree of clogging encountered with the waste. Normally, wastes with high grease content can be screened better with either the disk or rotating drum screen since high-pressure sprays can keep the screening continuously clean, although vibrating screens can be equipped with cleaning arrangements. (See Fig. 8 for typical pictures of the disk, rotating drum, and vibrating screens.)

A number of other types of screens are available which are variations of the above types and are too numerous to comment upon here. A new basic concept in screening has been introduced with a fixed nonmoving screen made with triangular bars on a parabolic curve. This has the advantage of no moving parts but requires a higher head loss than with most other screens and may be subject to clogging with grease bearing wastes, although it can be equipped with mechanical cleaners.

Where regulatory authorities specify the type of mesh screening, it varies normally from 10 to 40 mesh. The solids recovered from the screening operation will vary appreciably depending upon the type of operation and the type of waste. The volume of solids collected and dependent upon the aforementioned factors may vary from 15 to 80 lb/1,000 gal of waste screened. A 10-mesh rotating drum screen 4 ft in diameter and 8 ft long may be able to handle peak loads of 3,500 gpm, whereas that same size rotating drum screen with a 40-mesh screen is likely to handle only approximately 2,000 gpm. A 4- by 7-ft vibrating screen with 20-mesh stainless steel wire is reported able to handle 600 gpm of paunch manure waste water. The same size vibrating screen with 30-mesh cloth has handled 700 gpm of beet sugar pulp waste. As previously indicated, vibrating screens produce the driest solids, and the moisture content of these solids may vary between 70 and 95 percent, depending upon the product handled. The disposal of screening is discussed under the separate heading of solids disposal.

Grease Recovery

Food processors who handle a significant quantity of meat or poultry are faced with the problem of grease in their waste waters. When the food processor treats his own plant waste for discharge to a stream, separate grease recovery facilities are not always necessary since it is possible to remove grease readily with surface skimmers in the primary clarifiers of a conventional sewage treatment plant. Most municipalities require some form of pretreatment for grease recovery prior to discharge of industrial waste to the community sanitary sewer systems. The main objection to too much free grease in a municipal sewer is that it coats the sewer, thus reducing its flow capacity, and becomes a problem in pumping stations because of forming a heavy sludge blanket on the surface or creating large grease balls, which are a disposal nuisance.

Grease is not normally a problem in a sewage treatment plant that is adequately designed to handle a high greasebearing waste. Many times municipal sewage treatment plants are poorly designed for handling of grease, and therefore they get into operating problems such as poor clarifier performance because of inadequate scum skimmer systems and especially poor digester performance. The latter can be overcome with good mixers, since grease will readily digest. However, if the grease is not kept adequately mixed, it will form a heavy blanket at the top and never digest, therefore utilizing valuable storage space and reducing the efficiency of the overall digester operation.

There is an incentive for most food processors to recover grease, since grease has a market value and can be sold as a by-product. Normally, the sale of this grease not only will pay toward the installation and operation of the equipment but frequently is capable of netting a profit.

In order to recover grease properly, it is desirable to avoid pumping the waste so that it does not become emulsified. Diluting it with other non-grease-bearing wastes should be avoided so that the maximum detention time in a holding facility is achieved. Normally, it is best to recover the grease as close as possible to its source in order to avoid clogging problems with in-plant sewers. There are a number of different types of grease recovery units, from simple trap to air flotation units with chemical additions.

Grease from food processing operations falls into two main categories: free floating and emulsified. As the name indicates, the former is readily recovered. The free floating grease can be recovered in short detention periods of 5 or 10 min with flow rates of 3 to 6 fpm. This grease has the greatest economic advantage to the food processor, since it is normally collected in a relatively clean state and is high in grease content and readily processed for by-products recovery. Emulsified grease is created by process operations such as cooking and pumping. This grease tends to stay in suspension and requires long detention periods of quiescent conditions to rise to the surface. Highly emulsified grease may require 30 min to over 1 hr to be recovered, and even then the degree of recovery is not nearly as high as with the free floating grease. In the case of free floating grease, 90 percent recovery can be readily achieved. Emulsified grease recovery by gravity separation rarely becomes greater than 50 percent.

Improved grease recovery on emulsified fats can be achieved with air flotation units and vacuum flotation systems. Improvements can be achieved by the use of chemicals such as alum, activated silica, chlorine, and other coagulant aids. Although higher grease recoveries can be achieved with more sophisticated equipment and the use of chemicals, frequently such programs are not justified and seriously curtail the economics involved. Grease recovered under such conditions frequently is tied up and is not as readily usable as a by-product. Also, depending upon local municipal conditions, an emulsified grease may not be too significant in the community's sewer system, and therefore the need for recovery in more sophisticated systems becomes questionable. All these systems have their places, however, and each must be judged on the particular local conditions existing and what must be achieved to meet specific criteria.

Biological Treatment

Almost without exception, food wastes are amenable to biological treatment. These organic wastes demand oxygen to be stabilized. Microorganisms in a biological treatment system remove the organic materials by adsorption and direct metabolism. The end products of such biological oxidation systems are carbon dioxide and water. The key to the successful treatment of these organic wastes is to supply sufficient oxygen to maintain the growth of the organisms. In the treatment process, the pH and alkalinity of these wastes is altered because of the biological activity. Not all wastes contain sufficient nitrogen, and therefore in some treatment programs it is necessary to add a form of nitrogen such as ammonia. Biological treatment is achieved in such processes as trickling filters, activated sludge with contact stabilization, oxidation ditch, and other arrangements as variations, lagooning, anaerobic digestion, and spray irrigation. Wastes may be treated in any of these systems as strictly an industrial waste or in combination with municipal wastes. There is no simple expedient method for a selection of a waste treatment method. Many factors have a bearing on which system should be utilized or what is actually required to achieve a certain goal or criterion. Frequently, expert advice from consultants is not only desirable but necessary. The National Canners Association Laboratories have suggested a guide of some of the factors which should be considered in the selection of disposal methods and about which information should be obtained.

1. The size of the pack and packing periods for all products
2. The amount of soluble, suspended, and settleable solids in the liquid waste stream
3. The range and average of pH and BOD or COD values for the composite waste flow
4. The average and peak flow of liquid waste
5. The potential for solids removable by screening, if screens are not already in use
6. Characterization of waste flow from major unit operations as to the volume, strength, etc.
7. The completeness of treatment being required by regulatory authorities
8. The potential for land or lagoon disposal: (*a*) proximity of dwellings to the site, (*b*) availability and cost of land, (*c*) ability of the land to accept water and the nature of the cover crop, (*d*) water table level, (*e*) climatic conditions
9. Potential for use of municipal treatment, including sewer charges

Trickling Filters

Trickling filters are frequently used as a means of treating organic wastes because of their adaptability to handle peak shock loads and the ability to function satisfactorily after shutdown periods over weekends, etc. Conventional trickling filters normally consist of rock 4 to 7 ft deep over an underdrainage system. The continuous application of organic wastes develops on the filter bed stones a gelatinous film of microorganisms. The soluble colloidal organics in the waste as it trickles over the rocks are adsorbed by the film where it is stabilized by the microorganisms. The filters slough, and the microbial cells and slimes readily settle out in settling tanks for removal of the major portion of the BOD. Conventional rock filter beds have been designed on a basis of either standard-rate or high-rate trickling filters. The

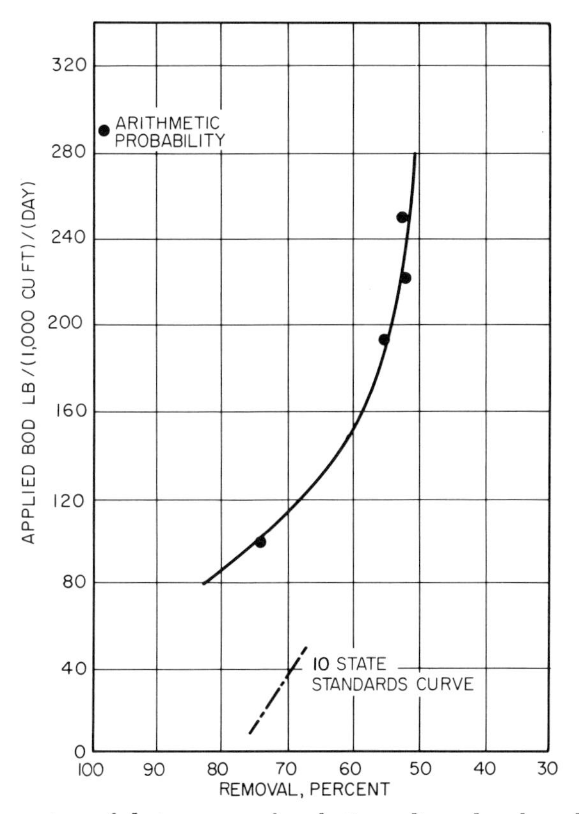

Fig. 9 A comparison of design curves for plastic media and rock media shows the improved ability of plastic media to treat heavy loadings of soluble organic wastes. The solid line indicates plastic media.

load on a standard-rate filter normally does not exceed 600 lb of BOD per acre-foot, whereas high-rate trickling filters have loadings in excess of 1,500 lb of BOD per acre-foot.

More recently, plastic filter media have been used for trickling filters. The advantage of this type of construction is that the lightweight plastic pack can be stacked many times higher than the conventional rock filter beds since there is no significant weight. Assuming a conventional rock bed of 6 ft, it is reported that a plastic filter media bed containing the same cubic foot content as a rock filter, but three times as deep—18 ft—would achieve three times the pounds of BOD removal for the same degree of efficiency. Figure 9 is the curve comparison.

Although plastic filter media is more costly than rock for the same cubic yard capacity, it has many advantages which overcome the higher unit cost rate. These include a much higher degree of treatment capability because of the increased surface area available, reduced odor potentials at higher loadings, and the capability of using lighter-weight structures for supporting the entire system. The deep pack beds also lead to a saving in land area utilized. Costly tiles and underdrain systems are avoided. The plastic media filters are reported capable of loadings over twice that of high-rate trickling filters for the same degree of treatment. Table 8 shows typical treatment of fruit and vegetable processing waste on plastic media 16 ft high with variable loading rates.

TABLE 8 Treatment of Fruit and Vegetable Processing Wastes on Plastic Filter Media*

Feed to plant, ppm BOD	Treated effluent, ppm BOD	Load, lb BOD (cu yd)(day)	Removal, lb BOD (cu yd)(day)	Removal efficiency, %
2,160	690	3.64	2.48	68.2
1,440	390	2.42	1.77	73.2
1,080	360	1.81	1.21	66.6
2,420	660	4.06	2.97	73.2
2,160	270	3.64	3.18	87.3
1,860	248	3.14	2.73	86.9
1,620	90	2.73	2.58	94.5
1,740	390	2.93	2.28	78.0
1,620	480	2.73	1.92	70.4
1,500	400	2.52	1.85	73.4
2,220	360	3.74	3.13	83.6
1,920	420	3.24	2.53	78.1
2,400	620	4.05	3.0	74.2
1,320	420	2.22	1.52	68.4
2,160	510	3.64	2.77	76.0
1,620	665	2.72	1.61	59.2
1,380	660	2.32	1.21	52.2
2,140	330	3.6	3.05	84.6
840	270	1.41	0.96	68.1
1,440	270	2.42	1.97	81.5
2,340	330	3.94	3.38	86.0
1,140	80	1.92	1.78	92.0
1,080	260	1.82	1.38	75.9
1,620	60	2.73	2.63	96.4
1,620	90	2.73	2.58	94.5
1,500	390	2.53	1.85	73.0
1,140	90	1.92	1.77	92.0
1,740	50	2.93	2.84	96.9
1,080	70	1.82	1.7	93.5
840	100	1.41	1.25	88.0
1,140	120	1.92	1.72	89.5
1,680	100	2.83	2.66	94.0
1,920	150	3.24	2.98	92.0
1,920	210	3.24	2.88	89.0
Mean 1,641	309	2.76	2.24	81.1

* From M. W. Askew, Plastics in Waste Treatment, *Process Biochemistry*, December, 1966, and January, 1967.

With either type of filter, knowledge of the waste is necessary in order to assure good bacteriological treatment. Many fruit and vegetable wastes are deficient in available nitrogen, and in order to have sufficient microbial growth on the filter, it is necessary to add nitrogen bearing chemicals. Conversely, meat packing wastes frequently are extremely high in concentrations of protein which produce heavy biological growth that can tend to clog, especially rock trickling filter media. Where high pH is encountered because of lye peeling or caustic cleaning compounds, it is necessary at times to adjust the pH in order not to destroy the biological growth on the filter. Some types of fruit wastes are sufficiently acidic as to warrant adjustment to a more neutral pH for optimum treatment on trickling filters.

Trickling filters normally have a greater flexibility in handling the inconsistencies in character sometimes involved with food processing wastes. Their one disadvantage over some other systems is the potential for odors, especially on rock beds if they tend to clog or not have sufficient air circulation.

Activated Sludge

Most food processing wastes are readily treated in a conventional activated sludge system. This process is carried out in tanks whereby a biologically active sludge is added to the waste water to be treated in predetermined concentrations and aerated vigorously to supply the organisms with sufficient oxygen. After sufficient detention time, which may be 2 to 5 hr, the mixture of waste and aerated sludge is separated in settling tanks, with the settled sludge returned for remixing with the raw waste. The required degree of detention time is dictated by the organics in the waste and the time required to be able to adsorb the organics on the sludge floc created. High efficiencies are obtainable with this type of system. However, secondary problems are frequently associated when food processing wastes are treated because the system is not normally amenable to shock loads or to extended shutdown periods such as over weekends. The activated sludge operation is not readily adaptable to wide variations in character of the waste. As in the case of trickling filters, an increasing of the loading factor will decrease the efficiency of BOD removal. The activated sludge process is normally more costly than other treatment systems, but if land is at a premium and the plant is located in an urban area where objectionable odors cannot be tolerated, this treatment system may provide the most practical approach.

There are a number of additional types of fluid bed treatment systems such as the activated sludge process. There are a variety of activated sludge processes, each one varying in detail but all having in common the contact of raw liquid wastes with preformed biological flocs in an aerated system. Part of the organic wastes which are susceptible to biological decomposition is changed to inorganics and part is converted to activated sludge.

In the *step aeration process,* the raw waste flow is split into equal increments and enters the system at separate points in the aeration basin, spaced equal distances apart. The organic load is therefore spread over the entire length of the basin, and accelerated growth and oxidation take place throughout most of the basin rather than at just one end.

In the *contact stabilization process,* settled waste is combined and mixed with returned activated sludge and aerated in a contact basin. The flow then goes to a thickening tank where the activated sludge is separated from the mixed liquor. A small amount of sludge is wasted while the major portion flows to a stabilization basin. Here the sludge is aerated and returned to the contact basin, where it is mixed with the incoming waste.

The *modified aeration process* is similar to the conventional process except that the volume of return sludge is smaller and aeration time is shorter. The result is a lower removal of organics.

In the *completely mixed* processes the raw waste is distributed uniformly throughout the aeration basin. The flow then goes to a thickening tank. In this process there is an option of wasting considerable quantities of activated sludge or returning almost all the sludge to the system. Complete sludge return is practiced where the organics in the waste are completely soluble and is particularly suited to the treatment of some industrial wastes.

The *oxidation ditch* process is a modified form of the activated sludge process which offers simplicity and economy. Screened sewage enters directly into an open ditch which serves as the aeration basin. The aeration equipment is in the form of a rotor aerator which extends across the width of the ditch. The rotor furnishes the necessary oxygen and keeps the liquid waste mixed and moving through the ditch. Mixed liquor flows from the oxidation ditch to a clarifier. Clarified liquid is discharged to a receiving stream while most of the sludge is returned to be mixed with the raw waste in the ditch. The oxidation ditch is a shallow, easy to construct, earthen ditch which can be either lined or unlined depending on local soil conditions. The shape of the ditch can be made to suit local requirements. Particularly on a small to intermediate scale, the oxidation ditch process is an economical treatment method capable of BOD reductions greater than 90 percent.

Lagoons

Lagooning is perhaps one of the oldest methods of biological treatment known to man but until recently was little understood and was not studied in depth to permit

TABLE 9 Middle West Storage Lagoon Operation, Seasonal Tomato Operation

Date	COD, ppm	BOD, ppm	Suspended solids, ppm
Waste strength to lagoon:			
9/6/61	1,980	760	664
9/22/61	2,430	800	806
9/28/61	1,194	1,127	1,374
Lagoon samples:			
12/18/61 (60 days after pack)	123	150	144
1/25/62 (90 days after pack)	127	135	122
4/12/62 (final lagoon drainage)	55	40	52

intelligent utilization with adequate design parameters. There are storage lagoons and flow-through lagoons. An example of the latter would be an aerobic lagoon because of low loading or mechanical aeration. Another type of flow-through lagoon is a combination anaerobic-aerobic system.

Storage lagoons are utilized primarily for small operations of a seasonal nature where it is possible to store the waste from an entire season's pack in a lagoon. Normally, these lagoons operate anaerobically until the BOD load is stabilized to a point where aerobic conditions take over. Frequently, the lagoon is aerobic on the top surface and anaerobic underneath. Most lagoons of this nature will sufficiently stabilize the waste in 90 to 120 days so that the effluent can be released to a stream after a sufficient storage period. The efficiency of these systems over an extended period of time is reported to achieve BOD reductions of 90 percent or better, depending upon the type of waste. A seasonal tomato operation in the Middle West utilizing a storage lagoon had peak daily loadings of 600 lb of BOD/acre/day. Table 9 indicates the strength of the waste to a Middle Western storage lagoon and the strength of the waste in the lagoon after storage for 60 days, 90 days, and final lagoon drainage approximately 6 months later.

The aerobic lagoon, when relying strictly on natural conditions for reoxygenation, is limited in loading. Some state health departments will permit only 20 lb of BOD/acre/day. Where there is sufficient sunlight and shallow depths are practical, loads of 1,000 lb of BOD/acre/day have been possible. Most food processing wastes in lagoons that experience winter conditions of freezing over normally will not function satisfactorily with BOD loads in excess of 100 lb/acre/day. Increased loadings can be achieved with mechanical aeration.

Dr. C. D. Parker's work in Australia has indicated that in a relatively mild climate it is possible to permit 400 lb BOD/acre/day for tomato waste and 200 lb BOD/acre/day for citrus waste by careful biological nutrient control of the pond so that the top surface area is aerobic but the major depth of the pond is under anaerobic conditions.

The major factor in higher lagoon loadings is the reduction of solids. Dr. A. J. Kaplovsky's work in Delaware on a poultry waste indicated that average BOD loadings of 214 lb/day/acre could be treated obtaining high BOD reduction if the waste was presettled and had flow equalization.

For the food processing industry, the aerobic lagoon is of considerable importance where land areas are not critical or costly. It has the advantage of maximum efficiency during the dry summer months when most local receiving streams have the least potential for assimilating a treated effluent.

The combination of anaerobic and aerobic lagoons is a recent development which is utilized primarily in the meat packing and food processing plants. It has the advantage of digesting the waste anaerobically for a high BOD reduction in a relatively short detention period in a small pond, with the effluent passing to larger aerobic ponds for final stabilization. Loadings in the anaerobic pond are normally from 12 to 15 lb of BOD per 1,000 cu ft, which is many times the permissible loading in an aerobic lagoon. With meat and poultry wastes, heavy scum layers on top of the pond reduce the potential for offensive odors. Most food processing wastes are high in volatile solids, and these are readily digested in the anaerobic portion. With anaerobic pretreatment ahead of the aerobic lagoons, the final loading to the aerobic lagoons is greatly reduced since where warm wastes are involved there is a BOD reduction of 70 to 90 percent and at least a 60 percent reduction at colder temperatures.

One of the advantages of the anaerobic-aerobic lagoons is that there is practically no raw solids load taken to the aerobic ponds and therefore very little solids are digested anaerobically on the bottom of the aerobic ponds. Thus, there is little odor potential during turnover periods in the spring and fall.

An interesting case is a small community in southeast Nebraska which had a relatively large poultry operation. Initially, the combined wastes were treated in aerobic

TABLE 10 Nebraska Lagoon System (ppm)

Characteristic	Influent	Anaerobic pond effluent	Aerobic* pond effluent
BOD:			
Avg.	840	350	67
Max.	1,260	675	99
Min.	450	218	24
Grease:			
Avg.	220	70	9
Max.	350	110	12
Min.	90	44	4
Suspended solids:			
Avg.	610	215	130
Max.	984	342	183
Min.	304	120	52
pH:			
Avg.	7.4	7.5	8.1
Max.	8.2	8.1	8.9
Min.	6.5	7.2	7.3

* Samples not filtered.

lagoons. Odor problems occurred during winter weather when the pond would alternately freeze and thaw, thus creating a turnover. The pond had an average loading of approximately 100 lb/acre/day and peaks of 160 lb BOD were reached. In order to overcome the odor problem, an anaerobic cell was installed immediately ahead of three 15-acre aerobic cells. The anaerobic cell was approximately 0.65 acre and had a loading of 12 lb of BOD per 1,000 cu ft. This cell soon became completely coated with grease, and odors have not been reported at a nuisance level. Table 10 indicates the effectiveness of this type of treatment on a system whose total load is approximately 85 to 90 percent from the poultry operation.

A factor not often considered in lagoon discharges is that the remaining BOD in the effluent as indicated in Table 11 consists primarily of algae. This is shown very dramatically on samples from an aerobic lagoon system containing a poultry and domestic sewage in western Ontario as shown in Table 11.

TABLE 11 Western Ontario Lagoon Effluent (ppm BOD)

Lagoon cell	Avg	Max	Min
No. 1	41	160	24
No. 2	35	69	11
No. 1 (algae filtered out)	4	8.4	1.2
No. 2 (algae filtered out)	7.8	20.0	1.2

Sodium nitrate will eliminate or reduce lagoon odors, based on the amount of nitrate added. Past work has indicated that feeding nitrate at the rate of 20 to 25 percent of the oxygen required for the 5-day BOD demand is a good guide as a suitable control criterion. The main function of the sodium nitrate is to furnish oxygen for aerobic bacterial decomposition, to stimulate growth of algae and other organisms which produce through photosynthesis additional oxygen, and finally to maintain an alkaline reaction. Therefore, sodium nitrate promotes aerobic decomposition with less odor by supplying oxygen both directly and indirectly.

Spray Irrigation

Spray irrigation of waste water should not be confused with farm irrigation. Waste water disposal by natural filtration encompasses a great deal more than the irrigation of waste water and its infiltration into the soil. Actually, it makes little difference if the water is purified while flowing overland, or if it percolates into the soil. Purification, in either case, is accomplished biologically and is dependent upon the biota and organic litter on and in the soil. Pure sand, without organic debris, will provide only mechanical filtration without reduction of the soluble solids.

The vegetative cover of a disposal field provides a protected habitat for soil microorganisms and a vast surface area for the adsorption of organic impurities. Almost any species of grass is satisfactory provided it produces abundantly, is water tolerant, and forms a turf. Grass serves a secondary function by protecting the soil from erosion and compaction.

Conventional spray irrigation systems with farm irrigation pipe relying primarily upon infiltration are well documented in literature. These systems require good site location in order to assure a high degree of infiltration into the ground. Therefore, the knowledge of local soil conditions is of paramount importance. Such systems on sandy loams that are well drained can tolerate in excess of 1 in. of water a day. Of perhaps greater importance, however, is the ability to spray irrigate on almost any type of soil regardless of its infiltration capability. The following discussion is limited primarily to concepts of spray irrigation on tight soils with overland flow and discharge of treated effluent to the receiving stream.

The "natural filtration" process of waste treatment using the "overland flow" method purifies water by flowing it over the soil surface in a thin, uniform sheet. The system is laid out on land that is contoured so that the wastes flow in a thin sheet across the surface of the land and are collected in terraces to be conducted from the field. As many as four or five sprinkler lines are laid out on a hillside, as shown on the diagram of Fig. 10. The waste is applied approximately $\frac{6}{10}$ in./day and flows in a sheet where it is purified in its travel across the field. The application is made approximately 4 hr/day, although on better lines the application is applied over an 8-hr period.

Of all the interacting phenomena in the natural filtration system, microbiological activities and the adsorptive capacity of the vegetation are the most important. Microbiology in this instance refers to the various forms of microscopic life usually present in an agricultural soil. It includes all the molds, fungi, bacteria, earthworms, and insects which feed directly or indirectly upon the organic waste. Adsorption is an electrochemical phenomenon which causes molecules of one sub-

Fig. 10 Spray irrigation surface filtration treatment.

stance to become attached to the surface of another. The process of adsorption continues during cold weather when the activity of microorganisms is low. Therefore, it is of particular importance to a system operating 12 months a year.

In any natural filtration system it is important that the treated water is allowed to escape either overland or underground. It must not collect in pools on the soil surface. If this occurs, vegetation will be killed and there is serious risk of creating odor problems and other undesirable nuisances.

The solids drop out almost immediately beneath the sprinklers and the soluble BOD is adsorbed on the organic litter that exists on the entire hillside slope. The efficiency is extremely high (Table 12). The waste applied averages 850 ppm BOD and the effluent as it runs off the property averages less than 10 ppm BOD. The slopes vary in length, depending upon the peculiar conditions of each hillside, but in general, most slopes are from 200 to 250 ft long.

The picture of the irrigation slopes in Fig. 10 was taken in the fall, and the outline of the terraces can be seen clearly. The dark area represents the northern grass species, primarily reed canary with some red top and fescue. The other area is primarily local Bermuda grass, which is dormant at this time of the year. The Bermuda grass has taken over where water has not reached an area during the hot, dry summer months. Towards the end of a dry season, some lines have such a high evapotranspiration rate that water does not reach the bottom of the slope.

The hydrological characteristics of the soil will always dictate the feasible method of disposal, and the cost of installation is always in favor of subsurface flow. On the one hand, suspended solids impose a limiting factor on an overland flow system—more so than on the infiltration system. The reason is, of course, that most suspended solids will be dropped from the liquid as soon as the velocity is reduced, mainly in the area where they are applied by sprinklers. Thus, in the overland flow system a heavy concentration will be built up directly under the sprinkler line at the top of the slope; whereas in the subsurface flow system, solids will be distributed equally by sprinklers over the entire disposal area.

TABLE 12 Northeast Texas Spray Irrigation Effluent (ppm)

pH			Suspended solids			Dissolved solids		
Max	Avg	Min	Max	Avg	Min	Max	Avg	Min
8.2	7.4	6.9	210	65	13	465	262	40
Dissolved oxygen			BOD			COD		
Max	Avg	Min	Max	Avg	Min	Max	Avg	Min
11.1	6.1	2.3	9.2	5.3	2	51	22.3	9.2

Where food processing plants are located in rural areas, attempts should be made to utilize the natural environment. At a site in South Carolina the soil was of high infiltration for the first 5 to 7 ft, and then encountered a denser clay structure. The entire field was underdrained at 200-ft spacing with perforated pipe wrapped in fiber-glass matting. This pipe conducted the drainage to a natural lagoon which is known locally as a Carolina Bay or savannah. The concept was to have complete treatment with the spray system and to follow up with a polishing pond. In actual practice, the infiltration rate of the subsoil is sufficiently high to keep the drainage to the lagoon at less than 10 percent of the applied flow. Part of this is attributable to the fact that the drainage field acts as a French drain system so that, although in the immediate area of application the perforated pipe is quite active in carrying the water to the lagoon, the flow travels through so much unsaturated area it infiltrates into the subsoil (see Fig. 11).

The system has been highly efficient. The sprays are laid out on a permanent underground system; there is no aboveground piping. The sprays are on a 120-ft triangular spacing. The application rate reaches 1 in./day. The efficiency of the system is 99 percent on a total pound BOD removal basis because of having practically no effluent. The efficiency on the effluent going to the lagoon is in excess of 95 percent. The waste applied averages 635 ppm BOD with all samples on the effluent less than 10 ppm BOD (Table 13).

One thing of interest is the application of these land treatment operations during the winter. For short periods of time, the ice has built up to several feet in thickness. Tests under these conditions and the following thaw indicate that the system still functions satisfactorily throughout the winter because of adsorption, and the system is not completely dependent upon the total oxidation of the organic load during cold weather. It was determined from field studies that the number of organisms increased during the cold weather, thus compensating to some degree for the decrease in rate of metabolism. Therefore, even under conditions of greatly reduced microbiological activity excellent BOD reductions are still obtained.

During cold weather, organic matter builds up under the sprinklers faster than it can be digested. When warm weather returns, the population of microorganisms builds up rapidly to feed on the accumulated material. However, if the accumulation of solid matter is too great, it may not have sufficient oxygen and odors will result. When the system is properly designed and operated, there will be no odors of putrefaction.

Fig. 11 Spray irrigation infiltration system treatment.

TABLE 13 South Carolina System (ppm)

Characteristic	Raw waste	Underground drain	Lagoon discharge
BOD			
Avg.	635	6.8	3.5
Max.	1,490	10	6.8
Min.	300	3	2
COD			
Avg.	1,030	40	23
Max.	3,270	68	47
Min.	320	6	8

Food processors who are faced with seasonal operations can, if located in rural areas, rely upon spray irrigation to accept extremely high loads for short periods of time. This use of the natural environment is extremely valuable for conventional systems such as activated sludge, which cannot be started up in a few days and cannot tolerate tremendous overloads. An example of the value of spray irrigation for peak seasonal loads is described below concerning a tomato operation in northwest Ohio.

Table 14 represents a summary of waste treatment practice by spray irrigation for the years 1961 through 1965, at a plant in northwest Ohio. During these 5 years,

TABLE 14 Northwest Ohio Tomato Plant Characteristics of Spray System, 1961–1965

	Volume sprayed*		Volume of runoff		COD applied		COD of runoff			
	million gal	gal/acre/day	million gal	% flor	ppm	1,000 lb	ppm	reduction	1,000 lb	reduction
1961	152.3	19,700	86.5	57 (29)†	576	732.6	94	84	67.7	91
1962	158.2	18,800	99.9	63 (36)	523	691.5	81	85	67.6	90
1963	203.0	27,900	86.6	43 (36)	369	625.5	94	75	67.8	89
1964	208.1	23,400	84.0	40 (29)	661	1,208.0	136	79	95.3	92
1965	316.7	32,500	152.6	48 (26)	500	1,319.3	94	81	119.9	91
Avg.	207.7	24,460	101.9	49 (30)	526	915.4	100	81	83.7	91

* Average season was 50 days.
† Parentheses indicate percentage after correction for rainfall.

TABLE 15 Characteristics of Raw Spray Waste, 1964–1965

Characteristic	Avg ppm	Max ppm	Min ppm
COD	916	2,000	352
BOD	548	791	401
Total nitrogen	30.0	48.0	12.4
Phosphates	10.7	21.3	2.6
Suspended solids	274	481	18
Volatile solids	161	268	11
pH	.....	9.3	4.9

Treatment performance on a mass basis

Constituent	Watershed I % reduction	Watersheds II and III % reduction
COD	95	81
BOD	...	85
Total nitrogen	93	73
Phosphates	84	65
Suspended solids	97	89

Hydraulic loads and runoff coefficients, 1964–1965

	Volume sprayed		Runoff	
	million gal	gal/acre/day	million gal	%
Watershed I, 1965	23.8	37,200	6.3	26 (16)*
Watershed II, 1965	49.6	23,900	39.6	80 (44)
Watershed II, 1964	38.4	21,211	20.1	52 (39)
Watershed III, 1965	19.8	13,300	7.4	39 (32)

* Parentheses indicate percentage after correction for rainfall.

the tomato season averaged 50 days/year, starting in August and ending by Oct. 1. During 1965, the average waste applied reached a total of 1.2 in./day. Individual spray lines in the field reach peaks of 4 to 5 in./day. Table 15 gives the characteristics of the waste sprayed in 1964 and 1965 for different watershed areas studied by T. W. Bendixen and R. D. Hill of the Taft Sanitary Engineering Center in Cincinnati. During this period, one entire watershed reached a maximum average of 1.4 in./day for the entire season. The results show that on a mass basis the percent reduction is extremely high, not only for COD and BOD but also for total nitrogen and phosphates. The strength of the effluent in terms of parts per million BOD is considerably higher than that of the Texas system (Table 12). However, the Texas system operates on a unit loading basis of $\frac{6}{10}$ in./day, whereas the Ohio system averages approximately three times this application rate. This shows the ability of the land disposal system to take peak loads for short periods of time and the advantage gained during summer or dry weather periods over wintertime application. The Texas system can have much higher dosage rates in summer, but the total system was designed to achieve maximum reduction under winter operating conditions.

The importance of nutrient removal has not been studied in depth. The data in the various analyses shown in Table 15 indicate a good control of nitrogen and phosphates with a spray irrigation system. We know from experience that it takes a number of years for the grasses to develop to their maximum growth potential. During this period of time, the nutrients in the waste are being utilized in the growth and development of the plants. We are presently contemplating a detailed study to ascertain if harvesting these grasses will be a factor in a continuing removal of nutrients.

Anaerobic Digestion Treatment Process

Anaerobic treatment of waste is as old as septic tanks, but the full impact of understanding an efficiently designed treatment system was not developed until recently. Dr. R. E. McKinney, in his book "Microbiology for Sanitary Engineers," wrote that "Anaerobic digestion is the uncharted wilderness in sanitary engineering. . . . The future is so full that it is impossible to state where the end lies and what the full potentialities are."

Development of a modified anaerobic contact process was facilitated by research work supported by the American Meat Institute acting through its committee on Meat Packing Plant Waste Disposal, and in collaboration with the program of the National Technical Task Committee on Industrial Waste. The first full-scale treatment plant was installed at Albert Lea, Minn.

The process consists of heating the entire flow of waste to 95°F and holding it in a digester for approximately 12 or 13 hr. In the digester the anaerobic organisms are mixed with the waste to digest the organic material to produce methane, carbon dioxide, and bacterial cells. This is similar to an activated sludge process wherein the treatment relies upon the contact between the organisms and the organics and nutrients in a favorable environment. The sludge, which consists of organisms and agglomerated organic matter, is separated from the treated liquid after degasification by drawing a vacuum in a cascade degasifier and settling in a sludge separator. The settled sludge is returned to the digester as seed. Figure 12 is a flow plan of the anaerobic contact process at Albert Lea, Minn. This plant is loaded at approximately 0.20 lb BOD/cu ft of digester capacity with 95 percent BOD removal. Table 16 shows the basic operating data through the anaerobic treatment system and following polishing ponds.

Another plant located at Austin, Minn., is reported to remove 96 percent of the BOD at a loading of 0.059 lb/(cu ft) (day) treating a raw waste of 1,400 ppm BOD.

The anaerobic treatment system is especially well suited to destruction of wastes of high organic material. The system has been utilized primarily for meat packing wastes, although other food processing wastes can be similarly digested. Table 17 summarizes one month's pilot plant operation on a tomato waste and a combination canning plant waste including soups, beans, and spaghetti. These data show the amenability of treatment by this method for wastes that heretofore have been possibly considered too weak.

This type of treatment process is less costly than conventional activated sludge and has the potential for less odor. It can accept shock loads and is capable of being shut down on weekends and for extended periods without loss of treatment efficiency.

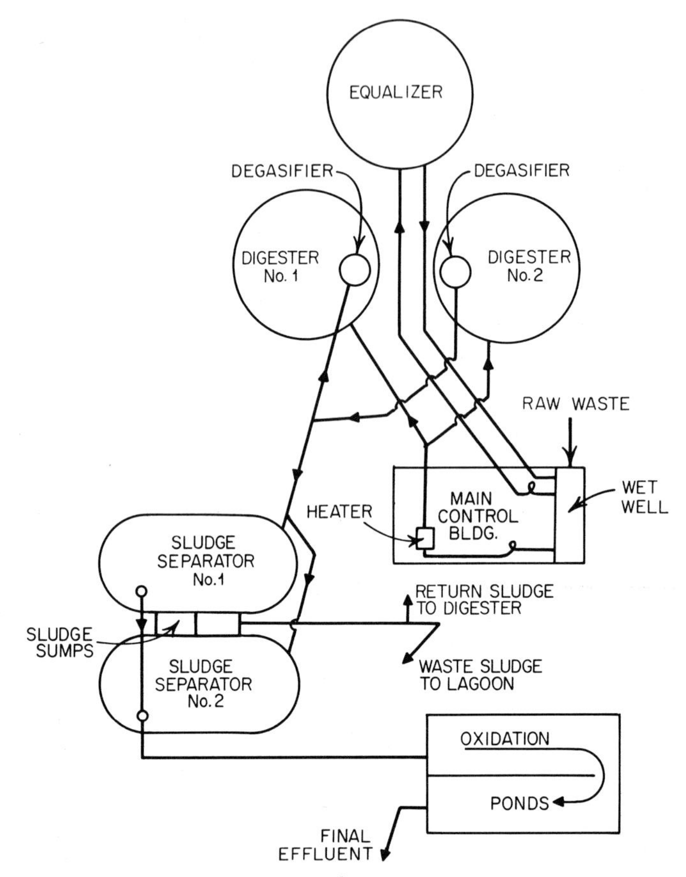

Fig. 12 Anaerobic contact process.

Solids Disposal and By-product Recovery

In some of the earlier sections, brief reference is made to solids disposal and by-product recovery. This is becoming more of a problem from several aspects. For those items that have been a valuable by-product in the past, there seems to be a greater tendency toward less economic return because other products have taken their places on the competitive market. For those materials that have no economic value, there is a constant cost escalation for disposal. In many cases, the food processor seems to be caught in an ever-tightening vise because burning wastes and solids creates air pollution problems, and sanitary landfill in urban areas faces the frightful prospect of having no disposal place at all.

Perhaps one of the most highly developed by-products recovery programs is employed in the meat packing industry, where we often hear the expression "everything is used except the squeal." By-products recovery means conservation. A. J. Steffen has pointed out that improvements in waste conservation practices yield a higher degree of sale-valued products in the meat packing industry and usually help reduce the strength of the waste water. Steffen has recommended the following seven basic conservation methods, which although they are stressed for meat packing operations, apply in general concept in the entire food processing industry.

1. Dry cleaning pens and floors before washing down.
2. Collecting kill blood in a separate blood tank, and squeegeeing the blood flow to the blood sewer before hosing during cleanup. Dual floor drains are provided to permit washing the residual blood to the sewer during cleanup.
3. Retaining casing slimes in casing washing operations. These slimes can be dried and added to animal feed products.

TABLE 16 Anaerobic Contact Process Treating Meat Packing Wastes, Albert Lea, Minn.
Average operating data—all killing days in 1960

	Raw waste	Anaerobic process effluent	Pond effluent	Loss in ponds
Flow gal.	1,410,000	1,410,000	772,000	638,000

	Raw waste		Anaerobic process effluent		Pond effluent corrected for seepage	
	ppm	lb	ppm	lb	ppm	lb
BOD.	1,381	16,220	129	1,517	26	304
Suspended solids.	988	11,610	198	2,325	23	268

	% removal			Digester loading, lb/(day)(cu ft)
	Through anaerobic unit	Through ponds	Through entire plant	
BOD.	90.8	79.8	98.2	0.156
Suspended solids.	80.2	88.4	97.6	0.112

4. Screening paunch manure and hog stomach contents with either rotary or vibrating screens.
5. Separating grease from all grease bearing waste waters by means of efficient gravity or air flotation separators.
6. Evaporating tank water, the liquid residue from wet rendering. After evaporation to about 35 percent moisture, the liquid "stock" is mixed with tankage and dried.
7. Evaporating blood water if blood is coagulated rather than dried directly.

Figure 13 indicates the degree of by-product recovery from an average 1,000-lb steer and the potential yield of various types of by-products. Paunch manure is usually segregated from the liquid waste and disposed of separately either by dumping on the land as a potential soil conditioner or in city plants by being trucked away with the garbage. Blood, casing slimes from the stripping operations, and tank

TABLE 17 Pilot Plant Studies, Anaerobic Digestion Cannery Wastes

				Raw					Effluent						
Type of waste	Total flow, gal	Digester deten-tion, days	Digester pH	Total solids, ppm	Vola-tile solids, ppm	Sus-pended solids, ppm	Total nitrogen, ppm	BOD ppm	Total solids, ppm	Vola-tile solids, ppm	Sus-pended solids, ppm	Total nitrogen, ppm	BOD, ppm	Mixed liquor sus-pended solids, ppm	Returned sludge sus-pended solids, ppm
Tomato:															
Avg.	35.92	1.13	7.0	1,411	858	591	41.21	859	269	269	9.5	21.7	20	72,096	86,194
Max.	44.4	1.7	7.2	2,466	1,502	1,088	94.8	1,673	1,712	384	33.3	40.5	36	8,352	10,768
Min.	23.5	0.95	6.8	740	352	200	18.0	419	450	140	0	11.6	13	6,528	7,208
Combination meat, poultry, vegetable, spaghetti, etc.:															
Avg.	59.46	0.82	6.75	1,191	693	370	36.94	737	639	282	183	40.37	46	10,427	15,263
Max.	100.0	2.3	7.6	1,466	1,066	578	64	937	900	590	378	67	84	15,050	20,856
Min.	17.5	0.40	6.6	972	538	142	7.2	526	284	122	0	15.6	16	7,335	9,067

water can be concentrated or dried and utilized in feeds and fertilizers. If the protein values of the feed and fertilizer ingredients produced at the plant are higher than the requirements of the particular fertilizer being produced, some quantities of lower-grade materials are sometimes blended with these materials. The larger packinghouses do both edible and inedible rendering using various processes including steam and dry rendering which produces no tank water, but skimmings from grease

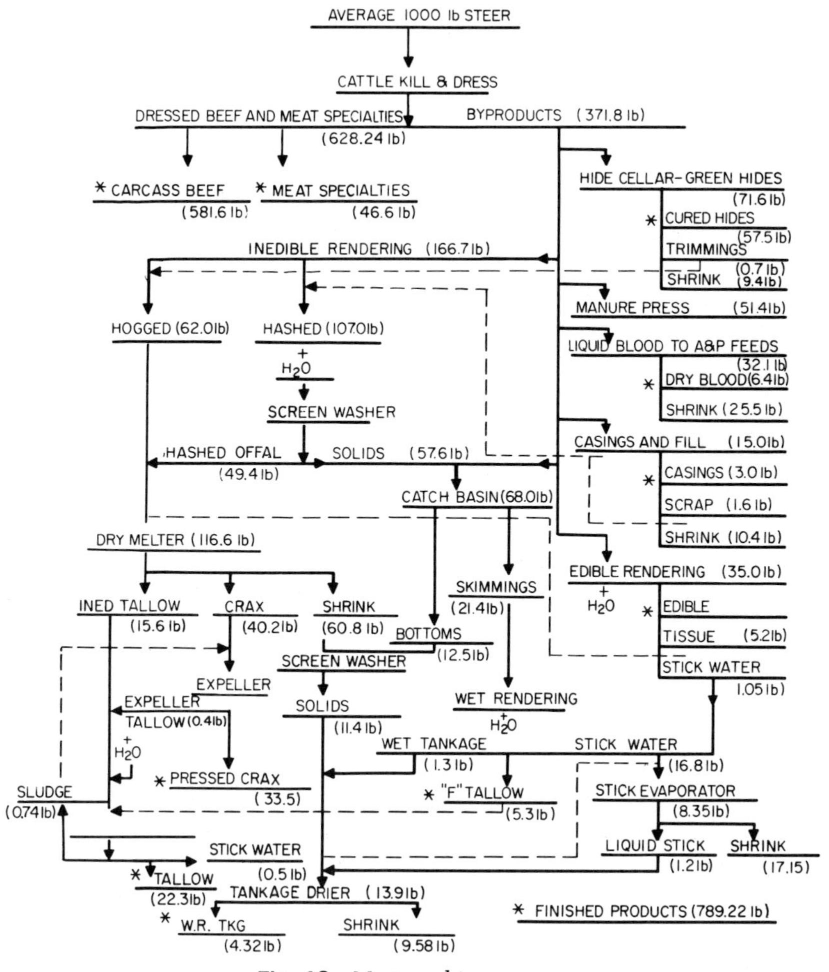

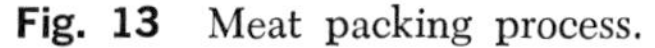
Fig. 13 Meat packing process.

tanks are wet-rendered because of water tied up with the grease sludge. In the larger plants, hair removal is accomplished by a mechanical scraper. The hair is removed, washed, and sold unprocessed, or is further processed in boiling water, dried, and baled for sale. Hair may also be hydrolyzed by steam rendering with the addition of lye. The rendered product then is dried to produce a powdered material. Tanning, wool pulling, manufacturing glues, soaps, and fertilizer are normally processed in separate plants. Glue is a collagenous matter extracted from head, feet, bones, tendons, and hide trimmings by rendering them in water at different temperatures. The inedible tallow is used in soap manufacturing. With dry rendering, a higher-grade material normally is reclaimed which can be used for stock feed rather than for fertilizer. The chemical industry has been gradually taking over the area of fertilizers, originally supplied from by-products of packinghouses.

Not all food processing operations have residual solids of as high an economic value as some of those from the meat processing industry. Many solids have in the past been disposed of by feeding to animals. This has run into difficulty in recent years in that the place of feeding hogs and other animals normally has to be some distance from the food processing, and more and more restrictions are being placed on what can be fed, how the food material is handled, cooking in transit, and other difficult problems. This type of solids disposal has become greatly reduced for most food processors.

A few cities permit disposal of solids by grinding up the material and passing it to the city sewers. This, however, creates additional sewage load and is a practice that has been generally frowned upon. It is not likely to gain wide acceptance in the future.

Incineration of waste solids has been practiced on comparatively dry solids. A few materials are even claimed to have some return benefit from heat values. This process does not appear to have long-range benefits because of the secondary problem of creating air pollution.

Land spraying of waste on top of fields for either air drying in thin layers or by plowing under has been practiced in a few areas. This, again, has the difficulty of requiring long hauling of materials and limitations by some county health departments. Long-range development of this appears to be limited because of our rapid urbanization around most food processing plants. Sanitary landfill falls somewhat into the same category. More and more food processors are being faced with no sites available for filling in. Under these circumstances there have been resultant prohibitive hauling costs for many operations.

More recently a great deal of work has been done by the National Canners Association in quick composting of solid wastes. This can be achieved with many materials, resulting in a humus that provides a satisfactory soil conditioner. In many places, other materials have to be mixed in, in order to obtain satisfactory composting or to start out with a material of sufficient dryness. The main difficulty with this approach is the finding of suitable markets for the final end product.

On the West Coast, efforts have been made to dispose of solids out in the ocean approximately 26 miles beyond the nearest point to the mainland. Material is disposed of from barges after it has been completely ground up. The disposal is done at the direction of authorities in charge of pollution control. Although this method appears to be satisfactory, it has been done only on a limited basis and, of course, is available only to those food processors near a coast.

The growing complexity involving disposal of solids in urban and semiurban areas indicates a need for a great deal more research on this problem. As indicated earlier, almost all these materials can be dried; however, the economic value of the subsequent product created has not in most cases been very beneficial. There is certainly a challenge to both the agricultural research chemist and the engineer and to industry to create new by-products from food processing. However, until better methods are obtained in creating economic by-products, it appears that most of the solids waste from the great majority of the food processing industry will be hauled away at considerable expense and treated as garbage.

GENERAL

No reference has been made to the costs of the various systems. In general, the land development systems for spray irrigation and for lagooning are considerably less than conventional systems of activated sludge or trickling filters. Each system represents to certain degrees the best adaptation to conditions in the local environment, and therefore, costs of installation never appear to be repeatable at some other site.

At the Fifth Annual Short Course for the Food Industry at the University of Florida, J. C. Dietz, P. W. Clinebell, and T. H. Patton, consulting engineers, provided a cost comparison for handling of meat packing wastes on three types of treatment: trickling filters, anaerobic contact process, and anaerobic-aerobic lagoons for flows of 0.5, 1.5 and 2.5 mgd. Figure 14 lists capital costs including land, with land valued at $500 per acre. The costs also included 2,000 ft of piping. Figure 15 lists an annual cost comparison including fixed charges using 5 percent interest and operating cost.

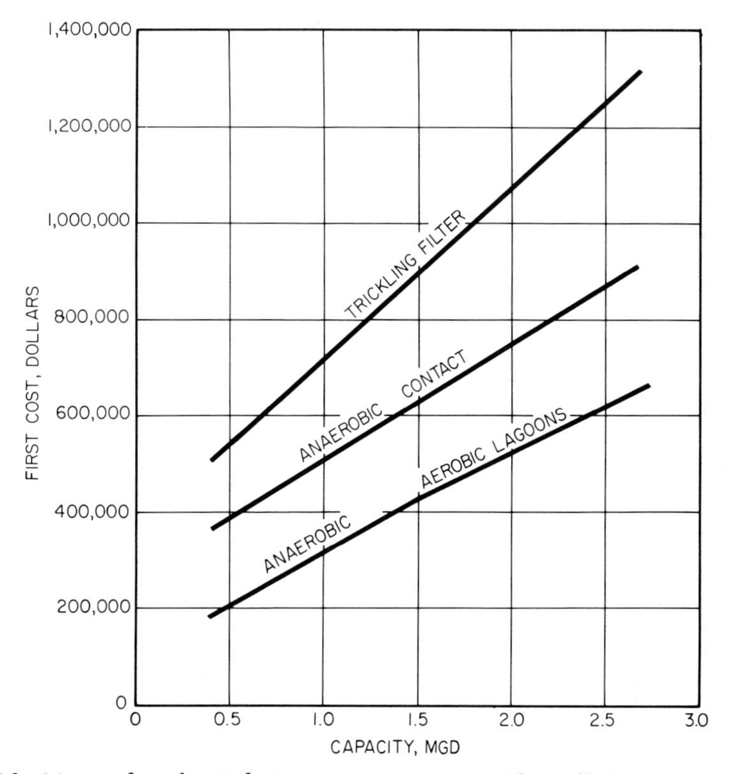

Fig. 14 Meat and poultry industry waste treatment initial installation costs. (*From J. C. Dietz, P. W. Clinebell, and T. H. Patton, "Solving Waste Problems of the Meat and Poultry Industry."*)

Table 18 provides a cost data evaluation for a meat packing operation in Cherokee, Iowa. In the cost comparison, the first three figures are taken from the U.S. Department of Health, Education and Welfare Public Health Service Bulletin 1229. These figures are compared with the actual cost for the Cherokee plant, which consists of an anaerobic lagoon followed by mechanical aeration and finally by aerobic lagoons. These data were presented at the 49th Annual Conference, Iowa Water Pollution Control Association, Fort Dodge, Iowa, June, 1967, by Dewild, Grant, Reckert & Associates, Consulting Engineers.

The Food Machinery Corporation of California has made an engineering study of seven different systems for disposal of solids wastes. Figures 16 and 17 provide information on the estimated operating costs of different types of systems for solids waste disposal after the amortization period. One of the major problems in the food industry is the fact that so-called solids may contain 50 to 95 percent water.

Therefore, there is a need to develop systems of either segregating the water from the solids and then treat the highly concentrated liquor or handling the solids with the water. Figures 16 and 17 show various methods of handling both the liquor and the solids.

Over the past decade or more, the food industry—which is very much water-oriented—has come to realize that the location of new plant sites is more and more being dictated by the source of a good-quality water supply and, perhaps even more

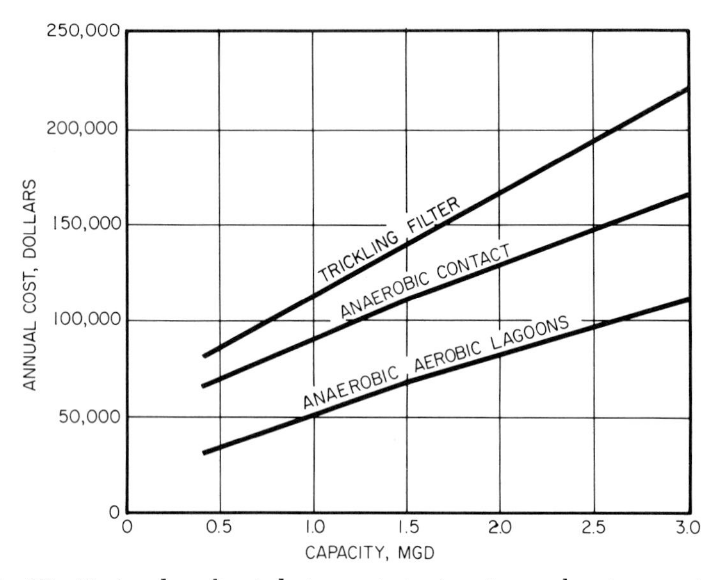

Fig. 15 Meat and poultry industry waste treatment annual cost comparisons.

TABLE 18 Construction Costs

Plant type	Design (population range)	Cost/population equipment (lower limit range)	Cost per capita (lower limit range)
Activated sludge	Up to 100,000	\$8.75†	\$12.35†
Primary—digestion	Up to 150,000	\$7.89§	\$ 8.88†
Stabilization ponds	Up to 60,000	......	\$ 1.75†
Cherokee plant	103,500*	\$3.26‡	No design population served

* Based on maximum flow and maximum BOD (0.17 lb BOD per capita).
† Based on 1957–1959 dollars.
‡ Based on 1964 dollars.
§ Based on trickling filter, separate sludge digestion treatment, 1957–1959 dollars.

important, the means and capability of sewage treatment and disposal. Many of the previous installations discussed in this chapter reflect different means of adapting to the environment so that the type of sewage treatment system—although it varies from site to site—normally represents the least-cost approach utilizing the natural environment in order to get maximum treatment and efficiency for each particular plant.

Pollution is an industry problem, a community problem, and a people problem.

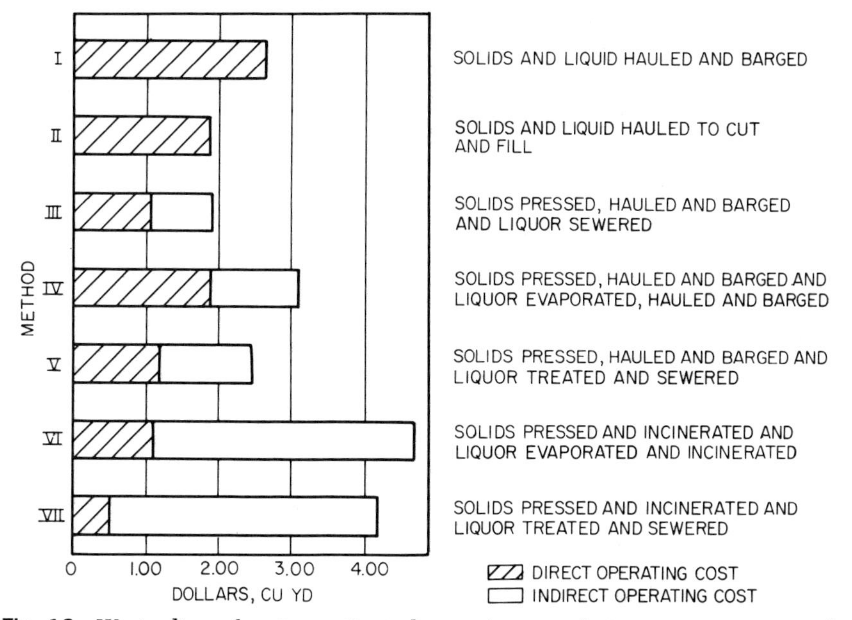

Fig. 16 Waste disposal system estimated operating cost during amortization period. (*Information developed by Food Machinery Corp.*)

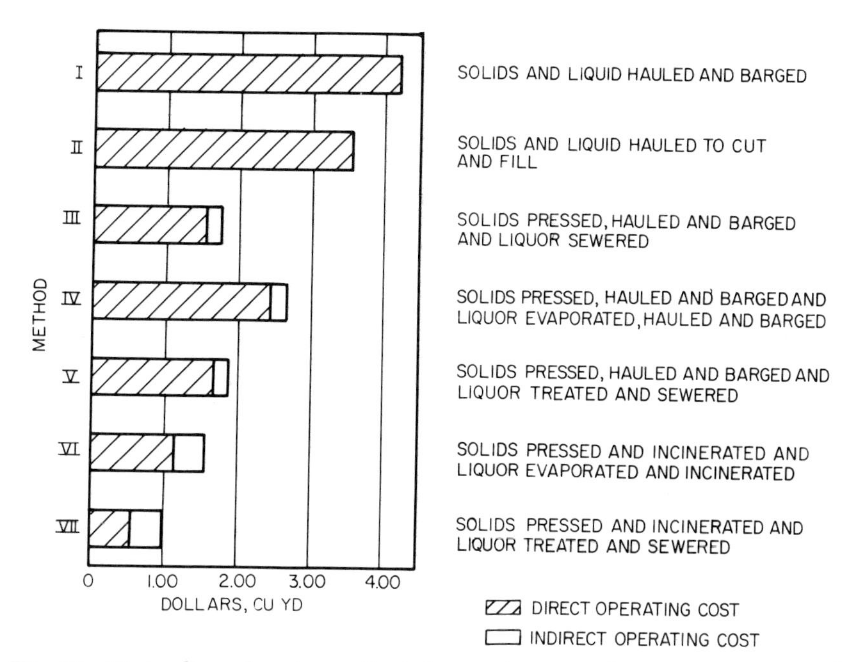

Fig. 17 Waste disposal system estimated operating cost after amortization period. (*Information developed by Food Machinery Corp.*)

Everyone has responsibilities, and there is a need to share responsibilties. The food consuming public is receiving high-quality food, convenience of use, and generally a lower-cost product. Organizations such as the National Canners Association and the American Meat Institute have developed valuable information and conducted excellent research toward the goal of obtaining reduced pollution, improved treatment techniques, and maximizing conservation efforts. The research of these organizations should be supplemented on a much greater scale in order to develop technologically practical and economically feasible methods of waste handling. It cannot be expected that each producer of industrial waste has a trained scientist or resources for waste treatment research, and therefore the local problems have to be considered by the community in complete cooperation. The complexities of the problem indicate a need for more investigators to become interested in research applicable to food waste treatment.

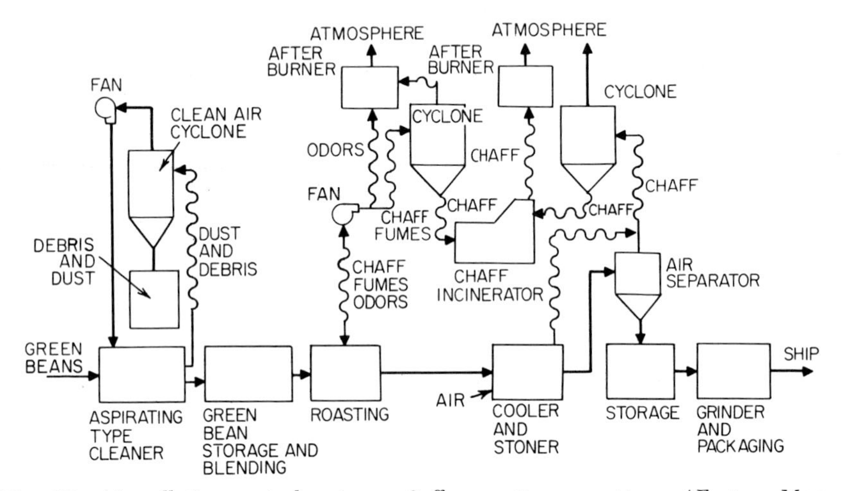

Fig. 18 Air pollution control system. Coffee roasting operation. (*Factory Magazine, October, 1967.*)

AIR

Carried to the extreme, each of us "pollutes" the air with heat, moisture, odors, germs, carbon dioxide, etc., with every breath; and every user of a home kitchen exhaust fan or backyard barbecue is guilty of air pollution in his neighborhood. An answer to the question of whether or not process exhausts from a food industry, or any industry for that matter, are to be labeled as air pollution depends on the definition of the term. Up to the present time, the term "air pollution" has usually been considered to mean the discharge of large quantities of obnoxious solids and gases into the air, to the extent that positive, measurable damage results. Powerhouses, steel mills, and chemical plants come to mind as potential "air polluters."

Most people like the aroma of roasting coffee. However, in a large built-up city such operations have come under criticism, and the processor has had to provide rather extensive air pollution control equipment. An example of this is shown in Fig. 18. The coffee roasting process consists of cleaning the green coffee beans, blending, roasting, cooling, and final cleaning of the roasted beans and grinding for packaging. The exhaust from various operations contains dust, chaff, odors, and fumes consisting of volatiles created during roasting. Roasting forms esters of chlorogenic acid, furfural derivatives, and other aromatic substances. Volatiles of tars and condensable and noncondensable organics are emitted as well as large quantities of chaff, or the outer covering of the bean. These emissions produce a very dense, odoriferous, bluish-white plume.

As shown in Fig. 18, the coffee industry uses mechanical collectors such as cylones and multicyclones followed by secondary devices, the most dependable of which is the direct fired afterburner operating in the range of 1200 to 1400°F. Other devices used in some cases are the indirect fired roaster and the catalytic afterburner.

Much of the grain and other dry ingredient food processing operations have problems similar to those of the coffee industry because of their conveying systems and cleaning systems which are capable of discharging dust in various quantities. In other food industries cooking odors may be objectionable from such operations as meat packing, onion preparation, cabbage cooking, and rendering operations.

An integral part of the food processing industry is the manufacture of cans. This activity contributes to air pollution through the operation of lacquer ovens which discharge solvent vapors and other by-products to the atmosphere. The majority of complaints against metal decorating operations have concerned odors, but some complaints have been lodged against smoke emissions.

The particulates consist primarily of breakdown by-products of the resin or drying oil in the coating or ink during the bake. These by-products are driven off as a gas which condenses when coming in contact with cooler air or surfaces. In some instances a noticeable plume or smoke is emitted. The major portion of the odor is also due to the above-mentioned by-products. Although solvent vapors are also emitted, the odor threshold is of the order of one thousand times that of the resin breakdown products. Bodymaker exhausts which contain solder (lead) and flux fumes also have to be controlled in the can manufacturing operation.

A number of different types of control devices are used in the food industry to curb air pollution. In the metal decorating processes, high-temperature incineration of the stack gases seems to be the most effective method of odor elimination. Other methods which have been tried are catalytic oxidation, scrubbing systems using water plus various additives, oil plus various additives, charcoal and silica gel adsorption systems, and low-temperature incineration.

In addition to having the proper equipment necessary for combating air pollution, industry must also educate its personnel. It is important that the employees of each plant become educated in the area of the importance and significance of controlling our environment. It is even more important that the operators of all process equipment and all pollution control equipment have complete knowledge of their equipment and know what controls are necessary to avoid pollution. Each operator should be trained so that he knows the capability of the equipment and understands his role in prevention of air pollution. Significant emissions to the atmosphere must be monitored and the equipment should always be maintained in good working order and kept in proper adjustment. The operator must be able to keep accurate records and understand them thoroughly so that he knows which changes in the recordings are significant and require remedial measures. When processes are modified or new equipment installed, it is necessary to be sure that increased venting or changes in venting materials will not cause an air pollution problem.

Major food processors have shown interest in improving the quality of our environment because they incorporate pollution control as an integral part of manufacturing activities and maintain people knowledgeable in the effects of air pollution, such as human health, nuisance, and the effects on vegetation, and corrosion problems. Many companies provide education in this area to their employees and stockholders as well as to the general public and newspapers.

Although most of the pollution present in our air today could be prevented, the fact remains that it is still present. The major reason for this is that prevention of air pollution is expensive. In past years, and in many instances today, air pollution codes are lenient enough so that the question of how far an industry should go in preventing air pollution is answered by evaluating how much money the industry is willing to spend on the problem. The problem is generally one of community relations, and the answer is usually geared to (1) how far the governing codes say we have to go; (2) the extent to which the other fellow goes to prevent pollution; (3) how much concern we have for our neighbors' reactions; and (4) how much we can afford to spend. Depending on the locality of a particular plant, industry takes various views on the air pollution problem. A plant in or near a residential area must normally make more effort to avoid offending than one which is relatively isolated.

The growing seriousness of air pollution, however, is forcing government regulations to become more specific and enforceable. In the future, therefore, there will be fewer decisions for company and plant management to make about control of air pollution. Plants will have to invest money for the prevention of air pollution with the option not of offending their neighbors but rather of being forced to curtail all process activities until the codes are satisfied.

The food industry, as a whole, is very cognizant of the effects of air pollution on our environment and the importance of preventing and controlling pollution. Perhaps as much as any other industry, the food industry has a responsibility to the community and values good community relations. An example of how an industry recognizes its responsibility to the community is Campbell Soup Company. It covered its sewage treatment plant with air-inflated plastic domes to control odors.

Just as important as educating plant personnel is the education of the public in general. The public, as a whole, has many misconceptions about air pollution which hinder a thorough understanding of the problem. It is difficult, however, to familiarize the public with such a thing as air pollution and its effects on our environment because of the many complexities involved and the technical nature of the subject. Too frequently, people are confused by harmless plumes of steam coming from various venting operations, or even steam condensation off cooling towers. These are perfectly harmless; yet inconspicuous stacks may be releasing odorless and colorless potentially harmful emissions. There have been times when newspapers have made the implication that the plumes of steam coming from a plant exhaust stack on a cold winter day are visible evidence of air pollution. Another example of the confusion which exists is the aesthetically obnoxious spewing of smoke from diesel bus and truck engines. This has, in general, annoyed people tremendously, but from a pollution standpoint it is far less significant than the unseen fumes from the average automobile exhaust.

There are, in general, two types of public complaints regarding air pollution. The first concerns plant emissions that actually endanger public health or property. The second, and probably the more frequent complaint, is that concerning odors which are not harmful from a health or property standpoint but which are considered a general public nuisance. Some plant emissions fall into both categories simultaneously, but others can be classified as either harmful or just nuisances. Given his choice, the average citizen, therefore, is likely to demand action where it may not be necessary, because the public appreciates the significance of what it can see, smell, and feel. There are cases on record where citizens have gone further than this and have registered complaints on days when a plant was not even operating.

To be sure, there are instances in the food industry when plant emissions are both harmful and nuisances. The argument in point, however, is that we must be knowledgeable enough to determine which particular emissions are objectionable and which are tolerable and then to be able to do something about the objectionable ones. Industry at large has to dispel as best it can the existing confusion in order to avoid illogical solutions demanded by the public and the press, which sometimes reach an emotional pitch indicating a crisis.

CROP DAMAGE FROM AIR POLLUTION

One fact that is often overlooked when considering air pollution in the food industry is that the damage inflicted upon the industry by air pollution is greater than the damage caused by air pollution originating from food processing. In 1967, for example, crop damage caused by air pollution in the United States alone was just under half a billion dollars. There are numerous cases on record where emissions from an industry other than food processing have been identified as the cause of crop damage.

The nature of air pollution damage to agricultural production is determined by genetic and environmental factors as well as by level and duration of exposure to pollutants. A suggested definition of plant damage is the creation of an economic loss and/or a change in usability of the plant. It is very difficult, however, to estimate the economic loss due to poor growth, lower yields, delayed crop maturity, and reduced product quality as an effect of air pollution.

The history of air pollution in the United States extends back well over a century. The earlier instances of pollution were associated primarily with smelting of ores. Present instances are less severe locally but are more frequent and of a broader nature. There is a rising trend of air pollution injury to crops and ornamental plants all across the United States. Chronic injury is obvious, but the major losses may be due to photochemical oxidants. Patterns of injury are becoming more clearly defined by various pollutants.

The nature of contaminants causing injury to plants may be either particulate or gaseous in nature. Particles exert their effect by causing tissue damage to exposed leaves and fruits, soiling of fruits and vegetables, as well as possibly being a toxic factor to forage crops. The gaseous air contaminants can directly injure plants by their toxic effects. The chemicals that are particularly damaging are fluorides, sulfur dioxide, ethylene, ozone, and peroxyacetyl nitrate. Some damage is due to the amount of the materials accumulated in the leaves, whereas damage to plants by other contaminants is related to the concentration of the toxicants in the air and the length of exposure of the plant to these toxicants. Ozone and peroxyacetyl nitrate are known to cause growth reduction in the absence of visible injury. These compounds result when hydrocarbons and nitrous oxides from automobile exhausts and other combustion exhausts react in the presence of sunlight. With the tremendous increase in population and industrialization, the potential damage to agriculture from this source will undoubtedly increase in the future.

Pollution has been severe enough to cause chlorosis or necrosis of leaves of plants such as lettuce, spinach, and tobacco. Texas A&M University reports that sulfur dioxide, especially in the presence of ozone, has been a factor affecting peanut crops. Similar detrimental effects have been reported in New Jersey for beans due to ozone-sulfur. Other plants damaged by sulfur dioxide include many vegetables, cereal crops, forage, forest, and fiber. Ozone has been reported as being particularly detrimental to cotton, grapes, and tobacco. Citrus fruits, corn, and many flowers have been severely affected by fluoride pollution. Ethylene has created major damage to cotton crops along the Gulf Coast of Texas. The buildup of contaminants in pasture grasses near highways is emphasized by an evaluation of two United States highways near Denver, in which grasses close to the roadway contain 3,000 ppm lead, while grasses next to a less traveled roadway contain only 700 ppm lead. Foliar damage to potato plants has been observed this year, consisting of necrotic spots that first appear on the underside of leaves. The ultimate size of the spots depends upon the susceptibility of the variety. In some cases, damage has been extensive. The cultivation of orchids in certain parts of California had to be discontinued because of the damage inflicted by air pollution.

H. E. Heggestad[1] sums up his evaluation of the increasing air pollution damage to vegetation by stating:

> Although more effort will be made to control air pollution at the source, we can anticipate continuing and serious problems. We may ask what can be done to minimize losses to food, fiber, and ornamental plants. Air pollution injury to vegetation would be more apparent and serious now, except for the selection

[1] H. E. Heggestad, Principal Pathologist, Leader, Plant Air Pollution Laboratory, Crops Research Division, ARS, U.S. Department of Agriculture, Beltsville, Md., Diseases of Crops and Ornamental Plants Incited by Air Pollutants, Phytopathology, vol. 58 no. 8, August, 1968.

and propagation of resistant plants showing least injury; that is, injured plants or species that grow poorly are automatically eliminated from culture by breeders, horticulturists, home owners, etc., even though the cause of injury is unknown. For example, cigar-wrapper tobacco breeders in the Connecticut Valley selected plants with least weather fleck before knowing the cause was ozone. The older, susceptible varieties that were used 10–15 years ago are no longer grown. Plant breeding has made possible the continued economic production of the crop. Air pollution losses can be reduced and production increased by greater application of plant breeding procedures to crop and ornamental species known to be damaged by pollutants.

Several research needs are evident: (i) More information is needed about effects of the various toxicants on different plant species, including bacterial and fungal pathogens; (ii) methods must be developed to more accurately evaluate the economic effects of pollutants; (iii) studies are needed on the effects on plants of two or more toxicants in mixed air; (iv) better biologic indicators of pollutants are needed; (v) the interaction of environmental factors, including nutrition and plant response to pollutants, requires more study; (vi) basic biochemical studies are needed to determine the mode of action of pollutants and the nature of plant resistance to pollutants; and (vii) the identification and development of pollution-resistant varieties of many species should be pursued.

With the constant pressure of increased size of urban communities, factories, and highways, we have a need for more recreation areas consisting of parks, flowers, and green belts. We are already experiencing significant damage to agricultural crops as well as trees and ornamentals, and with the population explosion we will be less able to afford in the future to reduce the green acres of the country within or in close proximity to the urban areas.

Not discounting the fact that attempts to curb air pollution should be made, it should be pointed out that pollution control in the food industry centers chiefly on water pollution control. Being a water-oriented industry, the disposal of used water is the main problem involved in food processing pollution control. Admittedly, there are instances when air pollution control poses a problem to the food industry, but on an industry-wide basis, air pollution created by the food industry is only a minor problem when compared with the difficulties involved in water pollution control. As indicated in this section, the food industry appears to be more and more adversely affected by air pollution due to crop damage.

Chapter **17**

Pollution Control in the Pharmaceutical Industry

BRUCE S. LANE

Director of Engineering and Maintenance,
The Upjohn Company, Kalamazoo, Michigan

INTRODUCTION

It is important at the outset to define a point of nomenclature. The term "pharmaceutical industry" will frequently be used throughout this chapter; this refers to the companies that are principally engaged in the manufacture of prescription drugs.

The pharmaceutical industry, although a strong and important entity in itself, is frequently considered to be part of the chemical industry. Indeed, the size and character of the larger pharmaceutical companies confirm this relationship. Figure 1, an aerial view of The Upjohn Company's plant in Kalamazoo, Mich., illustrates a typical manufacturing complex of one of the major companies.

The pollution problems of the pharmaceutical industry, particularly those of the major companies, parallel those of the chemical industry. Because the problems of the industry are not unique, this chapter will be concerned with emphasizing the problems, where they should be looked for, and some of the more interesting solutions that members of the industry have developed to solve these problems. There is a minimum of tabular data for the reason that much of this material is covered very adequately in other sections of this handbook.

Although there are over 1,300 producers of ethical pharmaceuticals in the United States, fewer than 150 of these are responsible for 95 percent of the industry's products. Obviously, the great majority of the companies producing drugs are relatively small, and the pollution problems of the smaller companies are probably not especially critical.

It is an interesting commentary, if not an axiomatic principle, that the pharmaceutical companies facing major problems in connection with both air and water pol-

Fig. 1 Aerial view of The Upjohn Company's manufacturing facility in Kalamazoo, Mich.

TABLE 1 Composition of the Pharmaceutical Industry

Based upon 1964 Data

Assets, Thousands	*No. of Companies*
From $100,000 to $250,000	14
From $ 50,000 to $100,000	5
From $ 5,000 to $ 50,000	47
From $ 50 to $ 5,000	759
From $ 000 to $ 50	488
Zero assets	12
All companies	1,325

lution are the ones that operate fermentation plants, principally in the production of antibiotics (penicillin, streptomycin, etc.).

WATER POLLUTION

Pharmaceutical Manufacturing

Pharmaceutical manufacturing represents all the various operations that are involved in producing a packaged product suitable for administering as a finished, usable drug. It would include such things as mixing of ingredients, drying of gran-

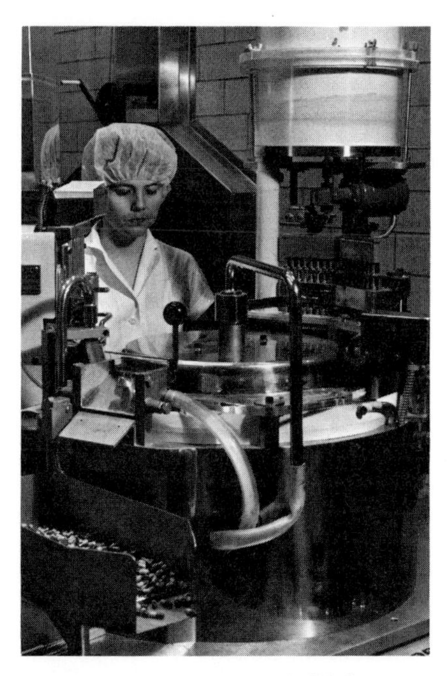

Fig. 2 Hofliger-Karg hard filled capsulating machine.

Fig. 3 Manesty rotary tablet press.

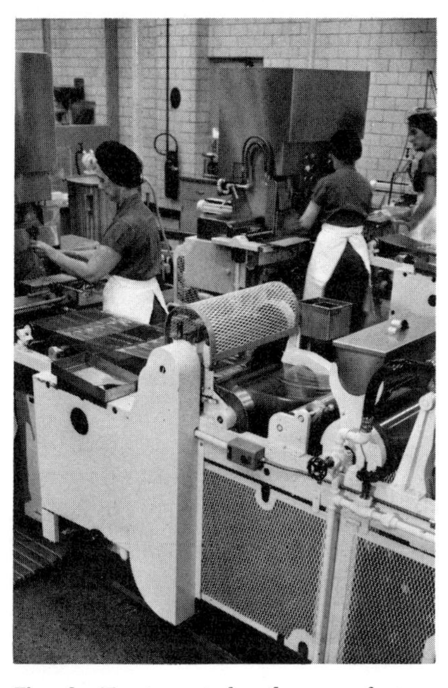

Fig. 4 Equipment for the manufacture of soft elastic vitamin capsules.

Fig. 5 High-speed tablet–counting-filling machine.

Fig. 6 An industrial vacuum cleaner being used for dry cleaning of a powder mixer.

Fig. 7 House vacuum system being used for dry cleaning a hard filled capsulating machine.

Fig. 8 Transporting a portable 1,000-gal mixing tank.

Fig. 9 Mixing tank attached to a package line filling machine.

ules, tableting, capsulating, coating of pills and tablets, preparation of sterile products, and finally the packaging of the finished product. Figures 2 through 5 illustrate typical pharmaceutical manufacturing procedures.

In general, none of these operations would be considered to be serious water polluters, for the simple reason that they do not use water on any basis that would cause pollution. In spite of this, however, there are a number of places where water pollution can be expected.

Washup operations are always suspect. The too generous application of too much water to too great an area can flush unusual materials, in terms of both quantity and concentration, into a sewer.

Fig. 10 Washup of a mixing tank at a central washup area.

Much of the routine cleanup in the pharmaceutical plant can be carried out most effectively by vacuum cleaning. Usually industrial vacuum cleaners are used for this job (Fig. 6). Frequently it is possible to install a "house" vacuum cleaning system (Fig. 7) so that this service is immediately and conveniently available at each machine, particularly if it is a dust generating operation.

The use of portable equipment, even large portable tanks that can be removed to a central washup area, provides better control to the possibility of haphazard dumping of "tail ends" of possibly harmful polluting material to the sewer. Figure 8 illustrates a portable 1,000-gal mixing tank. Such tanks are used for the initial preparation of a fluid or liquid pharmaceutical; following preparation, the tank may be moved to a filling or packaging line (Fig. 9); finally, when the tank is empty, it is moved to a central wash area for cleaning (Fig. 10). If the material in the tank or its wash water might be harmful as a water pollutant, special attention can be given at this point to prevent such material from entering the sewer.

The central wash area is usable for all types of portable equipment requiring this type of cleaning.

Dust and fume scrubbers used in connection with building ventilation systems or, more directly, on dust and fume generating equipment, can be a source of water pollution, depending on the nature of the material being removed from the air-stream.

In the interest of reducing the amount of waste water or in controlling specific materials that might be harmful as water pollutants, the use of baghouses may be called for. Figure 11 illustrates a dust collecting system used by Schering Corporation in connection with their central weigh room where active chemicals are handled.

Fig. 11 "Pulsaire" bag filter at Schering Corporation, Union, N.J., filtering the exhaust from the central weigh room where active chemicals are handled.

Waste Treatment

The adequacy of waste treatment facilities in connection with the pharmaceutical plant has an important relationship to water pollution.

In most instances biological treatment with trickling filters or an activated sludge process will meet a pharmaceutical plant's requirements, but the choice must be made carefully, in terms of the type of wastes that are to be treated, their volume, and of particular importance, the completeness of the treatment that is required.

Septic tanks, or somewhat more sophisticated anaerobic processes, may suit the smaller plant; trickling filters are flexible in the sense that they can be sized to most plants, from medium to large; whereas an activated sludge facility tends to operate more satisfactorily at larger volumes and should, in general, be restricted to larger plants or larger loadings. Figure 12 is a picture of a trickling filter at Lederle Laboratories, Pearl River, N.Y. Figure 13 is an activated sludge treatment facility at

Fig. 12 Trickling filters at Lederle Laboratories, Pearl River, N.Y.

Fig. 13 Activated sludge treatment plant at McNeil Laboratories, Fort Washington, Pa.

McNeil Laboratories, Fort Washington, Pa. Merck Sharp & Dohme, at West Point, Pa., have pioneered the use of a "Surfpac" treatment tank. It was the first unit of this kind in the state of Pennsylvania. Figure 14 shows the type of plastic packing that is used.

Many pharmaceutical plants are so situated that a municipal sewer and treatment system are adequate to their waste disposal problem. Such plants frequently have a distinct advantage over competitors that find it necessary to operate their own waste treatment facilities. However, even those fortunate ones that can function with simply a connection to a municipal sewer must recognize their responsibilities to the municipal systems that serve them. High-acid wastes can seriously damage a sewer, generally by destroying the material used to seal the sewer joints. It may be necessary to monitor and record the pH of sewage entering a municipal sewer to establish your innocence if a sewer is damaged or, if you are disposing of low-pH materials, to maintain and control the pH by neutralization. Figure 15 shows the neutralizing facility of Schering Corporation at Union, N.J. All plant wastes are controlled to pH 6 before entering the municipal sewer.

Another potential hazard in dumping materials to any sewer is the possible presence of flammable solvents in these wastes. Explosions and fires have occurred because of such carelessness. This can seriously damage and disrupt your own

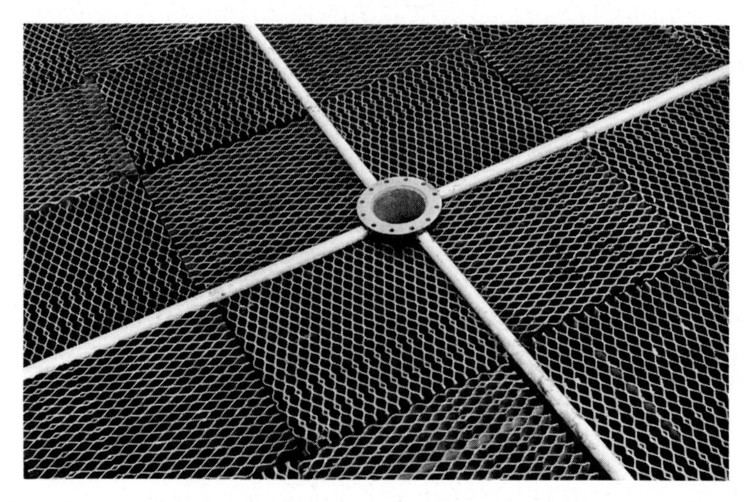

Fig. 14 Plastic packing in the "Surfpac" waste treatment unit at Merck Sharp & Dohme, West Point, Pa.

sewer system; when it involves a municipal sewer, the cumulative damage and your liability can be staggering. Figure 16 illustrates a recording explosion meter used at Hoffmann-LaRoche to monitor the sewer atmosphere for flammable concentration of vapors at the point where the effluent enters the public sewer. This instrument uses a carefully regulated propane flame to sense the presence of combustible vapor in a sample stream drawn continually from the atmosphere above the liquid level flowing in the sewer. The use of a flame instead of an incandescent platinum filament eliminates the problem of certain sewer vapor contaminants poisoning the more sensitive filament. Daily scrutiny of the recorder charts makes it possible to detect any deterioration in plant process control to prevent solvents from being discharged into the sewer.

Fig. 15 Neutralizing basin at Schering Corporation, Union, N.J., where all plant wastes are automatically maintained at pH 6.

Fig. 16 Recording explosion meter unit at Hoffman-LaRoche, Nutley, N.J. The meter monitors flammables entering a municipal sewer.

Chemical Production

Problems in this area might be classified as either normal or exceptional.

Normal problems include the possible pollution of water due to wastes, washup, ejector condensate, etc.; whereas exceptional problems are the unplanned situations, like spills, that find their way to floor drains, and leakage through a heat exchanger that results in chemicals entering a cooling water system.

Frequently an extreme problem, a fire or explosion, results in extensive, unanticipated water pollution.

The use of wet scrubbing of vapor and dust should be restricted to material that will not create a water pollution problem. Figure 17 shows the use of a dry dust collection unit at Wyeth Laboratories in connection with their chemical manufacturing operations.

Condensing steam jets and water sealed vacuum pumps can result in water pollution, depending upon the nature of the materials entering the discharge water stream.

Fig. 17 Central dust collecting facilities for chemical manufacturing at Wyeth Laboratories, West Chester, Pa.

A satisfactory steam ejector installation must give consideration to the possible pollution of the discharge water. If this is apt to be a serious problem, then the problem must be solved in some other way, including the use of surface condensers on the jet discharge. Figure 18 shows the installation of a surface condenser installed on the vapor line in front of a steam jet.

Water sealed vacuum pumps sometimes create pollution problems. A recirculation system of the type used in Fig. 19 is a means of greatly reducing the amount of water being discharged. By this means, it is sometimes possible to get the quantity of discharge seal water, containing pollutants, down to a more manageable volume. Figure 19 shows a large Nash pump. The tank for the recirculating water is in the background. Heat exchangers for cooling the recirculating water are at the right but do not show in this picture.

Fig. 18 Installation of a surface condenser on a vapor line in front of the steam jet to prevent pollutants from getting into the condenser water of the jet's after condenser.

Fig. 19 A recirculating seal-water system being used in connection with a large Nash vacuum pump.

Wastes and wash water from chemical manufacture are not always compatible with the biological system of waste treatment plants, and although it is sometimes possible to acclimate sewage bacteria to certain chemicals, there are many instances where chemical wastes are too concentrated or too toxic to make this feasible.

In dealing with wastes of this kind, a number of possible techniques are available. Further chemical treatment of the wastes may be called for. This may simply be a pH adjustment to make the waste more suitable to both the bacteria and the construction materials of the waste treatment plant.

Chemical treatment may, on the other hand, require more extensive efforts involving the precipitation of heavy metals, the elimination of cyanides, or the otherwise altering or removal of toxic materials prior to waste treatment.

In many instances it is not practical to try to change a chemical waste to accommodate a biological treatment system, and some more practical solution must be found. Waste solvents can frequently be recovered or incinerated. Figure 20

Fig. 20 The solvent recovery unit at The Upjohn Company, Kalamazoo, Mich., used for the recovery of a wide range of solvents necessary in the production of pharmaceutical chemicals.

shows the solvent recovery unit of one of the larger pharmaceutical plants. Figure 21 shows the solvent incinerator unit of this same plant. It is used to burn solvents that cannot be recovered economically, in addition to the "bottoms" from this same recovery unit.

When working with wastes that are not to be biologically treated, it is usually desirable to keep them as concentrated as possible, because the cost of treatment will be affected critically by the volume of material that must be handled. Handling great quantities of water in a nonbiological process can be expensive.

If a large amount of diluting water is present, it may be necessary to reduce the volume by evaporation prior to further treatment. This is costly, particularly where waste materials are involved, but it may be the only alternative. Figure 22 is a tri-

Fig. 21 A waste solvent incinerator used for burning solvents and still bottoms that cannot be economically recovered.

Fig. 22 A triple-effect evaporator used for concentrating wastes that are not suited to biological treatment.

ple-effect evaporator used at one of the pharmaceutical plants for reducing the volume of a waste material.

The concentrate from an evaporation step may possibly be disposed of by incineration or by landfill. Sometimes concentration will give the waste material an economic value that it did not have in its dilute form.

Deep-well disposal is a very practical method for disposing of many chemical wastes that are unsuited to biological treatment. The basic consideration here is the geology of the area in which the plant is located. Michigan is especially suited to deep-well disposal. Figure 23 shows the wellhouse and waste tanks at Parke-Davis' deep disposal well at Holland, Mich.

Fig. 23 Disposal well and waste tanks at Parke-Davis & Company, Holland, Mich.

The deep-well system used by Upjohn, in Kalamazoo, has been in operation about 16 years. There are two wells about 1,400 ft apart. They are 1,300 ft deep. The casing is 7 in. in diameter. Each well has a capacity of 75,000 gal of liquid wastes in a 24-hr period. In this particular operation, the only treatment of the waste is filtration to high clarity with diatomaceous earth. The waste is acid, about pH 5, and is not neutralized. The pressure at the wellhead is 200 to 300 psig. This is developed by high-pressure piston pumps. The cost of disposing of these wastes is about $2.00 to $2.50 per 1,000 gal.

Figure 24 shows the high-pressure pumps used in connection with the Upjohn Company disposal wells.

Fermentation Processes

Fermentation is an important production process in the pharmaceutical industry. This is the basic method used for producing most antibiotics (penicillin, streptomycin, etc.) and many of the steroids (cortisone, etc.). Fermentation facilities are limited to a few of the larger companies; in no sense is every pharmaceutical manufacturer involved in the fermentation business or burdened with the special pollution problems, both water and air, that characterize fermentation processes.

The most troublesome waste of the fermentation process, and the one most likely to be involved in water pollution problems, is spent beer. This is the fermented broth from which the valuable fraction, antibiotic or steroid, has been extracted. Spent beer contains a large amount of organic material, protein, and other nutrients. Discharging it to a stream or other body of water without eliminating or reducing these materials can only result in a serious water pollution problem.

Methods of treating liquid fermentation waste (spent beer and wash water) are generally biological in nature, i.e., trickling filters or activated sludge. Figure 25 shows the activated sludge part of Lederle Laboratories' sewage treatment plant at

Fig. 24 High-pressure piston pumps being used to pump chemical wastes into disposal wells at The Upjohn Company's plant, Kalamazoo, Mich.

Pearl River, N.Y. The activated sludge unit operates in parallel with a trickling filter plant. Figure 26 is a picture of Wyeth Laboratories' treatment plant at West Chester, Pa.

Although fermentation wastes, in rather concentrated form, can and are satisfactorily treated by biological methods, it is unquestionably a much less difficult job if these wastes have the benefit of dilution with large volumes of sanitary wastes produced by a large municipal sewage system.

Some spent beer can be evaporated to dryness, and the resulting dry product may have sufficient economic value to offset the drying costs. In most instances this dry material has been used as an animal feed. This is a highly desirable solution to both the waste treatment and water pollution problem when it is possible.

Another method of disposing of spent beer has been spray irrigation. Figure 27 illustrates an actual application of this method. Although seemingly a simple solution, spray irrigation has a number of limitations; large land areas are needed in the order of 125 acres for 100,000 gal of spent beer sprayed per day. The land should be reasonably flat so that runoff from the spraying does not result in erosion or

Fig. 25 Activated sludge waste treatment plant, Lederle Laboratories, Pearl River, N.Y.

Fig. 26 Waste treatment plant at Wyeth Laboratories, West Chester, Pa.

Fig. 27 Spray irrigation of fermented spent beers.

"puddling" in low spots. The latter, i.e., "puddling," will result in odors, and then the entire operation is likely to become a public nuisance.

Since, under the best conditions, spray irrigation is apt to have some odor associated with it, the land used for spraying should be as remote as possible from neighbors, particularly residences. Also, the amount of odor resulting from this spraying will depend on the attention that the operation receives. The irrigation system must be moved frequently to avoid saturation of the soil, because this condition, like "puddling," will result in an odor nuisance.

Although spent beer frequently contains high amounts of nitrogen, phosphate, and other plant growth factors, it is also likely to contain salts, like sodium chloride and sodium sulfate, as a result of the extraction process. The presence of such salts, depending, of course, on their concentration, can cancel out the value of the spent beer as a fertilizer. Land used for spray irrigation may not be very satisfactory for future use in producing agricultural crops, or at least such land may have only a limited use in this connection, because of a buildup of these undesirable salts.

Radiological Products

This is a highly specialized area of pharmaceutical manufacturing. Where nuclear materials are involved, the standards are largely established by the Atomic Energy Commission. The matter of water pollution necessarily gets special attention. In simple terms, all effluent from such a unit, including sanitary waste, is piped to large, closed, underground tanks. When one tank is full, the wastes are directed into another tank. The full tank is monitored until the radiation level falls to a safe limit, at which time it is pumped to the regular sewer system. Rotating two or more tanks in this manner prevents unsafe materials from entering the sewer.

Research Facility

Generally, quantities of materials being discharged by a research operation are relatively small as compared with the volumes generated by production facilities. To this extent, at least, the problem is generally a less important one. However, the problem cannot be measured entirely by volume of material going to the sewer. Research operations are frequently erratic as to quantity, quality, and time schedule when dumping occurs.

The most common problem is flammable solvents, especially low-boiling-point solvents like ethyl ether, that too frequently result in explosions and fires. The most effective approach to this one is to require laboratory personnel to dispose of all waste solvents in special waste solvent containers available in the laboratories.

Figure 28 illustrates a typical laboratory waste solvent container. Note the metal container is grounded to prevent static sparks when material is being poured.

Miscellaneous

Spills of both liquid and solid chemicals, not only inside production areas but in general plant areas such as roads and loading docks, have been an entirely too frequent cause of water pollution. Often such spills are washed down the nearest drain. Also, too regularly, the cleanup of a spill is done very quickly and expeditiously by the worker responsible in order to get rid of the evidence.

It is necessary to instruct workers properly so that they appreciate the problems that may result when chemicals get into a sewer system. In many cases, serious consideration should be given to the elimination of storm sewers and drains in a plant area. If possible, it may be a worthwhile safeguard to take all open drains to the sanitary sewer.

A common source of waste pollution, particularly with manufacturing plants in the northern part of our country, is the large amount of sodium chloride that finds its way into the storm sewer. Whenever snow and ice occur, salt is usually used very generously on roads and sidewalks. If storm sewer drains are provided, most of the salt ends up in what was planned as an unpolluted waterway.

Fig. 28 Waste solvent container for laboratory use. Note grounding strap.

AIR POLLUTION

As previously indicated under the subject of water pollution, the average, small pharmaceutical plant is not a serious polluter of the atmosphere. Air pollution problems are more likely to be the problems of the larger companies in which many of the industrial operations parallel those of the chemical industry.

With very few exceptions, the most serious problems relate to pharmaceutical manufacturers who operate fermentation plants in the production of antibiotics, steroids, etc. However, in no sense is air pollution exclusive to those that ferment. Others also have their problems, if only in the operation of their boiler plants.

Air pollution being the emission of undesirable materials into the atmosphere, we can usually expect to pinpoint the problem wherever processes or space ventilation are exhausted to the outside.

Pharmaceutical Manufacturing

Dust is one of the industry's greatest concerns, and whenever it presents itself, the matter is given serious consideration by members of the industry. Dust of any kind, either outside or inside of a pharmaceutical plant, is likely to cause serious repercussions. The drug manufacturer is very conscious of his public image; he does not want to alienate his neighbor by causing either a dust or odor nuisance; but perhaps of even greater concern is the presence of dust inside the pharmaceutical manufacturing plant. This leads to a completely different situation and one the manufacturer is becoming increasingly sensitive to as a result of the government controls relating to manufacturing practices in the industry. Dust inside the plant may cause "cross contamination," that is, contamination of one drug by another one.

There are a few materials that are capable of causing extremely toxic reactions, even when present in trace quantities. Unfortunately, penicillin is one of these materials. To an individual sensitive to penicillin, its presence in the most minute amount in, for instance, an aspirin tablet, is capable of causing a reaction of very serious proportions, a reaction which, in fact, might result in death. Therefore, it is absolutely essential that penicillin dust is not present in areas where other pharmaceuticals are manufactured.

Penicillin dust has been a particularly serious matter. Many of the manufacturers who produce penicillin containing products have isolated this production in buildings that are physically separated, even remote, from the other pharmaceutical operations. Nevertheless, all the manufacturers are concerned about any type of dust and go to great effort to eliminate it. The presence of dust inside the plant is an obvious problem and almost as obvious is the presence of dust outside of the plant, because this has a great tendency to find its way inside. The only way it can be kept outside is by extra care in filtering intake air going into the plant. The more dust that must be removed at the intakes, the more expensive is this job—hence the great and very much related concern with dust anywhere in the vicinity of pharmaceutical manufacturing.

Fig. 29 Typical Rotoclone wet scrubber installed at Mead Johnson & Company, Evansville, Ind.

Systems for removing pharmaceutical dusts are not unique. They are the same methods used throughout industry in general.

A very useful method for removing many pollutants that result from process operations has been the Rotoclone. Figure 29 is a typical air wash installation of this kind at Mead Johnson and Company, Evansville, Ind.

The use of water and the possible resulting problems of water pollution may dictate the use of a dry filter system rather than a scrubber or Rotoclone.

McNeil Laboratories exhausts all the air from most manufacturing operations through an extremely large Pangborn baghouse type dust collector. Figure 30 shows this unit. It is 33 ft long by 17 ft wide by 20 ft high. The inlet duct is 44 in. in diameter. This single unit has a capacity of 36,000 cfm. In the case of McNeil Laboratories, the pharmaceutical manufacturing areas are supplied with 100 percent outside air.

Fig. 30 Pangborn bag dust collector installed at McNeil Laboratories, Inc., Fort Washington, Pa.

Fig. 31 Large Rotoclone installed in the chemical finishing area at Schering Corporation, Union, N.J.

Chemical Production

Here, again, neither the problems nor their solutions are particularly unusual. Figure 31 is a large size Rotoclone used for wet scrubbing the exhaust from Schering Corporation's chemical finishing area at Union, N.J. Figure 32 is an HCl and HBr fume scrubber installed at Parke-Davis and Company, Holland, Mich. Figure 33 is a fume scrubber installed in the chemical manufacturing area of Wyeth Laboratories in West Chester, Pa.; this is an interesting installation because the scrubbing liquor is recirculated through the scrubber from a large holding tank into which the scrubber discharges.

Fig. 32 HCl and HBr vacuum scrubber installed at Parke-Davis & Co., Holland, Mich.

Fig. 33 Vacuum scrubber using recirculating scrubbing water installed at Wyeth Laboratories, West Chester, Pa.

Fermentation Processes

Most of the fermentations carried out in the pharmaceutical industry are aerobic; that is, air must be supplied to the fermentation organism. Compressed air is injected, or sparged, into the lower end of the fermentor, which is simply a large, vertical, circular tank. Capacity of such vessels may range between 5,000 and 100,000 gal.

Supplying fresh air to the fermentation vessel on a constant basis makes it necessary to vent or discharge an equal volume of what we might term "used" air from the top of the fermentation vessel. The so-called "used" air, or vent gas, has scrubbed a number of materials from the fermentation as it moves up through the fermenting mass. Included in these are carbon dioxide, in addition to many other more complex organic materials. It is the latter that results in the odor so charac-

Fig. 34 Stainless steel duct used for transporting fermentor vent gases from the fermentation areas to the boiler house to be used as combustion air, at Abbott Laboratories, North Chicago, Ill.

teristic of fermentation operations. These odors vary with the material being fermented and will vary somewhat between different fermentors of the same material. The odors may or may not be objectionable to all individuals, although the former is the most probable reaction.

This "used" air, or vent gas, from the fermentor is the principal air pollutant that we must be concerned with.

Wet scrubbing of the vent gases has never been particularly successful. On large fermentors, the volume of vent gases is so great that the water needed to do a scrubbing job, if water alone will do the job, is so large that, in order to eliminate or even partially reduce air pollution, water pollution is created that may have even larger dimensions.

Absorbing the odor of the vent gas on activated carbon can be done, but aside from demonstrating a theory, this is not practical in the case of large fermentors which would require an uneconomical amount of carbon to accomplish a satisfactory end point.

Incineration of the vent gases is a satisfactory solution, but if fuel must be supplied to raise the vent gas temperature from fermentation temperature (generally well below 100°F) to an incineration level, at which the odor is destroyed, this method may prove uneconomical.

There is, however, an interesting alliance of processes that can be accomplished to attain a more agreeable, economic result. It is possible to pipe the vent gas from the fermentor to a boiler house and use it for combustion air in the boiler. This system does work and is being used on a large scale. Figure 34 shows the large, stainless steel ducts carrying vent gases from the fermentors to the boilers at Abbott Laboratories in North Chicago, Ill. Figure 35 is a similar installation at Eli Lilly and Co., Lafayette, Ind. The large duct is shown transporting vent gases between the fermentation area and the boiler house, shown at the left side of the picture.

It should be noted that vent gases, since they are saturated with water vapor, will result in a slight lowering of boiler efficiency. Also, Abbott has noted some tube fouling that is probably the result of burning vent gases in their boilers.

Vent gases are not the only cause of odor in a fermentation plant. The waste mycelium, or filter cake, that results from the initial separation of solids from the fermented beer is a frequent source of odor. This is living cell material and is quite perishable. If it is not handled promptly, and if housekeeping standards are

Fig. 35 Galvanized steel duct used for transporting fermentor vent gases from fermentor areas to boiler house at Eli Lilly & Company, Lafayette, Ind.

not maintained at a high level, this part of the operation is also very likely to contribute to the odor problem. As a matter of fact, good housekeeping throughout the entire fermentation plant area will do much to improve an odor situation; even if this does not eliminate the odor, cleanliness throughout the plant somehow makes the odor more acceptable.

Radiological Products

As in the case of water pollution, the Atomic Energy Commission has established most of the standards that relate to operations of this kind. In general, the principle is to keep the process as free of dust as possible. Air intakes are equipped with high-efficiency, or absolute, filters to prevent dust from entering the facility via the air supply. Air exhausts are likewise equipped with the same high-efficiency filters to prevent any dust that may have somehow entered the facility from being emitted to the atmosphere. The exhaust systems are also monitored constantly for radiation levels that might be harmful.

Special precautions are taken to prevent dust from entering the facility via shoes, supplies, packaging materials, etc.

Fig. 36 Gas heated sterilizer used for destroying dangerous bacteria exhausted from a bacteriological laboratory hood.

Fig. 37 Plenum for removing harmful bacteria by absolute filters from the exhaust of a bacteriological hood.

Research Facilities

The problems of air pollution in connection with research laboratories tend to be matters of quality rather than quantity. How do you prevent pathological bacteria from being exhausted from a bacteriological laboratory? How do you safely control the exhaust from a radiation laboratory? Quantities of materials we are concerned with in research operations may be quite small but extremely hazardous. Consequently, this phase of air pollution in connection with pharmaceutical plants has received a great deal of consideration.

A common and older approach to the control of laboratory exhaust containing

Fig. 38 Use of absolute filter in a radiology laboratory hood to prevent radioactive materials from entering the atmosphere.

potentially dangerous living organisms has been incineration or sterilization of the entire exhaust airstream. This is done with either gas or electric heat.

Figure 36 shows the installation of a fairly typical laboratory exhaust, gas heater sterilizer. The temperature of the airstream is heated and held at 500°F. for approximately 13 sec. The burner develops about 600,000 Btu/hr when supplied with 1,000 Btu gas at 5 in. W.G. This unit takes care of a single hood in a bacteriological laboratory. By means of controls and pilot lights in the laboratory, the laboratory worker can start the burner and is advised by an indicating light when the temperature of the exhaust reaches sterilization temperature.

A more recent approach to this matter of handling dangerous organisms is the use of absolute filters that have an efficiency of 99.97 percent. Figure 37 shows one installation of this kind. The laboratory that it serves is held under a negative pressure; exhaust from the room is first filtered through a roughing filter and then through American Air Filter Astrocel. Following the final filtration, the exhaust can be safely exhausted to the atmosphere.

The increasing amount of radiation work in research laboratories has required special techniques in air handling to prevent dangerous radiation material from getting into the atmosphere. Passing the exhaust from such laboratories through absolute filters is the generally accepted solution to this problem. Figure 38 shows a hood in a radiation laboratory in which the absolute filter is mounted in an exposed position at the rear of the hood. In this location, the condition of the filter media can readily be assessed and monitored for radioactive materials.

Animal rooms can create air pollution problems. Odors are largely a matter of

Fig. 39 Special three-stage air filtration plenum used in connection with laboratory animal rooms. The final filtration stage is through an absolute filter.

housekeeping, but even with the highest sanitation standards, the exhaust from animal rooms can become a nuisance by spreading dust, including hair, animal food, dandruff, etc., in nearby research facilities. Figure 39 is the exhaust control plenum serving a group of 11 animal rooms at The Upjohn Company; 11,000 cfm is filtered through three stages of filtration. The first stage is an AAF Rollomatic Filter using an automatic. renewable-media roll for roughing filtration; the second stage is an AAF Dri-Pak No. 90 with a 50 percent efficiency (this stage gives a longer life to the final stage); and finally, the third stage is an AAF Astrocel absolute filter. When the exhaust from these animal quarters is finally discharged to the atmosphere, it is at much higher quality than most of the air we breathe.

Increasing concern in the industry with the toxicity of drugs, with the effect of dust of its own products and raw materials, on its own employees and customers and the public at large, has resulted in a large amount of investigation in this field. Inhalation chambers have been developed to study the effect of dust particles on animals.

Figure 40 shows one type of inhalation chamber. The exhaust from the chamber is filtered through three stages of filtration before being discharged to the atmosphere. A roughing filter is used to remove hair and larger particles; an AAF Dri-Pak No. 100 with an efficiency of 80 to 88 percent is used to give longer life to the final stage, an AAF Astrocel filter with an efficiency of 99.97 percent. The exhaust from the final filter can be safely discharged to the atmosphere.

Fig. 40 Inhalation chamber used for laboratory animals is on the left. A three-stage exhaust filter system in which the final stage is through an absolute filter permits air to be discharged safely to the atmosphere. The last two stages of filtration are in the portable plenum shown at the right.

Boiler Plants

Smoke and sulfur dioxide emission are the chief pollutants with which the pharmaceutical manufacturer's boiler plant is concerned. In this regard, it has exactly the same problems as other boiler plants.

Choice of fuel is usually a matter of economics. Pharmaceutical plants in the East tend to burn fuel oil; in the Middle West they are likely to burn coal, gas, or oil; in the Far West gas or fuel oil. Local ordinances and state laws have forced an upgrading of fuel standards requiring the use of higher-cost low sulfur coals and oils.

Boiler stacks in themselves do not eliminate air pollution, but improper design of stacks may compound the pollution problem by permitting smoke and flue gas to get back to ground level too easily. Adjacency to an airport necessitated The Upjohn Company's installing induced draft and relatively low stacks on coal-fired boilers at its plant in Kalamazoo. These were a constant source of irritation, particularly to employees who had to park their cars in the boiler plant area. Under some conditions of wind, temperature, and humidity these cars were covered with fly ash and condensation from the stacks. Eventually, because of this problem, these stacks were replaced by new ones that were designed with a smaller-diameter velocity tube at the discharge end. Figure 41 shows how the plume now leaves the stack at high velocity instead of, as previously, drifting downward and actually touching ground under some conditions.

Fig. 41 Specially designed boiler stacks at The Upjohn Company, Kalamazoo, Mich. The velocity tube at the top of the stack discharges the plume at high velocity to prevent its drifting back to earth.

In boiler house operation, periodic blowdown is required. Blowdown is characterized by the emission of black smoke from the boiler stack. Smoke ordinances recognize the need for blowdown and permit a smoke condition at stated periods. In order to control the smoke resulting during blowdown, the CIBA Pharmaceutical Company at Summit, N.J., found an interesting solution. Figure 42 shows a large Rotoclone installed on the boiler house roof. When, because of blowdown or other reasons, a high-smoke-density condition occurs, the gases from the boiler are passed through the Rotoclone. Figure 43 shows the black smoke emitting from the boiler without the benefit of the Rotoclone. Figure 44 shows the result of scrubber action, simply white water vapor from the Rotoclone.

Fig. 42 Large Rotoclone installed at the CIBA Pharmaceutical Company, Summit, N.J., is for handling high smoke densities during boiler blowdown.

Fig. 43 Black smoke being discharged at the CIBA boiler house during a blowdown, without the benefit of Rotoclone operation.

Fig. 44 The same blowdown being discharged through the Rotoclone, evidenced by the elimination of black smoke and the production of water vapor.

Incinerators

Although incinerators generally represent an auxiliary facility related to a major division or function of the company, they are a special type of problem and are recognized as such by most codes, ordinances, and laws relating to air pollution.

Large quantities of waste paper must be disposed of by most pharmaceutical manufacturers. Much of this material can be shredded and/or baled and disposed of as salable paper waste, but many types of paper are not salable; a lot of paper must be burned to ensure its being destroyed, such as correspondence, documents,

Fig. 45 A typical waste paper incinerator installed in a pharmaceutical manufacturing plant.

surplus labels, and cartons. Figure 45 is a picture of a typical waste paper incinerator installed in a pharmaceutical plant.

Although waste incinerators have improved a great deal in their ability to handle paper with a minimum of fly ash and smoke, under some conditions and locations their use must be questioned. In such cases other methods of disposal should be considered. Paper can be compacted and used in landfill. Another solution is to use a Somat system. This involves wetting the paper and reducing it to a pulplike mass which is then disposed of by landfill. In either case, air pollution is entirely eliminated. Figure 46 shows a Somat unit installed at Mead Johnson and Company in Evansville, Ind.

Sometimes there is both practical and economic justification for using one incinerator for two or more quite different wastes. Figure 47 shows an incinerator originally installed at The Upjohn Company for burning only waste solvents. The waste solvent burners are indicated at the right side of the picture. Because the quantity of solvent that could be burned was limited by the temperature of the furnace, a hatchway was added to the furnace through which animal litter and laboratory animals could be charged into the furnace. This opening is at the left side of the furnace. In this way, waste solvents are disposed of, at the same time producing fuel for incinerating large quantities of laboratory wastes. Figure 48 is a picture of a special incinerator designed to dispose of large laboratory animals. This is accomplished with no pollution of the atmosphere. The incinerator uses natural gas.

Fig. 46 A Somat system unit installed at Mead Johnson & Company, Evansville, Ind., for producing a wet, disposable paper pulp from waste paper.

Materials Reclamation and Regeneration

Wherever large quantities of activated carbon, filter aid, ion exchange resins, and catalysts are used in processing, there may be a strong economic incentive to reclaiming or regenerating these materials in order that they can be used again. Frequently, the regeneration process involves heating the materials at elevated temperatures simply to burn off the contaminants they have picked up in the process. Regeneration at high temperature is often an odoriferous process, because the contaminants are likely to be organic materials. Unless care is used in handling the waste vapor discharged from a regenerating unit, an air pollution problem is apt to result. Fortunately, water scrubbers are frequently a practical solution.

Fig. 47 A combination incinerator at The Upjohn Company, Kalamazoo, Mich., used for burning waste solvents and waste laboratory materials.

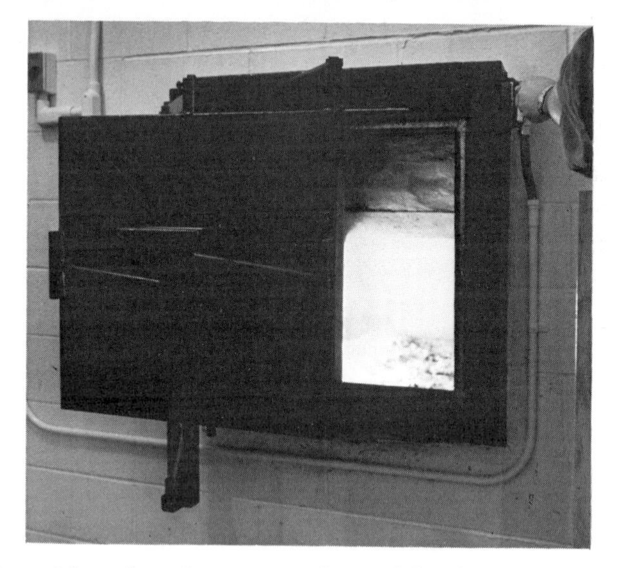

Fig. 48 A special gas heated incinerator designed for the disposal of large laboratory animals.

Figure 49 is a regeneration unit in which the waste vapors are removed by water scrubbing. The scrubber is shown at the left-hand side of the picture.

Waste Treatment Facilities

The waste treatment facility of a pharmaceutical plant can be a source of air pollution, usually as an odor producer.

Biological treatment plants, both trickling filter and activated sludge plants, are based on the oxidation of sewage by aerobic bacteria. Such plants, when they are functioning properly, do not generally result in a particularly odoriferous condition. If they are a source of nuisance odor, it is probably because they are not functioning properly and that a septic or anaerobic condition prevails.

Sometimes, however, characteristic odors can result from waste treatment facilities that are caused not by a septic condition but by the materials being treated. Abbott Laboratories experienced a situation of this kind. Their solution was somewhat unique. They enclosed the aeration tanks of their activated sludge plant with a relatively airtight housing of asbestos board. The housing is vented to the boiler house through a stainless steel duct, where the vented gases are incinerated as com-

Fig. 49 A water scrubber used for removing odor from the vent gases of a carbon regeneration unit.

bustion air. Figure 50 shows how the activated sludge tanks have been housed to prevent the escape of odors.

PUBLIC RELATIONS

Public relations problems in the pharmaceutical industry, insofar as water and air pollution are concerned, are undoubtedly the same as they are for any other industry that is likely, in the normal course of its operations, to generate water or air pollutants.

Fortunately, today there is a greater awareness and concern on the part of industry as to its responsibilities in this area. Unquestionably, stricter laws and more active enforcement of these laws have been factors in this awareness, but this has not been the only impetus the problem has received. Most of us have become increasingly concerned at the rate and the extent that our civilization is dirtying our environment. It is doubtful if there is any industry or many individual members of industry as a whole who are ignoring this problem. Certainly, to do so maliciously would be flirting with the wrath of public opinion.

There are two basic principles that will affect the public relations problems: (1) there must be honesty; problems cannot be hidden in the closet or swept under a rug; and (2) employees must be trained to understand what constitutes pollution and what their responsibility is in controlling pollution.

When a pollution situation exists, it is necessary that good internal communication exist between the various areas of the company that may be drawn into the problem. Public relations people should be the only ones in contact with news media. When there is a question of law, company lawyers must participate. Matters of technology must involve qualified engineers and/or production personnel.

Fig. 50 Airtight housings over activated sludge tanks to restrict the odor coming from these tanks, and permitting its being burned in the boiler plant at Abbott Laboratories, North Chicago, Ill.

None of this is to hide anything, but simply to avoid confusion and "muddying" of the problem waters.

LEGAL IMPLICATIONS

There is no evidence that the pharmaceutical industry has any uncommon problem in its relation to antipollution laws that is not faced by industry as a whole.

Laws, at the moment, may be somewhat more stringent along the Eastern seaboard than in other parts of the country. This results largely from high population densities in the East. Because a preponderance of pharmaceutical companies are located in the East, a large segment of the industry has become well acquainted with the stringencies of these laws.

In any case, the fact that antipollution laws may be stricter in one area of the country than in another is undoubtedly a very temporary situation.

Chapter 18

Pollution Control in the Pulp and Paper Industry

R. M. BILLINGS

Director of Environmental Control,
Kimberly-Clark Corporation, Neenah, Wisconsin

and

G. G. DeHAAS

Manager, Chemicals Development Department,
Weyerhaeuser Company, Longview, Washington

WATER POLLUTION

The pulp and paper industry is the fifth largest industry in the United States in value of assets, but is only tenth largest in value of shipments of all manufactured goods, a fact that is indicative of the relatively high capital investment required for the production of pulp and paper. In 1966, this production amounted to 35.6 million tons of pulp and 46.5 million tons of paper and paper products. Growth in production of paper products is relatively steady, averaging approximately 5 percent per year. Pollution from the pulp and paper industry, despite expanded output of product, is steadily decreasing. This improvement is achieved by the rapid increase in the rate of installation of industrial waste treatment facilities at old paper mills, by the phasing out of pulping processes which cause gross pollution, and by the incorporation of provision for maximum water recirculation and fiber recovery, followed by effluent treatment, as a fundamental part of the design of all new mills.

The problem of reducing pollution from old pulp and paper mills is one of the most vexing problems confronting the industry. These mills were constructed originally during a period when pollution was not considered a problem, and they grew as demands for their products increased. The resulting sprawling complexes with their multiple points of effluent discharge are a far cry from the compact, efficient, carefully designed units that the industry now constructs. Today, sites for new mills are carefully selected with the aim of keeping to a minimum the amount and effects of pollution. While each old mill can be improved by conventional abatement equipment, the degree of attainable improvement depends upon the specific mill, its location, and its technological age. Each such mill thus requires a solution of pollution essentially tailor-made to its particular situation. If technology can provide no

Fig. 1 Pollution abatement facilities at Kimberly-Clark's pulp and paper mill at Anderson, Calif., on the Sacramento River. Primary and secondary clarifiers in center foreground; stabilization and aeration basins right center; final storage lagoon top right.

tailor-made solution that is feasible, the mill will have to shut down and a new one must be constructed to replace its output. In this day of intense competition, the new mill will necessarily be constructed at a site that is optimum for markets as well as for labor, raw material, water, and power supplies, and also for waste disposal. This has caused some shutdowns at old locations and moves to new regions, with all the attendant expense and the resultant disruption of the local economies. The industry thus frequently finds itself on the horns of the dilemma of old mill inflexibility and extremely high capital investment.

Further complications arise from the almost total dependence of the pulp and papermaking process upon enormous quantities of water (see Table 1). Water is used as a vehicle for transporting wood within the mill, plays an integral part in the cooking and grinding processes, and then is used for carrying the separated fibers through the bleaching, refining, and sheet forming phases of manufacture. Pollutants thus occur in a highly dilute form, the ratio of water to pollutant varying from a few hundred to one, to several thousand to one. The removal of these dissolved or highly dispersed materials economically is a problem which defies simple solution.

Nature of Pulp and Paper Mill Pollution and Methods of Evaluation of Each

Care is usually taken to differentiate between wastes hazardous to human health —designated as contaminated—and wastes which may render the receiving waters undesirable or unsuitable for other legitimate purposes such as drinking, fishing, or swimming; these latter are said to be polluted. Wastes from the pulp and paper manufacturing process (see Fig. 2), since they are nontoxic to human beings and contain no pathogenic organisms, fall in the category of pollution. Five classifications cover the pollutional characteristics of pulp and paper mill wastes. By far the majority of the pollution problems of the industry, however, are included in the first three.

Suspended solids The solids found in pulp and paper mill wastes consist of (1) fine bark particles and silt from the wood room, (2) fibers and fiber particles from both the pulp and papermaking operations, and (3) coating and filling materials such as talc, clays, calcium carbonate, and titanium dioxide from the papermaking process. The settleable solids are the objectionable portion of the suspended solids and comprise from 75 percent to 90 percent of the total suspended load. They

TABLE 1 Effluent Volumes from the Manufacturing of Pulp and Paper Products

Category	Range, gal/ton	Avg representing superior performance, gal/ton
Pulp manufacturing process:		
Unbleached kraft	15,000–40,000	20,000
Kraft bleaching	15,000–35,000	20,000
Unbleached sulfite	15,000–50,000	25,000
Sulfite bleaching	30,000–50,000	40,000
Semichemical	8,000–40,000	10,000
Deinked	20,000–35,000	25,000
Groundwood	3,000–48,000	4,000
Soda pulp	60,000–80,000	65,000
Paper manufacture:		
Fine paper	8,000–40,000	10,000
Book or publication grades	10,000–35,000	12,000
Tissue	7,000–45,000	15,000
Kraft papers	2,000–10,000	5,000
Paperboard	2,000–15,000	8,000

settle to the bottom of the receiving stream to form sludge deposits or sludge banks which interfere with fish propagation and with other desirable forms of aquatic life. The deposits are unsightly and may become odorous as they decay, particularly where the sludge banks become partially exposed during low river flow. Submerged deposits in the process of decomposition exert an oxygen demand on the overlaying water, causing shallow quiescent areas to become anaerobic during hot weather. Much of the nonsettleable portion is colloidal in particle size, and is objectionable mainly from the aesthetic standpoint. It consists chiefly of wood fines, titanium dioxide, clay, and other finely divided coating or filler material. While it may impart opalescence to the receiving water and thus reduce the beneficial effect of sunlight penetration in the immediate area of discharge, effect upon the river's capacity for self-purification is generally negligible.

Total suspended solids are determined gravimetrically by filtering a measured amount through an asbestos mat in a Gooch crucible.[1] Since the preparation by this

[1] Procedures described in Am. Public Health Assoc., "Standard Methods for the Examination of Water and Wastewater," 12th ed., are used unless otherwise specified.

method is rather time-consuming, glass fiber-filter paper is generally used and gives results that correlate well with the Gooch crucible method. The use of rapid filtering papers can lead to erroneous results and conclusions. Such filtration should be used only for operating control purposes where speed is essential. The determination of settleable solids as a quick check on the efficiency of solids removal equipment is usually done volumetrically. Settleable solids content determined by the gravimetric method is considered in some quarters as more meaningful than total suspended solids, since it is indicative of the behavior of the waste in the receiving streams.

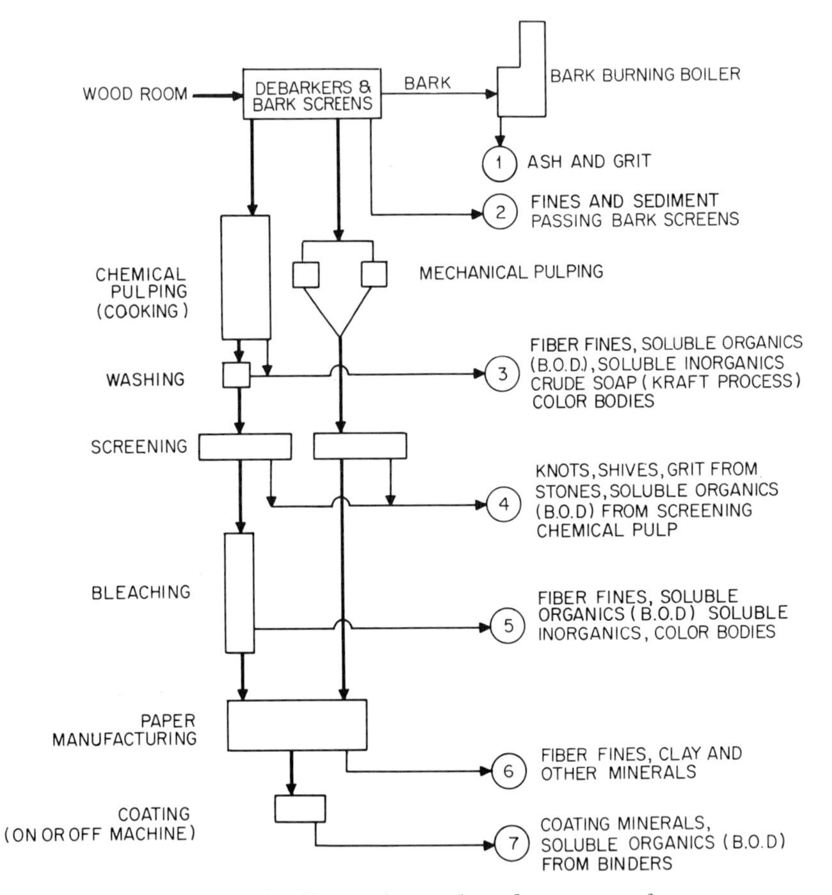

Fig. 2 Sources of pollutants from pulp and paper manufacture.

Soluble organics The second principal area of pollution concern in the pulp and paper industry is soluble organics. Cellulosic fiber is the only material desired from the pulpwood log, yet approximately half of the log is nonfibrous in nature. Therefore, depending upon the pulping process used, as much as 50 percent of the organic material may be discarded. Substantial portions of this discard are wood sugars, carbohydrates, and related compounds. These materials serve as food for the ubiquitous water microorganisms that form the base of the aquatic food complex or pyramid which ultimately results in fish. If the discard is controlled to a small and relatively uniform amount, it thus can be beneficial. Excessive amounts, however, result in the abnormally rapid growth of water microorganisms. Oxygen is consumed in this growth process, and the removal of oxygen from the surrounding waters can make them unsuitable for fish and certain other forms of aquatic life. The complete removal of dissolved oxygen from a river also activates the anaerobic bacteria which are always present, and the stream then develops the unpleasant odors associated with decay. The presence of high concentrations of organic material may also enhance the growth of river slimes, chief among which is *Sphaerotilus natans.* Biological imbalance resulting in slime growth, however, appears to be the result of a number of factors, of which availability of food is only one.

Five-day biochemical oxygen demand, or BOD,[1] is the universally accepted evaluation of the effect of organic wastes in the pulp and paper industry. Extreme attention to all details of the BOD test procedure and selection of the proper modifications is essential, for even under the best of conditions the precision is poor. Probably no other test which is used so frequently, and upon the results of which so many important decisions are made, is so lacking in precision as the 5-day BOD determination. It evaluates a characteristic, however, that to date can be obtained in no other way.

Chemical oxygen demand (COD) furnishes information regarding the total oxygen demand including that portion not subject to biochemical degradation. This test is quicker and simpler and is often used for industrial control, where the correlation between BOD and COD has been determined. Unfortunately, this correlation is specific to almost every process and even for the same process is subject to change and therefore must be checked repeatedly.

The continuous carbon analyzer is another control tool of recent development. It yields an answer in minutes rather than in days and, for an effluent of fairly constant composition, gives a good correlation with 5-day BOD. Here also, however, the correlation must be checked regularly.

Aesthetic pollution This type of problem of the pulp and paper industry can usually be held to a minimum by adequate primary treatment to remove settleable and floating solids, and scum and oil slicks, and by making sure that some dissolved oxygen is always present to prevent the stream from going anaerobic. There are three types of aesthetic pollution which at present can be only partially taken care of, in spite of extensive research to date. These are the dark color of effluents from the kraft pulping and bleaching processes, the regeneration of foam in certain waterways, and the highly visible streaks of finely divided clay and titanium dioxide.

Pollution toxic to aquatic life Pollution which is toxic to aquatic life occurs only infrequently in consequence of pulp and paper mill operations. A significant exception is crude soap, which is produced as a by-product in the manufacture of kraft pulp and must be collected. Where bioassays have shown some specific additive to be toxic, the user mill makes special provision to prevent its escape in objectionable concentrations. Fish kills have sometimes occurred, usually during mill startups, because of equipment failure or pipe breakage discharging toxic slugs of materials which are regularly used in harmless quantities for process purposes.

An effluent's degree of toxicity to fish can be determined only by bioassay methods requiring considerable knowledge of factors affecting results. Therefore, when toxicity is suspected or when mills contemplate using extensively a new agent suspected of possible toxicity, they normally employ a laboratory or agency experienced in bioassay to determine safe concentration limits.

Soluble inorganics Soluble inorganics in effluents from the pulp and paper industry have created very little problem to date. Increase in chloride content of receiving waters has been watched from the standpoint of possible increases in corrosion experienced by downstream users. At present low levels of chlorides in almost all the streams, however, no such effect has been demonstrated.

Variants of new membrane process technologies, including electrodialysis and reverse osmosis, are being researched and developed as advanced methods of treating industrial waste. These may eventually be found capable of separating and con-

[1] Where the term BOD is used in this chapter, 5-day BOD is always meant.

centrating dissolved organics and inorganics and releasing clean water for reuse in the mill. Proving the practical worth and feasibility of these new unit operations now in advanced stages of engineering development will depend on solving problems of utilizing or disposing of the membrane process concentrates. The present general answer to the problem of dissolved inorganics is dilution of the mill effluent.

The increasing problem of the growth of algae in our streams and lakes appears to be associated with the increasing presence of certain dissolved nutrients, chief among which is phosphorus. For this reason, reduction of phosphates from all sources—municipal, individual, and industrial—is being emphasized. Phosphates are not used in pulping processes and hence make up no part of the soluble wastes originating from this source.

Some phosphates are used in the paper mill for such purposes as the preparation of coatings and the dispersing of pitch. Recent laboratory and pilot plant developments in chemical treatment in conjunction with primary clarification have indicated that reductions in phosphates ranging from 70 to over 90 percent may be possible. Similar results have been indicated for treatment in conjunction with or following regular secondary treatment. Industrial application of these findings is under study.

Types of Pulp and Nature of Pollutants Produced

Kraft, sulfite, and semichemical are classed as chemical pulps. These, along with deinked pulp and groundwood, are the most common types of pulp produced today.

Bleached and unbleached kraft pulp The ratio of bleached and unbleached kraft pulp to total pulp production is increasing steadily, chiefly because of the successful application of the kraft process to a wide variety of wood species. Chemical recovery is an integral part of the kraft process, and hence the BOD discharged to the stream per ton of pulp produced is the lowest of the chemical pulps (see Table 2). Kraft pulp may be used unbleached, or bleached to a high whiteness.

Pollution parameters to be considered in design are suspended solids, biodegradable organics, color, foam, and materials potentially toxic to aquatic life.

Sulfite pulping The first pulping process used predominantly on the American continent, was sulfite pulping, primarily because of the availability of spruce and balsam, which are readily cooked by this process. Also, sulfite pulping used low-cost chemicals such as lime and sulfur in preparing the cooking liquor. These could be discarded after use, and expensive recovery systems were therefore unnecessary. The fact that approximately half of the log is thus discarded in the mill effluent rather than being burned during a process of chemical recovery explains the high BOD per ton of sulfite pulp produced. While unbleached sulfite is used as an ingredient in newsprint and liner stock, most of it, like kraft, is bleached to varying degrees.

Pollution parameters to be considered in design of sulfite pulping facilities are very high BOD, suspended solids, and color (although the color problem is less than for bleached kraft).

Semichemical pulps Midway between kraft and sulfite in pollutional characteristics are semichemical pulps. Because of the much lower degree of chemical treatment, the BOD per ton of product is less than that for sulfite pulp. However, most semichemical mills have no chemical recovery; so the BOD released to the stream is much higher than for kraft pulping. A few mills operate semichemical recovery processes for stream improvement purposes, since the value of the chemicals recovered is insufficient to support the cost of the recovery operation. Recent technical developments have demonstrated good chemical recovery through the application of fluidized bed incineration. BOD has been reduced by this process to the level obtained in kraft pulping. The chemicals recovered, however, are in a form not directly reusable in semichemical pulping. They are usually shipped to kraft pulp mills or otherwise marketed.

Pollution parameters are BOD, suspended solids, and to a lesser degree, color of effluent.

Deinked pulp This type of pulp may be used in almost every type of paper from news to high-grade business papers. The newer deinking processes are more efficient and are broadening the range of papers that can be deinked. The normal 20 to 40 percent overall shrinkage encountered in deinking results in a very heavy discharge of suspended solids. BOD of deinking wastes ranges from 60 to 160 lb/ton of deinked pulp produced, depending upon the grade of paper and the deinking process itself. Installations of the newer flotation process fall in the lower half of the range.

Pollution parameters to be considered are BOD and suspended solids.

TABLE 2 Untreated Effluent Loads from Pulp and Paper Manufacture

Effluent	lb/ton of product	
	Suspended solids, range of design values*	5-day BOD, range of design values*
Pulps:		
Unbleached sulfite	20–40	400–700
Bleached sulfite	25–60	450–800
Unbleached kraft and soda	20–30	25–50
Bleached kraft and soda	25–55	45–80
Unbleached groundwood	30–80	15–25
Bleached groundwood	45–85	25–60
Neutral sulfite semichemical	80–180	250–500
Textile fiber	300–500	200–300
Straw	400–500	400–500
Deinked	400–800	60–160
Fine papers:		
Bond—mimeo	50–100	15–40
Glassine	10–15	15–25
Book or publication papers	50–100	20–50
Tissue papers	30–100	10–30
Coarse papers:		
Boxboard	50–70	20–40
Corrugating brand	50–70	25–60
Kraft wrapping	15–25	5–15
Newsprint	20–60	10–20
Insulating board	50–100	150–250
Specialty papers:		
Asbestos	300–400	20–40
Roofing felt	50–100	40–60
Cigarette papers	100–800	20–30

* Design value is dependent upon yield.

Groundwood Groundwood waste pollution problems are chiefly those of suspended solids removal. The BOD resulting is low (see Table 2), and aesthetic problems are easily corrected. Bleaching of groundwood may more than double the BOD per ton, depending upon the bleaching process used and the degree of whiteness required. Where zinc hydrosulfite is employed as the bleaching agent, concentrations of zinc in receiving waters must be carefully controlled. The ratio of bleached to unbleached groundwood produced is certain to increase as demands for whiter paper continue.

Other pulps Besides the many modifications and combinations of the conventional pulping processes listed above, lesser quantities of a variety of pulps are made for use in specialty papers or because of local availability of a specific raw material. For example, a limited amount of bagasse pulp is produced from sugarcane in areas of sugar production; this pulp is used in a wide spectrum of paper and paper products. Jute, flax, cotton, and rag pulp all have certain properties particularly suited to given products.

Types of Paper and Nature of Pollutants Produced

Paper products are usually classified from the standpoint of pollution under five headings: fine, book, tissue, coarse, and specialties.

Fine papers Examples of fine papers are the free-sheet grades such as bond, writing, and business papers, together with mimeo and white papers in general. Necessary pollution abatement for mills producing these products consists chiefly in removing suspended solids from the effluent.

Book or publication papers These grades are used primarily in magazines, brochures, catalogs, etc. They consist of paper manufactured from mixtures of groundwood and chemical pulp, with a suitable filler, which is usually clay. To give varying degrees of gloss, or "snap," the sheet is heavily coated, usually with clay or calcium carbonate to which has been added titanium dioxide for increased whiteness. This coating material is bound to the sheet by means of starch, protein, polyvinyl alcohol, or some similar adhesive material. The major pollution problem is again one of suspended solids, which in addition to the fiber particles may contain as much as 50 percent mineral from the filling and coating process. Increasing demand for whiter sheets with higher gloss has resulted in double coating processes which put the finest and whitest materials into the surface coat. The binders that hold these coatings to the sheet are the chief reason that the BOD from producing a ton of book paper is nearly double that from a ton of newsprint.

Tissue papers Tissues, in addition to the facial, toilet, and toweling grades, include wrapping and decorative paper. They contain groundwood or reclaimed fiber in varying amounts, but the major portion usually consists of chemical pulp. Settleable solids removal is generally sufficient to accomplish the necessary stream improvement since the BOD imparted to waste waters in tissue manufacture is very low. Chemical aids to flocculation employed in clarification of tissue mill wastes are more effective in reducing BOD simultaneously with settleable solids content then when used with wastes from other types of paper. This is because the manufacture of tissue results in very little soluble matter in the waste waters.

Coarse papers This grade includes, folding carton and paperboard, coarse wrapping paper, building paper and board, and newsprint. Suspended solids are the chief cause of pollution from coarse paper mills. This problem is less than might be expected, however, since shives, specks, and color variations are not critical for most grades with the exception of newsprint. Water can be reused, and the system can thus be closed up to a higher degree for these coarse papers than with other types of paper manufacture—again except newsprint.

Specialty papers For some special uses such as for currency or cigarette paper, nothing else but paper can serve. But paper at one time or another has been used as a substitute for a wide variety of other materials. When a mill goes into producing a paper for a specialized use, it invariably creates a new variation on one or more of the familiar old problems. These variations are so many and so specific that space limitations preclude any attempt to enumerate them here. The multiplicity of processes, products, and pollution problems emphasizes the fact that the pulp and paper industry is not a single industry, but rather a large number of industries. Several or many mills may produce the same product. But few do it in the same way, and few can apply identical procedures to solve their pollution problems.

Methods of Abating Pollution in the Pulp and Paper Industry

Suspended solids reduction (See Table 3). First in order of importance in pollution abatement is suspended solids reduction not only because pollution of this nature is common to all sections of the pulp and paper industry but also because the removal of suspended solids is a prerequisite for almost any subsequent treatment. Attempts have been made to bypass the clarification stage with portions of dilute mill effluent containing low concentrations of suspended solids. Such bypassing, however, almost invariably results in expensive problems of operation, since the suspended solids so bypassed may, in certain subsequent processes, end up as a relatively inert circulating load. This in turn increases the difficulty of maintaining the desired relationship between food (BOD) and the active mixed liquor solids in any bio-oxidation process that follows. Effluent streams so bypassed,

TABLE 3 Pollution from the Pulp and Paper Industry—Sources, Types, and Methods of Abatement

Source	Nature of pollutant	Method of abatement
Wood room debarkers and screens	Bark, bark fines passing through screens, grit from logs	Bark pressed and burned in bark boiler, ash and grit collected for land disposal fines removed in primary clarifier
Mechanical pulping	Fiber fines, grit from stones, BOD (large installations)	Fines removed in primary clarifier, grit settled out in settling chambers, BOD reduced through biological treatment
Chemical pulping (or cooking) and washing	Fiber fines, soluble organics (BOD), crude soaps, color bodies, soluble inorganics	Fines removed by primary clarifier, clarified effluent treated biologically to reduce BOD by means of aerated lagoons or some modification of activated sludge process, crude soaps collected during liquor evaporation and shipped as tall oil or burned, concentrations of color bodies and soluble inorganics reduced by dilution in receiving waters
Screening........	Knots, shives, coarse fiber, soluble organics (BOD)	Refined, cleaned, and returned to system; water recirculated; dirt, shives, and fines rejected are then removed in primary clarifier, BOD reduced by biological treatment
Bleaching........	Fiber fines, soluble organics (BOD), color bodies, soluble inorganics	Fiber fines removed in primary clarifier or settling lagoon, BOD of clarified effluent reduced by biological treatment (lagoons, activated sludge process, or modification), color bodies and soluble inorganics reduced in concentration by dilution
Paper manufacturing	Fiber fines, clays and other minerals	Filler clays and fiber fines removed in primary clarifier—may require special chemical treatment with alum, lime, or ferric compounds
Coating (on or off machine)	Coating minerals, binders such as starch have a BOD	Primary clarification, employed to remove coating minerals in suspension, may require special treatment; BOD of clarified effluent reduced in biological treatment

although dilute, contain both settleable and colloidal suspended solids. The settleable solids eventually drop to the bottom in quiescent areas and create an added oxygen demand as they decompose. This necessitates expensive cleanouts of ponds and lagoons. Removal of settleable solids from the entire mill effluent as a first step in treatment is therefore recommended.

Maximum collection and reuse of water and solids within the mill is obviously a first step in reducing pollution. Flotation save-alls are frequently used to concentrate suspended solids. A large proportion of the fibers and minerals reclaimed through recirculation or flotation can be used in the process, and thus offset a portion of the cost of pollution abatement. All mills use the recirculation practice to some degree. However, as the process is progressively closed up and recirculation increased, certain existing problems are magnified and new problems develop. For example, the recovered solids no longer possess their original desirable characteristics, and become gelatinous or otherwise deteriorate in usefulness. They also have greatly increased capacity to hold water, frequently many times that of the original fiber. Recirculation of water and fines when carried too far can lead to such difficulties as reduced drainage rates on the wire, increase in foaming, aggravated slime problems, and concentration of dirt in the finished product. The degree to which a mill system can be closed up is dependent on the importance of the various offsetting factors to the particular process and product. The ultimate pulp or paper mill, operating with a closed system, is yet to be developed.

Sedimentation is the principal method used to reduce suspended solids in mill effluents. Materials so removed in most instances are not reusable.

Fiber traps or settling lagoons are used where land is available and initial capital cost is of primary importance. Such installations are extremely effective in reducing settleable solids, but initial saving in construction may eventually be offset by the subsequent cost and unpleasant problems of cleanout. Cleanout is facilitated by building the lagoons shallow and only wide enough to ensure that flow rate is well below the dropping velocity so that fibers will settle out. Even small lagoons should have the length at least four times the width and should not exceed 5 ft. in depth. This construction minimizes channeling. The usual practice is to construct two lagoons in parallel, the first for regular use and the second to be used during cleanout of the first.

Deep lagoons are extremely difficult if not impossible to dewater. Frequent cleanouts of the shallow lagoons is preferable to allowing them to fill until efficiency of settling falls off. The difficulty of cleanout appears to increase markedly as the depth of sludge after decantation exceeds 3 ft.

Mechanically cleaned circular clarifiers, usually center driven, have become the generally accepted facility for the removal of suspended solids. Efficiencies of the units in removing total suspended solids from a pulp or paper mill usually range from 70 to 90 percent, while settleable solids are reduced by 95 percent or more. Each mill effluent, depending upon water source used, raw material supply, process employed, and product produced, has its own specific settling rate. An examination of the Zeta Potential by means of the Zeta Meter and the streaming current detector sometimes proves helpful in predicting the settleability of suspended waste streams. While the test results are subject to considerable variation, settling results are usually very good in the range of 0 ± 5 mv. Poor settling is almost invariably encountered at $\pm$ 20 mv and beyond.

Primary clarifiers which are giving settleable solids removal of 95 percent or more in present operating practice are chiefly designed for an overflow rate of 600 to 700 gal/(sq ft)(day). In designing a pulp or paper mill clarifier, it has been found to be good practice to size it 20 or 25 percent larger than laboratory settling tests indicate if it is intended that the centrate from the centrifuges or belt filters used in sludge thickening is to be returned to the clarifier. Centrates contain a disproportionately large amount of fines and therefore place a heavy additional load upon the settler. Bottom area requirements are not too critical, ranging from 200 to 800 sq ft/ton of solids, but the collecting mechanism should be of the thickener or heavy-duty type. Average retention time should be 4 hr. Where gravity feed to clarifiers is contemplated, care must be taken to prevent the entrainment of air, which can reduce the efficiency of the unit considerably, particularly if some long fiber is present.

Sludge pumps used to remove the sludge normally at a consistency of 2 to 6 percent are usually of the positive displacement type, although centrifugal pumps have been successfully used below 5 percent consistency. A most important consideration in sludge removal is to place the pump as near to the sludge well as possible, desirably adjacent to it under the clarifier itself. Sludge suction lines should be a minimum of 6 in. in diameter for clarifiers 100 ft in diameter or less and proportionately larger up to 10 or 12 in. in diameter for units over 200 ft.

Thickening or dewatering of primary sludge for eventual disposal is accomplished by filtration using drum, disk, or belt filters, or by centrifuging. Thickening is followed in some installations by pressing. Most efficient thickening is obtained when feed consistency to the filter or centrifuge is at a practical maximum. General practice in the industry is to use the clarifier as a thickener within the mechanical limits imposed by the torque, with resulting sludge consistencies usually falling between 3 and 6 percent. Intermediate thickeners are little used in the industry except for specific situations.

Filtration of clarifier sludge is the most widely used method of thickening, yielding a dry cake solid content in the range of 20 to 35 percent. The majority of filters have been designed to fit the individual processes, but even so loading rates vary widely, ranging from 2 to 10 lb/(sq ft)(hr). This is not surprising, considering that primary sludges from the pulp and paper process may vary from 10 to 70 percent ash. The variety of filters and filter cloths available together with the use of precoats or conditioning agents have given broad application to filter thickening. Filter clothing can usually be selected to limit the consistency of the filtrate. This is an important consideration, since the filtrate must be disposed of, and its return to the clarifier can result in clarifier upset if the filtrate contains too heavy a load of fines. Chief objections to filters are the necessity to enclose them from the weather and the operating attention required.

Centrifuges have given satisfactory results in a number of applications. with resultant cake dryness averaging between 16 and 45 percent. Centrifuges occupy less space than filters, require less protection from the weather, and need less operator attention. But centrifuges are less flexible for change in nature of the sludge, they can suffer abrasion damage from effluents containing sand or grit, and they yield a high concentration of sludge fines in the centrate. The most successful centrifuge installations to date have been on effluents from long-fiber processes. In one instance centrifuges are being used on a sludge from kraft-groundwood-newsprint production, but in this case, the clarifier was sized with the return of centrate to the clarifier in mind. Size of centrifuge, specific design, and *g*'s required for any particular effluent must be determined by pilot plant studies in the field.

Disposal of the thickened sludge is a major problem to which there are at present only two general answers:

Land disposal is always used whenever land is available. A great deal of land, however, may be required since volumes of sludge become truly enormous. A plant producing 25 tons/day of dry sludge in the moisture and density ranges usually encountered in pulp and paper mill wastes would have approximately 5,000 cu ft to dispose of each day, or enough to cover 3½ acres to a depth of 1 ft each month. Thickened sludge can be used for filling depressions and then covered over with 2 to 3 ft of earth. It can also be used to raise the level of the ground by trenching. If sludge used in landfill stands in water, its decomposition may create a serious odor problem. There is no record of primary sludge disposed of on dry ground causing odor problems even when rained on, where original solids content of the sludge has been 15 percent or greater.

Incineration is a last choice because of high operating costs and recurring problems of operation. In some instances, however, it is the only solution available. For incineration, sludge must be further dewatered to a degree dependent upon its volatile content and whether or not it is to be burned alone or in conjunction with some other material such as bark or sawdust. To support combustion without supplementary fuel, the ordinary pulp and paper sludge must have a solids content in excess of 45 percent. To dewater the sludge mechanically from the initial thickening consistency to that suitable for the type of incineration employed is presently accomplished by either V type or screw type presses. Many times, neither yields a product with a solids content in excess of 45 percent, and so it is necessary to provide either supplementary fuel addition or else incineration in bark or power boilers.

Reduction of soluble organics (BOD) The soluble organic material expressed as BOD is the second most significant source of pollution in the industry. The lignin compounds in wastes from the pulping processes decompose very slowly. The influence that they exert is therefore gradual and usually absorbed by the normal reaeration characteristics of the stream. Other organic compounds resulting such as carbohydrates, however, have a high and rapid BOD. This type of BOD can deliver a shock loading to the waterway which exceeds the stream's capacity to handle. The 5-day BOD test is considered the most suitable for evaluation of this property, and reduction of 5-day BOD in pulp and paper mill effluents is one of the primary objectives of pollution abatement.

Clarification that removes suspended solids also reduces the 5-day BOD of pulp and paper mill wastes. Absorption plays a major role, but the reduction is also theorized as being due to the carrying down of the finely divided organic particles along with the larger particles in the settling process. Portions of these fines verge on being soluble and thus support bacterial action. The effectiveness of clarification on BOD removal varies widely, from practically zero for the sulfite wastes to 10 percent for kraft, 15 to 20 percent for newsprint, 20 to 25 percent for book mill, and 35 to 65 percent for tissue.

Stabilization basins or naturally aerated lagoons have been widely used to satisfy BOD. This is particularly true of mills situated in the warmer Southern states and in rural areas. Effectiveness of the ponds in removing BOD is dependent upon ambient temperature and exposed surface area. Reductions of BOD range from 50 lb/(acre)(day) during warm seasons in the southern United States to as low as 15 to 20 lb/(acre)(day) in some parts of Canada. Stabilization ponds are usually quite shallow, preferably averaging 5 ft or less, surface area being the chief consideration. With a 5-ft depth a reaeration coefficient k_2 of 0.15 is used. Storage time to accomplish a high percent removal (85 to 90 percent) should be 25 days or more. Where sufficient surface areas are not available, increase in loadings per acre of 10 percent with an increase of storage time of 50 percent have been employed.

Optimum operating requirements must be determined for each particular set of conditions. Prevailing winds and distance from surrounding habitations become of increasing concern as loading and depth of lagoons are increased. The bottom of almost every naturally aerated basin is anaerobic, and this condition intensifies with depth of the lagoon and degree of loading. In some instances the anaerobic zone must be kept at a minimum; in others the combination of anaerobic and aerobic action is beneficial. The advent of the floating aerator has made it possible to correct an odor problem rather quickly in critical areas.

Geological appraisal of probable direction of underground water flow and location of wells in the area is important. While soil bacteria acting on the dissolved organics will grow and within a few years will plug off the interstices even in gravel formations, dissolved material meanwhile can adversely affect the taste of water in neighboring wells. If this appears to have a strong possibility, the bottom of the basin should be compacted or sealed and the dikes core-walled to prevent seepage. It is extremely important that wastes flowing to stabilization basins have been subjected to some form of treatment to remove the settleable solids. If not removed, these solids settle out in the bottom and decompose and thus increase the area of the anaerobic zone and add to the BOD loading. Eventually, if allowed to accumulate, they can destroy the function of the basin altogether. Multiple basins, wing dams, or underwater fences should be installed in large basins to assure a uniform passage through the system. Stabilization basins have the distinct advantages of dependable performance, of being able to absorb wide variations in BOD load, and of inexpensive operation.

Mechanically aerated basins (see Fig. 3) are being looked to more and more as the solution to the BOD problem of pulp and paper mills. They have the advantages of stability of operation, six to ten times the BOD loading capacity per acre of a naturally aerated basin, and they avoid the extremely difficult problem of secondary sludge encountered in the activated sludge process. The oxidation rate indicates a coefficient k of 0.1 at 20°C and a k of 0.16 at 30°C. Most common designs provide a retention time of from 6 to 10 days with provision included for supplementary feeding as needed with nitrogen and phosphorus. Reductions of 60 to 75 percent have been reported with 4 days' retention without supplementary feeding where ambient temperatures were high. In another instance, 90 to 94 percent reduction in 6 days' retention with supplementary feeding was obtained. Here again the variable nature of waste streams requires a tailor-made answer based on local experimentation.

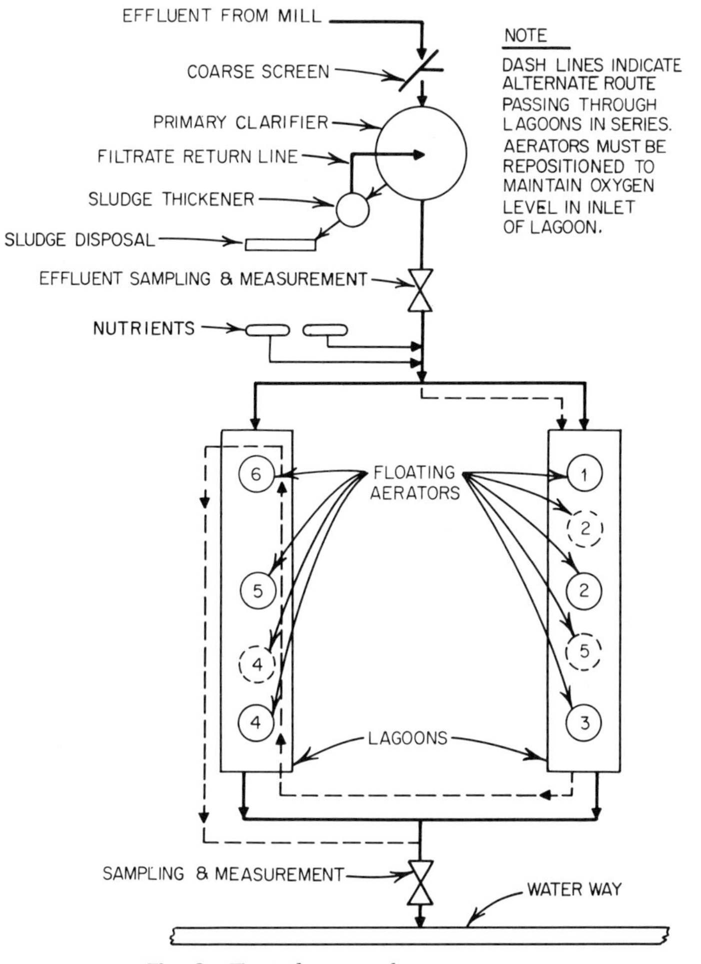

Fig. 3 Typical aerator lagoon system.

In the last analysis, the weight in pounds of BOD removed per horsepower-day is the only real criterion of any aerator for any specific job. In designing an installation, the figure of 50 lb BOD removed per horsepower-day is normally used, and experience has shown this to be a good figure. Both in the laboratory and in the field, experience indicates that less than 0.2 lb of new insoluble sludge is produced per lb BOD destroyed in extended aeration of this type. Moreover, the sludge is relatively inert, light, finely divided, and has little oxygen demand. Hence it does not stimulate *Sphaerotilus* or other slime growths, creates no oxygen demands on the waterway, and does not settle out to form sludge banks or sludge bars. Beyond slightly increased turbidity (which is quickly reduced by dilution) no effect, either positive or negative, due to its presence in the stream has ever been demonstrated.

Activated sludge treatment or one of its many modifications is being utilized for the biological treatment of pulp and paper mill wastes in those installations where space is of major concern. An activated sludge basin will accomplish in 4 to 6 hr the same 85 percent reduction as a natural aeration basin achieves in 25 days' retention time or mechanically aerated basins in 6 days. With clarification of 90 percent or better, activated sludge installations are currently removing 50 lb of BOD per 1,000 cu ft aeration basin capacity, with higher removals indicated for specific installations. As efficiency of initial clarification decreases, however, the efficiency of BOD satisfaction in biological treatment likewise decreases.

The above is only one example of the extreme sensitivity of the activated sludge process. Another is its susceptibility to upset due to shock loads which occur from time to time in every mill manufacturing pulp or paper. Oxygen content is also critical in waste sent to the secondary clarifier. Too little oxygen may cause the clarifier to go anaerobic; too much oxygen may result in sludge bulking and sludge carryover. Good operation calls for temperature not exceeding 110°F and pH normally maintained near the neutral point. Activated sludge usually tolerates short-term fluctuations to 6 pH or to 9 pH since the return sludge has a buffering capacity, but efficiency falls off rapidly as these levels are maintained or exceeded. If excessive fluctuation or some accidental spill of chemicals within the mill kills the sludge, there is nothing to do but discharge the sludge and wait weeks for a new population to develop. There is no proof that seeding from another source hastens the process of sludge development.

This extreme sensitivity at so many points requires that the process be under the supervision and guidance of a professional engineer or chemist, since conditions arise repeatedly which require a high degree of technical knowledge and skill.

Nutrient ammonia and phosphoric acid are essential to this process. At startup, nitrogen is usually added at approximately 1 lb/20 lb of BOD, and phosphorus at 1 lb/100 lb. After the operation has settled down, it will usually proceed successfully with a small reduction in nitrogen requirement and a much larger reduction in phosphates, since these may be present to some extent in the effluent.

Besides sensitivity, the big disadvantage of the activated sludge process is the problem of disposal of the secondary sludge formed—0.35 to 0.55 lb/lb of BOD satisfied, compared with 0.2 lb or less in mechanically aerated stabilization basins. The sludge contains active as well as inert matter and is in a flocculated rather than a finely divided form and cannot be discharged to a waterway. It is gelatinous and filters very poorly if at all. When centrifuged, it takes on the consistency and appearance of tapioca pudding. It can be filtered when 1 part is mixed with 3 parts of the proper primary sludge. The cake so obtained is usually disposed of in the same manner as primary sludge. Not all primary sludges, however, possess the good drainage properties required, and this limits applicability of the method. In areas where rainfall occurs only 2 or 3 months of the year, secondary sludge is being sprayed on level land at the rate of one application per week, and advantage is thus taken of the high natural evaporation rate. Spraying creates many problems during the rainfall months. In other circumstances, secondary sludge has been pumped to large holding lagoons as a temporary expedient until satisfactory methods of disposal are found. Such lagoons almost invariably cause odor problems during hot weather and require costly applications of lime. Probably no single phase of pollution abatement in the industry is subject to such intensive research as is secondary sludge thickening and disposal. Further application of the activated sludge process to the pulp and paper industry will probably be limited until satisfactory solutions to the secondary sludge problem are forthcoming. Cost of the activated sludge process is high.

Trickling filters, when operated at a high rate, give reductions in BOD of from 40 to 60 percent. Their chief advantage lies in the ability to handle large volumes of wastes over a wide range of BOD concentrations. The development of the plastic media as a filter at one time encouraged consideration of the trickling filter to pulp mill wastes, since plastic media largely eliminated the problem of clogging that was experienced with stone media. Expectations have as yet not materialized. With trickling filters as in other methods, best BOD reduction is obtained on wastes from which most of the settleable solids have been removed, although this is not so critical as with activated sludge. Trickling filters are not so widely used in the pulp and paper industry as other bio-oxidation processes. They do furnish satisfactory treatment to mill wastes which do not require a high degree of reduction, or where it is desirable to reduce the BOD load ahead of a final treatment process. Trickling filters occasionally serve, also, as cooling towers in situations where waste temperatures must be reduced before subsequent treatment or discharge. Such applications yield a BOD reduction of low magnitude.

Spray irrigation or land disposal is one of the few known methods in which BOD reduction can approach 100 percent. Where weak wastes are sprayed on the soil it is important to maintain a good crop of vegetation to keep the soil open or porous and facilitate evaporation of the water. Soil pH of 6 to 9 is satisfactory, and the sodium absorption ratio should be less than 8.0 to avoid plugging. Spray disposal requires an average of 40 to 50 acres/mgd, and care must be taken to avoid flooding that can kill the vegetation. Most soils can handle no more than 200 lb of BOD/(acre)(day).

Irrigation with weak wastes is particularly effective in areas where water is at a premium. This should not be undertaken without chemical analysis of both soil and water. For example, well waters high in boron may not be objectionable in the manufacturing process but may be unsuitable for certain crops.

Land disposal is also effective for strong wastes, if certain precautions are followed. Wastes of high specific gravity can lead to serious seepage problems, since they may go down through the groundwater substantially undiluted and then move sidewise along rock strata. For this reason spray disposal is usually preferred. Digester-strength spent sulfite liquor, for example, can be effectively sprayed at a daily rate no greater than 1 lb of solids/sq yd of surface. Soil bacteria in a 10-ft depth of sandy soil reduce the BOD by 80 percent on the average, and greater depths increase the reduction. Millions of gallons of spent sulfite liquor at a concentration of 10 percent and higher are used each year as roadbinder on gravel roads. The sulfite solids quickly become insoluble and will not wash out even during heavy rains. The binding properties significantly reduce road maintenance.

The development of by-products, particularly from spent sulfite liquors, is reducing the BOD of a number of waste streams from the pulping operation. Torula yeast is a by-product used as a food ingredient for a wide range of consumers ranging from babies to cattle. Ethyl alcohol is another by-product, but it usually requires government subsidy since the economics are not favorable. Lignosulfonates are used in both binders and dispersants—end products with highly differing properties. The binders are used in such things as animal feed pelletizing, linoleum paste, and wood product adhesives; the dispersants are of importance in oil well drilling muds and numerous other applications. The lignosulfonates reduce the need for water in preparing ready-mix concrete and increase concrete strength and durability. They serve as chelating agents, can be modified for agricultural use, and are the starting point in the production of a wide variety of chemical compounds.

To date most of these uses are specific, and markets of even the most successful are limited. For example, vanillin, the flavoring compound widely used in place of natural vanilla extract, is now produced from spent cooking liquor, and this source has largely replaced all others. Two mills, however, supply the entire market demand and bid fair to do so for the foreseeable future. Chemical by-product production is highly diversified, and the search continues for better uses for the other half of the log. By-products of the kraft pulping process are turpentine and tall oil, which do not enter the waste picture.

Control of aesthetic pollution *The color problem* resulting from the production and bleaching of kraft pulp still lacks a solution that is both technically and practically applicable. The massive lime treatment developed by the National Council for

Air and Stream Improvement is technically effective, but to date the economic problem of practical application to existing installations has not been solved. Clarification and biochemical treatment of mill wastes have relatively little effect upon color of effluent, seldom reducing it by more than 20 to 25 percent. Good brown stock washing practices along with adequate evaporator and recovery boiler capacities can reduce black liquor losses and thus reduce color pollution to a minimum. The degree of dilution in the waterway, however, is what usually determines whether or not the discoloration is objectionable. The hunt for solutions to this color problem is an extremely active area of pollution abatement research.

Foam from a pulp or paper mill can be almost completely controlled at the outfall by proper outfall design or by the use of foam fences such as log booms or canvas curtains, if necessary supplemented with knockdown sprays. Regeneration of foam in a waterway, however, can be a serious problem. The foam regeneration rate is determined by the nature of the effluent, the stream turbulence, and the dilution afforded by the waterway. Foam regeneration is a problem specific to each location and must be carefully considered in selecting a new plant site. In a stream where foam regeneration occurs, the mill can do little to control it, since even the most complete clarification and bio-oxidation process have only a minor effect upon foam regenerating properties.

White streaks appearing in streams below paper mills are best eliminated by dilution and dispersion, through diffuser pipes which release effluents at multiple points in the stream. The particles involved are usually so fine that they defy separation through flotation, sedimentation, or other conventional methods. These fines do not form sludge banks. The clay is of essentially the same composition as that found in the river, and extensive bioassays indicate that titanium dioxide is harmless to fish even at many times the concentration normally found below paper mills.

Elimination of pollutants toxic to aquatic life Fish kills traceable to pulp and paper mills have almost always been due to temporary discharges abnormally high in BOD. Oxygen depletion and resultant fish suffocation, not the release of any poisonous materials, have caused the kills. A relatively few preventive measures will keep down the discharge of materials actually toxic to aquatic life. In the kraft process, fatty acids and resin acids may become a problem if losses are excessive from the soap, tall oil, or black liquor system. Usually, however, a mill has no difficulty in meeting even such tight standards as 1 ppm fatty acids and 0.2 ppm resin acids in the receiving water after mixing with mill effluent. In the paper manufacturing process, zinc hydrosulfite is often used for bleaching groundwood. While it is a well-documented fact that zinc toxicity decreases as water hardness increases, published data on toxic levels are quite contradictory. For this reason and until the toxic level is definitely established, the paper mills have held to less than 0.05 ppm and thus have kept on the safe side even for spawning waters. Clarification has almost no effect upon zinc concentration, and bio-oxidation usually reduces it by only 10 percent.

Some of the new conductive coatings exhibit strong toxic properties which can be greatly reduced by the addition of preparations which act as toxic neutralizers. Bio-oxidation also reduces the toxic effect where the material can be introduced into that treatment process at a rate low enough not to interfere with normal biological activity.

Progress in Abatement of Water Pollution

Studies made by the National Council for Air and Stream Improvement have revealed that in the most recent period of years during which the production of pulp and paper products more than doubled, this industry's total waste discharge to surface waters was actually reduced. This has been accomplished both through increased emphasis on pollution abatement and through major technological advances.

New pulp mills now going into production in both the United States and Canada are almost all kraft mills, and these produce only one-tenth as much BOD per ton of pulp manufactured as do the older sulfite mills which they supplement or supplant (see Table 2). Water recirculation and solids recovery is considered to be an integral part of the fundamental design of new mills, whether pulp or paper. New abatement facilities are coming on line almost monthly at old mills to reduce their pollution.

Thus pollution abatement is now very definitely a cost of doing business. Usually it is a cost upon which there is little return on the money invested, since in the majority of cases the industry cannot use the materials removed. Capital costs of abatement facilities for new mills are averaging from 2 to 4 percent of the total capital investment. Yearly cost of operating these facilities is in the neighborhood of 15 percent of the original capital cost for the average installation. For a highly sophisticated installation, such as an activated sludge plant with solids disposal facilities, this figure may approach 30 percent.

Pollution abatement cannot be accomplished overnight, but it can and is being accomplished. At least 3 years, and more often 5, is required from the time that an idea is conceived in the laboratory until it can be carried through the pilot plant, full-scale design, construction, and shake-down stages, and eventuate in an operating unit. Stream improvement gains made by the paper industry during the last 10 years, therefore, are the result of long-continued effort and concern. Ten years ago the pollution problem was more serious than the general public knew or cared. Today, public awareness is acute, but extensive as the problem still may be, the tide has been turned in the pulp and paper industry.

ACKNOWLEDGMENTS

To the National Council of the Paper Industry for Air and Stream Improvement, the Pulp Manufacturers Research League, and Arthur Van Vlissingen.

AIR POLLUTION

Nature of Pollution

Most air pollution problems, specific for the pulp and paper industry, are associated with the production of kraft pulp and, to a lesser degree, with sulfite pulp. The emissions of the paper mills cause far fewer complaints.

Kraft pulping Wood chips, cooked in batch or continuous digesters with a solution of sodium hydroxide and sodium sulfide, produce volatile constituents, inerts, and water vapor. After these vapors are passed through a condensing and decanting system, separate streams of turpentine, condensate, and noncondensables are obtained (see Fig. 4).

The black liquor, washed out of the pulp, passes through a multieffect evaporator and a direct contact evaporator. The concentrated black liquor is sprayed into a recovery furnace. The recovery furnace flue gases flow through the direct contact evaporator and dust collecting equipment and then to the stack.

Molten salts, withdrawn from the bottom of the recovery furnace, are dissolved in a smelt tank. Vapors and entrainment arise from the boiling solution. The solution from the smelt tank, reacted with slaked lime, provides new liquor for reuse in subsequent cooks. Kilns are used in the lime recovery loop to prepare the spent lime for recycling. Flue gases from the lime kiln pass through a dust collector and then to the stack.

Sulfite pulping The sulfite process involves a sequence that is similar to kraft when recovery of chemicals is practiced. However, some of the older sulfite mills, still operating, were built prior to the time when recovery of the spent sulfite cooking chemicals was possible. The significant gaseous escapement during sulfite pulping is sulfur dioxide. Contrary to the kraft process, few objectionable volatile organic sulfur compounds are formed during the pulping.

Effects of Pollution and Abatement Objectives

Kraft mill odors have resulted in many complaints. The data indicated in Table 4 generally correspond to an intensity level of "barely perceptible" as far as the sulfur compounds are concerned. The reported threshold values vary considerably.[1]

Sulfur compounds, present in pulp mill emissions, can be detected by the human nose at concentration levels of less than 5 ppb. If the pulp mill is to operate without detectable odor, ground level concentrations will have to be maintained so as not to exceed that low level of concentration.

Fig. 4 Kraft pulp mill.

Sulfur dioxide in sufficiently high concentrations can cause corrosion, damage to metals, and leaf damage to vegetation. Some regulations indicate maximum allowable ground level concentrations of 0.2 ppm for an averaging time of 24 hr and higher concentrations over shorter periods of time. More recent proposed ordinances list sulfur dioxide concentrations of 0.05 ppm.

TABLE 4 Odor Threshold Values

Substance	*Volumes per billion volumes of air, ppb*
Hydrogen sulfide	20
Methyl mercaptan	40
Ethyl mercaptan	3
Methyl sulfide	15
Ethyl sulfide	3
Butyric acid	0.6

The plumes and particulates from both kraft and sulfite mills can result in corrosion and a loss of visibility. The ordinances in force now differ considerably. However, it is expected that before long new mills will be required to maintain the particulate losses at less than 0.05 grain/scf.

The removal of the particulates reduces the plume significantly. It seems unnec-

[1] H. F. Droege, APCA 60th Annual Meeting, 1967.

essary to establish a maximum allowable concentration in respect to water vapor, a natural component of the atmosphere.

Analysis and Monitoring

Analytical methods In pulp mill air pollution work, the rates of emission can vary substantially, a 10 percent variation being insignificant and a 50 percent range not being unusual. Most of the methods used by mill laboratories are adequate to compare abatement performances on a day-to-day basis. Nevertheless, it is advantageous to arrive at more uniformity for performance comparisons at the various mills. The National Council for Stream Improvement is taking an active role in carrying out investigations and advising on standard methods of analysis. These methods often originated in the research and mill laboratories of various companies.

The odorous sulfur compounds have presented the most difficulties. However, portable instrumentation with a short setup time has recently become available.[1] The system is based on techniques very sensitive to sulfur compounds. The instrument is built around a Mikro titration cell using electrolytically generated bromine. The response to volatile sulfur compounds approaches the odor threshold values. For example, concentrations of 10 ppb can be determined in the case of hydrogen sulfide and methyl mercaptan. However, the instrument has been used primarily in abatement work involving concentrations ranging from 0.1 to 100,000 ppm.

A gas chromatograph is recommended for laboratory investigations involving the volatile sulfur compounds and other organic compounds. For example, an investigation of the ambient air revealed, in addition to the various sulfur compounds, trimethylamine, methanol, α pinene, camphene, limonene, and butyric acid.

Monitoring of the ambient air requires an adequate network of samplers around the mill. A sampling method based on the tarnishing of silver membrane filters has recently been described by Falgout and Harding.[2] The method is sensitive and not time-consuming. Air is metered first through a cellulose filter and through the filter membrane. The change in reflectance is a measure of the silver sulfide formed by reaction of the silver and the hydrogen sulfide in the air.

Subjective odor tests are generally not suitable as a developmental or control method. However, the subjective odor tests should be included occasionally to confirm an improvement. An ASTM method expressing the odor strength as a dilution factor (odor units) is available.

Plant instrumentation Consistently good performance regarding pollution abatement with respect to recovery furnace and lime kiln flue gases requires continuous monitoring instrumentation. It is advantageous for most companies to use commercially available instrumentation of the same type in each of their plants in order to reduce the servicing costs and obtain comparable data.

Recording instrumentation for residual oxygen and combustibles has been used for years. Walther and Amberg have described a gas chromatograph to monitor odorous sulfur compounds.[3] Sensitive instrumentation, based on the electrolytical bromine technique, is now being used to record the wide range of concentrations of the odorous sulfur compounds normally encountered in the flue gases. This information is required to guide the operators.

A continuous recording in respect to particulate emissions is also necessary for maintaining the operation at an optimum level. The light from a fixed spot lamp passes through the gas. Depending on the dust load, more or less of the light energy is scattered before reaching the detecting element in the so-called "bolometer."

Abatement of Odor, Particulates, Other Pollutants

Recovery and control in kraft mills The reaction of the pulping chemicals, in particular sodium sulfide, with the methoxy groups of the lignin molecules consti-

[1] G. N. Thoen, G. G. DeHaas, and R. R. Austin, 21st Alkaline Pulping Conference, 1967.

[2] *Journal of the Air Pollution Control Association,* vol. 18, no. 1, p. 15, 1968.

[3] *TAPPI,* vol. 50, no. 10, p. 108A, 1967.

tutes the major source of the odorous organic sulfur compounds.[1] The reactions may be presented as follows:

$$2\text{ lignin} - OCH_3 + S^{=} \rightarrow 2\text{ lignin} - O^- + CH_3SCH_3$$
$$\text{Lignin} - OCH_3 + SH^- \rightarrow \text{lignin} - O^- + CH_3SH$$
$$2CH_3SH + O \rightarrow CH_3SSCH_3 + H_2O$$

Although the rate at which these compounds are formed can be affected by the pulping process, this technique is severely limited because of various operating requirements.

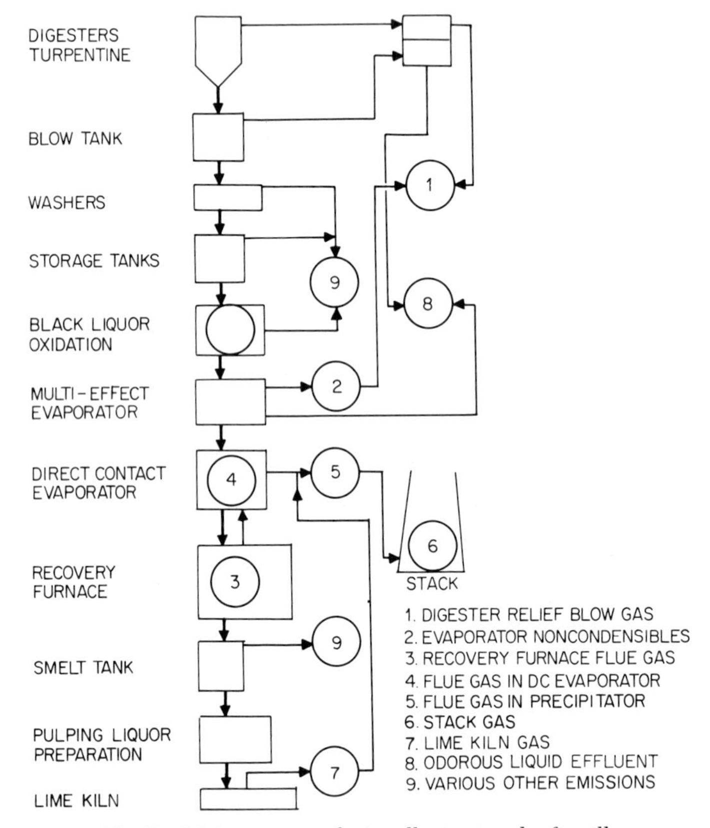

Fig. 5 Major sources of air pollution in a kraft mill.

The other major source of odor, hydrogen sulfide, originates from the reaction of sodium sulfide and carbon dioxide.

$$Na_2S + CO_2 + H_2O \rightarrow NaHCO_3 + NaHS$$
$$2NaHS + CO_2 + H_2O \rightarrow Na_2CO_3 + 2H_2S$$

The rate of release of hydrogen sulfide can be reduced during some of the operations in the mill by raising the pH of the black liquor.

The major sources of air pollution emissions from a kraft mill are shown in Fig. 5.

Digester Relief and Blow Gases. Methyl mercaptan, methyl sulfide, methyl disulfide, and some hydrogen sulfide are released with the gases and condensates.

[1] I. B. Douglas and L. Price, *TAPPI*, vol. 49, no. 8, p. 335, 1966.

Safety is a major consideration in the design of any large-scale digester gas handling installation. The limits of inflammability are indicated in Table 5. It has been determined that the digester relief gases require more than 50 volumes of air to render them nonexplosive.[1]

The concentrations found for volatile sulfur compounds are dependent upon the method of venting the digester, the degree of cooling of the vapors, the cooking conditions, and the types of wood used. The concentration ranges from 1 to 15 percent for sulfur compounds and two to three times that much for total volatile organics and hydrogen sulfide.

Although the average flow rates of noncondensables vary from 20 to 120 cu ft/ton of ovendry wood, peak flows of more than 5,000 cfm can occur during the blow of batch digesters. A floating cover, or diaphragm type of gasholder, appears to be the most satisfactory equipment to contain these peak flows (see Fig. 6).

Fig. 6 Gasholder for digester relief and blow gas.

In some large-scale installations the relief and blow gases, containing no oxygen, are pulled from the gasholder through a de-entrainer to eliminate the turpentine mist. It then passes through a flame arrester followed by in-line mixing with a flow of air that exceeds 100 times the flow of gas in order to dilute the gas below the limits of inflammability. Finally, the diluted gas enters the lime kiln for complete combustion, as indicated in Fig. 7. Installations using this system have been in operation without a single explosion for more than 10 years.

The gasholder and combustion system require an investment of approximately $100,000. The costs are lower for a system using continuous digesters because a gasholder is not required.

Evaporator Noncondensables. The noncondensables from the multieffect evaporator are, as in the case of the digester relief gas, relatively concentrated.

[1] G. M. Ginodman, *Chemical Abstracts*, vol. 42, p. 6111, 1948.
[2] A. A. Coleman, *TAPPI*, vol. 41, no. 10, p. 166A, 1958.

Some mills react the evaporator noncondensables, consisting mostly of hydrogen sulfide, with white liquor to obtain noticeable increases in the sulfidity, especially if unoxidized black liquor is evaporated. The residual gas is piped to the digester gas combustion system.

Recovery Furnance Flue Gas. Flue gas from kraft recovery furnaces can contain as much as 200 ppm of volatile sulfur compounds, mostly hydrogen sulfide, and 20,000

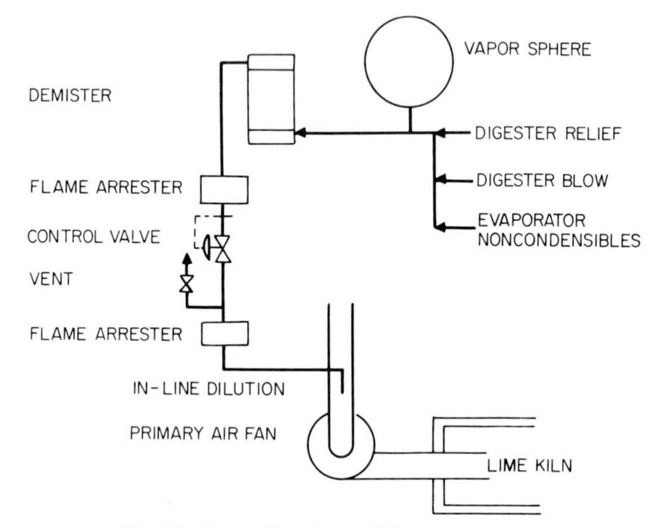

Fig. 7 Gas collection and burning system.

ppm total combustibles, of which about one-third consists of carbon monoxide. The flue gas, including the water vapor, amounts to approximately 15,000 lb/ton of oven-dry wood, assuming a 50 percent pulp yield.

Investigations involving many recovery furnaces have demonstrated that, in newer units, the odorous sulfur compounds can be reduced to less than 1 ppm. The results in Table 6 and Fig. 8 indicated that sufficient secondary air should be introduced into the furnace to maintain the residual oxygen at about 2 percent. In addition, sufficient turbulence in the fire zone is required in order to prevent the existence of "pockets" containing no oxygen.

TABLE 5 Limits of Inflammability in Air

Substance	Lower %	Upper %
CH_3OH	6.7	36.5
H_2S	4.3	45.5
CH_3SH	3.9	21.8
CH_3SCH_3	2.8	
Turpentine	0.8	
Digester relief gases	2.0	

It has been demonstrated that the SO_2 concentrations are determined primarily by the droplet size of the liquor sprayed into the furnace. The liquor temperature as well as the type of nozzle determine the diameter of the droplets. The larger droplets will fall into the drying and reducing zone and will create the least amount of

SO_2 contained in the flue gases leaving the furnace. The SO_2 concentrations can be kept below the 10 ppm range, which is insignificant from the air pollution point of view.

Although black liquor oxidation has a beneficial effect on the performance of

TABLE 6 Recovery Furnace Flue Gases

Oxygen in flue gas, %	Secondary air per total air, %	Sulfur compounds in flue gas				
		H_2S, ppm	RSH, ppm	RSR, ppm	RSSR, ppm	SO_2, ppm
0.5	35	135	20	...	20	
2.5	35	0.5	0.5	0.5	0.1	0.5
2.5	30	12	0.8	...	0.9	12
2.5	25	25	1	2	1.2	

direct contact evaporators, it is of little consequence as far as the recovery furnace is concerned.

Some furnaces can be operated at 20 percent or more over designed firing rate; however, overloading of the furnace, resulting in incomplete combustion, can contribute significantly to the emission of reduced sulfur compounds.

Adequate instrumentation to guide the operators and supervisory personnel in

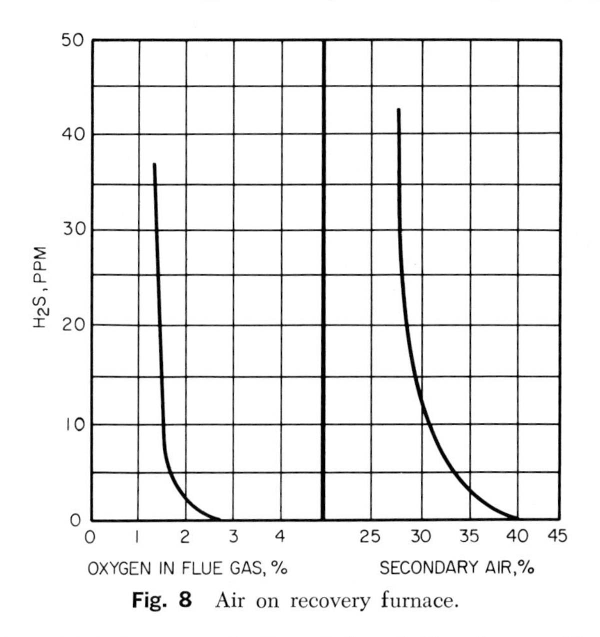

Fig. 8 Air on recovery furnace.

their efforts to maintain low concentrations of the volatile odorous sulfur compounds is necessary. The costs for oxygen and hydrogen sulfide recorders amount to less than a total of $15,000.

Flue Gas in Direct Contact Evaporators. As much as 200 ppm of hydrogen sulfide can be released to the flue gas by reaction of the carbon dioxide with the sodium sulfide in the black liquor.

Although raising the pH of the black liquor results in a noticeable reduction of the hydrogen sulfide released, the most effective means of abatement is either the reduction of the sodium sulfide concentration to less than 0.2g/1 by oxidation, or the elimination of the direct contact evaporator.

BLACK LIQUOR OXIDATION. Some kraft mills have operated a tower type black liquor oxidation system for more than a decade. The dilute black liquor is passed concurrently with air through two towers in series, packed with point welded corrugated stainless steel sheets. The rate of oxidation amounts to 15 to 20 lb Na_2S/(1,000 sq ft)(hr) at an airflow ratio of about 5 to 8 O_2/Na_2S. Another type of tower, using a packing of plates with diamond shaped imprints for better liquor distribution, achieves a somewhat higher rate of oxidation per square foot of packing.[1]

The strong foaming tendencies of black liquor have prevented the use of oxidation in many mills. As an example, the pulping of pine results in more foaming difficulties than the pulping of fir. The foaming can be reduced by concentrating the liquor first.

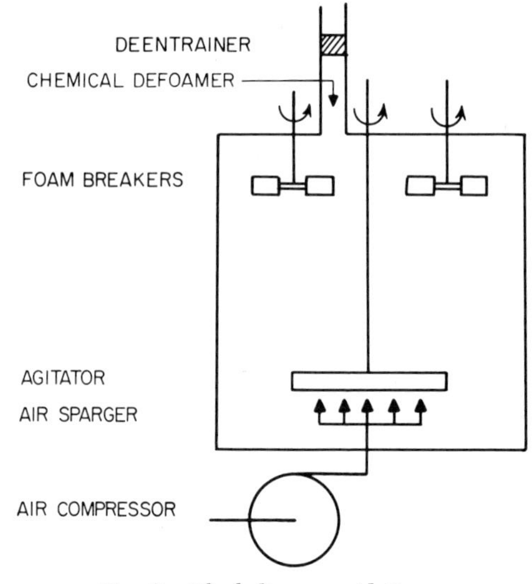

Fig. 9 Black liquor oxidation.

The newer systems, capable of handling either weak or strong black liquors, introduce the air under pressure through a diffuser below the surface of the black liquor in a tank. The air is distributed through perforated plates, or through an air sparger below an agitator, as indicated in Fig. 9. The liquor retention time ranges from 10 min to 2 hr, depending on the degree of air distribution and the Na_2S concentration. The horsepower requirements are higher than in the case of the tower system. Most installations are equipped with mechanical defoamers, and sometimes a few use antifoam.

The air leaving the black liquor oxidation system, containing 5 to 100 ppm of sulfur compounds, should be piped to the air intake of the recovery furnace or lime kiln.

Black liquor oxidation helps to contain more of the sulfur in the mill, as indicated in Table 7. Generally, as is the case with all improvements of the recovery system in respect to sulfur, an adjustment of the makeup ratio should be made in order to

[1] E. T. Guest, *Pulp and Paper Magazine of Canada,* vol. 66, no. 12, p. T-617, 1965.

prevent raising the sulfidity and the overall sulfur losses correspondingly. The cost of a black liquor oxidation system for a 500 ton/day mill amounts to about $150,000.

ELIMINATION OF DIRECT CONTACT EVAPORATORS. Although black liquor oxidation reduces the losses of volatile sulfur compounds to a low level, it does not prevent the escape of significant quantities of organic material in the flue gas during the direct contact evaporation. These compounds are similar in nature to the organic materials found in the condensate from the multieffect evaporator section containing the more concentrated liquor.

TABLE 7 Effect of Black Liquor Oxidation

Black liquor	Chemicals makeup		Mole ratio in makeup, S/Na	White liquor sulfidity, %
	Sulfur content, lb/ton pulp	Sodium content, lb/ton pulp		
Oxidized	18	34	0.38	27
Unoxidized	26	36	0.52	21

Most Swedish mills eliminate the need for direct contact evaporators by going to higher concentrations in the multieffect evaporators and by providing additional heat exchange capacity in the flue gas stream.

A new mill in the United States is installing an air contact evaporator. As indicated in Fig. 10, the combustion air passes through a direct contact evaporator and then to the recovery furnace. It has been demonstrated that the unoxidized liquor from the air cascade can be burned to a degree of completion corresponding to less than 1 ppm of odorous sulfur compounds. Although this system utilizes contact evaporators, there is no direct contact between the flue gas and the black liquor, and consequently this source of air pollution would be eliminated.

Flue Gas in Precipitators and Scrubbers. The gas leaving the direct contact evaporator contains 50 to 100 lb of fly ash/ton of ovendry wood. Generally, more than 50 percent consists of sodium sulfate with the remainder carbonate, sulfite, and other sodium compounds. Sodium chloride, when introduced with saltwater-borne logs, tends to accumulate preferentially in the fly ash. Sometimes unburned carbon particles will turn the fly ash gray. The bulk of the particles are in the 0.1- to 2-micron range.

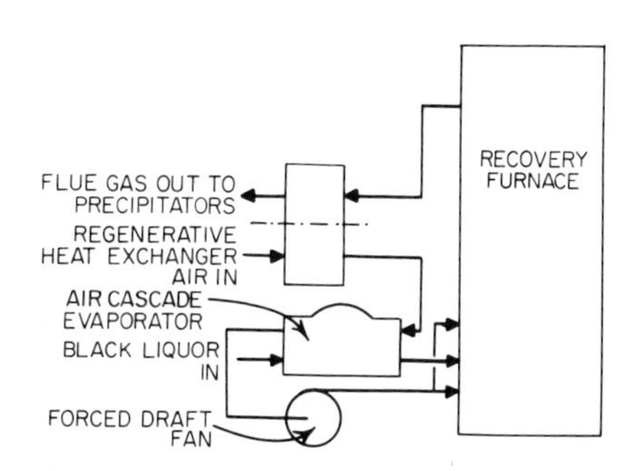

Fig. 10 Air cascade recovery system.

ELECTROSTATIC PRECIPITATORS. The dust particles in the flue gas are first exposed to electron emission from the discharge electrode. The particles move then to the collecting electrode. Periodic rapping of the electrodes shakes the dust down into the hopper bottom. The accumulated fly ash is either picked up by black liquor in a so-called wet bottom or removed by screw conveyors. The precipitators are well insulated to prevent condensation of moisture in the flue gases which could result in corrosion.

Almost all the kraft mills operate the precipitators at 250 to 350°F, and a few at 600 to 750°F. These temperature ranges give the fastest discharge.

Newer kraft mills are now being equipped with precipitators with rated efficiencies of 99+ percent. This should result in a residual dust loading of 0.03 to 0.04

grain/scf. The cost of such electrostatic precipitators, for a 500 ton/day recovery unit, is approximately $450,000.

VENTURI RECOVERY UNIT. Most venturi recovery units either present a higher pressure drop for the flue gas or use large quantities of water. The operating costs are higher than for electrostatic precipitators. However, the treatment with water or a solution allows the simultaneous recovery of some of the volatiles and additional heat.

One proposal involves a venturi-type direct contact evaporator using concentrated black liquor, followed by a venturi scrubber operated with dilute liquor, and then a separator to recover the entrainment. Finally, the gas passes through a cooling tower to produce hot water.[1] A good black liquor oxidation system is necessary in this case.

Older mills with overloaded or low-efficiency precipitators sometime install a venturi scrubber in series with the electrostatic precipitator. Pulp mills that have a salt problem will find that the sodium chloride tends to accumulate in the secondary scrubbing system.

Some venturi recovery units use solutions containing caustic, black liquor, chlorine, and other oxidizing agents to trap some of the volatile pollutants. Sequestered iron compounds are also suitable to oxidize hydrogen sulfide. However, none of these chemical treatments have been as effective as good combustion in the furnace.

Many different types of scrubbers have been proposed and tried. Most commercial units installed are of the venturi type. The costs vary considerably from $100,000 to $300,000 for a 500 ton/day pulp mill.

Stack Gas. The degree of dilution obtained with a well-designed stack is seldom less than 100 times and often considerably more than 1,000 times. This reduces a tolerable stack exit concentration of 0.5 ppm of odorous compounds to the objective of less than 5 ppb at ground level. As a rule, the hotter the plume, the higher the plume rises into the air. Superior dilution effects take place with the use of higher stacks located so as to be least affected by surrounding high terraces or buildings. Kraft mill stacks have been described by Cancea.[2]

Lime Kiln Flue Gas. The gas leaving the lime kiln can contain more than 100 ppm of hydrogen sulfide in addition to particulates consisting mostly of lime dust and sodium salts. Thorough washing of the lime mud and operation of the lime kiln with at least 2 percent residual oxygen effectively eliminates hydrogen sulfide.

Most modern lime kiln scrubbers obtain 99 percent removal of lime dust. Some, especially those equipped with venturi scrubbers, also achieve a high degree of removal of sodium salts. Piping of the lime kiln flue gas to the venturi recovery scrubber of the recovery furnace, if one is installed, could prove to be helpful.

Odorous Liquid Effluents. Untreated digester and evaporator condensates released to a river can be noticeable for several miles below the outfall. Lindberg described condensate steam stripping in a Swedish plant.[3] The use of air as a stripping medium is less expensive but not as effective. Chlorination of the stripped effluent is advantageous. The effluent air should be treated by thermal oxidation, preferably in the lime kiln or recovery furnace. The prevention of explosions has to be considered in the design of this system. As indicated earlier, turpentine in particular has lower limits of inflammability that extend to less than 1 percent.

Part of the condensate can be eliminated by reuse in the liquor making and pulp washing systems, or in some locations, it can be used for irrigation purposes. However, in most cases, reuse without treatment is not advisable since a major part of the volatiles are then lost through various vents.

Aerated stabilization ponds are effective in eliminating the odorous sulfur compounds. A slight, musty odor remains, but this odor does not carry very far. Anaerobic conditions, created either by overloading or by insufficient aeration, result in the

[1] I. S. Shah, *Chemical Engineering (New York)*, vol. 74, no. 7, p. 84, 1967.

[2] C. L. Cancea, *Pulp and Paper Magazine of Canada*, vol. 65, no. 7, p. T-309, 1964.

[3] S. Lindberg, *Svensk Papperstidning*, vol. 69, no. 15, 484, 1966.

development of significant quantities of odorous sulfur compounds and butyric acid.

Various Other Sources of Kraft Mill Pollution. The smelt tank requires a de-entrainer to recover droplets that are carried along in the vapor. The entrainment losses often amount to several pounds per ton of pulp produced. Periodically, the scale of calcium salts has to be removed by cleaning with hydrochloric acid. A Teflon mesh pad de-entrainer can be used. One mill has a short tower packed with 3-in. carbon rings. The vapors are passed countercurrently to weak pulping liquor.

The gas from tanks, roof vents, knotters, brown stock washers, and various other vents is minor but should be collected, condensed, and piped to the air intake of the recovery furnace or lime kiln.

TABLE 8 Kraft Mill Air Pollution Abatement

Source	Major air pollutants	Abatement procedure
Digesters....	Methyl mercaptan, methyl sulfide, methyl disulfide, turpenes	Collection of digester relief and blow gases, combustion in lime kiln
Multieffect evaporators	Hydrogen sulfide, methyl mercaptan, methyl disulfide, methyl sulfide	Scrubbing of the noncondensables with alkaline pulping liquor, combustion of remainder in lime kiln
Recovery furnace	Sodium sulfide, sodium carbonate, hydrogen sulfide, sulfur dioxide, various organic compounds, carbon monoxide	Adequate instrumentation to record volatile sulfur compounds and oxygen, sufficient total air to maintain 2–3% oxygen, sufficient secondary to primary air ratio, right size nozzle, air ports designed to maintain good turbulence, no excessive overload
Direct contact evaporator	Hydrogen sulfide, various organic compounds	Black liquor oxidation to reduce sulfide content to 0.2 g/l or less, or eliminate direct contact of flue gas and black liquor completely
Fly ash recovery	....	High-efficiency electrostatic precipitator or a combination of an electrostatic precipitator followed by a venturi recovery unit
Stack.......	....	The selection of the location and design to include the recommendations of a meteorologist
Lime kiln	Calcium carbonate, sodium salts, hydrogen sulfide	Adequate instrumentation to record volatile sulfur compounds and oxygen, efficient lime mud washer, venturi scrubber
Aqueous effluents	Hydrogen sulfide, methyl mercaptan, methyl sulfide, methyl disulfide, various volatile organics	Steam stripping and combustion of distillates or primary-secondary aerobic treatment, or irrigation
Smelt tank...	Sodium sulfate, sodium carbonate	De-entrainer
Other sources	Various pollutants	Collecting, cooling, and burning of various vent gases in recovery furnace, efficient incinerators

Systematic searches with the portable analyzer, described earlier, will help to find the remaining minor sources and maintain the continuity of the odor abatement program.

Condensation of the water vapor in the plume can result in an improvement in visibility.

Containment in sulfite pulp mills Sulfite pulp mills are subjected to fewer complaints than kraft mills.

A "tight" mill requires a good collection system for all the vent and flue gases. Conventional scrubbers or packed towers can be used to absorb the SO_2 effectively in a solution containing the base used in the pulping liquor. It is recommended that provisions be made for automatic control to regulate the flow of base according to the surges of various vent gases that are relieved into the scrubber system. A Teflon mesh de-entrainer should be provided to recover the carryover and sulfuric acid mist.

Elimination of other sources. Sulfur dioxide contained in the flue gas of power boilers using fuels with a high sulfuric content has been suggested as an economical source of makeup chemicals by Galeano and Harding.[1]

Pulp and paper mills sometimes have to handle substantial quantities of waste material, for example, bark from the wood chipping department and sludge from the water pollution abatement section. If land disposal is not possible, well-designed incinerators should be provided.

Conclusions

One of the newer mills, located in a sensitive resort area, has been operating without serious complaints. It is anticipated that some of the kraft mills now under construction will operate so as to be almost free from complaints. Table 8 lists the abatement measures that are recommended for kraft mills, the major source of air pollution in the pulp and paper industry.

[1] S. F. Galeano and C. I. Harding, *Journal of the Air Pollution Control Association,* vol. 17, no. 8, p. 536, 1967.

Chapter **19**

Pollution Control in the Aerospace and Electronics Industries

ROBERT G. REICHMANN

Plant Engineer, Electronics Division,
Aerojet-General Corporation,
Azusa, California

THE INDUSTRY'S PROBLEM

The items produced by the aerospace and electronics industry are as diversified as the processes that are required to complete production. A single company may produce at one location all or part of an airplane, liquid or solid rocket systems, biological life support systems, advanced weapon systems, and atomic energy systems as well as commercial products. Because of this diversification and the rapid rate at which the technology is advancing, many new techniques and processes are being developed. This process development must include adequate controls for waste products. Though most plants in this field employ many processes similar to those used in other industries, certain special situations continually consume their time and energy. They must contend with "military specs" and their limitations. Since they are in the public "glamour" limelight both locally and nationally, they must present a stronger public image than most other industries. And they tend to deal with relatively new, exotic metals, some of which are significantly tough pollution problems. Testing of solid rocket fuels, machining of beryllium or other rare metals, and plating rinses contaminated with rare metal particles are only a few of the pollution problem areas being challenged by the aerospace and electronics industry.

COMMUNITY CONTROLS AND CODES

In each community there will be found various control agencies responsible for the safekeeping of our environment. The names and exact charters may differ from area to area, as these regulations are city, county, and state controlled. Some duplication in each area may be expected. Table 1 lists some of the types of agencies involved and their probable areas of interest.

TABLE 1 Pollution Control Agencies

Agency	*Areas of Interest*
Sanitation districts	Uncontaminated waste, contaminated waste, domestic sewerage
Building departments	Liquid waste systems, water treatment facilities, process equipment, ventilation
Fire codes	Flammables, explosives fire sprinkler systems
Air pollution districts	Vapor emission, dust control, process equipment, smoke
Health	Odors, toxic vapors and dusts, industrial water, domestic water, waste systems
Safety	Explosive conditions and hazards, unsafe environments, toxicity, heat, pressure, ventilation
Water and flood control districts	Groundwater, storm drain systems, cooling waters, uncontaminated liquid wastes, contaminated liquid wastes, leach beds, settling beds, treatment plants

When air and water pollution limits are defined in succeeding sections, they will represent actual limits imposed by one or more communities. Air pollution regulations have been largely taken from the Los Angeles County air pollution rules, as they generally represent the strictest laws over the broadest fields.

FEDERAL CONTROLS AND CODES

In addition to local regulations that may define effluent restrictions, personnel and equipment, safety codes, or facility requirements, the aerospace companies may be governed by federal regulations of product acceptance or specific process requirements. The majority of government contracts issued to aerospace companies contain a certain number of process requirements. These are referred to as Military Specifications, or Mil Specs. These Mil Specs call out specific materials which can be used during a process. The exact ingredients of the materials are, in many cases, also spelled out in the Mil Specs. Mil Specs cover a wide range of materials, including such items as paints, cleaning solvents, lubricating oils, and protective coatings. Years of testing and research go into these materials. They have been tried and tested, and their general protection abilities are usually well known.

How are Mil Spec materials involved in air pollution? The Mil Spec materials specified on a contract may contain photochemically reactive solvents which are restricted as to the amount that may be released to the atmosphere. A typical example of Mil Spec materials that could be used to process a piece of metal is:

Process	Mil Spec	Material
Cleaning.	. . . .	Trichloroethylene
Priming.	Mil P-8585	Zinc chromate
Topcoat or finish.	Mil L-19537	Lacquer acrylic nitrocellulose

In certain parts of the country these Mil Spec materials violate solvent emission regulations. If the solvent emissions cannot be controlled within the limits prescribed by local laws, then the process would be in violation and would be stopped.

Substitution is not always possible unless sufficient testing has been accomplished to prove the durability of the new product.

Conflict at the Community Level

What happens when a community adopts local regulations that restrict their use of Mil Spec materials? This was the case in 1966 when the Los Angeles Air Pollution Control District adopted an air pollution regulation restricting the amounts and types of organic solvents that could be emitted into the atmosphere from any single source. By a unique cooperative effort between local industry and the Air Pollution Control District, a study of reactive vs. nonreactive solvents was initiated. Based on laboratory tests, a family of solvents was found that would contribute little or nothing to the formation of photochemical smog. The regulation of solvent emissions, known as Rule 66, limits to 40 lb/24 hr the amount of nonexempt (reactive) solvent that could be emitted from a single piece of equipment. Any piece of equipment exceeding that amount would require control equipment to reduce the solvent emission by 85 percent overall efficiency or not to exceed 40 lb/day. Approximately 18 months for complete compliance was allowed industry. A cooperative program involving the federal government and industry has been establish to review and reformulate Mil Specs in conflict with Rule 66.

The aerospace industries have embarked on a program to find nonphotochemically reactive solvents that could be substituted in Mil Spec materials and still give a satisfactory end product. At the request of the aerospace companies, local chemical manufacturers have been reformulating Mil Spec materials. When the chemical manufacturers have formulated what they feel is an adequate substitute, the materials are turned over to the aerospace company for further testing. These new materials are then tested against the Mil Spec requirement. When the test proves, based on short-term test, that they will perform satisfactorily, the test results and samples of the materials are forwarded to the appropriate government agency for approval. Federal government approval now allows the companies to continue to operate by conforming to local emission laws and Mil Specs required by the contract.

In general Mil Spec materials used in manufacturing processes are unaffected by local rules and regulations concerning water pollution. Those that pose pollution problems usually are purchased and handled as a liquid, and are already self-contained. The manner in which they may be disposed of depends on their nature and the local restrictions on water pollution. Recommendations for control of water pollution wastes restricted by local water pollution regulations will be covered in another section.

DUST CONTROL

Although dust particles contribute nothing to photochemical smog, they can produce an irritating visibility, health, and potential explosive hazard. State, county, or city safety laws, the National Electrical Code, local fire codes, and state health codes will regulate the level of control and the minimum requirements for the equipment. General industrial safety orders will set maximum limits for dust concentrations for the workers' environment.

Safe Working Levels

Typical safe working levels are listed in Table 2. These levels were designed as guidelines for continuous operations—100 ppm for 1 hr cannot be substituted for

TABLE 2 Dust Limits

Mineral Dusts	*Million Particles per cu ft*
Total dust	50
Asbestos	5
Silica (free or uncombined):	
Less than 5 %	50
5–25 %	20
25–50 %	10
Over 50 %	5
Talc (commercial or industrial)	20
Talc (chemically pure steatite)	50

Metallic Dusts, Fumes, and Vapors	*Mg/cu meter*
Antimony	0.5
Arsenic and compounds (except volatile hydrides)	0.5
Barium	0.5
Cadmium	0.1
Chromium, chromic acid, and chromates (as CrO_3)	0.1
Lead	0.15
Manganese	0.0
Mercury	0.1
Selenium	0.1
Tellurium	0.1
Zinc oxide	15.0
Total metal fumes	30.0

NOTE: The maximum acceptable concentrations for dusts contemplate the estimate of dust exposures by the method recognized by the U.S. Public Health Service.

800 ppm for 8 hr. Several typical operations that produce dust particles are sanding, grinding, sandblasting, and pulverizing of plastics, metals, or propellants. Each operation must be evaluated to determine the potential hazard of fire, toxicity, and/or explosion. Local operator and personnel comfort may require dust collection at a concentration below that established for health or safety. Once the dust particles have been removed from the work area, the exhaust from the collection equipment may be the next problem. Air pollution regulations have been adopted in some communities to limit the maximum particulate matter discharged from a single source or piece of collection equipment to not more than 0.4 grain/cu ft and a maximum allowable dust discharge rate based on the process weight per hour. Table 3 shows a typical process weight with corresponding maximum allow-

able discharge. These regulations control the amount of dust at the operation and the amount of dust that may be emitted from the collection equipment. Figure 1 shows four typical operations that range from no control to complete control.

TABLE 3 Process Dust Weight Limits

Process wt/hr, lb	Max wt discharge/hr, lb	Process wt/hr, lb	Max wt discharge/hr, lb
50	0.24	6,000	7.37
100	0.46	7,000	8.05
500	1.77	8,000	8.71
1,000	2.80	9,000	9.36
1,500	3.54	10,000	10.0
2,000	4.14	15,000	13.13
2,500	4.64	20,000	16.19
3,000	5.10	30,000	22.22
3,500	5.52	40,000	28.3
4,000	5.93	50,000	34.3
4,500	6.30	60,000 or more	40.0
5,000	6.67		

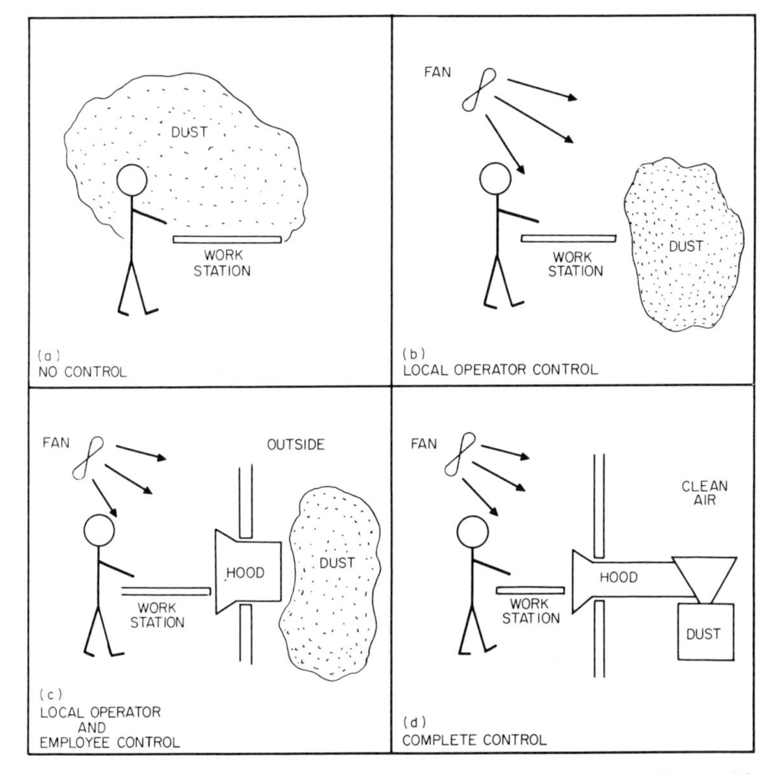

Fig. 1 Typical dust problems. (*a*) and (*b*) may require control by health and safety regulations. (*c*) and (*d*) may require control by local air pollution regulations.

DUST COLLECTION METHODS

There are two basic steps in the selection of dust collection systems. The first step is the effective capture of the dust particle; the second is the retention of the dust in a suitable container or medium. Local or adjacent exhaust ventilation is the most common method employed to prevent dust from escaping into surrounding areas. The necessary quantity of air and the proper way to implement the ventilation depend on characteristics of the surrounding air movement, the direction and velocity of the dust particle, and the size of the dust particle. The improperly directed discharge from the exhaust of a pneumatic tool can change an efficient exhaust system to a marginal one.

Once the dust laden air has been captured, the choice of proper air cleaning equipment depends on (1) the dust loading, (2) the nature of the material to be collected, and (3) the cleaning efficiency required. There are several dust collection systems available. A brief description of these systems can be found on Table 4. Because of the wide range of equipment available, the cost can vary from about $0.10 to about $1 per cfm.

TABLE 4 Dust Collection Methods

Class	Type	Contaminants removal	Particle size, microns
Air washer	Spray chamber	Lint, dust, pollens, smoke	20
	Wet cell	....	5
Electrostatic precipitator	Single stage	Lint, dust, pollens, smoke	2
	Two stage	....	1
Filters	Viscous	Lint, dust, pollens, smoke	5
	Dry	....	0.5–3
	Absolute	Bacteria, dust, radioactive	0.3
	Cloth bag	Dust, fumes, smoke	1
	Cloth envelope	Dust, fumes, smoke	1
Inertial dry collectors	Settling chamber	Dust, fumes, smoke	50
	Skimming chamber	Dust, fumes, smoke	20
	Cyclone	Dust, fumes, smoke	5–10
	Impingement	Dust, fumes, smoke	10
	Dynamic	Dust, fumes, smoke	
Solubles	Cyclone	Dust, fumes, smoke, gases, vapors	10
Wet	Impingement	Dust, fumes, smoke, gases, vapors	5
	Dynamic	Dust, fumes, smoke, gases, vapors	10

Optimum particle size recommended by equipment manufacturer.

One of the most critical operations involving toxic dust is metal removal using beryllium. Such dust collection systems must be highly efficient and require that stringent maintenance practices be followed.

SMOKE AND FUMES

Sources of visible smoke pollution can be found in most industrial plants today. Existing in some communities are regulations which limit both the visible and the invisible effluent emissions. Regulations are applied to such equipment as oil fired boilers and heaters, incinerators, and smoke producing processes.

Community Controls and Codes

In general the regulations limit stack emissions as follows:

1. No atmospheric discharges are permitted from any single source for a period or periods aggregating more than three minutes in any one hour which is

a. As dark or darker in shade as that designated as No. 2 on the Ringelman chart (Published by the U.S. Bureau of Mines), or

b. Of such opacity as to obscure an observer's view to a degree equal to or greater than does smoke described in Subsection (a).

2. No atmospheric discharges are permitted from any single source, one or more of the following contaminates in any state or combination exceeding in concentration at the point of discharge:

a. Sulfur compounds calculated as sulfur dioxide (SO_2) 0.2 percent by volume.

b. 0.3 grain per cubic foot of gas calculated to 12 percent of carbon dioxide (CO_2 at standard conditions).

In addition to stack emission restrictions, fuels may be regulated to the following:

3. A person shall not burn any gaseous fuel containing sulfur compounds in excess of 50 grains per 100 ft^3 of gaseous fuel calculated as hydrogen sulfides at standard conditions, or any liquid fuel or solid fuel having a sulfur content in excess of 0.5 percent by weight.

In addition to these regulations, a law usually exists covering nuisance emissions which will cause injury or annoyance to considerable persons or the public (odor).

Each community, as soon as low sulfur control levels are adopted, starts searches for a suitable supply of low sulfur fuels. Alternate sources of fuel are manufactured gas, natural gas, low sulfur fuel oils, and low sulfur coal. An industry that is presently using a nonconforming fuel oil or coal can expect an increase in fuel costs when it is required to make the conversion. How much additional cost will depend on the area and the converted type of fuel. Much investigation is being done on economical and practical methods to control sulfur emission from stacks. It can be assumed that suitable control equipment will be developed in the near future or that low sulfur fuels will become more readily available.

Incinerator Controls

Of prime interest to aerospace companies was the restriction imposed on standard trash burning incinerators. Trash incinerators were one of several methods used to dispose of the classified material generated by defense projects. All companies doing classified work with the federal government are required to conform to DOD 5220, 22-M, "Industrial Security Manual for Safeguarding Classified Information." The method recommended for destruction is burning. The purpose of burning is to destroy classified material beyond recognition. Each contractor is responsible for ensuring destruction completeness. In a standard incinerator this is a problem. After burning, the ashes have to be raked over to make sure all material has been destroyed. Incinerators are available, however, which will conform to these air pollution regulations and do more than an adequate job of material destruction. They are called multiple-chamber incinerators and are described as a piece of equipment used to dispose of combustible refuse by burning, consisting of three or more refractory lined combustion furnaces in series, physically separated by refractory wall, interconnected by gas passage ports or ducts employing adequate design parameters necessary for maximum combustion. These multiple-chamber incinerators leave a fine ash which more than satisfies security destruction requirements. Table 5 shows some typical sizes of multiple-chamber incinerators available and their fuel requirements. Figure 2 shows a typical multiple-chamber incinerator. The multiple-chamber incinerators can be designed to handle such other wastes as sawdust, wood scraps, garbage, and human and animal remains.

The approximate cost of a 500 lb/hr multiple-chamber incinerator not including gas service is $5,000. Operating cost per hour can be calculated by current natural or manufactured gas cost per 100 cu ft and the cost of wages for one man per hour. No attempt will be made to discuss effective control of rocket exhaust fumes, as no practical method exists. There are, however, federal regulations covering tables of distances based on toxicity, explosive hazard, and fire hazard for the various fuel combinations used.

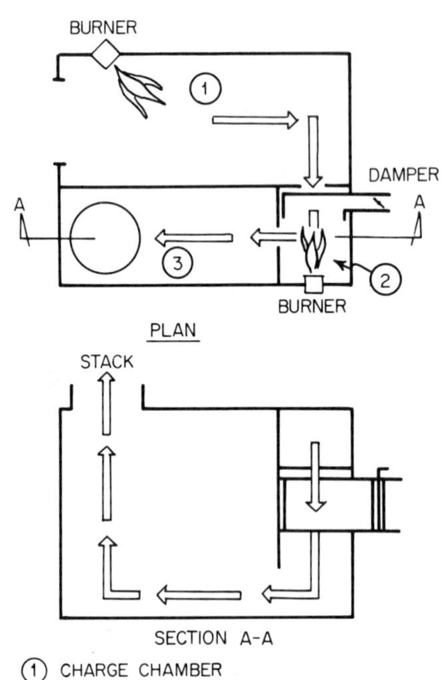

Fig. 2 Multiple-chamber incinerator.

HYDROCARBON POLLUTION CONTROL

In 1966, the nation's first regulation for the control of solvent emissions was adopted by Los Angeles County. Under the regulation the amount of nonexempt solvent that can be emitted into the atmosphere from any single source is restricted to 40 lb/24 hr or must be reduced by control equipment by at least 85 percent overall. A second condition states that *any* solvent in contact with flame, heat cured, or heat polymerized must be controlled to a maximum emission of 15 lb/day or be reduced by control equipment to at least 85 percent efficiency.

TABLE 5 Incinerator Sizes

Size Capacity, lb/hr	*Fuel Requirements, Approx cu ft gas/hr* (1,050 *Btu/cu ft*)
50	250
100	250
150	375
250	500
350	750
500	1,200

DEFINITION OF SOLVENT POLLUTION

Laboratory tests were conducted on certain solvent classifications to determine if they could be segregated into different reactive vs. nonreactive status. All the solvents tested indicated some degree of reactivity. The solvent classifications tested are listed in Table 6. All solvents listed above the naphthenes are considered as reactive in Los Angeles County, and all those listed below, including naphthene, are

TABLE 6 Solvent Classification

Solvents classified by their order of decreasing reactivity:
- Xylenes and heavy aromatics
- Isophorone
- Toluene
- Methyl isobutyl ketone
- Trichloroethylene
- Naphthenes
- Mineral spirits
- VM&P naphtha
- Stoddard solvent
- Isoparaffin mixtures
- *n*-Paraffin mixtures

considered as nonreactive. Based on these experimental results, Section K of Rule 66 was developed to define a photoreactive solvent and is described as follows:

> Section K. For the purposes of this rule, a photochemically reactive solvent is any solvent with an aggregate of more than 20 percent of its to al volume composed of the chemical compounds classified below or which exceeds any of the following individual percentage composition limitations, referred to the total volume of solvent:
>
> (1) A combination of hydrocarbons, alcohols, aldehydes, esters, ethers or ketones having an olefinic or cycloolefinic type of unsaturation: 5 percent;
>
> (2) A combination of aromatic compounds with eight or more carbon atoms to the molecule except ethylbenzene: 8 percent;
>
> (3) A combination of ethylbenzene, ketones having branches, hydrocarbon structures, trichloroethylene or toluene: 20 percent.
>
> Whenever any organic solvent or any constituent of an organic solvent may be classified from its chemical structure into more than one of the above groups of organic compounds, it shall be considered as a member of the most reactive chemical group, that is, that group having the least allowable percent of the total volume of solvents.

OPERATIONS REQUIRING CONTROL

A survey of aerospace-electronics manufacturing processes was conducted by a number of companies to determine which processes use hydrocarbon materials that may be affected by Rule 66. The major problems were divided into the following four areas:

- Cleaning and degreasing
- Primers and topcoats
- Chem Mil maskants and strippable coating
- Plastic adhesives (investigation revealed little or no problem existed in this field because of the small amounts of material used)

The surveys indicated that all companies used trichloroethylene as a cleaning material. This material was listed as photochemically reactive and therefore would require investigation. Some 15 Mil Spec primers and topcoat finishes were currently being used by various aerospace companies. Not all companies involved were using all 15 Mil Spec materials, but these represented the most commonly used paints. These were:

- Mil P8585 Primer coating Zinc Chromate
- Mil P23377 Epoxy Polyamide
- Mil C22750 Epoxy Polyamide
- Mil P7962 Primer Coating Intermediate N/C Modified Alkyd Type
- Mil L19537 Lacquer Acrylic-Nitrocellulose Gloss
- Mil L19538 Lacquer Acrylic-Nitrocellulose Camouflage
- Mil L52043 Lacquer Semi-gloss Cellulose Nitrate
- TTE 529 Enamel Alkyd Semigloss
- TTL 32 Lacquer Aircraft
- TTP664 Primer Corrosion Inhibiting
- Mil C27725 Integral Fuel Tank Coating
- TTE 516 Enamel Lustreless Styrenated Alkyd Type
- Mil P22332 Paint Priming Exterior and Interior
- TTL 20 Lacquer Camouflage
- Mil P27316 Primer Epoxy

The Chem Mil maskants and strippable coatings were not Mil Spec materials but were used extensively in the manufacture of airplane and missile components. If the solvents could be changed to conform to solvent rules, little or no change would be required in the production facilities.

Plastics and adhesives are commonly used by all aerospace companies but were not classed as a Mil Spec item. These materials are used as components, and any change could require extensive testing to assure equal reliability of the assembled end product.

In all four areas processes were found in which the solvents used were in excess of the total premissible emission and the solvents were classed as photochemically reactive.

Each aerospace company was assigned responsibility for one of the major areas of investigation. In addition each participating company would have a working member on each committee. By doing this, each company would be able to contribute its own problem.

Control Methods

The first approach method of compliance would be to install control equipment to collect, neutralize, or destroy the illegal portion of the effluent.

The following is a brief description of such control equipment:

Carbon absorption units (see Fig. 3) Solvent laden fumes are taken from the process unit and passed through a bed of activated carbon. The solvent is adsorbed by the activated carbon and held until removed by steam. Generally two systems are employed to allow continuous operations. One unit is active while the second unit is being stripped of the solvent by steam.

Advantages. Cost of recovered solvent helps to reduce payoff time. Highly efficient and good for low air volume, high solvent concentration conditions.

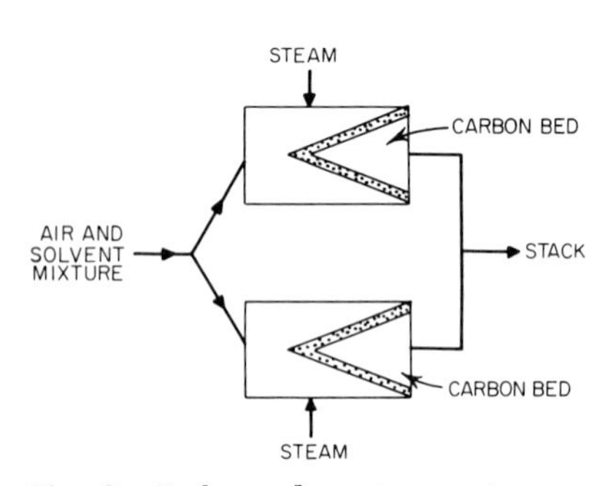

Fig. 3 Carbon adsorption equipment. Operating cycle: one unit is on line collecting solvents while second unit is being stripped by steam to remove collected solvents.

Disadvantages. Requires a steam supply. Some materials are corrosive, requiring expensive materials for construction. Materials other than solvents may poison the carbon bed. Increased plant operating cost. Approximate cost $70,000 to $120,000 (installed on a vapor degreaser approximately 12 by 4 ft).

Incineration (see Fig. 4) There are several types of fume incinerators available on the market. They include both the straight pass-through, the heat exchanger type, and effluent/air supply units. Solvent laden fumes are passed through an incinerator and the mixture is brought up to approximately 1200°F and held for a period of time. The hold time will depend on the type of solvent being consumed, the solvent concentration, and the temperature of the incinerator.

Advantages. Good for low volume of air. Small space requirement, low-maintenance operation; high efficiency can provide waste boiler heat for other operations.

Disadvantages. Expensive to operate in areas of large air requirements. Requires a volume fuel supply; cannot be used if solvents or effluent products of combustion are toxic. Approximate cost: $0.50 to $4 per cfm.

Other methods of vapor collection, such as fume scrubbers or refrigeration condensers (see Fig. 5), may be used if their efficiency can be demonstrated to local air pollution officials to be at least 85 percent overall and they are found to be economically feasible. Presently no equipment of these types is readily available for organic solvent vapor collection applications.

Processes most commonly affected by the solvent control rule will now be reviewed to determine if any of the available control equipments are feasible for the particular operation.

Primers and top coating Both dry filter and water spray paint booths are generally used when applying a primer or topcoat. State laws require that a minimum velocity of 100 cfm/sq ft of booth opening be maintained. A booth with an opening 10 ft high by 15 ft long would require 15,000 cfm of air to satisfy state health requirements. With this large amount of air, a small amount of solvent is added. In addition, small amounts of paint may be carried up the stack. Both the carbon adsorption equipment and the incinerator proved to be too large and costly to be considered as suitable control equipment. (Example: At 15,000 cfm approximately $13,500 would be required for the incinerator and approximately $10.50 per hr would be required for fuel costs.) See Fig. 6 for equipment and operating costs.

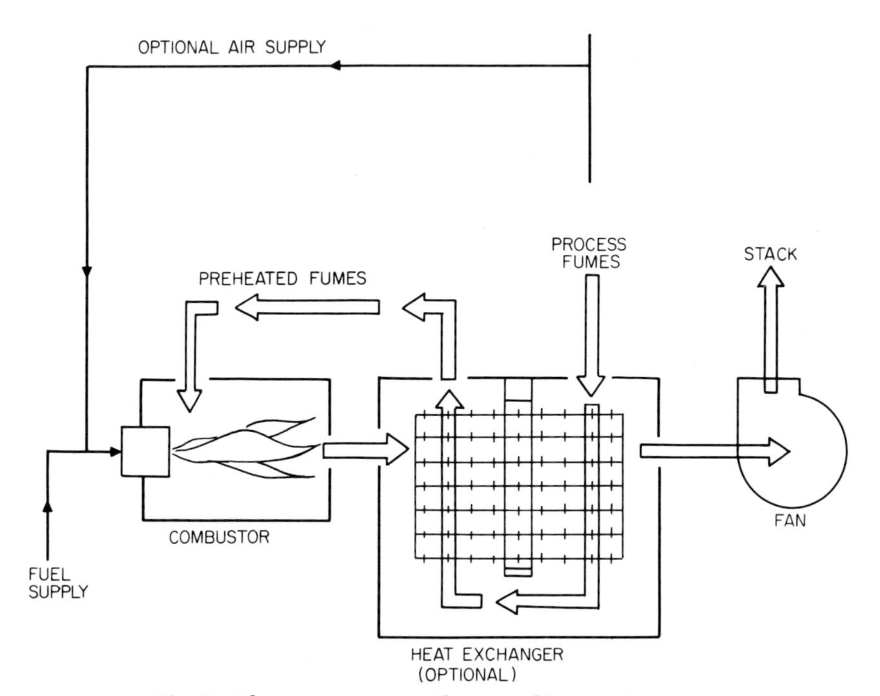

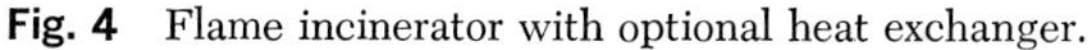

Fig. 4 Flame incinerator with optional heat exchanger.

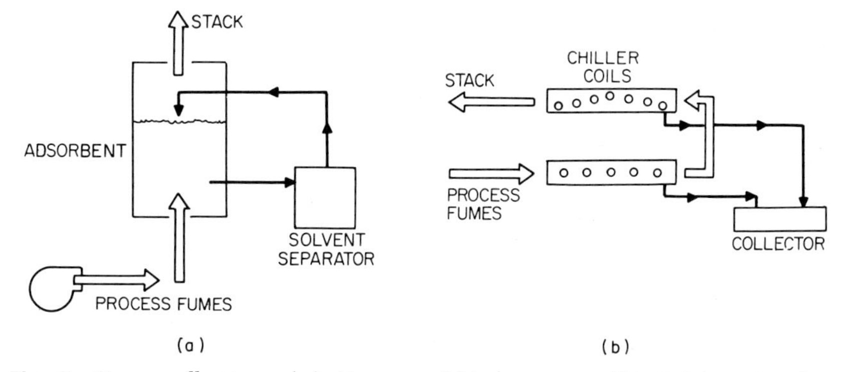

Fig. 5 Vapor collection. (*a*) Fume scrubbing system. (*b*) Refrigeration fume collection system.

Cleaning and degreasing There are a variety of exempt cold cleaning solvents available on the market. The type of cold cleaning solvent depends on the material to be cleaned, the contaminant, and the toxicity of the cleaning fluid.

Trichloroethylene used in vapor degreasing has been classed as a nonexempt solvent in Los Angeles County. Perchloroethylene (another Mil Spec approved exempt degreasing solvent) may be substituted for trichloroethylene provided the plant is equipped with high-pressure steam. Trichloroethylene requires 15 psig steam (low pressure) and perchloroethylene requires 50 to 60 psig steam (high pressure).

Vapor degreasers are common in almost every aerospace operation. Those using trichloroethylene may require control of emissions in excess of 40 lb/24 hr.

The vapor degreasers are equipped with automatic heat controls, water separator, and a cooling coil for vapor control, and may be fitted with covers which can remain closed when not in operation. Lip exhausters are generally used to remove vapors that escape above the cooling coils because of sudden drafts or the piston effect of loading or unloading. In properly operating degreasers the lip exhaust system may collect up to 70 percent of all the vapors escaping above the cooling coils. The remaining vapors will be dissipated into the room. If control equipment is added to the existing lip exhaust ventilator and this control equipment is 100 percent efficient, it still does not capture sufficient vapor to comply with Rule 66 (85 percent is required). Elaborate steps can be taken to enclose the degreaser totally within a room, and all vapors can be fed into a vapor collection system. This approach has several obvious drawbacks. Production rate would be affected, the operator would be in the stream of slightly toxic vapor, and the collection equipment would be very large. Incineration was ruled out, as incomplete combustion could result in the manufacture of highly toxic gases. Carbon adsorption equipment was also eliminated from consideration because of its size and expense. Losses from a degreaser may be reduced by careful supervision of its operation. The following are some common rules for carrying out proper degreasing operations:

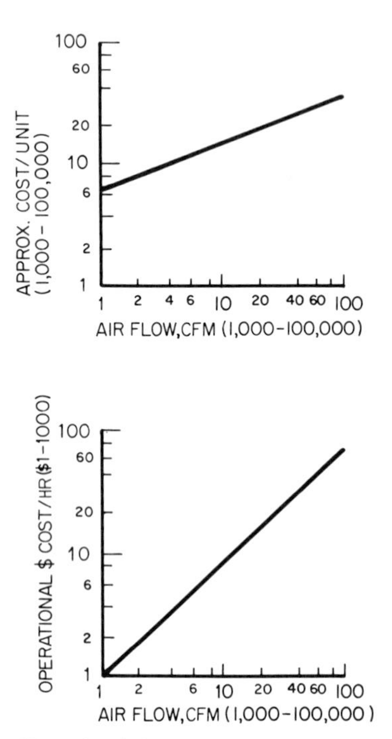

Fig. 6 (*a*) Incineration equipment costs. (*b*) Incinerator operating costs.

1. Load parts for proper drainage.
2. Lower and remove baskets slowly (11 fpm recommended maximum speed).
3. Leave work in vapor zone until vapor stops condensing.
4. When removing basket, hold it above the vapor zone but within the freeboard area long enough to permit complete drainage and drying.
5. Never use a spray nozzle above the vapor zone.
6. Close degreaser when not loading or unloading.
7. Do not use plant air for drying.
8. Notify supervision whenever operation conditions indicate something is wrong with the equipment.
9. Replace covers when not in use.

Chem Mil maskants and strippable coatings Coating may be applied by spray, dip, or flow coating. Hoods, when used, are similar to the exhaust system employed for paint booths. The large areas that may be involved require large amounts of air and make the installation of control equipment impractical.

Control by Solvent Substitution

In reviewing the three operations affected in the aerospace industry, it appeared that existing vapor control equipment was not practical except in the case of curing ovens, where incinerators could be installed to match the low volume of exhaust air.

The rule in itself provides a second method of compliance—the substitution of exempt solvents to replace existing nonexempt solvents. From a technical standpoint this can and is being done. Many exempt architectural and industrial finishes have already been developed and are offered to the public. Chem Mil maskants and shop strippable coatings using exempt solvents have been developed whose performance is equal to or better than previously used materials. Test data indicate that Mil Spec materials may perform equally well using exempt solvents. Samples of exempt manufacturing coating materials and supporting test data are being submitted to the federal government for review and approval. Presently of 14 Mil Spec coatings being tested by aerospace companies, 6 have completed tests and the data have been submitted to the cognizant government agency for approval. Initial tests on the remaining coatings are most favorable. After permission is received from federal agencies to use a substitute solvent material, some form of quality receiving inspection may be desirable to provide background evidence that a standard grade of materials is being received.

A new vapor degreasing fluid has been developed to replace trichloroethylene. The material 1,1,1-trichloroethane has been used for many years as a cold solvent cleaning agent. The addition of inhibitors has made this material suitable for vapor degreasing, and a special Mil Spec has been issued to allow its use, Mil Spec MIL-T-81533.

Generally the cost of exempt material is slightly higher than that of the nonexempt. Early reports indicate increases from 0 to 20 percent. It can be expected that the prices will be reduced as the usage of materials increase.

INDUSTRIAL WATER, CONTAMINATED WASTES AND CONTROL

Contaminations resulting in water pollution by the aerospace industry are, with few exceptions, generally similar to contaminant problems found in the waste water product of other industries. The marked distinction in the control of contaminated water as a by-product of the aerospace industry as opposed to other manufacturing and fabricating processes is the great divergence of aerospace waste products. A pulp or paper manufacturing plant, as an example, may have as a result of its manufacturing procedures a constant and continuing water pollution handling problem over which limits and controls may be determined and proper safeguards established. The aerospace industry and, in particular, that activity having to do with space exploration and space vehicle propulsion, presents water pollution problems of varying magnitude and sources because of the experimental and research development procedures and changes.

Closely allied with water pollution and its efficient control is the ethics of water conservation. This chapter, whose primary purpose is to deal with contamination problems, their control, and their final efficient disposition in accordance with governing ordnances, will also present concepts for the more efficient use of industrial water and highlighting the resulting economics.

Local Codes and Regulations

In general six different systems are available for the disposal of liquid wastes generated by industry. They are:

1. Domestic sewage systems
2. Industrial treated systems
3. Leach beds
4. Storm drain systems
5. Deep-well injection
6. Contract hauling

Uniform codes or regulations do not exist to control the permissible amount of water contaminants allowed to flow to local waste water disposal systems. Such regulations as now exist are dictated by many factors such as population and industrial density, municipal or public sewage disposal systems, ultimate discharge points of waste water effluent, subsoil water tables, and types of resultant pollution, to name but a few. Since the disposal of contaminated waste water is accomplished by such means as its introduction to storm and surface drainage systems, sewerage disposal complexes, ground percolation, and deep injection wells, it is apparent that disposal regulation is a local or community problem and does not lend itself to general compliance to other than broad concepts.

Furthermore it is also true that the limits of tolerable contamination have been accomplished in many areas. New industry or modifications of the manufacturing procedures within existing plants cannot make use of the established municipal disposal facilities usually because they are generally working under overloaded conditions. Under these circumstances plant engineers must use other means of contaminated water disposal. Some of these methods are discussed as follows:

Contaminated water disposal methods

Haulage. For small plants or plants having a small quantity of effluent, it is generally found cheaper and more convenient to pay haulage and disposal charges.

Deep-well Injection. When concentrations, toxicity, and complexity of treatment and excess quantity of polluted waste water preclude economical disposal to surface systems, the effluent may be delivered to an acceptable geological stratum below the earth's surface. Before such a method is considered practical, research is required to determine the depth of the wells required. The depth of injection is determined by several factors:

1. The ability of the subsoil strata to accept the anticipated flow rate
2. The nature of the subsoil layers to preclude the possibility of plugging
3. Determination of waste pretreatment required, if any, to prevent scaling and corrosion
4. A geographical survey to eliminate the possibility of contamination of existing potable water tables in the vicinity of the injection well and controls for constant surveillance of the adjacent potable water tables

Injection wells may vary in depth from 500 to 2,500 ft. The economics of this disposal method may take into account the initial cost of the drilling required, piping from the receiving reservoir to the well entrance, and pump operating costs.

Artificial Lagoons. The use of artificial lagoons as settling ponds may be divided into two types: (1) those which permit percolation of liquid contents and dissolved solids into the ground and (2) those used as settling ponds for gravity precipitation of suspended matter with no resulting ground percolation. In the first instance care and study will be required to prevent contamination to subsoil water tables. This method of waste water disposal assumes that the nature of the soil is such that it will accept and dispose of waste liquid at a practical rate. Consideration must also be given applications where the frost line would make the pond of little use during extended subfreezing ground temperatures. The suspended solid matter in waste water disposed of in this manner must not be of a type which would prevent percolation under extended usage.

In line ponds or those constructed on impervious ground, other problems must be considered. Such ponds require frequent clearing of precipitated or suspended solids and are practical only if the clear liquid has practical reuse. A common fault of these ponds is that the settling out of solid matter is usually at an extremely slow rate after apparent water clarity has been obtained. If the dissolved solids remaining make the reuse of the clarified water practical, it is drawn off by controlled overflow.

If the volume of water delivered to ponds or lagoons is on a large scale and the retention time required is of long duration, the size of settling areas required may become a major economic problem. Often it is possible to dilute lagoon water for reuse by the addition of surface or rainwater runoff. If the diluted water is still not acceptable for reuse, the degree of dilution obtained by this method may make the effluent acceptable to local storm or sewerage drainage systems. Again, dependent on the nature of the pollutants, ponds and lagoons lend themselves to controlled chemcial dosing to render the waste liquid disposable by other means.

Precipitators and Clarifiers. The aerospace industry frequently has use for water flow rates at elevated capacity and pressures in hydrodynamic test facilities. A rocket engine plant in central California utilizes two such facilites, the larger of the two being capable of circulating water at 18,000 gpm at 650 psig. The water is continuously reused and the quality maintained by using flocculant precipitators and a series of diatomaceous earth and activated charcoal filters. The volumetric water capacity of the facility is approximately 200,000 gal. Less than 1 percent per month of makeup water is required to replace that lost by evaporation and waste drawoff from the precipitator and filters. The only source of contamination is chiefly small metal filings and sulfated cutting oils. It is estimated that the entire reservoir contents of 200,000 gal would need replacing every 2 weeks during periods of peak usage if the above equipment were not installed to maintain water purity. Thus a considerable saving results in the purchase of potable water, but what is more important, the degree of purity is maintained in the pump sumps and the cost of waste water treatment and disposal is held to a minimum.

WATER CONSERVATION AND RECLAMATION

Many methods suggest themselves to restore water (otherwise wasted) to a state of reusability. The method selected will depend on the type and concentration of the contaminants and the degree of purification required for the specific reuse of the reclaimed water. A survey of plant water requirements may disclose the reuse possibilities of treated or diluted waste water where the need for high-quality water becomes progressively less. The proper use, reuse, and application of water conservation methods is desirable in any plant operation. Therefore, attention is called to the need for volumetric reduction of waste water disposal.

Even in areas where the cost of municipal water is a minor factor, it is worthwhile to consider and study the means available to reduce potable plant water intake as a step in the cost reduction of waste water disposal. Only a very small fraction of the potable water demanded in the aerospace industry is actually consumed. Consumed water is that which finds its way into the final product, and that which is lost by evaporation and outflow from wash and rinse tanks. The bulk of the water used is discharged as polluted water in varying degrees of contamination. Therefore, the cost of disposal of effluent water is closely linked with plant economics. It may be demonstrated that the cost per gallon to treat and dispose of contaminated water exceeds many times over the initial water charges.

Measuring Water Usage

Each industry and each process in each industry presents an individual problem of water conservation. An overall study of entire plant operations will undoubtedly reveal many areas where corrective steps may be initiated toward the desirable end of water conservation. Potable water outflow from compressor jacket cooling may be reused as makeup and rinse water in one or more metal etching, stripping, and anodizing processes. Final rinse water overflow in turn may be used in process heat exchangers or may be diverted for use in air scrubbing operations. The foregoing are but a few of the many reuse possibilities of partially contaminated water. Obviously a quantity of water used in several manufacturing processes not only reduces the quantity of overall potable water requirements but (more importantly) reduces problems and costs of effluent treatment, handling, and ultimate disposal.

Consideration should also be given to pretreatment of potable raw water for purposes of scale prevention, oxidation, and corrosition in engine jackets and heat exchangers, etc., with the view toward the elimination of the reduction of the quantity of additives otherwise required. This is true in the case of water used for boiler makeup purposes. The economical approach to pollution reduction is good water management in all its phases. A survey of any manufacturing plant would unquestionably reveal areas where corrective steps could be initiated and costs reduced.

Total Water Management

Following is a description of a well-integrated program of waste water control and ultimate disposal as practiced by a large aerospace complex operating in central California. It illustrates the close liaison existing between plant management and the State Water Quality Control Board.

The raw water supplied to the plant is from two sources: (1) from open ditches, supplied by a public utility company, and (2) from deep wells located on the property. Only sufficient water is pumped from the deep wells when an additional source is required as in drought periods or during peak demand periods.

A major use of water is during periods of rocket engine test operations, when it is used principally as cooling water for flame deflector plate protection. The overflow is temporarily impounded by a dam downstream of the test stand operations area where it is continually and automatically sampled, analyzed, and treated if necessary before release to a natural creek which runs through the property. During this retention time the thermal level of the effluent has an opportunity to stabilize. In most cases this would reduce any problems of meeting federal thermal pollution levels. Downstream of the point of entry of the waste water into the creek an aquarium has been constructed and is supplied with creek water. The aquarium is stocked with fish supplied by the State Department of Fish and Game. The growth and health of the fish is checked periodically to assure that no material harmful to aquatic life is present in the water. The overflow from a completely integrated acti-

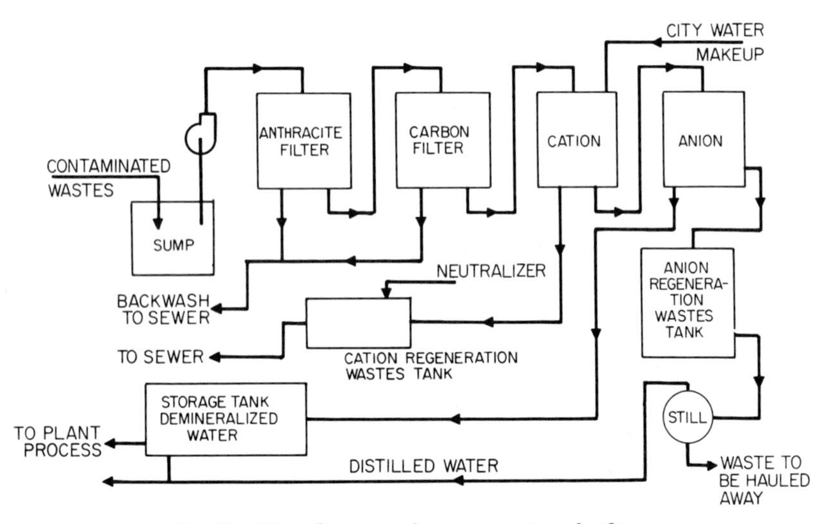

Fig. 7 Flow diagram of water treating facility.

vated sludge sewage disposal plant is also drained to the creek. The liquid waste is aerated, clarified, and chlorinated. Continuous monitoring of the liquid discharge from the plant, prior to release to the creek, is maintained to determine the efficiency of the plant in reducing BOD and to evaluate the degree of disinfection achieved.

After all waste waters have entered the creek, all water is pumped into one of two retention ponds. The waters are retained in the ponds until chemical analysis confirms compliance with discharge standards.

Waste effluent from operations which preclude treatment for discharge into the creek is pumped into deep subsurface injection wells. Sampling wells at several locations on the property are maintained in order to eliminate the possibility of contamination of subsurface streams by the material delivered to the injection wells.

This rather sophisticated system, while not typical, illustrates the extent of control which the industry may be required to exercise in the near future.

Water Pretreatment and Recirculation

Another large aerospace plant located in the southern California area producing fabricated airplane components has solved its waste water disposal in a unique manner by eliminating for all practical purposes the need to dispose of any contaminated water.

All water used in process work other than that used in cooling towers is first demineralized. Waste process water is returned to a common sump, which serves as a collection basin for the entire plant operation. From the sump it is pumped to a storage tank and then delivered to anthracite and charcoal filters. The water leaving the filter banks passes through ion exchangers and is stored in a 50,000-gal holding tank, from which it is again introduced to the process water system as required.

All anion waste from the ion exchangers is fed to a holding tank. From that tank it is delivered to a barometric still. After leaving the still in vapor form, it is condensed and the outflow commingles with other process water (see Fig. 7).

Prior to the system installation, approximately 160,000 gal of polluted process water per day was dosed in treatment ponds and, when considered acceptable, pumped to the municipal sanitary water system. At the present time only 5,000 gal/week is treated and delivered to the sewer system.

It is estimated that the cost of the plant may be written off after approximately 5 years because of the reduced intake of potable water and the cost of waste effluent treatment.

Water Conservation Applied to Metal Surface Treatment

In the overall picture of the aerospace industry, the greatest problem of contaminated water treatment and disposal is in the area of metal surface treatment. This water use embraces some of the following processes:

1. Passivation
2. Pickling
3. Corrosion prevention
4. Stripping
5. Rust removal
6. Etching
7. Coating
8. Plating
9. Anodizing

Table 7 shows the types of operation being performed on various metals and the chemicals being used. While these processes are in themselves users of water in usually large volumes, the major pollution and disposal problem is found in the several stages of rinsing required as a follow-up procedure from these processes.

TABLE 7 Typical Process Material

	Material
1. Cleaning	
a. Aluminum	Solvent
	Alkaline
	Phosphoric acid
	Nitric acid
	Chromic acid
	Sulfuric acid
	Sodium phosphate
	Altrex
	Oakite
	Diversey
b. Steel	"Electro" cleaner
	Solvent
	Muriatic acid
	Sodium phosphate
	Chromic acid
	Hydrochloric acid
	Sulfuric acid
	Silver cyanide
	Potassium cyanide

TABLE 7 Typical Process Material (Continued)

	Material
1. Cleaning	
b. Steel	Potassium carbonate
	Sodium cyanide
	Sodium hydroxide
	Cadmium oxide
	Sodium carbonate
	Cuprous cyanide
	Rochelle salts
	Sodium polysulfide
	Potassium stannate
	Hydrogen peroxide
	Chromic acid
	Sulfuric acid
	Chromium 110 salts
	Nickel sulfide
	Nickel chloride
	Boric acid
c. Stainless steel	Potassium permanganate
	Sodium carbonate
	Sodium hydroxide
	Nitric acid
	Hydrofluoric acid
	Solvents
d. Titanium	Alkaline
	Solvent
	Sodium dichromate
	Sodium hydroxide
e. Magnesium	Oakite
	Alkaline
	Solvents
2. Plating	
a. Aluminum	Sodium dichromate
	Chromic acid
	Sulfuric acid
	Oxalic acid
	Nickel acetate
	Aluminum sulfate
3. Passivating	
a. Aluminum	Chromic acid
	Nitric acid
b. Steel	
c. Stainless steel	Nitric acid
	Sodium dichromate
	Hydrofluoric acid
d. Titanium	Nitric acid
4. Coating	
a. Aluminum	Alozinc 0500
	Chromic acid
	Nitric acid
	Sulfuric acid
	Iridite H14
b. Steel	Sodium dichromate
	Sulfuric acid
c. Stainless steel	
d. Titanium	
e. Magnesium	

In rinsing procedures, considerable economy may be effected by cascading the effluents of multistage rinsing requirements. This is illustrated in Fig. 8, which shows a four-stage rinsing operation.

Makeup water enters the final (fourth) rinse tank, and the overflow is delivered progressively to the third, second and first tanks. This method adds makeup water to each tank by admitting water with progressively lower contamination to the preceding rinse water. As a consequence only 25 percent makeup water is required as opposed to adding raw water to each tank.

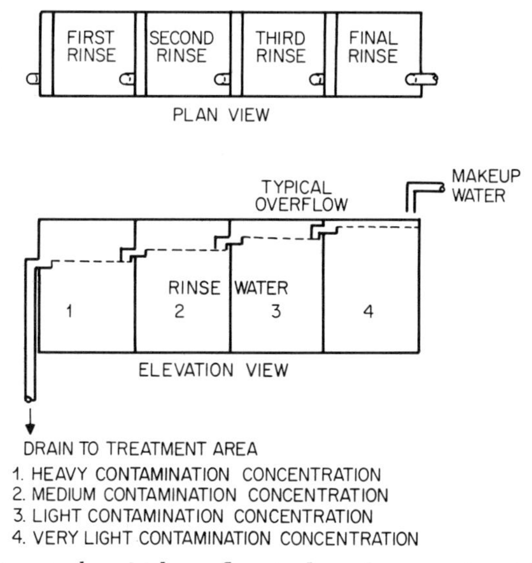

Fig. 8 *Top:* Rinse tanks. Makeup flow is dependent on allowable contamination concentration in each tank and parts loading. *Bottom:* Cascading rinse system.

"Once-through" Systems versus Recirculating Systems

Basically "once-through" systems are more costly to operate than using recirculation of cooling water. Water used in "once-through" processes is generally not acceptable for discharge into municipal or plant operated sewage disposal plants. Heated water into which no oxygen has been added produces a volumetric load on activated sludge disposal plants. Lack of oxygen content in the fluids and the elevated temperature of the water start growth of sulfate producing bacteria, creating dilute concentrations of acids detrimental to the sewage collection system.

A recirculating system usually involves the use of a cooling tower, and only the bleedoff waste is discharged to the sanitary sewer system. Further savings in operating costs result in the use of cooling coils in the cooling tower for purposes of heat removal from other heat exchangers.

Figure 9 shows the basic differences in the two systems and demonstrates the additional cooling capacity available and the reduction in volume of polluted water discharge, which could be as much as 90 percent less than that resulting from a "once-through" system.

LIMITATIONS GOVERNING USE OF SANITARY SEWER SYSTEMS

All industrialized areas have governing codes intended to prevent the overload of sanitary sewer systems and to prevent the discharge of excess quantities of material that would be harmful to the action in sewage disposal plants, cause damage to the piping or sewage disposal equipment, or result in a fire hazard.

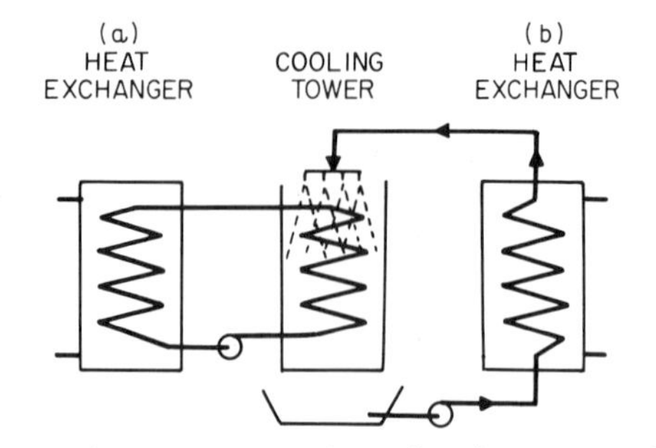

Fig. 9 Recirculating cooling system. (*a*) A closed system cooled by the sensible heat transfer to the water of an open recirculating system which, in turn, is absorbed as the latent heat of evaporation of water transferred to the atmosphere. The closed system minimizes the problems of sludge, scale, and bacterial growth within the heat exchanger. (b) An open recirculating system in which the heat is absorbed as the latent heat of evaporation of water transferred to the atmosphere. The open system increases the probability of problems of sludge, scale, and bacterial growth within the heat exchanger.

The following is a listing of materials usually not considered acceptable for discharge into sewage systems:

1. Excess quantity of water
2. Flammable materials
3. Insoluble solids
4. Toxic materials
5. Strongly odorous wastes
6. Oil and grease
7. Wastes less than 6 pH
8. Hydrazine derivatives
9. Nitrogen tetroxide
10. Fluorine and boron compounds
11. Most chromates
12. Liquids otherwise acceptable except that their temperatures are above 140°F
13. Wastes and runoffs that would be acceptable to the storm drain system
14. Clean water

Special limitations may be imposed when necessary. The governing criterion for acceptability is essentially whether the industrial waste discharge will increase plant costs and is generally considered appropriate for sewer disposal.

In the Los Angeles area it is estimated that a flow rate of 1 gpm requires $600 worth of sewage disposal facility.

RADIOACTIVE WASTES

Plants handling or using radioactive materials are required under federal law to be licensed to permit their use. The license may be issued and controlled by a state provided they have established radiation control regulations satisfactory to the Atomic Energy Commission. They are known as agreement states. In requiring licenses, for most radioactive materials, the services of a qualified radiological officer are assured. The necessity of having a qualified radiological officer also reduces the possibility of improper disposal of radioactive wastes. The following are excerpts from a typical state regulation for radioactive waste disposal.

1. General Requirement—No user shall dispose of any radioactive material as waste except:

a. by transfer to a person holding a specific license to receive the radioactive waste, or

b. as authorized pursuant to other Sections XXX* of this document.
2. Disposal by Release into Sanitary Sewerage Systems
a. No user shall discharge radioactive material into a sanitary sewerage system unless:
(1) it is readily soluble or dispersible in water; and
(2) radioactive material released into the system by the user in any one day does not exceed the larger of:
(a) the quantity which, if diluted by the average daily quantity of sewage released by the user, will result in an average concentration not exceeding the limits specified in Section XXX* of this document or
(b) ten times the quantity of such material specified in Section XXX* of this document
(3) the quantity of any radioactive material released into the system by the user in any one month, if diluted by the average monthly quantity of sewage released by the user, will not result in an average concentration exceeding the limits specified in Section XXX* of this document.
(4) the total quantity of radioactive material released into the sewerage system by the user does not exceed one curie per year.
3. Disposal by Burial in Soil
a. No user shall dispose of radioactive material by burial in soil unless:
(1) the total quantity of radioactive material buried at one location and time does not exceed, at the time of burial, 1,000 times the amounts specified in Section XXX* of this document; and
(2) burial is at a minimum depth of four feet; and
(3) successive burials are separated by distances of at least six feet and not more than 12 burials are made in any year.
b. the department will not approve any application for license to receive radioactive material from other persons for disposal on land not owned by the Federal or State Government.
4. Concentrations in Effluents to Uncontrolled Areas
a. No user shall release or cause to be released into air or water in any uncontrolled area any concentration of radioactive material which, when averaged over any year, exceeds the limits specified in Section XXX* of this document. Such concentration limits shall apply at the boundary of the controlled area. The concentration of radioactive material discharged through a stack, pipe, or similar conduit may be determined with respect to the point where the material leaves the conduit. If the conduit discharges within the controlled area, the concentration at the boundary may be determined by applying appropriate factors for dilution, dispersion, or decay between the point of discharge and the boundary.

In addition to radioactive materials requiring the services of a licensed radiological officer, there are many materials of low radioactivity which require control only from a local plant health and safety standpoint. Most of these materials are sold complete with disposal instructions which, in many cases, suggest they be returned to the manufacturer. There is little or no guarantee that the user of these materials will follow the suggested disposal methods. This is where the radiological safety officer can be utilized. Procedures should be established requiring his approval on all purchase orders for radioactive materials. This will allow him to review what is being purchased, where it is to be used, and the disposal requirements. When the materials arrive, he can review with the user the required disposal methods and any special handling techniques involved.

The bulk of disposal of radioactive materials falls into the low-level class. Licensed disposal firms will haul the waste material to an approved dump site. Dump sites are under the jurisdiction of the state if they are an agreement state or, if not, under the Atomic Energy Commission.

In general, deep-sea disposal is not used because of its high cost. Deep-sea disposal is under the jurisdiction of the Atomic Energy Commission.

* These sections will be found in the General Rules; because of their size, they have not been duplicated in this report.

Some typical costs for material disposal are:

1. Low-level liquid wastes—ICC approved containers provided by the hauling contractor, $3 per gal
2. Low-level solid wastes—ICC approved containers provided by the hauling contractor $6.50 per cu ft

Costs for high-level radioactive waste materials must be figured individually because of the many variables involved.

Chapter 20

European Industrial Pollution Control Practices

STANLEY K. SMITH

Supervisor of Facilities Engineering,
Honeywell Inc., St. Petersburg, Florida

GENERAL COMPARISONS

It is always dangerous to generalize the pollution control practices of so many political and ethnic divisions as exist in Europe, for it is always possible to point up an exception. Nonetheless a general concept of how pollution control practices in Europe differ from those in the United States will be made. Keep in mind, however, that Europeans process and control their pollutants using the same basic methods we do. They have not developed any major "breakthroughs" in pollution control processes. What they have developed are unique applications to their economic and social situations and a more sophisticated working relationship between industry and government enforcing agencies.

This is as it should be, because the Europeans have been practicing pollution control for a long time. In fact, most of the laws under which they operate today stem from the old "nuisance" or "public health" laws that were legislated around or before the turn of the centrury. This is being changed in most countries, and new pollution control oriented legislation will be forthcoming in the next several years. This does not mean standards are loose but that they are high where they exist, and if they do not exist, there is a high moral obligation on the part of industry, particularly large industries, to give the public the greatest possible protection.

Air pollution is controlled on a national basis and water pollution by local authorities. Control is generally effected through a system of permits which industry must obtain before they can operate. Generally, the "best practical means" attitude is adopted. This may entail alterations in plant and method as new "means" become available. Legal sanction is rare for enforcement; instead education and persuasion are used to ensure compliance. There is an exception in Germany, where the VDI (Association of German Engineers), whose Commission for Clear Air sets the standards, have adopted the "generally accepted technical rule." That is a rule that has been tried and tested in practice.

Research for pollution control is generally carried out by industry, although in England and Holland the government has research laboratories where industry can purchase research capability at nominal cost. The larger industries, particularly in Germany, have large pollution control research staffs that do research not only for the industry but for the government on a commission basis.

Since this is true, one finds a tendency for these staffs either to develop their own monitoring devices or to modify off-the-shelf devices so that they are tailored to the particular process. There is also a tendency to run large-scale models of a new process for pollution control for at least a year before it is utilized on a full scale. Very often, if two or more processes are in question, they will be operated under like conditions for a year to obtain economic operating data before a selection of process is made. This usually results in some modification to the unit that is accepted, so that it is quite rare to find unmodified purchased equipment.

To illustrate the characteristics of European pollution control practices, a series of industrial operations have been chosen as examples. These examples are generally well-known ones and best illustrate the differences in American and European operating techniques. Of necessity, all countries could not be included, and some other practical examples have been omitted for various reasons. The entire pollution operation for the particular plant representing a certain industry will be given in sufficient detail so that a reasonable comparison can be made with similar American operations.

Chemical Industry

Bayer, Leverkusen, Germany This plant of 40,000 employees is considered a model of integrated pollution control. Its pollution control problem is one of the most involved, as the wastes from manufacturing several thousand chemical products run the gamut of industrial contaminants. From 1956 to 1967, Bayer had invested over $62.5 million in capital equipment for water and pollution control. Operating costs have amounted to $25 million. The most important statistic of all is that they have a planned program to spend $6 million a year on continued research and improved control equipments through the year 1974. This is a large investment but amounts to only 5 percent of total capital outlay and is considered well within feasible economic operating limits.

The most important aspect of the dollar figures is not how many but how they are spent, and this has been with a view of integrating not only the needs of the entire complex but also the needs of the surrounding communities.

They have a complete pollution monitoring center; a very complete and efficient waste combustion plant; a new sewage plant for their laboratory wastes that is also designed for research studies; a joint sewage purification and waste disposal project with local governmental authorities; and very importantly, a separate monitoring section for pollution control.

In their pollution monitoring section, they have 38 assigned persons, of whom 8 have doctor's degrees, 20 additional are college graduates, and the remaining 10 are technicians. This high degree of qualification of the personnel leads to a high quality of pollution organization and control.

The pollution monitoring station was located in the northeast corner of the complex in the path of the prevailing winds. This station has two functions: one, detecting failure of controlled air emissions and initiating immediate remedial action; two, continuous sampling to evaluate individual sources of emission, developing model results for various hypothetical control measures, and evaluation of results of actual control measures taken.

To accomplish these tasks, the control center has a network of 50 measuring points inside and outside the plant boundaries. In the most important processes, if possible, the waste gases are controlled and monitored continuously—automatically. At other points, the atmosphere is monitored continuously but the results are funneled to the center periodically and evaluated by computer. In addition to the 50 monitoring points, three technicians rove inside the plant and make instantaneous analysis with special portable equipment. There is also a mobile laboratory operated outside the plant that has continuous recording instruments and reports unusual concentrations back to the control center. The items monitored are SO_2, H_2S, CL_2, nitrous gases, CO_2, CO, hydrocarbons, dust, and aerosols.

Fig. 1 In the center of the works area a swiveling and tilting TV camera with zoom lens is mounted in a height of 60 meters. It transmits visible emission to the monitor in the measuring center.

There is a separate SO_2 network of 12 continuous SO_2 recorders in the area up to 6 miles distance to obtain information about SO_2 ground level concentrations. These data, in conjunction with simultaneous weather data, are evaluated by a computer program which reveals gas distribution and dilution in the atmosphere under different conditions. This computer program also determines the proportion of contribution by each source, industrial and nonindustrial, taking into consideration factors of wind conditions and other weather data.

In addition to the recording stations, the center also has a visual monitor in a television camera (Fig. 1) with a zoom lens and remote position control. This is used to make routine checks of the area for visible undesirable emissions. The camera makes a complete 360° circuit every 15 min.

Many of the measuring systems used are not off-the-shelf items but have been developed by the engineering department at Bayer. New measuring processes were developed so that parts of 1 in 1 billion could be accurately measured and recorded in the atmosphere. Some of the analytical instruments developed have been licensed to manufacturers of measuring instruments for commercial availability. This is an example of how income has been derived from a function considered an expense, helping to defray pollution control costs.

Bayer has developed a patented double contact sulfuric acid production process that reduces SO_2 emission from 2,000 to 200 ppm.

The *combustion waste plant* was developed to dispose of chemical wastes that are difficult to render harmless by conventional methods.

This waste differs from garbage in three different aspects: one, they are not only solid but liquid and pasty; two, their mean heating value is two to three times that of conventional municipal garbage; three, when burned, they develop much more soot, smoke, and odor than normal garbage. This is why Bayer developed a large pilot plant in 1961 to provide data for the construction of a full-scale combustion waste plant in 1967.

From Fig. 2, the plant's operation is as follows: The solid and liquid industrial wastes are transported from the plant to the combustion plant about 1.9 miles away by trough-tipping vehicles or in movable containers, via an access road especially provided for this purpose. Solids, difficult to ignite, are stored intermediately in a bunker (1); bulky material is previously reduced in size in an impact pulverizer (11). A grab crane hoists the material to the feed hopper provided at a height of 43 ft from where it reaches the rotary kiln (2). At temperatures of 2200°F and a residence time of 30 min in the kiln the wastes are burned completely. Even materials having an offensive odor are reliably destroyed at this temperature. The resultant liquid slag flows into a trough filled with water (10) and, upon discharge by means of a chain conveyor, is deposited on the dump. The volume of the wastes is reduced by this procedure to such an extent that only one-twentieth of the quantity charged need be carried to the dump in slag form. Combustible liquids are heated up to pumping consistency in tanks heated by steam (12) and are pumped via pipes into the steam pressure type special evaporation burners, which are installed at various places (9) of the plant.

The fumes, which result from the combustion of solid and liquid wastes and which have a temperature of 2200°F, are cooled to about 840°F in a steam boiler (4) and to 570°F in a subsequent air preheater and freed in a high-efficiency electrostatic gas precipitator (5) from 99.5 percent of the entrained particles. A suction-pressure blower passes the waste gases still at about 540°F into a smokestack 390 ft high, at a rate of about 39 fps. This ensures good dispersion of the waste gases in the atmosphere, and the environments are not affected. Steam is produced while the fumes are cooled down in the boiler. This steam is fed to the plant system via a pipeline 1.7 miles long. Wastes which, when burned, produce fumes of intensive odor, e.g., hydrochloric acid, are destroyed in a combustion chamber (8) of special design. Because of the corrosion risk involved for the heating surfaces, the heat of these fumes cannot be utilized in the steam boiler. The fumes are scrubbed out with water, are subsequently freed from dust, and are added at a temperature of about 158°F to the main stream.

Steam produced is approximately 25 tons/hr and is pumped back to the main plant 1.7 miles away at 570 psi and 750°F. In order to start the plant, 24 tons of solids and 10 tons of liquids are accumulated. Daily combined capacity is 100 tons. Supplemental oil is used at the rate of 0.6 percent of the total heat for the boiler. Cost of the plant was $4.5 million, and operating costs are estimated at $750,000 a year, but one-third of this ($250,000) is returned in value of the steam. The plant also had an auxiliary stack to be used if the main stack was down for any reason.

The *domestic sewage plant,* completed in May, 1966, is a pilot plant operating on the activated sludge principle and will be used as the basis for design of the inte-

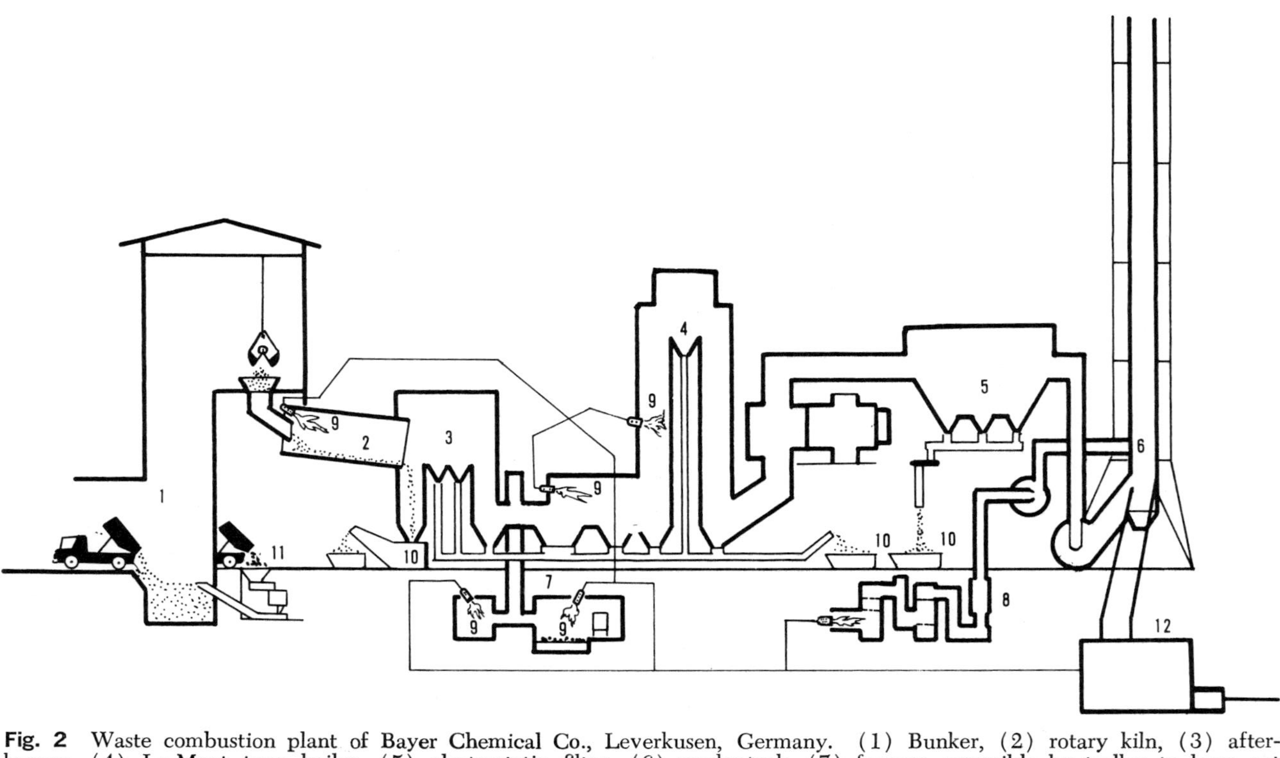

Fig. 2 Waste combustion plant of Bayer Chemical Co., Leverkusen, Germany. (1) Bunker, (2) rotary kiln, (3) afterburner, (4) La-Mont type boiler, (5) electrostatic filter, (6) smokestack, (7) furnace accessible by trolley to burn out combustibles in equipment, (8) combustion chamber with scrubber, (9) burners using liquid wastes, (10) slag trough, (11) impact pulverizer, and (12) combustible liquid tank.

grated plant which is described below. It is designed for 60,000 persons and processes the sewage from Bayer laboratories and some offices. It employes mechanical air agitators and has an archimedes screw with a lift of 20 ft which feeds effluent to the plant. The reason for this type of lift is that it can tolerate varying loads with less power expended and does not separate the activated sludge from the incoming sewage as would be the case with centrifugals. One unusual feature is in the first stage in which pH control values are kept between 6.0 and 8.5. Two-thirds of the settled sludge is returned to the agitation tank and one-third to the archimedes screw.

A *joint waste project* is being undertaken with the local and Wuppertal River Authority for a sewage purification plant and landfill. It will employ 20 chemists when in full operation in 1972. Total cost of the project is estimated at $50 million, of which Bayer pays $36.5 million and governmental agencies the rest. This project involves 247 acres of land and the rerouting of two rivers, the construction of new roads, dikes, dams, and bridges. When the project is finished, it will provide a sewage plant with a daily capacity of 4,768 million cu ft. Industrial and municipal sewage will be mixed on a 1:1 ratio basis. Bayer uses approximately 600,000 cu meters of water/day, and 20 percent of this contains organic and inorganic waste. BOD of the effluent will be approximately 20 as it is fed into the Rhine River. In addition, a landfill of 148 acres will be used for noncombustible wastes, and disposition will be under rigid controls.

Refuse Incineration

There are many municipal incineration plants in Europe, and the one at Munich is chosen as one of the more successful. The city of Munich was running out of area for refuse burial and decided to incinerate it.

To compound the problem, European refuse by American standards is inferior in several significant aspects.

For one thing, European refuse has a lower calorific value, 2,700 Btu/lb vs. 4,500 Btu/lb. It also has a higher moisture content, 28 vs. 17 percent by weight, which makes it more difficult to burn. It also has a greater content of dust and fines, which tends to increase the amount of particulates emitted. To compound this problem, European standards of particulate emission have been four to five times more stringent than our own until recently. Upon viewing these facts, one would think that the burning of the refuse was going to be very expensive and hardly worth the efforts. Yet they did burn the refuse and on a sound economic basis.

Since 45 percent of the heat from incineration is wasted, they decided to use this heat to (1) generate steam to run electric generators and (2) generate hot water for domestic space heating, both of which are salable items.

The result was that electricity is generated for $01.125 per kwhr, and the average domestic heating bill is $90 per year per customer. Both these rates are very competitive by European standards. The rates would be much lower, but coal, which is burned in conjunction with the refuse, costs $20 a ton.

Coal must be used to maintain a constant operating efficiency. Refuse varies greatly in calorific value and its burning rate is difficult to control; so the plant is designed to burn refuse at the minimum demand rate of 40 percent, and pulverized coal is used to meet the rest of the load.

The key to their efficient operation is the refuse burning system, which utilizes the Martin grate (shown in Fig. 3). This system permits burning rates of up to 95 psf/hr, or approximately 50 percent higher than United States systems, and requires only one grate per furnace. Dry burning, burning, and afterburning are done on a very wide but short stoker. Heavy serrated cast steel bars make up the grate surface. A reciprocating device feeds the grate, which is sloped about 30°. Under the grate are zonal air compartments which control the air from about 4 in. W.G. at the top to ¼ in. W.G. at the lower end. This air control compensates for changing moisture content and the thickness of the fuel bed as the refuse progresses.

There are many other innovations that account for an efficient operation like having a negative air pressure in the storage pit so that dust laden air and odors are drawn through the grates as primary air for combustion and using electronic automatic weighing equipment that weighs each bucket load. The hopper is designed to feed refuse into the furnace continuously. Arching and bridging are prevented by having two sides of the hopper vertical and two sides sloping. At the discharge end of the hopper, a swing gate prevents backfires and backdrafts. Continuous water flow through nonpressurized jackets handles cooling, and oscillating pushers regulate feed rate into the furnace. The furnace itself is of water wall construction, and only the lower portion is refractory lined.

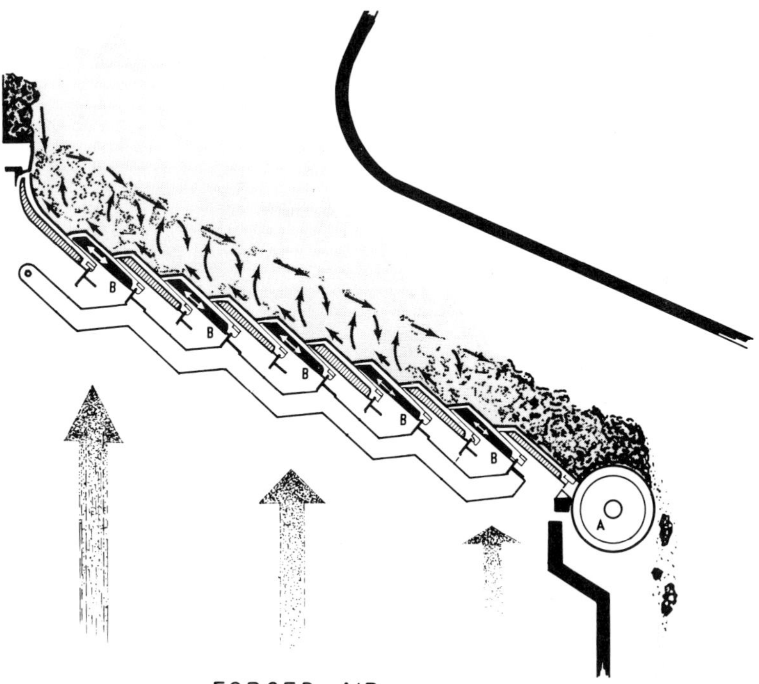

Fig. 3 Martin gate system. Bars marked *B* are movable and push uphill in a slow reciprocating motion against the downward refuse flow. Each stroke, part of the burning material, is pushed against and under the unignited fuel. Relative motion of about 1 in. between adjoining bars will free any material lodged in the approximately 1/16-in. interbar gap. The drum (*A*) on the end is of variable speed to control discharge. Residue is 2 to 3 percent of the original refuse.

Even with this efficient refuse burning, they would not be allowed to operate unless they had effective emission control. In Germany, stack discharges from large-sized incinerators must be limited to 150 mg/cu meter NTP for CO_2 content of 7 percent, and this is equivalent to a residual in the cleaned gas of 0.065 grain/cu ft NTP. Sulfur is not a problem, and they do not have SO_2 control. Sulfur content of refuse is 0.3 percent and of the coal 1 percent. Emission of SO_2 is on the order of 0.15 to 0.17 grain/cu meter. What little SO_2 is emitted is dispersed by the high stack.

When firing 40 percent refuse and 60 percent coal, the total dust collecting efficiency of the precipitator is 99.75 percent. When only refuse is fired, the total dust collecting efficiency is 99.9 percent and the residual only 0.0078 grain/cu ft NTP. It should be pointed out that paper particles are no problem as special electrodes with capturing profiles do not let light ash particles become reentrained by the gas flow once captured. It should be emphasized that no additional mechanical collectors are used in the cleaning of the stack gases; only the elctrostatic precipitator is used.

The secret to obtaining these high efficiencies is the careful control of two things, the temperature and the velocity of the gas. Cooling is accomplished by injection of water into the gas stream in such a way that evaporation time of the water is short. Velocity of the gas, of course, is regulated by the design of the system.

In order to decrease down time by arc-over in the precipitator, they use two series arranged and electrically independent precipitation fields. There is one power pack per precipitator. Each series precipitation field has a different voltage imposed upon it. The leading field has approximately 33,000 volts and the leaving field 40,000 volts. They also have complete parallel operation of the precipitator so that it can never be completely inactive unless arc-overs occur in both parallel branches at once. Arc-over is rare, by the way, and total full effective operating time is

Fig. 4 Waste oil incinerator, Munich.

quite high. The inlet of the precipitator has three perforated plates in series installed in the gas inlet transition to ensure even gas distribution.

The total operation is very efficient, clean, and quite odor free because of the high operating temperature of the system. A total of 250 megawatts of electricity is produced from three generators. Total cost of the plant is $37.5 million, of which $6.25 million is assigned to the refuse burning equipment and $1.25 million to the district recirculating heating system. The question remains, is this type of operation applicable to the United States? The answers lie in the economics of the local community's ability to dispose of trash, and its need to produce electricity. Certainly, we can always use the higher operating efficiencies of its grates and precipitators in our incineration, especially as our standards approach theirs.

Incineration of oil and chemical wastes, Munich, Germany This is a private venture in which several large companies cooperated.

The experimental incinerator (Fig. 4) is a horizontal barrel type about 20 ft long and 3 ft in diameter. The operating temperature is kept at 1300 to 1400°C to ensure that no explosions occur when burning hydrocarbons. It has a capacity 2,000 tons/month.

The average oil content of the residues received is about 10 percent. When the residue arrives at the plant, it is placed in a tank and the oil fraction is separated by appropriate methods. These include the addition of water, ferrous sulfate, caustic soda, or other chemicals. Flotation is used once in a while. The furnace cannot handle large quantities of ash, and for this reason, soil containing oil must be sent to another experimental plant in the vicinity—which from information given is a rotary kiln in which oil is burned from the soil.

Initially, great difficulty was experienced in blending the various materials received, but now they have by trial and error discovered appropriate blends for burning. A 30 percent cyanide solution was sprayed into the furnace after all other materials have been removed, and it was found that the cyanide is completely oxidized after 6 hr of burning. Phenols and solvents were also burned along with the oils and were successfully disposed of.

The unit is experimental, but results were highly encouraging, and it seems reasonably certain that this method will be adopted generally, particularly along the Rhine and in similar highly industrialized areas. Not only did it remove noxious oils and chemicals, but it did not create any water or air pollution problems.

This is an example of how European industry has taken the initial step in solving pollution problems. These first solutions will not be the last, and the cooperative effort of German industry will continue to discover more effective methods. The keynote here is that several industries banded together to find a joint solution by which they could all profit. The German government encourages this action but does not directly contribute.

Cement Industry, Holderbank, Switzerland

Cement production is normally a very dirty business, but the Holderbank plant has a clean, dustless operation and in doing so has remained highly competitive.

The Swiss Society of Manufacturers of Cement adopted severe dust emission standards in 1964. They did this to rid themselves of claims for dust damage, but there was no law to force them to reduce dust. They enforced their standards within the framework of the society.

The standards laid down were that dust emission shall not exceed 100 mg/cu meter in all new installations, and within 10 years it shall reach a level of less than 150 mg in old plants. Most of the old plants have reached their assigned level already. They enforce these standards themselves and have a standards section that has free access to all plants. It measures dust emission with the same instrument at each plant.

In 1964, the Holderbank plant had emitted 2,000 mg/sq meter/day, dropping to about 100 mg a mile away. This has been reduced to 100 mg maximum in the immediate vicinity of the plant and 50 mg 1 mile away, which is approximately the background level.

They achieved this level using electrostatic filters. They chose the electrostatic filter for two reasons; one was they eliminated the human element needed to correct a bag fault while the electrostatic filter had automatic controls that were virtually foolproof. The other reason was that the maintenance cost of the electrostatic filters was much less. They conducted a 1-year test with an electrostatic filter on one kiln and a baghouse on the other, and the results favored the electrostatic filter.

Each electrostatic filter is sectionalized and utilizes parallel operation much the same as at the Munich incinerator. Each section is fed and controlled independently. Down time for a section is rare, and this does not close down the operation. From the records, a section was down about once every 3 months for 8 hr. Efficiency of the precipitators was 99.75 percent.

The secret of this efficiency is the careful control of gas temperature and velocity. Gas feed velocity for the wet kilns is about 200 fpm and for the dry kilns about 150 fpm. The most effective operating temperature in the precipitator is 150 to 200°C, the actual value depending upon local conditions. When dry kilns are used, it is always economical to cool the hot gas by water spray to this temperature to decrease the volume. At this temperature, all the water vaporizes and no sludge problems arise. Continuous recorders are installed but do not work satisfactorily.

Cost of precipitators runs between 8 and 14 percent of capital costs depending on size of plant and location.

They have also realized a 40 percent reduction in man-hours to produce a ton of cement compared with similar United States plants. Some of the increased efficiency must be attributed to superior working conditions. The area around the plant was green and lush. Flowers grew within 15 ft of the kilns. The plant floor was exceptionally clean.

Pulp and Paper Industry, Morrums, Sweden

When AB Morrums Bruk decided to build a pulp mill near the Swedish resort town of Morrums on the Baltic Sea, they knew that the control of pollution would be extremely important if they were to receive a permit to build the mill.

They sought expert advice to control the problem of odor release and settled on a system designed by the British Columbia Research Council of Vancouver, Canada. Even so, nobody knew for certain whether the proposed blending measures would work in their particular application. But it was decided to go ahead and build the mill. It was quite a gamble, as the additional cost of the pollution equipment was $4 million, or 20 percent of the total capital investment for this 400 ton/day facility.

Briefly, the pollution aspects are: (1) Steps are taken to remove the noncondensable sulfur compounds such as H_2S and mercaptans from the gas streams. These are scrubbed out, and the gas, which still contains trace quantities of these constituents, is then blended with chlorine dioxide and chlorine water from the bleaching stage of the plant so as to result in neutralization. (2) The effluent streams are blended and all H_2S is either recovered as sulfur or oxidized. The solid material is settled out in ponds. (3) Cyclones and an electrostatic precipitator are provided for the furnaces.

In the early stages of design, the following steps were taken to reduce odors:

1. They eliminated a direct contact evaporator. This reduced large quantities of hydrogen sulfide released from this source and also provided additional heat economy.
2. An odor control system was installed that included a two-stage scrubber and a black liquor oxidation tower for deodorizing the gases from various sources in the mill.
3. The exhaust from the washer hoods is mixed with the chlorine containing exhaust from the bleach process, and the brown stock washers use contaminated condensate.
4. The recovery furnace was designed of ample capacity and with maximum control of combustion to prevent H_2S from being emitted.

The key to the reduction of odors in the mill is the oxidation tower and the scrubbers. The tower consists of two packaged units operated in series with respect to black liquor flow and in parallel to airflow. Gases from the digester and noncondensable gases from the evaporators are mixed with the airstream entering the towers. These gases then pass through the oxidation towers, where the hydrogen sulfide is absorbed in the alkaline black liquor. The methyl mercaptan is absorbed and oxidized to methyl disulfide. The exhaust gases, containing a residual of methyl sulfide and methyl disulfide, then go to the two-stage scrubber.

The first stage of the scrubber contains chlorine water piped from the seal box of the chlorination washer, and the exhaust gases are passed through the deluge. The methyl disulfide is oxidized to methyl sulforyl chloride or methyl sulfuric acid. The methyl sulfide is converted to methyl sulfoxide.

In the second stage of the scrubber, the gases containing some chlorine and small amounts of acids are exposed to an alkaline spray, which comes from the cold caustic extra action stage in the bleach plant.

The gases released from the second stage are almost odorless and smell similar to a laundry bleach solution. Tests taken with a gas chromatograph show that 99.9 percent of the methyl disulfide, methyl sulfide, and methyl mercaptan are removed from the scrubber exhaust gas. There is no detectable trace of hydrogen sulfide.

As stated previously, the effluent goes to settling ponds. Rosin floats to the surface is recovered for use in other products such as lacquers. The clean effluent is discharged at a distance of 3 miles into the Baltic. It has a low pH of 3.5, but the Swedish authorities stated tests had proved that diffusion at the end of the line rendered the concentrations so low that absolutely no harm was detected to fish or other forms of marine life.

This is an example of what can be accomplished to reduce obnoxious odors in pulp mills. True, first costs were high; but this facility has remained competitive, primarily because it is highly automated and the pollution control aspect was planned in the beginning.

Under adverse temperature inversion conditions, a faint odor is detectable approximately 1 mile from the mill. Normally, the odor is not noticeable.

Metal Industries

Hoesch Steel Works, Duisburg, Germany Hoesch has a research laboratory for pollution control with over a hundred personnel. They work on numerous problems, but the one broad area of concentration is development of reliable continuous sampling apparatus. They produce a unit for manufacturing artificial dust for test purposes. Dust is fed into a blower equipped with an oil burner where it is then passed up a stack. Several test instruments can be mounted in the stack to measure properties of the dust. They test various instruments of different manufacturers, then modify them and determine if improvement in performance is possible. They use oscillating light beams to try to improve performance. The relationship between dust particle size and light value varies with different dusts, and it is difficult to integrate the two functions.

On the roof of the laboratory is a continuous SO_2 recorder (Fig. 5*a, b*) utilizing the principle of peroxide oxidation of sulfur dioxide. The sulfuric acid resulting is then measured by electric conductivity to give a value. The instrument is pivoted so that it points always into the wind. It takes samples every second and records the results; it also automatically integrates the recorded results every ½ hr. The instrument has good reliability, as it operates without attention for about 1 year.

The plant contained six bessemers of 170 tons capacity that utilized a wet gas cleaning plant to reduce pollution emission. The gas was drawn off the furnaces by a fan of 150,000 cu meters/min capacity. The gas temperature was then reduced from 1600 to 800°C, then passed through a spray chamber, where the temperature was further reduced to 70°C, i.e., just above the dew point. Most of the dust would fall out of the gas at this point in the process. The gas was then passed through a cyclone, where a small additional quantity of dust was extracted as well as any water droplets that remained. From there it went out the stack. It was originally intended to burn the gas, which contained 90 percent CO, but with six furnaces going the quality of the gas varied so much that the flame often went out. Originally, the plant had a cooling tower, but explosions frequently occurred at the air duct connection between the bessemer furnace and the dust take-off.

To correct this condition, they designed a venture chamber of 0.65 meter in diameter which was equipped with an intense water spray at the throat. This has proved very effective in elimination of explosions and as an added benefit proved to be an excellent collector. This system was not in full-time operation but looked very promising.

The director of the laboratory stated that some of the criteria under which they operated were as follows:

1. The emission limit for SO_2 is 0.7 mg/cu meter at 20°C.

2. The federal government monitors the area for SO_2 emission, and if the above value is exceeded, they must change to a low-sulfur fuel. If conditions do not improve, they then must stop production. Emission is monitored in 1 sq km increments in industrial areas.

3. For existing factories, emission values permitted are lowered from time to time as technology improves for individual industries.

4. New factories must obtain permits before they are allowed to build. They must prove that emission will not be greater than 0.35 mg/cu meter before they are allowed to start production.

5. There is the right of appeal for any disagreements in the courts.
6. There is no regulation concerning size and height of stacks.
7. Coke oven effluents must be treated by the individual factory. The levels of dissolved solids permitted are controlled by the local river authorities.

Violations and court cases are extremely rare. Industry, through the national engineering societies, has a self-policing policy which works quite well.

(*a*)

(*b*)

Fig. 5 (*a*) Hoesch steel works, Duisburg, Germany. Continuous SO_2 recorder on plant roof. (*b*) SO_2 recorder on plant roof.

Sulzer Works, Oberwinterthur, Switzerland The Sulzer foundry is a clean, noise reduced operation in a normally dirty and noisy business. This they accomplished without legal action being taken concerning pollution control. They considered their actions good business judgment to foster good will with the townspeople and to provide improved working conditions to obtain more efficient production.

To obtain better working conditions in the foundry, they use artificial ventilation. The air changes at the working position vary between 6.5 and 10 changes per hour, the more frequent changes being for the more arduous working conditions. The overall rate of change for the foundry is 2.5 changes per hour. To obtain local variance in cfm, the air is ducted to the location needed. At high dust points the area is completely enclosed.

The foundry building is 450 ft wide, 668 ft long, and 49 ft high, except in the heavy foundry section, which is 61 ft high. The capacity of the foundry is 24,000 tons/year, one-quarter being cast steel and the remainder cast iron. To produce this tonnage, they have two hot-blast cupolas of 12 tons/hr each, four electric arc furnaces from 3 to 14 tons each, six medium-frequency induction furnaces of ½ to 10 tons each, and a large annealing and an elevator annealing furnace.

To cool the foundry, they pass 40.67 million cu ft of air through the building per hour, of which approximately 28.1 million cu ft is makeup air and the rest recirculated air. The incoming air is filtered electrostatically before use. The expelled air is cleaned by multivane water scrubbers. The air is then dewatered and passed through woolen dust bags. Approximately 1.5 tons of dust each hour is removed in this matter.

Water to operate the foundry comes from city mains (two-thirds) and from a deep well (one-third). The water from the well is chlorinated before use. The lime content of the water is high, and precipitation occurs at 50°C. It therefore requires treatment to prevent precipitation of lime when heated as cooling water. In winter, this water is used to supplement the heat of the foundry by passing it through a heat-exchange system. In summer, the excess heat is dissipated through three roof heat exchangers. There is a third industrial water circuit that is nonpotable. This water is used for dust settlement, dust control spraying, and sand washing. It is treated in a settler with flocculents. Seventy percent of the water is returned to the circuit, and 30 percent goes to the municipal sewage system. The settlement sludge, approximately 35 tons dry weight/hr, is filtered and then resettled, dried, and transported to burying pits. Candle filters are used for filtration.

The cost of the operation for the air and water pollution equipment was a little over $6,000 per month.

Certainly, this cost has proved well worth the investment by providing better working conditions and admirable community relations. They have a noise abatement program that is very effective. The result is a clean, odorless, and very low noise foundry that is extremely competitive.

The town homes abut one wall of the foundry and farms the others. The area surrounding the plant is dust free, and there is hardly a trace of foundry noise. The immediate area is a very pleasant place to live.

Duisburg Kupferhutte (copper refinery), Germany The Duisburg copper refinery was formed originally to process pyrite for the production of sulfuric acid, which it does not do today at all. Today it receives pyrite from all over Europe and extracts copper, silver, gold, zinc, lead, nickel, cadmium, cobalt, and thallium from the pyrite. To produce the extracts, they give the pyrite (after adding rock salt) a chloridizing roast and extract the majority of the materials by leaching. The residue from the leaching is a very high grade of iron ore, which is put through blast furnaces from which silver bearing lead is tapped.

The chloridizing roast produces a product containing 6 g/cu meter of hydrochloric acid and sulfur dioxide. The SO_2 is washed out and then also the hydrochloric acid in towers equipped with Pauli rings. This yields a 7 percent hydrochloric acid solution which is used for extracting metals during the leaching stage. Sixty percent of the residual sulfur in the pyrite and in the coal is converted to sodium sulfate. Zinc chloride is recovered at 85°C by a plastic lined electrostatic precipitator.

All metals are recovered stagewise from the leached solutions, and at the end of the processing a calcium chloride solution is discharged into the Rhine River. A salt furnace is used for chloride roasting. The dust is not recovered by bags because no material has been found which will withstand the gases. A spray extractor is used, and cuprous and cupric chloride have to be cleaned weekly from the extractor area.

They have a requirement for large quantities of Rhine River water for cooling, and this is cleaned by a thorough water treatment system. After the cooling use, the water is used for processes and is re-treated between stages and reused as often as possible. The complex processes involved will not permit purification at the final stage. The aluminum sulfate sludge from the water treatment is dumped downstream back into the Rhine River. The levels of metal discharge into the Rhine River are not disclosed. SO_2 and HCl were extracted from the roasters at considerable cost. SO_2 emission from the stacks was monitored continuously by an optical apparatus using infrared rays. Water cleaning processes were considered equal to or better than comparable United States plants. It was stated that 10 percent of their capital outlay was for pollution control equipment.

To illustrate the extent to which European industry will go to compare products, Kupferhutte ran a comparison test for 2½ years using four different makers of SO_2 recorders. Two were of United States make and two German. They were operated under near identical conditions, and a careful analysis was made of operating characteristics, breakdown time, accuracy, and operating costs. This practice is not unusual, and the usual result is a modified off-the-shelf item that meets the specific requirements of the industrial operation. There was no legal obligation to measure SO_2 and certainly not to the degree of conducting research on the best method for reliability and accuracy. But industry does feel a moral obligation to protect the public interest to the best of its ability.

Electrical Manufacturing

Siemens, Munich, Germany The total integrated system approach was used at the Siemens plant to control its pollution problems. In their automatic circuit board plating plant, water was recovered from spent solutions and metal from plating solutions. In the water conditioning plant, they produced and circulated 25-megohm deionized water. But the most impressive pollution control contribution was in the basic and applied research area.

Out of the 230,000 employees in Siemens, over one in four is directly concerned with research. This is a staggering percentage, and though, of course, only a small percentage of this is concerned with pollution control research, they are certainly doing their share, as illustrated by the following case.

Since 1954, Siemens has been working on a new mechanical dust collecting system that they call the "tornado dust collector." The system separates dust from 4.5 microns upward and, depending upon the specific application, fractional efficiencies of 97 of 99.7 percent. The operating principle is given in Fig. 6. The collector is very versatile and can be readily adapted to many applications. Here are listed some of the features to illustrate this versatility:

1. Efficiency is constant: (*a*) up to 600°F, (*b*) up to a grain loading of 90 grains/cu ft, (*c*) with a dirty variance of —30 to +15 percent of nominal flow.
2. If dirty gas flow varies, the secondary gas flow can be regulated to yield a constant separating efficiency over the whole load range.
3. There is no abrasion of the collector walls, as the particles are prevented from reaching the walls by the secondary gas flow.
4. There is no particulate buildup in the collector.
5. The collector can be mounted vertically or horizontally.
6. It can be operated in parallel banks without undue regard to distribution of flow rates to individual units.
7. The collector is virtually maintenance free except for fans.
8. The size is small, 2.5 to 40 in. diameter for 20 to 4,000 cfm capacities.
9. With a high-temperature primary dirty gas, it can be cooled with a cold sec-

ondary gas, and conversely with heated secondary gas, it is possible to dry wet primary gas.

The collector has been used to collect lapping powder used to clean silicon plates. The collection efficiency in this application has been 99.7 percent and the lapping powder is recovered to be reused. As the powder does not reach the walls of the collector, there is no maintenance to it. The unit is also used as a classifier in the production of aluminum and copper powder. With this device, it is also possible to remove water from a water and air mixture. Special units have been designed to remove high dust loads. One such unit essentially consists of two units in tandem for the collection of flour with an efficiency of 99.6 percent.

Operating costs for a single 1,000-mm unit, assuming 75 percent fan efficiency and $0.1 per kwhr, are $0.7 per 1,000 cu meters of gas cleaned. Of course, parallel operation of two cells would lower energy costs but increase capital costs. This cost lies between that of the cyclone and that of the wet washer. The above cost assumed atmospheric air as secondary air. If dirty gas is used, the cost can be reduced one-half, but efficiencies are lower.

There are still other applications in the laboratory stage of development. One such application still using the "tornado" principle is the protection of the SO_2 probe in measuring stack gases. This probe is continually being clogged by the particulate matter in the stack. At Siemens they have developed two miniature collectors and placed them in tandem. The SO_2 probe samples the stack gas in the second collector and does not become clogged.

Another development has been in the field of burners for boilers. Still using the same principle, it is possible by regulating the secondary airflow to have the dust not spiral down to the bottom of the collector but form a ring like a doughnut around the wall. One ring can be formed for each secondary tangential air source. The ring can also be lowered to the next lower level by momentarily interrupting its secondary air source. Instead of dust, a fuel such as coal dust or oil or a mixture of both is fed into the collector, formed into a ring, and ignited. It can be seen that the rate of burning can be controlled by both the primary and secondary source of air. As burning progresses, the ring is lowered to the next level until it reaches the bottom, where the residue is deposited in a hopper. It is reported that radiant

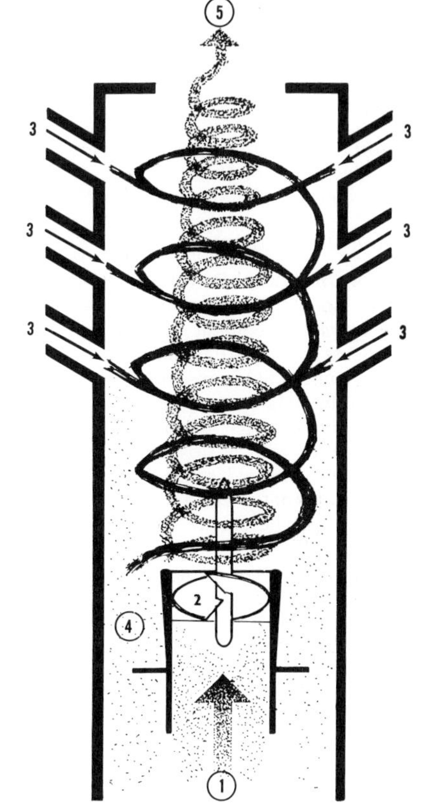

Fig. 6 The tornado dust collector. The collector consists of a cylindrical tube into the center of which the dust laden gas stream enters (1). The dirty gas stream flows past a stationary turning vane (2). In this primary flow, centrifugal forces direct the particles toward the wall of the collector. Near the wall, the particles are assisted by a secondary gas stream (3) from several tangential nozzles which are directed downward. The particles are carried into the gaseous stream by centrifugal force plus drag forces at the vortex. The secondary gas stream then transports the particles (4) into a hopper by the collector. The cleaned gas leaves the collector at the top (5). The secondary airstream may consist of the cleaned gas (5), dirty gas (1), or atmospheric air depending upon particular operating requirements.

burning efficiencies are increased 15 to 20 percent. It can be readily understood that particulate matter in the stack is greatly reduced.

In conducting these studies at Siemens, they also discovered that SO_3 was a better indicator of burner efficiency than SO_2. So they developed an SO_3 recorder that takes about 1 sec to indicate the amount of SO_3 in the stack gas. They do this by using alpha-iron oxide to absorb SO_2, which changes to SO_3 at 650°C. The SO_3 is absorbed by water droplets. This is heated and condensed in a second flask, where the electrical conductivity is measured to yield SO_3 concentration. These are but a few examples of industrial research being conducted in Europe.

Petroleum Industry

Shell Refinery, Pernis, Holland This refinery has a capacity of between 16 and 18 millions tons of crude oil per year. They burn approximately 3,000 tons of 3 percent sulfur content fuel a day. This had led to problems in controlling H_2S gas, SO_2 CO, and dust emissions.

To control SO_2 emission, they replaced twenty-five 250-ft high stacks by one stack 700 ft in height so as to have greater dispersion of stack gases. The stack is 15 meters in diameter at the bottom and 10 meters in diameter at the top. It is made of concrete with a 1.5-in. steel inner liner. There are three main ducts to the new stack from the old stacks which contain manually operated butterfly valves. There is one automatic valve to compensate for varying wind velocities across the top of the stack. The exit velocity of the gas is 50 mph, and this provides enough velocity for the gas to penetrate through any normal temperature inversion. This provides a ground level concentration of between 0.1 and 0.15 ppm SO_2. Cost of the stack system is $4.8 million.

Catalytic dust is extracted by cyclones and is a small problem. Carbon monoxide is piped to the boilers and burned by a special CO burner. H_2S is burned or used for sulfur recovery. All pop valves open into a closed system; so there is no discharge of pollutants. The H_2S is absorbed by the saltwater stripper, is degassed by steam, and is then diluted with cooling water and discharged into the Rhine River. There are no towns below the plant; so this practice is acceptable. However, plans are in the making to provide an ocean outfall in the North Sea.

The refinery recovers oil with an elaborate set of tilted plastic separators. The effluent is presently 50 to 80 ppm, but this is being reduced to 30 ppm. Even lower effluent values will be possible when a new chemical treatment plant is built.

Spent caustic, phenols, and thiophenols are collected in a closed system and stored in a tank. The contents of the tank are then pumped into oil tankers as ballast, and this is discharged in mid-Atlantic Ocean.

SEWAGE TREATMENT

Two methods of sewage treatment that have been developed in Europe to a greater extent than in the United States are the oxidation ditch and the stainless-steel cloth microstrainer.

The microstrainer has been employed since 1950, principally in England. It was developed as a mechanical "tertiary" treatment for the removal of algol growths before sand filtering. It can also be used as a substitute for sand filters, humus tanks, for the treatment of industrial water supplies, and as step in the "micellisation-demicellisation" (M.D.) process which employs microstraining, ozonization, and sand filtration.

In principle, the system consists of a revolving drum which has micropore stainless-steel panels attached to it. The pores are 60, 35, or 23 microns in diameter. The upstream side of the drum is open and the downstream side is closed. As the drum rotates, it collects impurities in the mesh; these impurities are backwashed out of the fabric at the top of its rotation cycle. The waste water is discharged into a hopper attached to the hollow axle of the drum.

The backwash water is drawn from the downstream side of the drum and pumped through a row of self-cleaning, adjustable jet nozzles. The debris-laden backwash water is then disposed of in an appropriate manner. In sewage treatment it would be returned to the inlet of the plant to be mixed with raw sewage.

In the tertiary treatment of effluents, the typical removal rate of solids is 70 to 80 percent and BOD 60 to 70 percent with a 23-micron mesh. With a 35-micron mesh, solid removal is 50 to 60 percent and BOD removal is 40 to 50 percent. In the removal of microorganisms, it can be over 97 percent efficient.

Total installed costs for the system are in the neighborhood of $30,000 to $60,000 per mgd capacity. The lower figure represents a capacity up to 50 mgd for the 35-micron mesh and the higher figure a capacity up to 3 mgd for the 23-micron mesh. Total operating costs are approximately from $4.75 per mg treated to $6.30. This is considerably less than with sand filters of equal capacity.

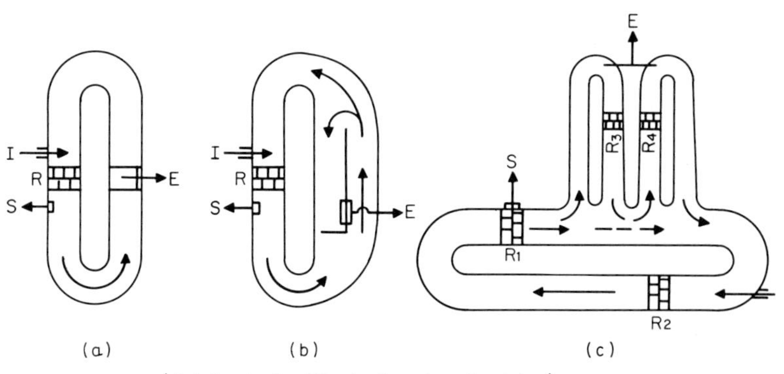

Fig. 7 Types of oxidation ditches. (*a*) In the intermittent system, the single ditch functions alternately as an aeration and clarification basin. To function as a clarification basin, the rotor is switched off and the sludge allowed to settle. Raw sewage is then pumped into the ditch and, as the water level rises, the effluent is siphoned off. When the proper amount of effluent is siphoned off, the rotor starts again and the ditch becomes an aeration basin. (*b*) In the semi-intermittent system, an additional parallel channel is used as a clarification basin and operation may be nearly continuous depending upon load conditions. (*c*) In the continuous system, two additional channels are used, one of which is always acting as a sedimentation basin. The effluent is siphoned off from the channel that is inoperative. In this system, rotors R_1 and R_2 operate continuously and R_3 and R_4 alternate their operating periods. Speed of rotation of the rotor is about 75 rpm. The sludge is normally pumped into drying beds. Sludge load is extremely low, from one-tenth to one-twentieth that of the conventional activated-sludge process.

Microstraining has proved to be an effective and economical means of providing tertiary treatment of sewage effluents and should be considered as a method of meeting higher effluent quality.

The "Pasveer" oxidation ditch has been used successfully in Europe since 1954. Experiments have been conducted since then by TNO (Research Institute for Public Health Engineering) Delft, the Netherlands. This method was first developed for small communities of less than 10,000 inhabitants where conventional systems are quite expensive. Today it has proved to be an economical method for treating industrial wastes of up to 250,000 inhabitants capacity.

The system basically consists of an oval ditch which is shaped like a race track and has a rotor capable of supplying an oxygenation capacity-to-load ratio of 2. The ditch may or may not be lined depending upon local soil conditions. The oxidation of the fresh sludge and the bacterial floc formed during purification is so complete that open drying of the sludge does not produce objectionable odor. Thus, neither a digestion tank nor a primary sedimentation or humus tank is needed.

The aeration rotor has several functions. It acts as a comminutor, keeps the floc in suspension and supplies oxygen to the mixture of sewage and sludge. The sludge may be separated from the effluent either by intermittent, semi-intermittent or continuous operation. The single ditch is used in the intermittent operation (Fig. 7). In the semi-intermittent system, one additional parallel channel is added to the ditch, and in the continuous system, two additional channels perpendicular to the ditch are added.

The raw sewage is pumped directly into the ditch through a bar screen. For domestic sewage, the volume of the ditch is calculated at 300 liters per 54 g of 5-day BOD load disregarding the volume in which the quantity of BOD is present. In order to keep the floc in suspension, a velocity of 30 cm/sec is maintained at the center of the ditch. The suspended dry solids content of the mixed liquor is maintained at about 4,000 mg/l, thus maintaining a low BOD per floc ratio compared to conventional systems. The resulting surplus sludge may amount to 30 g per inhabitant per day of dry solids. Normally, this can be dried in several days under good weather conditions. Cost of operation is between 18 and 20 kwhr per inhabitant per year. This is higher than conventional systems but is offset by lower initial construction costs.

Experiments have been conducted by TNO in the treatment of industrial wastes using the oxidation ditch. By mixing industrial waste with domestic sewage, effluences of 99 percent for phenol, 100 percent for CNS, and 86 percent for COD have been reported. In addition, peak loads of sulfuric acid (pH2) and HCN (30 ppm) have been treated without noticeable effect on the system. Thus, the oxidation ditch can tolerate peak loads of some toxic materials better than the conventional systems.

In winter operation, the oxidation ditch functions very well if the rotor and weir are protected. This usually consists of a cover and some source of heat to provide for normal operation. Ice forming on top of the ditch acts as an insulator and biological reactions continue at a lower rate. Only in extremely cold weather does the process cease.

Both the oxidation ditch and the microstrainer are two European sewage treatment devices that have industrial applications and should be examined for possible use elsewhere.

Summary

This is but a scant cross section of European industrial practices. All countries of Europe, including those behind the Iron Curtain, are aware of the pollution problem and are doing something about it. Most countries, including Russia, in practice, have pollution criteria that are more restrictive than our own.

Prosecution of violators is rare, but compliance with best "practical means" is followed because of diplomatic persuasion. Industry considers it in their best business and political interest to be the leader in adopting the most practical antipollution methods without government intervention.

Industry either conducts research itself or commissions government laboratories to do it for them. There is a general interplay of knowledge among these associated in the pollution control field. This exchange of knowledge crosses industrial and political boundaries.

There is a general spirit of independence and of finding a better way to accomplish the task at hand. Off-the-shelf items are not considered good enough for their particular application. This leads to a large amount of modification of instrumentation apparatus and equipment. Large-scale pilot plants are not unusual to determine the adequacy and economics of a system.

European industry is practicing pollution control with a spirit of "let's get on with it," and they want to accomplish it in the most pleasant manner possible.

ACKNOWLEDGMENTS

The information for this article was gathered primarily during the American Institute of Plant Engineers' Pollution Study Tour of Europe during September and October, 1967. Appreciation is extented to the following for the information provided:

Bavarian government, H. Orchem. RAT Dr. Von Ammon, Water Pollution Control Official, Munich, Germany

VDI (Verein Deutscher Ingenieure), German Society of Engineers—Air Pollution Committee, Herr Ing. Kramer

Bayer, Leverkusen, Germany, Dr. Hans H. Weber, air and water pollution; Dr. T. Wolff, water pollution, dumping; Dr. Hamburg, biological problems: Dr. K. Winkler, air pollution; Dr. Breuer, applied physics, instrument development; Dr. I. Mersch, planning of treatment plants, engineering; Dr. Fabian, incineration plant

Munich Incinerator, Germany, Dr. M. Andritzky, consultant designer; Dipl.-Ing. Gunther Schmolling, Lurgi Apparatebau Gesellschaft mbH; Dr. Bachl, H. Lindner, and I. A. Kollmannsberger of the Nord-Kraftwerke

Oily waste incinerator, Munich, Germany, Dr. Bachl, H. Lindner, and Mr. Mayer of the Nord-Kraftwerke

Oxidation ditch, TNO Research Institute Laboratory, Delft, the Netherlands

Holderbank cement plant, W. W. Thut, Chairman Swiss Cement Manufacturers Association; Dr. Walter Wieland, Director Holderbank Technical Center; P. Lenzin, Manager of the Holderbank Cement Plant; E. L. Kappeler, Assistant Manager, Holderbank cement plant

Pulp and paper plant, Morrums, Sweden, G. Emil Haeger, Technical Manager; Bo Rasback, Plant Manager; Thorsten Peterson, Public Relations Manager

Hoesch Steel Works, Duisburg, Germany, Herr Eickel Pasch, Dipl.-Ing., Director, Air Pollution Control Laboratory, Eugen Gulker, Dipl.-Ing.

Sulzer Works, Oberwintertur, Switzerland, Arnaldo Gianotti, Manager, Production Planning, Foundries

Duisburg Kupferhutte, Germany, Dr. Ing. Wilhelm M. Teworte, Prokurist; Dr. Ing. Habil

Siemens, Munich, Germany, Dr. K. R. Schmidt, Diplomphysiker; Dr. Schuster, Plant Engineer; H. Dr. Schniederman Z. K.; Dr. Robert Hoffman, Consultant

Shell Refinery, Pernis, Holland, Mr. Andrea, Technical Advisor

England, Arnold Marsh, Director, National Society for Clean Air; Dr. S. R. Craxford, Head, Air Pollution Division, Warren Springs Laboratory; H. B. Berridge, Water Pollution Consultant

Tour Members, D. O. Marsden, Group Water Treatment Engineer, Anglo American Corporation, Johannesburg, South Africa; Carmen F. Guarino, Deputy Commissioner-Water Department, City of Philadelphia, Pa.; Herbert F. Lund, Editor-in-chief, Modern Manufacturing magazine, McGraw-Hill Publishing Co.

Section 3

Pollution Control Equipment and Operation

Chapter **21**

Organizing and Planning a Pollution Control Department

HUGH P. GRUBB

WED Enterprises, Inc.,
Glendale, California

ORGANIZATION

The industrial pollution control program deals with the abatement and control of pollutant generating influences caused by the interaction of processes which involve men, machines, materials, and methods. Its purpose is to reduce pollution effects to harmless levels by identifying, measuring, reporting, and controlling the pollution source. Industrial pollution control management is being practiced with an increasing scope and range of application. Programs produce simple and complex, small and large, collection and treatment facilities to protect our environment.

The management of these programs is influenced by the Water Quality Act of 1965, enacted for the establishment of water quality standards which encompass three things:

1. Determination of the uses to be made of a body of water.
2. Assignment of criteria of water quality necessary to support those uses.
3. Development of a compliance plan to achieve and maintain these criteria. It essentially provides for the upgrading of polluted waters and the protection of clean waters.

Three basic considerations are necessary to a concomitant air pollution control program:

1. The air pollution saturation potential, be it liquid, solid, or gaseous, of the industrial environments
2. The industrial area characteristics which are subject to the causes and effects (chemical or photochemical changes) of the pollutant and its economic impact
3. The capacity of the industry to generate air pollutants in terms of types, emission rates, and atmospheric reactions

Air pollutants are generally classified in two basic categories: (1) soiling, corrosive, deposition of which may be injurious to people or property; (2) adhesive particulates which have possible physiological impairments, toxic, carcinogens, and radioactive materials. It is readily evident that science and law are highly interdependent in the industrial pollution control activities because of legal and enforcement problems as well as the scientific analysis of pollution problems.

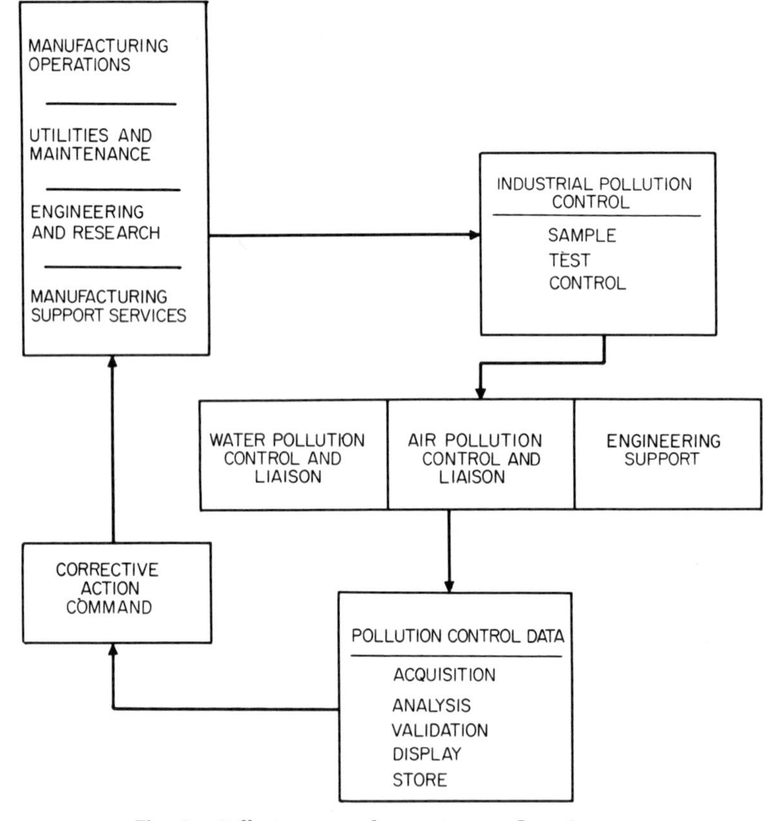

Fig. 1 Pollution control organization flow diagram.

The methodology for achieving these objectives requires the management to (1) organize the activities into a system, examine the data, alter the activities where required, and develop a feedback loop; (2) perform an aggressive empirical study of all pollutants, their sources and behavior, in a systems framework to portray clearly the criteria for adequate pollution control technology and management; (3) develop program objectives, expressed in quantifiable terms, to measure accomplishment and aid the decision making process in the presentation of the objective (see Fig. 1).

Planning concepts are important. Developing and promulgating creative management techniques and practices, which can be adopted into a sound practical industrial pollution control program, requires they be founded on strong technical structure but oriented to communicate to the industrial personnel served as well as the pollution control staff. The manager must move decisively into these areas which will underwrite and promote increasing competence and effectiveness of the program. The aim is to bring about an organizational framework within the plant, whereby supervisors and operators can perform expeditiously in a stable administrative environment that is favorable to productive pollution control and quickly responsive to the plant's needs. Initially management should define the functions, responsibilities, and authority of the pollution control program. The industrial environment must be considered as a total package encompassing water and air pollution, solid waste disposal, occupational health, and radiological health. These considerations, about which we develop considerable expertise, cannot be implemented effectively and independently of economic, social, and ecological factors.

Functions

1. Maintain liaison with the various research, manufacturing, and facilities operations, equipment operations, and processes, both new and revised
2. Investigate these operations prior to their implementation so as to recommend administrative or operational safeguards that may be required to abate and/or control the pollutant
3. Review and approve or disapprove manufacturing, research, and facilities operations governing pollutant generating activities
4. Initiate and implement new or revised procedures to cover new or modified operations
5. Monitor these operations for compliance with the industrial pollution control plan
6. Identify pollutants at their source, measure process loading and environmental loading from pollution emission, and conduct tests needed to assure compliance with appropriate pollution control agencies
7. Initiate corrective action to operations or areas of noncompliance
8. Report the industrial pollution control plan activities to management and such data as are required to the controlling agencies

Responsibility

1. Prevent, in accordance with the controlling authority, the release of any material which is defined as containing pollutants
2. Review administrative and operational procedures covering the use of pollution generating equipment
3. Certify instrumentation, test pollutants in all environmental media affected, and report the results

Authority

1. Direct, as authorized by the company's functional charter of responsibility, the abatement and control of all pollutants which could result in the violation of the local, state, or federal pollution ordinances
2. Approve all operations and practices within the company having pollution generating implications
3. Identify all equipment, and the specific pollution emission, and determine if the equipment requires a permit to operate from the responsible pollution control agency
4. Obtain the necessary operating permits, when required, from the controlling agencies

PROCESS EVALUATION

Establish a procedure for evaluating and certifying process and/or facility equipment to comply with the statutory regulations for pollution control. Manufacturing process documents generally contain a list of all materials used including manufacturing items such as cleaning agents and solvents with a notation of dangerous materials coupled with a list of safety precautions. Large equipment is generally listed separately, together with secondary devices pertaining to its use. In certain states it is necessary to register all sources of pollution and identify the related equipment. The industrial pollution control plan must compile data detailed enough to allow

POLLUTION CONTROL CERTIFICATION

PERMIT NUMBER________ DATE ISSUED ________ BY ________

DECERTIFICATION DATE ________ RECERTIFICATION DATE ________

This document certifies the equipment listed below complies with statutory and pollution control requirements when operated in accordance to the process documents listed below.

EQUIPMENT: ASSET NUMBER________ DESCRIPTION ________

LOCATION________ DEPARTMENT________

MANUFACTURING PROCESS:

NUMBER	TITLE	POLLUTANT			REMARKS
		WATER	AIR	SOLID	

OPERATIONAL CHARACTERISTICS:

PROCESS INPUT (LB/HR)________ NORMAL________ MAXIMUM

PROCESS OUTPUT (LB/HR)________ NORMAL________ MAXIMUM

WASTE WATER (GPM)________ POLLUTANTS________ PPM________

VENTILATION (CFM)________ POLLUTANTS________ PPM________

OPERATING SCHEDULE________ HR/DAY________ DAYS/WEEK________ TOTAL HR

SAMPLING METHOD________ LOCATION________

COMMENTS:________

APPROVAL________ TITLE________ DATE________

Fig. 2 Pollution control certification.

analysis and engineering review so the pollution potential can be weighed against the statutory requirements. The inventory data should indicate equipment and component descriptions, and their locations, which affect the emission of pollutants. It should include engineering information relevant to pollution potential such as a description of the process(es) rate of emissions, size of outlets, and composition of the effluents.

The effective identification and control of all sources of pollutants requires that the industrial pollution control plan functional be aware of all new or revised manufacturing processes and other operations which will alter the pollution loading level of our surrounding environment. This process evaluation should be accomplished immediately during the manufacturing process development stage and prior to the first production run of a new or revised pollutant generating process. The review should evaluate pollution potential inherent to process chemical properties, equipment ventilation, and drain systems. The review of the process documents must consider the physiological effects, cost burdens, and nuisance factors of the process interactions having pollution generating implications.

The plan should determine where the process will be used and when production is scheduled to start, sample the effluent from the process drains and/or stack emissions, and test for concentration and evaluate the results against the pollution control requirements. The characteristics of the pollutants and their emission rates should serve as the basis for scheduling surveillance after certification. Higher concentrations of critical pollutants will necessitate more frequent review. All new pollutant sources should immediately be incorporated into the certification schedule and specific pollutants identified.

New or revised equipment required in the process is generally registered with the controlling agency when application is first made for approval to construct or install, after the design is complete and prior to procurement. Equipment performance tests should be executed by the manufacturer in accordance with standard practices and witnessed by the owner. "First-run" performance evaluation should be mandatory so as to assure compliance with statutory and industrial pollution control plan requirements under normal operation. All pollutants should be identified and measured during the "first-run" tests (Fig. 2). The reference datum plane for each source of emission should reflect normal maximum operating conditions for the source involved. For example, emissions from a plating process exhaust should be initially measured under full process loading where a maximum surface area is being plated at the maximum current density and at the maximum temperature permissible to the process.

After a review of all existing process documentation and establishing of a procedure for reviewing the initial drafts of the future processes, to determine if pollution controls are necessary, methods should be developed for identification, measurement, and control and should be incorporated into the industrial pollution control manual. The technical ability and experience of analysts vary greatly. Success in using the methods outlined in the manual will depend not only on the detailed compliance with specified procedures but on proper preparation, general practices, and interpretation. The appropriate pollution control technique applicable to the process document should be recorded by terms of reference to a specific section of the pollution control manual. Example when pollution controls are required:

Pollution Control Requirement

Operational practices of this process shall be in compliance with section____of the Pollution Control Manual.

Example when pollution controls are not required:

Pollution Control Requirement
Not Applicable

The decision of the process review should be recorded in the process review register indicating the type of pollutant involved (Fig. 3).

POLLUTANT SOURCE

The industrial pollution sources are usually consigned to one of three categories: (1) manufacturing process operations, (2) utilities operations and plant maintenance, or (3) support services ranging from photographic laboratories to research and development laboratories. The pollution control program assigns all the manufacturing processes having pollution implications (Fig. 4) and the related equipment into two classifications, "permit required" and "permit exempt." Major sources of pollution, such as industrial power plants, incinerators, and metal finishing facilities, and those which require extensive design are designated "permit required" pollution sources. "Permit exempt" sources of pollution encompass those items of equipment or processes which generate minor sources of pollutants but are legally and administratively exempt from permit requirements as defined by the responsible pollution control agency.

All equipment units, mobile or stationary, have been surveyed, evaluated, and the pollutants they generate recorded in the pollution control review register. The pollution control administrator should prepare applications for permits, on forms and in a manner prescribed by the responsible pollution control agency, for all construction, alteration, expansion, or operations which, through its operation or maintenance, could possibly discharge or emit a pollutant to the plant environment. It is generally necessary to obtain a "permit to construct" in accordance with plans and specifications which have been approved by the responsible agency. Upon satisfactory completion of the construction, a "permit to operate" may be required by the agency based on the degree of compliance evidenced in the first-run test. The permit to operate may be revocable should continuing operating practices fail to meet the pollution control criteria. The permit number should be indicated on the industrial pollution control certification decal, which is affixed in a conspicuous place on the equipment (Fig. 5). The certification decal is to be displayed on all equipment, whether "permit required" or "permit exempt," under the surveillance and control of the industrial pollution control program.

PROCESS REVIEW REGISTER

PROCESS NUMBER	REVISION LEVEL	DESCRIPTION	DATE OF REVIEW	ACTION		EQUIPMENT		POLLUTANT	
				MEMO	DATE	NUMBER	DESCRIPTION	HYDROPOL	AEROPOL

Fig. 3 Process review register.

POLLUTION RECEPTORS

The inital stream survey establishes the conditions of the receptor prior to the activation of any plan processes or plants involving pollutants. The survey should take into consideration such factors as BOD, solids concentrations, pH, temperatures, flow, depth, and flora or fauna. The sampling, measuring, and test methods should be approved by the appropriate controlling agency or as defined in the American Water Works Association "Standard Methods of Analysis for Water and Wastewater." Other documents from which suitable methods may be drawn are listed by preferential rank:

U.S. Public Health Service Publications, W P Series

Ohio River Valley Sanitation Commission, Water and Wastewater Analytical Procedures

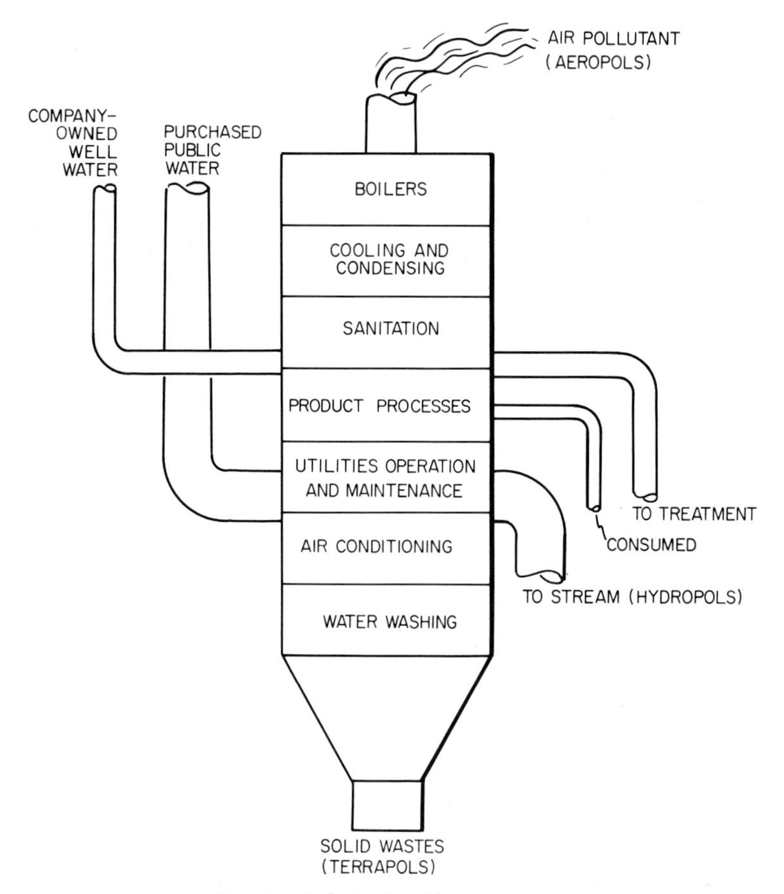

Fig. 4 Industrial pollution sources.

American Society for Testing Materials, D-19, Water and Waste Water Committee Publication

Scott's Standard Method of Analysis

Streams and lakes which receive plant effluents should generally be examined every 6 months to 1 year or when circumstances dictate, e.g., accidental spills.

Initial air samples, taken prior to the beginning process or operation, will determine somewhat the extent of air pollution existing in the plant area. Sampling methods should equal those specified in the U.S. Public Health Service "Field Operation Manual." The sample size selection depends upon the type, emission concentrations, and source of the pollutant. Air pollutants should be dissolved in appropriate media and the subsequent analysis conducted accordingly, for example, passing sampled air through a dilute aqueous sulfuric acid solution containing hydrogen peroxide to measure sulfur dioxide. Fallout jars half filled with distilled water can be used to collect particulates. Sampling stations shall be contiguous to the sources as well as on the property boundaries.

TRAINING

It behooves us to be continuously on the alert for new or improved methods of pollution abatement, to make better use of our personnel, and to establish better training and motivational techniques.

Most pollution control training programs are geared to sewage and waste water treatment. Therefore, industry develops its own quasi-training programs suited to its particular situation. These are often not sufficiently related to the total industrial pollution control concept, and because of the overhead cost constraint, are not likely to be so.

The development of modern pollution control training and operating manuals, which would also embody the Water Pollution Control Federation's "Manuals of Practice," provides the best technique to the solution of the decision making process of the operating problem situations. The manual must be developed by those people who are knowledgeable in manufacturing process and equipment operational practices. They must know the pollution abatement processes, equipment, instrumentation, construction practices, maintenance and repairs, sampling, testing, and how to interpret test results. A looseleaf notebook series provides for revisions and insertions on a practical-need basis. The development of the operating manual lends support to a review of the more burdensome routine operations, which could be automated, thus freeing the operator for the profitable activity of testing and analysis. This helps to close the incremental gap between actual operation and achievable operation. The prognosis of adequacy ascertained from the design and installation drawings and specifications could be applied by the same approving authority to the industrial pollution control manual.

POLLUTION CONTROL

CERTIFICATION

EQUIP. No. ____________

DEPT. ____________

PERMIT No. ____________

CERTIFICATION DATE ____________

APPROVED ____________

Fig. 5 Industrial pollution control decal.

Certification of the industrial pollution control plant operators should be restricted to the practices defined in the operating manual. The certification could be nontransferable, restricted to the industrial plant for which the operation manual was written and approved. Basic mathematics and chemistry courses, presently required in most sewage and water treatment plant operator's certification courses, would likewise be included as a prerequisite to receiving an industrial pollution control plant operator's certificate. Industrial plant operator certification could also be a condition of approval for issuing a "permit to operate" the major industrial pollution generating processes.

Training of the pollution control operator guarantees accuracy and precision in his work and confidence in the results obtained. This includes technical as well as housekeeping and safety training. A well-informed operator understands the importance and objectives of his work. The philosophy of the pollution control plan is one of individualistic responsibility. Analytical methods may represent sound scientific principles and years of development, but if they are misused or misinterpreted the effort is wasted. Likewise, the most modern facilities and equipment do not ensure analytical success if improperly used. Expect the pollution control operator to contribute more than a cookbook application of analytical schemes.

CONTROL METHODOLOGY

A continuing audit of manufacturing and facilities operations is performed by the pollution control program to assure compliance with requirements. Each audit

POLLUTION CONTROL FACILITIES SCHEDULE

PROCES No.	EQUIPMENT NUMBER	EQUIPMENT NOMENCLATURE	OPERATIONS				CERTI-FICATION DATE	DECERTI-FICATION DATE	RECERTI-FICATION DATE	CRITICAL POLLUTANT			STATUS		EST HRS	YEAR	WEEK NUMBER OF YEAR											
			NUMBER FANS	CFM EACH	NUMBER PUMPS	GPM EACH				NAME	CONCENTRATION		EXEMPT	PERMIT			J	F	MAR	APR	MAY	JUN	JUL	AUG	SEP	OCT	NOV	DEC
											SPEC.	ACT.																

Fig. 6 Pollution control facilities schedule.

should consist of a statistical sampling and analysis of the air and/or water effluents from the operation. The audit is a scheduled repetition of the "first-run" test to assure continuing compliance. A typical schedule is indicated on the pollution control facilities schedule (Fig. 6). The accuracy sought will not be the highest order, and tests will be performed to give the greatest economy of time and motion. Preparation of the program of analysis should, if possible, precede sampling. Reasonable forethought must be applied to achieve effective sampling. Exact locations and flow patterns of all drains, ditches, canals, exhaust stacks, and solid wastes sources, including their transport and discharge methods, should be known. The chemical and physical nature of all pollutants and their relationship to the system that carries them should also be known so that scientific disciplines can be imposed on a realistic and meaningful basis.

Certain restraints are necessary when production and/or facilities operations are not conducted in compliance with the established procedures. A pollution control action memo is released to the operating department identifying the process in a noncompliance status and recommending corrective action. It explains in detail the conditions of noncompliance and establishes a date for completion of the corrective action. Serializing the action memos improves the follow-up and recording functions. The pollution control program assures prompt and suitable reaction to the action memos. If suitable corrective action is not taken, a second action memo will direct the withdrawal of the noncomplying process and/or equipment from production until corrective action obviates the decertification. When the pollution control standards are again satisfactorily complied with, a statement of approval is issued. The pollution control procedure must be controlled to current requirements. Each time reference is made to a specific pollution control procedural memorandum in the manual, it should be reviewed for its exactitude and application. If the memorandum does not reflect actual procedures because of new or revised processes or practices, the procedural memorandum should be revised appropriately, identifying the process and the reason for the changes or cancellation.

Emergency operating procedures may be necessary, dependent upon the nature and criticality of the specific system component. The pollution control procedure must be adaptable to compensate for random occurrences and respond quickly. Failure analysis techniques should be used in all cases of random failure, to determine the mode of failure, cause and effect of failure, and corrective action necessary to prevent recurrence of the failure. Each condition must be evaluated on the basis of legal and technical considerations. A copy of the report should be distributed to all those affected directly by the emergency. Emergency operating procedures should serve to prevent violation of pollution control requirements, prevent production schedule interference, and establish corrective action and systems defect prevention.

COST CONTROL

Pollution control is a multidimensional phenomenon. Therefore, the cost associated with operating pollution control programs is difficult to estimate for planning purposes, but they should be budgeted. The cost estimates for achieving the water treatment goals established by the federal government are shown in Fig. 7. The total production of domestic and manufacturing wastes is shown in Fig. 8. Note the anticipated increase in manufacturing wastes. We also need to recognize the bias in expressing total costs due to the pollution effects evidenced in the social interactions. It has been estimated, in certain situations, a game fish has been valued at $30 to $40 as represented by the money the fisherman spends to catch it. The psychological costs, those causing disappointment or anguish, must also be considered a part of the total costs. The pollution effect on property values appeared to be linearly related to the geometeric mean sulfation rates, in a housing market model study, which explained 90 percent of the variance in property values.

The cost of industrial pollution control and abatement must be related to the effectiveness in removing or controlling the pollutant. It should be expressed in quantified terms, i.e., pounds of BOD (biochemical oxygen demand) removal vs. pounds of product produced. Another important indicator of the management system effectiveness is how often it fails to satisfy the selected quality standards. Cost controls should relate to these assigned quality standards in terms of minimized

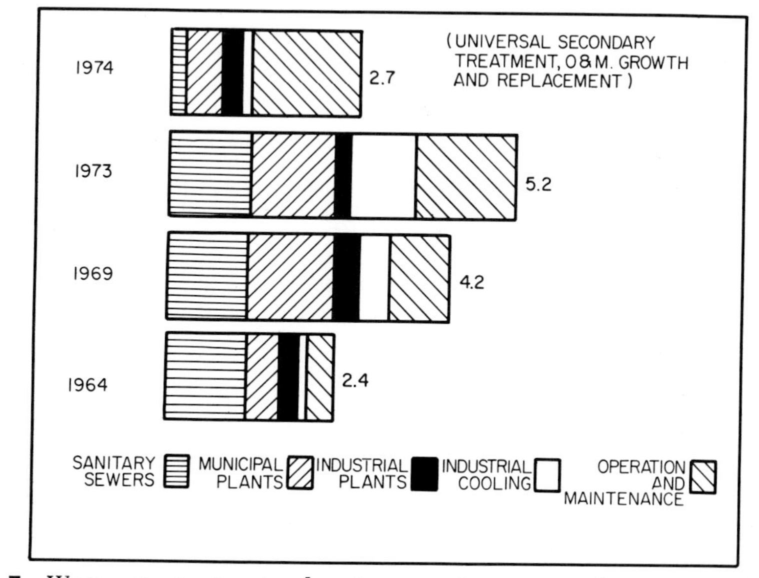

Fig. 7 Waste water treatment and sanitary sewering costs to achieve treatment goal by 1973. (*FWPCA, 1968.*)

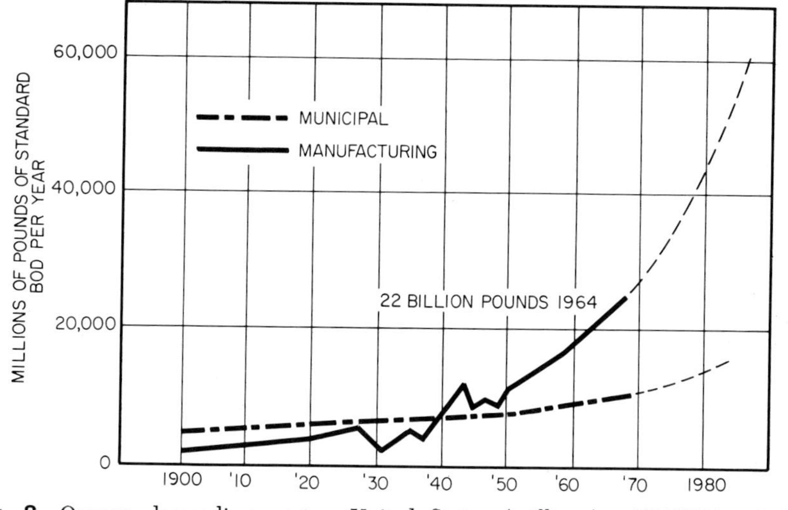

Fig. 8 Oxygen demanding wastes, United States (millions). (*FWPCA,* 1968.)

costs. The pollution control plan operation can be simulated by synthetically producing input data to obtain monthly operation costs and yearly average costs. The assumption that the costs of pollution control increase linearly with the degree of control (treatment) makes it possible to apply linear programming techniques to the problem. The optimization technique requires a careful evaluation of the following parameters:

1. Process pollution identity and the minimum and maximum quantity, e.g., paunch refuse in meat packing plants, hair, chicken feathers, strong acids and alkalies, certain gases and toxic substances
2. Pollutant loading received by the pollution control facilities, e.g., BOD, pH, suspended solids, chlorine requirements
3. Operating cost of each pollution control system or facility based on volume of flow and strength of waste, e.g., baghouse, afterburner, scrubber, vapor control system
4. Stream conditions prior to receiving the plant's effluent and the stream quality criteria
5. The merits of a unified system serving a number of industries in the area

The optimal level is attained when the combined costs for all the pollution control processes are minimized in maintaining the standards of the final effluent to the receptor.

Industrial pollution control plants are generally operated without regard for efficiencies or for the temporal or spatial variations of the pollution generating processes. Systematic planning of work schedules can result in better utilization of personnel and thus reduce costs. Work measurement of contiguous data can be determined by the development and application of engineered performance standards. A review of work planning and measurement often will result in a second look at tools and equipment used which could net a further cost reduction. A strong maintenance program maximizes equipment availability and minimizes deterioration economically over a long run.

Perfect information is not always available, nor for that matter, are clearly defined objectives in industrial pollution control activities. We operate in a situation of uncertainty and risk, for all actions have levels of risk and uncertainty associated with achieving the objectives, but using a systems approach directly implies comprehensiveness for it insists on quantitative answers. It is essential to learn how to use quantified information, coupled with maturity, intelligence, and judgment, so that the resulting decisions maximize benefits to the industry.

Chapter 22

Plant Operation and Training Personnel for Pollution Control

DAVID H. REEVE

President, Effluent Controls, Inc.,
Lakeland, Florida

and

JOSEPH SCRUGGS

Project Engineer, D. J. Stark and Associates, Inc.,
New York, New York

OPERATING ABATEMENT EQUIPMENT

The effect of poor equipment operation in pollution control work is generally more detrimental to efficiency than poor operation of other process control equipment. To be specific, the plugging of 10 percent of the aeration nozzles in an aerator, used to reduce the BOD in the liquid effluent, will decrease efficiency as much as 50 percent or greater, because the poor operation can set into effect a chain reaction. The lower level of oxygen in the effluent will kill or discourage the growth of sufficient bacteria to complete the biological reaction; if bacteria are killed, the BOD is increased by the weight of the bacteria killed. As the chain reaction continues, the retention time in the aeration pond becomes critical, allowing untreated wastes to go into the settling basins, where the continued biological reaction hinders settling.

Similarly, much scrubbing is done by condensation. The inlet gas temperatures are reduced by water sprays. If these sprays begin to plug off, the gas temperature following these sprays is raised, evaporating less water into the gas stream, and therefore reducing the amount of condensation which can take place in the particle separation section of the scrubber. Since gases hold limited quantities of water vapor, a few degrees of temperature rise reduces efficiency rapidly.

These malfunctions of equipment are costly to the plant through fines by state control agencies and poor public relations. Since pollution control equipment is usually installed where the gases enter the atmosphere or the waters enter a river, there is no second chance to recover a mistake or the effect of poor operation. There is no recovery sump or next process where corrections can be made.

In 1966 in the state of Florida, a phosphate slimes dike broke. The slimes entered the river where all the fish were killed. The penalty for this pollution was $200,000 paid to the state to restock and clean up the river.

Air pollution in Garrison, Mont., fomented a request by the governor to have a federal hearing on the problem. The plant was forced to shut down until adequate control equipment and auxiliary backup equipment were installed to guarantee a safe level of pollution in the area.

These fines and penalties are the exception rather than the rule, in the year 1967, but all present legislation shows that it is going to be costly to allow pollution to continue and that penalties will be enacted to ensure adequate operation of equipment which is in pollution control service.

Therefore, training personnel to operate the equipment properly becomes a very important part of selecting, installing, and operating control equipment.

TYPES OF EQUIPMENT AND SPECIFIC TRAINING ITEMS

Numerous types of equipment are used for industrial pollution control. In fact, many total plants are constructed for the sole purpose of eliminating a major pollution problem and controlling pollutants such as sulfur dioxide roaster gases, utilizing a sulfuric acid plant as the pollution control device. Another example would be the incineration of black liquor in the pulp and paper industry for the recovery of sodium sulfate to be reused in the process.

But generally speaking, industrial pollution control equipment can be classified into the following simple categories:

1. Dry particle removal
2. Gas vapor removal
3. Liquid entrainment removal or mist removal such as evaporator processing equipment
4. Combination gas, liquid, and particulate removal
5. Settling ponds for liquid effluent neutralization
6. Biological aeration ponds and/or chemical fermentation ponds

Various types of equipment are used in each and every one of these categories, and in many instances, several types of equipment are used in conjunction with each other to provide the desired emission control, the primary combination being product recovery control equipment followed by effluent gas emission control equipment. The following section is written primarily to summarize and give examples of specific control points which operators are required to know to maintain continuous, trouble-free operation of the emission control equipment.

Bag Dust Collectors

Operators need to be thoroughly informed of the type of bag or sleeve being used and the materials of fabrication. Their understanding of this equipment can decrease fire hazards or bag failures by improper handling during installation and/or improper cleaning when the units must be mechanically cleaned. An understanding of the bag-shaking mechanism should be pointed out in order for them to be able to troubleshoot and determine the causes of blockage, plugging, and so on. Where multiple units are used, it is advisable to indicate to the operators the methods of flow; that is, parallel, series, or series and parallel flows in the event of maintenance or mechanical problems (see Fig. 1).

Cyclones

Cyclones are relatively simple apparatus. Operators need to understand fully the principle that is being used to separate the suspended particles from the gas stream. Gas streams laden with particles are directed through the tangential entries and enter a funnel-shaped area where gas flow is reversed, making it impossible for the dry solids to follow the gas stream (see Fig. 2). The method by which this reversal throws out solids is an important principle for operating personnel to understand. The effects of moisture in these systems and how and where the buildup occurs when excessive moisture has reached a cyclone will aid the operator in cleaning the discharge of the unit. The effects of high-moisture gas streams on baghouses and cyclones should be thoroughly understood by the operating personnel, and they should also be given a course in methods to control moisture from the primary process equipment such as driers and evaporators. Generally a system is

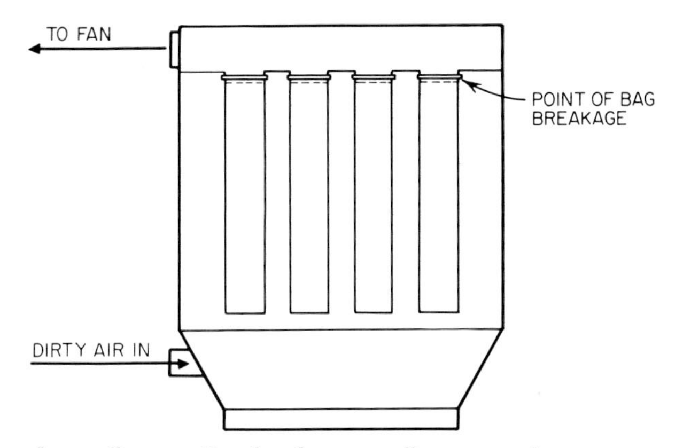

Fig. 1 Bag dust collector. Bag breakage usually occurs when system operates on wet gas or when dust plugs the filters because of lack of attention.

designed for a drier and the moisture of the gas stream is controlled arbitrarily by sizing the fans to give an excess airflow, but when plants are pushed beyond the design capacity, the moisture-to-dry-air ratio changes, creating operating problems that did not exist in the original design, and only by a thorough understanding of these basic principles can an operator troubleshoot the system and correct the problem before moisture has created solids buildup and blockage or ineffectual operation.

Electrostatic Precipitators

Again, moisture control is one of the primary prerequisites of an operator's knowledge in handling the unit. He should also be thoroughly informed of the mechanism by which the particles are attracted to the precipitator nodes and the mechanism by which these particles are released and returned to the product system. A fundamental understanding of the electric currents involved, the amount of voltage being used, and static electricity pickup is necessary to train the operators in handling the system safely.

Wet Scrubbers

The wet scrubber is a relatively simple piece of equipment. The prime reason that wet scrubbers have trouble in maintenance or decreased efficiency is blockage in the spray nozzles, discharge pipes, or other places where the slurry deposits solids, blocking the flow of the gas or liquids. Changing the sizes of the nozzles to eliminate buildup problems completely violates the basic design of the wet scrubber, since particle size or surface area of the liquid droplets is the prime determinant for contaminant removal. The surface area of water droplets is also a prerequisite for noxious gas absorption; therefore, the last corrective alternate for plugged nozzles should be arbitrarily increasing the nozzle sizes, which effectively increases water droplet size.

If the operators are instructed to go to the source of the problem, to determine the source of the contaminants which are causing the nozzles to plug, they may readily eliminate the problem by installing a screen or settling basin in the liquid circulating system to prevent solids from getting to the nozzle orifices. If nozzles continue to plug under these circumstances, a qualified engineer should be asked to investigate the cause of the problem, and if necessary to substitute larger nozzles,

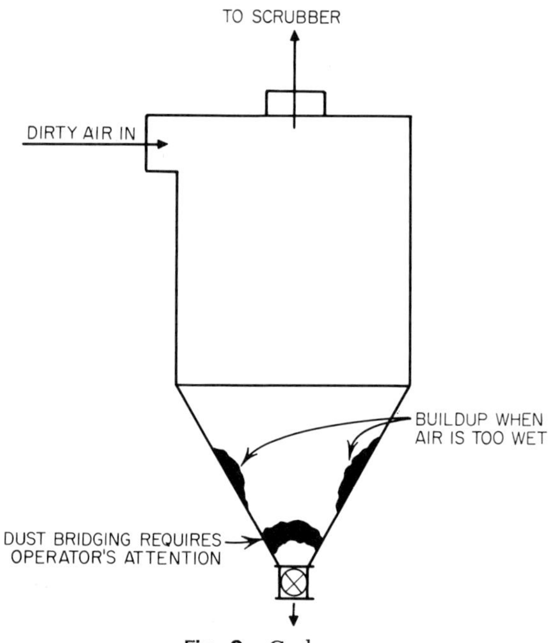

Fig. 2 Cyclones.

being specific to obtain a nozzle which will give the same flow and liquid particle size as the smaller nozzle by using a slightly different principle of atomization or by using additional power through air or higher pump pressures.

Entrainment Removal

There are many different types of apparatus designed to do nothing but remove liquid particles from the gas stream. The basic cyclonic scrubber is designed so the gas enters tangentially, throwing the water against the wall by centrifugal force and allowing the coalescence of the water to stop reentrainment of the liquid particles (see Fig. 3).

Fixed vane separators such as a Centrifix or propeller type unit depend upon the gas flow entering the vanes and imparting a swirling motion to the gas stream, creating a centrifugal force where liquid particles are separated on the walls above the main separator, very similar to a tangential entry cyclonic scrubber. The operators should be informed of the critical velocities of operation through these units and the points of discharge of the liquid entrainment so that they can quickly pinpoint buildup areas. For example, when the liquid stops flowing, the entrainment separator outlet line is probably the cause of the problem.

Mist eliminators such as fine woven mesh materials or baffle type eliminators, utilizing gravity separation for the larger particles, are used on many wet scrubber systems for straight entrainment separation from evaporators and similar process services. Again, operating and maintenance personnel must be made to understand the critical velocities required within the system, the design velocities, and the tolerance above and below design velocities for proper operation of the entrainment separators. Where a combination of gravity and baffle entrainment separation is taking place, two parameters of critical design velocities should be given to the operators in their operating manuals for easy reference so that they readily understand the effects of increased capacity of the production unit, or the effects of surges from the production unit, on the actual entrainment separation operation. Without the understanding that gravity settling would be eliminated by sudden surges or increases in design capacity which would overload the baffle entrainment separator, and without understanding the effects of increased rates on the entrainment separation, an operator cannot understand the cause for mists rising out of the stack when it has never occurred before.

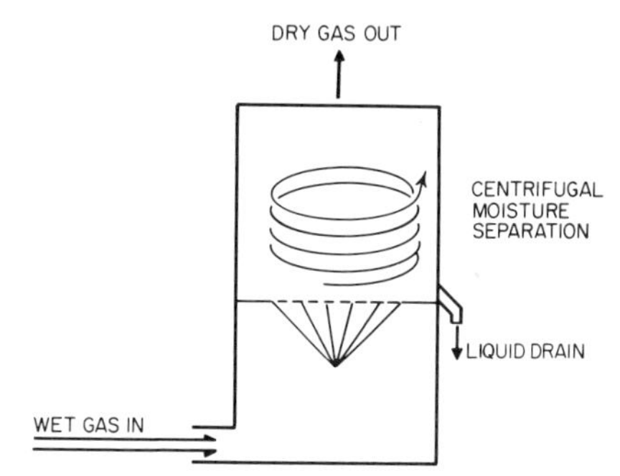

Fig. 3 Vane-type entrainment separator.

Settling Ponds

Settling ponds generally have two weaknesses which are troublesome in the normal course of using them as pollution control devices, the first being that poor maintenance of dams which have been standing for several years will allow the footings of the dam to be weakened, creating potential failure points which can create massive pollution problems if the dam fails. Therefore, the fundamental approach to most settling ponds would be a routine weekly inspection from top to bottom to

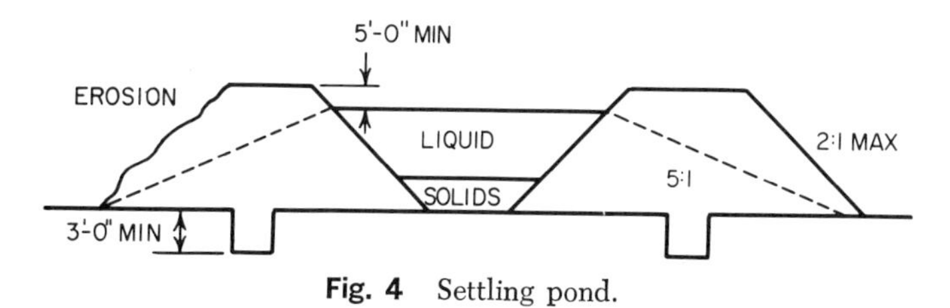

Fig. 4 Settling pond.

be sure that they are maintained in proper repair. The second area requiring attention is bypassing or short-circuiting the liquid flows within the settling areas. When improper inspection of the settling ponds has allowed the solids to build up in specific areas in the basin, many bypass currents can be created, which make the settling basin completely ineffective, therefore allowing the finer particles to be carried into the overflow or discharge to the river. Operators should be shown the need for continuous surveillance of equipment of this type, which normally does not give operating problems but needs to be continually inspected in order to operate for several years in a trouble-free manner (see Fig. 4). By informing operators of the surface requirements needed, the particle settling velocities in the solution, they can better understand the problems which they must look for during their inspection tours of normal settling basins. If the settling pond is used as a neutralization operation, they can readily understand the buildup of solids at the inlet or point of neutralizing media addition; therefore, this would be one of the major trouble spots that they inspect thoroughly and most frequently. If it is primarily a fine particle sedimentation basin, then several areas could build up because of eddy currents or irregular flows in the system. More than likely, a short distance from the inlet, where the liquid is rapidly cooled by mixing with the other liquids being retained, will be the point of maximum buildup and a place where solids should be raked level or removed periodically to allow the basin to operate properly.

Combination Systems

When a combination of pollution control circuits are added together, an understanding of the basic principles of operation of each unit is mandatory for the operator to know the effects of an upset or failure of one unit upon the other units. With proper training he has a basis for making judgments as to whether to continue operation or to shut down the operation until the first control device is brought into control, repaired, or replaced. If the second control system can handle surges of capacity for short periods of time, it may be possible to repair the first control device without shutting down the plant and without materially affecting overall recoveries within the system. Therefore, on combination systems, we can only reemphasize that training is an essential component to maintaining good operation of the equipment, and that inspection plus training is as necessary a function for pollution equipment as it is for production equipment.

ECONOMICS OF OPERATION

There are generally five areas which affect the cost of pollution abatement. They are:

1. Capital cost
2. Maintenance cost
3. Power cost
4. Reagent cost
5. Operating labor

Many pieces of equipment such as cyclones, bag collectors, and settling ponds have only three cost control areas: (1) capital, (2) maintenance, and (3) power.

Capital costs for normal installations range as follows:

1. Cyclones, $0.10 to $0.50 per cfm handled
2. Bag collectors, $1.10 to $2 per cfm handled
3. Wet scrubbers, $0.20 to $2 per cfm handled
4. Aeration or settling ponds, $0.30 to $0.70 per cu yd of earth moved

Maintenance costs will run from 8 to 15 percent of the installed capital cost per year if dredging, nozzle cleaning, and other cleanup services are considered as maintenance for good operation.

Power costs will vary widely depending on the complexity of operation and energy requirements needed to meet a specific state code. In the removal of particulate material from gases, the following power uses are typical:

hp per 1,000 *cfm*	*Particle Size, Microns*
⅕–⅓	15–40
⅓–½	5–15
½–¾	2–5
¾–1½	Less than 2

In processing of liquid wastes in ponds adjacent to the plant, power costs fall in the range of 15 to 30 hp/1,000 gal where the liquid is returned for reuse. If the liquid is discharged to the river after entering a settling pond, the power cost will be 7 to 15 hp/1,000 gal of liquid processed.

Aeration costs for BOD and COD reduction are based on the load and retention time required. No typical figures appear to be representative.

Reagent costs are also highly variable, dependent upon type of reagent used, transport costs, and quantity of reagent required. Each case must be evaluated separately using the lowest-cost reagent (i.e., limestone for acid neutralization) which can accomplish the work.

Operating labor figures are confusing for most pollution control equipment, because this equipment becomes part of an integrated plant and the work load is therefore distributed to the present operating force. As a general rule for pollution equipment, one-third man/shift is required for each $100,000 capital cost of installed equipment. At $500,000 the labor is reduced to one-half man/shift for each $500,000 of installed capital equipment. Further reductions in labor can be accomplished with automation, but the above figures are averages of 1967 requirements.

Specific examples of pollution control costs by industry are as follows:

A. Apartment house and hotel incinerators, wet scrubber
   1. Capital cost installed.................................... $2–$3/cfm
   2. Power requirement.................................... 1.3 hp/1,000 cfm
   3. Maintenance cost when including cleaning................. 8% of capacity/year
   4. Reagent cost for cleaning................................ $0.15/week
   5. Operating labor, automated system required............... 2 hr/week

B. Phosphate industry fluorine, phosphate, and gypsum disposal
   1. Capital cost
      a. Gypsum settling pond............... $0.20–$0.40/ton solids
      b. Clear water cooling pond............. $4,000–$6,000/acre of cooling area
   2. Operating cost
      a. Reagent cost, limestone neutralization for fluorine, and P_2O_5 neutralization... $0.30–$0.50/ton P_2O_5 processed in plant
      b. Labor cost to handle inspection, neutralization, and analysis.............. ⅓ man/shift

Controlling cost of operation is limited to two areas: initial purchase and training of operators and maintenance men. When considering the purchase of centrifuges, gas scrubbers, settling ponds, or other equipment, the following general rules will apply:

1. Be sure the equipment is properly applied. Laboratory data on settling rates or particle size analysis of airborne dusts are costly but necessary to proper design of equipment.

2. Where possible leave a 10 percent margin of safety in design. This is an expensive safety factor, but upset operations at a future date will not create pollution with this margin of safety and may save lawsuits, penalties, or neighborhood complaints.

3. Eliminate buildup areas and small orifices which can clog, upsetting the control equipment. If small nozzles or aerators must be used, then adequate strainers and air filters should be installed with the initial equipment.

4. Be highly selective in the materials of construction. Corrosion has been a major maintenance item in pollution equipment.

5. Do not let price be your only guide in the selection of equipment. Price must be secondary if the selection is to be a good one.

Since operator and maintenance training is covered under training personnel below, no further comment is required here.

POLLUTION CONTROL OPERATING COSTS

The costs of pollution control and plant product recovery are usually directly related, as can be seen by the foregoing example of air pollution. A similar situation exists in water pollution. In a phosphoric acid and fertilizer complex, phosphoric acid mist and fluorine fumes are scrubbed from the gas stream using fresh or recirculated pond water. The gypsum cake is discharged to the pond using the scrubber solutions, as shown in Fig. 5.

If water usage in the plant is not controlled, the neutralization costs rise, if the reactor or evaporator lose excessive quantities of phosphoric acid because of upset conditions. The loss of product reflects in higher operating costs and lower product recoveries, while neutralization costs increase simultaneously. Costs of pollution control in this system can easily double by poor operation and/or poor supervision of the operation.

Proper training of operators and supervisors is an intregal part of production cost control in a fertilizer complex. A 1 percent loss of 100 percent phosphoric acid can be worth $75,000 per year in a phosphoric acid manufacturing plant.

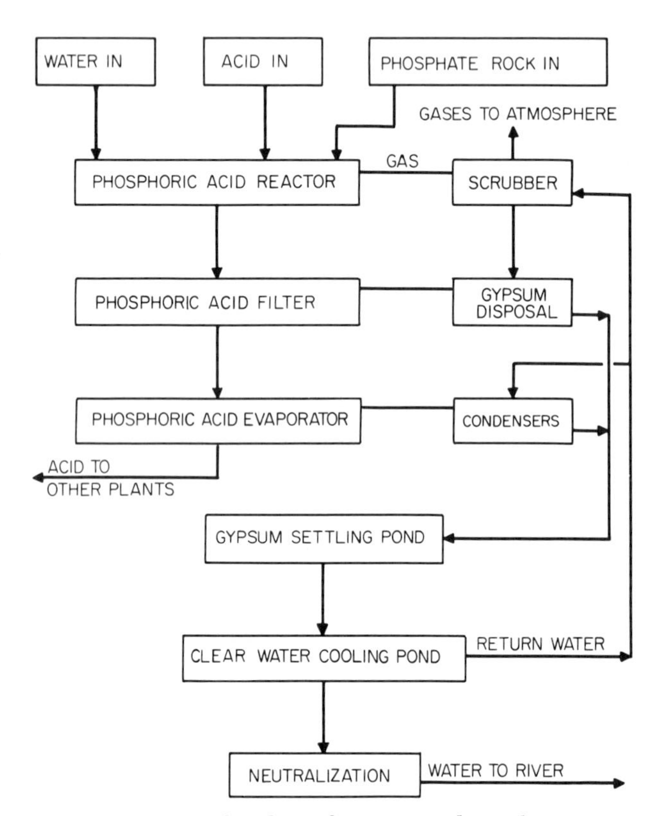

Fig. 5 Phosphate plant—waste disposal.

In the manufacture of fluosilicic acid, gases containing silicon tetrafluoride enter a two-stage water scrubber (see Fig. 6). The scrubber has been staged to effect a high absorption of the acid. The first stage of scrubbing is accomplished with 25 percent fluosilicic acid.

$$3\ SiF_4 + 2\ H_2O \rightarrow 2\ H_2SiF_6 + SiO_2$$

When the gases go to the second-stage scrubber, they are washed with 10 to 11 percent H_2SiF_6. The concentrations are maintained by vapor pressure equilibrium, temperature of operation, and balanced operating efficiency in both scrubbers.

Silicon dioxide precipitates cause blockage in pumps, lines, and scrubbers which creates upset or unbalanced operating conditions.

Describing the balances required, setting up 10-day cleaning cycles, and training the operators to monitor internal and external symptoms of change made this process successful.

With untrained supervisors and operators, production was 50 percent of the design rate.

Trained operators and supervisors had to know the effects of

1. Vapor pressures in relation to temperature and acid strength
2. Pressure losses as they affected recovery
3. Temperature and its relation to acid strength

After 4 months of 50 percent capacity operation, management realized the need for training. After 2 weeks of intensive classroom training and 1 month of in-plant

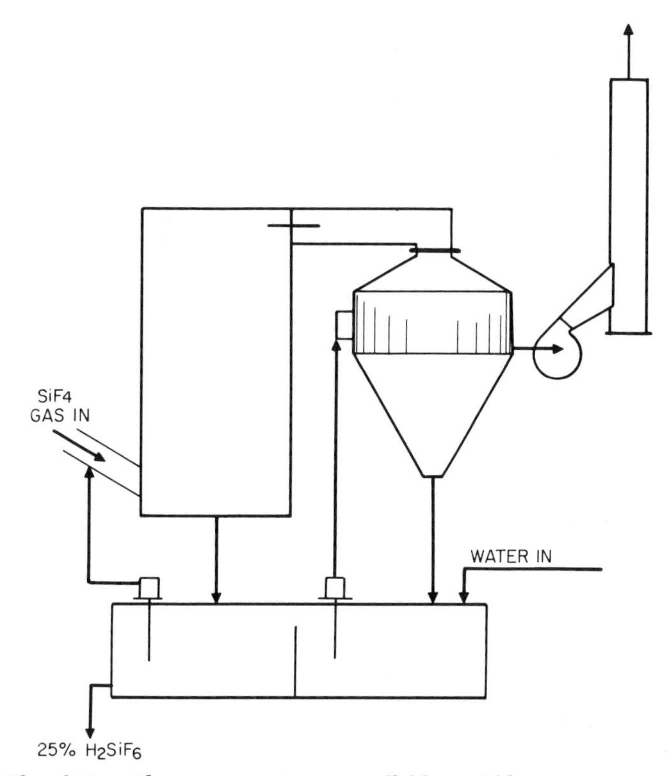

Fig. 6 Fluosilicic acid recovery system controllable variables: temperature of gases, rate of water feed, liquid recirculation rates to each scrubber.

training, recoveries of H_2SiF_6 began to rise. In 2 months 90 percent of design rates were reached, and 1 month later the plant exceeded the design capacity.

TRAINING PERSONNEL

Many of the control devices in operation have been criticized for two major problems. One has been high maintenance cost and the second is the low operating factors. A really close look at most training manuals will indicate that the recovery devices in many plants are treated as being secondary to the production equipment, even though the effluent control equipment will often shut down a plant operation as easily as one of the basic production unit failures. Therefore, we can only stress that training and retraining of operators to handle pollution control equipment should be handled in no way less than basic process equipment. Continuous preventive maintenance inspections are necessary to maintain this equipment in good operating condition, to maximize product recovery and minimize damage to plants or property outside the production system. At no time should an effluent control device be an auxiliary piece of equipment appended to the tail gas stream.

Training of supervision can do much to eliminate both the above problems with control equipment. The emphasis in supervisor training needs to be stressed in

1. Downtime analysis
2. Preventive maintenance
3. Analysis of operating efficiency
4. Integrated system control

Although downtime analysis appears to be a logical operation for each supervisor to log, there are many examples where the logged individual data are never summarized and analyzed to point out the frequency of a recurring problem or to summarize many problems occurring over several months which may point to extensive erosion or corrosion problems. The plant manager does not want to know the problems in each shutdown, but he must be informed of trends, specific problem areas, and the suggested corrective action which is being or needs to be taken.

A proper preventive maintenance program is also based on a review of the downtime analysis logs, pinpointing frequency of cleaning, analysis of corrosion, and scheduled repair or replacement of worn parts. A typical downtime log sheet might be as follows:

Component	Plant	Unit	Date	Inspector
Dikes				
Aerator nozzles				
Influent piping				
Sludge pond settled solids				
Effluent pipe				
Pump shafts and bearing inspection				
Pump impeller erosion				
Aerator nozzle replacement				

A visual and a measured analysis should be made of each point of wear, corrosion, or buildup and a note made indicating the condition and/or corrective action taken. The analysis of these log sheets then becomes the basis for all preventive maintenance work and scheduled repair programs.

When the aerator nozzle plugging forces the shutdown before a scheduled shutdown, the cause of buildup should be inspected and analyzed to determine if more frequent shutdowns are necessary for proper operation or if the problem is a temporary one which can be repaired now and eliminated.

The third area of supervisor training which needs to be stressed is operating efficiency. Most pollution control devices are not monitored continuously, and therefore other indicators of effective control must be set up so the supervisor can measure his success or failure to maintain optimum efficiency. Such control indicators could be:

1. On a scrubber the flow of water or recirculated liquid is one of the controlling factors of operating efficiency. This along with a clear stack may be sufficient to gage operating control continuously. These indicators matched with in and out gas samples once or twice a month can prove very reliable.

2. Pressure drop across air diffusers in pond aerators indicates the flow of oxygen into the water and therefore is an indirect measure of BOD reduction. The pressure drop manometer readings on the fans and periodic pitot tube readings on the air ducts will be effective operating controls on a system which has a constant inlet

BOD content. If the inlet BOD varies widely, only BOD or COD analysis can be used to measure effectiveness of operation.

Independent variable indicators are available on every system which can be used to analyze control problems and/or measure operating efficiency. Training supervisors who know and use these controls will save many laboratory analyses and improve operating efficiency.

Integrated system control is difficult for many operators or supervisors to understand. This understanding usually is proportional to their special abilities because it requires integrating two or more components to control a third component, for example, a cyclone, a primary recovery scrubber, and a tail gas scrubber in series. All are dependent upon one fan which controls the flow of air through the systems but are dependent upon no buildup or restriction in any one of the three systems.

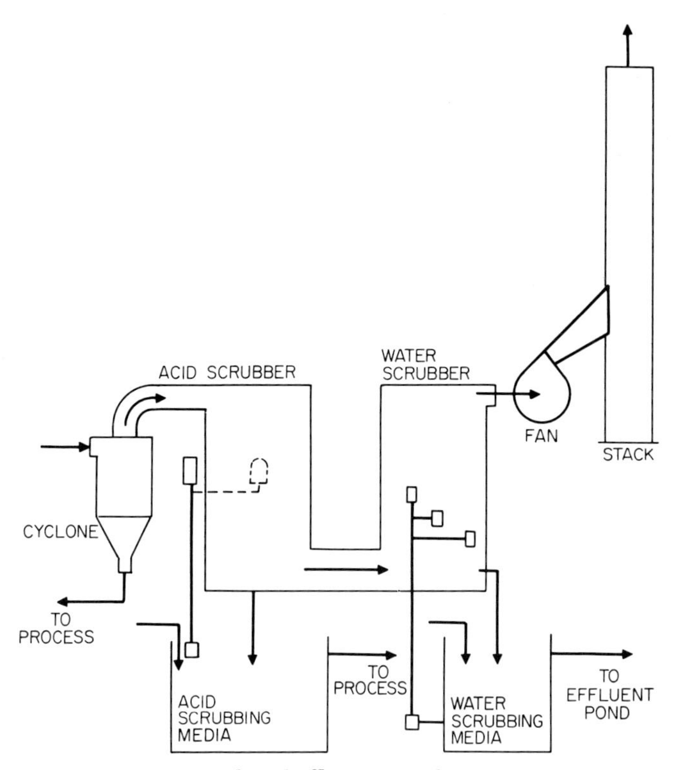

Fig. 7 Combined effluent control equipment.

If a drier before the cyclone produces a saturated gas which creates a buildup in the entrance to the cyclone, the cyclone efficiency will decrease because the decreased flow of air will decrease centrifugal force, allowing more solids to go to the primary recovery scrubber (see Fig. 7).

If increased solids continue to reach the primary scrubber, an increase in scrubber acid will be necessary to stop buildup and plugging problems occurring within the primary scrubber. If the problem is not analyzed correctly and corrected, then product from the primary scrubber will carry over to the tail gas scrubber being lost in the effluent ponds. A supervisor who knows the normal operating pressures and who uses periodic analysis of the scrubber solutions to control the system can usually pinpoint an approaching problem and take corrective action.

Take a look at your plant to see how many integrated pollution control systems have evolved over the past years, where no effective integrated training course has been given to familiarize supervisors or operators with the interrelated operation of each component.

GENERAL TRAINING TOPICS

The following is an outline of the major subjects which should be covered during a training course:

1. Introduction
2. Process theory
3. Equipment familiarization
4. Equipment operation
5. Process recoveries
6. Testing procedures
7. Operating maintenance
8. Safety

The order in which these items are listed has no bearing upon the importance of the subject matter.

In the case of training mechanics in the repair of equipment, emphasis should be placed on the introduction, equipment familiarization, maintenance, and safety. Operating recoveries and methods of operation would be only lightly touched on when training maintenance personnel. When training operating personnel, test methods and operating maintenance would probably be the least important of the groups set forth above.

Introduction

As discussed previously, the introduction should be primarily used to set a relaxing atmosphere and for the operators to become familiar with the instructor and their fellow students. Listening to gripes and/or problems would be part of the teacher's introduction to the group and will allow him to relieve any unnecessary worry that the group is concerned about. The instructor should briefly outline the objectives of the course, the reason for training, and the methods that will be used.

At this time, he also would give the students their training manuals so that they can become familiar with the subject which they must learn.

The instructor must transmit an enthusiastic approach to the topic in hand and set a light, joking atmosphere concerning errors which the operators might make and relieve their concern that they might be asking foolish questions throughout the course.

Process Theory

Process theory must be thoroughly integrated into the course so that the operator can understand the reasons behind the unit operation and each of the individual components within the unit. If possible, it would be useful to obtain the process engineer for an hour or an hour and a half's discussion with the operators so that he can explain the theory behind the components that have been installed for the operator to handle.

From these general discussions, the operator will learn the boundaries of good operation and the definitions of upset conditions. With a thorough knowledge of the process theory, he can become something more than a pair of hands. He will have the ability to detect changing operations and can relate this to the theoretical design aspects to troubleshoot approaching problems. He can predict, with a thorough knowledge of process theory, when maintenance will be required in given areas. A thinking operator more than pays his way by increased recoveries and reduced downtime.

An example of the process know-how which the operator should be given is best demonstrated by the following example of training on a wet scrubber. The main points to be stressed for the theory of wet scrubber operation are:

1. The theory of airflow and how it creates pressure drop through the system.
2. The type of nozzles and degree of atomization which are required for the unit to operate properly. High and low limits should be designated.
3. The size of nozzle openings and pipelines to determine the size of particles which will plug nozzles or restrict the flow of liquid to the scrubber.
4. The method or equipment which is used for entrainment separation. If entrainment separation is accomplished by a combination of centrifugal and gravitational forces, the students need to understand the basic principles underlying this process operation and their effects when one of the laws is violated. Again, minimum and maximum operating ranges should be noted in the operating manual.
5. The effect of velocities on the liquid-gas contacting efficiency and how the contaminant removal is actually accomplished.

As this process theory is reviewed in the classroom, the limits of operation and specific control points that are being discussed need to be referred to in the operating manual so the operator knows quickly where to reobtain the data when necessary.

Equipment Familiarization

It is desirable during the training course for the students to inspect the equipment in the plant and to have the individual pieces of equipment pointed out and named. The operator should be allowed to climb over and in any equipment which is open so that he can ask questions concerning internal structural designs. Often he will point out to the instructor shortcomings in access platforms and sample points and other areas with which he is familiar. Many times these operator comments have allowed changes to be made before the unit was completed, correcting an oversight in design.

Continuing with the equipment familiarization in the classroom, every piece and portion of the unit to be operated should be drilled with the students until they can rapidly name each piece of equipment without hesitation. In the field during start-up, a thorough understanding of the names of equipment will expedite corrective action being taken when an operator is requested to close the valve on the recirculating tank. If he does not know what a recirculating tank is, he is then nothing but a pair of hands which are more in the way than helpful during the initial operation of the unit and/or during emergency operations at a future date.

Since equipment manufacturers and engineers use many long-sounding names, describing equipment by function or trade name, the operating manuals must have a glossary of terms to explain to the novice the meaning of chemical formulas, technical terminology, and trade names.

Equipment Operation

After process theory and equipment familiarization, operation becomes a combination of the previous two training units and therefore should be conducted in a conference type atmosphere, where operators and instructor are freely exchanging methods by which the system can be operated. The students by this time have learned why and how the system runs and what is to be accomplished by the unit operation, and they should know the names of all pieces of equipment in the unit. Therefore, the instructor can break down the unit operations into discussions of when to turn what valve, how soon a pump should be turned on, and how to start up a fan. If he directs the questions to the students in this simplified manner and has set the proper atmosphere, where they are free of the fear of criticism, the operators will respond, indicating the methods by which they think this unit should be operated.

At this point in the course, the instructor can evaluate the progress of each and every individual in his class and readily pick up mistaken concepts which have been inadvertently transmitted to the students.

The entire operation should be reviewed at least three times for normal operating conditions, and then a series of upset conditions can be simulated on the blackboard or on a process trainer, if one is available. Operators using their instruction manual can determine what changes should be made to bring the unit back to normal operation.

Process Recoveries

Recoveries are being used in this category in the broad sense of the word. Since a pollution control device may primarily be concerned with reducing BOD in the streams, no effective material recovery has been obtained. But the operator must be oriented so that he understands his objectives as if they were in process recoveries or valuable materials to be recovered. Again, an open forum type of discussion with the students will allow the instructor to pose operating problems which cause reduced recoveries, and the instructor can readily determine how well the process theory has been absorbed by the class.

Recoveries and understanding of the chemical reactions which take place within the unit should be transmitted to the students. They do not need to know all the fundamental chemistry but must know how and why a chemical is added or why a material is aerated to be able to integrate this into their process theory and unit operation knowledge.

Testing Procedures

Performance tests are becoming more of a norm in checking pollution control equipment. Therefore, operators should understand the best basic methods and test procedures used to determine the efficiencies of their operation. Depending upon the circumstances, they may never run these tests themselves, but by understanding the test procedures, they can cooperate with the people performing the tests, such as holding the unit in static operating conditions during performance of efficiency tests.

Explanations in this area should be made on how the stream is sampled, what is being measured, the time period required and the expected accuracy of the test procedure. Operators should have sufficient knowledge, after reviewing the training manual, so that they could perform these tests if necessary.

Operating Maintenance

Operators and maintenance personnel can often be joined together for instructions in maintenance procedures and a briefing on materials of construction and expected areas where erosion, corrosion, or plugging can take place within the system. Engineers will tend to gloss over expected problem areas within the production unit by saying, "Erosion, corrosion, and plugging have been eliminated by effective engineering." Instead of taking the engineer's word for it, all materials of construction should be reviewed with the operators, all wear points in mechanical moving equipment should be reviewed, and where possible manufacturers' representatives for pumps, air compressors, and blowers should be brought in to train the class in proper repair procedures, lubrication, and potential trouble areas.

When the operators are informed of these problems, they can often prevent future downtime by making a routine preventive inspection of the equipment prior to a planned shutdown which will detect overheating bearings, improper lubrication, noisy fan drives, and many other obvious trouble spots. Then a preventive maintenance inspection during the shutdown can review these real or imaginary problem areas.

Maintenance instruction manuals should be given to the maintenance man assigned to a unit which would include operating and maintenance manuals from individual equipment manufacturers. If these documents are pulled together in one booklet, they tend to get lost less easily and will often remain on the shelf above the maintenance foreman's desk for ready reference. Some additional training should take place with the maintenance men in the field in which the operators do not have to participate.

The maintenance people should have the opportunity to see all equipment inside and out prior to startup of this equipment.

Safety

Maintenance and operating personnel should again be brought together for instruction in safety practices. The safety engineer may do part of the instructing and may take the class through the unit a second time to pinpoint areas where slips and falls, due to chemical spillage or obstacles, could create hazardous conditions. A complete indoctrination on hazardous chemicals being used, and methods to prevent fires, treat burns, and safely administer first aid should be included in this course.

The above outline is detailed to cover major equipment installations, but no individual section should be deleted, no matter how small the pollution equipment installation might be. A total training period, encompassed by the outline above, would take approximately 10 days for something as complex as a sulfuric acid complex startup. This can be reduced to 4 to 5 days for a simple scrubber installation, but a reduction below this time limit will generally make the training ineffectual.

SPECIFIC TRAINING PROGRAM OUTLINES

A. Supervisory training outline. Subjects—general pollution control equipment
   1. Glossary of terms
   2. Object of control device
   3. Process description
      a. Types of equipment used
      b. Theory of operation
      c. Designed control limits
      d. Standard operating procedures
      e. Upset operating procedures
      f. Corrosion problems
      g. Chemistry of operation
   4. Troubleshooting
      a. Practice in nonstandard operation
      b. In-plant inspection
      c. Problem solving, simulated conditions
      d. Maintaining equipment
   5. Safety program
      a. Limitations of safe operation
      b. Preventive safety
      c. Caring for injured
      d. Special precautions for particular chemicals and equipment
   6. Maintenance
      a. Preventive program
      b. Corrosion problems
      c. Limits of materials of construction for repairs and replacement parts

B. Operator training outline. Subject—general pollution control equipment
   1. Glossary of terms
   2. Object of control device
      a. Recoveries expected
      b. Capacities of equipment
      c. Effect of extreme conditions on plant and neighborhood
   3. Process training
      a. Standard operating procedures
      b. Equipment in system
      c. Control limits
      d. Chemistry of operation
      e. Theory of equipment operation
      f. Corrosion problems
      g. Upset operating procedures
   4. Troubleshooting
      a. Simulated upset operation
      b. Maintenance equipment

  c. Cleaning requirements
  d. Dependence of interlocked equipment on operation
  e. Electrical interlocks and use
5. Safety program
  a. Safe methods of chemical handling
  b. Safety devices installed
  c. Maintenance of safety equipment
  d. Caring for injured
6. Maintenance
  a. Expected requirements
  b. Preventive maintenance program
  c. When to request maintenance checks

SETTING UP A TRAINING MANUAL

Teaching and training has been a process which has progressed along with the increased technology of our civilization. There is little that can be added to existing known methods, but it is our hope that we can rearrange and organize the training program more effectively to meet the needs of the pollution control industry.

One of the first objectives in planning a training program is to write a training manual. This manual, after being written, should be available to all operators on the unit, providing a ready reference for operation control limits, terminology, and theory of operation. It must be simplified to the point where an existing operator can use the manual to train a replacement and yet sufficiently detailed that the replacement can do his job effectively.

A recommended outline of the topics and subject matter is as follows:

Pollution Control Training Manual

1. Glossary of terms
  a. Explanation of technical terminology
  b. Definition of chemical terminology
2. Abbreviations and symbols glossary
  a. Drawing symbols for the unit so the operator can refer to blueprints
  b. Chemical symbols
  c. Abbreviations such as in. mercury pressure, in. water gage, ΔP, gal per in., or gal per ft
3. General description of the unit
  a. Its purpose
  b. Its design characteristics
  c. Its physical size
  d. Its location in relation to other unit processes
4. Theory of equipment operation
  a. Special design characteristics
  b. Limits of operability
  c. What reaction or method is utilized to accomplish its purpose
5. Chemistry of operation
  a. Chemical compounds
  b. Properties of chemicals
  c. Physical and chemical reactions
6. Detailed operating procedures to be outlined by progressive steps for startup and shutdown
7. Safety
8. Maintenance procedures
  a. Operators' duties before maintenance personnel take over
  b. Safety precautions needed during maintenance
  c. Wear points or expected trouble areas

If this training manual can be bound in a corrosion-resistant binder, such as a plastic-covered notebook which can be easily cleaned, and sections can be easily replaced, it can effectively keep operators well informed and decrease the need for retraining. Emphasis should be placed on brief 2- to 3-day retraining sessions each year, with a 3-month checkup on operating manuals to be sure they are available to all operators at all times.

Training to handle pollution control equipment has not been emphasized because of the secondary nature of the equipment in relation to the producing equipment within a plant. Since pollution is now a matter of public interest, every effort to stalled equipment. At $500,000 the labor is reduced to one-half man/shift for each train personnel to operate pollution control equipment as effectively as possible must be made. Training personnel to handle pollution control equipment is different from training people to operate manufacturing equipment in the following ways:

1. Effective operation of production equipment can be measured by how much product has been produced. Effective measures of production efficiency for pollution control equipment are not nearly so clear-cut.

2. The importance of pollution control is an intangible object to a shift operator. It must be made real in a black-and-white manner.

TRAINING—HOW TO TRAIN SUBORDINATES

In recent years much attention has been given in industry to training personnel. Very few people, even the professional specialists, come into a company fully prepared for immediate employment. The required training can range from a "shakedown" on company policy to the learning of a new trade or skills.

There is a definite economical point where it is no longer practical to "pirate" trained employees from competitors. This practice does not increase the number of skilled employees, and ultimately the point is reached where people must be trained to fill jobs created by an expanding economy. Today's industrial workers are often hired without specific skills and are trained on the job to meet the requirements of the company. The educational level ranges from the recent chemical engineering graduate being trained as a process man to a fifty-five-year-old man with a seventh grade education learning the duties of office janitor. As technology and company objectives are changed, old employees must be retrained in new methods or skills.

The supervisor is the person closest to the problem of training the people who work under his supervision. His success as a supervisor is directly tied to the success or failure of the people to learn their new jobs. He is also in an excellent position to judge and evaluate the training needed. He can be assisted and guided by a "company training director," but in the final analysis, the success or failure of any training is his responsibility.

Training committees may help select trainees, adapt training to company needs, teach classes, and evaluate results. Another method is to train the supervisor, and he in turn trains each of his subordinates. But in either case, the supervisor bears a heavy primary responsibility for personnel training.

A number of training methods are in use by industry. Most of the more successful ones are based on Charles R. Allen's four-step teaching method. Mr. Allen's method has been modified to meet various company needs, but no one will be far from his objectives if he follows each of the four steps in sequence. The four-step teaching method is

1. Preparation
2. Presentation
3. Application
4. Examination

Preparation is divided into two parts, what to teach and preparing the worker's mind for the new ideas it is to receive during training. Each of these two parts is of equal importance. The results wanted must be clearly determined, and course of action to achieve these results must be outlined. The subjects to be taught must be investigated and complete lesson plans written, even down to the details of questions to ask to cause the learner to think along certain predetermined lines. A good method to use in preparation is the card file system. The instructor lists all the operations that make up the trade to be taught. Each operation has a card made listing

1. Name of operation
2. Methods that are to be used in teaching
3. Instruction materials

The instructor then takes the pile of file cards listing the various operations to be taught and arranges them into an instructional sequence. The simpler operations are to be taught first regardless of the order in which these operations follow each other in normal production routine. The student must learn to crawl, stand, walk; then he may learn to run. In the same manner, the operation cards must be arranged in sequence from the easiest to the most difficult operations. After the sequence has been determined, the time required to teach each operation is assigned to each card. With this step, the basic outline of the course is completed. The instructor is now free to start preparation of his lesson plan and training aids for teaching each operation.

The second step is presentation of information to the student. There are basically four methods of presenting information:

1. Tell-lecture method
2. Demonstrate-conference
3. Illustrate
4. Experiment

The tell-lecture method is one of the fastest though less efficient methods to use. The instructor tells the student what he wants him to learn, and there is no two-way exchange of thoughts or ideas.

The demonstrate-conference method is one of the more efficient though slower methods to use. There is an exchange of ideas and information between student and instructor, and the instructor is able to evaluate the progress in learning each student is making prior to any final examination. Where the conferees lack information and experience and have a weak instructor, there is a danger of "sharing the ignorance."

The illustrated method uses simulation as an approach to learning. This can be either the case method or business game. In practice it consists of two parts: the case history and the case discussion. The two parts may be combined into a "programmed teaching machine." It is essentially a self-administered multiple-choice or true-false examination. The student reads a section, then tests himself by comparing his answers to questions to the correct answers. If his answers are wrong, he will be able to see why they are wrong. Two or three times through the program and the student knows the right answers as well as why the answers are right. This method is very good in helping a person to learn the many details of his job.

Experiment is the most expensive method to use. The student is self-taught through trial and error. This method should be used only when exploring an area where there is no trained instructor and knowledge on the subject is not available.

In the application step the learner performs the operation in question himself with assistance if needed. Application is a very important step in the teaching process and may take more time than all the other steps combined. Usually the student will have to perform his task or new duties several times, each time making fewer mistakes that the instructor needs to correct. The third teaching step is completed when the student can perform the entire operation without error.

Examination or testing is the final step and may occur any time after the application step, though it usually occurs right after the completion of the application step. The examination usually requires the worker to perform the entire operation satisfactorily without assistance. When the student can do this, the learning process has been completed for this operation, and the student can advance to more difficult lessons or the completion of the training.

If you are teaching new skills or upgrading an older worker, you wish only to teach the worker certain information. You are not concerned with the knowledge the worker already possesses. When the worker learns what you want him to know, his training is complete. Either the tell-lecture or visual instruction lends itself to this need. When you are dealing with a group of people who are experienced in the field but are not equal in knowledge in all areas, it is desirable to utilize knowledge already possessed by the worker. This can best be done by using the conference method of instruction. This training method was developed for industry by Charles R. Allen. It has gained widespread acceptance throughout industry in recent years since the business conference has largely superseded oratory as a device to develop new ideas. The training conference is conducted in exactly the same manner as the business conference. This gives the supervisor a training method he should be very familiar with. Any supervisor who aspires to advance to executive ranks will find a knowledge of conference-leadership techniques of greater value than the ability to make a good speech; so if he does not already have this ability, he should acquire this knowledge as soon as possible.

The major advantage of the conference method is that the students can be better informed on the subject than the instructor and will still have a successful and meaningful training session. The instructor only guides the conference and withdraws every bit of information he can from the group. Convinced of this, there can be no reason or justification for the instructor ever to lecture the conference group. The leader must remain impartial and limit his discussion to carefully worded questions designed to develop further ideas from the group.

It is extremely important to prepare for any conferences as carefully as for any other teaching method. Again use the four-step method to prepare a guide for your conference. An experienced leader always develops a detailed plan for the meeting, questions to ask, and an opening statement outlining the object of the conference and rules by which the conference will be conducted. It is of equal importance to close the conference with a summary of what has been discussed and accomplished in the conference.

The most common error in job training is to omit the preparation step from the teaching process. Neither the instructor's nor students' minds are in such a state as to give and receive new ideas. Another common mistake is to attempt to teach too much at one time. The student can absorb only a certain amount, depending on his past exposure to education and his mental capabilities. The instructor many times forgets or does not realize how difficult it is to learn the operation he is teaching. He knows the job and feels his students must be stupid if they cannot grasp new ideas as fast as they are presented to them.

An area of training often neglected is the third step, application. The student is not given enough time and assistance to practice what he has been taught. The result is a lack of confidence in ability to perform his new duties.

Last but not least is testing. Every training session should be followed up with an examination, and the student should be informed of his strong and weak areas. When the students have weak areas, the fault invariably lies with the instructor. When the student fails to learn, the instructor failed to teach!

Many people place much emphasis on "visual training" and believe it to be an excellent and easy training method to use. What must not be forgotten is that this is only one of four steps in the teaching method, presentation. The other three steps (preparation, application, and examination) must not be omitted. The motion picture is not a training method; it is only a training aid or tool and must be used as such if the student is to get the maximum benefit from the training. Visual instruction is a great aid and tool when the instructor uses it correctly.

The only justification for any training program is to solve a specific problem. Training is expensive to conduct and must be justified in terms of dollars saved or increased profit for the company. Very little training in industry is conducted just for the public relations. When the problem has been solved and results shown, it should be shortly discontinued.

TRAINING TO OPERATE

Operation of effluent control equipment is similar to usual plant process equipment. Once it is in operation, certain target operating objectives are used as guides for satisfactory control.

Uusually, new facilities, expansions, and additions are constructed by contractors or construction groups that are separate from the operating departments. During construction and the final phases of completion of pollution control equipment is a crucial time for operating and maintenance personnel to become acquainted and familiar with the new facilities. When the equipment installation, painting, etc., have been completed, a semiformal planned program is necessary to facilitate transfer and acceptance of the equipment, equipment records, etc., to the operation departments.

It is wise for an inspection of each piece of equipment to made before written acceptance is made. Visual inspection of equipment includes thorough inspection of all components. Listed are some examples of items requiring visual inspection:

1. Bag dust collector.
   a. See that bags are installed properly.
   b. Check the bag securing.
   c. All doors and joints are dusttight
   d. All mechanical items locked or fastened for shipping are commissioned and ready for operation.
   e. Manometers are installed, connected, and filled.
   f. All surfaces are protected against corrosion.
   g. Unit is free of foreign material.
2. Wet scrubber.
   a. All nozzles are installed and in the proper places.
   b. Protective lining is installed and no damaged places evident.
   c. Piping is connected.
   d. Packing is in place.
   e. Baffles are securely in place.
   f. Blinds are removed or in place as specified.
   g. Unit is free from all foreign material.

Equipment with running gear should be "run in" if at all possible to ensure that all components are in order. Initial run-in of equipment is intended only for rotation checks, and brief runs to ensure equipment is not noisy, belts are trained properly, switch gear and electrical components are in order, and generally that equipment operates mechanically. This check has no bearing on the equipment meeting process requirements. These checks would be as listed:

1. Dust collectors.
   a. Bag shakers shake.
   b. Fans run and in correct rotation.
   c. Reclaim screws rotate quietly.
2. Wet scrubber systems.
   a. Pumps turn in correct rotation when bumped over electrically (these should not be run but a few seconds since seal damage usually occurs when dry).
   b. Valves will cycle.

This inspection would include mechanical, electrical, lubrication, correct rotation, and any other phases that would be relative to the equipment and its acceptance. Experts or qualified personnel quite familiar with each phase of inspection should represent the accepting group. All discrepancies should be *immediately* covered in writing to the parties concerned so as to expedite corrective action. If disagreement arises as to whether a piece of equipment is acceptable or in proper operating order, this should be expressed in writing to the parties concerned for a later reference.

Equipment acceptance and the covering paperwork should be completed on a daily basis. When equipment checks out, initial run and general mechanical acceptance is completed, a new phase of planning and philosophy is entered.

The second phase of "getting the equipment on the line" incorporates planned operation of the equipment under simulated operating conditions, usually beginning with initial operation of a few single circuits or individual pieces of equipment and finally completing the "phase-in" with as much equipment on the line as is possible at one time.

First, let us consider the initial operation of a wet scrubber system. The scrubber liquid system would be filled with an inexpensive liquid. This liquid would be of correct viscosity and have other characteristics suitable for the purpose of washing out, cleaning and general check-out of the piping, pumping, control valving, and spray systems. Pumps would be started to facilitate circulating within the liquid system. Circulation of the gas stream can be simulated with air so that the complete scrubber would be in simulated operation.

Precautions are usually required, according to the circumstances, for removal of foreign material from the system. Various types of strainers are used temporarily in piping, sumps, pump suctions, and any other places as required in liquid service. If the gas stream ducts could contain foreign material, ice snow, etc., and are inaccessible to cleaning, then temporary wire screens should be installed in order to facilitate the removal of these unwanted materials, which may damage the lining and plug pumps, strainers, and nozzles.

A special effort should be made to record locations of all leaks that cannot be stopped while in operation which can be repaired immediately after shutdown.

Manometer and pressure readings should be taken as if in normal operation. Since we have simulated operation of a wet scrubber unit, let us now review the simulated check-out of a cyclone and bag dust collector system.

The equipment and ducts should be first inspected again for foreign material. Temporary screens should be placed as thought needed. An airstream should be directed through the cyclones and into the dust collector. Manometer readings should be noted for checking pressure drops, particularly across the dust collector bags. All ducts and openings should be checked for air leaks. Stacks and ducts should be traversed to determine gas velocities at critical points. Check cyclone belts to determine if chains are swinging properly. These preliminary runs serve two good purposes. First, the equipment is given a good check and second, the operating personnel become familiar with internals, components, and simplified operation of the units.

Inspections after Water Run. Equipment should be reinspected after the equipment is shut down, drained, and opened. All foreign material should be removed. Strainers and screens should be cleaned. If the systems appear clean, the strainers and screens could be removed, but if there is evidence that foreign materials are still circulating, the strainers and screens should be replaced. Smaller-mesh screens can be used if the temporary screens did not properly clean up the system.

In dry systems, careful inspection should be made to ensure no moisture accumulations occurred, especially in bags of dust collectors. All shaking and air pump mechanisms should be rechecked to see that the operation runs and is satisfactory. Manometers that did not appear to operate satisfactorily should be rechecked for plugging, etc. It is quite common that the wrong weight fluids have been installed, giving misleading readings.

During initial water washes and water runs, considerable useful information can be compiled which can be very useful for startup operations. Power readings and flow rates would be of particular importance for both wet and dry type control equipment. Correlation with initial equipment operation where temperatures, viscosities, specific gravities, etc., can be scaled up or down may give indications of serious problems before the actual process is under way. Any means of catching design problems before major upsets occur usually saves money, time, and frustrations. Pump curves, fan curves, and pipeline sizes can be roughly checked for proper sizing by observing and recording data during preliminary water runs.

Actual Operation. Now that preliminary inspection, equipment checks, and water runs have been completed and the parties concerned are satisfied that the facilities are ready and personnel are well trained, the actual operation is the next step.

Startup of actual operation should be well planned and operating personnel completely familiar with the control equipment and its operation by this time. Initial use of equipment under operating conditions should not be such a big step for supervision and operating personnel if preliminary work and training have been successful.

Personnel should pay very close attention to all equipment during initial operation. They should be particularly on their toes to locate straightforward mechanical problems, those from lubrication, process irregularities, and design errors. Sometimes it is quite hard to determine the source of problems during upsets and failures since they become so involved, but recording data and evaluating the data during the first shutdown can resolve most problems.

As equipment and systems can be run, even for short periods, the operating personnel should be closely observing both equipment and process phases, so that a complete familiarity can be gained in as short a time as possible.

As soon as our control equipment has begun functioning, immediate steps should be made to determine efficiencies and losses. After all, control equipment of all types is used to control losses of one type or another. Efficiencies are usually very important to control equipment operation.

Measurement of losses is most generally accomplished through sampling of one type or another. Types of sampling vary from spot samples to continuous samples which are fed through analyzers registering continuous recorded data. Some examples of samples can be as follows:

1. Dust in gas stream.
   a. A specific volume of dust laden gas is drawn off from the main flow through a porous permeable tared filtering container which collects the dust but lets the gases through. The filtering container is dried and reweighed to determine the dust loading.
   b. A specific volume of dust laden gas is drawn off from the main flow and directed through a wet collecting apparatus which washes and collects the dust as a slurry. The slurry is dried, or filtered and dried, leaving the solids to be weighed for determination of dust loading.
2. Gases in airstreams. A specific volume of gas is removed from the main flow and passed through a series of absorptive solutions that will capture the gas. The solutions are then weighed and analyzed to determine the quantities of gas absorbed. Calculations then determine the particular gas density in the original gas stream.
3. Moisture in gas streams. A specific volume of gas is removed from the main flow and passed through liquids or desiccants which completely absorb the moisture. When the quantities of moisture captured are ratioed with the gas quantities bled off the main stream, the moisture content of the main gas stream is determined.

By using the available sampling techniques, measurement of losses can be determined for all phases of the effluent control field.

For best results, all data should be gathered under similar conditions. Detailed notes should be made with each sample, calculations and correlations to be used as future reference. All data should be correlated on a continuous basis and referenced so that they can be useful to all parties responsible.

Conclusions that are the results of or are drawn from these data usually are the basis for changes which are required to improve efficiencies and recoveries within a control unit.

Safety should be included in all training and operating philosophy. It should also be included in everyday work and refresher training until it automatically becomes part of the operations. Safety is a means of developing an acute awareness and respect for all operating equipment, processes, and materials used in operation. It also can be used to create discipline which is needed to execute operating functions in a satisfactory manner.

Maintenance is sure to be a major component in economical, satisfactory operation of control equipment. Sloppy maintenance will cause any number of problems, including poor recoveries, impossible operating conditions, and extremely high costs for equipment operation, all of which can be avoided.

Maintenance personnel require detailed training and familiarization with the equipment and processes which are involved.

Maintenance must be performed on a planned and scheduled basis. Preventive maintenance plays a very important role in high operating factors and efficiencies. It must have a high priority, with proper direction and administration. Good records and accumulated data can be most useful for sustaining economical, aggressive maintenance.

CONCLUSIONS

To summarize the total desired training program, the following outline may be helpful:

1. Determine clear-cut objectives for operators to measure their efficiency of operation, making it clear that upset operations cause harm to the plant, internally or externally, because a pollution device is not operating properly.
2. Write an operating manual, in a clear, concise manner, and bind it for rough use in the operator control room. The binder should facilitate page changes or replacement.
3. Do a thorough training job, including classroom and on the site training.
4. Do preliminary water check-out of equipment with the operators.
5. Follow through startup and maintenance, refining the training and evaluated performance.
6. Retrain operators at least once a year, and be sure operating and maintenance manuals are always available for their use.

Chapter 23

Air Pollution Control Equipment

RALPH R. CALACETO

Vice President, Airetron Engineering Division, Pulverizing Machinery, a unit of the Slick Corporation, Midland Park, New Jersey

THE NATURE OF AIR POLLUTION

Before we can discuss the nature of the equipment used to control air pollution, we must first define what constitutes air pollution, in order to understand the nature of the specific problem which control apparatus must be designed to meet.

For our purposes, air pollution consists of particulates or impurities that are suspended in or conveyed by a moving stream of gas or air. This pollutant material may exist in liquid form (commonly described as mist), as gaseous fume, or as solid particulate matter, including dust. It will generally result in at least one of four detrimental effects: loss of valuable products, an atmospheric nuisance or safety hazard, damage to the quality of the manufactured product, or mostly plant and equipment maintenance.

Particulates are formed and classified in either of two size categories: submicron or micron. Equipment design and collection efficiency are directly related to particle size. Particles 1 micron or larger in size (1 micron equals1/25,400 in.) are generally easily collectable; smaller (submicron) sizes are more difficult. One exception is the gas molecule. Although extremely small (0.005 micron in size), it is usually easily removed by packed tower type methods discussed at length later in this chapter.

The first consideration in the proper selection of a collector is the nature of the pollutant materials to be collected.

Origin of Air Pollutants

Both the chemical and physical characteristics of all particles must be known in order to choose the proper collector(s). The origin of the particle may furnish a clue as to whether the pollutant constitutes a fume, dust, mist, or just an undesirable gas.

Particles formed by mechanical disintegration may generally be classed in the micron size group. These may be formed by pulverizing, crushing, and grinding. Generally, these particles are larger than 1 or 2 microns. But there are always exceptions to the rules. For example, a particle of dust which is subject to continuous "self-disruption," as in a roller grinding mill, may produce particles with distribution into the submicron size range.

Particles created by coarse disintegration such as sawing, jaw crushing, or tumbling (as in slow-moving rotary equipment) may be above the 15-micron range.

Particles created by a physical change of state (such as sublimation or chemical reformation) or from an intensive heating and melting operation are generally submicron in size.

Furnace operations, in both the steel and nonferrous industries, usually produce exhaust or combustion gases containing metallurgical submicron fumes. The generation of finer fumes is directly related to the heat intensity.

Two current areas which are known producers of extremely submicron metallurgical fumes are:

1. Oxygen injection—to raise melting temperature at increased speed
2. Electric furnace—to reach temperatures above 2000°F.

Pollutants having a composite combination of particles may be found in air which has been swept through the inside of a plant having various processes, or which has been trapped in the hooding around operating equipment.

The secondary consideration in selecting a collector is the feasibility of available collectors to suit the specific needs of the user. Factors such as particulate or gas recovery, the area which is to be occupied by the collector, power usage, disposal problems, maintenance, replacement parts, and situations governed by the overall process operation weigh heavily on the best choice for the type of collector. Often a combination of these collectors will offer many advantages.

INFORMATION REQUIRED PRIOR TO EQUIPMENT DESIGN

Because each air pollution problem is unique, exact preliminary knowledge of pollutants is required in order to design compatible equipment. Particle size, gas temperature, and corrosive constituents are all factors influencing equipment selection and design. To acquire this knowledge, samples of dust and mist must first be collected.

Sulfuric acid mist (SO_3) illustrates how the chemical properties of a pollutant can cause complications. Appearing as a visible fume with little odor, SO_3 is a submicron aerosol mist which forms sulfuric acid upon combining with moisture. When this occurs on moist skin tissues, a tingling feeling or burning sensation results. To be scrubbed, SO_3 must first be hydrated to convert it into the removable form of sulfuric acid. Wet conditioning prior to scrubber entry performs the hydration. This requires a dual operation at the scrubber—conversion plus energy of impaction.

Metallurgical plants have generally collected sulfuric acid with an electrostatic precipitator. More recently, plants manufacturing sulfuric acid have utilized venturi scrubbers (operating at 30 in. W.G. pressure drop) to make the characteristic whitish cloud of SO_3 mist essentially disappear.

A similar collection problem is involved with the phosphorus pentoxide fumes,

P_2O_5. Although simple collection methods are not sufficient for this submicron aerosol mist, venturi scrubbers operating at 30 in. W.G. pressure drop have proved efficient.

As a rule, the submicron mist must be reduced to approximately 1 or 1½ mg/scf to approach invisibility.

Analyzing Constituents of Polluted Airstreams

Particle size distribution Distribution of particles is commonly measured in total airborne mass concentration, of grains per standard cubic foot (60°F). This can be misleading, since fractions of airborne material in the large micron range may appear as a small quantity by count, although heavily affecting the weight of the mass concentration. Conversely, smaller particles that have little effect on the mass weight offer greater dispersion (visual disturbance) and a more serious potential inhalation hazard. Comparisons of percentage by count and percentage by weight must therefore be carefully interpreted to avoid inadequate equipment selection. Percentage by count reveals particle distribution and the size ranges of counted particles (often by a visual microscopic examination). The influence of counted particles on the weight percentage varies according to the volumes of the particles or spheres. For example, where the volume of a sphere equals $\pi d^3/6$, count and weight would vary as follows:

Particle size	% count	Volume*	% by weight
1 micron..........	50	1	0.1
10 microns.........	50	1,000	99.9
		1,001	100.0

* Volume is interchangeable with weight since the particles all have the same density.

Techniques for particle size analysis *Photoelectron microscopy* although most commonly accepted, is a time-consuming method of measuring particle size and requires an experienced interpretative eye to determine whether particles are discrete (separate) or agglomerates. After gas samples have been bottled for dispersion in the reflection chamber of the microscope, it is virtually impossible to know whether they are dispersed in the same way that they originally traveled in the airstream. Thus, the true size of particles in flight is extremely doubtful. Although accuracy improves with increased particle size (results are satisfactory for particles 5 microns or larger in size), the method is inadequate for the smaller particles that constitute the majority of true pollution problems.

Cascade impactor analysis is an inherently simple method that utilizes processes in inertial impact similar to those of the respiratory system. Recently made practical for field survey sampling, it is advantageous in that it requires only a series of mass determinations, rather than individual sizing of several hundred particles.

Cascade impactors are also used efficiently to determine the size distribution of one component of an aerosol in situations where the presence of background particulates would prevent accurate sizing by microscopic or other methods. They can also determine the size distribution of dusts which are altered by the process of collection, or which decompose or decay following collection. By grading an aerosol on the basis of its airborne particle size, the size analysis will be the same whether particles swell or shrivel or droplets spread, coalesce, or partially evaporate—as long as some fraction of derivative of the original material remains on the slides. The particles are actually graded by their effective aerodynamic diameter in the cascade impactor.

Analyzing dust, fumes, and particulate matter *The thimble method,* assembled as in Fig. 1*a*, utilizes a holder equipped with disks or cylinders made from Whattman paper or alundum (aluminum oxide). The alundum has the advantage of withstanding temperatures of over 1000°F. Although the standard thimble holder is ideal for collecting samples of 1 to 20 g in weight, minute samples necessitate the use of the disk in a retainer ring. Note that use of this apparatus at low temperatures may cause condensation of moisture, clogging, and eventual breakdown of the

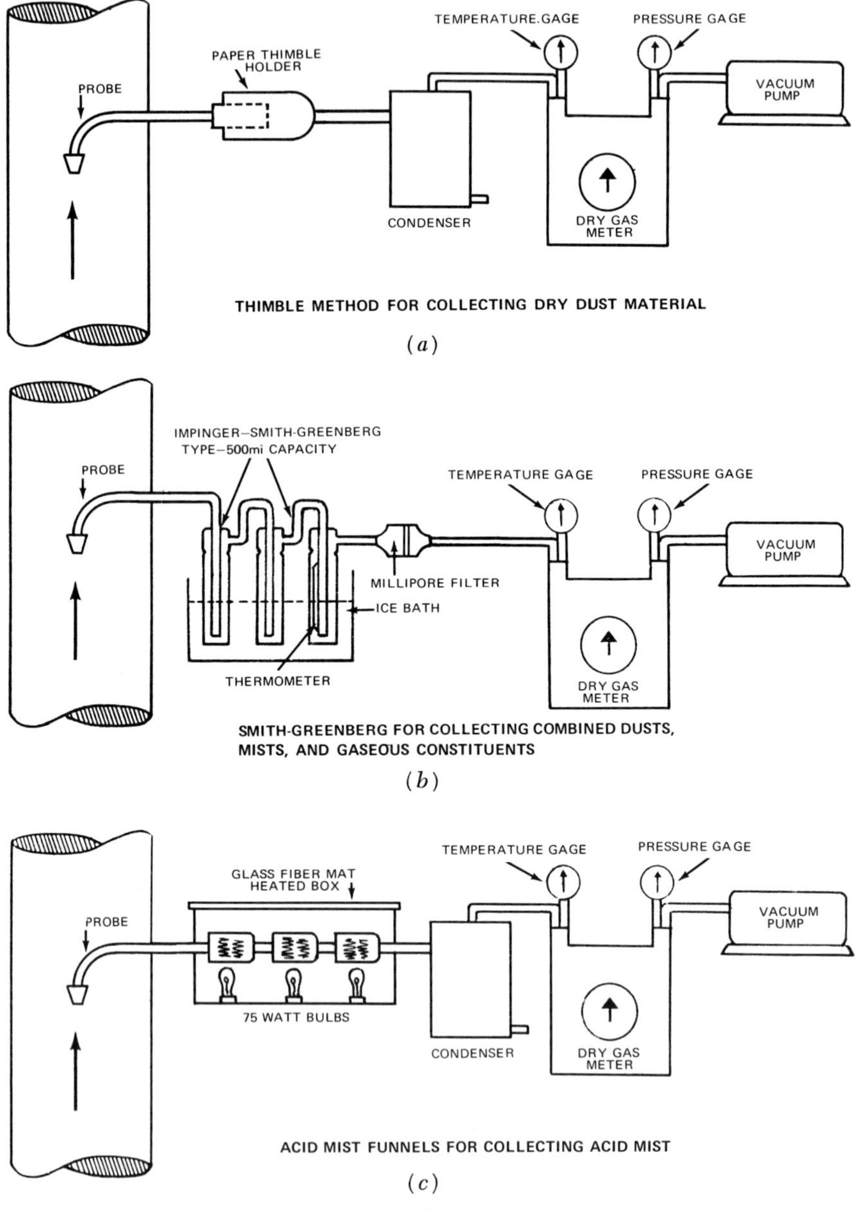

Fig. 1 Preliminary procedures include collecting and analyzing particles for size and density with this apparatus.

disk or thimble. Accordingly, provision for heat should be made in those cases where temperatures are low and condensation may occur. Usually a stripheater or wraparound heating element must be installed to keep temperature above the dew point. After this, the gas enters a condenser or ice bath where water vapor is trapped out, proceeding then through a meter which takes pressure-temperature readings. Movement of the gases follows this order whether a vacuum pump or ejector is used.

A Smith-Greenberg Impinger (see Fig. 1b) may be installed in similar fashion following a thimble holder if other contaminants such as SO_2 are present. Since condensation causes an increase in the liquid volume in the impinger, this amount is measured, calculated back to gaseous volume, and added to the metered air volume. This corrected volume is then used in the calculation of grains per cubic foot or milligrams per cubic foot of contaminants.

Determining gaseous contaminants Constituents which are strictly true gases (SO_2, HF, NH_3, CO_2, H_2S, etc.) are relatively simple to collect. As a rule, Smith-Greenberg impingers, possibly two or three in series, give excellent service, with NaOH or dilute H_2SO_4 added to collect basic or acidic constituents, respectively. The preceding apparatus should be used in a similar manner.

Determining the gas volume An important factor in ultimate design, gas volume determination requires two precautions:

1. Determine gas velocity at a point in the duct where gas moves in a uniform straight line pattern. Any circular or curvature interference in the gas flow will yield erroneous results as to actual volume. Uniform gas flow is generally indicated as 5 to 10 diameters upstream from a bend in the gas duct; otherwise, straightening vanes mounted within the duct may be necessary.
2. Choose a location convenient to the setting up of the test apparatus to assure accuracy in technique in handling the equipment.

Pitot tube analysis with an incline manometer is relatively simple and well established. Data should be taken in a flue duct using a cross-sectional arrangement, thus equalizing readings taken over a number of equivalent areas—the center of each area being the point where the pitot tube measurement is taken. Usually, the areas are taken in a manner designating annular areas from the center portion of the duct out to the circumference. Measurement should also provide for taking data at right angles. Any large deviation in measurement from the average figure will indicate that an uneven flow pattern exists in the duct—introducing some doubt as to the accuracy of the velocity measurement.

If the gases are conveyed by use of a fan or blower, a fan curve will be useful. Knowledge of the horsepower consumed (verifying static pressure) and the gas temperature is requisite.

Determining gas density This involves questions of whether moisture is present or whether the gas molecular weight is other than standard air. Unless a proper density correction is provided, the velocity measurement can be erroneous. Variations in density measurement may be due to large quantities of CO_2 and SO_2.

Determining mists, dusts (including fumes), and gaseous constituents Each of these three categories requires special apparatus. A multiple test apparatus assembly is necessary where more than one type of constituent is present in the gas stream.

Basic Technique for Sampling Aerosol Mists. Apparatus as shown in Fig. 1*c* is required to collect SO_3, P_2O_5, and monoethanolamine (MEA). Filter funnels, three in series, are filled with a 1-in. layer of glass fiber. Gas should have a pressure drop of approximately 10 in. of mercury across the funnels for excellent mist collection. In the presence of water vapor, the funnels are mounted in a wooden box fitted with light bulbs to prevent water vapor condensation. From this, the gas proceeds into Smith-Greenberg impingers to remove the gaseous constituents such as SO_2. The impingers may contain 100 cu cm of 0.5*N* NaOH. Standard laboratory titration procedures are then used to determine the SO_3 washed from the glass fiber as H_2SO_4, as well as the SO_2 absorption liquid taken from the impingers. Gases, which are moved by means of a vacuum pump or an ejector, are then metered, and pressure-temperatures recorded.

Purpose of Control Equipment

Recovery is economically desirable in many instances, as in the fertilizer industry, which utilizes recovered ammonia. In the steel industry, blast furnaces produce iron oxide which is collected and returned to sintering plants for reuse, while the clean gas containing high CO is also returned for heating value.

Factors in the recovery of particulate matter First, determine whether it can be returned to process in wet or dry form. With excessive moisture content, the material may have to be filtered and dried. Overly fine dry particles may be subject to continuous re-entrainment.

If contaminants are not returnable to process, determine whether they are readily marketable. This may necessitate their being in a fine dry powder or in a liquid container.

There are three basic methods of recovery: wet scrubbers, cloth type collectors, and electrostatic precipitators (all described at length later in this chapter). Reuse of recovered material normally dictates the ultimate method of collection. Therefore, specifications for the proper collection equipment must take every element of your problem into account.

If conveying the recovered material presents a problem, for example, you may have to choose a wet scrubber for relatively easy recirculation of the recoverable slurry. Wet scrubbers should also be chosen where dry materials present a fire hazard, as in the case of aluminum, bronze, magnesium, wheat, or flour dust. Certainly, the danger of explosion from solvent vapors would preclude the choice of a precipitator.

Factors in disposal Where organic constituents are not recoverable, they may be disposed of by combustion. Combustion offers a positive method for deodorizing materials, since it results in a harmless exhaust gas of carbon dioxide and water vapor. One variation is use of a catalytic device which permits lower, and in some cases no fuel consumption to raise gas temperature to combustion levels.

Disposal of nonrecoverable collected solids may be a simple matter of pumping to a pond. In some cases a sanitary sewer is permissible. This depends on the community facilities and regulations, however, and may present a problem.

When the collected material is pumped into a settling pond, the sludge must be removed from the pond at intervals, depending on the site of the settling basin and the quantity of emission. Collected solids may also be settled out and used for landfill.

Disposal of dry solids In applications that are an integral part of an industrial process, pneumatic conveying has numerous economic advantages over mechanical conveying of granular, flaked, or powdered materials.

In both high- and low-pressure types of pneumatic conveying, compressibility of air is the major consideration in flow calculation.

Positive pressure conveying systems employ high air pressure created by a blower at the material loading station. These systems are determined by an air-to-material ratio, with a figure of 12 cu ft/lb recommended for operational efficiency.

Negative (low-pressure) conveying systems, on the other hand, employ suction, attaining virtually 100 percent separation of particulate matter from the air or gas. In these systems, pressure in the conveyor system is below that of the surrounding atmosphere from material pickup station to point of discharge.

These suction air conveyors utilize four basic components: pickup nozzle or feeder device, conveyor piping, receiver-separator, and exhauster. Although pickup nozzle designs vary with the application, they employ basic aerodynamic principles to accelerate the feed to the conveyor without interim settling out of the material.

Unless abrasion is a factor, 16-gage wall thickness is sufficient to withstand the light pressures in conveyor piping. It is vital that joints do not leak, since virtually 100 percent recovery or separation of conveyed material is required. Receivers consist of a primary mechanical separator followed by the cloth filter unit or a filter type dust arrester. To maintain constant pipeline velocity, a constant cleaning filter, cloth type, should be used (the reverse jet type, for example). Strong housing construction is necessary to withstand design pressure.

Single-stage centrifugal type exhausters are normally used, unless flow rates below 250 cfm necessitate choice of a positive displacement type. Exhausters with constant pressure should be avoided, or conveyer lines will plug with slight overloads.

Flexibility of the Air Conveyor. Flexibility in the suction system is limited to the material pickup end, since the receiver is a stationary installation consisting of a dust collector and exhauster. Thus, material can be delivered to a single station, although switching arrangements can make possible pickup from several locations. The reverse is true in positive pressure conveying, which permanently locates the blower at the material loading station. More complex systems are also available, offering increased flexibility at both ends via combinations of pressure and vacuum systems.

Limitations of pneumatic conveying Transporting materials via pneumatic conveying requires more power than that needed for mechanical conveying. Also, pneumatic systems are sensitive to overload conditions that are easily handled by mechanical systems.

Advantages of pneumatic conveying Maintenance of pneumatic systems is generally limited to the dust collector—an ordinary filter type operated on an equivalent hourly basis. This simple filter is easily interchanged. In some instances, power-driven devices at the feed end must be treated as machinery, or pipe bends transporting abrasive materials may require periodic replacement.

Stainless Steel Specifications. Light gage conveyor ducting is far less costly than the stainless steel frequently required in a conveyor system. Obviously, the advantage of the pneumatic conveyor increases in proportion to the transport distance. Also, the cost of the dust collector-receiver is increased 12 to 15 percent by specifying stainless steel.

System Cleaning. Thorough cleaning is a problem in mechanical conveying systems which process several types of a similar product. Yet cleaning a pneumatic system is a simple matter of conveying relatively small quantities of three inexpensive materials: leather dust, pearlite, and sawdust. Although filters in the receiver may require frequent changing, this is generally quite simple.

Contamination of Product Handled. The unique exhaust system of the suction type pneumatic conveyor is beneficial to cleanliness, safety, and the well-being of operating personnel. In addition, virtually no product contamination results from material transportation. This is due to five inherent characteristics of the suction type conveyor: (1) the path of conveyance is a gastight closed duct from pickup point to discharge station; (2) filtered and/or conditioned air may be supplied at the feed end; (3) the lack of mechanical parts eliminates the chance of lubricants contaminating the product: (4) lower allowable transport velocities permit use of traps to separate tramp; (5) permanent magnets may be inserted into the duct to separate magnetic particles.

SPECIFYING APPROPRIATE TYPES OF COLLECTION EQUIPMENT

Manufacturers must take many factors into account when selecting the correct method of treating a specific air pollution problem.

Each type of control equipment embodies particular design peculiarities which influence installation, maintenance, and efficiency.

A manufacturer must also weigh the interrelated process and environmental factors. Frequently, the preferred solution to one or more side problems may indicate a choice other than that first apparent. Thus, the collection equipment eventually selected often represents the most desirable compromise. Alternately, a combination of equipment may be the answer.

Factors Affecting Equipment Specification

Location. Proximity of the gas collector to the gas source can determine the treatment required. As a general rule, a reasonably close location minimizes many side problems. Among problems alleviated are conveyance of dust particles which could settle in the ducts, and cool spots which cause condensation and ultimate corrosion. Close placement also minimizes pressure drop losses which occur when gas is drawn through a long distance.

Location is also influenced by the nature of the dust loading. If the dust is micron size or larger, in an essentially dry and ambient condition, the collector may be installed as close to the source as possible. However, if the gas loading is submicron in size, factors such as nuclei growth should be considered. Collection efficiency is somewhat enhanced when submicron particles have an opportunity to agglomerate with one another while traveling a long distance to a collector.

Space Requirements. The feasibility of a particular type of equipment may hinge on space factors. If the source is on the roof, for example, considerations of roof loading will affect collector design. These can involve size, as in extremely large bag collectors or precipitators. In the case of a wet scrubber, on the other hand,

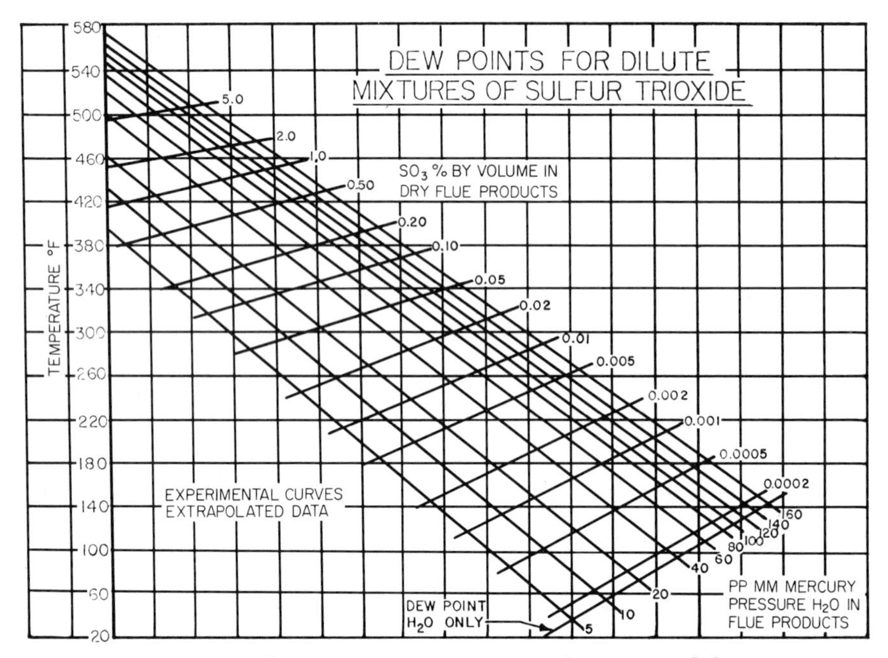

Fig. 2 Corrosion in electrostatic precipitators can be prevented by maintaining temperatures above the SO_3 dew point.

weight may be the determining factor, when recycle tanks are requisite. Foundation factors of support materials and design are therefore involved whether on the roof, near the ground, indoors, or outdoors.

Complicated Fumes. Fumes consisting of heavy particles, tar vapors, or organic solvents may create side problems varying from fire hazards to dust accumulation on the walls. In this case, short duct systems are preferred. If impractical, consideration should be given to spraying the ducts, necessitating sloping the ducts (possibly 5°) to direct drainage toward the scrubber.

Temperature. When handling extremely hot gas, the collector must be placed at some distance from the source to allow cooling in transit. This is especially true in the case of bag collectors, where pretreatment of the gas, by cooling, improves collection efficiency and protects the filter fabrics. These fabrics affect the overall operation and cost of the equipment, and require specific temperature ranges for maximum performance and longevity (see Table 1, page 23-11).

Excessive temperature may contribute to duct warpage.

Sudden cooling often causes a "wet-dry line" condition, in which solids accumulate at the cooled duct walls.

Optimum temperatures for electrostatic precipitators range from 350 to 500°F. The presence of such acid compounds as SO_2 and SO_3 requires serious consideration. Mixtures of sulfur trioxide have a high dew point which can create a corrosive atmosphere.

The SO_3 dew point (illustrated in Fig. 2) should be considered at all times.

Generally, extremely hot gases are best handled by wet collectors, which effectively utilize several cooling techniques. Adiabatic cooling to saturation is one means of reducing the temperature. For best scrubbing results, gas temperatures above 450°F should be saturated prior to entry into any wet scrubber.

Taking the following points into consideration, typical choices for handling various constituents are shown in Fig. 3.

CLOTH FILTERS

The oldest known method for removing dust from an airstream is the cloth filter. It is versatile and highly efficient for collecting solid particulate matter in a wide range of sizes.

As a general rule, efficiency increases in direct proportion to the amount of cloth area in the filter. Maximized cloth area delivers lower pressure drop requirements, greater reserve capacity for surges or expansion, and longer media (fabric) life. Thus, efficient design of fabric dust collectors involves a compromise between the ideal of large bag filters and their considerable cost. To maximize filter surface per unit volume, designers frequently determine size by selecting the highest filter rate (velocity through the media expressed in feet per minute) consistent with good operation.

Filter fabrics The actual fabric of the filter is chosen on the basis of its ability to withstand temperatures and stresses inherent in the process, as well as its compatibility with the pollutants to be collected. The fabrics are fibrous material, either natural or man made, in woven or felt form. In the case of woven fabric, the material serves as a base for the accumulation of a porous layer known as a filter cake. This cake heightens the filter's cleaning efficiency by screening out submicron particles. With felt materials, cleaning is accomplished as the gas travels through the maze of fibers in the fabric itself.

Factors affecting fabric filter performance These include the fineness and size distribution of particulate matter, particle shape, agglomeration tendencies, static charge or tendency, physical properties such as adhesion or sublimation, and chemical properties such as crystallization and polymerization reactivity. System factors include gas constituents, loading, media limitations, temperature, humidity, desired differential pressure, turbulence, and dust origin.

Increased temperature, for example, requires more cloth, probably because of an attendant gas viscosity increase. This is eventually counterreacted by reduced density. Dust load is also a major consideration, requiring more cloth area at higher loadings (more frequent cleaning or precleaning). Certain continuous automatic collectors can handle more material without lowering cfm per square foot when loading surpasses a certain point, normally in excess of 100 grains/cu ft. This is probably due to total saturation of the air with dust, thus giving bag surfaces a saturation-limited rate of accumulation per unit time.

Specialized application factors fall into three basic categories. The first of these is nuisance venting, which includes relief of transfer points, conveyors, and packing stations. The second, product collection, may involve air conveying-venting mills, flash driers, and classifiers. The third, process gas filtering, ranges from spray driers to kilns and reactors.

Equipment Basic equipment in the filterhouse (baghouse) includes a number of

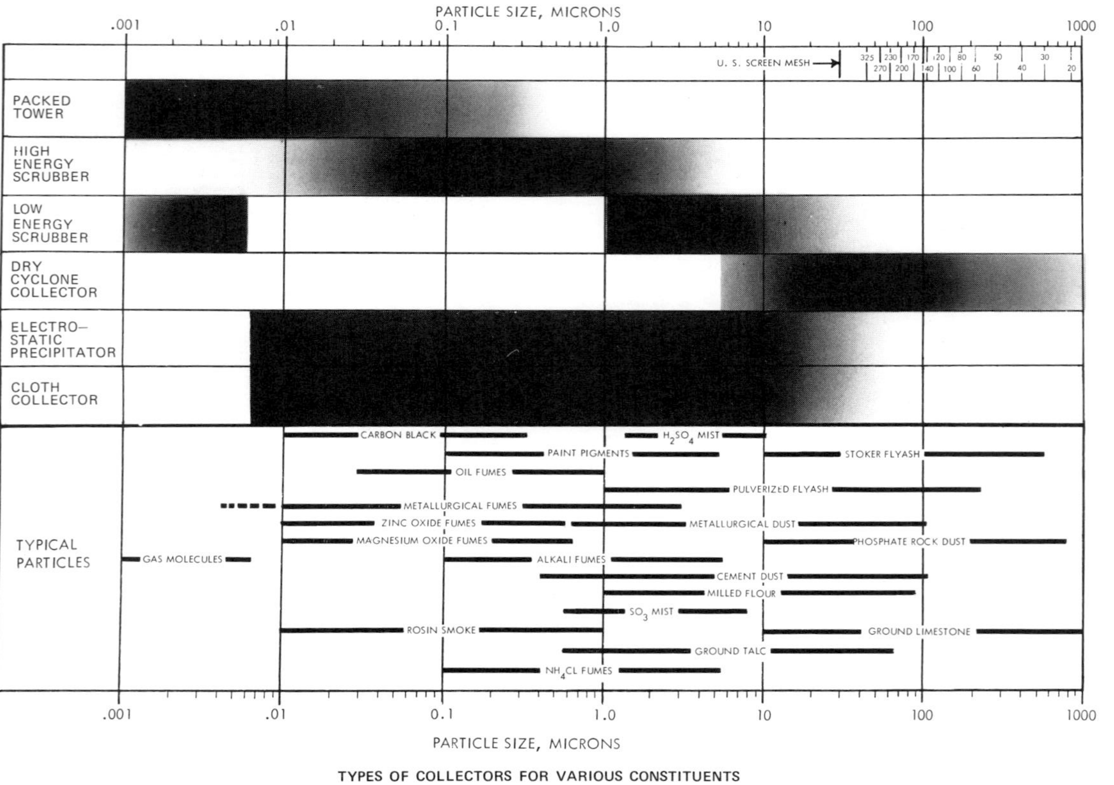

TYPES OF COLLECTORS FOR VARIOUS CONSTITUENTS

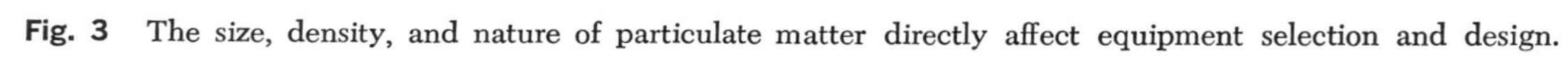

Fig. 3 The size, density, and nature of particulate matter directly affect equipment selection and design.

TABLE 1 Recommended Dust Collector Fabrics

Fabric	Structure	Max operating temp, °F*	Mechanical resistance to abrasion	Chemical resistance to: Acid	Chemical resistance to: Alkali	Other	Remarks
Cotton	Woven	180	Fair	Poor	Excellent		
Wool	Woven	190	Good	Good	Poor	Degraded by hot sulfuric acid	
Wool	Felt	190	Good	Good	Poor	Degraded by hot sulfuric acid	
Glass	Woven	550	Poor	Good	Good	Attacked by hydrofluoric and hot phosphoric acids	Recommended silicon treatment for fiber lubrication. Will pass gaseous acid radicals at operating temperatures above dew point without degradation to fabric
Nylon	Woven	200	Excellent	Poor	Excellent		
Nylon	Felt	200	Excellent	Poor	Excellent		
Dacron	Woven	275	Good or excellent	Excellent	Good	Subject to hydrolysis attack during vapor phrase transition temperatures of water in presence of SO_3	
Dacron	Felt	275	Good or excellent				
Orlon	Woven	275	Good	Excellent	Fair	Seriously attacked by high concentration of metallic chlorides	Available in copolymer staple form only
Orlon	Felt	250	Good	Excellent	Fair		Available in copolymer staple form only
Dynel	Woven	160	Fair	Excellent	Excellent		
Teflon	Woven	425	Fair	Excellent	Excellent		
Teflon	Felt	425	Fair	Excellent	Excellent		
Polypropylene	Woven	200	Excellent	Excellent	Excellent		Exhibits very high elongation characteristics as maximum recommended temperatures are approached
Polypropylene	Felt	200	Excellent	Excellent	Excellent		
Metallic	Woven						Limitations and chemical properties dependent on metals used
Metallic	Felt						
Nomex	Woven	425	Excellent	Good	Good		
Nomex	Felt	425	Excellent	Good	Good		

* For surge temperatures of short duration, add 5 to 10 percent.

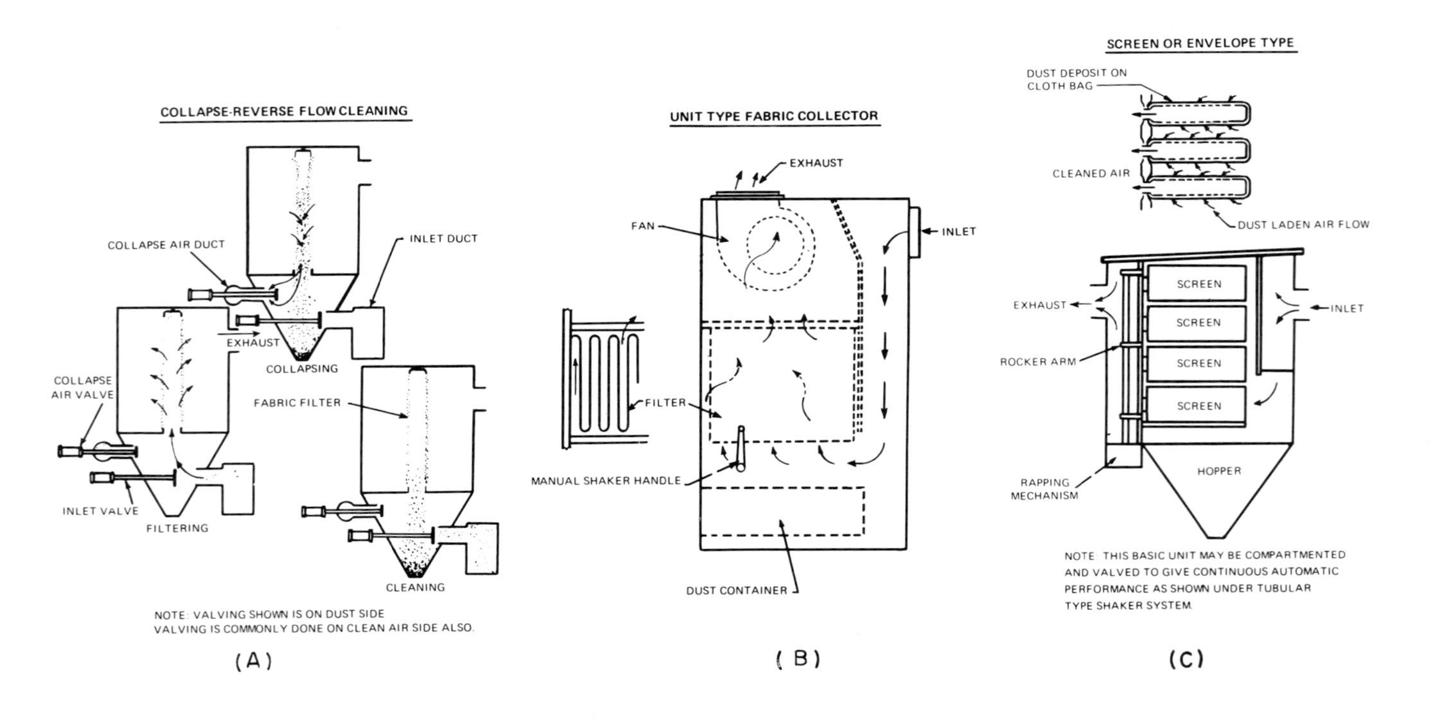
COLLAPSE-REVERSE FLOW CLEANING
COLLAPSE AIR DUCT
INLET DUCT
EXHAUST
COLLAPSING
COLLAPSE
AIR VALVE
FABRIC FILTER
INLET VALVE
FILTERING
CLEANING
NOTE: VALVING SHOWN IS ON DUST SIDE
VALVING IS COMMONLY DONE ON CLEAN AIR SIDE ALSO.
(A)
UNIT TYPE FABRIC COLLECTOR
EXHAUST
FAN
INLET
FILTER
MANUAL SHAKER HANDLE
DUST CONTAINER
(B)
SCREEN OR ENVELOPE TYPE
DUST DEPOSIT ON
CLOTH BAG
CLEANED AIR
DUST LADEN AIR FLOW
SCREEN
SCREEN
SCREEN
SCREEN
EXHAUST
INLET
ROCKER ARM
HOPPER
RAPPING
MECHANISM
NOTE: THIS BASIC UNIT MAY BE COMPARTMENTED
AND VALVED TO GIVE CONTINUOUS AUTOMATIC
PERFORMANCE AS SHOWN UNDER TUBULAR
TYPE SHAKER SYSTEM.
(C)

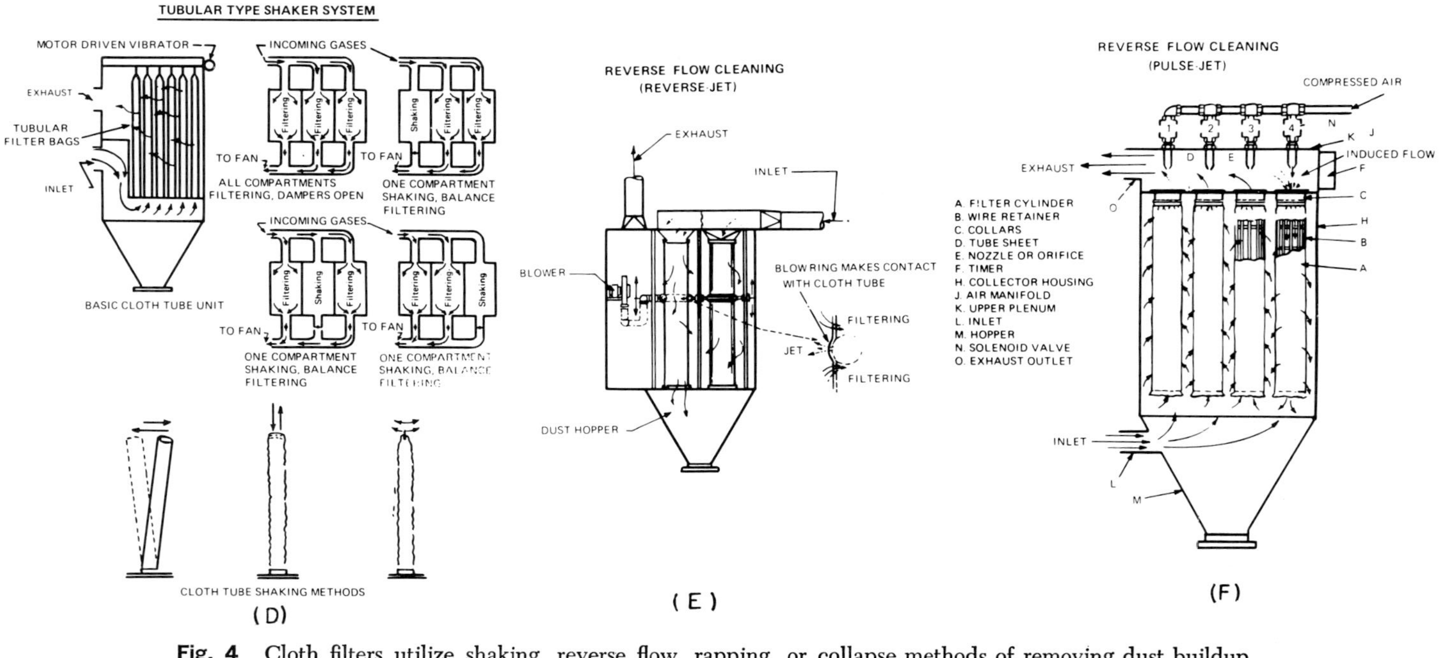

Fig. 4 Cloth filters utilize shaking, reverse flow, rapping, or collapse methods of removing dust buildup.

tubular cloth filter bags which are supported top and bottom with gas entering from one end. The dust bearing gas is moved by either suction (negative pressure) or propulsion (positive pressure) through the filter, so that dust particles are trapped on the approach side of the fabric, while clean gas passes through. Typical examples are shown in Fig. 4.

The filters themselves are supported in a housing or frame structure. (Where gas is moved through the process by suction, the housing must be gastight.) The housing includes accommodations for a dust laden gas inlet, a clean gas outlet, and a discharge for collected particulates.

In systems which collect dust on the outside surface of the filter element, a support frame (frequently a wire cage) safeguards it against collapse. The open ends of the elements are mounted in clusters on a tube sheet or cell plate. These act as dirty gas entries for unsupported filters, or as clean gas exits for supported filters.

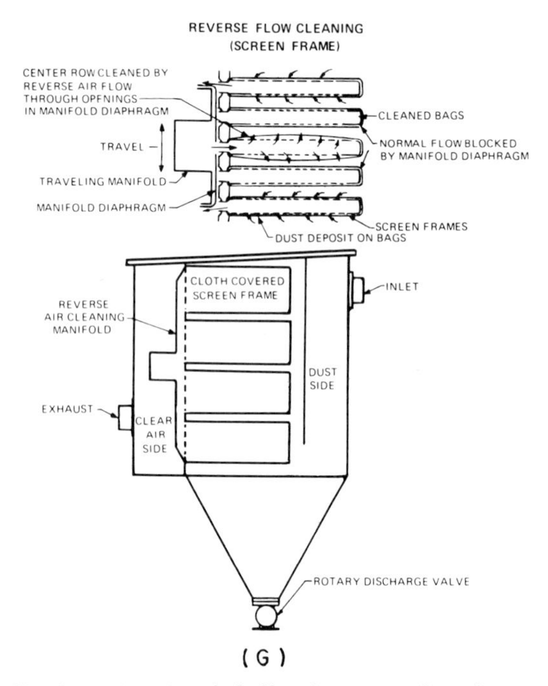

Fig. 4 continued Cloth filter for reverse flow cleaning.

Fabric filter systems are available in preassembled units, or in modules for field erection on a larger scale. In either form, fabric filters are extremely flexible and can accommodate a wide variation in capacity rate.

Accessories. Among the accessories developed for glass filters are aluminum caps with eyebolts and rods for hanging, preferably with large 2-in. washers to distribute the load at rod or eyebolt. *Steel thimbles* in the cell plate clamp at the top. Stainless steel clamps attach the bag to both cap and thimble.

Compression or tension-type springs tension the bags for collapse cleaning (bags should have 25, 50, or 75 lb tension when in operation). These must be compressed (or extended until this loading, added to that obtained from the structural growth, equals the desired total poundage).

Construction Materials. Materials vary according to the type of filter, climatic conditions, and the degree of standardization in the plant. Outer shielding may range from steel plate to corrugated asbestos. Insulation depends on the climatic condition. For example, bags stiffen and break easily if gases are permitted to cool to the point of condensation. Conversely, glass collectors must be designed with consideration for the higher temperatures to which they will be subjected. Dust from the hoppers is handled by both screw and pneumatic conveyors.

Kiln draft is virtually unaffected by a properly designed baghouse, although an excessive build of pressure drop across the bags will decrease the amount of available repressuring air proportionately.

Clean gas can be emitted from the collector via a stack or through louvered exhaust ports which can open and close automatically via a temperature controller.

Dampers used in glass collectors are controversial because of a multiplicity of requirements: clean seating, stability under temperature, ability to modulate rate of gas flow, and low maintenance needs. Standard butterfly dampers are commonly used, although problems can arise from the high temperatures and difficulty in modulating rates of flow. The new "popper" type—a flat plate with a 20° collar which seats inside the duct end—is preferred for its self-aligning and self-cleaning characteristics. Louvered and slide dampers are also in use.

The normal ease of operating a glass baghouse can be disrupted by improper design, equipment malfunction, poor maintenance, or carelessness. Excessive pressure drop, for example, is usually a symptom of insufficient bag cleaning.

Filter Materials. A variety of filter fabrics are available to help contemporary designers meet the problems inherent in each pollution problem. Table 1 indicates their respective superiorities.

Fiber-glass filter bags Among the most frequently chosen are silicon processed fiber-glass filter bags, which were first developed and marketed in 1950. In addition to the regular advantages of fiber-glass fabrics (temperature and chemical resistance, high tensile strength, etc.), the filter bags employ silicons bonded to the individual fibers at very high temperatures. This process provides a lubrication which prevents the fibers from self-abrasion during flexing, and acts as a release agent, assisting dust removal.

First utilized in the critical Los Angeles area, fiber-glass filter bags are indicated for both air pollution and products collection where high temperatures and high efficiencies are required, and have proved especially applicable to the needs of the cement industry.

Advantages. The glass bag does not require shaking, making possible the use of a "top inlet" design, where incoming gas flow tends to flush dust into the hopper. This, in turn, permits use of longer bags (single bags 44 ft long and 50-ft bags with an inner cell plate), and results in reduced ground space requirements, better bag cleaning, and more efficient use of the entire bag length than were possible with the "bottom inlet" bags used prior to glass.

Design Variations. Two systems are in common use. "Suction" system collectors, which employ the fan on the clean gas side, are completely enclosed and compartmented, thereby improving temperature controls and eliminating mechanical collectors. "Pressure" systems have the fan on the dirty air side, permitting relatively open housing and no compartments within the bag area.

Glass bag filters are available in varying designs. The most common design in the cement industry utilizes continuous filament woven in a "crowfoot satin" construction for permeability from 12 (used on finish grinds) to 17 (on kilns). New fabrics using "bulked" or "textured" yarns offer higher permeabilities (60 to 80) and excellent collection efficiencies. Bags are usually 11½ in. in diameter, with greatest variation in length (top inlet bags go to almost any length).

Method of bag cleaning When dust buildup on the filter surface becomes excessive, the unit may be cleaned by one of four methods: reverse flow (backwash); shaking, rapping, or vibrating the filter element; complete or partial collapse of the elements; and combination methods. The collapse method is shown in Fig. 5.

Bags can be cleaned by shaking if caution is taken to reduce flex wear by adjusting the shakers to minimal frequency and amplitude. Generally, the systems of reverse air and collapse are preferred because they provide minimum bag wear and completely eliminate the cost and maintenance of shakers. The collapse method, however, requires careful balance of tension, damper action, and cleaning cycle in order to avoid high pressure drops or short bag life. The ultimate cleaning cycle must be planned with adequate consideration of the characteristics of the dust being handled, the design of the equipment, and the operational peculiarities of the plant.

Sonic cleaning is another method of cleaning bags. Another, the use of "flutter valves" in the ducts, utilizes a rotating "stove" damper which sets up a pulsation in either inlet or repressuring gases. Various techniques of gas pulsation are now being offered. Showing even greater promise is a new technique employing a series of annular rings which are formed of a curved stainless steel section and connected to each other with a stainless steel bar. Since the rings hang inside, independent of the bag, the bag can be held open during the reverse cycle for thorough cleaning via a free reverse flow of air through the fabric. The fact that the bag is not collapsed reduces the need for carful tensioning, speeds damper action, and eliminates null periods.

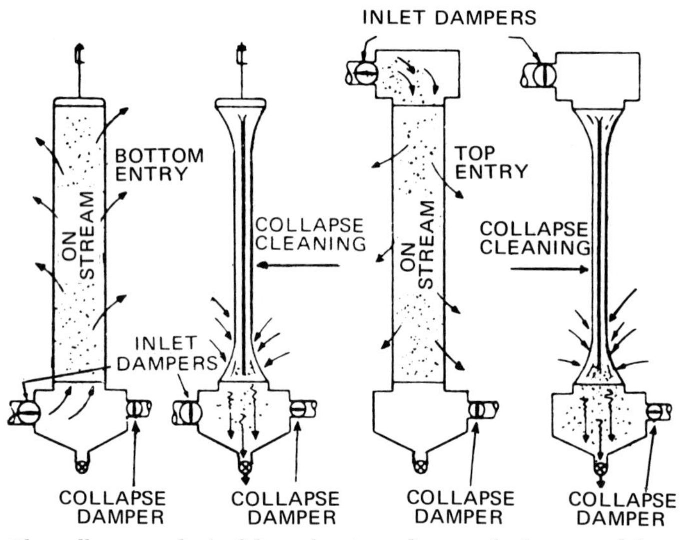

Fig. 5 The collapse method of bag cleaning adapts to both top and bottom inlet variations.

Limitations of cloth filters Basic to fabric filters is their necessity for periodic maintenance and repair. When this requires work stoppage, it can be quite costly. The filter bags themselves necessitate constant protection against the stiffening breakage caused by condensation. Conversely, excessive temperature may require a preliminary cooling spray. Short bag life can also result from flex wear during shaking or inadequate attention in the collapse method to balance off tension, damper action, and the cleaning cycle.

MECHANICAL COLLECTORS

Cyclones

One of the earliest types of dust collectors used in industry is the mechanical cyclone collector. More cyclone collectors are used in commercial applications today than any type of collector.

Design variations Basically there are two types of mechanical collectors in commercial applications—tangential inlet cyclones, and axial inlet collectors—as shown in Fig. 6*a* and *b*.

Axial inlet cyclones are frequently referred to as multicyclones. Generally of small diameter (from 3 to 24 in.), they are grouped together in a common housing with a common inlet and a series of gas outlet tubes that discharge at a common plenum. Because of their small diameter, they are usually of higher efficiency than the large-diameter tangential cyclones. However, the smaller size of the axial cyclones usually results in axial inlets which are proportionately small and prone to plugging. This is especially true where there is moisture and/or a high inlet dust loading. Designed for large-scale operations. multitube type cyclones can handle multitudes of gas volume at about 50 percent of the cost of a large-diameter tangential cyclone.

In principle, however, both types of collectors are alike. Whether the dust laden gas enters tangentially through a duct or is directed by "turning" vanes, it is caused to spin in a downward spiral.

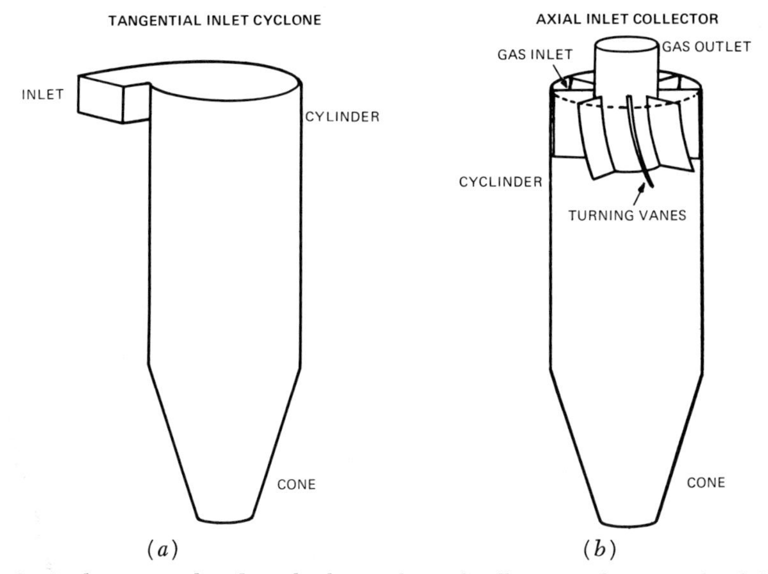

Fig. 6 Both tangential and axial inlet mechanical collectors utilize centrifugal force for particle collection.

General principle In both types of collectors the dust laden gas is caused to spin in a downward spiral. The spinning action causes the suspended particles to migrate toward the inner wall of the cyclone. The vertical force component of the downward spiral causes the particulate to be carried down through the cone section and into the hopper. As the gases move downward, the cyclonic phenomenon takes place. An ascending vortex is developed by gas migrating along the entire length of the cyclone. This vortex has its terminal vertex at the apex of the cone. Because of centrifugal and inertial force the migrating gas is generally particulate free. The ascending vortex rises through the center of the cyclone and is finally exhausted through the gas outlet tip and/or vortex finder, as shown in Fig. 7.

Forces affecting particles Disregarding gravitational forces, each dust particle is subjected to three different forces:

1. *Inertial*—this is the tendency to move in a straight line and is a function of particle mass and velocity.

2. *Centrifugal force*—this force is created as a result of the shape of the cyclone. It is a function of the particle's velocity, radius of curvature within the cyclone walls, and mass.

3. *Viscous drag*—this force is created as a result of the resistance to flow within the gas medium. Shown in Fig. 8, viscous drag is a function of particle size and shape and the viscosity of the gas.

The essential difference between an ordinary and a high-efficiency cyclone is the ability to create, within the cyclone and by its design, sufficient centrifugal force to affect particle concentration without developing any currents to reduce its efficiency.

Cyclone operation Basically there are four steps required for good cyclone separation:

1. *Concentration of the dust*—in a well-designed cyclone, most of the collectable dust will be concentrated on the cyclone's inner walls by the time the particles have traveled less than one revolution.

2. *Dust conveying*—as the gases enter the cyclone, they are caused to spin in a downward spiral. By the time one revolution has transpired, concentration has been effected. The concentrated dust must now be conveyed through the outlet of the cone and into the hopper. As the gas is forced downward, a gas migration (cyclonic phenomena) takes place toward the center of the cyclone. This migration of gas forms what is known as the ascending vortex of clean gas. Since the ascending gas flow has been reduced by the migration toward the center of the cyclone, an increase in spin velocity must take place in order to maintain particle concentration along the entire length of the cyclone. This is accomplished by a taper in the cone section of the collectors.

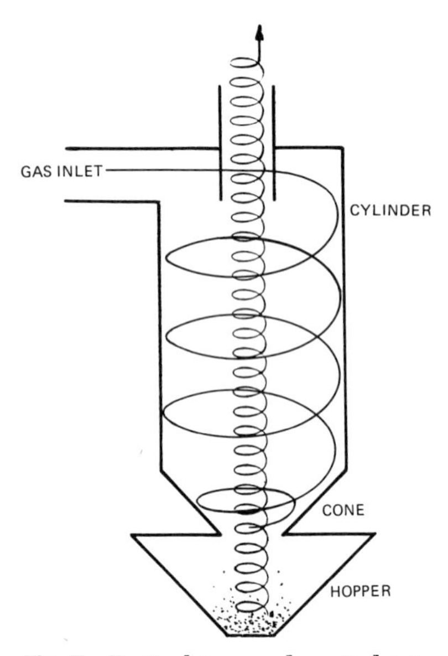

Fig. 7 Particulates are deposited into the hopper as gas is redirected from a downward spiral into an ascending vortex.

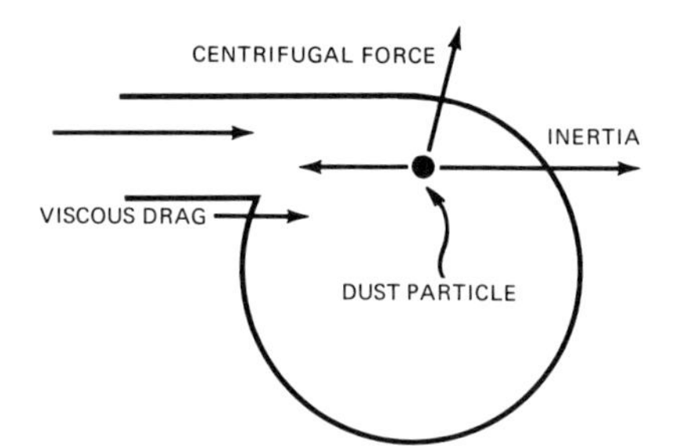

Fig. 8 Efficient cyclone design effects a balance between the inertia, centrifugal force, and viscous drag forces which interact on the particles.

3. *Discharge of particulate*–at the cone outlet, the dust which is being conveyed has high rotational and downward force components. This causes the dust to discharge downward and away from the ascending vortex at the scrubber section.

4. *Ascending vortex*—the ascending vortex is comprised of particulate free gas which migrates radially and inwardly the entire length of the ascending vortex.

Being a simple device, the cyclone collector is composed of several stationary parts, or components—each of which contributes to the performance of the collector. The relationship of each component to the other determines the collector's ultimate separation efficiency. Since determining the correct component relationships is an exacting science based on empirical observation and laboratory work, all cyclones actually constitute a compromise of sorts. For example, a high-efficiency cyclone approaching the efficiencies of a high-energy scrubber can be designed. However, its component parts would be of such tremendous proportion, and the pressure drop so high, that its use would be economically undesirable. The alternative would thus be a compromise between size, pressure drop, and efficiency, for the design of a collector of acceptable cost.

In a given cyclone design, changing the proportions of any of its component parts will affect the performance characteristics of that collector as follows:

Relationship of Components

Parts to be changed	Effect on efficiency	Effect on pressure drop	Cost per cfm
Increase diameter....................	Reduced	Reduced	Up
Increase the length of the cone........	Increased	Increased	Increased
Increase the inlet arc..................	Reduced	Reduced	Reduced
Reduce cone length...................	Plus or minus	Plus or minus	Little change

Figure 9 shows the effects of *temperature inlet velocity on* ΔP.

A dust that would not eventually settle in air (still air) cannot be collected in a cyclone collector. The longer the time required for a given type of dust to settle in still air, the harder it is to collect in a cyclone.

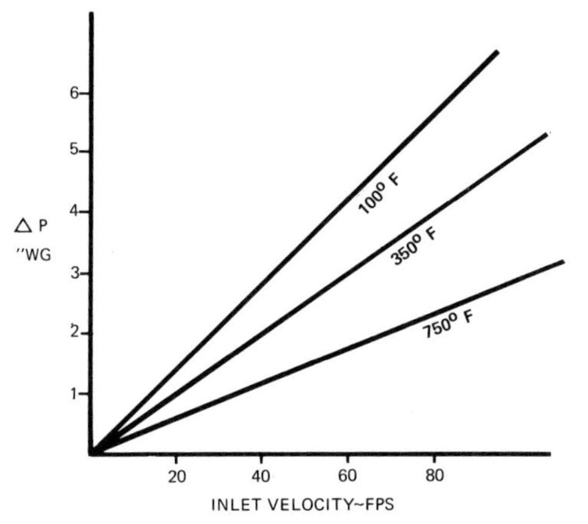

Fig. 9 Cyclone component relationships depend on the effect of temperature inlet velocity on pressure drop.

Operational factors Since there are no moving parts in a cyclone collector, maintenance is minimal. Mechanical collectors, however, are susceptible to factors such as erosion and corrosion. However, this can be minimized by giving consideration to the gas velocity at the inlet of the cyclone (not to exceed 3,600 fpm) and to the materials of construction.

Temperature limitations of cyclone collectors involve only their materials of construction. In many cases cyclone collectors can be lined with insulated bricks for temperature protection.

High dust loading will generally increase the efficiency of cyclone collectors. Generally, the efficiency of mechanical collectors is expressed in graphic form, plotting the percent collection vs. particle size. A typical cyclone efficiency curve is shown in Fig. 10.

Typical applications and percentage collection efficiencies In essence, the efficiency of a mechanical collector is dependent on the particle size of the dust to be collected. These efficiencies can vary from 65 to 99.9 percent, depending on the dust and the cyclone design.

Application	Particle size	% collection
Cement kiln	35 %–10 microns	75–85
Lightweight aggregate	25 %–10 microns	80–95
Fertilizer plant	40 %–10 microns	80–95
Calcium chemical process	10–50 %–10 microns	80–97
Petroleum (catalytic cracking)	10–20 %–10 microns	99 plus
General industrial applications	10–60 %–10 microns	65–98

Cost variations of cyclone collectors are primarily due to variations in the materials of construction or in the required efficiency.

Some typical unerected equipment costs are:

1. Process kiln (mild steel cyclones) is $0.20 to $0.35 per cfm
2. Process kiln (alloy cyclones) is $0.75 to $1.50 per cfm
3. Industrial cyclones (mild steel) are $0.35 to $0.60 per cfm

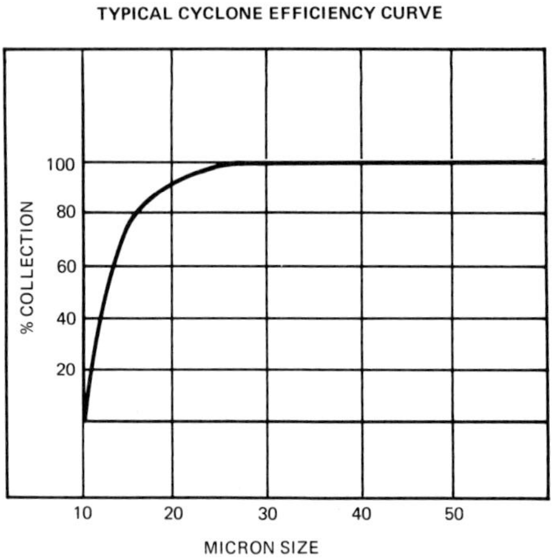

TYPICAL CYCLONE EFFICIENCY CURVE

Fig. 10 Particle size is a prime factor in the collection efficiency of mechanical collectors.

Typical installed costs for cyclone installations can vary from 75 to over 350 percent of equipment price. Operating costs are difficult to estimate, but they can vary anywhere from $0.1 to $0.7 per cfm of gas handled on a per year basis.

Other Mechanical Collectors

The gravitational settling chamber Although infrequently used today, this is the most elemental of the mechanical collectors. Used primarily for collecting large particles (43 microns or more in diameter), a settling chamber may be merely a simple enlargement of a flue or duct. As the gas stream enters the enlarged chamber, velocity is reduced, and the heavy dust settles to the bottom for discharge into a hopper. Extremely low pressure drops are required, since collection efficiency requires low velocity and minimized gas turbulence. Relatively inexpensive because of their simplicity, settling chambers sometimes contain suspended curtains, rods, or screens, which simultaneously reduce turbulence and increase collection by impingement.

Impingement separators For the collection of micron-sized particles, impingement separators employ impaction by permitting the gas to impinge on a plate, vane, or scroll. In essence, impingement separators utilize the inertia of particles in the dust laden gas stream, which is directed through a virtual obstacle course. As the gas is deflected around each obstruction, the inertia of the particles causes them to maintain their original course, thus impinging on the obstacle's surface.

The baffle chamber is the simplest form of impingement separator. It effectively collects particles 20 microns or larger in size. Baffle chambers are designed so that the gas flows around zigzag plates or a staggered pattern of variously shaped obstacles. Material collected on these obstacles is flushed off by water. Baffle chambers are frequently used to remove suspended particles carried over from boiling liquids.

ELECTROSTATIC PRECIPITATORS

The most widely used high-efficiency dust collector today, the electrostatic precipitator is based on the simple principle that a charged body attracts one oppositely charged.

General principle Electrostatic attraction and repulsion are accomplished in the electrostatic precipitator via a set of electrodes. Between these is a voltage potential of magnitude sufficient to create a negative-corona effect on the negatively charged electrode so that ionization occurs. Both positive and negative ions are generated. Of these, the positive ions remain on the negatively charged electrode, while the negative ions pass over the grounded (collector) electrode along the force lines of the electrostatic field between the electrodes.

As dust laden gas passes between the electrodes, the dust in the suspended particles intercepts the negative ions and is electrostatically charged, thereupon being attracted to (collected by) the grounded electrode. The dust remains there until removed, and the gas which has thus been cleaned moves on to recovery or exhaust. The grounded electrode which collects the precipitate has the secondary function of discharging it without permitting it to be reentrained in the gas stream. Movement of gas or air is accomplished by a fan or blower stationed at either end of the cleaning system, depending upon application.

Fields may be arranged in series and/or parallel. The width of the field is determined by aerodynamic considerations and the volume of gas being cleaned. The length of the field is governed by these factors, as well as the specific particle characteristics, temperature and humidity considerations, and the desired degree of cleaning efficiency. In handling corrosive atmospheres, field elements may be made of corrosion resistant material.

Electrostatic precipitators can achieve cleaning efficiencies of 99.9+ percent, and may be combined with another type of collector under special operating conditions. Size and cost increase in proportion to efficiency desired.

Highly selective particle removal can be accomplished by manipulating the temperature, velocity, and/or collector potential of a precipitator, thus affecting the size or mass of the particle precipitated. Moreover, potential can be manipulated differently in various fields throughout the precipitator, according to the electrical characteristics determined by the chemical composition of a particular substance.

Variations in design Electrostatic precipitators are available in either of two basic design variations.

The dry or plate type of precipitator utilizes a number of grounded electrode plates suspended parallel to each other within a rectangular shell or casing. The spacing between the plates acts as a gas duct and contains vertically suspended discharge electrodes. Deposits of accumulated material are cleaned from the collecting electrode plates by a continuous automatic rapping device. Actuated by electric, pneumatic, or hydraulic power, the rappers dislodge the precipitated particles into hoppers below.

The pipe or tube type of precipitator (shown in Fig. 11) is usually a circular shell or casing divided into two separate gas compartments by a header plate or tube sheet. This type of precipitator is generally more expensive. The gas is moved through pipes in the header plate from the lower compartment to the upper compartment, with the actual work of cleaning accomplished by discharge electrodes suspended within the pipes. Collected material is removed by water flushed through the collecting pipes, and is carried off as a dilute slurry.

Basic advantages When feasible for use, the electrostatic precipitator is desirable for its economy and operational simplicity. Operating cost is measurably below that

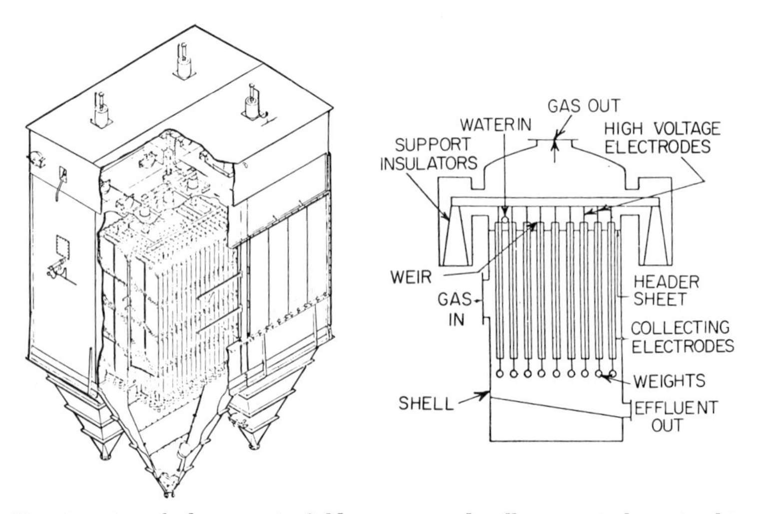

Fig. 11 A series of electrostatic fields attract and collect particulates in this pipe type electrostatic precipitator.

of other systems because of the low fan power consumption derived from the low-pressure drop (typically, ½ to 1 in. W.G.) across the cleaner.

Low Current Drain. Current drain is virtually negligible, usually between 0.05 and 1.5 amp. Occasionally, the pressure drop available from the flue itself is sufficient to dispense with added air moving equipment. This combination of reduced pressure drop requirement and low current drain yields a system extremely economical both in size of blower/fan equipment and in power consumption. Maintenance is also low, since there are essentially no moving parts.

High Efficiency. One of the precipitator's foremost advantages is its highly selective ability to remove a broader spectrum of particles, ranging from submicron to greater than micron size. Also, it can be easily designed to handle moisture and oil.

Because of this higher cleaning efficiency, the precipitator has the added advantage of eliminating consequent water cleaning and effluent problems.

Operating is extremely flexible, since the pressure drop through the unit shows little change for increases in volume handled. Thus, a significant overload can occur for a short period without materially altering the appearance of the gases leaving the stack. There is rarely a steam plume—a factor of high psychological significance in community relations.

The high recovery value of the dust collected offsets the cost of the equipment. In steelmaking, for example, dust can be collected and added to the sinter mix or pelletized in a "flying saucer" and returned directly to the steelmaking vessel without treatment.

Basic limitations Two primary limitations of the electrostatic precipitator from the manufacturer's point of view are its high initial cost and its extremely large space requirements.

Operational Limitations. Technically, a predominant disadvantage is that precipitate particles can escape from the discharge collector electrodes during the rapping cycle. Since these are generally coarser particles, solution of this problem can necessitate installing a secondary mechanical cleaner. More recently, adding electrically charged baffles to the precipitator unit has proved to increase the collection efficiency significantly. The charged baffle system is a minor cost added to the overall precipitator.

The danger of incomplete combustion prohibits the use of a precipitator with an explosive gas stream. To combat this, excess air in the form of a higher gas volume is required.

Variations in flow can spell trouble for the electrostatic precipitator. Also, sensitivity to the amount of SO_3 in the gas stream presents a problem in that efficient operation may require maintaining a minimum amount of sulfur in the fuel. This may result in a conflict with pollution control regulations.

WET SCRUBBERS

Perhaps most frequently chosen for industrial applications requiring the cleaning, cooling, and deodorizing of air and gas emissions, wet scrubbers generally use water as the scrubbing agent. Variations include caustic additives for SO_2 and H_2S collection, use of phosphoric acid for the recovery of ammonia, etc. These variations can pose related problems in construction materials and auxiliary treatment and/or disposal operations.

Variations in design Since wet scrubbers are available in an almost limitless variety of geometric shapes, the layman's lack of basic technical understanding places a burden and risk on the user. Several variations are shown in Fig. 12.

Each of the three basic types of wet scrubber design–spray towers, packed towers, and venturi type scrubbers—offers a unique capability.

General principle Except for the packed towers, all operate on the basic aerodynamic principle, which may be illustrated by the simple analogy of a basketball and a BB. Water droplets (the basketballs) are projected so that they collide countercurrent (crossflow) with the particles (BBs) carried by the gas stream. If the droplets and particles are of comparable size, collision will take place, with the result that the particles will adhere to the droplet and be easily collected. Studies have shown that a surface film surrounding a water droplet has an approximate thickness 1/200 of its diameter. Accordingly, the BB (the particle in flight), having a diameter less than 1/200 diameter of the "basketball," will flow through the streamline film around the basketball without collision, as shown in Fig. 13. Therefore, a fume particle 0.5 micron in diameter requires droplets less than 100 microns (200×0.5) for adequate collection. Efficient scrubbing therefore requires atomizing the liquid to a fineness which will afford maximum contact with the particles to be caught.

Dust concentration in the gas is another consideration. The probability of a droplet's hitting the particles decreases in relation to the dust concentration, just as a ball would be less likely to hit a single BB than a dense mass. To equalize these factors, scrubbers are regulated as to the volume of gas to be scrubbed (pressure drop) and water to be sprayed (pressure).

The scrubbing chamber's height and diameter must also be tailored to the known characteristics of the gas. The effective path of the water through the gas must equal in length the distance traveled by the water, and the pressure drop.

Pressure drop as a function of collection efficiency It is well established in wet scrubbers that pressure drop is directly related to collection efficiency. How-

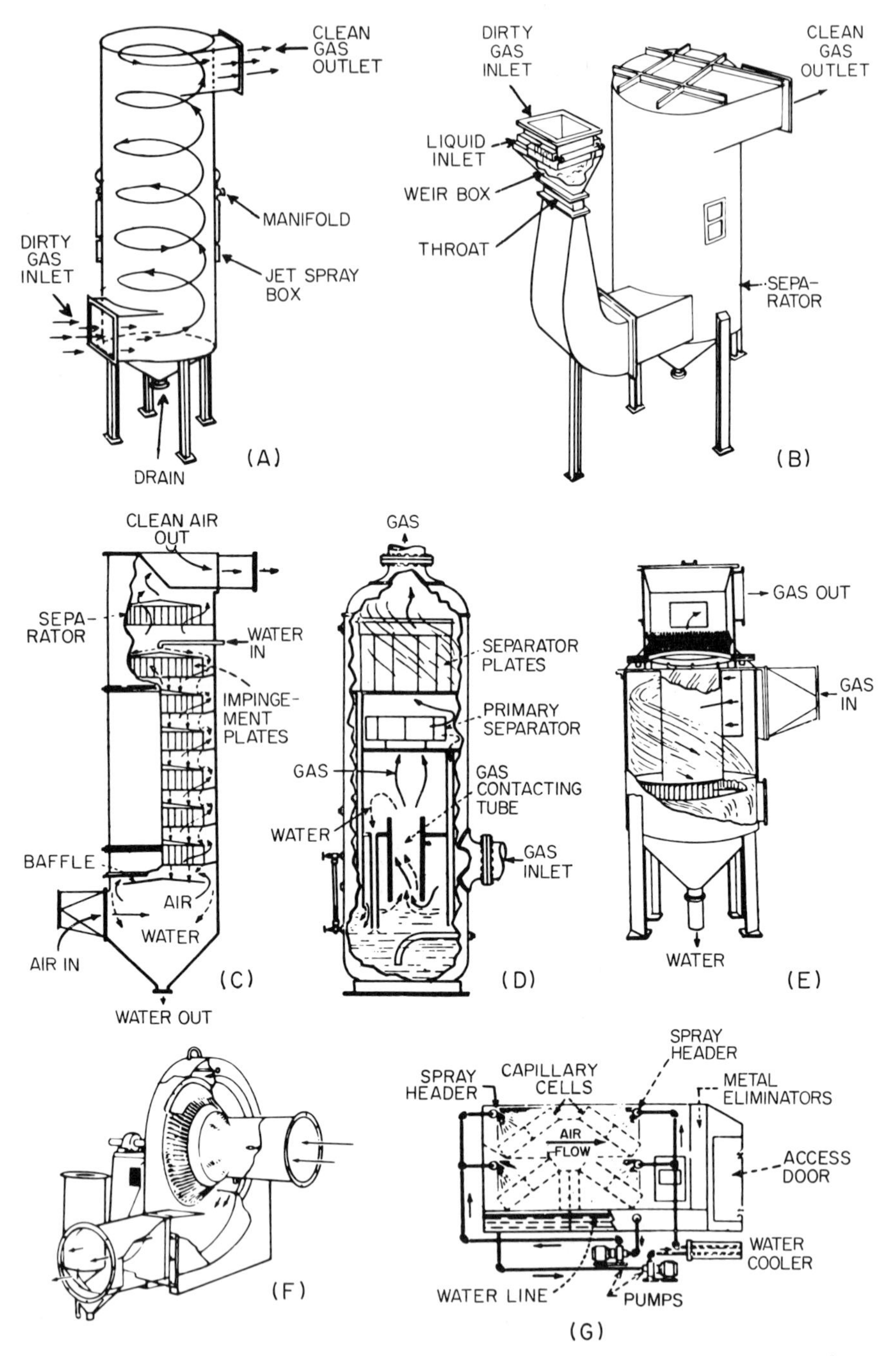

Fig. 12 Scrubbers and washers. (*a*) Airetron/Mikro cyclonic scrubber. (*Pulverizing Machinery.*) (*b*) Airetron/Mikro venturi scrubber. (*Pulverizing Machinery.*) (*c*) Multiwash scrubber. (*Claude B. Schneible Co.*) (*d*) Liquid vortex contractor. (*Blaw Knox Co.*) (*e*) Hydrocyclone type scrubber. (*Whiting Corp.*) (*f*) Centrimerge rotor scrubber. (*Schmieg Industries.*) (*g*) Wet filter. (*Air and Refrigeration Corp.*)

ever, it must be determined whether all or most of the pressure drop is consumed for scrubbing.

In the first case, energy is consumed by any scrubbing device to which the dirty gas is subjected, in the region where water is contacted. These include water sprays, curtains, flooding beds, etc., which cause greater resistance to gas flow.

Compensating for Pressure Loss. Energy can also be lost in miscellaneous resistance encountered in other areas. Once compensation has been made for these resistances, the work of scrubbing is equivalent in any device using the same fumes and gas density.

Simple provisions in scrubber design, such as the "well-rounded corner," lead to minor losses in horsepower (pressure drop), and ultimately allow higher horsepower for scrubbing.

Among the causes of pressure loss in gases moving through a collector are sudden enlargement, sudden contraction, the entrance resistance of projecting pipes, or sharp corner turns, as shown in Fig. 14*a* to *c*.

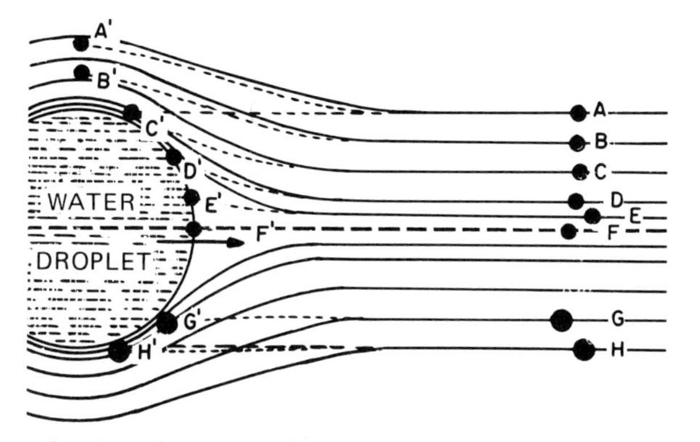

Fig. 13 Particles less than 1/200th the diameter of a water droplet will flow around it with colliding.

Entrance loss Pressure is lost when a gas or fluid flows from a large vessel into a small pipe. This loss of pressure or head, as seen in the case of valves and fittings, can be expressed by the formula

$$h_L = K \frac{v^2}{2g}$$

in which h_L = loss of head, ft of liquid
K = resistance coefficient
v = mean velocity in the pipe, fps
g = acceleration of gravity, 32.2 ft/sec^2

Gases entering well-rounded ducts are subjected to a lower pressure loss than that caused by entry into a sharp corner duct. These resistances to flow in piping systems are diagrammatically shown in Fig. 14*a*.

Sudden enlargement The consuming of pressure drop by sudden enlargement (shown in Fig. 14*b*) is seen in the formula

$$h_L = \left(1 - \frac{a_1}{a_2}\right)^2 \frac{v_1^2}{2g}$$

in which a_1 and a_2 = cross-sectional areas of smaller and larger pipes, respectively
v_1 = velocity in the smaller pipe, fps

Sudden contraction Less pressure is lost in a sudden contraction (see Fig. 14c) than in a sudden enlargement, as seen in the formula

$$h_L = \frac{K_c v_2^2}{2g}$$

in which K_c is a function of the ratio of the cross-sectional area of the smaller to the larger pipe. The following table gives the value of K_c for various area ratios:

Values of K_c for Sudden Contraction*

a_2/a_1	0.1	0.2	0.3	0.4	0.5	0.6	0.7	0.8	0.9
K_c	0.362	0.338	0.308	0.267	0.221	0.164	0.105	0.053	0.015

* O'Brien and Hickox.

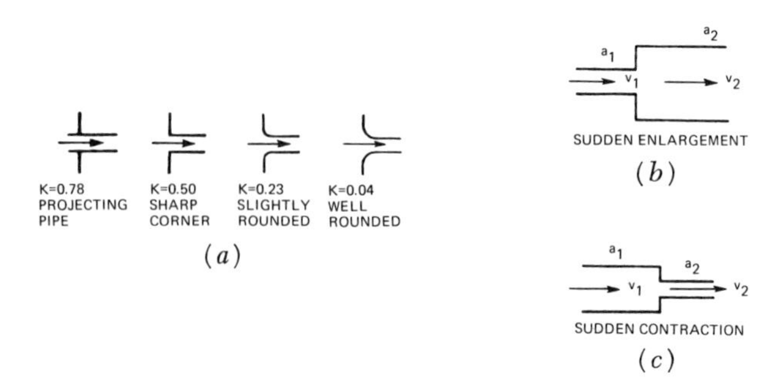

Fig. 14 Various resistances to flow within the collector cause a loss of pressure.

Venturi Scrubbers

Venturi or orifice type scrubbers are employed in installations requiring high-energy collection of submicron particles.

The venturi's basic construction and principles of operation are noncomplex in nature. Since there are no internals, workings are accessible from the unit's exterior.

The venturi is the most accurate of fluid meters, since it contains no moving parts to impede the airflow. The unique shape of the venturi offers 98 percent velocity head (power consumption) recovery, thereby allowing efficient introduction of fluid to meet the gas crossflow in the throat region. Atomization and impaction occur as the injected liquid is shattered by the unscrubbed gas into minute droplets which collide with and carry away minute particles.

Construction materials Because of the equipment's simplicity, designers can choose from a full range of construction materials to handle problems of corrosion, abrasion, or both, depending on the nature of the emission. Among materials frequently specified are stainless steel or Hastelloy. Lead linings may be employed for protection against concentrations of sulfuric acid. Brick (ceramic) linings are used for protection against high temperatures and/or excessively abrasive particles. Rubber linings provide protection against fluorine and phosphoric acid fume emissions. To minimize deterioration and eliminate maintenance, current designs frequently use plastics such as fiber glass and PVC (for nitric acid).

General principle Atomization of the scrubbing liquid takes place in the throat of the venturi. Here, the liquid is introduced at relatively low pressure and is shattered into minute droplets by the onrushing gas flow. For coarse particles, such as those found in the lime kiln gases and flue gases from the powdered coal furnaces, efficient collection may be attained with lower velocities and water rates than those needed for the collection of submicron particles.

As a general rule, a higher liquid rate is usually more advantageous to use than a higher gas velocity. Figuring 3 gal/1,000 cfm, to take an extreme example, the combination of high velocity and low liquid rate would probably create an unwetted void through the middle of the venturi. If we were to look down the venturi, we would see an area in the middle where wetting action does not take place. Hence, it is always preferable to lean toward a higher liquid rate to ascertain a proper liquid-to-gas impaction level. At the other extreme, a problem can occur where a low gas velocity exists with too high a liquid rate (12 gal/1,000 cfm, for example). Under these conditions, liquid shattering may be reduced to such a point that the scrubber starts to operate like an "ejector," which means that very poor collection efficiencies are achieved. In summary, as a general rule, the ideal operating range of the venturi type scrubber varies from 5 to 8 gal of liquid/1,000 cu ft, so that maximum and minimum scrubber contact velocities of 140 fps at 300 fps are attained.

Basic advantages Even though venturi scrubbers occupy less ground space than most bag collectors and precipitators, they can also clean gas volumes within a minimal equipment area. The venturi's lack of internals, however, eliminates the need for work stopping checks for plugging and deterioration.

An effective firestop in applications where extremely fine, dry, or combustible materials are involved, the venturi is capable of handling hot gas, some noxious gases, sticky dust, and moisture, with no secondary dust problems.

Venturi operation costs may be low if the collected sludge can be disposed of without clarification. However, sludge that requires clarification and collection can raise the capital cost of the venturi plant to that of the electrostatic system.

Basic limitations The basic disadvantage of the venturi scrubber is the rise in operating cost—usually 50 to 60 percent above that of electrostatic cleaning—primarily because of the high power cost for fans and water pumping. As mentioned above, further processing of the wet sludge poses a costly auxiliary problem.

Continuing overhead also includes fan cleaning, which must take place at frequent intervals to prevent mud from collecting on the blades, causing the fan to go "out of balance." In systems handling saturated gases, equipment design must make full allowance for corrosive gas handling to avoid costly maintenance.

Steam Plume. In terms of psychological community relations, the venturi (as well as any wet scrubber) is under a disadvantage in that the stack emits a steam plume, especially in cold weather. Although recent innovations in cooling towers have made it possible practically to eliminate the plume, it may give the impression that cleaning is less satisfactory than in actuality.

Other Types of Wet Scrubbers

Spray chambers are the most elemental of wet scrubbers. These are empty towers utilizing liquid introduced via a bank of spray nozzles at the top. Gases passing countercurrent to the falling drops are scrubbed clean of particulate matter of larger than micron size. These towers may also be used as coolers or as primary cleaners. Though these units were acceptable in prior years, recent developments have surpassed their performance.

Cyclonic spray scrubbers combine the spray technique with the mechanical principle of centrifugal force. The cylindrical tower contains a central manifold, from which droplets are sprayed into the airstream, which enters through a tangential duct. Centrifugal force created by the gas rotates the droplets at high velocities, enabling them to collide with and carry particulate matter to the scrubber walls.

Normally designed for specific installations, cyclonic spray scrubbers can accommodate gas entrance velocities up to 200 fps, with low-energy pressure drops ranging from 2 to 6 in. W.G. They have been utilized for cleaning micron-sized particles created from mechanical disintegration.

Packed Towers

The packed tower type of gas scrubber is a prime method of scrubbing a true gas. In use since the 1800s, packed towers are used for the removal of gaseous fumes, noxious gases, and entrained droplets in gas cleaning installations for various chemical processes.

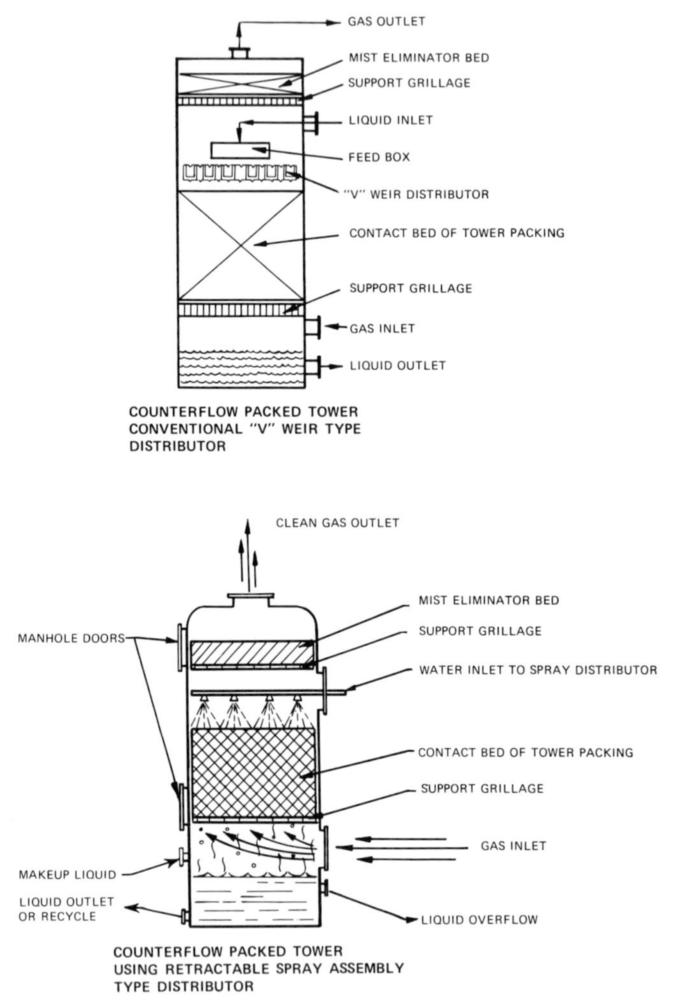

Fig. 15 The weir box and spray assemblies are used for liquid distribution in counterflow packed towers.

General principle Basically, the packed tower is a vertical vessel in which various fill material is wetted. Surface area provided by the various packings offers a basis for inducing interaction between the liquid and gas phases. The air or gas enters the bottom of the tower and receives a preliminary washing as the scrubbing liquid drains in an opposing flow from the packed, irrigated bed. This liquid, which is pumped into the top of the tower, flows down over the packing bed. En route, it covers the surface areas of the packing with a liquid film that accomplishes the major work of collection. Finally, the airstream passes through a mist eliminator section (a dry packing pad) before it is permitted to exhaust.

Variations in design Spray towers with baffles or slotted plates are also used to produce a circuitous path for the contaminated airstream. As in the case of packing, these cause the contaminant to contact the water film. High-pressure nozzles maximize air and water contact.

Figure 15 shows two typical examples of packed towers. Because of the complexities involved in packed towers, air pollution control specialists normally design

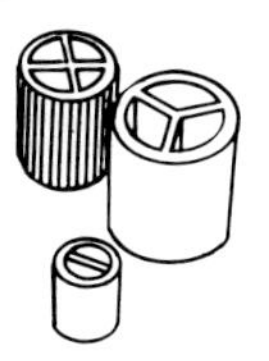

PARTITION TRICKLERS MADE OF ACIDUR CHEMICAL STOEWARE AND CHEMICAL TECHNICAL PORCELAIN

PACKING RINGS MADE OF GLASS

PACKING RINGS MADE OF CARBON AND PLASTIC(PVC)

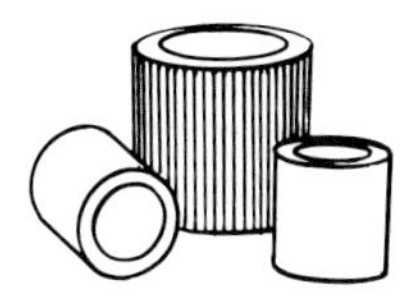

CYLINDRICAL RINGS MADE OF ACIDUR STONEWARE AND CHEMICAL TECHNICAL PORCELAIN

INTALOX-SADDLE-BODIES

INTOS RINGS

PALL–RINGS

MASPAC

TELLERETTES

WIRE SPIRALS AND WIRE MESH RINGS

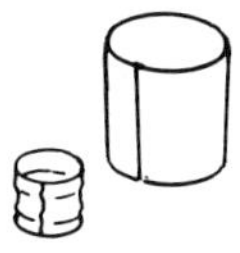

PACKING RINGS MADE OF SHEET STEEL, ALUMINUM, COPPER, STAINLESS STEELS

LURGI SPIRAL SPRAYER

LETSCHERT REFORMER RINGS

SADDLE BODIES ACCORDING TO PROF. BERL

Fig. 16 Material and shape determine the collection efficiency of various packings for different particulates.

equipment for each specific industrial situation, extrapolating data from laboratory and/or pilot plant studies to accommodate the variables inherent in the particular gas flow.

Factors in specifications The following guidelines will help to determine the best uses of the packed tower.

Velocity. To obtain maximum throughput in a vessel so that there will be sufficient contact between the gas to be collected and the liquid media, the gas flow, or velocity, should range between 3 and 6 fps.

Selection of Packing. Packing should be specified to provide maximum surface area along with minimum void area.

Maximum surface is a prerequisite for covering the packing with a liquid film sufficient to contact and react with the gas film. Obviously, if the packing is not wettable during the absorption cycle, this vital interface reaction is reduced. In this case, more packing is required. Similarly, if a tower has a very large packing bed with minimal surface exposure, the liquid film created may be minimal. Thus contact will be limited to the gas and droplets which exist between the packing, thereby reducing interaction of the gas and the liquid film on the packing surface.

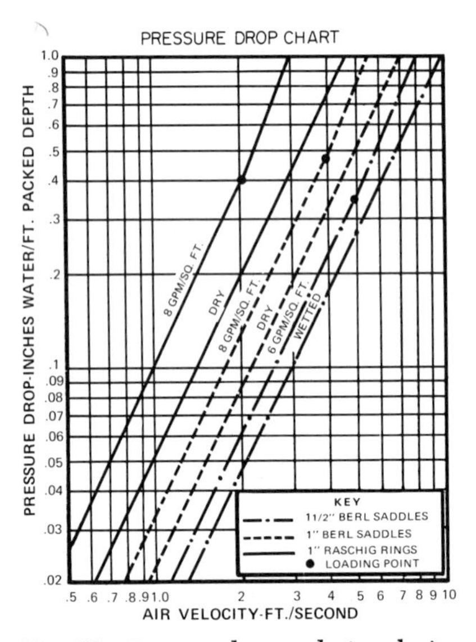

Fig. 17 Pressure drop and air velocity interact differently for various packing materials.

Generally, the smaller the packing size, the more contact offered, thereby requiring more energy (pressure drop) to pass the gas through the tower. At the same time, the smaller packing requires less height in the tower.

However, larger quantities of packing with large void space have proved to be equal in contact efficiency to smaller packing with greater surface area. Experience has shown that a nominal packing size ranging from 1 to 2 in. is optimum for the capacity of gas handled as well as for the factors of grillage availability, water distribution, and pressure drop consumption.

Packing Materials. The tower's packed "bed" may contain any of a variety of materials in an assortment of shapes, each specifically chosen for its compatibility with the constituents in the particular gas stream. Conventional packings include Berl saddles or Raschig rings, which are commonly constructed in chemical stoneware. Porcelain and carbon are also frequently used. Specialized packings marketed today include those made of polyethylene and polypropylene. These packings are commercially known as Maspac, Pall rings, or Tellerettes. Figure 16 illustrates a variety of commonly used packings. Different packing materials will have different pressure drop which in turn result in various horsepower consumption (see Fig. 17).

Packing Height. Calculation depends on such factors as the absorption coefficients, the partial pressure of the gaseous contaminant, and the cross-sectional area of the tower. Usually effective irrigated packing heights range between 4 and 8 ft with a 9 in. high mist eliminator bed located above the spray region.

Pressure Drop. Pressure drop is normally described in terms of inches water gage (W.G.). Increased water gage refers to the power consumed in order to move the gases through the equipment. This power is generally consumed by the air moving device, which may be the blower or the fan. As a rule of thumb, a 64 percent mechanically efficient blower consumes 1 hp/1,000 cfm of gas when developing approximately 4.3 in. W.G. Approximately 408 in. W.G. is equivalent to 14.7 psi, or 27.4 in. W.G. is equivalent to 1 psi. As a general guide a low pressure drop through a packed bed is associated with a high percentage of free or void space.

Table 2 offers variables of the many packings that may be used in the general equation for pressure drops.

TABLE 2 Pressure Drop Correlations for Irrigated Towers

General equation:

$$\Delta p = a \times 10^{bL} \frac{G^2}{P_g}$$

where Δp = pressure drop, in. H_2O/ft of packing height
G = gas mass velocity, lb/(sec)(sq ft)
P_g = gas density, lb/cu ft
L = liquid mass velocity, lb/(sec)(sq ft)
a and b = constants given in the table, valid for H_2O − gas system

Packing type	Packing size	a	b	L (range)	% void
Stoneware construction:					
Raschig rings	½	3.10	0.41	0.08–2.4	58.6
	¾	1.34	0.26	0.5–3.0	72.0
	1	0.97	0.25	0.1–7.5	69.2
	1½	0.39	0.23	0.2–4.0	75.9
(Dumped)	2	0.24	0.17	0.2–5.5	81.6
Berl saddles	½	1.20	0.21	0.08–6.0	63.2
	¾	0.26	0.17	0.1–4.0	71.0
	1	0.39	0.17	0.2–8.0	70.9
	1½	0.21	0.13	0.2–6.0	70.0
Intalox saddles	½	0.82	0.20	0.15–4.0	77.0
	¾	0.28	0.16	0.1–4.0	75.0
	1	0.31	0.16	0.7–4.0	74.0
	1½	0.14	0.14	0.2–8.0	78.0
Polypropylene/polyethylene construction:					
Maspac FN-200	2	0.089	0.133	1.11–2.78	90.0
Pall rings	1½	0.110	0.119	0.417–2.50	90.5
Tellerettes	1	0.099	0.095	0.555–1.665	92.0

Purpose of application Packed towers are greatly influenced by the solubility of the gas to be collected.

Simple gases are defined as having "gas film controlling" whereas difficult gases are defined as having "liquid film controlling." Simple gases which are readily water soluble (gas film controlling) are as follows: NH_3, HCl, SO_3 in strong H_2SO_4, SO_2 by alkali and NH_3 solution, H_2S in caustic solution, also the evaporation and condensation of liquids. Those gases which are not readily soluble (liquid film controlling) are CO_2, O_2, H_2 by water, the absorption of CO_2 in alkali solution, and the absorption of chlorine in water.

Difficult, or readily insoluble, gases require special considerations which are beyond the scope of this discussion. Furthermore, their products are not generally associated with the air pollution nuisance purpose.

The general design of packed towers is dependent upon whether the gas contains any solid particles. Where the solid particles are in concentrations higher than 5 mg/cu ft, a packed tower is not recommended.

Treating water soluble gas In a very simplified manner, we have listed those gases which are readily removed using water on a once-through basis. There are also times when the concentrations in water allow recirculation with maintained optimum removal. Such a concentration is largely dependent upon the gas temperature leaving the tower as well as the concentration of the contaminant in the gas stream. At high temperatures, gases have a tendency to revaporize from the surface. Accordingly, the recirculation cycle becomes limited as the gas vapor pressure above the liquid surface increases. Those gases which are ideal for water contact are as follows: ammonia (NH_3), fluorine (F_2), hydrogen fluoride (HF), hydrochloric acid (HCl). Gaseous tetrafluoride (S_iF_4) is also effectively removed, provided the packing does not receive gas concentrations above 5 mg/scf as F. Upon initial wetting S_iF_4 converts partially to the gelatinous S_iO_2, which will foul and plug the packing. Where the gases need additional treatment to convert them to the soluble form, solubility is influenced by a neutralizing additive. Gases which need such additional neutralizing treatment are sulfur dioxide (SO_2), hydrogen sulfide (H_2S), and nitric–nitrous oxides (NO—NO_2).

The packed tower type scrubber further lends itself to the handling of gases which may have entrained droplets as from plating or various metallurgical operations. These droplets may consist of corrosives such as sulfuric, phosphoric, and nitric acids. Caution must be exercised here (particularly when dealing with sulfuric and phosphoric acid) to ensure that the droplets are true, and not a decomposition product existing as an aerosol mist. These "aerosol mists," as they are termed, occur as a white cloud and cannot be effectively removed by a packed tower scrubber.

Description of a packed tower may be further clarified in that the tower uses a support grid to hold the packing or fill. The grid has an open pattern to permit an upflow of gas through it with minor resistance to gas flow.

Methods of liquid distribution *The "weir box"* is a unique liquid distribution vehicle with V notches which permit uniform liquid distribution. It is subject to reasonable leveling within the tower. The primary disadvantage of the weir box device is that the operator cannot tell whether a reasonably uniform distribution exists and is effectively compensated for with variations in gas flow. V notches in the weir box having spacing beyond 3 in. do not offer full coverage at the upper edge of the packing, resulting in loss of effective packing height. V notches spaced at 3 in. may lose up to 2 ft of packing height. Also, weir boxes are very costly and add excessive weight to the tower.

Spray Assemblies. In place of weir boxes, the spray pipe assembly method has recently been demonstrated to be superior as it allows positive and adequate distribution throughout, and fuller use of packing height, since distribution exists at the upper edge of the packing. Further, removal of the spray headers from the exterior is allowed. Disadvantage of the spray assembly occurs when recirculated water contains particles that may plug the spray nozzles. But careful choice of nozzles will help prevent such a happening.

Use of plastics—such as FRP or polypropylene for the shell housing—has increased in recent years. The packing made of polypropylene and the pipe headers of PVC and spray jets of Teflon has proved to be ideal construction. Plastic is also advantageous as it is lightweight, noncorrosive to any of the fluids mentioned above, easily repaired and assembled, and requires no outside painting. Of course, there are some temperature limitations where PVC is employed, but this situation is primarily restricted to the pipe header and does not prove to be a problem in most instances.

Basic limitations Although small quantities of fine micron sized particles can be scrubbed by the packed tower, it is not generally recommended as a primary apparatus for the collection of micron particles, because such particles can close the space between the tower's packing and cause it to become clogged, rendering the equipment inoperative. A good rule of thumb is to allow a maximum of 0.2 grain/scf of dust to enter a packed tower. Otherwise a serious problem of blocked packing may result.

STEAM PLUME

The steam plume is a very important psychological consideration which may affect the overall equipment selection and design.

A wet type scrubber or wet type electrostatic precipitator which is doing an excellent job will show a steam discharge plume upon becoming saturated with moisture. Although the steam will have no deleterious effects on the surrounding area, its appearance may cause concern with the novice that the air cleaning equipment is inadequate or malfunctioning.

Cause Steam plumes are the result of rapid cooling of gases or air carrying moisture to below the saturation temperature. Obviously, gases from wet scrubbers will quickly show a steam plume at the stack discharge. During the colder winter months, the gases are subjected to a colder atmosphere. They are thus air-condensed sooner and have shorter plume trails compared with equivalent gases cooled by the summer atmosphere.

As a guide, saturated gases which discharge from the stack below 105°F will have a negligible appearance and will not create a questionable steam plume. At 105°F the volume of moisture content is less than 7 percent, whereas at higher saturation temperatures, the percentage of moisture by volume is as follows: 130°F, 15.0; 160°F, 32.5; 180°F, 51.0.

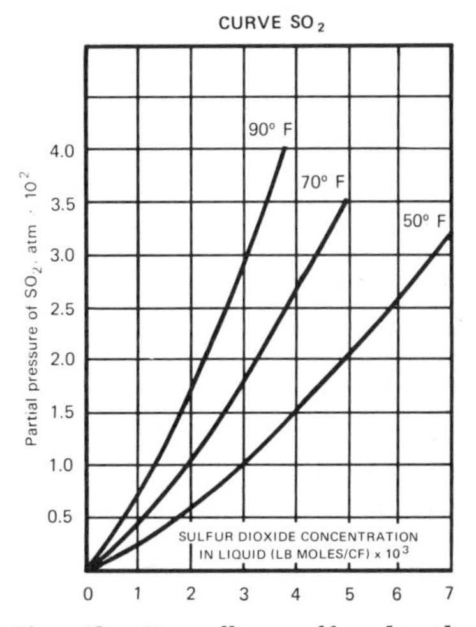

Fig. 18 Controlling sulfur dioxide concentrations by manipulating temperature levels can minimize steam plume side effects. (*Data from R. V. Parkison, Journal of Technical Association of Pulp and Paper Industry* (*TAPPI*).)

Side effects In addition to appearance, steam plumes have other side effects which include:

1. SO_2 (or other corrosive gases) may become aggravated through steam plume condensation when SO_2 is absorbed by the newly formed droplets into sulfurous acid and then falls on homes and industrial sites (see Fig. 18).

2. In some cases odoriferous constituents may be entrapped by falling droplets to increase odor at ground elevation.

Methods for steam plume minimization

Indirect Cooling of Hot Gases. A quick inspection of the psychrometric chart in Fig. 19 reveals that the saturation, easily reached by all wet type scrubbers, is dependent upon the initial moisture content (lb moisture/lb dry air) and the initial hot gas temperature. Reduction of either or both of these conditions ultimately decreases the saturation temperature.

In the example in the chart, point A is at an initial temperature of 500°F and mois-

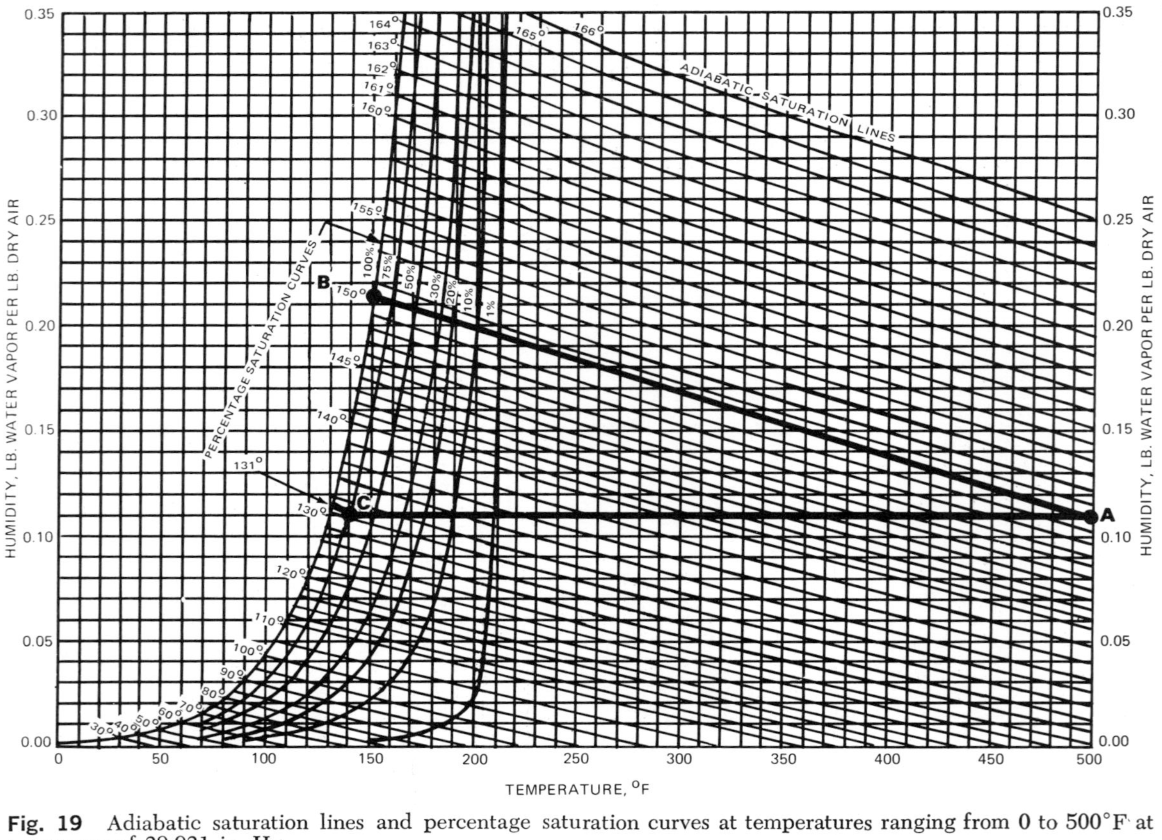

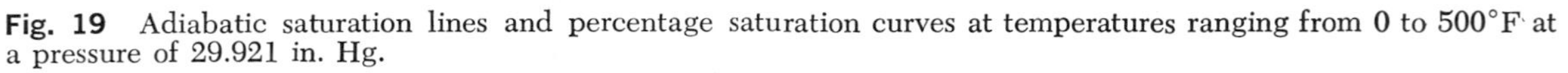

Fig. 19 Adiabatic saturation lines and percentage saturation curves at temperatures ranging from 0 to 500°F at a pressure of 29.921 in. Hg.

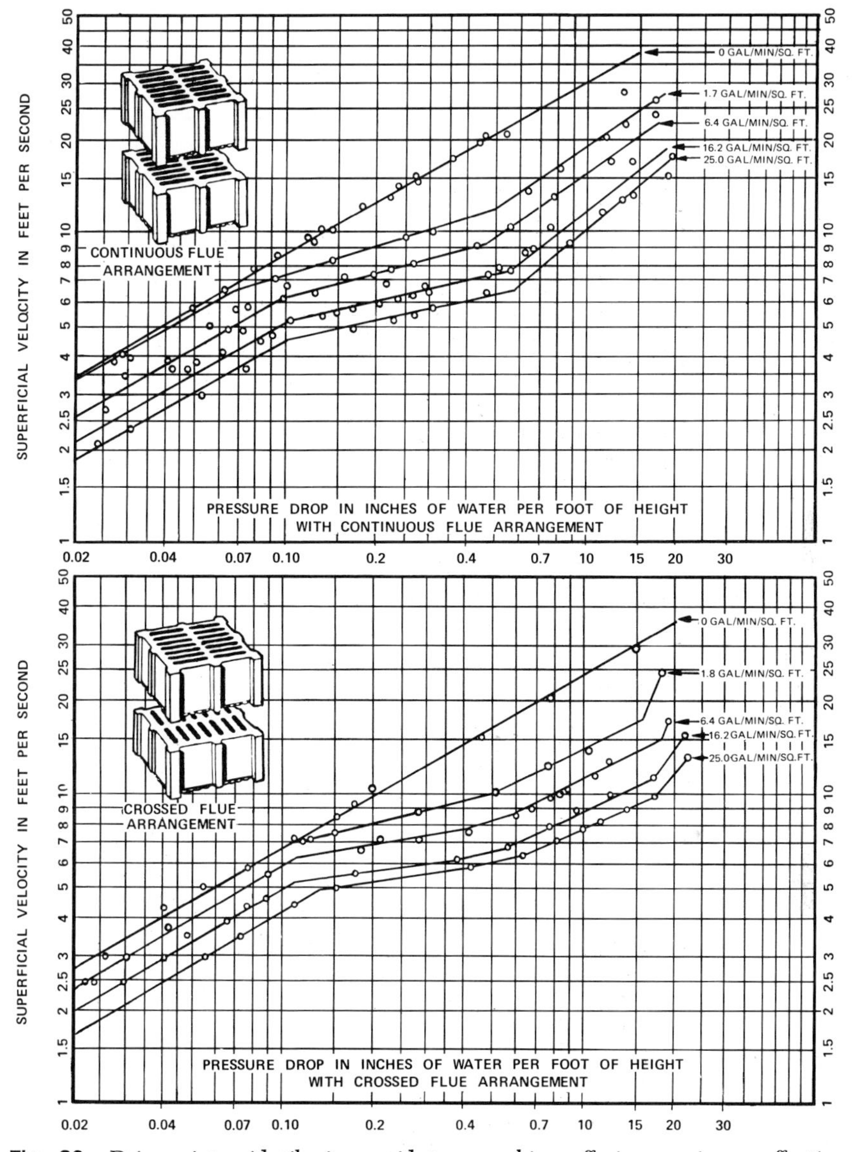

Fig. 20 Drip point grid tile is a grid type packing offering maximum effective surface via straight slots. Spray chambers in a series utilize a patented number of drip points for liquid distribution. Alternate grid tiles and straights provide more effective open surface area than possible with the checker arrangement.

ture content of 0.11 lb moisture/lb dry air; therefore, the saturation temperature will be 150°F at point *B*. Point *C* is designated a reduced initial temperature at 140°F and the same moisture content, 0.11 lb moisture/lb dry air. Now the saturation temperature will be 131°F.

The 17°F reduction at the final saturation represents a reduction from 25.2 percent water vapor to 15.3 percent water vapor content (approximately a 40 percent water vapor content reduction).

Cooling is effected within a continuous S shaped duct with sufficient surface exposure for radiation and cooling by atmospheric air surrounding the ducts. Often the ducts may be arranged essentially vertically with U bends returning upward and downward. The bottoms of these U bends should be furnished with cleanout doors and hoppers to permit intermittent dust removal.

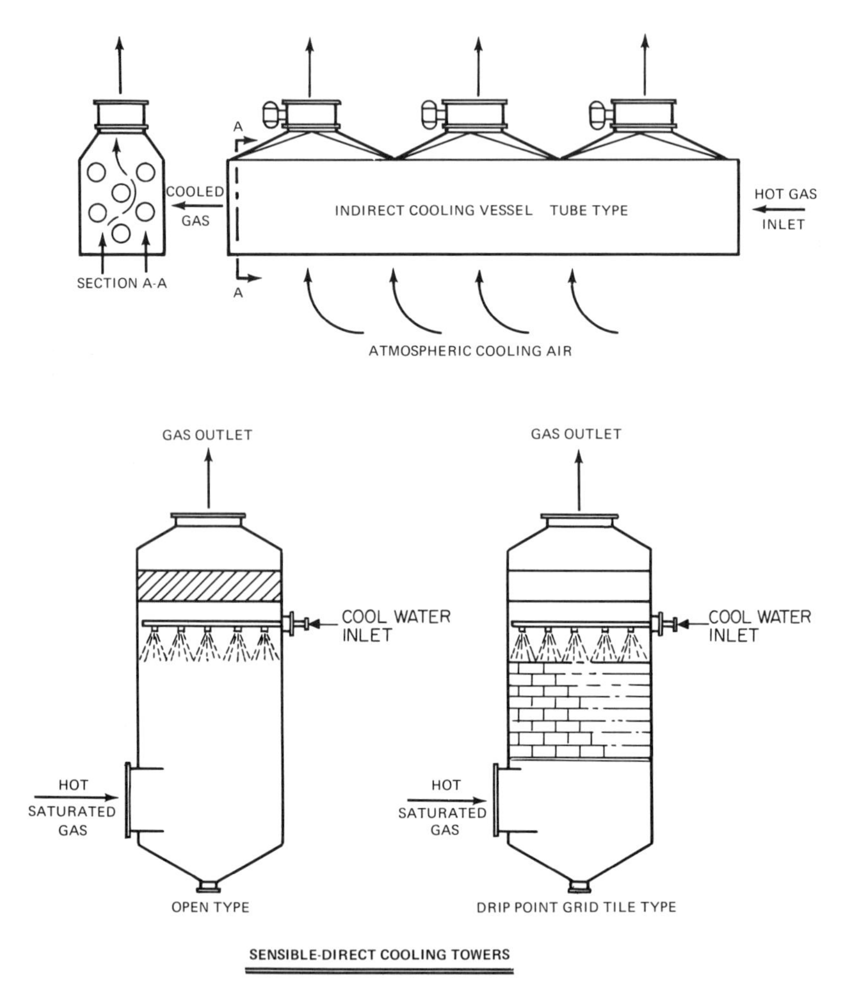

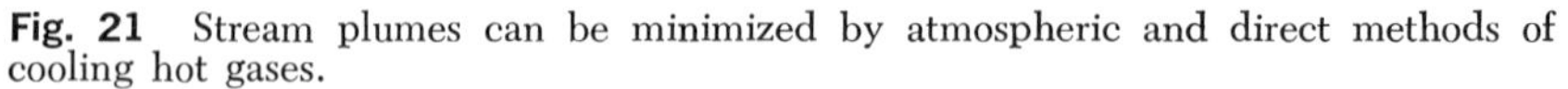

Fig. 21 Stream plumes can be minimized by atmospheric and direct methods of cooling hot gases.

The addition of fans, handling large quantities of atmospheric air, blowing or inducing air across banks of ducts which transport the uncleansed gas, has recently been introduced.

Sensible or Direct Cooling of Saturated Gases. This technique cools the already 100 percent saturated gases with cool water. Sufficient coolness is required to dehumidify or condense water vapor down to the desirable lower saturation.

Thus we see, in Fig. 19, point *B* at 150°F saturated has a heat content of 267.0

Btu/lb dry air cooling to 105°F saturated; the heat content is 73.6 Btu/lb air. Heat to be removed equals 267.0 — 73.6, or 193.4 Btu/lb dry air.

By using a standard spray tower where available, cooling water at 70 to 85°F may be introduced at 25 psig. Droplets in the range 500 microns in diameter will fall counterflow to the gas passage and carry away latent heat of the water vapor and sensible heat of the dry gas. As a general rule, the cool water leaving a properly designed tower will approach within 10 or 15°F of the entrance gas temperature. Therefore, pounds of 80°F water needed equals 193.4/(150) — (80 + 15) or 3.51 lb water/lb dry gas or 0.42 gal/lb dry gas.

Use of a tower filled with drip-point grid tiles (see Fig. 20) offers a method to obtain benefit of maximum heat transfer with an approach of approximately 5°F or less (between the gas and liquid discharging).

Cooling by Mixing with Atmospheric Air. In some special cases sensible cooling may be obtained by the addition of atmospheric air having a low dew point temperature. However, this method becomes impractical where already large saturated volumes of gas having high saturation temperatures are involved.

Figure 21 shows both direct and atmospheric methods of cooling.

REGULATIONS

The United States has established varying air pollution control codes through regulations by municipalities, counties, or control districts (which comprise the combining of areas under one jurisdiction). State regulations in a few cases govern where no controls have been established by the cities or villages.

While the movement of air or gaseous exhausts recognizes no boundary lines, very few of the regulations across the country are identical. In some cases mountains aided by hot and cold air levels limit the movement and replacement of air. These areas, typically in California, have created a concentration of suspended particulate matter. Further, the situation has prompted early passage of air pollution control measures which have served as a guide or pattern for regulations in other areas in the United States.

Variations in control criteria There are essentially five general types of pollution control measures setting the maximum allowable emission. These are based upon the following criteria:

1. *Ringelman smoke chart.* Usually permitting a Ringelman number 2 as the allowable shade of smoke corresponding to the grid pattern of the chart. Refer to the Bureau of Mines *Information Circular* 7718 of August, 1955.
2. *Process or plant production weight.* The "process weight" is the total weight of all materials introduced into any specific process which may cause a discharge to atmosphere. Refer to the Los Angles Air Pollution Control District Table.
3. *Exhaust location of coarse or fine particles.* Emission governed by distance from property line and discharge height from ground level. Refer to the New Jersey Air Pollution Control Code.
4. *Test code for dust separating apparatus* of the American Society of Mechanical Engineers, which states that dust in a gas stream is not to exceed 0.85 lb/1,000 lb of gas, adjusted to 12 percent CO_2 content for products of combustion.
5. *Fixed emission schedule* for various processes and industry. Refer to the Air Pollution Control Code Ordinance No. 167, City of Detroit, Mich.

Comparative calculations There are numerous units of measurement in the various codes and literature. To simplify their interpretation, the most commonly used terms, abbreviations, and conversion factors are tabulated:

pound (lb) = 7,000 grains
= 453.6 grams (g)
= 453,600 milligrams (mg)
grains = 64.8 mg
g = 15.43 grains = 1,000 mg
microgram (μg) = 0.000001 g
cubic feet/minute = cfm = cu ft

35.21 cfm = cubic meter/minute (cu meter/min)
cfm = cu meter/hr × 0.57, h
scfm = cfm corrected to basis of 32°F at 1 atm pressure (referred to as standard conditions)
acfm = actual cfm at the temperature and pressure of operation
grains/cu ft = 64.8 mg/cu ft
g/cu meter = 0.437 grain/cu ft
29 lb of dry air = 359 scf
1,000 lb of dry air = 12.390 scf
0.85 lb of dust/1,000 lb of exhaust = 1.0 lb of dust/14,600 scf

Correcting for excess air Often a correction for excess air, as associated with a combustion process, is required. Accordingly, the exhaust, or flue, gases are analyzed for CO_2, CO, and O_2 using an Orsat analyzer. Nitrogen, N_2, is determined by subtraction.

For example, it is determined that flue gas from a source contained 0.346 lb dust/1,000 lb gas; also the Orsat revealed 10.1 percent CO_2, 11.1 percent O_2, and 0.08 percent CO. Based upon an allowable 50 percent excess air, the flue gas dust loading is then corrected as follows:

$$\text{Ratio of actual to theoretical air} = \frac{N_2}{N_2 - 3.782(O_2 - \frac{1}{2}CO)} = \frac{79.7}{79.7 - 3.78(11.1 - 0.04)} = 2.09$$

Allowing 50 percent excess air or 150 percent total air,

$$\frac{2.09}{150/100}(0.346) = 0.481 \text{ lb dust/1,000 lb gas}$$

Percent total air = 100 percent + percent excess air
Ratio of actual to theoretical air = percent total air/100

Sampling Programs

Industry can protect itself against variation and change in codes with a routine environmental hygiene monitoring program. This sampling should be undertaken prior to plant construction. [If the plant is already in existence, sampling necessitates suspending plant operations to establish a (less defensible) comparison between conditions with and without emission.]

By determining the normal load of emission in the surrounding air, vegetation, and water, an industry can plan for and evaluate the effectiveness of pollution control equipment, making certain that effluent discharge is maintained at a level which is safe for human, agricultural, and recreational activities in the area. The study would also chart wind directions and velocity patterns prior to plant startup. Perhaps most important, it provides management with documentary evidence that can be used as protection against claims of pollution damage or violation.

A recommended initial sampling program should cover these six major areas:

1. *Sampling of dynamic and static air,* including detailed engineering sketches for construction of the equipment, as well as order information for component parts
2. *Water sampling,* including a detailed engineering sketch for equipment to be constructed and order information for component parts
3. *Vegetation and soil sampling,* providing detailed information as to location and type, plus frequency of sample collection
4. *Training* of laboratory personnel and validating analytical techniques
5. *Formulating a brief* for the state board of health on steps taken to control and prevent air pollution

6. *Formulating a reporting procedure* for the data in question to facilitate the program's objective.

Protection of this nature is especially valuable in light of the 1967 Air Quality Act, which gave emergency powers to the Secretary of Health, Education and Welfare. In the event that states do not enforce appropriate standards, the Secretary may request the Attorney General to bring suit to stop any emission creating a hazard. For violators, this could mean shutdown of all plant incinerators, a shift to low-sulfur fuels or natural gas, or complete shutdown of plant facilities.

The 1967 law also increases pressure on industry via designated air quality control regions. This new regional control structure uses federal funds to organize and publish guidelines for regions sharing a common air pollution problem encompassing two or more communities, a part of a state, or several states.

Pitfalls of current codes The new control unification is aimed at alleviating fallacies prevalent in some current codes.

Misleading criteria for satisfactory operation are often based on a stack plume measured for density by a reading on a Ringelman chart, or by equivalent opacity. This method is actually a subjective measurement based on individual interpretation of stack effluent appearance and does not basically relate to the emission's actual amount, by weight. The various factors which affect stack appearance (and thus Ringelman reading) are particle size, color of dust, background against which the emission is seen, light, stack diameter, and plume velocity. An extremely fine particulate will cause more visual density in an emission than would a more highly dispersed large particulate of equal weight. Thus the opacity of the plume is not an accurate basis for design and cannot be reasonably held as evidence of non compliance with a code unless it is correlated to measurement by weight.

Much of the confusion in current air pollution control can also be traced to the fallacy that codes should be as strict as possible. In reality, codes cannot be excessively stringent if they are to be operable and enforceable. When no equipment can possibly meet the demands imposed by regulations, no polluter can be held responsible for violations.

Chapter 24

Water Pollution Control Equipment

W. ALLEN DARBY

Consultant, Environmental Equipment and Systems Division, Dorr-Oliver Incorporated, Stamford, Connecticut

Treatment of waste water, whether from industrial or domestic sources, is accomplished with certain basic processes. Application varies with the characteristics of the waste water that govern the design criteria. After the treatment problem has been analyzed, processes should be selected and equipment chosen that are best suited to assure the carrying out of those processes in the most practical and efficient manner.

Equipment available reflects the concepts of the manufacturer based on his experience and is necessarily specialized. In other words, it is designed and marketed for specific processes. Consequently the designing engineer must first select the processes needed, survey the market for the equipment available to do a satisfactory job, and build his specifications around the equipment. This frequently involves designing structures of the right dimensions to fit the machinery. The equipment may then be placed in the structure as a turnkey operation. It is not often feasible, for example, to design a flocculation basin with specific dimensions and then find the paddles or stirrers that will operate in it efficiently. The design stage requires close collaboration between the designing engineer and the equipment manufacturer to ensure that the facility will function as intended when built.

TRANSPORTATION OF WASTE WATER

Piping Systems

The characteristics of the waste water to be handled have to be considered when the piping or conduit system is selected. If the waste water is on the acid side, some materials will be affected. If concrete, steel, cast iron, or aluminum are used, coatings such as epoxy paints or coal-tar enamels should be specified. Polyvinyl chloride or similar plastics, clay, and to some degree, asbestos cement are resistant to deterioration. On the other hand, PVC cannot be employed where high temperatures and pressures are encountered. An epoxy lined asbestos cement pipe is manufactured specifically for industrial applications, as is a bituminous fiber pipe. Glass lined or plastic lined pipes minimize grease buildup if this is likely to be a problem.

Quantity of flow and size of conduit are also considerations in selection of the material. Cast-iron, clay, asbestos cement, and PVC pipe have upper size limitations. The trench load has a bearing on selection. Materials that lack strength should not be used indiscriminately under structures and roadways. Ability of the pipe to withstand trench loads and pressure imposed is spelled out in handbooks distributed by manufacturers' associations and institutes and can materially aid designers in preparing specifications.

Pumps

Pumps employed in industrial waste applications are of two general types: centrifugal for the main flow and reciprocal, plunger, screw, or diaphragm types for the underflow solids or sludge. However, some centrifugal pumps are designed to handle sludges, grit, and other high-solid streams. In these and all other centrifugal pumps used in waste water applications, the critical feature is the impeller design. Economy and freedom from clogging depend largely upon the shape of the impeller, its casing and clearance. Where untreated waste containing irregular and large-diameter solids or fibrous materials is handled, the impeller should have such clearance between vanes that any object entering the pump will pass into the discharge pipe. Some designs provide a hatch type cleanout in the impeller casing to facilitate removal of clogging materials. Others involve an integral screen and pump. One type of centrifugal pump has a recessed impeller which transmits power as in a fluid type torque converter. This will handle sludge—and the main stream flow, though with less efficiency than the conventional centrifugal pump. The more open the impeller, generally the less efficient the pump, and a balance should be struck between the relative economics of using open impeller pumps and more conventional designs plus screens upstream from the point of pumping.

Specifications should require that rotating parts be accurately machined to provide near perfect rotational balance and to avoid resonance at normal speeds. Centrifugal pump selection should be on the basis also of head capacity characteristics corresponding to the conditions encountered. Head capacity curves should be prepared to cover all conditions, including maximum and minimum friction losses. The maximum speed is determined by the net positive suction head available at the pump, the quantity handled, and the total head. The speed at which a pump will operate should be checked against the limiting suction requirements of the Hydraulic Institute Standards. Since waste water flows are highly variable, variable speed drives offer advantages, particularly economic, with respect to wet well construction, though use of multiple pump installations with on-off operation is most frequently employed.

Reciprocating pumps, used for handling high-solids flows have one to four plungers, in which the stroke is adjustable, Ball type valves are employed with construction such that the valves are easily accessible for removal of obstructions. Diaphragm pumps are also equipped with ball valves in single and duplex design, with adjustable stroke. Both reciprocating and diaphragm pumps are V belt or chain driven, usually from electric motors. Protection against leakage is important and is controlled by the selection of good packing material. Acid, alkali, or brine resistant packing is desirable, depending on waste characteristics.

SOLIDS REMOVAL

Discharge of effluents containing appreciable suspended materials to streams is a violation of most water quality standards, and such materials must be removed. Solids in industrial effluent may be classified as large floating, heavy (high specific gravity), or finely divided. The first two categories include materials that should be removed from the waste water to protect equipment used in further treatment steps as well as to avoid pollution.

Fig. 1 Yeomans pumps in one of the Brockton, Mass., pumping stations. Yeomans manufactures pumps in all sizes for all types of application in the handling of sewage and waste water. (*Yeomans.*)

Large floating solids are essentially materials that can be removed by screening and may vary from pieces of rag and wood to spillage from such operations as canning. Heavy solids are generally referred to as grit and may include sand, mill scale, fly ash, coke fines, and many abrasive materials. The finely divided solids are those that are slow-settling and too fine to be removed by screening.

Generally, if all such materials are present, the succession of steps is screening, grit removal, and sedimentation or flotation. If the solids are flocculant or can be precipitated with chemicals, flocculation may precede sedimentation.

Screens

Large floating solids may be removed from the waste flow by racks of vertically stacked bars known as "bar screens," which are mounted in a channel at angles from 10 to 45° with the horizontal, the acute angle on the downstream side. They may be hand or mechanically cleaned. Openings between the bars are ordinarily 1

Fig. 2 Model 10A comminutor manufactured by Chicago Pump, installed in an Army sewage treatment plant.

to 2 in. The channel should be of a configuration that will allow a minimum velocity of 1 fps to avoid sedimentation and a maximum of 3 fps to eliminate the possibility of forcing solids through the bars. Where the screen is mechanically cleaned,

(*a*)

Fig. 3 (*a*) Eight-foot north sewage screen at Waseco, Minn., sewage plant. Machine is also filtering raw intake primary sewage. Note water passing through wire mesh cloth filtering medium that covers cylinder.

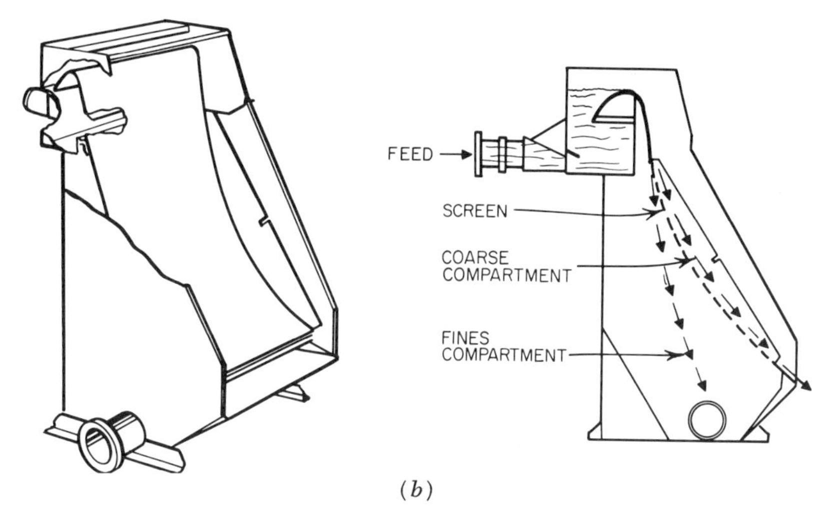

(*b*)

(*c*)

Fig. 3 cont. (*b*) Design views of liquid filtering machine. (*c*) Close-up of two NRM-148 liquid vibrating screens in waste water treatment plant at Lakeside Packing Co., Manitowoc, Wis. Screens remove fine solids from waste water, which then meets state board of health regulations and is discharged directly into city sewerage system. Each screen has a capacity of 750 gal of water/min. Other Link-Belt equipment includes Thru-Clean bar screen, Straightline grit collector, and flight conveyor. Lakeside packs peas, string beans, beets, kidney beans, and carrots at this plant.

toothed rakes move over and between the bars, pulling screenings to a platform at the top of the structure. The rakes are driven by electric motors, usually equipped with overload protection devices. Some screens are equipped with curved bars, which are cleaned by a revolving rake. The operation of rakes and consequently frequency of cleaning can be governed by a time clock or by a float control actuated by head differential.

For removal of fine, nonflocculant, or noncolloidal particles, disk type or drum screens may be used. In these the screen revolves on a horizontal axis and is cleaned continually by water sprays, cascading, breeches, etc. The screening medium is usually stainless steel, manganese bronze, or alloy wire cloth up to 60 mesh in disk screens and higher in some drum screens, which may have apertures as small as 20 microns.

Vibrating screens are also available and are particularly advantageous where screenings should be dewatered before disposal, or for recovery of materials for reuse or by-product manufacture.

Another type of screen available comprises a curved screening surface consisting of a series of spaced wedge wire bars set at right angles to the line of flow. The material to be screened is fed tangentially to the screen via a parabolic shaped weir. This screen is self-cleaning. The oversize material travels along the screen surface and is discharged at the end. Liquid together with fines passes through the screen to be discharged via a separate outlet (see Fig. 3*c*).

Grinders

In some cases it is desirable to employ grinders, which reduce large solids to smaller particles prior to subsequent treatment of the waste stream.

This may be accomplished by grinding the solids in the waste stream without first removing them from the stream or by grinding the solids after removal and then returning them to the stream. The first method involves the use of a rotating, horizontally slotted drum screen fitted with cutters that mesh with a fixed comb also with cutting members. The waste stream passes through the slotted drum and out the bottom of the unit. The large particles retained on the screen surface are cut into small pieces until they also pass through the slots. Alternate devices consist of a fixed horizontal semicircular bar screen with cutters. Cutting action on the large solids is caused by an oscillating comblike device also equipped with cutters.

The second method involves the use of mechanically cleaned screens with separate grinding devices such as cutter pumps or swing hammer mills. Some cutter pumps are capable of handling very heavy grinding loads and are also capable of pumping against a considerable head. One such unit uses a notched disk for an impeller mounted at an oblique angle. The notches mesh with serrations in liners. Such types of cutter pumps are of particular value where screenings are to be transferred to an anaerobic digester or where primary sludge is to be thickened ahead of a centrifuge for further dewatering.

Grit Removal

Grit is not a common problem in treatment of industrial waste waters. When it is encountered, it can be removed by differential settling. Reduction of the velocity of stream flow to about 1 fps with proper area will permit high specific gravity solids to settle and the remainder to be transported to other treatment units.

In the design of a grit chamber, area is the most important consideration, and the capacity is directly proportional to the area, regardless of width, depth, and velocity. In a square tank, for example, Table 1 can be used to determine the size of grit chamber to remove particles of a certain size (of particles at 2.65 specific gravity).

Rectangular channels or square or circular basins may be used. Where channels are employed, care must be taken to maintain the desired velocity, which can be

TABLE 1*

Mesh of Grit to Be Removed	*Overflow Rate, gal/(sq ft)(day)*
35	73,000
48	51,000
65	38,000
100	25,000

* W. A. Kivell and N. B. Lund, Grit Removal from Sewage, *Water Works and Sewerage*, April, 1940.

Where sludge from primary sedimentation tanks is dewatered by means of centrifuges, it is desirable to remove 100 to 150 mesh grit so as to minimize wear in the centrifuge.

done by providing integral Sutro weirs or Parshall flumes. Mechanical removal of the grit is desirable in a continuous flow operation, with endless chain flights of scraping bars or buckets.

In square or circular chambers, revolving scrapers move the grit to a collection well or trough in which a conveyor operates to remove the solids for disposal. Washing type collectors may be used. In these, conveyors or classifiers move countercurrent to the flow to remove the grit from the waste stream where sprays may be applied.

Grit chambers may be aerated, which action tends to clean the heavy solids and often can be applied for combination oil, grease, and grit removal, with oil and grease collecting at the surface. Another method for separating and cleaning grit is by means of a cyclone which has a tangential feed inlet. The centrifugal force imparted to the feed separates the grit from the organics.

Oil-Water Separation

Where floating oil is encountered, it can be removed by directing the flow into a basin having a retention period of at least 1 hr and a velocity not greater than 2 fpm. Oil is collected at the surface and is skimmed off by a conventional sludge scraper skimming mechanism, similar to that used in clarifiers and described later under Clarifiers. Two stages are frequently used.

If the oil is emulsified, the emulsion must be broken by physical or chemical means. The American Petroleum Institute provides information on the design of separators and emulsion breaking. The principle of the separator is basically that of flotation. Other procedures are described under Flotation Units.

Flocculation

This step is often combined with sedimentation when the solids are light and tend to coalesce or when chemicals are added, to produce flocculant particles. The process takes place in a basin equipped with a means of slow agitation or stirring, which causes the smaller solids to be attracted to the larger particles by mechanical collision or electrochemical phenomena. With the suspended matter concentrated into dense particles, the rate of settling is increased.

The means of agitation may be horizontally or vertically mounted paddles or baffles revolving through the waste stream or slats or bars moved horizontally or vertically by eccentric and reciprocating mechanisms. The moving elements are driven by electric motors operating through geared speed reducers. Baffles or paddles revolving on a vertical axis are direct driven from a power unit mounted on a bridge across the top of the tank. Those revolving about a horizontal axis may be direct, chain and sprocket, or V belt driven. Where they are direct driven, power units must be placed in an adjacent dry well. A single power unit may be used to drive flocculation mechanisms in adjacent parallel tanks, with a clutch to permit one to operate without the other. Peripheral speeds of rotating elements should normally not exceed 1.75 fps.

Horizontally mounted units are designed for use in rectangular tanks, vertically suspended elements in circular tanks. Some manufacturers provide a combination flocculator and clarifier or sedimentation basin. These are circular, with the flocculator in a central well at the top of the tank.

Care must be taken in design not to restrict flow passage from the flocculation basin to the sedimentation basin, because appreciable increase in velocity will tear the floc. Velocities greater than 1 to 2 fps should be avoided. If chemcials are employed, a preliminary "flash" mixing basin should be provided. In this, the mixing of chemicals with waste is accomplished by means of a high-speed turbine or propeller type mixer in a square or circular tank preceding the flocculation basin. A detention time of about 10 to 30 sec. with rapid stirring is adequate. Longer detentions may cause floc breakup. The ratio of depth to width is about 2:1. Chemicals are applied in a predetermined dosage, measured dry, and then dissolved or suspended, on the basis of volume or weight. Most flocculating chemicals can be obtained in liquid form, in which case they are fed to the waste stream by proportioning pumps.

Clarifiers

Removal of fine suspended or flocculant solids is accomplished by means of clarifiers (also called "sedimentation basins" and "settling tanks"). In these, the velocity of the waste stream is diminished to 1 fpm or less by dispersal into tanks of large

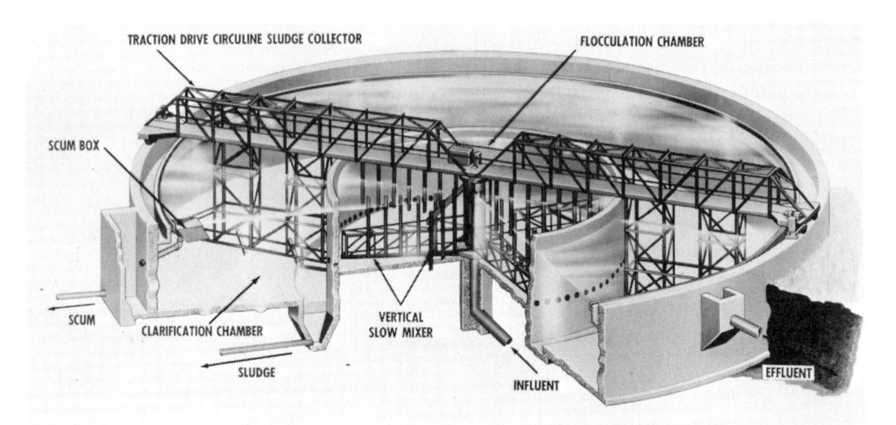

(*a*)

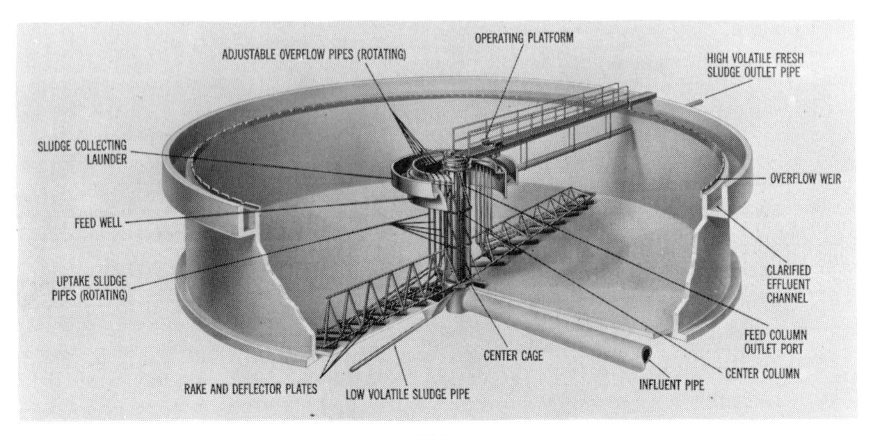

(*b*)

Fig. 4 (*a*) An unusual system for treating industrial wastes consists of a combination flocculator-clarifier, designed and built by Link-Belt, installed in a 125-ft-diameter tank. The flocculation chamber in the center has a Link-Belt vertical slow mixer, while the clarification chamber employs a Circuline sludge collector. The treatment plant is in operation at the Lock Haven, Pa., pulp and paper mill of New York and Pennsylvania Company, Inc. (*b*) The Dorr-Oliver RSR clarifier, a cross section view.

cross-sectional area and held for two or more hours, as a rule. Lower retention periods are possible with some wastes, and pilot plant or laboratory studies can determine this. The tanks may be circular with a central inlet and peripheral outlet or the reverse. If tanks are rectangular, the flow is from one narrow end to the other, though there are some types in which flow is vertical, with the tank divided into cells, each with its own take-off weir. Square clarifiers are also possible, with central inlet and peripheral outlet. Water depths are on the order of 12 to 14 ft, and with rectangular basins the length-to-width ratio is at least 5:1. Other controlling factors are the overflow rate and weir loading. Recommended overflow rates vary from 500 to 1,200 gpd/sq ft of surface area for domestic waste, and most industrial wastes should fall in this range.

Sludge removal is mechanized in modern plants. Rectangular tanks are equipped with a hopper bottom channel across the inlet end for deposit of sludge. The sludge is moved to this by endless chain and flight scrapers driven by electric motors equipped with gear reduction. The arrangement frequently calls for the flights to move along the surface as well as the bottom, to permit removal of scum. The collector moves very slowly, usually not over 2 fpm. The bottom of the basin is sloped to the collection channel, about 1 percent or more. In one design for rectangular tanks, a motorized carriage, from which a single blade is supported, moves back and forth along the tank rim.

In circular tanks, the sludge collection well is in the center of the tank, with scraper blades rotating from a central shaft, or being driven by a traction mechanism operating along the outer rim. In the latter, the scrapers are suspended from a truss. As in rectangular basins, the floor is sloped toward the collection well. Where sludge is heavy, requiring high torque, heavy-duty mechanisms are recommended.

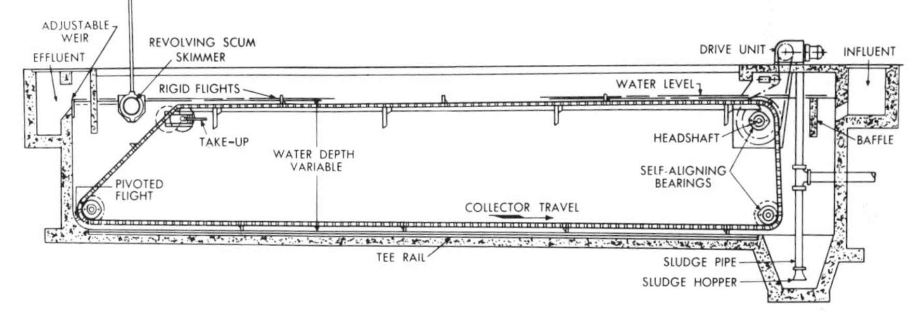

Fig. 5 Sludge collector cross section.

Sludge withdrawal is by means of pumps, and sludge concentration can be controlled by density meters to avoid pumping too much liquid with the solids.

One type of sludge collection mechanism, particularly adapted to the activated sludge process (used for organic wastes), involves a hydraulic principle. Conduits are suspended from a rotating truss, or a single tapered arm is used equipped with orifices.

Flotation Units

Some solids, which are floating because they have low specific gravity or are fibrous in nature, are best removed by flotation. The method also is applied to grease removal. It involves application of air, with the air bubbles carrying that material to the liquid surface, where it is removed by skimmers. Two methods are generally employed, pressurized aeration and vacuum. In the former, the waste stream is subjected to aeration from a blower under pressure which may exceed 25 psi. The pressure is then released in a flotation chamber and the dissolved air forms bubbles which carry the suspended material to the surface. In the vacuum units, aerated waste is passed into an enclosed chamber, where vacuum is applied to cause air to be released to the surface.

Sludge may be thickened by this method, with polyelectrolytes used to increase sludge yield.

Design criteria are frequently expressed in pounds per square foot of surface per 24 hr range and from 4 to 20.

Factory built steel tanks may be used, or the units may be field constructed of steel or concrete, either rectangular or circular, with mechanized skimmers. Combination clarifiers and flotation units are available.

SOLIDS HANDLING

Thickeners

Sludges pumped from clarifiers are likely to entrain a high percentage of water. For disposal by anaerobic digestion or for dewatering by filtration or centrifugation, the solids content should not be less than 5 percent. Sometimes this can be controlled by density gages actuating the pumps in the sludge withdrawal process. However, it is commonly done by thickeners.

Fig. 6 Overall view of a large Dorr-Oliver thickener in flue dust recovery system. Thickener prevents discharge of basic oxygen furnace flue dust and recovers usable iron.

Flotation may be applied for thickening, as described in the foregoing section, or gravity methods may be used. In the gravity units, sludge is passed to a tank, usually circular in configuration, where settling takes place. Rakes stir the sludge gently, release water, air, and gas and push the concentrated sludge to a central collection well. Centrifuges, which are described under the section on Sludge Dewatering, may be employed.

A type of rectangular basin thickener uses helicoid flights rotating on a transverse axis. Combination clarifiers and thickeners are also available.

Anaerobic Digesters

These have limited application to industrial effluents, but if the waste water is organic in character, the sludge may be stabilized anaerobically. Usually, if a waste has a BOD of 1,000 mg/l or more, the method may be applied. Toxic materials will seriously interfere, and the method should not be attempted if they are present in any significant concentration. The sludge is contained in enclosed tanks and held for a long period of time, ranging from days to several weeks, rate of digestion depending on the character of the waste. The tanks are circular and of concrete, equipped for continual mixing of the contents. Liquefaction of the organic solids occurs, and gas is produced. An optimum temperature of 80 to 100°F should be maintained by heating if required.

The gas produced is usually flammable and must be handled with care; it may be burned in a waste gas burner or be utilized for generation of power or heat. Collection lines are equipped with flame and condensate traps.

The digester may be equipped with coils, through which hot water is circulated. Submerged gas burners can also be used. External heaters can transfer more heat per unit of exhanger surface than can internal heaters. In these, sludge is recirculated from the digester through a heat exchanger.

Digester mixing may also be accomplished by electric motor driven propellers or turbines and gas mixing erected at multiple locations through the tank cover.

The diameter and depth of digesters are largely governed by available equipment and economics of construction. A 20-ft side water depth is common. They may be up to 100 ft or more in diameter.

Aerobic Digestion

Intensified aeration of sludge over an extended period of time will result in some thickening and mineralization of organic matter by oxidation of volatile solids. The detention time is about the same as for anaerobic digestion, but digestion can take place in open tanks and gas collection and heating equipment is not required.

The solid fraction is readily thickened and dewatered, and the liquid portion may or may not require further treatment. In the former case it may be returned to a biological treatment unit.

Air may be supplied by blowers and dispersed by diffusers placed along one side of the tank, which imparts a rolling motion to the tank contents as in activated sludge treatment. Mechanical aerators may also be used. These are turbine mixers or surface agitators driven by electric motors. The equipment used is the same as in the activated sludge process and is described more fully under that section.

Sludge Dewatering

Dewatering is usually an intermediate process between solids removal processes and final disposal by incineration or burial. This may be accomplished by air drying on sand beds or use of vacuum filters, filter presses, or centrifuges.

A vacuum filter is in the form of a drum suspended in a tank containing the sludge. It is covered with a filtering medium which may be natural fiber or synthetic fabric, woven stainless steel wire, or coil springs. Radial partitions divide the drum into compartments, which are subjected to vacuum. As the drum rotates, sludge accumulates on the filter surface and the vacuum removes water. As the drum approaches the submergence line, the vacuum is broken and the accumulated cake is discharged, sometimes by compressed air, aided by a scraper. In coil spring filters, the springs leave the drum, which causes the cake to discharge. Similarly in the "belt" filter the filter cloth leaves the drum and passes over a small auxiliary roll for cake discharge. Another type of filter uses cords placed around the drum to effect discharge. The vacuum is around 20 to 26 in. of mercury.

Submergence of the drum in the tank is usually about 25 percent of the drum diameter but is usually adjustable, and the degree of submergence affects cake formation. The tank is equipped with agitators to keep the sludge suspended and avoid concentration.

With some sludges it is necessary to add conditioning chemicals to effect efficient dewatering. These may be ferric chloride and lime or polymers.

The performance of a vacuum filter is measured by the ratio of dry solids produced per surface area and varies between 1 and 8 psf/hr.

Filter presses consist of a series of plates supported on a frame in a vertical position. Each face is covered with a cloth medium. Sludge is pumped into the spaces between the plates, forcing the liquid through the cloth. Other types employ bags placed in a frame face to face between drainage sheets. Water is forced through the bags by pressure applied by platens. In general, filter presses are used only in

Fig. 7 Dorr-Oliver vacuum filters dewatering sludge in a sewage treatment plant in northern New York State.

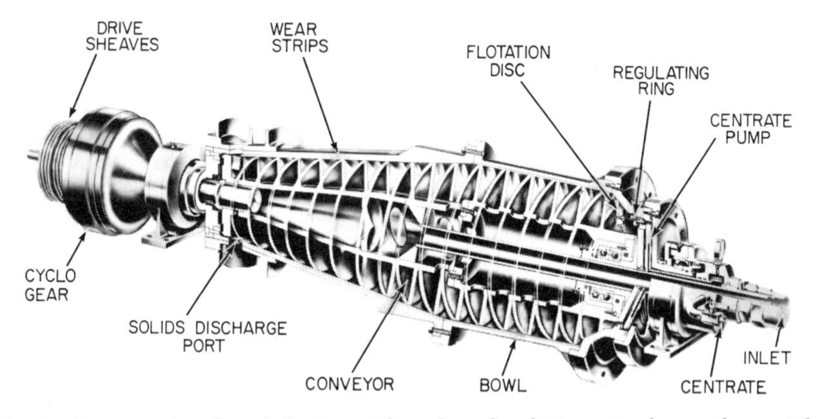

Fig. 8 Design principles of the Dorr-Oliver long bowl MercoBowl centrifuge. The long cylindrical section, attached to the conical section, provides greater clarification capacity and operating efficiencies.

the case of sludges that cannot be readily dewatered by means of discontinuous filters. They are suitable most generally for batch operation; however, at least one type for continuous operation is used in the waste treatment field. In it, thickened sludge is passed between rotors, the bottom one of which is equipped with a wedge slot screen.

Continuous flow centrifuges are employed for thickening as well as dewatering.

The solid bowl centrifuge has been found most suitable for waste sludge dewatering. It operates on a horizontal axis and consists of a solid bowl shell with an internal scroll or screw conveyor rotating at a speed slightly faster than the bowl. Sludge is fed into the hollow shaft of the conveyor, and the solids settle through a liquid pool and become compacted against the bowl wall. Liquid flows toward the end of the bowl, and that overflowing a regulating dam can be removed by gravity or by pumping. The frictional force of solids on the conveyor pushes solids from the pool across a drainage deck, where they are dewatered before discharge. Centrifugal forces developed vary from 3,000 to 5,000 G, in which

$$G = 1.42 \times 10^{-5}(2r) \qquad (\text{rpm})^2$$

and r is the radius of the bowl.

The pool is maintained by an internal weir and regulates the retention time, which is around 0.25 to 1.25 min. Increasing retention time results in greater clarity but higher moisture in the sludge. Increasing bowl speed increases product recovery and produces drier solids. Decreasing conveyor speed has like results. Recovery is also improved by decreasing feed rate, increasing feed consistency, and with the addition of flocculants. Drier solids result from increasing feed rate, decreasing feed consistency, and reducing the amount of flocculants. Increasing temperature improves both recovery and solids.

Sludge Disposal

Sludge cake from filtration or the discharge from centrifuges still has considerable moisture. Where the sludge is largely organic in nature and has value as a fertilizer, drying is necessary for further moisture reduction. This consists of heating the sludge sufficiently to evaporate the moisture. For every pound of water, about 1,122 Btu are required at a room temperature of 60°F. The sludge cake has to be broken up for fast, efficient drying, and a pug or cage mill is used for this. A rotary drier affords continuous operation and consists of a mechanically rotated drum through which sludge and hot air are passed from a furnace at one end with a fan moving the air. One type has a double shell arrangement, with the hot gases and sludge passing through the inner shell and gases returning through the annular space between the shells. In flash driers, sludge is passed through a cage type disintegrator and the solids are held in suspension by a stream of hot gases. A cyclone is used to separate water vapor from the sludge.

Complete destruction of sludge from organic wastes is the most desirable method of disposal. This is accomplished by incineration of the volatile solids, leaving only a sterile ash.

Two general types of incineration methods are commonly employed for this purpose. In one, the dewatered sludge is introduced to a fluidized bed consisting of sand suspended in air and heated to temperatures of 1400 to 1500°F. Combustion gases are scrubbed by plant effluent, and the entrained solids are removed and dewatered by the scrubber followed by a cyclone (see Fig. 9).

In the second, the dewatered sludge is introduced to a multiple-hearth incinerator consisting of a refractory lined steel shell containing four or more hearths or trays. Rotating "paddle arms" or plows move the sludge on each hearth to discharge ports so that the sludge passes through all hearths during incineration. Temperatures must be maintained at about 1350°F to avoid odor production. As in the case of the fluidized bed system, the exhaust gases are scrubbed to avoid air pollution.

Another method of sludge disposal in use is the Wet Oxidation process, wherein the wet sludge is oxidized in a reactor operating at about 500°F at relatively high pressures (350 to 3,000 psi). Heat exhangers are used to transfer heat from the processed material to the incoming sludge.

A fourth process uses a vertical reactor in which thickened sludge is atomized and heated to temperatures of 1000 to 1400°F. The entrained water flashes into steam, and the sludge particles fall through the reactor in contact with air introduced at midpoint, combustion taking place in the process.

REMOVAL OF DISSOLVED ORGANICS

Organic wastes, following solids separation, must be further treated to remove dissolved unstable constituents. This may be done by biological oxidation, in which cultures of microorganisms are maintained to feed upon the organic matter and thus destroy it. Three general processes are employed: (1) stabilization in large ponds, (2) passing the waste through beds of coarse media, known as trickling filters, and (3) subjecting the waste to intensive aeration, known as the activated sludge process.

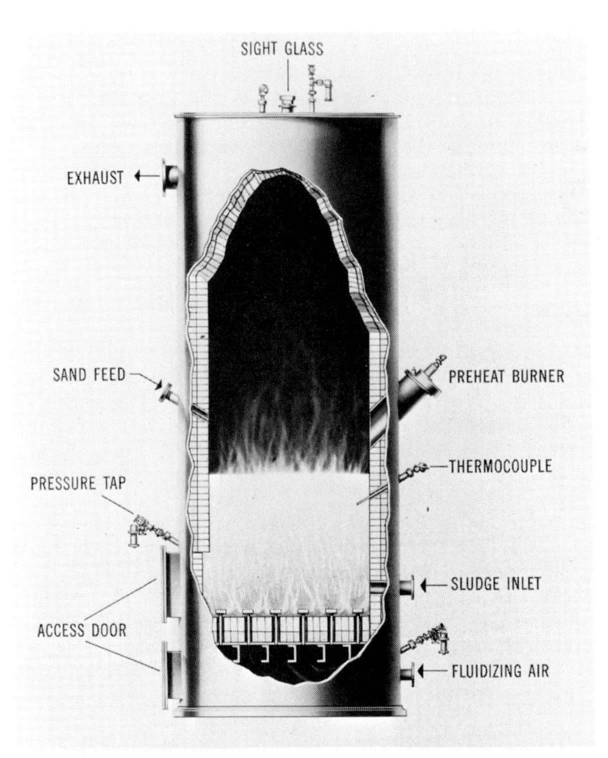

Fig. 9 Cross-sectional view of a Dorr-Oliver FluoSolids single-stage reactor. Fluidized air is pumped by a multistage fan into the bottom compartment. The air is distributed evenly into the bed through a constriction plate and percolates through the sand bed, expanding or fluidizing the bed. Temperature in the bed is maintained at 1200 to 1400°F. Stack gases are completely deodorized in the FluoSolid's reactor. Ash is stripped from the bed by the upflowing gases and is discharged out of the exhaust exit.

Stabilization Ponds

These are large shallow earthen lagoons in which the waste is retained for several weeks. The biological system involves algae which by photosynthesis produce oxygen to effect oxidation. The ponds are kept shallow (2½ to 5 ft) to maintain algae growth throughout their depth. The loading recommended is around 10 to 50 lb BOD per acre per day, for domestic wastes. Proper loadings for industrial effluents should be determined by experimentation. Raw wastes with a low suspended solids content may be treated in this manner without initial solids separation. In areas where percolation through soil is a problem, the ponds may be lined with asphalt or plastic materials.

Multiple cells in series and parallel are frequently used for flexibility of operation, to prevent solids spill and to accommodate the varying biota systems occurring with progression of oxidation. Freeboard should be about 3 ft plus allowance for frost heave. A beam of about 10 ft should be provided with an inside slope of 1:4 to 1:7; outside slope about 1:4. The dikes should be seeded with short-root grass. The shape of the ponds should avoid islands. peninsulas, and coves. Influent feed is usually less than 1 ft from the bottom.

The loading on the ponds and depth may be considerably increased by aerating them, so that one-fifth to one-tenth the land area will be required. Equipment for aeration includes both surface type mechanical aerators and diffused air systems. Mechanical aerators may be bridge or column supported or mounted on floats supported in place by cables. Conventional diffused air systems, discussed under Acti-

Fig. 10 Empty lagoon with installed aeration units on access platform, Laurel, Montana. (*Surperator, by Permutit Co.*)

vated Sludge, may be employed when the configuration of the pond permits baffling to achieve the rolling action required. One system designed specifically for pond installation involves the use of perforated plastic tubing snaked transversely across the bottom of the pond, thus forming a series of cells, separated by screens of air bubbles.

Trickling Filters

In these, bacteria and other microorganisms are cultured specifically for the waste being treated. The devices are beds containing rock or slag in a size range of 2 to 4 in., specially fabricated plastic media or redwood slats. The waste is distributed evenly over the surface of the bed by a rotary mechanism. The distributors may be motor driven or hydraulically actuated. The bed is supported by a floor designed to allow free drainage of the effluent and permit air passage at all times. The structure is usually concrete, but steel or tile shells can be used.

The microorganism culture, which assumes the form of a slime, becomes attached to the media. As the waste percolates through the media in thin films, the culture feeds on the organic matter, reducing it to stable compounds of carbon and nitrogen.

Rock, slag, and redwood media filters are 3 to 10 ft in depth. Plastic media filters, because of superior air passage qualities and the light weight of the media, can be much deeper, minimizing ground area requirements.

Fig. 11 Fallbrook, Calif. An example of Air-Aqua solving a severe odor problem. Originally the plant consisted of an Imhoff tank and trickling filter, without a final clarifier. Overloading and high sulfate content of the waste caused objectionable odors. Now, aeration ponds have solved the odor problem.

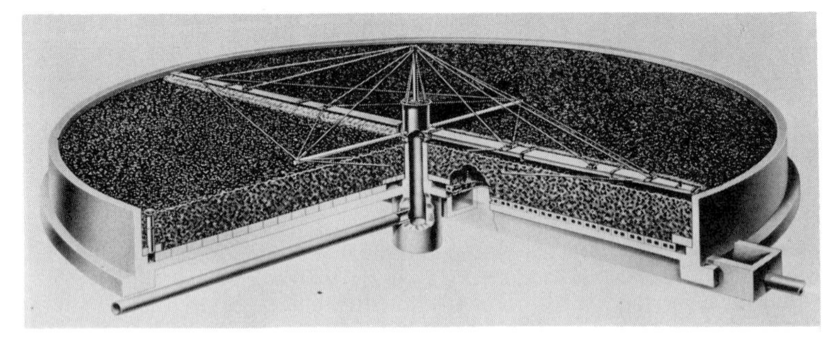

Fig. 12 Cutaway of a typical Dorr-Oliver biofilter unit, showing distributor mechanism details.

In the redwood filters, modules of sawed slats are assembled, with the space between slats about 0.7 in. Plastic media are made up of corrugated sheets of polystyrene or polyvinyl chloride, which are assembled into modules for placing.

Recommended loading is in terms of pounds of BOD per day per unit volume and ranges from 5 to 300 lb/1,000 cu ft. The effective loading varies to some extent with the waste to be treated. The media must be kept wet at all times, and provisions must be made for recirculation of the effluent to compensate for low flows. There is some virtue in recirculation under all conditions, thereby effecting a higher degree of removal of organics. Some humus which is discharged from the filter from the dead bacterial cells has to be removed by a final settling tank.

The Activated Sludge Process

This method produces the highest degree of treatment of all biological oxidation processes but is sensitive in its operation and requires careful, intelligent operation. For those reasons, trickling filters and stabilization ponds are sometimes preferred.

The reaction takes place in tanks equipped with diffusers, through which air is supplied by blowers, or the tanks may be equipped with mechanical surface or tur-

Fig. 13 Model 30 Swingfuser diffusers with precision diffuser tubes. Manufactured by Chicago Pump, the units are installed at Redlands, Calif.

bine aerators driven by electric motors. In diffused air aeration the diffusers are placed along one side of the tank, at or near the top or bottom, so that the release of the air imparts a rolling motion to the tank contents. This keeps the solids in suspension and permits oxygen absorption from the atmosphere. The diffusers may be of porous ceramic material, stainless steel, Saran wrapped tubes, headers with nozzles, or just perforated pipes. Blowers may be centrifugal or positive displacement types designed to operate at heads of 8 to 10 psi, though this is affected by the depth of the tank, usually 10 to 20 ft. Tanks may be as wide as 30 ft or more. When fine bubble diffusers are employed, the air should be filtered.

There are several types of mechanical aerator, the surface, draft tube with turbine, and turbine with air injection being most common. The surface aerator is a paddle wheel or brush which is vertically suspended and partially submerged. The draft tube type is a vertically suspended propeller, usually mounted in a vertical tube. The waste is pulled up or down in the tube, effecting mixing and aeration. A modification consists of a vertically suspended turbine equipped with a sparge ring for dispersing diffused air. Surface aerators may be of the updraft or downdraft type. Mechanical aerators are rated on the basis of pounds of oxygen input per horsepower-hour, ranging from 3.5 to 4.5 for aerators varying from 1 to 150 hp.

The microorganism culture is suspended in the waste, and since it is recycled, a final settling tank has to be provided. There is an optimum solids content to be maintained in the aeration tank, and consequently the rate of return of sludge from the final settling tank is controlled.

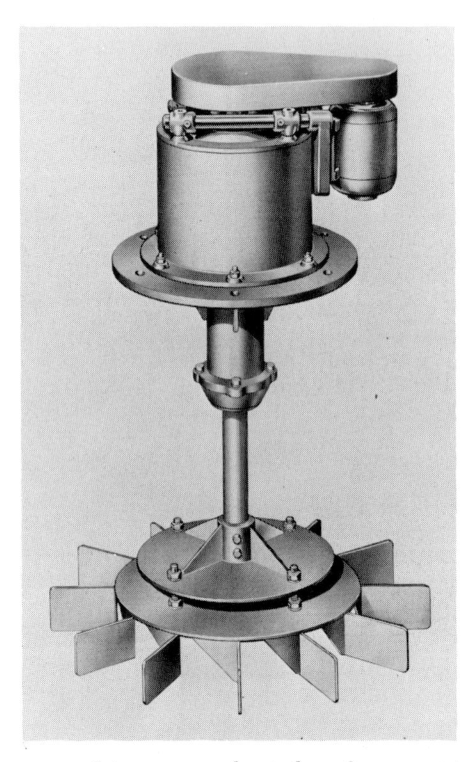

Fig. 14 Walker process Intensaer mechanical surface aerators are the result of intensive research and development. Full-scale tank operation, with scuba divers making underwater tests, has resulted in one of the most efficient units available. (*Walker Process Equipment.*)

There are three modifications of the activated sludge process, generally categorized as "conventional," "extended aeration," and "contact stabilization." In the conventional process, the aeration period is from 2 to 6 hr. In extended aeration, it is up to 24 hr.

In the contact stabilization process, the sludge is aerated for an extended period of time before it is mixed with the incoming waste. The main waste stream is aerated in the presence of the sludge about 30 min. The process requires an aeration tank for the main waste stream, a sludge aeration tank, and a final settling tank.

Conventional activated sludge and contact stabilization processes are generally applied to clarified wastes and the extended aeration process to raw wastes. The extended aeration process produces very little sludge, and that produced is organically stable. It therefore has advantages over the other two methods, which dispose of appreciable quantities of excess, unstable sludge.

The design criteria include relations between BOD loading, tank volume, and aeration provided. For conventional activated sludge, the loading ranges from 25 to 300 lb BOD/(day)(1,000 cu ft). Diffused air requirements are 900 to 1,100 cu ft/lb BOD. For contact stabilization, the air requirement is 1,600 to 1,700 cu ft/lb BOD. The sludge aeration tank in contact stabilization should have a holding capacity of 2.5 hr based on average flow and the mixing or main waste stream aeration tank 30 min. In extended aeration plants, the air requirement is 1,500 to 3,500 cu ft/lb BOD. Loading varies from 12.5 to 25 lb BOD/1,000 cu ft.

REMOVAL OF DISSOLVED SOLIDS

While developed for demineralization of saline or brackish waters, desalination processes may be employed to remove dissolved organics as well as inorganics. The methods are costly compared with other waste treatment procedures and would not

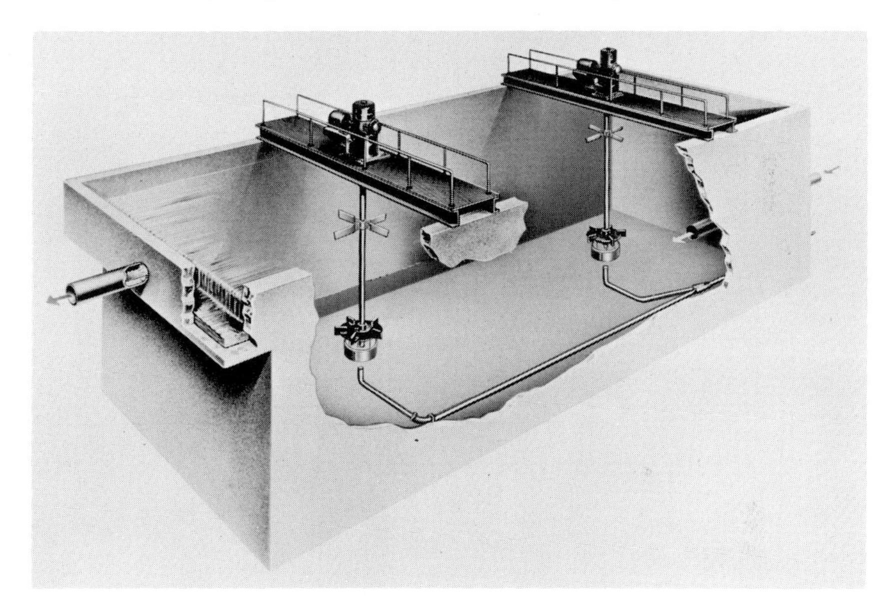

Fig. 15 The Dorr-Oliver aerator, a motor driven, high-capacity mixing unit that can be adapted to a variety of new or existing tank sizes and shapes.

be justified except where a very high degree of treatment would be required or if objectionable constituents could not be removed by any other method. An example of possible application would be to convert waste to potable or process water where no other source of usable water is available. It is necessary to have a solids free and—for some of the methods—bacteria free influent for efficient results.

The processes include distillation, electrodialysis, and reverse osmosis. In distillation, part or all of the water or waste is evaporated and subsequently recovered as usable water by condensation. The economics depend on an economical source of energy. The mechanics of the evaporation-distillation cycle are widely varied in practice, but three methods are in general use: multistage flash distillation, multiple-effect long-tube vertical evaporation, and vapor compression. To a limited extent, solar distillation has been employed to demineralize saline waters and could have application in waste treatment.

In multistage flash distillation the water is heated to about 200°F and flashed into an evaporator at a lower pressure. The steam produced is condensed to product water. The residual concentrated waste is flashed through additional evaporators, each at a lower pressure than the previous one, and each one produces product water.

The multiple-effect long-tube vertical distillation process involves a similar principle of successive evaporation. Each evaporator or effect contains long vertical tubes through which preheated influent falls, boiling as it falls. Vapor and hot water collect at the bottom of the evaporator. The hot water is pumped to the tubes in the second effect, where the process is repeated. A number of effects can be provided until efficiency reaches an optimum. Steam is used as the initial heat source, picking up the vapor, moving from one effect to the other. Condensate is collected from each evaporator.

In vapor compression distillation, the influent is evaporated under atmospheric pressure. The resulting vapor is compressed and is returned to heat the evaporator. The temperature difference between the compressed steam and influent in the evaporator makes this possible. The process has an advantage over other methods in that no cooling water is necessary to effect condensation.

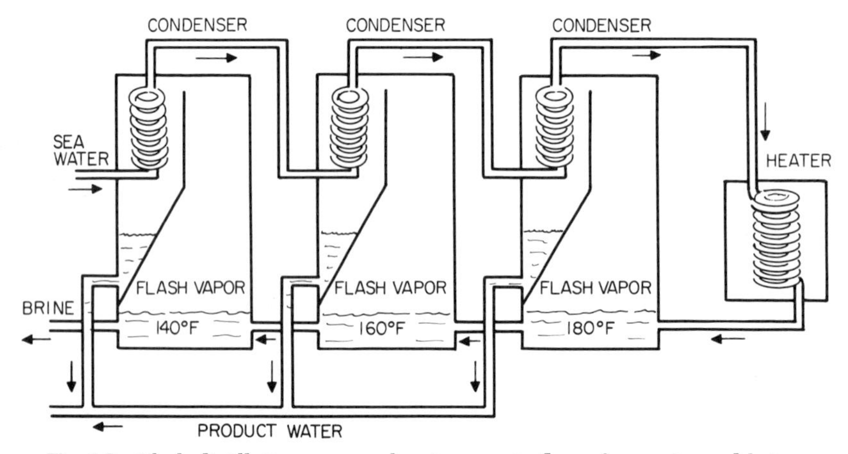

Fig. 16 Flash distillation process showing counterflow of seawater and brine.

The electrodialysis procedure consists of passing influent through alternating layers of cation permeable and anion permeable membranes that selectively remove dissolved solids from alternate compartments and concentrates from other compartments. The energy component is electricity, and high voltages are employed.

Reverse osmosis also uses membranes but of such character that only dissolved solids of certain molecular weights may pass through, depending on the "tightness" of the membrane material. This membrane "tightness" also controls the flow rate obtainable at a given operation pressure. Where membranes are used for secondary or tertiary treatment in waste treatment systems, capacities are of the order of 10 gal/(sq ft)(day) at feed pressures of about 50 psi. The membrane structure is such that it retains all solids having a molecular weight of over 10,000. This includes all oxygen consuming material (see Fig. 17).

In all reverse osmosis processes, there is a residual product of concentrated substance that requires disposal.

Ion Exchange

This is useful in recovery of process chemicals as well as for removal of objectionable constituents. The influent must be solids free. It is not competitive with other methods for demineralizing brines.

The mechanism is chemical, using resins that react with cations and anions. Two resins are employed which may be kept in separate receptacles or in the same (mixed bed). The receptacle is usually of steel, totally enclosed and operating under pressure. Water is passed through it as in a filter. The resin becomes exhausted after a period of time and must be regenerated or discarded. The cation resins are regenerated with acid and the anion type with caustic. The beds are operated at 6 to 8 gpm/sq ft of filter surface area.

The quantity of minerals that can be removed without regeneration is dependent on the resin capacity and the concentration of the influent. The units have to be controlled to be assured that the resin capacity has not been exceeded. This can be done automatically by employing a conductivity cell in the control circuit.

Fig. 17 A Dorr-Oliver ultrafiltration system at a Connecticut industrial plant treating sanitary waste flow to the tertiary level by removing 95 percent of the BOD and suspended solids. Unit shown in partial view has a designed capacity of 3,600 gpd.

Provision must be made for disposal of the products of regeneration, which will be more concentrated than the original waste. Rinse water may also offer a disposal problem.

Chemical Treatment

Many waste waters are too acid or too alkaline to be discharged directly to a stream, and neutralization is required. Others, such as chromium plating wastes, have constituents which may be removed by reaction with specific precipitants.

Neutralization or pH correction involves addition of acids or alkalies. This requires mixing of the added chemicals to the waste water stream.

Mixing for pH control requires a relatively small basin equipped with a means of agitation such as vertically suspended paddles or propellers. Air supplied by a compressor and dispersed through diffusers is sometimes used. The design of mixing basins is covered under Flocculation.

Acids, liquid caustics, or any solution may be proportioned into the waste stream by a diaphragm or other reciprocating pump. These are subject to control by a variable-speed motor drive or adjustment of the length of stroke.

Dry chemicals, such as lime, alum, and other coagulants, require gravimetric or volumetric feeders. These are usually equipped with a solution or slurry tank, so that fluidity is achieved. The gravimetric feeders involve a weighing device, by means of which the chemicals are added in proportions by weight. Volumetric feeders measure chemical addition by means of the bulk of the material. The former is, of course, the more desirable for precision in treatment.

CHLORINATION

Most industrial waste streams do not carry pathogenic microorganisms. However, chlorination of such waste waters is widely practiced for other reasons—including odor abatement, BOD reduction, control of slime organisms, color reduction, and as a specific reactant.

Chlorine may be added in the elemental form, which is stored and shipped under pressure as a liquid in steel cylinders. It may also be fed as a hypochlorite solution, either made up as it is used with calcium hypochlorite and water or purchased as a sodium hypochlorite solution. A third form used is chlorine dioxide.

Elemental liquid chlorine is fed by means of a device which meters and controls the amount of chlorine discharged from cylinders as a gas. Vacuum type feed is preferred to direct pressure feed for safety reasons. In both, chlorine flow is regulated by adjusting the pressure differential across a fixed orifice or by varying the size of the orifice. In vacuum feed, the vacuum is created by a water injector, which in addition to drawing the chemical from the cylinder places it in solution. At 20°C, 1 volume of water dissolves 2.3 volumes of chlorine gas.

The chemical is highly toxic, and since leaks may develop, the equipment should be housed in a closed-off separately ventilated room, preferably equipped with an alarm system. The room should be heated to a temperature of 70°F. Rate of withdrawal or evaporation of the liquid is limited by temperature. The rate may be increased by installing special heating devices.

Calcium hypochlorite is obtainable in powdered, granular, or tablet form and contains about 70 percent available chlorine. It is batch dissolved in solution tanks, usually of plastic. Sodium hypochlorite may be purchased in concentrations up to 15 percent available chlorine and may be fed in such concentrations directly. Both types require diaphragm or positive displacement type proportioning pumps for metering the solution into the waste stream.

Chlorine dioxide, prepared from sodium chlorite and chlorine, is a gas and is fed by gas metering equipment of the water injection type. Stable chlorine dioxide solution may be handled directly without dilution by chemical proportioning pumps.

PROCESS INSTRUMENTATION

Particularly where chemicals are applied, economies can be effected by installation of control equipment. Labor costs may be reduced, and chemicals are fed precisely in the desired proportions regardless of variation in flow or concentration of the waste stream. The basic control system consists of a primary device, sensor, transmitter, and receiver.

Primary Devices

These produce a pressure differential and afford a means of measuring flow. A venturi tube, venturi nozzle, Dall tube, concentric orifice, Parshall flume, weir, or pitot tube will suffice. For control of a plant operation, or a portion of it, the primary device is placed in the main waste stream and acts as a rate of flow sensor.

Sensors

Other than rate of flow, pressure, level, or the physical or chemical characteristics may need determination for controlling a process. Liquid level is sensed by floats, electrodes, temperature sensitive probes, or compressed air systems.

Floats involve a mechanical linkage to a cable and drum, the shaft of which is geared to a transmitter to actuate mercury switches for pump or valve operation or to produce a signal. Water containing any appreciable amount of dissolved minerals will transmit electric current. Hence, when an electrode becomes immersed in it, the flow of current is detected. In pump control, two are used, one at the bottom of a vessel for turning a pump on and the other at the top for turning it off. Temperature probes sense the differential between air and liquid temperatures. Compressed air systems involve producing bubbles which form an air pocket in a pipe or bell. Back pressure in the air line caused by level increase is sensed by bellows which trigger switches.

Pressure sensing devices are usually bourdon tubes, bellows, or diaphragms. These are placed directly on the line without an intervening primary device. Bourdon tubes may be used in gages or may involve a mechanical linkage to position the trip arm of a transmitter, producing a signal which varies directly with pressure. Diaphragms and bellows operate on a similar principle.

Turbidity, temperature, pH conductivity, dissolved oxygen, oxidation reduction potential, and other characteristics may be continually determined by means of meters designed for that purpose. In fact, any chemical characteristic which can be detected by colorimetric means may be measured automatically for recording and control purposes. Turbidity meters and color sensitive devices involve automatic sampling. Others use electrodes or probes.

Transmission and Reception

The sensor or primary device measures a variable quantity which has to be transmitted to the point of observation or use, where it is received by an appropriate component. Electric current, electronic signals, or pneumatic signals may be employed. Pneumatic systems are limited to a transmission distance of about 1,000 ft. These involve variation in air pressure. Pressures varying from 3 to 15 psi are maintained in a closed circuit. An integrating device proportions the air pressure within this range to the variation in the measured quantity. The pressure in the circuit actuates a receiver, usually an indicating or recording meter.

Electric control systems use a dc wire system, and signals are sent as pulses, the duration of which is proportional to the measured quantity. Radio waves may also be employed with electronic transmitters and receivers involved. Here, pulse variation is transmitted as tones by AM (amplitude modulation) equipment. In addition FS, or frequency or carrier shift, may be used. The latter has the advantage of being capable of transmitting two signals on the same channel and of providing highly reliable transmission under adverse communication circuit conditions.

Supervisory Control

When the signals are transmitted to a central point for indication and recording and it is desired to control operation from that point, "report back" features are incorporated in the system, thus achieving automation.

With electrical transmission of pulses, scanning switches driven by electric motors are used in the receivers, at both the dispatching point and the central control station. As many as 28 control or advisory functions can be handled by a single pair of wires. With electronic transmission, similar principles are involved. Control is accomplished by arranging to have pumps start or stop or valves open or close by predetermined conditions. An examples is pH control. The pH is measured by an electrode system, and signals are transmitted to the control point. If operating to correct an acid condition, the receiver senses a pH below the desired level and starts an alkali feeder which functions until the predetermined level is reached.

A further step is computer control, by which variations in flow and quality characteristics, for example, are logged. The computer is programmed to determine chemical dosage based on these variables, and it actuates the necessary controllers to make the adjustments necessary. To help the central station operator keep up with the plant functions, indicating and recording gages are employed, along with lights to indicate "on" and "off" conditions of the various pieces of equipment and alarms to indicate malfunction. Automatic control systems have to be designed on a "fail-safe" basis to avoid damage, and any automatic operation must have corresponding manual overriding switches.

The various indicating lights and gages may be mounted on a panel incorporating a flow diagram of the plant units. This helps the operator determine at a glance the operating condition of each component.

OTHER EQUIPMENT

Package Plants

Where a waste is amenable to biological oxidation, complete plants may be obtained, factory assembled and ready for installation, requiring that only piping and electrical connections be made. These employ extended aeration, contact stabilization, and trickling filter units. Comminutors, aeration and settling basins, or filters are contained in one cylindrical or rectangular shell, usually of cathodically protected steel. Around 100,000 gpd is the usual capacity limit for factory assembled units. However, field assembled units may be obtained in capacities up to 1,000,000 gpd. These are generally designed for application to domestic wastes but can be sized for industrial effluents, based on flow and BOD characteristics.

Tertiary Treatment

This is a term given to advanced waste treatment processes, derived from the premise that primary treatment consists of suspended solids removal and disposal and secondary treatment of oxidation of dissolved organic matter. A tertiary treatment unit would process the effluent from a secondary treatment unit. Biological oxidation and solids removal processes in current use ordinarily cannot produce effluents with less than 10 mg/l BOD and 20 mg/l suspended solids, and the reduction obtained from raw characteristics is seldom more than 95 percent. Thus a raw waste with values in these parameters on the order of 1,000 or more would still contain 50 mg/l in the effluent from these processes.

The most frequently used tertiary treatment device is the stabilization pond, possibly because it is the least expensive to construct. Others consist of intermittent sand filters, rapid sand filters, microstrainers, activated carbon or diatomite filters, membranes (electrodialysis or reverse osmosis) and ion exchangers. The effluent may also be applied to land by sprays. A number of industries use spray irrigation as the sole method of disposal, but appreciable quantities of land are required.

The design of stabilization ponds is covered under another section. As a tertiary treatment device the criteria are the same. However, there is a problem with the profuse algae growth that develops, so that immediate reuse of the effluent is not feasible. Intermittent sand filters are designed on the basis of about 0.1 million gal/(acre)(day) or 50 to 100 lb BOD/(acre)(day). They are shallow beds of sand, around 3 to 4 ft deep, equipped with drainage facilities. Their function is both biochemical and physical. The surface layers of the sand become coated with a gelatinous film of oxidizing and nitrifying bacteria, and the rest of the sand bed acts as a strainer. More than one bed has to be provided to permit alternating them. During periods of nondosage, air enters the spaces between the sand grains to maintain aerobic conditions. The filters are flooded by means of a dosing tank containing a siphon or by a valved distribution box. A concrete apron is provided to protect the bed area around the inlet from becoming eroded.

Design and operating criteria for intermittent sand filters are well established, but those for rapid sand filters as waste treatment devices are largely experimental. Coagulation of the waste before application is essential. Some success has been realized from the use of mixed media beds. These are filters in which the media layers are graded coarse to fine from the top of the filter to the bottom, or vice versa, with materials of selected specific gravity, so that when the filter is backwashed, they will assume the original gradation. Application rates can be higher than those recommended for filtration of water, up to 5 gpm/sq ft. The mechanism of purification is primarily solids removal, with little reduction of dissolved organics.

Likewise, solids may be removed by microstrainers. These are drum screens equipped with a stainless steel medium of low porosity, 10 to 40 mesh. They are washed with water sprays, so that operation is continuous. Diatomite filters may be employed. They consist of wire wrapped or ceramic elements, which are coated

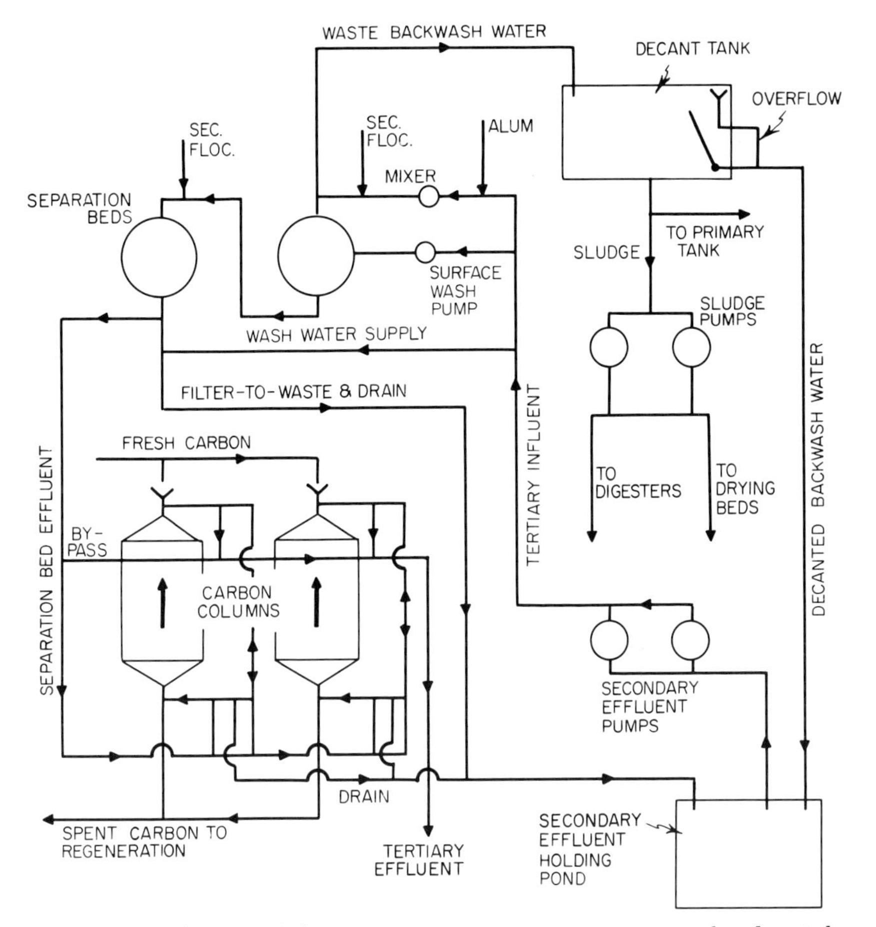

Fig. 18 Flow diagram of the tertiary sewage treatment process employed at Lake Tahoe.

with diatomaceous earth. They may be enclosed in pressure vessels or open tanks. In the latter case, a vacuum is applied. In operation, the elements are precoated with diatomite using a slurry feeder, and the filter aid may be added continually during filtration. Backwashing is accomplished by releasing pressure in pressure vessels or vacuum in open tanks. The design basis is on the order of 1 to 2 gpm/sq ft of filter surface.

Removal of dissolved organic matter from an effluent containing very little suspended solids is possible by adsorption in a granular activated carbon filter. The carbon may be periodically reactivated in a furnace.

SELECTION OF PROCESSES

The states have adopted stream quality standards which require at least the equivalent of secondary treatment of industrial effluents for most waters. These are based on criteria promulgated by the Federal Water Pollution Control Administration, U.S. Department of Interior, pursuant to the provisions of the Federal Water Quality Act of 1965.

The minimum standards adopted by Indiana for the Maumee River Basin are typical:

1. Free from all substances attributable to municipal, industrial, agricultural or other discharges that will settle to form putrescent or otherwise objectionable sludge deposits.
2. Free from floating debris, oil, scum and other floating materials attributable to municipal, industrial, agricultural or other discharges in amounts sufficient to be unsightly or deleterious.
3. Free from materials attributable to municipal, industrial, agricultural or other discharges producing color, odor or other conditions in such degree as to create a nuisance.
4. Free from substances attributable to municipal, industrial, agricultural or other discharges in concentrations or combinations which are toxic or harmful to human, animal, plant or aquatic life.

In addition, higher standards apply to streams at points where water is withdrawn for use as a potable water supply. These specify a maximum bacterial count, threshold odor number, content of dissolved solids, concentration of radioactive substances, and chemical constituents of a toxic nature. Standards applied to points of withdrawal of water for industrial processes indicate a minimum dissolved oxygen, give a pH range, and specify a maximum temperature as well as dissolved solids content. There are further standards for maintaining aquatic life, for recreational water quality, and for water used for agricultural and stock watering.

Means of implementation have also been adopted in all states. These usually consist of notification of polluters and establishing a timetable for compliance.

All states have classified waterways on the basis of use, but in general, the intent of the 1965 Act, as interpreted by the Federal Water Pollution Control Administration, is to require secondary treatment by municipalities and the equivalent by industries.

"Equivalent secondary treatment" has not been defined, but the term implies solids removal, removal of all floating material, and production of a clear effluent with little or no dissolved organics or toxic constituents. In addition the pH would be in the neutral range. Specific interpretation should be sought from the state regulatory body before facilities are planned. Most states require the submission and approval of plans before construction of a waste treatment facility, and many require a permit for waste water discharge to a stream.

In making any plans for treatment, an industry should take into consideration future as well as present stream uses. The FWPCA has insisted that an "antidegradation" provision be made a part of all state standards on the order of the following:

> Waters whose existing quality is better than the established standards as of the date on which such standards become effective will be maintained at their existing high quality. These and other waters of a State will not be lowered in quality unless and until it has been affirmatively demonstrated to the State water pollution control agency and the Department of the Interior that such change is justifiable as a result of necessary economic or social development and will not interfere with or become injurious to any assigned uses made of, or presently possible in, such waters. This will require that any industrial, public or private project or development which would constitute a new source of pollution or an increased source of pollution to high quality waters will be required, as part of the initial project design, to provide the highest and best degree of waste treatment available under existing technology, and, since these are also Federal standards, these waste treatment requirements will be developed cooperatively.

A strict interpretation of such a mandate would be that as the quality of a stream becomes improved, improvements are going to have to made in treatment plants. Thus, it is conceivable that additional treatment methods may have to be employed.

Using present technology, for example, an organic waste containing suspended solids might have to be treated with the following steps: clarification, solids disposal, biological oxidation, chemical precipitation, sand filtration, and activated carbon filtration.

ECONOMIC CONSIDERATIONS

Designing for Future Expansion

As manufacturing operations increase, so will the waste water flow. Furthermore, there is a likelihood that regulations on pollution control and their enforcement will become increasingly stringent. The treatment plant must therefore be designed initially with these factors considered, and the designer should develop a long-range plan that will permit orderly expansion of the plant.

One problem of the future is the availability of land; and land availability frequently limits manufacturing operations. By considering these factors jointly, it should be possible to arrive at the treatment plant area required to match the ultimate expansion of the manufacturing operation.

The treatment plant should be designed for construction in stages, for orderly development with anticipated increase in waste flow. If the quantity of waste water is relatively small, stage construction could be handled by package plant erection, for example, with each stage calling for one or more plants. Where field erected facilities are required, plans should be made to duplicate at least primary and secondary treatment units with each stage. With many wastes, the solids handling phase is the most expensive. Consequently, these facilities probably should initially be made large in proportion to the remainder of first-stage units.

One trick that has been used is to take advantage of the various modifications of the activated sludge process. The initial plant could be of the extended aeration type, in which the retention period for aeration is 24 hr. As waste flow increases, the process can be converted to other modifications that require about the same area and few structural changes. Parallel conversions are from plain settling to coagulation by chemicals, and starting with a trickling filter plant and later supplementing it with an activated sludge unit.

Plant design should incorporate site arrangement and hydraulic gradients that permit tertiary treatment devices to be added when and if needed.

Equipment Life and Materials of Construction

As with any facility used by industry, the investment in waste treatment equipment must be weighed against the anticipated useful life. As in-plant processes change, so will waste treatment procedures. Thus a treatment facility designed to last 30 years, as in the case of domestic waste treatment, may become obsolete. Knowledge of past histories of other industries using like manufacturing processes and of research likely to change manufacturing methods should help.

Equipment life also depends on the use to which it is put. An electric motor rated at 100 hp will last longer if operated at 95 rather than 115 hp. Some types of pumps operating continually will have a longer life than ones operated intermittently. Consideration should also be be given to the corrosiveness of the waste. Cast iron and steel are not as resistant as clay or plastic. Polyvinyl chloride and similar plastics, where they can be applied, are probably the materials of preference where corrosive substances are handled. The use of special coatings and application of cathodic protection on submerged surfaces will greatly extend the life of steel equipment and steel and concrete tanks. *Manual of Practice* 17, published by the Water Pollution Control Federation, describes in detail the application of various type of coatings.

TABLE 2 Estimated Life of Waste Water Treatment Facilities*

Item	*Useful Life, Years*
Buildings	50
Grit chambers:	50
Rotary grit collectors	30
Chains and flight collectors	10
Screens:	
Manually cleaned	40
Mechanically cleaned	30
Grit and screenings incinerators	20
Primary settling tanks:	50
Rotary sludge collectors	35
Chain and flight collectors	15
Trickling filters:	50
Rotary distributors	25
Activated sludge units:	
Aeration tanks	50
Blowers	25
Intermittent sand filters	40
Secondary settling tanks:	50
Rotary sludge collectors	35
Chain and flight collectors	15
Chlorinators	15
Sludge digesters	50
Sludge drying beds	20
Vacuum filters	25
Sludge incinerators:	
Multiple-hearth	35
Traveling grate	20
Sludge driers:	
Rotary drum type	20
Flash driers	20
Electrical equipment:	
Transformers	30
Switchboards and switchgear	25
Motors and generators	25
Underground conduits	75
Mechanical equipment:	
Centrifugal pumps	25
Vacuum pumps	25
Blowers and compressors	25
Gas engines (slow-speed)	25
Conveyors:	
Belt	15
Spiral	15
Channels and pipelines:	
Concrete	75
Cast iron	75
Pavements:	
Asphalt and concrete	20
Macadam	15
Gasholders	75
Chimneys, brick or concrete	50
Fences, chain link	25

*C. E. Keefer, *Public Works*, vol. 93, p. 7.

Table 2 prepared by C. E. Keefer gives the estimated life of facilities used in domestic waste treatment. Another source of information is the Internal Revenue Service, which has published tables of useful life of depreciable property.

Chapter **25**

Equipment Guarantees, Performance Testing, and Startup Procedures

ORLAN M. ARNOLD

Vice President, Ajem Laboratories, Inc., Livonia, Michigan, and Environmental Research Consultant, Grosse Pointe Park, Michigan

INTRODUCTION

The increasing concern and accelerated activities in air and water pollution control on federal, state, municipal, and public levels have been prompted by the realization of the belated beginning of seriously controlling the individual pollutants in air and water systems. At the same time there has been recognition of the greater concentrating of population in limited geographical areas, and of the fact that the higher level of economy in the country has produced a greater demand by the population for more worldly goods and that these demands result in greater industrial production and the accompanying evolution of more airborne and waste water by-products and pollutants. There has also been a recognition that many industrial by-product pollutants, formerly ignored, are of serious concern to the health and welfare of the people living in that environment as well as to the preservation of property and to the protection of the wildlife environment.

OBJECTIVES FOR AIR AND WATER POLLUTION CONTROL, EQUIPMENT GUARANTEES AND AGREEMENTS

From field experiences and observations in the rapidly expanding air and water pollution control fields, it is recognized that progress has often been impeded because of several factors and existing conditions, including:

1. Lack of understanding of requirements of new air and water pollution control regulations
2. Incomplete and misleading information on applicability and anticipated performances of air cleaning and water treatment equipment
3. Failure of purchaser to evaluate fully and properly the needs and magnitudes to consider for specific air and water pollution control problems
4. Frequent disappointments in performances of installed dust and chemical fume removal systems and of water handling and treatment equipment.
5. Misunderstandings between purchasers and suppliers on such equipment
6. Omissions from purchase contracts of necessary specifications on equipment, of process operating conditions, of guarantees on qualities of performances, of assurances for sustained control of air outlet loadings and quality of discharge water, of individual and joint responsibilities of supplier and purchaser, and of related stipulations
7. The ignoring or failure to recognize the paramount importance and use of accurate, reliable test data, particularly on performances of dust and fume collection equipment and of water treatment processes and facilities in the evaluations and careful selections of such methods and devices, in the proving of satisfactory performances to the purchaser's management on the merits of pollution control investments and to the enforcing agencies of different air and water pollution code regulations
8. Neglect of contract statements setting forth nature and extent of specific liabilities in relation to malfunctioning of equipment and to deviations in manufacturing and negligence in maintenance
9. Disregard of relations of processes and conditions to liabilities resulting from possible present and future personnel injuries or death and/or property damages
10. Limited attention to safety methods and equipment
11. Omission of contract statements on responsibilities of parties concerned, in possible patent infringement lawsuits involving manufacture and/or use of equipment bearing patent coverage

With the awareness of these pitfalls from lack of understanding, omissions in contracts, failures to obtain and use actual and reliable test data, and carelessness in defining required performances in purchase of equipment, the aims and intended uses of this chapter are to bridge these gaps by alerting those concerned in the air and water pollution control programs of these shortcomings so as to prompt the taking of proper steps and to ensure success in the air and water pollution control efforts with satisfaction to all involved parties.

GUIDELINES TO ESTABLISH CONTRACT RESPONSIBILITIES AND AGREEMENTS

1. To assist the purchaser and others concerned in recognizing and handling the numerous factors involved to attain the goals and set regulation requirements in air and water pollution control
2. To alert the parties of concern of the pitfalls in evaluating air and water pollution control equipment and in the determination of their true performance abilities
3. To provide a mode of understanding that will assure the purchaser a soundness and permanence of investment in air or water pollution control facilities
4. To point out the significances of the marked decreases in limits of individual pollutants in air emissions or waste water discharges with the new or revised code regulations

5. To enumerate and suggest the different elements that constitute a complete and realistic contractual agreement in respect to starting time, place, and dimensions and completion time for installation of air and water pollution control equipment.

6. To provide guidelines for establishing and defining guarantees and assurances between suppliers and purchasers

7. To emphasize the obligations of both the suppliers and the purchasers to realize and assure satisfactory equipment applications and operations

8. To demonstrate the importance of accurate, reliable test data on which to evaluate performance of air and water pollution control installation

9. To provide guides and suggested interrogations to evaluate and assure that a testing agency is capable and will procure dependable accurate test data on air emissions and water discharge pollutants

10. To encourage the inclusion of statements of liabilities carefully defined as to scope and extent in relation to supplier and to purchaser

11. To prompt the understanding, specifications, and use of adequate safety measures and facilities including interconnect control devices

12. To urge incorporation of contractual definitions of time limits of supplier's responsibilities in respect to components, materials, workmanship and performance of specific air and water pollution control equipment

13. To recommend inclusion of clauses in contract of responsibilities of purchaser bearing on performances and care of the air or water pollution control equipment conditional on purchaser maintaining prescribed adequate utilities in electrical power, water gallonage supply, proper removal of collected wastes, on providing of competent maintenance personnel, in preventing excessive overloads beyond stated capacities and in avoiding introduction of new dangerous, toxic, corrosive substances that interfere with the contracted pollution control system.

AIR POLLUTION CONTROL EQUIPMENT—AGREEMENTS, GUARANTEES, PERFORMANCES, TESTING, STARTUP, AND MAINTENANCE TRAINING

While there is a common objectivity in air and water pollution control to have clean air to breathe and clean, safe water to use, there are wide differences in the factors, methods, and facilities to realize the clean air and clean water so vital to life and to industrial functions.

Since air and water are different in their properties, in the nature of their respective pollutants and contaminants, in their requirements for handling and treatment, in the separate grouping of regulatory codes, in the type of equipment for purification, and in the modes of sampling and testing, it is necessary to treat air and water under two separate sections in this chapter.

Influences and Effects of Accelerated Air Pollution Control Programs

The accelerated air pollution control activities in the form of new requirements of more rigid regulation codes and of increased scope in recognition of individual pollutants have presented many new facets as well as demands of assurance from an increased number of concerned parties inclusive of the public. The influences and effects of the accelerated air pollution control programs have marked effects on the requirements, methods, guarantees, and procedures in air pollution control, as evidenced in being far-reaching in respect to:

1. Increased consciousness and concern of the public in air pollution control

2. Organization from civilian groups to government agencies for regulation and control of pollutants in the air

3. Use of different communication media, radio, television, newspapers, magazines, and technical journals, to stimulate the concern of the public in air pollution control, to bring pressure on government agencies for taking action, and to take measures against the industrial sources producing the major part of the polluted air.

4. Accelerated establishment of much more rigid regulations and codes in air pollution control through legislation on municipal, state, and federal level
5. Reversal of earlier nonresponsive attitudes of industry in taking steps to curtail air pollution emissions
6. Establishment of courses in universities in industrial hygiene and health for training personnel in the fields of air pollution control
7. More realistic and rigorous enforcement of air pollution control regulations
8. Recognition that selection, purchase, installation, and assurances of satisfactory performances of air pollution control equipment must be made on sound basic facts

Uncertainties in Limits of Specific Air Pollutant Outlet Loadings with Changing Code Regulations

The rapid adoptions of air pollution control regulatory measures have called for five- or more-fold decreases in the emission limits of airborne pollutants. This reduction is well illustrated in metallurgical processes, in which over a few years the outlet emissions have been reduced in code regulations from 0.5 or 0.3 grain particulate/scf dry air to 0.05 grain or less of particulate per standard cubic foot of dry air. A similar reduction, such as from 5 to 1 grain solid particulate matter/1,000 scf dry gas in allowable outlet loading of solid particulate matter from paint spray booths, has been experienced. Another dangerous pollutant, carbon monoxide, which has been allowable in the past up to 100 ppm or in undefined quantities, has been progressively reduced in limits of concentration to 50, to 30, and to 20 ppm in different regulations, with present consideration of reduction to 10 ppm in view of recent studies of physiological effects. This has been brought about by recognition of the seriousness of many of the pollutants formerly ignored or taken for granted and the realization of the increasing quantities of emissions with industrial expansions. These rapid changes in limits of specific pollutants have been accompanied by expressed uncertainties as to whether the new limits are permanent or will be further changed to lower numerical values. This in turn influences the thinking in terms of equipment selection, performances, adaptability, and assurances.

In the past a general knowledge of the presence, at least, of certain pollutants was the extent of concern. With the uncertainties as to where the limits of certain pollutants in regulations would eventually be established, the questioning of where the enforcement of limitation of pollutants would reside and primary lack of reliable technical information for positive control of the individual pollutants created the lack of understanding and clarity of the air pollution control requirements.

Why Certain Industries Resist Investing in New Air Pollution Control Equipment

A number of factors have been responsible for passiveness and delays of industries in installing air pollution control equipment. Among these factors are:

1. From the economical point of view the argument has been that the air pollution control equipment is not a direct productive investment item. This thinking has more recently been reversed, since in many cases it is found that the proper air pollution control equipment can make money such as in salvaging raw materials, protection of personnel, buildings, and equipment, and raising the corporate image with the public.

2. Another discouraging influence has been the unfortunate experiences in many industries with installation of inferior or inappropriately applied dust and fume collection equipment. Many times this is reflected in the selection of equipment based on the inaccurate claims of the seller of the air pollution control equipment that is found later in field testing not to give satisfactory and acceptable performances.

3. Another factor has been the lack of reliable test data in understanding the specific air pollution problems. This, many times, is because of failure to recognize the whole of the air pollution control requirement in the nature of the pollutants, their relative quantities, and the ranges in size of the particulate matter. When pollution particulate is to be removed, the definition "total particulate" means material, except

uncombined water, that exists in a finely divided form as a liquid or solid at standard conditions. It should be observed that there are distinct differences between total solid particulate matter and total particulate, as the latter includes the discrete solid particles and the finely divided liquid droplets that may be all of liquid phase or may contain submicron particles in the liquid finely divided form.

4. Lack of sound test data to evaluate the equipment accurately for specific air pollution control application. One of the more serious restraints and also causes for misleading judgment in selecting proper air pollution control equipment has been the lack of accurate, reliable test data on the performance of the individual air pollution control devices. This lack of sound facts and data has several reasons:

a. Attempted evaluations with inaccurate test methods and equipment

b. Careless performance of sampling and related pollutant measurements

c. Use of equipment and test methods that account for only a portion of pollutants such as failure to capture all the ultra-micro-sized particulate matter

d. False claims on air pollution control equipment performances based on capture of a portion of the pollutant

e. False conclusions and statements in respect to claims for performances of air pollution control equipment in new application as based on data obtained in another field in the removal of different types and particulate size pollutants

f. False claims based on sampling and tests that account for only a portion of the pollutants rather than the total

It can be easily seen how inaccurate and wrongly used data can result in confusion and also would be misleading to the purchaser of air pollution control equipment. Sound guarantees on air pollution control equipment performances can be made only with full factual information and data on the specific application.

5. Uncertainty of code regulation requirements. Another hindrance in the progress of air pollution control has been the uncertainty in minds of officials responsible for taking steps in air pollution control in industry as to what limits of individual pollutants will be in the ultimate code regulations. This uncertainty has been brought about by the rapid changing of these higher limits to lower values during this accelerated era for air pollution control. Besides being due to lack of accurate, reliable data, this uncertainty also comes from differences in municipalities and states and from questioning on federal levels as to where these limits will be set in magnitudes or whether they will be changed in the future. This has prompted many officials in industries to question, when they buy a certain piece of equipment, whether it will meet presently indicated or defined conditions for a period of years or if there is a way in which this same air pollution control equipment can be modified or improved to meet even more severe restrictions in the limits of emitted pollutants in the future.

6. Failure to recognize changing times and to accept responsibilities. Another set of influences retarding the accomplishment of air pollution control has been the failure to recognize that, with an increasing economy with its higher production and more concentrated population centers, the air pollution combined with effects of atmospheric conditions has grown more rapidly than had been anticipated. Another retarding influence has been resistance to accepting the responsibility of attacking the air pollution at its source rather than just let nature in its ambient way take care of the airborne pollutants without regard of effect on the neighboring environment. Fortunately the latter attitude in industry has been changing to acceptance of responsibilities to handle the pollutants at the source and to contribute to the attainment of a healthful environment.

7. Ambiguous citations by code regulation enforcement agencies for infraction of air and water pollution control laws. There have been a number of ambiguities in the citations given by regulatory air pollution enforcing agencies. In the transition from a lax state in control of air pollution to more rigid controls on limits, it is natural to expect some irregularities. Part of these ambiguities have been caused by a misunderstanding in several of the facets of the air pollution control programs. They will be treated in further sections of this chapter. At this point it may be stated that the context of the code regulations must be properly understood, that ac-

curate, reliable test methods are the only way by which to evaluate the emissions and to enforce their application. Furthermore, a basic requirement should be that all communications, particularly those dealing with tests, should be described and written in complete form for use by both enforcement agencies and the industrial organization receiving a citation.

8. Unqualified, untrained personnel assigned to handle pollution control needs. With accelerated programs for air pollution control, it is unfortunate that many persons are catapulted into jobs in air pollution control capacity in a company who, though they may be conscientious, are unqualified and untrained for such work. This has contributed to misunderstanding, retardation of the program, misjudgment in type of equipment best suited for an application, and lack of understanding of regulations and of dealing with enforcement agencies. This situation can be corrected. Any person assigned to such a responsible position should be encouraged or required to take special short courses, at least, in universities or schools offering such instruction and to attend meetings and seminars that are instructive in these fields.

Parties Concerned in Air Pollution Control Equipment Guarantees

In consideration of guarantees and assurances of performance of air pollution control equipment, several parties are actually concerned:

1. Purchaser of the air pollution control equipment. The purchaser is naturally concerned in having positive assurance that his investment will serve the purpose of removing the pollutant to satisfy a number of factors, not only of environmental health, recovery of materials, and handling of collected pollutants for disposal, but also of meeting regulations defined by the air pollution control code regulations for that area.

The purchaser is also involved in any agreement not only for payment for the equipment but for provision of space, facilities, and assurance of proper maintenance of the installed equipment.

2. Supplier of the air pollution control equipment. The supplier of the air pollution control equipment is deeply involved, because it is basically his responsibility to produce, install, and prove to the purchaser, enforcement agencies, and others concerned that the equipment will do the job in removing the individual specified pollutants from a specific industrial process. It is the supplier who must give positive assurances of making the equipment perform to the stated degree recognized in a purchase agreement.

3. The air pollution regulatory enforcement agencies. Health and industrial hygiene enforcement agencies are involved, as in most cases approvals or permits must be obtained from these authorities for the proposed dust or fume collection system to be installed as based on guarantees of performances for the specific equipment in the particular application. Likewise, the enforcement agencies are concerned with even field testing, if deemed necessary, to determine after installation of equipment that the air pollution emissions are being reduced in conformity with the code regulatory requirements and also at later times that the equipment is sustaining the stipulated performance in removal of the pollutants.

4. Interested observers and environmental inhabitants. Naturally the civilians living in the environment are concerned with whether the installation will give clean, healthful air in their environment. This concern of civilian groups is demonstrated by the numerous local residential committees organized to serve as vigilant watchers for observing and demanding the the pollution be brought under control. Some civilian organizations are equipped with apparatus to register evidences of violations on pollutant emissions.

It may be seen that a number of parties are involved when guarantees on air pollution control systems are written, accepted, observed, and checked for performances by accurate test methods.

Essential Phases to Establish Equipment Guarantees

In the past and through the present the pattern of agreements to satisfy business and legal counsels in the purchase and contract for installation of pollution control equipment has been the use of a standard purchase order and simple acknowledgment with the names and addresses of purchaser and supplier, the location of site for installation, the prescribed times for starting and completing installation, the terms of payment, and the inclusion of brief descriptions and specifications such as volume capacities and possibly a few layout drawings. Recently technical and legal counsel and others have recognized that in the fields of air and water pollution control there was greater need for depth and scope of assurances, of guarantees on performances based on trustworthy accurate data, and of defined responsibilities of both supplier and purchaser in installation and operation of equipment. Likewise, in the engagement of engineering or consulting firms more depth of assurance, definition of services, qualifications of personnel, and requirements that recommendations be backed by sound facts and technical data should be incorporated in the related contract.

The new trend for "turnkey" installations makes it all the more imperative that full understanding, specifications, guarantees, assurances, and responsibilities be set forth in the contract.

With the complexities and variations in factors involved in purchase and installation of air pollution control equipment it is impossible to prepare a universal contract format that would cover all the elements of guarantees, specifications, and liabilities for all cases. Furthermore, space in this handbook would not permit contract forms for all the cases and their individual conditions. It remains for the technical staff and legal counsel of parties concerned to incorporate the different elements that define and fit the specific case.

Several phases are concerned in guarantees for air pollution control equipment. In fact, agreements require assurances from both the purchaser and the supplier of the equipment. The different elements of guarantee and contract assurances involve the following phases.

Selection and quality of components Assurances are important and necessary on the selections and qualities of the components to be used in construction of the air pollution control equipment because of the variation of conditions of operation, environmental factors, and properties of air pollutants that are to be handled and removed from the air. The equipment is expected to perform from such an investment for long periods of time; hence full definitions of the specifications for components will result in a better understanding between the supplier and the purchaser. Examples could be cited where misunderstandings occur when certain types of pollutants such as corrosive chemically active substances to be experienced are not identified and given by the purchaser to the supplier, or changes in the industrial process result in the formation and presence of the new active or corrosive chemical pollutants. This has been costly and in certain cases has limited the life of the air pollution control equipment because proper structural materials resistant to the specific corrosion and abrasion were not used.

Workmanship While warranties on workmanship are common in most contracts, in the case of air pollution control equipment involving high investment expenditures this phase should be well defined in any contractual arrangement between a supplier and a purchaser of the equipment. This may involve careful definition of a procedure of installing as well as selection of certain components for resistance of temperature or active chemical environments or widely differing operating conditions. If there are mechanical parts such as fans, pumps, controls, or other moving components, the expected quality of workmanship should be defined to assure positive air and liquid movement and control to prevent leakage of air or water and to ensure protection against corrosion.

Time of delivery Another point of definition normal in the contractual arrangement is understanding and specification of time of delivery of the air pollution control equipment and specific time when the equipment will be ready for operation. This understanding and agreement on time of delivery becomes of great importance to the purchaser where time limits have been given by air pollution control enforcement agencies to bring an emission under control or when a time schedule for production is to be met. This also reflects understanding of arrangements committed by the purchaser for location, dimensions, and availability of space for installation and provision of facilities such as utilities for the installation proper.

Equipment installation Another phase which should be very carefully defined in the contractual statements bears on installation. This may involve the purchaser in preparing the space and the foundation work and bringing the necessary services to the site. If the supplier is to install the foundation, the purchaser should be committed in the contract to furnish all available information on ground structure and engineering data on neighboring structure. Along with this should be a definition of the responsibilities specifically of the supplier and the purchaser as to what individual preparations will be performed by each party and when the specific step or installation is to completed.

Guarantee on performances of the air pollution control equipment While this subject will be treated more extensively in a further section, one of the primary parts of the contract should be specific on the performance of the equipment in removing the specified air pollutants involved in the function of the installation. The specific pollutants to be captured should be individually named and the respective limits of outlet loadings from the equipment defined. For the assurance to the purchaser of the equipment, such guarantee should include specifications as to the employment of accurate test methods to be used in evaluation of the performance and also that the performance is to be sustained as verified by accurate tests taken over stated periods of time at a minimum of several months. This latter consideration is important inasmuch as some equipment is found to perform fairly well at the beginning but soon drops off rapidly so that the outlet loadings may become several times greater than what it was guaranteed to permit as an upper limit, and as a result the purchaser exceeds the code regulation limit of the outlet loading for the specific pollutant. When such decreases in efficiencies of performances occur, the purchaser becomes subject to citations of enforcement agencies for not meeting regulations of the air pollution control code.

Adequate and defined installation space The first five essential phases in contractual agreements as defined above involve primarily the supplier. There are several phases that involve more directly the obligations of the purchaser. One of these is the definition of available horizontal and vertical space and the relationships to other environmental equipment.

Adequate utilities Another condition to be included in such an agreement is a statement on specifications for adequate facilities including electricity, compressed air, and water where wet type dust or chemical fume collectors are to be utilized. Likewise, facilities must be defined for discharge of collected pollutants, whether it is of a dry collected solid or whether it is liquid or slurry form. While the supplier may not be contracting to handle the collected material, the installation and its functioning generally involve the supplier in respect to the way the collected pollutant is discharged that is agreeable to the purchaser for his planned further handling of the collected material.

Adequate maintenance One of the serious faults and difficulties in industry in recent years has been the lack of provision for adequately and properly trained personnel to maintain equipment. Since a limitation in industry has been the availability of enough persons assigned to do the maintenance work and frequent change of personnel occurs, a paragraph in the contract agreement on assurances by the purchaser or owner to maintain a proper number of knowledgeable servicemen for a specific installation is important. This will save, many times, breakdown of equipment, shutdown of equipment or of line production, and unnecessary expenses and ill relations with the supplier. Also, the proper maintenance is necessary in order that equipment, whether simple or complex, will be meeting the code regulations in respect to control of outlet emission. This phase also means that the purchaser agrees to follow the manufacturer's operation instruction. Proper maintenance of equipment also means longer life of the apparatus and savings to the owner.

Training of purchaser's personnel Another important part of the contract should be the agreement between purchaser and supplier that the supplier will give a training and instructional preparation to responsible personnel designated by the owner, particularly in the maintenance field, on the care, handling, and servicing of the equipment. The instructional training should include furnishing of pertinent drawings, circuit diagrams, and service manuals covering the specific equipment.

Adequate access platform, ladders, railings, etc. Since platforms and ladders are essential on air pollution control equipment, not only for servicing and inspecting the equipment but also for sampling of emission products in testing for performances, the contract should specify installation of such well-planned, well-engineered, and properly located and secured accessories. The provision for such accessibility has economical worth in that it allows easy checking and service as a preventative to costly maintenance and repair. It also simplifies the preparation for testing whether it be an agency of the purchaser, of the supplier, or of the enforcement agencies.

Use and installation of safety devices Another part of the agreement should specify and define the safety devices that are to be installed in different areas of the equipment for control of the system as a whole as well as of the components in case of malfunctioning, breakdown, accidents, fire, power failure, lack of water, changes in functioning of associate industrial equipment, and imposed unexpected abnormal conditons, particularly in respect to the specific types, specifications, and locations of each safety control device.

Responsibilities between purchaser and supplier in respect to steps in the installation The agreement should define in detail the steps for which the purchaser and the manufacturer or contractor of air pollution control equipment are responsible, jointly or individually, in respect to preparation, footings, fabrication, assembly, connection of electrical, water, and air facilities, startup in production line, trial runs, and testing.

Handling and disposal of collected pollutants Another point of understanding in the contract between the supplier and the purchaser of air pollution control equipment is in respect to the disposal methods and handling of the collected pollutant, whether they be of solid particulates, organic substances, or even soluble corrosive or chemically treated solutions. It is important to plan carefully and define this step in the contract between the supplier and the purchaser so that the supplier of the equipment can tailor his equipment to discharge the collected pollutants efficiently to the defined receiver, if the latter is to be furnished and handled by the purchaser. If the total air pollution control and residue handling systems are to be furnished by the supplier, further details as to the receiving of the eventual collected pollutant must be understood and defined in contract with the purchaser so that a smooth flow in the discharge of the captured pollutant can be effectively functioning with the air pollutant capturing equipment. This point is sometimes overlooked and causes delay in getting the equipment in operation and also results in lack of adequate capacity of the residue handling equipment for receiving whatever the discharges are from the air cleaning equipment. This has also been known to cause malfunctioning of air pollution control equipment or to cause a shutdown.

Guarantees on performances of dust and chemical fume collection systems The guarantee on the performance of the dust and fume collection system should specifically define the outlet load limits acceptable for each individual air pollutant that is to be collected, treated, and handled, preferably in relation to the regulations of air pollution control codes that are in force in that specific area. This can mean in relation not only to municipal but also to state and federal levels.

Purchasers determination in contract agreement on future changes in air capacities in industrial processes and in production rates Another primary understanding that must be in record, first for the engineering proposal and stated in documentary form, is that the supplier will be furnished information by the purchaser as to any anticipated changes in the cfm capacities required and any anticipated changes in rate of production or the nature of the industrial process before the equipment is designed and installed so that the equipment can give a performance in removing the specific pollutants to meet specified limits of outlet loadings in accord with the code regulations in force.

Acceptance of equipment performance based on accurate, reliable sampling and tests by qualified personnel One of the important sections of an agreement and the basis of the guarantee on air pollution control equipment is that the acceptance of equipment will be in part based on the results of reliable quantitative sampling and measurements of the specific particulates or fumes under normal process operation. This stipulation for acceptance is one of the most important statements in the contract. Many times in the past, unreliable, inaccurate methods have been used for testing, or testing has not been done in an accurate manner by qualified personnel or the testing has not been made during a normal industrial process operation. This has resulted frequently in the purchaser's finding that he has not received the equipment with a performance as possibly earlier claimed and understood. It may result in citations of violations when accurate, reliable testing is later done by the enforcement agencies. It can also result in bad relations, arguments, and even lawsuits between the supplier and the purchaser, if tests have not been made during normal production operations and by use of accurate test methods. The time has come that accurate, factual information on the performance of an installed air pollutant collection system in relation to a promised and stated guaranteed performance must be used as the basis for contractual selling, purchasing, and acceptance of air pollution control equipment.

Purchaser's cooperation to furnish data and information on pollutant sources Another stipulation that should be incorporated in the contract is that the purchaser shall provide adequate information on the process, particularly when it is cyclic in nature, during the operation of the primary pollutant generating system while sampling and testing is being done so as to correlate such information with the sampling and testing data. As a result of failure to make this correlation, disappointment of the purchaser and grief of the manufacturer of the air pollution control equipment can result, besides the difficulties with enforcement agencies. This correlation should be required as part of a full written report covering performance of the equipment, no matter what testing agency is employed.

All change requests and approvals in writing In most all major constructions and stages of installations there are reasons for changes, modifications, or substitutions, with the rapidly changing and improving field of air pollution control equipment being no exception. One of the causes of misunderstanding, possible irritation, bitter relations, and arguments, particularly in later financial adjustments and settlements that could end in lawsuits, is the lack of use of formal procedures with written records covering the changes. The contract should contain a clause to the effect that

> After the formal consummation of the purchase of air pollution control equipment, any change or deviation from original specifications instituted by either party should be presented in a written clearly defined request or proposal over the dated signature of an official. To become valid and binding, a reply to this request should be made in writing from the second party whether accepting, rejecting, or modifying the request and should also be signed and dated by an official or delegated agent of the second party. This change request may cover a portion or all of the basic components, their partial or complete installations as defined with the specifications in the supplier proposal and formally recognized in the purchaser's or his agent's purchase order. Further a change or deviation may be major or minor in nature for innovations, for changes of dimensions, locations, capacities, requirements in utilities, for accommodations to expected or unexpected changes in the industrial process, or for any modifications and substitutions.

Another portion of this clause in the contract pertaining to changes should read that it be required in these written communications on changes that the financial responsibilities of the respective parties for such changes be fully defined.

Cancellation settlement of engineering development equipment installation The agreements in offering or purchasing of air pollution control equipment should include a special clause that explicitly defines the financial obligations or liabilities of the individual parties or both parties jointly in accordance with the original contract to the specific reasons or causes for cancellation and in relation to the stages of the preparatory work and of installation. The wording of such a clause would have to fit the specific case and be in accord with the original arrangements.

Contract requirement for full written report on tests constituting evaluation of equipment performance Another stipulation should be included in the contract in relation to guarantees to the effect that there should be detailed recording and complete summary of information on method, procedures, and conditions of sampling and testing that are used to determine the efficiency in equipment performance during the normal process operation. Too many times in ignorance or intentionally, inaccurate, slipshod tests have been conducted that have been misrepresentative of the true performance of the equipment or there has been substitution for accurate, reliable methods of sampling and test procedures that account for only a portion of the pollutants. These deviations have been proved to be of high magnitude in relation to quantitative, true emission quantity determinations. It should be remembered that at the present time with the increasing knowledge of testing procedures and availability of better equipment in the hands of enforcement agencies, purchasers of such equipment that has been so misrepresented may face citations.

Contract requirement for proof of sustained guaranteed equipment performance In the contract a portion of the acceptance in respect to guarantees on performances should be based on proof of sustained performance of the air pollution control equipment to prove the dust and fume collecting system will give sustained performance, not only at the time of installation but at later periods such as 6 months or a year or at other specified times within a period of warranty of not less than 1 year. This demand for sustained performance efficiency tests as part of a guarantee cannot be overstressed, inasmuch as quantitative field tests have revealed outlet loadings from installed equipment that has been claimed by the manufacturer to be able to give low specific limiting outlet loading values deviate by several-fold increase in a matter of months or even less time.

Until the appropriate, reliable sampling and test methods are employed during the true normal industrial operation with the dust collecting equipment for proof of sustained performance, the customer cannot be sure that he has received equipment to do the defined job, nor will the supplier of such equipment have the satisfaction of knowing that his product has met the guarantee to which he has given his signature.

Statements in contract on equipment liabilities Another section of the contract should carefully define the liabilities assumed by the supplier or seller of the air pollution control equipment and the purchaser in respect to coverages by different types of insurances, service commitments, and replacements of components and defective parts in basic equipment, or expendable items.

Identities and extents of other liabilities affecting supplier and purchaser With the complexities in present industries, the uses of complicated interdependent equipment, the trends to place responsibilities on different parties, the occurrences of expensive lawsuits on property damage claims, personal injuries, deaths, interruptions of installation or of production, as well as on patent infringement and use suits, it becomes necessary in the field of air pollution control equipment for the legal counsels of the suppliers and purchasers to set forth very carefully the nature, extent, or exclusion of liabilities concerned with each party.

Protective clause for the supplier against misuse, malmaintenance, and damage by purchaser Carefully worded legal clause or clauses are important in the contract to cover protection of the supplier, his personnel, materials, and tools during construction and installation on site and protection of the supplier during warranty and after warranty following completion of installation against damages incurred by the purchaser or his agents through failure of the purchaser to maintain the equipment properly, negligence, sabotage, or results of fire, explosions, storm, riot, or war. A similar protection of the purchaser is required in writing.

It is not the intent nor within the scope of this chapter to prepare a format of these contractual elements of guarantee, agreements, responsibilities, and extent of liabilities for each air pollution control equipment installation, Each case requires its own contract containing the particular set of these elements defining the transactions with full understanding, specifications, committals of obligations, assurances, and protection to each party concerned. Furthermore, the preparation of the contract with the selection of the above applicable elements can best be presented in proper order and expressed in the desired legal phrasing individual to the legal counsel in charge.

Primary Steps in Evaluation, Setting Requirements, Prescribing Guaranteed Performance Specifications, and Installing Dust and Fume Collection Systems

In the past, planning air pollution control programs, writing guarantees on performance of equipment, selection of the air cleaning devices, and testing have been conducted too much on a casual basis without regard to primary steps to be taken in making critical choice of equipment and investments. The situation has been also complicated by the fact that there have been rapid changes in the regulatory air pollution control codes and many of these values of emission limits were not being properly recognized and taken into account in evaluation of equipment for air pollution control. There are a number of basic steps which may be followed as a guide and that are of assistance not only to the purchaser of the equipment for air pollution control but also to the supplier of the equipment and to the enforcement agencies responsible for permits and regulations. These steps may be as follows.

Determination of the individual pollutants from the process to be brought under emission control The casual and careless approach in the past has been somewhat like calling all the solid emissions smoke, overlooking the presence of organic aerosols, and ignoring the presence of invisible toxic or corrosive pollutants. This ignoring of the exact nature, identification, and quantities of the individual pollutants has caused, in many cases, failure of purchased and installed equipment to meet code regulations and also other difficulties in the selection of structural components in the dust or fume collection system. This careless, incomplete approach has been costly in many cases, requiring replacement of the dust collection system or extreme modifications in equipment. Another common oversight is that the liquid organic aerosols can be carrying solid particles in the microdroplet, though these droplets are of extreme small dimensions. Also, recognitions of the presence of invisible active chemical materials that are in the true solution phase in the air are ignored. Many times these invisible materials, unrecognized pollutants, are deadly in nature, such as carbon monoxide, or are highly corrosive acidic substances destructive to structural components.

Investigation and study of regulations in the air pollution codes governing particular areas An initial step is to obtain copies of the regulations from the municipal, state, regional, or even federal level in respect to limits of individual pollutants from specific industrial operations. It is imperative that the most recent of accepted code regulations that govern that area be obtained instead of depending upon old issues which may be outdated. If the municipality, state, or federal authority has not issued and accepted a code for a specific industry in a defined area, a check should be made as to what the air pollution control committees have in mind and what tentative emission limits may be prescribed.

Appraisal whether stricter regulations in air pollution control are in the offing or anticipated modifications in such regulations are to be made It is recognized from references to the present and the past that the next few years are a transition in respect to what the defined limits of air pollutants were and will be from individual processes. Before selection of equipment, a careful check should be made of the trend and the expected values of outlet loadings that will be recognized in code regulations and used by enforcement agencies. These rapid changes in outlet loadings which were carelessly defined in the past but have new sharp limitations in new code regulations are illustrated by the particulate matter permitted from cupola furnace operations. Not long ago the outlet loading of particulate matter was allowed up to 0.5 grain/scf dry air. This value not many years ago in certain areas was dropped 0.3 grain/scf dry air and rapidly to 0.15 grain/scf and still more recently to 0.05 grain/scf dry air for a production cupola furnace of medium or larger size.

The same can be said of other pollutants such as reduction of permissible carbon monoxide. These cases illustrate the transitions that have been occurring in establishing these new lower values, which create a severe challenge on abilities of methods and devices for the removal and control of specific pollutants.

Determination of relative quantities of individual pollutants from a specific industrial process One of the deficiencies in air pollution control planning and in recommendation and selection of air pollutant removal systems has been the lack of information on the relative quantities of the specific individual pollutants generated from the process. This deficiency is compounded by failure to recognize that these relative quantities *must* be known not only for the emission for average operating conditions but for the maximum generation of the individual pollutant during the specific industrial process before a supplier can properly recommend and guarantee the performance of this equipment for removal of the particular emission products. The same applies to the purchaser in that he cannot make a true decision in selection of equipment to remove the pollutants unless he has such specific basic information at hand.

Consideration of expansion of the industrial production operation Another factor that must be taken into account to ensure that the investment in the air pollution control equipment is adequate is to know whether an accelerated production program is contemplated or to be undertaken and whether the considered air pollution control equipment can accommodate such increase of emissions in industrial process whether it is with new process equipment or addition to present production equipment or acceleration of use of the present equipment. This is well illustrated in the case of certain metallurgical operations in which originally a certain tonnage of metal is scheduled per hour but with modification of the method the tonnage may be greatly increased to double or more. This expansion in production results in increased emissions and requires provision of air pollution control equipment with adequate capacities.

Check on anticipated environmental changes—proximity and increase of population in the area A point that is frequently overlooked where the regulations in an air pollution control code have not been established or extended to the area is a tendency to ignore the fact that where a plant is installed, for instance, in the open country, sooner or later it will be surrounded by the residences of the workers. If control of the emissions has not been duly planned and taken care of, the increasing population in the area is suffering from this pollution and may make the existence of the air pollution a focal point and cause considerable activity in demanding immediate action in providing high efficient emission control in the removal of the pollutants of concern.

Study of present and possible future relation to regional and interstate air pollution control Just as in the trend to treat water pollution control by regions governed by the lakes and river basins, so it may be in terms of air pollution control that regional regulations, probably from the federal level or agreed control between areas divided by state or municipal lines will be in effect. These regional regulations may be also extended between countries, as is well known with the International Air Pollution Control Commission operating between the United States and Canada.

Study whether industrial process will be changed with new or increased amount of individual pollutants In the planning for present and future handling of airborne pollutants, a study should be made by the industrial management and the results made known to the supplier of air pollution control equipment whether there is any anticipated change in the process or whether a new process will be put in operation that could change the relative amounts of individual pollutants or volumes of air. With the rapid increase of technology, as is well illustrated in the metallurgical fields, the new processes, let alone expansion of older methods, become of concern in the design and selection of air pollution control equipment. A study of these new processes must be made to determine, for instance, whether the pollutants have changed in their nature, whether a larger percentage of submicron particles is anticipated, or whether other combinations of pollutants have resulted from the new operation. All these factors influence the design and the selection of the equipment and the guarantees that are given.

Check for the variations of temperature and volumes of pollutants from each specific industrial process One of the reasons for disappointment in the performance of installed air pollution control equipment has been the failure to determine accurately the temperature and volume variations of the gases that must be handled in the equipment. The air pollution control equipment must be adequate to handle the maximum of emissions and to sustain the quality of performance with wide fluctuations in quantities of pollutants and volumes of gases. These factors are important in considering each specific type of air cleaning system. It has not been well recognized in the air pollution control field that many of the different types of systems for air cleaning have a relative high sensitivity in respect to the ability to remove pollutants when the temperatures fluctuate and the volumes of air change to a marked degree. An important criterion in judging air pollution control equipment, particularly when it is servicing more than one source of an industrial air pollution

TABLE 1 Inlet and Outlet Loadings of ASP 100 Test Pigment Dust Quantitatively Measured in Determination of Performance Characteristics of Centrifugal Spray Generated Dust Collectors*

Test No.	Inlet loadings of ASP 100 airborne pigment, grains/1,000 scf dry air	Outlet loadings of ASP 100 airborne pigment, grains/1,000 scf dry air
1	333	0.946
2	1,043	1.150
3	327	0.935
4	992	1.036

*Unpublished data from Quantitative Air Pollution Control Study Measurements, Industrial Hygiene Dept., Ajem Laboratories, Inc., Livonia, Mich.

emission, is whether the equipment has a wide accommodation factor to sudden and large fluctuation in quantities of pollutant per unit quantity of air. Some equipment not only has the high accommodation factor for temperature and volume changes of air but also has low dependence upon wide changes in the quantity of pollutant present and is able to sustain a uniform controlled low outlet loading of particulate matter. Quantitative performance testing has verified that the powerful centrifugally generated spray type dust and chemical fume collector systems have this high accommodation characteristic.

Table 1 shows the low variation in magnitudes of outlet loadings experienced with large changes of inlet loadings of a very fine particle size test dust introduced under controlled conditions into a modified design of the centrifugal spray generated dust collector illustrated in Fig. 1. The data presented in Table 1 were obtained in quantitative study measurements on the abilities of commercial size centrifugal generated spray collectors in removal of dusts such as the ASP 100 pigment from the air when the test dust is aspirated at metered rates into an air inlet duct not shown in Fig. 1. The ASP 100 test pigment has a particle size distribution of all particles under 5 microns and approximately 45 percent less than 0.5 micron. In the quantitative tests code regulation inlet and outlet ducts are attached to the collector and precision methods and apparatus were used in sampling and measurement of outlet loadings of test dust.

The first column in Table 1 gives identities of test runs in which the respective inlet and outlet loadings in grains per 1,000 scf dry air of the specific ASP 100 test dust pigment were quantitatively measured in the performance testing of two wet type centrifugal generated spray dust collectors of medium rated efficiency. The inlet and outlet loadings are given, respectively, in the second and third columns. Tests 1 and 2 are runs of 5 and 4½ hr, respectively, sampling time on one collector. Tests 3 and 4 were performed on another similar collector with sampling times of 3½ and 4 hr respectively.

It will be noted that the magnitudes of penetration of this test dust containing high percentage of submicron size particles through these wet type dust collectors are low. It can also be seen that the outlet loadings have experienced relatively small changes even if the inlet loadings change in a high ratio. The extended studies have also revealed that a repeat of these tests on the same dust collector under the same conditions at a later time will give almost identical performance data.

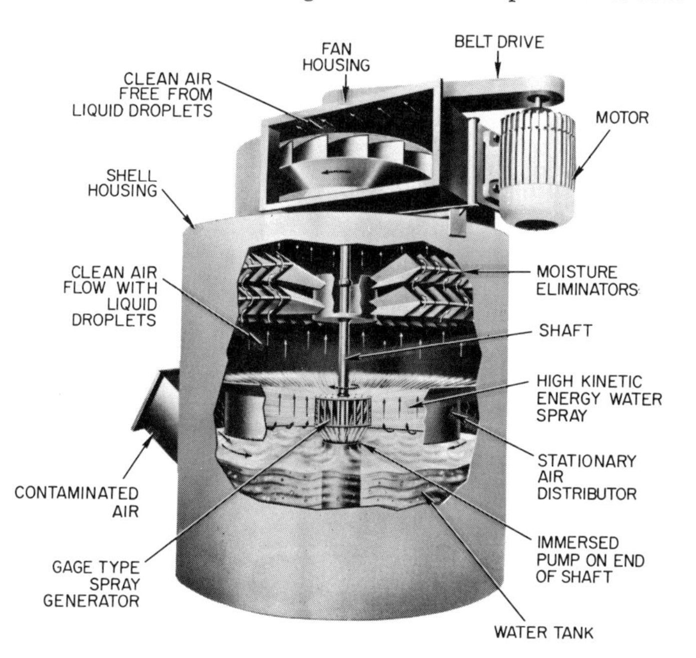

Fig. 1 Cutaway view of centrifugal high-kinetic-energy spray air scrubber. (*Centri-Spray Corporation.*)

The centrifugal spray generated dust collector shown in Fig. 1 is a self-contained wet type dust and chemical fume collector which has centrally located in its shell a vertical rotatable shaft equipped with drive sheave at top, a fan in the fan housing area, a spray generating distributor in the spray chamber area, and a feed pump for liquid to the distributor. Either the feed pump is in the liquid at the bottom of the collector housing and on the lower end of the shaft or it is a separate pump at a remote position. The contaminated air is brought in through the inclined side entrance duct attached to the housing. The contaminated air is directed up through the powerful 360° spray generated at the distributor elevation. The washed air with some airborne droplets moves upward through stationary moisture eliminators by action of the fan, and liquid droplets are removed, with clean air going through the fan and exhausted to the exterior or into a receiving air duct.

Study of climatic, meteorological, and atmospheric conditions in relation to seasons and to anticipated extremes in production emission quantities A number of the types of air pollution control equipment are sensitive to atmospheric conditions. Furthermore, in certain areas of the country, specific pollutants must be eliminated as much as possible in order to prevent the compound effect with the atmosphere in creating serious air pollution conditions. This means that any equipment considered must be a system that will adequately adapt itself to the wide range of conditions whether it is the temperature of the environment, humidity, or stagnation of air in a specific geographic location.

Evaluation of air pollution control methods and available dust and fume collector equipment For assurance of sound investment, economy in operation, and high efficiency in performance with sustained removal of the specified pollutant, quantitative evaluation of the equipment performance must be made for the removal of the particular pollutant. Care in this stage of evaluation and comparison can mean the difference between success and failure, between the meeting of regulations of code or facing citations, and between the enjoyment of efficient functioning equipment or experiencing of equipment that is erratic and inefficient in performance.

Necessity for quantitative sampling and measurements on quantities of specific pollutants as experienced at maximum production operation One of the causes for misjudgment in selection of air pollution control equipment has been the failure to have quantitative tests made for the individual pollutants of concern as experienced when the industrial process generating the emission is under maximum production operation. This point cannot be too strongly emphasized, inasmuch as many times estimates are made from poorly performed tests not correlated to the production rate or inaccuracies in tests resulting in estimates that do not give accurate emission evaluations. This frequently has resulted in lack of capacity of the selected system to perform with higher emission rates, with the higher volumes of gases, or at higher temperatures than anticipated. This has sometimes been followed by costly and unsuccessful attempts to modify the air pollution control equipment with improvements that still fell short of the goal in meeting the regulations.

Stipulation in guarantee for accuracy and reliability of all test data submitted A stipulation of importance in the contract is for check and verification that the procedures and instrumentation used in determining the emission quantities from a process or from the operation of a reference air pollution control system will give guaranteed reliable, accurate test results.

It should be added that even with the best instrumentation, without skilled and qualified test personnel, accurate quantitative results may not be forthcoming. Therefore, it is necessary to demand and require specifically that reliable, accurate, and guaranteed or certified test results will be submitted. The criteria for judging accurate, reliable test results will be discussed in a further section.

Check for versatility of the considered air pollution control equipment To ensure the sound investment in the air pollution control equipment, three further basic facts should be established:

1. The considered equipment should be checked as to whether the contemplated equipment can be applied with assurance and simplicity if a portion or all of the air pollution control equipment is applied on removal of other industrial source emissions.
2. This check of the equipment for versatility and adaptability should also be made in respect to the possible changes of pollution emission products resulting from modification of the industrial process, including increased production rate on the equipment for which the air pollution control system was installed.
3. Another reason for this check is also to determine whether the considered equipment could be modified in the future in a simple manner to upgrade the performance and meet a changed regulation in the code for a lower outlet loading. Consideration of this last point finds that many of the offered air cleaning equipment systems cannot be easily upgraded to meet a change in the quality of performance such as to meet a lowering of outlet loading limits in new revised regulations or when the emission products become more difficult to remove such as with higher percentage of submicron particulates present in the emissions.

Factors for Evaluation of Air Pollution Control Equipment

In the previous section, different steps in the general approach to solving an air pollution control condition have been discussed. In the present section more detail will be given to the individual factors that determine whether the equipment will be adequate and a sound investment.

Adaptability of considered air pollution control equipment for removal of each specified air pollutant issuing from the industrial process For an accurate appraisal of the adaptability of considered equipment several conditions and factors should be considered. Among these are the following:

Occurrence of Wide Range in Sizes of Particulate Matter. One of the basic understandings about air pollution equipment is to determine the degree of efficiency in removing the specific size ranges of particulate matter emanating from the industrial source. It is in this connection that many times the wrong evaluation is made on equipment performance because of use of data from inaccurate test methods that do not account for the total of particulate matter. There is also the correlating fact that the industrial air cleaning equipment in many cases does not have the capability of removing particularly the submicron particle sizes. The investor should have positive proof if he is confronted with ultrafine particles such as in the submicron size that he knows the relative quantities present in the emission before he undertakes the selection of equipment. Moreover, the investor should demand proof of the performance ability of the specific equipment under consideration to remove specific submicron particulate matter efficiently.

TABLE 2 Comparison of Quantitative Differences in Determined Magnitudes of Total Particulate Matter and of Solid Particulate Matter in Outlet Loading Emissions from Dust Collection System Serving Metallurgical Operation*

Test No.	Outlet loading of total particulate matter, grains/scf dry gas	Outlet loading of total solid particulate matter, grains/scf dry gas
1	0.121	0.053
2	0.150	0.052

* Unpublished data from Quantitative Air Pollution Control Study Measurements, Industrial Hygiene Dept., Ajem Laboratories, Inc., Livonia, Mich,

Presence of Solid Particulate and Organic Aerosols. A common error that has been made is ignoring the fact that particulate matter includes the solid as well as liquid particles or aerosols, the latter of which can be carrying still smaller size particles of solids. It should be recognized that certain air cleaning equipment may be efficient in removal of submicron solid particles but not efficient in removing organic aerosols. Furthermore, when certain metallurgical operations are involved, the amount of organic aerosols being emitted my be an appreciable percentage. This has been verified by quantitative feld testing.

In the first column of Table 2 the identities of the two comparison runs are given. In the second column the "total particulate matter" is recorded for each run in grains per standard cubic foot dry gas. In the third column the corresponding weights for the "total solid particulate matter" are given and expressed in the same units.

In Table 2 it will be noted that for this field case in test 1, the total particulate matter is 2.3 greater than the total solid particulate matter, while in the test 2 the corresponding ratio is 2.9. It will be recognized that increments of increases in the total particulate matter over the total solid particulate matter are accountable in the presence of organic liquid aerosols such as derived from oils and organic residues on metal scrap. Extensive quantitative field testing has shown wide ratios in the values of these two ways of recognizing particulate matter. In some cases the nominal values of total particulate matter and of solid particulate matter are practically the same, while in other instances the ratios may be higher or lower from the values cited in Table 2, depending on:

1. The relative quantities and species of organic oily matter on the parts or occurring in the industrial process
2. The operational conditions of the industrial process such as in a metallurgical furnace
3. The completeness and efficiency of the dust and fume collection system in removing particularly these organic aerosols

The seriousness in the occurrences of these varying differences in the magnitudes of total particulate and of total solid particulate should prompt careful thinking and caution in selection of air pollutant control systems in respect to:

1. Whether organic oily aerosols are being emitted from the specific industrial process.
2. Whether the particular emission from the specific industrial process is to be subject to code regulation limiting discharge on the basis of total particulate matter or of total solid particulate matter. This can be of major significant difference in respect to the selection of the air pollution control system and the efficiency required in the system
3. Importance of the ability of the dust removal system to remove the total particulate inclusive of these organic aerosols not only for avoiding possible code regulation violation citations but also as a safety measure. It should be emphasized that there are a number of types of air pollution control systems that do not have the ability to capture and handle such aerosols. In fact, there are a number of field cases in which these organic aerosols have been responsible for causing costly fires and resultant major loss or damage of the equipment. It should be pointed out that some of the air pollution control equipment on the market is not basically designed or adapted for removal of organic aerosols and also that some air cleaning systems are more vulnerable than others to malfunctioning or suffering from fires when such organic aerosols are present and are not properly handled.

Presence of Corrosive Chemicals with Solid Particulates or with Organic Aerosols. Air pollutants are not limited to solid particulate matter that is visible, for many times corrosive or toxic chemicals are present with solid particulate or with organic aerosols or with the mixture. If the corrosive or toxic chemicals are objectives for removal as well as the other particulate matter, it is necessary to know the identities and relative quantities of these individual corrosive or toxic pollutants which may be present and invisible. Another reason for determining whether these corrosive and toxic materials are present is not only to prepare the best ways for their removal and capture but knowledge of their presence which may be the cause of side chemical reactions and, particularly, of corrosion of the equipment for air cleaning. Many installations have been made on the assumption that only inert solid particulate matter is present but it is found that corrosive toxic materials are accompanying the solids in the emission and cause rapid corrosion even with the lower grades of stainless steel alloys. Therefore, the adaptability of the considered type of air cleaning equipment must be evaluated on the ability to collect the chemically active emission products if they are to be removed simultaneously with the solid and organic particulates and also on the chemical resistance of the air cleaning components to the possible chemical attack in the specific environment.

Chemical and Physical Properties of Each Corrosive or Toxic Material. In realizing successful removal and treatment of corrosive and toxic materials, not only their identifications must be obtained but also knowledge of their relative quantities occurring in ambient conditions and the individual chemical and physical properties. This latter information is imperative in order to use the proper chemical reagents to react and destroy the captured active chemical pollutants.

Properties of Mixtures of Corrosive Airborne Materials. It is not enough to know the properties of the individual pollutants, particularly of the active chemicals. It is important to determine whether there are chemical reactions between the different air pollutants when two or more chemically active species are present. This may form new emission products not suspected in the original process and as a result may require changes in the air pollution control system. Again, it is known that the presence of traces of certain corrosive materials with other acidic materials, for instance, can greatly accelerate the attack on structural materials like the ferrous metals. An example of this is the comparison of the rate of attack of a sulfuric acid solution on ferrous metal product and the accelerated rate of chemical attack at the same temperature when traces of halogen derivatives are present. The increased rates of chemcial attack may be many-fold and thus demand that a different structural confining material be used such as for wet type dust collectors.

Dependence of efficiency in performance on variation of cfm through equipment for removing the pollutants Another factor to consider in appraising air cleaning equipment is whether the particular air cleaning apparatus changes in performance efficiency when different magnitudes of cfm are experienced through the same unit. In this respect it should be warned that a number of air cleaning devices have high variations in performance efficiencies when the air volumes change. In seeking the best air cleaning equipment, the apparatus that has a low dependency of efficiency upon fluctuation of air volumes as influenced, for instance by the process or temperatures can be expected to give the best performance. Therefore, one should seek to find air cleaning equipment that will give sustained performance over a considerable range of variation of the magnitudes of the incoming cfm of polluted air from an industrial process.

Degree of fluctuation of efficiency as a function of temperature and environmental conditions While fluctuation of temperatures causes changes in the volumes of the gases from an industrial process, the results are like varying the cfm to the collector system. An examination should also be made of the industrial process to see whether the temperatures are high or low and of the relative effects for other reasons such as whether the emission products after formation change chemically or physically with the temperature and the activity or state of the gases or pollutants when received in the dust collection system. It should also be determined whether the temperatures are of such a high value that they will cause an increased activity of one or more of the airborne pollutants and result in severe corrosive attack and damage to the interior part of the ducts or fume collection system.

Potentialities for recovery of pollutants that are collected A factor to examine is whether any of the captured pollutants can be converted into useful forms and amount to a source of revenue that would pay for the air cleaning process in part or total.

Evaluation of components in dust collection system subject to rapid deterioration, high rate of wear, breakdown, and malfunctioning To ensure long life of air pollution control equipment, it is important to examine whether the proposed equipment and its individual components have adequate protection and resistance to possible high abrasion, chemical attack, and particularly corrosion.

Consistency of performance This factor should be carefully studied and answered with use of accurate, reliable data on measurements of the performance in removal of the specific pollutants present in the emission. Appraisal of air cleaning equipment on this condition alone would rule out a number of air cleaning systems for certain applications because of irregularities in functioning and of effects of variations in temperature, of changes in volumes of the air, of presence of contamination, and of possible breakdown or malfunctioning of components in the equipment. Consistent reproducible performance can be obtained with the proper equipment for removing a specific soil. Proof of ability to give consistent performance may be found in the data presented in Table 1 giving the quantitative values of inlet and outlet loadings of ASP 100 test dust pigment as determined during studies of the efficiencies of the centrifugal generated spray type dust collectors. The measurements reported for tests 1 and 2 were for performances of a large size commercial self-contained centrifugal spray generated dust collector as illustrated in the type shown in Fig. 1. At a later date with another similar wet type large commercial dust collector quantitative measurements were made while using the same species of test dust pigment. The results of the latter measurements shown in Table 1 for tests identified as 3 and 4 confirm excellent agreement in the corresponding inlet and outlet loadings.

Maintenance—time for servicing and repair and cost of replacements Many times the operating costs in terms of maintenance, repair, and replacement of components are far higher than anticipated for air cleaning equipment. In fact, some of the cheaper original installations have been responsible for high expenditures with limited operating efficiency. From the point of view of economics the purchaser should have concrete field record data on the maintenance requirements of different types of air cleaning equipment to evaluate properly and project the cost of operation of these air cleaning devices in the future.

Operating ability of wet type systems for functioning with captive water system When the wet type dust and chemical fume collectors are being considered, an important factor to determine and analyze is whether the air washing equipment can operate with recirculated liquid in a captive water solution arrangement. Several of the wet type dust collectors are critical and demand fresh solutions or fresh water, free from suspended material, or otherwise jets, slots, and orifices will become plugged and the efficiency will be reduced to an unsatisfactory level. In comparison, there are dust and chemical fume collectors of the wet type such as of the centrifugal generated spray design that are capable of operating on captive water systems to the extent of high concentrations of suspended material as well as of dissolved components. When a captive water system is employed, auxiliary equipment may be installed for filtering out solids, for removal of stratified oily-like materials by oil skimmers, or for chemical treatment with use of chemical feed devices to destroy corrosive and toxic materials or to inhibit rusting of ferrous metal surfaces. Such auxiliary equipment makes it possible to handle and remove collected pollutants by either manual or automatic control. The captured and isolated pollutants may be of monetary value to certain industrial companies. It should be noteworthy to industries generating airborne corrosive or toxic materials to know that certain wet type dust and chemical fume collectors are capable of operating with captive chemical reagent solutions and have the ability to take out solid and chemically active airborne materials simultaneously. Many field applications and quantitative performance tests of centrifugally generated spray wet type dust and fume collectors have shown that this type of air pollution control equipment, besides removing particulate matter, can function as a chemical reaction vessel with addition of chemical reagents specifically to convert active chemical airborne materials into inert captured states.

Compactness and flexibility in design for adaptation to available space One of the initial considerations is the adaptation of an efficient and applicable dust and fume collector to the available space near the source of the particular emission products. Some air cleaning equipment is exceedingly bulky and requires excessive space as compared with others which are available. Consideration should be given to efficient units that are flexible in their adaptability and arrangement. This flexibility is illustrated in recent installations of the centrifugally generated spray dust and fume collector type where the spray generator, the fan, and the moisture eliminators and other accessories can be self-contained or separated in their locations relative to each other with the necessary connecting ductwork to the pollutant source and to the stack. Each can therefore be adapted to the different industrial units.

Power and plant facility requirements Whether it is an industrial, chemical, metallurgical, or manufacturing process that is to be equipped for air pollution control, the electric power source and related and required plant facilities should be carefully evaluated when steps are taken in considering equipment to handle the air pollutants from a particular source.

Ability to realize sustained, efficient performance in removal of pollutants One of the paramount requirements that should be confirmed for a specific air cleaning device is that its functioning has been proved by accurate, reliable test measurements that it can give a sustained, efficient performance in removal of the specified pollutants. Field experience has established that a high efficiency may be claimed for a number of well-known air pollution control devices which may be shown and realized at the time of the installation but that considerable deviation occurs in a relatively short time. This has been observed for certain air cleaning equipment inclusive of dry type of electrostatic and of bag filter types and in certain cases of the wet types.

In a series of quantitative field measurements on the solid emission particulates from a centrifugally generated spray dust collector system installed on a large ferrous metal cupola furnace, observations were made on the performances of this type of wet dust collector over a period of time and its ability to sustain the removal of the airborne cupola solid emission products. The quantitative sampling and test methods, further described in a later section, as used in these field measurements were capable of collecting and measuring all the solid particulate matter inclusive of submicron particles down to a fraction of 2 percent of the total solid particulates.

Figure 2 is a photograph of this particular centrifugally generated spray dust collection system installed on the ferrous metal cupola furnace. This photograph was taken while the cupola was in full operation and during the time one of the sets of measurements reported in the data on Table 3 was being made.

The type and magnitude of the solid emission products from this ferrous metal cupola before the cupola furnace was equipped with the centrifugally generated spray dust collection system are shown in Fig. 3, a photograph of the emission of a similar size of cupola.

Fig. 2 Photograph of installation of centrifugally generated spray dust collection system on large ferrous metal cupola furnace, taken during normal production and during performance tests.

TABLE 3 Performance Data on Centrifugally Generated Spray Dust Collector Adapted to Collection of Emissions from Production Nodular Iron Cupola Furnace

Time of Performance Measurement	*Outlet Loadings of Total Solid Particulate Matter from Centrifugal Spray Dust Collector System, Grains/scf Dry Gas*
At time of installation	0.0503
3 weeks after installation	0.0473
5 months after installation	0.0470

Table 3 gives the performance data in this series of tests over a period of 5 months in the removal of the solid particulate by the centrifugally generated spray dust collector in the above installation. The first column gives the relative times of the three sets of tests referring to time of installation as starting time. In the second column the averages of the solid particulate outlet loadings from the identified type of collector system taken while the cupola was in normal operation are given for three sets of measurements. The first set was taken at the time of installation, the second set at 3 weeks, and the third at 5 months. Each value of outlet

Fig. 3 Emissions from large ferrous metal cupola furnace, similar to the cupola studied, before its conversion with installation of centrifugally generated spray dust collector system.

loadings in the second column is the average of three quantitative samplings and measurements expressed in grains of solid particulate per standard cubic foot dry gas. The decreases in the outlet loading values after the initial set of tests can be accounted for by improvement changes made by the manufacturer on this centifugally generated spray dust collector system. These quantitative performance determinations verify that sustained performances can be maintained. These findings also give proof that sustained control in removal of air pollutants can be attained with the assurance of continued operation to meet the specific code regulations.

Other field observations and quantitative field measurements have revealed that with some other types of dust collection systems considerable deviations from claimed performances are experienced and that large fluctuations with marked increases to the extent of several-fold occur in outlet loadings from these systems.

In other words, from these findings, the judgment of whether the equipment will meet the code regulation requirements with a sustained performance must be verified by actual field tests over a period of time on the equipment under consideration. These tests not only must be accurate and reliable but must encompass removal of total pollutants that are of concern in functioning of the equipment.

Expected life of air pollution control equipment There is considerable difference in the expected life of air cleaning equipment and its particular components. The life varies widely not only with the type of equipment serving to remove the air pollutants but also as influenced by the environment in which the equipment is employed. There are many areas in which complex mixtures of active corrosive materials or highly abrasive particulate matter are captured. Before the supplier designs or makes a guarantee on the life of air pollution control equipment or even predicts its life, a careful study and determination must be made of the environmental conditions, particularly the abrasive materials, that may be collected and the corrosive or chemically active materials that would act on some component in the system, such as in normal ferrous metal steels or casting production. If there is any anticipation of a high corrosion rate, a selection of more inert structural material is necessary in order to assure the customer of longer life of the purchased equipment.

In expecting a guarantee, the purchaser and the owner of the equipment must also take into account the environmental conditions and stipulate the conditions that must be adequately and properly maintained.

Modification of dust collection systems for upgrading performances with changes in air code regulations, for change in nature of the pollutants, or difficulties in removal of new pollutants At the present time there is a rapid change in the stipulated limits of specific pollutants with new code regulations in air pollution control. Serious attention should be given to whether the contemplated equipment could be modified in a simple way to upgrade its performance if the regulation becomes more severe in respect to demanding lower outlet loadings. Likewise, if there is a change in the nature of the process or emission products, the same serious attention should be given to whether the equipment can be modified to meet the new imposed conditions in removal of the new pollutants.

In a recent panel meeting on foundry air pollution control such a question was addressed from the audience, and it was exceedingly perplexing to many suppliers of air pollution control equipment as to what could be done. The answers ranged from total replacement of the old equipment with new equipment, addition of another section, or other undefined changes. This is in contrast to other cases such as the wet type centrifugally generated spray design of dust and chemical fume collectors where changes are necessary only in mechanical spray generator, the magnitude of the horsepower driving the distributor, and the pumping capacity of the water to the distributor, without any major change in the housing of the equipment or duct system with its fan and stack. This effective upgrading of air pollution control equipment is illustrated in the ability to modify and tailor the centrifugal spray generator wet type dust and fume collector to meet the regulation changes in allowable solid particulate outlet loading from a paint spray booth such as cited in the earlier section. Changes in code regulations with reduction to a fifth or less of the outlet loading have been met and quantitatively verified when the parameters of distributor design and speed, applied horsepower, and water supply were changed.

Figure 4 shows a cutaway view of the wet type centrifugal spray generator dust and fume collector adapted to a paint spray booth. With changes of the above parameters the solid particulate in heavy industrial painting was reduced in one case to 1.0 grain/1,000 scf dry gas. In other applications to the paint spray booth the solid particulates have been reduced to a fraction of a grain per 1,000 scf dry gas. This closely ties in with the importance of versatility of the equipment for adapting to different type of pollutants by simple modifications of the basic equipment.

Requirement for Accurate, Reliable Sampling, and Test Data

The high importance of reliable, accurate test data has been emphasized in connection with the writing of air pollution regulation codes, in establishing judgment on the control requirements for specific air pollution emission sources, in preparing agreements in terms of responsibilities and guarantees on performance, in verifying the performances of equipment considered or installed for control of these emissions, and in proving to code enforcement agencies that an installation will satisfactorily remove the pollutants. To be more specific, the importance may be recognized under:

Magnitudes of individual pollutants and appraisal of air pollution control requirements At the start in solving air pollution control problems one of the fundamental requirements is to make an accurate evaluation of quantities of the individual pollutants whether they be of solid, liquid aerosols, or chemically active toxic gaseous substances. A frequent error in the past has been that in mixture emissions certain components were either overlooked or not properly identified and not determined accurately in the range of concentrations in which they occur. This latter point is important to note, because many times one test in an ambient industrial process shows that a low quantity or even none of a certain pollutant such as a corrosive material may be appearing but a high quantity is present in other stages of the operation. This can be deceiving in the choice of air cleaning devices and in the writing of specifications and guarantees on the performances of the equipment. Reliable information is necessary not only to the owner of the industrial process but also to the supplier of equipment who is proposing, engineering, or installing equipment for handling the pollutant.

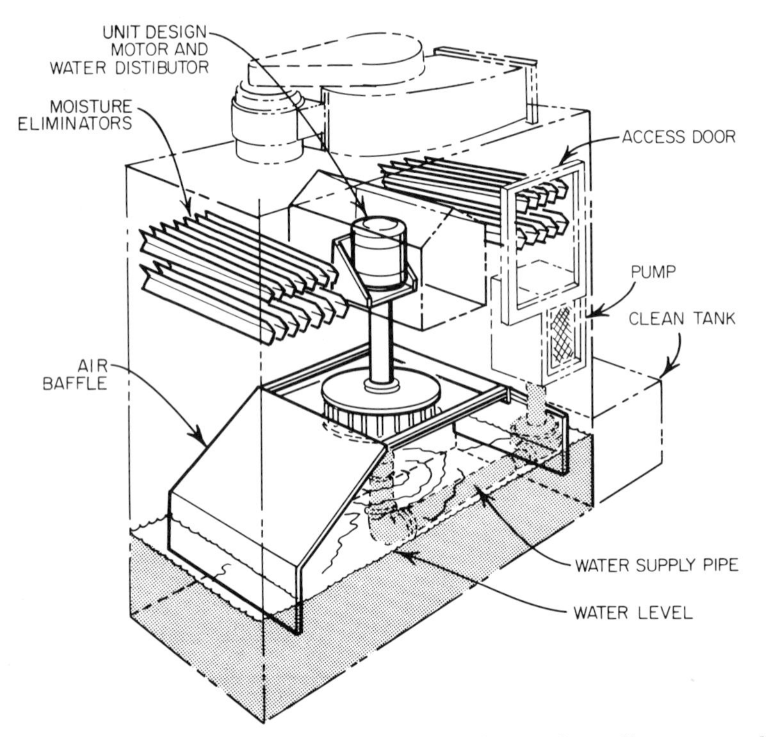

Fig. 4 Adaptation of versatile centrifugally generated spray dust collector principle to a thin module for attachment to spray booth chamber. (*Centri-Spray Corporation.*)

Significance of reliable data for evaluating existing equipment efficiencies With the present serious movement to enforce regulations on air pollution control, it is important to the owner as well as to the original supplier that accurately determined test data be used to obtain reliable information on the performance of the equipment presently in operation. This may determine whether the existing equipment is adequate or whether modifications or new equipment must be installed. When a new process is being installed in a plant, parallel tests should be made on a similar operation to evaluate the emission products in respect to their identities and relative quantities.

Establishing pollutant emission limits related to new air pollution codes Anyone who has served on air pollution regulation control code committees realizes the difficulty in arriving at fair and equitable limits of outlet loading of emissions for individual air pollutants. Such limits must be based on a number of factors including relation to health, safety, prevention of damage or destruction of property, and nuisances. Fundamentally accurate, reliable test results are necessary in order to arrive at these fair controlling regulations of individual pollutant limits.

Need of accurate data in establishing reliable performance guarantees At the heart of any air pollution control equipment purchase contract is the guarantee of the supplier to the purchaser in respect to the performance of the equipment that will be experienced under defined field operating conditions. There is no substitute for accurate, reliable test data in respect to the evaluation of the removal of specific pollutants as occurring in the field industrial process, in both representation of the equipment potential performance and also proof of its performance after installation. It is the agreement between these two sets of values that gives the satisfaction and assurance in performance and also satisfaction from the contractual agreement between the parties concerned.

Securing installation permits With the establishment of most of the new code regulations on air pollution control, permits are demanded by health departments, building commissions, and industrial hygiene departments and other responsible agencies for the installation of the equipment to serve in removal and handling certain pollutants. One of the primary concerns of the permits is the performance that will be experienced in removal of the pollutants by use of proposed equipment. This involves guarantees as the basis for issuing of a permit on the claimed performances that can be legitimately and equitably expressed only in terms of accurate measurements of pollutant outlet loadings from that specific design of air cleaning equipment on the particular pollutant or group of pollutants.

Lawsuits involving citations for excessive emissions Equitable judgment in lawsuits concerned with citations for excess emissions must be based on accurate quantitative measurements whether presented by the air pollution control health enforcing agency or in defense of the user or manufacturer of the equipment. Wrong judgments in lawsuits can occur if the results of inaccurate measurements are introduced as evidence. It can be said that if representation of equipment for air pollution control or knowledge of its performance by the purchaser is known and established by accurate tests, citations and lawsuits can be avoided in most cases.

Establishing new regulations and codes It cannot be too strongly emphasized that the recognition and use of quantitative reliable test measurements are at the heart of any air pollution control code whether in its formulation or in its enforcement and understanding. While it is not the objective in this chapter to describe or compare different test methods, it behooves the committees proposing the regulations in codes on air pollution control to specify that approved accurate sampling and test methods for detecting and determining quantities of the particular pollutant be employed.

Status of Air Pollution Control Testing and Resultant Data

It must be recognized that there has been and is at present a wide latitude in the quality and accuracy of data relating to emissions of air pollutants as reported and used today. Wrong methods of measurements, omission of correlated observations, poor test equipment, improper use of apparatus, lack of calibrations, and testing by unqualified and nonskilled personnel can account for high deviations and inaccuracies in test results. It behooves anyone concerned in use of data on the performance of air pollution control equipment to look carefully at the reliability of the method of determination of data that has been referred for study or obtained from former or present measurements. While it is not in the scope of this chapter to evaluate different data that have been used or published nor the background and methods that were used in procuring the data, still caution must be used in their acceptance and consideration. It is the objective in this section to alert those concerned in air pollution control evaluations to the importance of accurate data in formulating and accepting guarantees on proposed equipment and also on proofs of the performance.

Inaccuracies in sampling and test methods, apparatus, and procedures are well illustrated in a series of simultaneously performed quantitative comparison tests on emissions from a dust collection system serving a ferrous metal cupola furnace. A widely used test method for stack sampling was compared with the proved accurate quantitative sampling and test method in respect to efficiency in sampling and measuring the emission products coming from a dust collector on a cupola furnace.

The results of these tests, as summarized in Table 4, show considerable differences in the measure of total particulate matter by the two methods. Three simultaneous runs were made on the same discharge duct with the two sampling methods.

In Table 4 the first column identifies the individual runs. The second column gives the total particulate matter expressed in grains per standard cubic foot dry gas as collected and measured by the standard quantitative method. The third column gives the corresponding values of the total particulate matter as determined by the Ajem method and expressed in grains per standard cubic foot dry gas. The fourth column gives the percentage of total particulate collected and accounted for in terms of the total measured by the Ajem method which experiences only a small fraction

TABLE 4 Quantitative Comparison Determinations on the Abilities of the Standard Quantitative Method and the Accurate Ajem Method in Collecting and Measuring Total Particulate Matter from Dust Collector on a Cupola Furnace

Test No.	Total particulate emission product collected by standard quantitative method, grains/scf dry gas	Total particulate emission product collected and measured by Ajem method, grains/scf dry gas	% total particulate collected by standard quantitative method
1	0.0530	0.1210	44
2	0.0642	0.1500	43
3	0.0673	0.1610	42

of 1 percent penetration through the submicron range. It can be seen from these data that it is dangerous to judge and purchase air pollution control equipment that has been evaluated and rated on the basis of removing total particulate if the test method used in the claims for performance accounted for only a small fraction of the total particulates.

In this comparison test, the method was not capable of collecting the relatively high amounts of submicron aerosol particles originating from the cupola furnace. The author talked at a later date with the inventor of the method and was informed that it was not developed or intended for collecting small micron and submicron particles. This is an example of how a method that becomes widely used may be projected by others and extended in use such as in submicron range without proof of accurate performance in sampling and measurements.

It is imperative that the individual air pollutants be sampled and tested with methods and apparatus capable of accurate determinations of the individual emission species and of the total range of submicron particle size.

It is also the intent at this point to call attention to the fact that the quality of data varies and to alert those concerned to why these large deviations have and can occur. These factors include:

Existence of variety of test methods and procedures While effort is being directed to improving, selecting, and accepting accurate test methods, there is a wide variation in methods and processes for sampling and testing. It is recognized that test methods must be selected in relation to the nature of the individual specific pollutants, possibly their relative quantities present, the conditions in their formation, their changing properties, and the influence of other pollutants in the mixture.

Wide latitude in types and quality of measuring equipment Part of the cause of the wide variation in data reliability has been in the type of equipment used for sampling or measuring along with the associated techniques. It is strongly emphasized that these wide deviations in measurements of emission quantities are caused by different conditions or limitations such as:

1. Presence and influence of a high percentage of submicron solid particulate matter. A study of the data from many hundreds of quantitative measurements, with solid particulate sampling and testing equipment such as Greenberg-Smith impingers followed by effective backup filters on collection of airborne solid particulate emissions such as of test dust pigments and of metallurgical furnace particulates that are very concentrated with the submicron particles, has revealed that, even if the sampling is done isokinetically in respect to the flow of air carrying the submicron particulate matter, the Greenberg-Smith impingers may pass from a few percent to near 75 percent of these very finely divided particles. The relative penetration of submicron particles through the Greenberg-Smith impinger is influenced by the rate of flow of air with submicron particles through the Greenberg-Smith impinger train, by the nature of the solid particulates, and by the relative concentrations of submicron particle size particularly of the smaller micron dimensions. These findings give proof that an efficient backup filter must follow the Greenberg-Smith impingers to account for the total of solid particulates. It may be seen that the evaluation of emission or performances of air pollution control equipment by use of apparatus or portion of test equipment that is capable of sampling or measuring only part of the particulate matter may lead to serious error and wrong judgments, and such erroneous data are not acceptable in agreements or guarantees.

2. Use of particulate air sampling devices for collecting submicron particles that were not intended or developed for capture of submicron particulate matter. A serious error has been made in the use of sampling equipment which was not capable of collecting all the airborne particulate matter, since the devices were not designed or developed for effectively capturing particulate matter in the low or submicron ranges.

3. Lack of proof of performance of sampling and measuring equipment. Another serious situation in past and present sampling in respect to particulate measuring devices is that the particular device has not been evaluated or calibrated for performance in capture of particulate matter of the type in which it may have been or is being employed in the field.

Measurements performed by inadequately trained personnel To have modern up-to-date equipment does not assure accurate, reliable test data. Many tests are performed by personnel who are not skilled in the art and are not aware of the special problems in sampling and testing. They may lack experience for recognition of conditions that may involve an abnormal situation, as it is necessary to record all the pertinent facts inclusive of correlations with the industrial process. Though schools and training courses are available, still many of the testing personnel have not had the experience and are not fully qualified for performing the required accurate sampling and testing.

Variation in the responsibility and reliability of testing agencies Since the beginning of accelerated programs in air pollution control, many agencies have sprung up as well as individuals offering service in testing for emissions. These agencies for testing vary widely in the degree of accuracy in which they determine the air pollutants. At present there is no listing of qualified and accredited test agencies. This present status in the field of testing resulted from a number of factors, including lack of defined accuracy required in testing, use of old test methods and measuring equipment capable of limited accuracy, oversimplification and carelessness in procedures, short cuts in sampling and measurements to reduce cost of testing, failure to recognize the new rigid code regulations that require accurate and reliable test data, and failure of the contractor in demanding accurate determination of each pollutant of concern in the emission. It is believed in the future that there will be an attempt to evaluate and have on reference qualified test agencies and personnel. In a later section the particular questions will be discussed that will determine in some measure whether an agency is qualified to the degree required for conducting specific tests.

Careless use of older methods and equipment formerly employed when the emission limits were high or not defined Another condition that exists in relation to obtaining accurate reliable test data is that in many cases older methods are used that were not designed for sampling and testing accurately the submicron particle.

Assurance of accurate, reliable test results from qualified personnel and adequately equipped sampling and testing apparatus It is encouraging to note that there are laboratories that are well equipped and have understanding, well-qualified personnel to do accurate, reliable testing of emissions.

Reasons for Inaccurate, Unreliable Data on Emissions

If may be more helpful to those concerned in obtaining services in testing for emission to summarize more specifically the reasons that many determinations of emission products are inaccurate and unreliable. These reasons include:

1. Failure to recognize the presence of different pollutants and their individual chemical and physical properties.
2. Use of wrong or inapplicable sampling and measuring apparatus for a specific or combination of airborne pollutants.
3. Failure to recognize that with submicron size particles, different instrumentation is necessary.
4. Tests performed by unqualified, inexperienced personnel.
5. Poor condition of apparatus.
6. Failure to calibrate specific pieces of measuring equipment.
7. Failure to follow testing procedures.
8. Lack of quantitative evaluation of airborne products.
9. Poor handling and transferring of samples.
10. Improper cleaning or lack of cleaning of test equipment.
11. Lack of proper corrections for temperature, pressure, humidity.
12. Improper sampling positions. A serious cause of poor test results is improper positioning of sampling component, particularly the sampling nozzle.
13. Too small sample collected. Inaccuracies can occur if the sampling does not involve large enough volume of air or collection of large enough sample of the particulate matter or chemical pollutants.

Contractual Guarantees of Performance of Air Cleaning Equipment

While a proposal or purchase contract must give the specifications, capacity, size, dimensions, power required, and the price of the equipment under the arrangements with which it is to be installed, several other basic elements must be included to give full and detailed understanding between the supplier and the purchaser of air pollution control equipment. One of these phases of guarantees is important and necessary on the performance of the equipment. While this section of the chapter is not attempting to state the conditions of contract in the legal terminology that may be individually used by the legal counsel in different ways to express the guarantees of a supplier and the expected requirements and conditions of the purchaser, there are a number of fundamental elements that should be recognized and included in the legal contractual statements relating to performance of the air cleaning device. These include:

1. The performance of the installed air pollution control equipment should be evaluated, determined, and accepted or rejected by the results of accurate, reliable sampling and quantitative measurements of pollutants. These conditions are paramount in order to satisfy the stipulations given in an earlier section.
2. Recognition and definition of pollutants to be removed to prescribed limits. Another portion of the guarantee must recognize, identify, and define the individual pollutants to be removed and their prescribed outlet limits from discharge ducts or stacks. A condition in writing a guarantee for performance of air cleaning equipment is that the individual pollutants must be specifically named and the individual limits of these pollutants in outlet loadings.

3. Requirement that test sampling and measurements in determination of performance are to be conducted only with the field installation of equipment and with the production process under normal operation. A stipulation that should be included in the guarantee is that the test sampling and measuring must be performed in the field on the equipment while the industrial operation is in full production. An example would be that the sampling of the emission products from the collector system on a cupola furnace must be taken during the melting operation.

4. Requirement for guarantee on accuracy and reliability of quantitative sampling and measurement of pollutants. A statement should be included that the sampling and measurement shall account, for instance, for the total solid particulate matter to a fraction of a percent and other pollutants specified as to the accuracy to which they are to be collected and measured. This will establish in one way that the evaluation of performance of the equipment and acceptance has been based on reliable test results.

5. Requirement of submission of full data on sampling and measurements. There has been a tendency to report results such as in the performance of equipment in simple values or merely by saying that the equipment is satisfactory or passed in respect to a certain reference value. For protection of the supplier as well as the purchaser and to have documentary evidence to show to enforcement agencies, complete results must be presented including:

1. Operating conditions of industrial process.
2. Periods of sampling in relation to status of industrial process.
3. Description of test method and equipment used in sampling.
4. Procedures for:
   a. Detailed steps in sampling.
   b. Transfer of collected air pollutants.
   c. Concentrating for drying for preparing and weighing samples.
   d. Size of samples of individual pollutants. It is important that the tests be conducted in a manner over a period of time that sensible quantities of pollutants are obtained for weighing or for other methods of determination.
   e. Application of corrections. Another requirement is that adequate measurements have been made, for instance, on temperatures, gas pressures, and moisture content so that the appropriate corrections can be made and shown in the report.
   f. Calculations. It should be a requirement that reported performance results be accompanied with the indicated calculations for arriving at the reported values of outlet loadings.
   g. Expression for relative accuracy of determinations. A point should be made clear in the requirements for the tests as well as in the reporting of results, no matter who are the performers in making the samplings and tests, that the accuracy of the tests should be stated in the report.

6. Quantitative tests after periods of time following installation of air pollution control equipment. It is known that many air cleaning devices deviate widely in their performance efficiency as influenced by a number of factors. The owner or purchaser of the equipment should demand in the contract that the equipment for a specific air pollution control installation should give a sustained performance over a period of time. It is known that a number of types of equipment will give a good performance characteristic with a low outlet loading at time of installation, but in a short time the quality of pollutant removal changes rapidly. A criterion that should be introduced into the contracts is guarantees of stated performance not only at time of installation but at periods of time following the startup of the equipment. For such a criterion, quantitative sampling and evaluation of performance should be performed in such a schedule as

1. Time of installation
2. At 6 months or so after installation
3. At 12 or more months after installation

This would be a protection to the purchaser of the equipment and also confirm to the supplier that his equipment is meeting a certain guarantee that he has given to the purchaser of the devices. This guarantee on results of good performance would better satisfy the enforcement agencies, who are naturally going to follow and observe the performances. A requirement in the contract should specify that full complete reports of tests on the performance of air pollution control equipment must be submitted in writing. Such reports should include descriptions of test

AJEM LABORATORIES
INDUSTRIAL HYGIENE DEPARTMENT
LIVONIA, MICHICAN

AIR POLLUTION EMISSION TEST SUMMARY SHEET

TEST IDENTIFICATION NO. ______ DATE ______ AUTH. BY ______

LOCATION ______ TYPE OF EQUIPMENT ______

TEST SERVICE P.O. NO. ______ AIR POLLUTANT SOURCE ______

CFM @ STACK CONDITIONS	
CFM @ 70°F 29:92" "HG" DRY	
CFM @ INLET CONDITIONS	
DISTRIBUTOR – NO. AMPS PRESS GPM	NO. 2 AMPS PRESS GPM
FAN AMPERAGES 1	FAN NO. 1 FAN NO. 2
SATURATOR–	AMPS PRESS GPM
SATURATOR SUPPLY–	AMPS PRESS GPM
COLLECTOR SUPPLY–	AMPS PRESS GPM
MOIST. ELIMINATOR	AMPS PRESS GPM
SYSTEM SUPPLY	AMPS PRESS GPM
TEMPERATURE – ABOVE SATURATORS °F	
TEMPERATURE – SATURATOR WATER	
TEMPERATURE – COLLECTOR INLET AIR	
TEMPERATURE – COLLECTOR WATER	
TEMPERATURE – FAN INLET AIR	
TEMPERATURE – ELIMINATOR INLET AIR	
TEMPERATURE – ELIMINATOR WATER	
TEMPERATURE – DISCHARGE AIR	
TEMPERATURE – AMBIENT AIR	
STATIC AIR PRESS WG – ABOVE SATURATOR	
AIR SPWG – COLLECTOR INLET	
AIR SPWG – ABOVE SPRAY PATTERN	
AIR SPWG – BEFORE FAN	
AIR SPWG – BEFORE DAMPER	
AIR SPWG – BEFORE ELIMINATOR	
AIR SPWG – DISCHARGE	
TOTAL NO. CHARGES	
DAILY AVG. MELT-RATE – TONS – IRON/HR	
NO. OF CHARGES DURING TEST	
TEST AVG. MELT RATE – TONS/HR	
CUPOLA TUYERE AIR VOLUME – CFM	
CUPOLA TUYERE AIR TEMPERATURE – °F	
CUPOLA TUYERE AIR BACK PRESSURE	
AFTERBURNERS OPERATING/GAS DATA	
% VOLUME DRY GAS BASIS DISCG. GAS ANALYSIS	CO_2 O_2 % CO %
THIMBLE NO. BAROMETER "HG"	
CONDENSATE-CC VAPOR TRAP °F	
SOIL WEIGHT IN THIMBLE – MILLIGRAMS	
SOIL WEIGHT IN IMPINGERS – MILLIGRAMS	
SAMPLED GAS VOLUME @ METER CONDITION	
SAMPLED GAS VOLUME @ STD. CONDITIONS	
GAS METER VACUUM "HG" TEMP. °F	
NOZZLE SIZE NOZZLE SAMPLING RATE	
SAMPLING TIME – TOTAL / PERIOD	
OUTLET LOADING – TOTAL PARTICULATE	
OUTLET LOADING – SOLID PARTICULATE	
GRAINS / SCF DRY – TOTAL PARTICULATE	
GRAINS / SCF DRY – SOLID PARTICULATE	

NOTES ______

TEST SUMMARY SHEET NO. ______

Fig. 5 Air pollution emission test summary sheet.

methods, modes and positions of sampling, correlation with process operation during production, observed data and measurements, calculations, and test summary sheets. An example of an air pollution emission test summary sheet is shown in Fig. 5 as adapted for recording and presenting test data from measurements on the performance of dust collection systems on ferrous metal cupola furnaces. Modifications of such summary test data sheets are necessary for measurements involving other emission sources and evaluating specific air pollution control systems. Such air pollution emission test summary sheets are not only helpful in preparing the data and their correlation but are of help to all parties concerned including the design engineering organization, the supplier and installer of equipment, the purchaser of the air pollution control system, and the air pollution control enforcement agencies.

Startup Procedures for Air Pollution Control Equipment

Another portion of the contractual agreement should be committed by the supplier to the purchaser for furnishing the following:

1. Full descriptive, instructional operation manual. This manual should be a complete description in respect to mechanical components, fans, dampers, rotational parts, electrical devices with specification on voltage, amperages, air pressure controls, defining pressures, volumes of air required, water facilities specifying volume and pressure required in different sections, pump capacities, safety controls, and other air handling regulation devices and rating of volume at referenced conditions of air to be processed in the system.

2. Provisions for technical, engineering help from supplier. The contract in purchase of equipment should also define a provision of the supplier for furnishing technical engineering instructional help on the site at startup.

3. Interconnect controls with process equipment. A check should be made whether the interconnection controls to synchronize the air cleaning equipment with the process operation have been satisfactorily installed and are operating properly.

4. Special instructional material for startup. A special, carefully written set of instructions specific for the startup of the dust and chemical fume control equipment can be of great help in assuring good operation and understanding between the supplier and the user of the equipment.

5. Maintenance and service manual. One of the best insurance steps for continued operation and sustained high efficiency, besides good relation between purchaser and supplier, is the use of a well-written maintenance and service manual. This manual may encompass the content instructional manual described above, identifying the individual components of the system and requirements for their maintenance. Another reason for a good maintenance and service manual is that the personnel assigned in many companies for keeping equipment in operation or cleaning it change rapidly, and new personnel must become acquainted with the equipment. One of the serious complaints and findings in industry is the lack of adequate maintenance and service on equipment, particularly new apparatus. Industry can realize great savings not only in capital equipment investment but also in the cost of service if the maintenance and service are maintained on a better quality level.

6. Presence of supervisory personnel from equipment manufacturer. To assure satisfactory startup and continued performance of the equipment, the agreement in purchase of equipment for air pollution control equipment should stipulate that qualified supervisory personnel from the equipment manufacturer will be available not only at time of startup but also during preceding stages of installation and also in training of the owner's personnel who will service the equipment.

7. Check on installation of sampling ports at proper positions. At time of startup, the installation of ports in outlet stacks, ducts, and other positions should be accurately checked to assure meeting the conditions for proper sampling of emissions, determination of temperatures, measurements of air volumes, or other correlated measurements.

8. Provision in installation of safe platforms and ladders for test sampling. One of the common omissions on installation of air cleaning equipment is failure to put wide safe platforms, railings, and ladders not only for servicing of the equipment but also for access to the proper positions for test sampling in determining the efficiency of the system. In fact, lack of proper platforms and ladders has been the reason for not performing tests that should have been made to establish whether the equipment is functioning to the specifications appearing in the purchase agreement

and also has been the cause of doing poor testing, resulting sometimes in erroneous test data.

9. Shutdown controls. At the time of startup one of the particular steps is to assure the proper shutdown and interconnect controls have been installed and are in operation. This is important not only in relation to the operation in the future but particularly at startup, when some malfunction or interconnect controls may not be operating properly.

10. Check on auxiliary equipment. In many air pollution control systems, auxiliary equipment is incorporated. A careful check should be made on all auxiliary equipment such as for the handling of the collected pollutants, pumping facilities where wet type units are operating, and proper direction of rotation of power equipment. A stipulation in the original contract should be to the effect that adequate description and instructions on operation of each auxiliary combination of equipment must be provided by the supplier. At time of startup, careful check should be made of each component of these accessory units supplementing the major air pollution control equipment.

11. Protection of control wiring from heat, chemical fumes, and environment. A step in preparation for startup is that a careful check should be made not only that the wiring of different electrical devices meets the industrial electrical specifications but also whether the control wiring is adequately protected or shielded from heat, chemical fumes, or other severe environmental conditions. In this connection it should be observed whether any equipment has been jumpered out which would prevent safety devices from operating or cause some other imbalance.

12. Specific measurements at time of startup. Several specific sets of measurements should be made qualitatively if not quantitatively at the time of startup, with a written recording of the respective values. These include:

a. Static air pressure across specified sections of equipment

b. Evaluation of airflow referred to a reference temperature and position in the system

c. Water pressure where water is used in the wet system

d. Volume of water supplied or being used in different components

e. Voltage of electrical power sources

f. Operating amperages of individual motors or other electricity consuming equipment

g. Fluctuation in any of these utility services supplied by the industrial company

A correlation should be made from these different steps in preparation of the startup or from initial observations to determine not only whether the equipment is operating properly but also whether it is meeting the basic conditions in the purchase agreement of the system and the related guarantees in its functioning. This prompt correlation enables quick adjustments to be made if necessary at the time of startup to get the system functioning.

13. At startup a check should be made whether the equipment at its maximum air volume setting is handling all the emission products from the industrial process operating at normal production or is handling the stipulated air volume of air pollutants.

Examples of Erroneous Determinations of Pollutants

Lack of accuracy in testing in past In earlier times, accuracy of testing was not taken seriously or challenged. This accounts for the wide range of values that have been reported. The reasons for these inaccuracies have been reviewed in an earlier section.

Wide latitude in range of outlet loading In many localities there was no definition of limits of outlet loadings, and in other cases wide ranges of emission particulate matter were permitted.

Earlier laxity in enforcement of air pollution control There was a time when mere installation of some type of air pollution removal system was enough to satisfy the control groups. In many cases no tests were made by the enforcement agencies or by representatives of the supplier or the purchaser. Such is not the present case, with the advent of a multitude of code regulations in air pollution control from municipalities to the federal level. Now there must be positive proof in performance of equipment in meeting a regulation.

Early lack of identity or description of testing procedures or test equipment There was a time when, particularly before the initiation of the new code regulations, if a statement was made that tests had been performed, this was accepted as satisfying the condition of air pollution control. Such laxity cannot be tolerated in view of the new regulations and ability of enforcing agencies to test emissions from different sources accurately.

Lack of dependability in reported test data At present as well as in the past in reporting tests on air pollution emissions, reference is made sometimes that a certain test method was being used. Though it nominally was being used, actually it has been repeatedly observed that there was wide deviation in the actual performance of the test procedure. This deviation occurs to different degrees, depending on what variation is performed and the laxity or carelessness displayed by the personnel performing the tests. This lack of rigorous performance of the named test method greatly discounted the worth in accuracy of the reported results.

Recognition of value of new accurate, reliable test methods and their performance It is recognized and proved that new accurate, reliable test methods are available as well as the equipment with which to perform such tests. It can be emphasized that these methods will be used by responsible skilled personnel in the enforcement of air pollution control regulations. These reliable test results will be recognized and used in court prosecution cases. This again means that such test methods should be required and specified in contracts and accordingly used in relation to the evaluation and selection of air pollution control equipment and the associated apparatus as well as proof of efficient performance after installation.

Qualifications of Organizations for Sampling and Testing of Emissions

The frequent and serious question that is raised at air pollution control conferences in different industries is to the effect that there is no list of qualified agencies for sampling and measuring emissions originating from different industrial processes. With the realization that code regulations must be realistically met, the concern is expressed for obtaining accurate, reliable test results. With the accelerated programs in air pollution control, many personnel and organizations offer their services. Since there is no registration of qualified testing agencies at the present time, the owner and supplier are dependent upon what they can improvise with their own personnel for testing or frequently, without proof of the accuracy and capabilities in testing performance of an agency, engage such an applicant for the testing of certain emissions. This has placed the supplier as well as the purchaser at the mercy of testing by poorly equipped, poorly trained persons with limited skill in making the tests. This can be serious and misleading not only in the representation of equipment to the customer before selection but also in the verification of performance of the equipment after installation to satisfy guarantees between a supplier and purchaser and eventually prove to enforcement officials that the code regulations are being met. Field observations have shown that references for the applicant to test do not adequately give the rating of the ability of the person or agency for testing.

In view of this quandary and uncertainty in the quality of sampling and testing that can be done, selection is left to the judgment of the organization engaging the testing agencies to evaluate the ability and expected quality of the testing. To overcome this difficulty at present, there is a set of questions to which the purchaser, for instance, can demand answers from the laboratory or agency proposing to do the tests. These questions may be put in a written questionnaire form and addressed to the agency or person proposing making application or being considered to do the tests. They may be enumerated as follows:

1. What emission products is your laboratory or organization equipped to sample and test quantitatively?

2. What percentage of submicron solid particles can your equipment collect in sampling?

3. What proof can you produce that submicron particles do not penetrate through the sampling train that would be used? If penetration does occur through the sampling device, what percentage of the finely divided particulate is not captured?

4. How do you collect the organic aerosol particulate matter, and what percentage of the liquid aerosol particulate penetrates the sampling train What proof of the degree of collection of the liquid particulate or liquid aerosol can you give or have established?

5. How do you express the outlet loadings of particulate matter?

6. What percentages of total emission particulate and of total solid particulate can you guarantee to collect in the sample?

7. How do you measure the volume and quantities of air in relation to the samples collected and in relation to total gas emissions?

8. What corrections are applied in calculating the true outlet emission loadings?

9. Do you correct to a standard gas reference condition?

10. How do you test for other pollutants such as carbon monoxide, sulfur dioxide, sulfur trioxide, halogens, or other specific corrosive toxic materials involved in the particular emission?

11. What accuracy do you guarantee in respect to determination of the total particulate matter?

12. Will you submit all the recorded data?

13. How do you perform this sampling in relation to the industrial process?

14. How do you judge and decide the period of sampling and testing?

15. What volumes of contaminated air do you consider minimum for quantitative testing?

16. What quantities of solid particulate do you consider minimum in collection for quantitative evaluation of emission particulate matter?

If the person or organization proposing to test does not answer these questions satisfactorily or, for instance, the responses to these questions in writing demonstrate the agency is not capable in their procedure and use of specific equipment to account for all the solid particulate matter to a fraction of 1 percent, or if corrections are not being used in calculation of data, or if the tests are not correlated with the industrial process, then there is grave doubt as to the reliability and value of the data that would be obtained.

17. How are solid particulates transferred and concentrated?

18. What equipment is used for testing the specific emission product?

19. What are your criteria for positioning of sampling stations in a stack or in a duct?

These questions will give a good idea as to the abilities and the reliability in guarantees that the testing personnel or organization will give in respect to the accuracy of the data to the company for which the testing is being done.

If guarantees on performances of air pollution control equipment between the supplier and the purchaser are to be a part of the contractual agreement for such installations, then another element of the guarantee must be that only competent testing personnel and reliable, accurate test methods will be employed in evaluations of emission products. In turn it will be required of the testing agency that it guarantees that the sampling and testing for specific pollutants will be performed with use of reliable methods and with assurance of specified accuracies.

Costs in Sampling and Measuring Airborne Pollutants and Testing Equipment Performances

A few statements should be made in relation to costs of sampling and analyses of airborne pollutants and also in the determination of the performances of installations for air pollution control. The accelerated programs for air pollution control, the extension in the demand for control of an increased number of pollutants, the complexities encountered particularly in industrial airborne emissions, the transitional changes in code regulations on limits of individual air contaminants, the uncertainties in efficiencies of equipment offered for air pollution control, and the lack of definitions in the degree of control of specific pollutants are prompting the incorporation of guarantees and stipulated conditions between supplier and purchaser in contractual agreements.

There is no substitution for sound, factual, reliable information in a defined and equitable contract such as between supplier and the purchaser of the air pollution control equipment. A significant portion of this sound information must be based on accurate test data on the relative and varying quantities of individual air pollutants to be controlled in arriving at the stipulations and conditions contained in the contractual agreement. The factual information and assurances serve not only as the basis of understanding between a supplier and a purchaser but also in relation to the air pollution control enforcement agencies in granting permits, from the time before and in issuing approvals or rejections after the installation is in operation and has been functioning over a period of time. It is recognized that the testing and quantitative determinations of the air pollution cost in terms of effort, preparation instrumentation, measurements, time, and services of qualified personnel. The employment of accurate and reliable test methods, the use of required associated precision equipment, and the conducting of tests by skilled personnel is the more economical choice and step in the long run in the analysis of air pollution control problems, in recommending and evaluation of equipment to control the particular emissions, in arriving at understandings between parties concerned with assurances and confidences and the satisfaction in bringing the pollution under control. The cost of obtaining accurate data on the control of these airborne emissions by the field measurements and laboratory determinations is far cheaper than making errors in judgment in producing or selecting the air pollution control equipment with possible failures in performances to meet code regulations and the possibilities of facing citations, shutdown, or lawsuits.

The complexity and variations in sampling, measuring, and determining of individual pollutants make it impossible to publish so-called flat rate costs for quantitative testing such as in the cases of metallurgical furnaces. The effluents vary to a great degree and physical conditions differ to large extremes. In arriving at a just cost on a proposal for testing air pollutants for a specific case and in fairness to all concerned, a preliminary survey on the site with all available information at the time taken into account should be accepted as a first-step proceeding before writing and offering any proposal on details of measurement services, on guarantees of accuracies in determination of specified pollutants, and in evaluating the costs for the tests.

The requirements in making the tests, the preparation, the scope, the precision to be realized, and the imposed conditions in the determination of contaminants include a number of factors:

1. Number and identities of individual air pollutants to be sampled and tested
2. Relative quantities of specific pollutants present in the emission and the related method and time required for sampling.
3. Specific requirement and instrumentation for precision determination of particular pollutants
4. Number of samplings to be made for each pollutant
5. Possible requirement of monitoring in testing for specified pollutants over stated period or periods of time
6. Nature of the pollution producing source or process and adaptation of sampling or tests to production cycles such as in metallurgical operations
7. Availability or requirements of ladders, platforms, railings, and safety devices to gain access for personnel and instruments to measure emissions correctly.
8. Effects of weather conditions on continuity of the testing
9. Occurrences of production breakdown or shutdown
10. Relative location of site for field testing
11. Requirement for travel time and expenses, and transport of testing equipment
12. Possible requirement of special new instrumentation
13. Insurance coverages for property damages and liabilities
14. Requirements for analysis of field samples at testing laboratories
15. Requests and requirements for particle size distribution evaluations
16. Number and type of personnel required

17. Requirements for different degrees of skill at corresponding rates of pay, differing from one area of the country to another and in foreign countries

It should be pointed out that field tests can require all the way from two to five skilled persons to obtain dependable accurate data. At no time should there be fewer than two competent personnel at the site for field testing, not only to accomplish the sampling and correlated field measurements, but also as a safety measure.

It can be seen from these enumerated factors that preparation, provision of quantitative test equipment, planned and controlled performance of testing, and employment of skilled personnel are essential for assuring and providing meaningful accurate air pollutant measurements.

Summary

A review of the influences of accelerated air pollution control programs, of recognition of the necessity of control over an increasing number of air pollutants, of legislative establishment of stricter but varying code regulations on air pollution control by all levels of government, of the offering of wide ranges of air pollution control equipment, and of the complexities inherent in air pollution control is presented in relation to the designing, offering, selecting, installation, operating, testing air pollution control equipment, and satisfying enforcement agencies.

Evidences are presented of factors and conditions that have impeded the progress in air pollution control such as lack of understanding of new air pollution control regulations, misjudgment in applicability of equipment due to incomplete and misleading information, misunderstandings between purchaser and supplier, absence from contracts of definitions of responsibilities and liabilities of parties concerned, failures to require and use accurate, reliable test data, omissions of requirements for safety measures, omissions of full specifications and of guarantees for sustained quality performance of equipment, and failure to include protective and conditional clauses in agreements.

The aims and objectives of the chapter are set forth for steps to take in overcoming the impedances to progress in air pollution control, for establishing better understandings in the specific requirements for air pollution control and in the relations between supplier and purchaser, for instituting and using contractual agreements that are more applicable and specific to air pollution control installations by inclusion of guarantees for sustained equipment performances, of explicit definitions in responsibilities, obligations, and liabilities in supplying, installing, purchasing, and maintaining of air polluton control equipment, in specifying that accurate, reliable test data be used in claims for performances of proposed equipment and in proofs for the performances after installations of air pollution control equipment, and for confirmation in soundness and economy of investment with satisfaction to purchaser, supplier, enforcement agencies, and the environmental citizenry.

The reasons and conditions in the air pollution control field necessitating full and specific detailed written contractual agreements are analyzed, and the elements constituting equitable, meaningful contracts covering guarantees on performances and responsibilities of parties concerned before, during, and after installations are described and defined. Strong emphasis is placed on the requirement, as part of the agreement and stipulations in the contract, the incorporation of reliable, accurate test methods with a stated precision in determination of quantities of specified pollutants as the basis of representing equipment as claimed by supplier and as proof of performance. Quantitative test data showing small to large unexpected deviation in accuracy and abilities of sampling and test methods to account for all the pollutants are numerically shown for specific cases.

Proof of accuracy and reproducibility of proper test methods in determining quantities of very finely divided particulates are shown in test data.

A review is presented of the points to consider and use in the evaluation of proposed air pollution control equipment and their abilities to remove specific pollutants from the air.

The importance is emphasized for inclusion of stipulations in a purchase contract that the air pollution control equipment should not only meet the code regulation requirements at time of installation but as part of the guarantee the equipment should give sustained performance in removing the defined pollutants as proved by tests at later designated times.

The present quantitative field test data in the performance of a wet type centrifugal generated spray dust collector system in removal of solid particulate from a ferrous metal cupola furnace over a period of 5 months verifies that sustained performance of air pollution control equipment can be experienced, though marked increase of outlet loadings over a period of time has been observed with a number of air contaminant removal systems.

The present status of methods and apparatus for sampling and measuring air pollutants is discussed in relation to accuracy and reliability of resultant test data with illustration of wide deviations in quantities of total emission particulate matter as determined simultaneously by the Ajem method, which is capable of accounting for all total particulate matter to a fraction of 1 percent, and by a method widely used in stack sampling.

Recommendations are made that testing agencies must submit a description of their proposed test method and measuring apparatus to be used and give assurance of the accuracy with which the quantity of each pollutant will be measured.

A type of questionnaire to be used by the supplier or purchaser of air pollution control equipment is recommended in determining the qualifications of the testing agency in obtaining accurate, reliable test rata.

Since an equitable contract agreement has a clause on guarantee of performance of equipment, and since the true performance of air pollution control systems has to be evaluated with accurate, reliable test data, it should be a requirement in engaging a testing agency that it give a guarantee or certification on the accuracy of the test results in determination of each specified pollutant.

Special attention is directed to the coverage in the agreement for responsibities of supplier and purchaser at startup time of new or modified equipment and at times of testing.

(Field test data assistance from Victor W. Hanson.)

WATER, POLLUTION CONTROL EQUIPMENT—AGREEMENTS, GUARANTEES, PERFORMANCES, TESTING, STARTUP, AND MAINTENANCE

The belated concern and activities through the past years on water pollution control, the recent and sudden realization of the magnitudes of pollutants entering the streams, rivers, lakes, and adjacent bodies of water, the recognition of an ever increasing number and variety of contaminants being discharged into these waterways, the acknowledgment of the existence of so many uncontrolled pollutant sources and lack of proper waste treating facilities, and the roused alarm at the damages to plant life and at the hazards and seriousness in the effects of the pollutants to fish, animal, and human life have been responsible for the recent accelerated programs and efforts in water pollution control. This acceleration is reflected in the enactment of numerous legislative control measures at all levels from federal and state to municipal, in the wide publicity directed toward stimulating public consciousness of water pollution and the urgency of remedies, in the enlistment and sometimes coercion of industries to treat and control liquid waste discharges, in the extensive and improved quality of literature on the related subjects, and in the educational efforts of universities and interested agencies in conducting formal or special training for personnel in the water pollution control field.

Accelerated programs such as in water pollution control can be hampered by a number of conditions and factors such as the lack of correlated information, the failure to consider all the water pollutants and their relative amounts that will be received by the treatment system, the too limited planning in respect to volume capacity and modes of treatment, the appearances of new and unexpected contaminants, the occurrences of side effects such as flooding or poor timing in treatments, and the selection of equipment and the structural components. Also, impedances and sometimes disappointments in the steps of water pollution control code regulations are due to misunderstandings between a contractor or supplier and purchaser, due to omissions of adequate specifications and of definitive statements on responsibilities and liabilities of parties concerned, due to absence of assurances and guarantees in performances of installations due to failures to specify in agreements the use of reliable analytical test methods and procedures, and due to failure to monitor the polluted water for a period of time to determine the variations in type and quantities present.

Several objectives of this presentation relating to the above phases of water pollution control are:

1. To help eradicate a number of these existing impediments and to enable a more rapid realization of successful prevention of contamination of water sources.

2. To enable better planning and engineering in each project, whether large or small, to control water pollution by directing the effort to basic facts, requirements, and goals in the specified treatment and conditioning of water, and by use of the best technological and engineering skill.

3. To focus attention on the necessity of considering each and every pollutant present in the water to be treated because of possible interchemical and physical influences of contaminants on each other.

4. To alert those concerned with installation of water pollution control equipment that widely varying conditions and compositions of liquid wastes can occur and that versatility in equipment is required to meet these changes.

5. To emphasize that before contracting for water treatment and handling facilities all facets of the particular water pollution case must be fully examined and the requirements completely defined.

6. To stress that all engineering estimates and evaluations of pollutants in the water should be based on reliable data determined by accurate methods and instrumentation appropriate for measurements of the specific contaminant. The obtaining of full data may require monitoring a polluted water to determine the variations in quantities of individual pollutants and the influences of the environment.

7. To insist that a thorough study be made of all present and possible changes in water pollution control code regulations pertaining to that location, inclusive of municipal, state, regional, and federal level.

8. To enable a thorough understanding between the supplier or manufacturer of water pollution control or treatment equipment and the sponsor or purchaser of the installation.

9. To prompt an early evaluation and committal for all power and utilities requirements and their location for the installation.

10. To point out that a greater economy will be experienced and a more successful solution of the water pollution will be realized by formulating the plans on established facts.

11. To urge that greater use be made of the contract by including more detailed descriptions of equipment specifications in respect to structural components, capacities, and rates of treatment under certain conditions for stated pollutants.

12. To include guarantees in contract for the efficiencies of a water pollution control process and associate equipment for the treating or removal of individually specified contaminants in the water.

13. To urge the incorporation in the agreement of definitive statements of the responsibilities and liabilities of the supplier or installer of the water pollution control equipment, as well as of the purchaser.

14. To stress the importance of stating in the agreement the obligations of the supplier to furnish engineering and qualified technical personnel help in the startup of equipment and to furnish adequate instructional material and service manuals.

15. To stress simultaneously the inclusion of statements of obligations of the purchaser to furnish competent personnel to operate and maintain the installation.

16. To prompt that the agreement be understood between seller and purchaser that all necessary permits and approvals for processes and installations are obtained before making the official contract final or beginning the installation.

17. To recommend the inclusion of protective clauses to the supplier in the contract that relieves the supplier of responsibilities in cases of excessive overloading of stated capacity, in case of introduction of new interfering pollutants not named in the contract, in case of poor maintenance, and in cases of sabotage or other unfavorable imposed conditions.

Similarities in Air and Water Pollution

There are a few similarities between air pollution and water pollution and the control of the respective contaminants. In the first place, both air and water are classified as fluids. It is also true that both air and water can contain pollutants in the different states of matter, namely, gas, liquid, and solids. In both air and water media, pollutants may be forming true solutions. In many cases they are in solid-liquid suspension or in aerosols in terms of air pollution or in the water as a class of colloidal dispersions. The contaminants in the air and water media may range from simple to very complex systems. Though both media, air and water, are inorganic, the contaminants may be inorganic or organic or mixtures.

Dissimilarities between Air Pollution and Its Control and Water Pollution and Its Control

The differences between air and water pollution may be reviewed as follows:

1. Air is an expandable medium in volume since it is composed of a mixture of gases subject to the temperature and pressure gas laws. In contrast, water is a medium that in the liquid state is near constancy in volume.
2. In air there is much less buoyancy for suspending partiiculate matter as compared with water with a much higher density.
3. In water the presence of solids, liquids, and gases can be at larger magnitudes per unit volume than normally found in air media.
4. The larger quantities of material per unit volume, whether capable of appearing as true solution or a suspension phase in liquid water media, present more complex systems as compared with the occurrence in equal volume of air.
5. The range of concentrations per unit volume may vary from low to very high in the water media, depending on the nature of the pollutant and its individual characteristics.
6. Water as a medium has a high specific heat as compared with a unit weight of air.
7. Because of the structure and characteristics of the water molecule, it displays a high dielectric constant compared with air. This characteristic can have marked influence on behavior and characteristics of pollutants that are contained in water and its liquid state.
8. With air as medium, there is rapid geographical motion due to winds, convection, and thermal effects, as compared with the motions experienced in water as a medium for contaminants.
9. Water has a distinct characteristic of freezing at 0°C, or 32°F, while air as a medium is far from liquefication or solidification.
10. Because of the unique characteristics of water, it provides an environment in which the water not only has strong influence on the dissolved or suspended materials but also permits a strong influence of solutes and dispersions on each other in the bulk water liquid media.
11. It may be stated that water serves as an influential environment for inducing, controlling, and participating in physical changes and chemical reactions. This has significant importance in relation to the control of pollution in water media. There are chemical reactions that will take place in water media which will not occur in air and many other media. This presents different problems as well as methods required for treating the individual pollutants present such as in the water media.
12. It is also obvious that different sampling and testing techniques are required for identification and measuring quantities of the pollutants in the water media as compared with those contaminants in air.

13. There are distinctly different sets of laws and regulations in the codes for air pollution and water pollution control. The code regulations also vary between air pollution and water pollution in respect to the geographical area controlled in the individual regulations. An example is that water pollution is defined in terms of bodies of water such as lakes or river basins.

14. Another difference between air and water pollution control is that the governing enforcement is by different agencies or sets of authorities.

15. In comparison of the code regulations for control of pollution of air and water, the objectives, procedures, and requirements differ greatly in their nature and specific requirements.

16. In air pollution control, most steps are taken to collect, treat, or convert the pollutant from its source. In the past and a great deal in the present, practice in water pollution is to dump the waste into a sewer and let it be subjected to a common treatment at the end. The fallacy of this general procedure for water treatment is obvious with the increasing variety and seriousness of many pollutants entering into the waste discharge systems.

From this review it is obvious that the dissimilarities between air pollution and water pollution control far outweigh the similarities.

With the recognition of the reasons for differences in laws and code regulations for air and water pollution control, it is clearly evident that a different set of guarantees and assurances must be employed between the supplier of water pollution control or handling equipment and the owner or purchaser of the equipment.

17. Another distinct difference in air and water pollution control processes is in the timing not only of the collection but also of the treatment. In the case of air pollution control immediate treatment is required at the source.

These differences in air and water pollution control necessitate also the writing of a different set of definitions on responsibilities and liabilities as compared with those specific for air pollution control.

Phases of Water Treatment and Handling

It is not in the scope of this chapter to review in detail different phases and modes of water treatment as adapted for controlling different pollutants. A number of the major activities and processes in cleaning and purification of water include the following:

Desalination of water With the increasing use of water, particularly in expanding industrial societies and areas, the importance of providing water from the sea has received a great deal of attention and effort. This has been the approach from the necessity of getting more water and also from the economics ot converting seawater to potable and industrial usable water.

Removal of solid suspensions One of the major fields of water pollution control is in the use of different methods for removal of suspensions which may range from the organic to the inorganic and in particle size from submicron to appreciable dimensions. The new code regulations vary in their maximum allowable quantities of the solid suspensions, and the present trend indicates that the allowable parts per million will still be decreased in the future.

Removal of organic pollutant content appearing in soluble and in suspension forms With the increasing concentration of population and the accelerated and expanded industrial processes, a high amount of organic matter is entering streams and other bodies of water. The organic materials range in all degrees of chemical stabilities, molecular weights, inertness, toxic properties, and different degrees of chemical acitivity response to environment or even to treatment. The presence of the organic matter has been seriously condemned in view of their effects on marine life, on vegetation, on the aesthetic quality of the water, on the safety in human consumption, on general usefulness of the water, and in the high consumption of dissolved oxygen.

Removal of odorous substances, both inorganic and organic Because of the sensitivity of the public in the use of water slightly tainted with odor and from extremely odorous bodies of water from waste and industrial operations, the control and removal of contaminants and preventing of the odor have been of paramount importance in many cases. The sequence of chemical changes, especially in the presence of bacteria and fungus, has presented very complicated problems for the controlling of varied sources of the odorous materials. In this connection, a number of newer methods and better testing have made better appraisal possible and at least a start on the prevention or elimination of the odorous content in water.

Removal of corrosive chemically active substances In industrial areas, particularly where heavy manufacturing of different products is involved, large quantities of chemically active materials such as acids and strong alkalies, oxydizing agents, and reducing agents have been discharged into the waterways. This has been of great concern not only to municipalities from the destruction of sewer systems and water treatment equipment but also in regard to damage to marine life and vegetation and exposed structures in the path of the corrosive water. Some of these same chemicals have relatively high toxicity and chemical activities which prevent the reuse of this water.

Removal of toxic substances Many of the suspensions mentioned, organic and inorganic, have other objectionable properties, and some may be highly toxic. They may be toxic not only to marine life but to the bacteria normally responsible in digesting for conversion of organic to a more simplified decomposable material in secondary stage waste treatment plants, or they may be deleterious in certain ways to plant and animal life or destructive to some industrial processes as the water is reused from the flowing streams.

Control of fungi and bacteria There has been an increasing concern and effort in the study and evaluation of the influences of the increasing list of bacteria and fungi species. These studies range from detection, identification, and determination of individual physical and chemical properties, to the evaluation of harmful effects or usefulness in treating particularly organic waste. One of the phases concerned with bacteria and fungi is in methods and processes for destroying hazardous and dangerous species. Some bacteria are cultured and successfully used in steps particularly of decomposing and chemically breaking down organic waste materials.

Removal of radioactive materials Since discovery and utilitarian concern in the use of radioactive and fissionable materials and elements, a great deal of quantitative effort has been devoted to methods of keeping radioactive materials isolated to prevent their gaining entrance into the main waterways and to find ways for disposing of the mixtures of radioactive materials with safety in handling and permanence in isolation.

Removal of toxic metals It is recognized that in many industrial processes, not only of the metallurgical type but also those involved in making products of well-known and exotic metals, high toxicity conditions can be experienced from these derivatives in the water supplies. This has necessitated special care in detecting their presence and ways to keep them from entering into, for instance, the waste treatment plants of municipalities.

It is not in the scope of this chapter to review individually these different phases in water pollution control and others not enumerated above.

The activities in water pollution control of these different phases and others not mentioned are well reviewed in discussions and documented by extensive bibliographies of presentations appearing in literature over a number of years. A number of these of more recent date well describe the methods, present status, and direction in which each of these phases are of concern in the water pollution control field.

Responsibilities of Specialists or Engineering Firms in Surveys, Planning, Design Engineering, and Installing of Water Pollution Control Processes and Equipment

The urgency, complexity, and crucial status of future programs in assuring healthful environments in terms of availability of clean, safe water has been repeatedly stressed by leaders in government and in water pollution control.

In such exhortations, the directives are that "The discharge of polluted waste into waters of the Nation *must* be controlled. The word should or encourage must be substituted by the dynamic present directive of *must.*"

Another recognition is that "Decisions on the type and degree of treatment and control of wastes and disposal in use of adequate treated waste water must be based on thorough consideration of *all* the technical and related factors in each portion of each drainage basin."

Different summaries and formal statements of policy of water pollution control have set forth different conceptions of the policy in the United States and also in other countries. A step forward in this direction is found in the 14 points of statement of policy of water pollution control as revised by the Board of Control of Water Pollution Control Federation.

Because of the complexities, magnitudes, and acceleration of water pollution control for private industries, small to large municipalities, states, regional river basins, and federal undertaking, the engagements of responsible and qualified specialists in engineering firms are required for assuring successful control of the contaminants in the discharges to the waterways.

To date very limited definitions of the responsibilities and the requirements of assurances from survey or engineering firms engaged to solve the particular water pollution problems have appeared in contractual agreements. Far more should be incorporated in the contract agreement between the engineering and consulting firms to be charged with the responsibility of study, recommendation of methods, equipment, and details of an installation than merely appearing in a simple drafted business contract with signatures of officials in the engineering company with a brief statement of services expected and statements of amount of fees and time of payments.

While steps have been taken in some areas and states for assuring qualified waste water treatment plant operators, and qualified personnel in engineering undertakings, there has been a lack in demanding these qualifications and certifications in many cities and municipalities and also in contracting with engineering firms to require assurances of qualified personnel to work on the assignments.

The contract with an enginneering firm responsible for any phase in the study, development, recommendation, or building of a water pollution control system should contain a number of important stipulations beyond the normal business points of agreement. This more complete contract would contain many of the stipulations that would be normal between a contractor for installation and the owner or purchaser of the installation. There are, however, other stipulations and assurances in a written contract with the engaged survey, engineering, or consulting firm which include the following:

1. Statement defining the scope of the investigation, study, field survey, type of measurements, and recommendations. One of the best insurances of a successful program is careful definition of all facets for which the engineering or consulting firm is responsible. In view of the complexity and diversity in water pollution control, no single set of points or stipulations can be selected to cover all cases. Each set must be individually composed.

2. The statement should be included for requiring proof of qualifications of personnel who will be available and assigned by the engineering or consulting organization to the specific water pollution control investigation.

3. Another important requirement is that the engineering or scientific survey agency will investigate, study, and base their recommendations and report on consideration of *all* pollutants present or anticipated in the water systems to be treated. This is well recognized as a partial responsibility of industry to be worked out jointly with those concerned in waste treatment and water pollution control.

4. The contract should be ample in provision that full testing for pollutants and their variations will be made including the use, if necessary, of monitoring systems by the engineering firm engaged for the work.

5. For economy and future planning, a stipulation should be that the engineering firm make a study and give a projected recommendation for future expansion or modifications for future anticipated needs.

6. Another stipulation is that the engineering organization or testing agency should give assurance of the reliability and accuracy of tests performed in the project.

7. A requirement should be included in the contract that the report of the engineering or testing agency be submitted in full, identifying test methods of sampling and positions of taking samples for tests.

8. A further stipulation should be incorporated in the contract that test data should have correlated information in respect to locations of the sampling, time of year, and timing relationship in respect to ambient manufacturing operations that industrial wastes are involved.

9. Full detailed reasons should be demanded of the engineering firm is respect to the selection of the methods of treatment of the waste or sewer discharge, such as for organic matter whether an anaerobic or aerobic process would be preferred to handle the organic waste material characteristic of the waste polluted water to be subjected to treatment.

10. In evaluating different processes for treatment of polluted water, assurances should be demanded of the organization making the study, survey, testing, and selection of equipment that defined efficiencies in treatment of the individual specified classes of contaminants will be realized and sustained in performances of the recommended installation.

11. A protective clause should be included in the contract that with the understanding of the defined pollutants present at the time the contract was consummated the engineering firm will not be held responsible if other unanticipated impurities or toxic materials are introduced in the waste water at a later date.

12. An additional stipulation to aid in the success of the study and recommendations is that the engineering firm will not be responsible for capacities of liquid waste to be treated beyond the maximum stated in the contract.

13. It should be a requirement that as part of the recommendations the engineering report will include specifications in respect to materials capable of resisting corrosion, chemical attack, and thermal effects in the selected and recommended processes.

14. It should be also stipulated that the sponsor, owner, or purchaser of the installation will make all information available to the organization doing the investigating and testing.

15. The requirement in the contract should be to the effect that all recommendations for removal of individual water pollutants and the assurance for performance of the equipment will be correlated with the water pollution control regulations that are in force in an area, whether they are municipal, state, regional, or federal.

16. The contract should be so written that if there are to be any responsibilities for control of the taste of water by treatment of waste water, the limits of specific tastes will be clearly identified and defined.

17. When any plant, petroleum operation, or waste treatment plant whether private or municipal is or will be concerned with loss or release of oily or organic substances, a full investigation of requirements, written report, recommendations, and guarantee of pollutant control should be demanded of any engineering or survey agency engaged for such work. Assurances must be given with documentation that methods and equipment meet all levels of such oily waste control regulations.

18. When radioactive wastes are concerned in waste treatment planning, the engineering firm should be required to study and recommend positive, safe methods and equipment for isolated handling, treatment, and disposition of the radioactive wastes in keeping with the quantity and radioactive characteristics of the particular radioactive element or elements present. Such recommendations must include full safety measures and devices with continual monitoring detection and measuring instruments through all steps.

19. In cases where thermal conditions in waste water discharge occur, it should be required that the engineering firm make studies of temperature, time, and volume of water into which discharge is made over an adequate period of time including climatic seasonal changes.

20. The engineering organization making the survey and recommendations, whether it is separate or a part of the eventual contracting company for installing selected water pollution control equipment, should be defined in terms of its responsibilities and obligations during the further step of installation, testing, or cooperative effort with the contractor in the construction.

Criteria and Factors in Selecting Water Pollution Treatment Methods and Water Pollution Control Equipment

In the planning stage operation for water pollution control, modification of older systems, or innovation of new and expanded facilities, certain criteria that fit the individual case should be included in the formulating and consummation of contracts for installation. These standards of judgment would include:

1. Stated present maximum volume of liquid to be processed and estimated quantities in the future.

2. Consideration of all the pollutants, whether solids in suspension or dissolved state, liquids in suspension or in solution with water, or gases dissolved in the water media.

3. The judgment in selection of components in a water pollution control treatment system must be based on the composite varieties of pollutants present from their minimum to their higher concentrations. The basis for decision must be on the results of accurate dependable data accumulated over a period of time. Full consideration also must be given to the influence of one pollutant on another at different concentrations, as it may affect how one pollutant removal system may be interfered with or nullified by changing concentrations and the presence of other pollutants. Also to be considered is the interchemical influence of different pollutants such as the influence of halides present in acid systems with resulting higher corrosive rates.

4. The evaluation of good water pollution control methods and equipment must be determined on the basis of proved sustained performances in reducing the pollutant to a specified level of concentration.

5. Another required point of evaluation must be in respect to the requirements of the code or code regulations that are enforceable on the specific type of pollutant inclusive of the code regulations of city, municipality, state, regional or federal level.

6. Consideration must be given in the correlation of performance of recommended methods and proposed equipment with water pollution control regulations that can be expected in the future.

7. Another criterion for judging water pollution control equipment is the relative power consumption or chemical treatment demand.

8. An important point of judgment in selection of methods and equipment for water pollution treatment is the flexibility and variation in ability to remove the individual pollutants from low concentrations to high concentrations and in different mixture combinations.

9. Another important factor in judgment of the method and equipment is in the relative safety or the ability to provide safety equipment for the particular water pollution treatment system.

10. A judgment on the qualifications of the method and equipment for removal of specific pollutants should be based on field performance of the equipment in question, though laboratory testing may have been directive and valuable in the development and application of the method.

11. Each component in the total water pollution treatment handling facility should be judged on its ability to give a sustained performance through the climatic range of environment and also through the wide variations of individual pollutants.

12. A thorough investigation should be made of the presence of inorganic and organic toxic materials or the possible occurrence in the future of such materials in contaminated water discharges to the waste treatment plant. Field experiences have shown that well-engineered waste treatment plants with primary and secondary stages have been seriously affected by the unexpected appearance of significant quantities of toxic substances. In one recent case of a modern municipal plant the fermentation process in a secondary stage was completely stopped.

13. In cases where radioactive substances could appear from any source, the planning should include installation of radioactive detection monitoring devices, and provisions should be made that the source can be quickly determined and steps taken as quickly as possible to isolate the radioactive source. The plans should always be to handle, isolate, and treat radioactive and toxic bearing wastes at the sources.

Reuse of Water

Though the title of this section is short, it contains one of the greater challenges for understanding of its fullest meaning, requirements for preparation and treatment of the water in water pollution control activities and its evaluation in respect to the following use for which the water will be employed. It brings to realization that at the present rate of water consumption there is necessity for frequent reuse of the same water. This large consumption does not permit nature to use its processes in a slower manner. The penalty for greater use of water is that man must assist nature and must recondition the water at a faster rate. Every step considered or taken in water pollution control reflects a phase of the reconditioning of water in anticipation of reuse.

Until recently the public thought of water primarily as the one-time immediate use which they were witnessing. With the new consciousness from federal to public citizen of the necessities of cleaning up a water system, a clearer picture has come to mind of the past, present, and future role water plays in the activities of man.

A more successful water pollution control will be realized with a conscious effort in study, planning, and treatment of water in respect to conditioning of the water for reuse.

A more vivid understanding of the reuse of water and the frequency of its reuse comes if one projects one's imagination to a biography of a molecule of water or group of water molecules in their cycle from the oceans and back to the seas. Such a biography of molecules would be informative and surprising to the average mind. It would make us realize the importance of water, the uniqueness of its properties and the multiplicity of its effects, the influences and participation in all types of physical and chemical processes. In imagination one can picture the water molecule evaporating from the ocean surface, its voyage in airborne gaseous state followed by condensation and precipitation at some place on land with the most circuitous and various sets of experiences before it gets to the sea. It would be found that the water molecule not only has lazily flowed from brook to larger stream but has been picked up by marine life, has experienced the environment in the waters controlling fish and aquatic plant growth, and has been the carrier of nutrient to be picked up by marine life. In this sequence of experiences it may have entered physically or chemically in plant life or been absorbed and impounded by the rocks and components of the earth surface. Its time of stay in these different conditions and states may vary from a short time to even thousands of years but it is still water in full circuit of travel. Further, as it collects with its associate molecules in the lakes and rivers, man in his enterprising activities takes it up for his vital use, which includes biological uses. Still, it is a water molecule and its associates which again continue on the way. In man's greater activity in industry, water is used in high volumes. It is here that the water molecule becomes exposed to its highly polluted state through exposure of corrosive materials, toxic reagents, salts, alkalies, organic matter, and suspensions that cause it to be highly contaminated and lose its purity in bulk solution. It is here that it is realized that water has a good many physical and chemical properties that are vital not only to life but also to industrial processes. In most cases the water is in its chemical stable state, ranging from the gas in steam engines or from evaporated causes to liquid to ice or in all three states at the same time. Here the importance of water is recognized in terms of pleasure in boating, fishing, skating, and skiing that man so enjoys. In a few cases the water enters into chemical reactions and loses its identity with release of hydrogen or oxygen, but eventually the oxygen and hydrogen will in most cases be back again at some time as the little water molecule H_2O. The repetitive reuse of water becomes very obvious from the experiences of water in even one cycle from ocean through its experiences and back to the sea.

If the water molecule could write such a biography, it would probably complain and would register its greatest complaint in the unfair severe environmental conditions imposed on it by man's activities such as in industrial operations with exposure to conditions of corrosive toxic materials, dissolved organic and inorganic substances, suspensions, decaying matter, and chemically active products which water is forced to associate with and carry in its bulk. It is in this contaminated state that the water molecule would realize that it is perilous for its use in this polluted state in terms of danger to life processes.

Since it is the carrier of so many foreign materials such as active chemicals, suspended solids, bacteria, fungi, and disease germs that may be damaging or fatal to animal life as well as to plant life, it is also carrying the contaminate that can interfere with, destroy, or even stop chemical reaction processes.

This strongly brings the realization that human, animal, and plant lives are dependent upon maintaining a chemical composition with water as the medium and that the water molecules must be in a high degree of purity so that no pollutants that are detrimental to the biochemical processes or plant cell function and growth will be present. At the same time water is the supporter and carrier of other nutrients and is the base medium in fluid in life streams whether plant life or the blood in the circulation in the living body.

With vivid recognition of the characteristics and participation of water whether or not in living organisms, all study, all planning, all legislation, all engineering, and all installations must be focused to the particular problem at hand for rescuing this precious water molecule from these contaminants and selecting the treatment to condition the water for the next anticipated use or storage of the water for a delayed use, whether it be for industrial processes, for protection of marine life, or for its functioning in a biological process.

With the recognition that water should be in its proper degree of purity for its function in the vital biochemical, botanical, and industrial processes, water should be receiving higher evaluation in its unique roles. Full consideration and action are essential to make it possible to have the water in the right condition of purity for its next reuse.

It is encouraging to note that, whether one considers the challenge from a point of view of conservation or from the ability to keep water in a particular impure state separated from discharge into lakes and streams, if simple steps are taken, captive recirculated water systems in industry can serve the purpose in the industrial processes, and such reuse with even partial purification can be of economical saving to the industry.

It has been too often the practice that when water is adequately available it is taken through a plant process and immediately discharged without consideration of reuse or even of purifying it to a satisfactory degree before discharging into the river, stream, or lake. This practice has been due to overoptimism about the continued availability of a supply of water and the inconsideration for those concerned in the necessary reuse of the water. Where there have been limited sources of water or where the natural water has been contaminated, greater consciousness has been developed toward reuse and captive recirculation of the water in industrial processes.

Several cases illustrate not only the possibilities but the desirability of using captive recirculated water systems. At one automotive plant millions of gallons of water were used per week in one process as a one-time use of this water. By the introduction of simple methods of filtering out small quantities of suspended matter and adding small amounts of chemical additives, the water was found to be more effective and could be used repeatedly with the only demand for water makeup for evaporation. Another case was in the washing of engine parts. The general habit in the industry had been to use a strong alkaline cleaner, run the power washer cleaning the parts a day or two, and then dump into the sewer and recharge with fresh water and new chemical additions. In this case it would seriously contaminate the water, which went into a small stream that in turn supplied water for a hospital a few miles down the stream. By confining the water solution in a captive system, and adding water to make up for evaporative losses in the circulated system, and using proper effective chemically stable cleaners introduced into this wash solution, it was found that the dumping of the solution every few days could be avoided and in one case the life of the solution before dumping was extended to 2 years and 2 months. At the same time a greater economy in the cleaning operation with less maintenance was required.

In another case the general habit in the particular automotive manufacturing operation was to use wet type paint spray booths to catch the overspray paint. This calls generally for a schedule of dumping the large bulk of liquid with a portion of the collected paint into the sewer every day or two. In this connection success in good air pollution control ties in many times with good planning and control in terms of water pollution. In this case, with high quantities of overspray in the magnitude of 400 gal of paint/day, the paint spray booth was redesigned with a centrifugal spray generator air wash chamber, and a specially developed chemical for killing, treating, and enabling the easy removal of the killed paint was employed. The new booth as contrasted to the older one, which was dumped every day or two, was not connected to the sewer and operated for 2 years and 7 months before it was finally dumped and cleaned. This also netted a high economic saving in production plus the prevention of paint waste discharges into the sewer and stream systems.

If ever the saying that one is his neighbor's keeper is true, it certainly is most applicable in the responsibilities including the effort required in taking the proper steps in treating and purifying the water for its reuse so that the water in its varied experience will go on to its next use in an unburdened state in respect to pollutants and contaminants but will be conditioned to serve its next unpredicted function with safety to all concerned.

Treatment of Polluted Water at Its Source

The attitudes in the past have been first to demand clean water and second to discharge the waste water with the responsibilities placed on others to clean the water to the degree necessary for discharge back into the main streams or lakes. This is reflected in the way municipalities have set up their domestic and industrial waste and storm sewer systems for handling waste and drainage water in large capacity by a general treatment.

In respect to the industrial waste discharge, serious situations arise in that with extreme variations of the nature of industrial wastes, their magnitudes and range of chemical activity and exotic properties, the normal single-stage treatment plant is not capable of coping with these varying quantities of pollutants in the water discharges.

When the second and third stages are introduced in the handling of waste, particularly from industry, still great difficulties and sometimes failure of the system to function properly have been caused by the introduction of chemicals and by-products that may nullify the processes for treatment of waste sewage material.

It is believed that it will be forthcoming in water code regulations as well as in the recognition of the responsibility of the industry that these dangerous chemically active pollutants must be localized and treated at their source. This is well illustrated in a recent case where a municipality with great pride and cost installed a modern waste water waste treatment plant. Plans were adequate for capacity, and the methods were modern and efficient for handling materials, particularly the organic, by bacterial fermentation and breakdown of these organic wastes. Along about the same time, permits were given to a large manufacturer to install a metal product plant and discharge the waste to a common sewer leading to this new municipal waste treatment plant. It so happened that the firm engineering the original planning as well as construction of the plant overlooked the facts that highly corrosive and toxic materials would be generated in this new plant. The municipal waste plant started out with good performance, but it was found that the bacterial inoculation had to be performed more frequently in the holding fermentation tank. Furthermore, the yield of the methane dropped and reduced self-made fuel to be used for maintaining the solutions at the proper temperatures. In a short time it was found that, no matter how much new strain bacteria was introduced, the fermentation was dropping to a very low rate and almost to zero. This was brought about by the fact that highly toxic metals and their derivatives in a widely fluctuating pH range of discharge were getting into the system and were inhibiting the normal fermentation action and in all probability destroying the bacteria in complex chemical reactions.

Repeated cases in the field point to these kinds of discharges that not only interrupt one of the important waste treatment processes but can interfere with the separation of solids subjected to standard or physical chemical treatment.

These findings should be fully taken into account and also prompt the incorporation into contracts of the responsibilities of the specific treatments at the sources of such chemically active substances. It should be specifically required that a number of sources of such dangerous toxic and detrimental materials will be ruled out from discharge into the main sewer systems, particularly from the industrial realm. Any study, planning, or contracting by organizations for water pollution control must take the possible presence of these substances into account, and there should be specific requirements covering at least the following and other similar cases. It behooves all regulatory agencies to see that the water pollution control codes have sections defining restrictions of discharge of such wastes and the responsibilities of those plants where they could originate.

The ability to handle solutions in a captive state has been cited above in the preparation and reuse with the high resulting economy. There is no reason why economy and safety and protection of a water treatment plant and process cannot be simultaneously realized. Therefore, it should be required:

1. All toxic and highly chemically active materials should be treated at the source and not be discharged in any active state into the sewers leading to the central plant or the municipal waste treatment. This may include removal of soluble and insoluble forms of metals such as zinc, copper, lead, or their derivatives and other similarly dangerous classified metals and their by-products.

2. It should be required that all highly toxic inorganic chemicals inclusive of cyanides, fluorides, and related objectionable anions must be treated and removed from the water at or near the source to the degree specified in the code regulations. This includes the chromates and other special complex anion derivatives.

3. It should be required that highly active metals inclusive of finely divided magnesium or aluminum alloys should not be discharged in the sewers but be treated and removed by special methods and equipment at the source. Many sewer explosions and fires are caused by finely divided active metals discharged into waste sewer systems, with resultant generation of combustible and explosive gas mixtures.

4. Another group for exclusion from discharge of waste in the sewers should include highly acting oxidizing agents, particularly peroxides of organic and of inorganic structures. Their proximity to other susceptible oxidizable materials can be of serious consequence. This group should also include the exclusion of other powerful oxidizing agents inclusive of chlorates, perchlorates, nitric acid, and other similar products. The relative small volumes of the liquid containing these generally does not necessitate the use of any large, bulky equipment for treating them at the source. Methods of treatment would be in accordance with the nature of the material, the presence of other contaminants, and the requirements of the chosen equipment and procedures.

5. Regulations should restrict the discharge of volatile organic materials into the waste. Again, these should be isolated and treated at the source. Sometimes reclaim of some of such materials, particularly of the solvent character, can be realized with some monetary value. This restriction is a *must*, inasmuch as some very disastrous explosions have occurred in sewer systems where volatilization of the organic matter creates an explosive mixture or some other conditions set off chemical reactions.

6. The restrictions should also apply for treating at the source toxic materials particularly of the organic family that are known to be dangerous to plant, animal, or human life. In this connection it should be kept in mind that some of these toxic

materials are chemically stable and are not easily destroyed by the common waste treatment methods. Likewise, sometimes these toxic materials have a vapor pressure that can be significant in terms of air pollution.

7. Restriction of discharge of any radioactive material into waste treatment systems. All solutions containing radioactive products must be kept isolated and treated at the source.

The argument is sometimes given that it takes too much maintenance, too much equipment, and too much space for treatments at the source. The requirements that the main central waste treatment, whether single or multiple stage, whether private industrial waste treatment plant or a municipal waste treatment installation, must function consistently rule out any exceptions of allowing these dangerous materials to go to the sewer and cause malfunctioning in large-scale treatment. These objectionable chemicals can be more satisfactorily treated at source in small bulk and actually with least cost in overall water pollution control programs.

Treatment of these different classes of dangerous hazardous materials should likewise be considered, studied, and treated by those responsible for their handling. Such planning and preparation for handling such hazardous substances requires the same thoroughness as defined in a further section in the responsibilities of the supplier and installer of the waste treatment plant and of the sponsor or purchaser of the system for handling and treating other wastes on a large scale.

Contractual Guarantees Required by the Purchaser-Owner of Water Pollution Control Systems

With the repeatedly expressed desires to realize a healthful and protected environment with clean water, with the directives and regulation enforcements to remove the pollutants from the water, and with the enormous appropriations and commitments of money to assure greater acceleration and success in water pollution control programs, the specifications, responsibilities, and liabilities of the parties concerned should be made a part of the contract. It is time that contractual agreements became of full stature in dealing with water pollution control engineering, development, escalation, and operation. With the astronomical quantities of water as estimated at 310 billion gal or greater per day, the handling and treatment of water has become a major activity and business with enormous investments of capital. Therefore, the contracts should contain all the elements that are related to full understanding and requirements in water pollution control.

With new systems, more severe requirements, and larger magnitudes of water for processing, water pollution programs will be implemented by using the contract to bring a common understanding of the requirements, specifications, and obligations of each party concerned.

The present section deals with the points in the contract that would be and should be required by the owner or purchaser. These include:

1. Assurance for the volume of water handling capacities of systems to be installed. This stimulates careful engineering so that there will be no step in which there is restriction in the treatment or handling capacity of the polluted water.

2. A primary requirement from the purchaser should be assurance from the supplier on the ability to meet the particular set of code regulations in removal of specified pollutants and contaminants.

3. A natural common section of a contract would contain the committal on completion and readiness for operation of equipment by a specified time.

4. Another primary stipulation of the purchaser in the contract with the supplier is that the method, equipment, and facilities will give a sustained performance in the removal of the water pollutants occurring in the ranges specified. The sustained performance applies not only in the removal of solid suspended particulates but also can include liquid dispersion strata, oil layers, dissolved solutes both inorganic and organic, and also odors.

5. Another must for incorporating in the contract is a stipulation that the supplier or manufacturer or engineering firm in charge shall supply a manual and instructions on the equipment and its operation.

6. Well worded and specifically applicable to the water pollution control equipment is the incorporation of a guarantee to be made by the supplier to the purchaser on workmanship and structural materials. The selection of structural materials as well as components is of high importance in view of the ambient pollutant occurrences and the chemical attack of some pollutants on certain structural materials. This is in another sense a guarantee of long life of the equipment that will be in contact with the solutions.

7. Another portion of the contract should specifically state a guarantee of the supplier to the owner or purchaser that the system is capable of removing individual and mixtures of pollutants to a certain loading. Sometimes it is easy to remove a pollutant that is in a water system by itself. The problem becomes more complicated and frequently more difficult from the interinfluence of other pollutants in a mixture. Still further to meet the code regulations with a stipulated maximum residual in the water, assurances should be given that the system is capable of such performance.

8. For assurance of safety in the operation of the equipment as well as to the operating personnel, the contract should have clauses that define and specify the necessary safety features and controls that will be installed. It is important to note the alarming increase in the frequency of disabling injuries occurring as reported in the New York State waste water plants. It is reported the frequency of occurrences in lost time accidents per million man-hours has increased from 1963 to 1966 in the above plants by over 32 percent. It is also alarming to note that the frequency of occurrence of lost time accidents per million man-hours in waste water treatment plants is 16½ times greater than that occurring in the explosives field in 1958. Also, in comparison with the National Safety Council average of 41 major industries for frequency of occurrence of lost time accidents per million man-hours in waste treatment plants it is nearly six times greater. This indicates that contracts in water pollution control activities must be more complete in specifying requirements to ensure safety. Different protective measures may be taken. It should be added that more monitoring systems may be of help in bringing better control and safety to the operators. Though effort has been directed from different agencies for safety in waste water works, there is much that can be done and should be stipulated in the contracts whether it involves the designer, the operator, the manager, the supplier of the equipment, or reviewing agencies inclusive of those for insurance coverages.

9. The contract should include provisions for specific monitoring equipment with the appropriate sensitivity and located at strategic positions. This is important in view of the widely varying conditions in which pollutants occur in the wastes that are being discharged and received by waste treatment plants. This ambient condition can greatly upset processes of water treatment and should have detection devices specific for certain pollutants so steps can be taken in regulation automatically or by manual methods.

10. Provision should be made in the contract so that the installation will be so designed and constructed that the owner may expand the capacities of the systems at a future date and with possible modifications for changes of pollutants and their quantities or requirements in the code regulations.

11. In the contract there should be definite statements of assurance from the engineering firm responsible for the design or the contractor constructing and installing the equipment that adequately trained supervision personnel should be furnished at startup time and during the initial operation for any adjustments.

12. Requirement should be made in the contract that after its signing, all changes in the installation whether requested by the supplier or by the contractor have to be in writing and receive approval in writing by responsible officials representing the other party.

13. Review by owner's engineers of soil survey data obtained by the contractor before installation is contracted or undertaken should be a stipulation in any preliminary agreement or conditional contract. Proper evaluation of the soil and its substructure are of paramount importance for long life of the installation without serious damage due to effects of the supporting foundation soil or other environmental conditions.

14. The owner or purchaser should stipulate that all structural materials and components of the equipment must have the approval in writing of a representative of the purchaser at the time of formulating the contract and during and after installation. Deviations from the specified materials would require written approvals.

15. The acceptance of the installation and system should be stated in the contract as conditional on proof of performance in removing the named pollutants as measured with accurate, reliable test methods and apparatus. The nature of these tests and the timing of these tests should be defined in the contract and later agreed upon in writing if changes are made between the supplier and the purchaser.

16. The owner or purchaser should specify in the contract that trained, qualified personnel and appropriate instrumentation be employed in any measurements evaluating the performances of the installed system.

17. It would be well to state in the original agreement the method of testing for specific pollutants and qualifications by which the test personnel will be selected.

18. It should be clearly stated in the contract that the acceptance of a portion or all of the equipment system will be based on the performance of that section of the equipment or total system as determined by reliable testing procedures and testing apparatus.

19. In view of the complication and nature of the installations, specific clauses should define the liabilities of the contractor or supplier in cases where damage is due to fault of the contractor or injuries or death due to carelessness or negligence of the contractor or his personnel. This should be well defined in terms of any insurance coverage that is designated to take care of part or all of such damages or injuries.

20. A statement of responsibility or insurance coverage by the contractor in view of damage or injuries to the purchaser personnel or to subcontractor personnel should be carefully defined and included in the contract.

21. Another portion of the contract must be carefully formulated in respect to the responsibilities of the contractor in case the industrial processes or municipal operations are stopped or interfered with by the negligence or failure of fulfillment of the contract in any of its portions. This must be well understood to define the limits of such responsibilities.

If the engineering firm designing the system is a part of the contracting and supplier firm, then the stipulations discussed in an earlier section as applied to the engineering firm would also be included in the responsibilities of the contractor as discussed in this section and required by the purchaser.

22. The engineering firm responsible for planning, or the contractor for installation if representing the purchaser, must guarantee and assure that all permits have been duly issued covering the particular installation.

23. A stipulation in the contract from the owner or purchaser should be to the effect that all test data on performances will be fully recorded and presented inclusive of methods of sampling, measurements, and correlation with the operating conditions of the water treatment plant as well as monitored at inlet sources of pollutants.

24. The new concern in regulation of toxicity or the poisoning effect from waste discharges upon fish and other aquatic life and also biostimulation or the effect of waterborne nutrients causing algae growth requires careful definition of the specificity of the functioning of the intended equipment and the guarantees associated with its effectiveness.

25. Where thermal water pollution, oily or organic wastes, odorous substances, or taste producing chemicals are to be controlled, the contract should identify the sources and species of contaminants and require of the contractor first the submission of detailed proposed methods and equipment for the particular control and second a statement of guarantee to what level the specific contaminant will be reduced.

Contractual Conditions and Guarantees Required by the Contractor

In previous sections the stipulations have been cited that are important to the purchaser or owner of the equipment for water pollution treatment and control. There are also a number of conditions and assurances that the supplier or contractor of the installation of the equipment should have included in the contract to assure the full understanding and cooperation through the installation and starting the system. These conditions and assurances needed and required by the contractor include the following:

1. Assurances should be given by the purchaser or owner to the contractor or supplier as well as to any engineering firm for design that, as a preliminary step, as completely detailed statements as possible of the specific water pollution control project should be made available whether the installation is to be completely or partially engineered and installed by the supplier.

2. Another stipulation is that the contractor will be given, in writing, explicit direction as to the location and exact definition of the allocated space for the installation.

3. Assurances should be given to the contractor or engineering firm that all information if available from any previous ground survey and structural footing conditions should be supplied.

4. The contractor should be given statements in respect to the responsibility of the contractor for making any soil survey and determination of footing requirements.

5. A section of the contract should have sections in respect to (*a*) Definition of volume capacities per units and time required for each treatment. (*b*) Identities of each pollutant, whether in small or large quantities, soluble or insoluble, with indication of any likelihood of new pollutants to be received by the system. If the contractor or engineering firm is responsible for exploring and testing for each pollutant identity, it should be so designated in the contract for this responsibility. (*c*) Statement of maximum loadings of each pollutant that would be discharged into the waste treatment system. (*d*) pH ranges in waste discharges as determined by monitoring for acid concentration or series of tests over a period of time. (*e*) Statement of notice in respect to the presence of specific corrosive or chemically active substances. (*f*) Presence of toxic organic or inorganic substances. (*g*) Temperature range of the water system at different seasons of the year, or with variations occurring in the industrial processes affecting water temperatures. (*h*) Presence of oil or diphase systems in waste water. (*i*) Presence of radioactive substances.

6. If known, the contractor should be informed of any data on the ambient ranges of individual pollutant concentrations or volumes so as to appraise the required facilities better.

7. A paragraph in the contract should state the precise location and area of any additional space available for construction purposes if plans for expansion are contemplated. If there are any limits in the use of the additional space, they should be carefully defined by the purchaser to the contractor.

8. Availability of electrical power and other utilities.

9. If there are any specific treatments and chemical additions or limitations placed in the processing of the polluted water they should be stated by the owner in the contract.

10. As one of the standard inclusions, the terms of the payment for services and the installation should be carefully worded and defined.

11. The contract should contain a written statement with references to the specific code regulations to be observed in limits for each pollutant in the discharge liquid from the treatment plant.

12. An important paragraph in a contract of this type is full definition in respect to disposition or reuse of any or all of the process water.

13. The contract should be specific in stating the limits of responsibility of the supplier-contractor in respect to the mode and nature of disposition of the captured pollutants and whether any step or portion of a step in the recovery, reclaiming or disposal of the specific pollutant is involved.

14. An assurance must be included in the contract that the purchaser will make an appointment of designated qualified operation personnel to whom instruction can be given on completion or near completion of the installation.

Steps have been taken for making assurance in state programs for examining and certifying waste water treatment plant operators. There are underway different steps further to assure qualified personnel. With the complicated water treatment systems, if qualified personnel and operators are not available, a continuous, sustained performance cannot be attained even though there was intent as expressed in guarantees. This means that cooperative effort between the contractor and the purchaser is necessary in the training program and conditional provision of qualified personnel.

15. Another assurance is deserved by the contractor from the owner in that the purchaser will give full and adequate maintenance to the installation. Negligence at this point can cost a great deal of money and bad relations between supplier and purchaser and malfunction of waste treatment systems that can affect many organizations and persons.

16. In the event that the information supplied originally from the owner in respect to pollutants is not available or adequate, the contractor should provide the owner with amendment to the contract with proposal to cover estimated time, cost, and steps for obtaining the necessary data from which to plan and construct the equipment for water treatment. Acceptance or modification of the proposed amendment by the owner should be made in written form agreement.

17. An important clause of protection to the supplier is to include in the contract a statement that in case there is an introduction of pollutants that are not specified in the contract or are appearing far in excess of the stated amount or if toxic materials that destroy and interfere with the process are introduced into the system or in case faulty maintenance occurs or if sabotage is proved or other calamities take place, the contractor is exempt from responsibilities.

18. The contract should also protect the contractor or engineering firm in respect to responsibilities if a plant, municipal practices, or other conditions prevent installation or proper operation of the plant water treatment handling system.

19. The contract should also define the limits of responsibility of the contractor with the appearance of new or different pollutants that can affect the selected water treatment process. This is well illustrated in the case previously cited in which a municipality with cooperation of a reliable contractor installed a first-class water treatment unit. Unexpectedly toxic metals and their derivatives were found to be introduced in high concentration and were capable of destroying the bacterial action in the fermenting stage. Such cases should leave the contractor exempt from any liabilities or modifications of the system.

20. It should be a requirement of the contract that the owner or his representative will furnish a climatic record of the environment as well as any other imposed conditions on the waste treatment system.

21. The contractor or supplier should have provision in the contract to the effect that the owner or his representative will supply a list of the types of industries and industrial operations being served.

22. If the treatment of water anticipates thermal water pollution control, removal of odor producing chemicals, elimination of oily and organic waste discharges, and the reduction of objectionable tastes in the water, the engineering agency and contractor should require the owner or sponsor of the waste water treatment installation to state explicitly in any request or contract order the presence or possible future occurrences of these type of pollutants, the maximum occurring quantities in each category, and the limits permitted in respect to code regulations. Furthermore, there should be a clause of exclusion of responsibilities if the particular pollutant exceeds the stated maximum amount per unit volume of water or the maximum stated capacity of waste volume to process in the original contract.

There are other special conditions and stipulations that should be included in contract agreements for the mutual benefit of the parties concerned in the water pollution control systems to assure the success of the operation and to materialize in the installation on the time scheduled. With such improved understanding, greater progress can be made and also in the future greater economy, speed in designing, and constructing of much needed water treatment systems can be attained. The contract and preliminary agreements give confidence to the parties concerned and assure acceleration to the program to be put into effect for combating water pollution.

Summary

The accelerated effort and programs for water pollution control are analyzed in terms of the need for better understandings, agreements, and assurances between all parties concerned in the realization of clean water, inclusive of engineering organizations, contractors for water treatment equipment, testing laboratories, sponsors or owners of the installation, and the enforcement agencies.

The nature and causes of past and present impedances to progress in water pollution control have been reviewed and cited for the basic reasons and needs in establishing new objectives to remedy water polluting problems.

The objectives of the presentation are set forth to enable better approaches in planning and engineering water pollution control projects, to focus attention on the necessity of recognizing and evaluating each and every pollutant, to examine all facets of a particular case, to recognize the value and use of accurate reliable test methods and instrumentation in any survey or evaluation of performance, to make fuller use of contracts for definition of responsibilities and guarantees of performances, and to define fully the specifications of the system.

The similarities and dissimilarities between air pollution and its control and water pollution and its control are reviewed. Arguments are presented for recognizing water pollution control as having its own requirements, its individual characteristics, different codes of regulations, individual methods and equipment for handling and treatment, specific analytical testing procedures, and types of contracts to cover responsibilities, guarantees, and assurances for performances of installations.

As a background, references are made to the different phases of water treatment and handling as correlated with the steps taken for individual methods of removing specific pollutants and conditioning the water for reuse. Emphasis is given to the responsibilities and required assurances in contracts from specialists and engineering firms serving in the role of planning, design, selection of methods, preparation of specifications. In view of the importance of making the contracts more inclusive and meaningful, the criteria and factors for requirements in selecting water pollution control methods and associate equipment are reviewed.

The reuse of water in so many facets of animal and plant life as well as industrial processes is emphasized in respect to the obligations and necessities of eliminating the different forms of pollution so as to maintain the quality of water for the next reuse in the cycle of the role of water in nature and man's activities.

Strong emphasis is placed on the importance of treatment of many of the polluted waters at the source, particularly where hazardous, toxic, odorous, or radioactive materials are present. Several specific instances of realizing economies in captive industrial water systems are cited.

In view of making the contracts more useful with assurances of realizing the elimination of pollutants, of keeping water at a high quality level, and assuring successful results in the water pollution control program, a detailed listing of the assurances, guarantees, understandings, and definitions of responsibilities as required by the purchaser or owner of the water pollution control systems is given. The facts that water pollution control is concerned with the processing and maintaining of quality water for reuse amounting to hundreds of billions of gallons in daily use and that tremendous investments of money are necessary to realize the handling, conditioning, and treatment of water subjected to different pollutants confirm the arguments that contracts bearing on water pollution control should become of full stature, with incorporation of the discussed elements that will enable positive returns from the investment and the realization of the objective goals in the elimination of the pollutants and will assure a sustained high quality of water.

Another section of the chapter is devoted to the parts of a contract that would state conditions, assurances, and guarantees as required by the contractor or supplier of the water treatment equipment.

In all cases where government sponsorships from federal to municipal level are involved with participation in part or total, the stipulations from the governmental agencies should be incorporated in proper form in the contract with engineering organizations and contractors for construction.

Emphasis is placed on the use of reliable test methods and testing equipment by competent personnel in appraising the nature and types of water pollution and in evaluating the performances of water pollution treatment methods and equipment.

Emphasis is given that, by employing contracts to their full potential use and incorporating individual clauses defining the specific responsibilities and obligations of the contractors and owners, there will be greater assurance in realizing the clean quality of water, the preservation of lakes and rivers, the protection of health in human, animal, and plant life, and the improvement in the industrial facilities with certain economies that can be experienced while conforming to the multiplicity of water pollution control regulations.

Chapter 26

Operating Costs and Procedures of Industrial Air Pollution Control Equipment*

HERBERT F. LUND†

President, Leadership Plus, Inc., Environmental Engineering, Manufacturing Analysis, and Motivation Training Consultants, Stamford, Conn.

A survey was conducted in 1967 among the plant engineering members of the American Institute of Plant Engineers (AIPE). It was felt that plant engineers should be the source of this vital information as they have the primary responsibility to operate and maintain industry's air pollution control equipment. Information requested covered (1) type of pollutant and source, whether "under control," "planned for control," or "not controlled"; (2) cost of control equipment; (3) installation costs; (4) operating costs (power, fuel, labor, replacement parts); (5) equipment breakdown frequency; (6) cost of collected waste disposal; (7) a listing of unsolved air pollution problems; (8) person or agency most helpful in air pollution control information and services; (9) percent cost related to manufacturing cost; and (10) suggested design improvements for air pollution control equipment. Air pollution control problems vary significantly by type of industry. For this

* This chapter first appeared as an article in the *American Institute of Plant Engineers* (AIPE) *Newsletter,* and in the *Journal of the Air Pollution Control Association,* vol. 19, pp. 315–321, May, 1969.

† Formerly editor-in-chief, *FACTORY* magazine, McGraw-Hill Publications Company, New York, N.Y.

reason, survey results have been studied and reported by two-digit Standard Industry Classifications (SIC).

While many estimates of the cost of air pollution control have been published, these figures have mainly dealt with capital first costs, consulting or engineering fees, and initial installation costs. None covered the daily, monthy, or yearly operating expense after control equipment is installed.

To fill this void, the American Institute of Plant Engineers conducted a survey of its membership. Plant engineers in American plants usually inherit the operating responsibility for air and water pollution control equipment. Their response to the survey makes the data particularly informative and significant.

In general, AIPE plant engineers were asked questions on (1) effectiveness; (2) capital, installation, and operating costs; and (3) improvements needed for air pollution control equipment in manufacturing plants. They were also asked to list unsolved air pollution problems, and who was most helpful to them. Data in this report have been tabulated by major standard industrial classifications. Of the 409

TABLE 1 Basic Response Data

	Survey plants and No. of pollutants				Control status industrial process air pollutants (boiler heating excluded), %			Plant location	
	Total plants reporting	Net total plants useful form	No. air pollutants	No. air pollutants per plant	No control	Plan control	Under control	States	Major city
Foods	30	19	46	2.4	28	12	60	17	18
Chemicals	36	32	94	2.9	18	7	75	16	16
Rubber, plastics	22	21	42	2.0	26	17	57	13	13
Stone, clay, glass	15	14	24	1.8	17	15	65	11	7
Primary metals	16	15	40	2.7	26	23	51	10	10
Fabricated metals	60	49	114	2.4	29	16	55	21	22
Powered machinery	20	20	35	1.75	15	27	58	13	11
Electrical machinery	17	15	47	3.1	18	10	72	11	13
Professional and scientific instruments	21	18	60	3.3	39	15	46	10	11
Aerospace manufacturing	26	22	63	2.9	18	25	57	11	13
Total 10 industries	263	224	565	2.52	24.1	16.7	59.1	42	41
Total survey	409	330	790	2.4	27.7	16.1	56.2	47	48
% 10 industries of total survey	65	68	71	105	91	104	105	89	86

total raw responses received, 330 contained useful material worth documentation. This 330 figure roughly represented 10 percent of the AIPE membership.

The plant engineers were asked to classify their plants by 25 major (two-digit) SIC industry classifications. Of the 25 classifications listed, only 10 industries contained sufficient representative information; the other 15 contained insignificant information. But we do have valuable data on 10. And in these times of information resistance, that is significant. The 10 are listed in Table 1 as:

SIC 20 Foods
SIC 28 Chemicals
SIC 30 Rubber, plastics
SIC 32 Stone, clay, glass
SIC 33 Primary metals
SIC 34 Fabricated metals
SIC 35 Powered machinery
SIC 36 Electrical machinery
SIC 38 Professional and scientific instruments
SIC 39 Aerospace manufacturing

IMPORTANT OBSERVATIONS

Table 2. Operating Costs vs. Capital and Installed Costs

Based on average cost figures, operations costs per year exceed capital and installed cost of equipment in certain industries: (1) Foods ($30,050 vs. $14,580 + $7,440); (2) powered machinery ($67,210 vs. $38,900 + $8,100); and (3) aerospace manufacturing ($47,400 vs. $34,900 + $5,750). In each instance the yearly cost of collected waste disposal was 27–30 percent of the yearly operating costs. Primary metals scored the highest average operating costs ($112,150). Yet compared with capital plus installed equipment cost, this high operating cost was only one-fifth. The greatest portion of this operating cost is from an unusually high power, fuel, and water cost ($86,000).

TABLE 2 Air Pollution Control Equipment Operating Cost per Plant

	Avg $/year					Avg capital and installation cost	
	Power, fuel, water	Material, spare parts	Maintenance labor	Collected waste disposal	Avg total operating cost	Capital Equipment	Installation
Foods	$15,100	$ 4,800	$ 1,250	$ 8,900 30 %	$ 36,050	$ 14,580	$ 7,440
Chemicals	10,125	3,670	2,650	33,200 67 %	49,645	49,400	40,000
Rubber, plastics	6,200	7,750	1,900	11,950 43 %	27,800	93,000	56,500
Stone, clay, glass	13,500	25,000	3,670	7,650 16 %	49,820	128,000	75,300
Primary metals	86,000	750	18,200	7,200 7 %	112,150	351,600	228,500
Fabricated metals	7,800	505	2,170	17,700 63 %	28,175	26,200	10,240
Powered machinery	13,900	910	34,300	18,100 27 %	67,210	38,900	8,100
Electrical machinery	18,200	2,360	1,750	2,520 10 %	24,830	123,000	79,400
Professional and scientific instruments	7,500	475	386	2,750 25 %	11,115	30,000	11,500
Aerospace manufacturing	16,100	2,700	4,700	13,100 28 %	47,400	34,900	5,750
Avg. cost/year	19,440	4,900	7,100	14,010	45,450		
% of total cost	43 %	10 %	16 %	31 %			

Other points: In chemical plants, operating costs were approximately one-half of the capital plus installed cost of equipment. Similarly, for fabricated metals plants the yearly cost of operating and maintaining air pollution control equipment was nearly 80 percent of Capital Equipment and Installation costs.

Thus, in one-half of the 10 type industries, the yearly operating costs has become a major item of expense. Most reports in the past have overlooked the yearly operating costs problems.

A breakdown of percent average operating and maintenance costs for the 10 industries shows 43 percent power, fuel, and water costs; 31 percent collected waste disposal costs; 16 percent maintenance labor costs; and 10 percent material and spare parts costs. The 31 percent figure, second highest, for collected waste disposal points out that plants have not paid much attention to this item of operation costs. In Table 3 on collected waste disposal, this fact is borne out by a more detailed question.

Nearly 60 percent stated "Cost is not above expected amount," and yet 57 percent also stated there was a need for a better method to dispose of collected wastes. Conversely, 67 percent stated "Waste handled properly" while 33 percent on the average expressed "Need of salvage or by-product" from these wastes. It appears that plants are burying the costs in other accounts. But the problem exists. What do you do with the solid particles or fines once you collect it? How do you get rid of it efficiently at low cost? Need for equipment manufacturers to consider this aspect seriously as an integral part of control system design is apparent. (For possible answers, see Chapter 7, Section 1 on "Pollution Waste Control.")

TABLE 3 Collected Waste Disposal

	Is cost above expected amount? No %	Are wastes handled properly? Yes %	Need better method? Yes %	Possible salvage by-product value? Yes %	Avg $/year collected waste disposal
Foods	N-57	Y-60	Y-82	Y-55	$ 8,900
Chemicals	N-42	Y-65	Y-61	Y-25	33,200
Rubber, plastics	N-73	Y-46	Y-49	Y-18	11,950
Stone, clay, glass	N-24	Y-62	Y-34	Y-71	7,650
Primary metals	N-87	Y-55	Y-88	Y-57	7,200
Fabricated metals	N-76	Y-71	Y-70	Y-32	17,700
Powered machinery	N-89	Y-91	Y-22	Y-9	18,100
Electrical machinery	N-95	Y-77	Y-86	Y-13	2,520
Professional and scientific instruments	N-56	Y-81	Y-27	Y-2	2,750
Aerospace manufacturing	N-57	Y-65	Y-51	Y-48	13,100
Avg. percentage	65.6% no	67.3% yes	57.2% yes	33.0% yes	

TABLE 4 Air Pollution Control Equipment Costs Capital and Installation

	Original equipment cost		Installation cost		Ratio, %, installation cost to original cost		
	Low	High	Low	High	Low	High	Avg
Foods	$ 1,200	$ 40,000	$ 100	$ 20,000	8	150	51
Chemicals	1,300	100,000	600	125,000	21	150	81
Rubber, plastics	480	800,000	170	300,000	18	79	39
Stone, clay, glass	19,000	300,000	2,000	200,000	11	67	43
Primary metals	8,000	1,500,000	2,500	1,000,000	29	67	49
Fabricated metals	7,000	60,000	600	30,000	7	60	33
Powered machinery	7,500	100,000	2,000	30,000	10	43	25
Electrical machinery	10,000	200,000	500	50,000	25	63	49
Professional and scientific instruments	70	200,000	20	50,000	16	80	31
Aerospace manufacturing	200	255,000	500	25,000	4	375	160
Miscellaneous plants	3,000	125,000	300	225,000	12	300	77

Table 4. Breakdown of Capital and Installed Equipment Costs

In one industry, aerospace manufacturing, the installation costs exceed the capital costs of equipment. In chemical plants the percentage ratio of installed to capital costs was 81 percent. Other relatively high percentages were (1) foods 51 percent, (2) primary metals 49 percent, and (3) electrical machinery 49 percent. The lowest

TABLE 5 Unsolved and Not Controlled Air Pollution

	Unsolved air pollution problems		% air pollutants, not under control	Typical air pollutants listed as "not under control" (manufacturing process)
	Yes	No		
Foods	23	77	28	Baking fumes, and vapors; shortening vapors; nondestructible gases from tankage cookers of animal by-products; cracker dust and salt, coffee fumes
Chemicals	46	54	18	Fumes—process oil and wax, carbonizer, acid, production acid tank storage. Dusts: residue from railroad carloading and cyclones; lime; catalyst; processing; production
Rubber, plastics	77	23	26	Fumes—resin from treating ovens; talc mixing; rubber curing exhaust stack; production line; carbon black production line dust; alcohol vapor from coating ovens; phenol odor
Stone, clay, glass	78	22	17	Toxic vapors from epoxy solvents; glass furnace combustion by-products; special bottling treatment releases SO_2 exhaust; ceramic dust, cement kiln dust; limestone dust
Primary metals	73	27	26	Fluxing chlorine gas; open hearth; oxyacid burning releases iron oxide; zinc furnace blowout gas; pickling acid mist; foundry silico dust; carbon black from coating; reclaim metallic dust; zinc oxide from casting; carbonate gas from burning oily aluminum scrap
Fabricated metals	52	48	29	Hydrocarbons from industrial ovens; acid fumes from plating; salt fumes; oil smoke from heat treating: paint spray and solvent; welding fumes; powdered graphite; molding gas fumes
Powered machinery	32	68	15	Spray from paint ovens; fork truck exhaust
Electrical machinery	53	47	18	Paint vapors, plating process fumes, exhaust from chemical cleaning; nitric acid fumes
Professional and scientific instruments	48	52	39	Plating process fumes—nitric, chromic, sulfuric acids; heat treating fumes; brazing, bronze melting smoke; lacquer thinner vapor; paint spray; research labs—miscellaneous chemicals; caustic salt fumes; Freon vapor; trichlorethylene vapor
Aerospace manufacturing	24	76	18	Oil mist; heat treating smoke; paint spray; toluene exhaust; degreasing and cleaning hydrocarbons; chromic acid from plating process; rocket propellent vapors
Avg. 10 industries	50.6	49.6	24.1	Other unsolved problems: pulp, paper—cooking gases; print, publish—ink solvent fumes; textile—nauseous odor from chemical and plastic finishing operations

percentage ratio, 25 percent, was found in powered machinery plants. The highest average costs were logged in primary metals (capital—$1,500,000 per plant, installation—$1,000,000). Generally, the lowest average costs were registered on the average food plant.

Table 5. Unsolved and Not Controlled Pollutants

In about half of the 10 industries, a high percentage of plants stated they have an unsolved air pollution problem. The situation seems most critical in three types: stone, clay, glass 78 percent; rubber and plastics 77 percent; and primary metals 73 percent. The two with least unsolved air pollution problems: (1) foods 23 percent and (2) aerospace 24 percent. It is interesting to note in answer to "percent air pollutants *not* under control," the three most critical industries, "stone, clay, glass," "rubber and plastics," and "primary metals" recorded 17, 26, and 26 percent respectively. Even though these three industries recorded highest unsolved air pollution problems, they have been most aggressive in control of air pollutants in their plants with known solutions.

For the two low-scoring industries on unsolved problems, their percent air pollution *not* under control about equals the unsolved problems. One must conclude, in foods and aerospace, that these plants are waiting for new technology to help solve their unsolved air pollution problems. The four industries in the middle are "chemicals," "fabricated metals," "electrical machinery," and "professional and scientific instruments," with nearly 50 percent of plants scoring "unsolved air pollution problems."

Under each of 10 industries, typical unsolved air pollution problems are listed. Most common are various acid plant vapors, dusts of various sources, hydrocarbons from heat treating, odors, and paint vapors.

Table 6. Needed Equipment Design Improvements

In order of scoring, (1) better corrosive wear resistant materials, (2) easier preventive maintenance, (3) better waste handling system, (4) improved capacity, (5) heavier-duty design, (6) improved controls, (7) less noisy, (8) less installation costs. By far, improved materials in fabricating equipment was the most recommended improvement.

Table 7. Who Is Most Helpful to Plant Engineer with Air Pollution Information and Solution?

Of the six different sources listed, the federal agencies still scored lowest, similarly in the pilot survey. But state agencies were picked as second most valuable compared with a significantly low rating in the pilot survey.

The most helpful on the average was the company staff specialist. Six of the ten industries picked the staff specialist as the primary information and solution source. Again, second, the state agencies; then third, consultants; fourth, plant managers; fifth, city agencies; and a distant last, federal agencies.

ANALYSIS BY INDUSTRY

SIC 20. Food Plants

On the average, food plants responding to the survey spent more on their yearly operation total costs than on dollars spent on capital plus installation cost of equipment. And the highest amounts were spent on power, fuel, and water ($15,100) and collected waste disposal ($8,900) out of $30,050 total cost. Concerning the waste costs, 82 percent of these plants stated a better method was needed compared with the all industry average of 57 percent. Also, food plants were high (55 percent) in calling for salvage or by-product use from collected air pollution wastes.

Typical air pollutants listed as "not under control": (1) baking fumes, (2) shortening vapors, (3) cracker dust and salt, (4) coffee fumes, (5) nondestructible gases from animal by-product tank cookers. Food plants state air pollution equipment improvements are needed in (1) better corrosive resistant materials, (2) easier preventive maintenance, (3) better waste handling systems, and (4) improved equipment controls. State agencies were most helpful to food plants, with staff specialists second.

TABLE 6 Needed Equipment Design Improvements Rated Four Most Important

	Better corrosive wear resistant material	Easier preventive maintenance	Better waste handling capacity	Improved capacity	Heavier-duty design	Improved controls	Less noisy	Less installation cost
Foods	1	2	3			4		
Chemicals	1	2	3	4				
Rubber, plastics	4	1	2	3				
Stone, clay, glass	1	2	4			3		
Primary metals	1			4	2			3
Fabricated metals	1	2	4					3
Powered machinery	1	3	2	4				
Electrical machinery	1	2			3		4	
Professional and scientific instruments	4		3	2			1	
Aerospace manufacturing	1	4		2		3		
All industry rating	1	2	3	4	5	6	7	8

TABLE 7 Most Helpful to Plant Engineer with Air Pollution Information

	In order by number preference					
	State agency	City agency	Federal agency	Staff specialist	Consultant	Plant manager
Foods	1	4	6	2	5	3
Chemicals	3	5	6	1	4	2
Rubber, plastics	4	5	6	1	2	3
Stone, clay, glass	4	6	5	2	1	3
Primary metals	3	5	6	1	2	4
Fabricated metals	2	5	6	1	3	4
Powered machinery	2	3	6	4	5	1
Electrical machinery	2	5	6	1	3	4
Professional and scientific instruments	1	5	6	2	3	4
Aerospace manufacturing	2	6	5	1	3	4
All industry preference	2	5	6	1	3	4

On maintenance practice, food plants have kept their breakdowns to insignificantly low occurrences with proper preventive maintenance procedure. They also stated that spare parts recommended for replacement by equipment manufacturers are greater than in practice. The annual cost of materials and parts is $4,800, which is below total industry average of $4,900.

Food plants covered 17 states and 18 major cities. There were 2.4 average air pollutants reported per plant. Of this, 60 percent were reported under control, 12 percent planned for control, and 28 percent not under control.

SIC 28. Chemical Plants

Average capital equipment cost and average total operating costs per year were about equal ($49,000). Of the total operating costs per year collected waste disposal costs account for 67 percent of the total ($33,200). Also, 65 percent of chemical plants stated a need for improved waste handling system but rated this third behind better fabrication materials and built-ins for easier preventive maintenance. Though this waste disposal cost is high, reports indicate it is not above the expected amount.

Concerning installation of control equipment, some of these plants report costs of installation exceed (high by 150 percent) the first or capital cost of control equipment. The average plant reports 81 percent for percent ratio of dollars installation to dollars original cost.

Chemical plants report a low 18 percent of air pollution not under control. Examples of these are generally in the fumes or dust category.

Fumes. Process oil and wax, carbonizer, acid, acid tanks, and storage

Dusts. Lime, catalyst, processing, residue from railroad carloading, cyclones

Most pressing design improvements: (1) better materials, (2) easier preventive maintenance, (3) better waste handling system, (4) improved equipment capacity.

Staff specialists were most helpful to plant engineers with information and solution to air pollution problems. Plant managers ranked second, then state agencies. There were 2.9 air pollutants per plant on average from 16 major cities and states.

Maintenance is always a problem in chemical plants, but air pollution breakdowns were kept relatively low. But spare parts were reported above manufacturing recommendations (42 percent).

SIC 30. Rubber and Plastics

Operation costs per year are relatively low compared with capital equipment and installation costs in a 1:5 ratio. But again, collected waste disposal costs accounted for 43 percent of the total annual operation cost figure of $27,800. Generally, this was not above expected costs nor do these plants feel a need for better methods or value of salvage or by-product (18 percent) compared with an average of 33 percent.

Original equipment costs ran as high as $800,000 for a plant and as low as $480. But installation costs ran respectfully low, only 39 percent of original equipment costs.

Though 77 percent of these plants stated they had unsolved air pollution problems, they had all but 26 percent of total air pollution problems under control. Fumes from treating ovens, talc mixing, rubber curing, carbon black, alcohol vapors from coating ovens, and phenol odor were examples of air pollution "not under control."

Easier preventive maintenance was most needed equipment design improvement; second "better waste handling system." The staff specialist rated as most helpful, with second to consultant and third to plant manager. Average number of air pollutants per plant was 2.0 from 13 major cities and 13 states.

No difficulty was reported in maintenance or spare parts or material costs.

SIC 32. Stone, Clay, and Glass Plants

Though average yearly operating cost of air pollution equipment is third highest ($49,820) of 10 industries, this figure is only a quarter of the capital plus installation cost averaging $200,000 per plant.

Of the operating costs, the material and replacement spare parts cost ($25,000) accounts for half the yearly figure. The second important factor is power, fuel, and water. This high replacement cost, above the other nine industries, reflects the extremely rough nature and atmosphere in this type of plant. However, maintenance labor costs ($3,620) are less than half of the all industry yearly average of $7,100. This high ratio of 7:1 replacement parts to maintenance labor speaks well for the craft efficiency.

Concerning their waste disposal collection problem, these plants call for possible salvage or by-product value (71 percent affirmative vs. 33 percent industry average). A low capital cost of $18,000 balances out a $300,000 high, with a normal ratio of installed to capital cost percentage of 43 percent.

Stone, clay, and glass plants recorded the highest percentage (78 percent) of unsolved air pollutant problems. Yet, also listed is one of the lowest percentages of air pollutants not under control (17 percent). These figures indicate these plants have gone ahead to control nearly all air pollutants but fully recognize the need to solve tough unsolved problems. Examples of unsolved air pollution problems: glass furnace combustion by-products; ceramic dust; toxic vapors from epoxy solvents; SO_2 exhaust from special bottle treatment. Better wear resistant materials and design for easier plant maintenance were first two voted as equipment design improvements.

Outside consultants are most helpful of six listed titles; staff specialists second.

Surveyed plants covered 11 states and 7 major city areas, with an average of 1.8 process type air pollutants per plant reporting.

SIC 33. Primary Metal Plants

The pattern of facts and figures for these giant complexes is justified—highest yearly operating cost ($112,150); highest capital equipment and installation cost savings ($351,600 and $228,500). Yet, percentagewise, of the big yearly operating costs, only 7 percent is spent on collected waste disposal. This is borne out as primary metal plants scored highest; 88 percent for the need for better methods to handle collected waste disposal. According to these plant reports, 87 percent report these costs are below expected amounts.

This is obviously a pretty safe assumption on their part, as they have allocated 7 percent of their yearly operating costs, compared with 31 percent for all of industry. Simply, these plants know of their need for better methods but will not spend or are not authorized to spend the proper amount in this category. Seventy-three percent primary metal plants admit to having unsolved air pollution problems but list only 26 percent of all their air pollution as not under control.

Typical air pollution "not under control" are: fluxing chlorine gases; zinc furnace blowout gas; pickling acid mist; carbonate gas from burning oily aluminum scrap, foundry silica dust.

For "needed equipment design improvements": (1) better corrosive, wear resistant materials, (2) heavier-duty design, (3) less installation cost, and (4) better waste handling. Most helpful: (1) staff specialist, (2) consultants, (3) state agency. Under maintenance, primary metal plants recorded most air pollution equipment breakdowns per year and highest preventive maintenance hours per year. Also, spare parts replacement was highest (51 percent) of all 10 industries in "above manufacturing recommendation." This reflected in the second highest maintenance labor cost. A major contributor to high operating cost was an extremely high "power, fuel, and water" bill. On the average, 2.7 air pollutants were reported per primary metal plant.

SIC 34. Fabricated Metal Plants

Cost of operating per year is above the cost of capital equipment ($28,000 to $26,200). *Collected waste disposal* 63 percent of annual operating cost and is considered below the expected amount (76 percent); 70 percent state "need better method"; 71 percent handled wastes properly. The average annual cost to collect and dispose of air wastes was $17,700, third highest of the 10 major industries.

Comparatively low ranges on capital and installation costs: $7,000 to $60,000 for capital; $600 to $30,000 for installation. Only food plants are lower.

Unsolved air pollution problem: about half report they have a problem; 29 percent air pollution not under control; example—hydrocarbons from dust ovens, acid fumes from plating, acid fumes from heat treating, paint solvents.

Design improvements: (1) better materials, easier plant maintenance.

Most helpful: (1) staff specialist, (2) state agency, (3) plant manager.

The fabricated metals category had the largest number of responses of all 10; sources 21 states and 22 major cities.

They reported 2.4 air pollutants per plant, exactly the average of all process air pollutants reported.

Maintenance—relatively low equipment breakdown, low plant maintenance and annual cost of labor ($2,120). Way low on spare parts $500 to $4,900 average.

SIC 35. Powered Machinery

Second highest operating annual cost, $67,000, in this category exceeds capital and installation costs by $20,000.

Maintenance labor is the highest operating cost ($34,300); second, collected waste handling ($18,100). For this industry, above normal equipment breakdowns of 5 per year were reported. They also do not spend enough on materials ($910 vs. $4,900 all average). The result is high maintenance labor from breakdown plus other maintenance labor costs.

Collected waste disposal is not a problem (91 percent handled properly) in spite of $18,100 average yearly expense.

Average capital and installation costs ranges: low claims on unsolved air pollution problems 32 percent, only 15 percent not controlled, 72 percent under control.

Observation: high operating costs, but claim no problem. Examples: paint spray, fork truck fumes.

Design improvement: (1) better materials, (2) better waste handling, (3) easier preventive maintenance.

Most helpful: (1) plant manager, (2) state agency.

Comment: The only industry listing the "plant manager" as most helpful to plant engineer. With highest operating costs, the plant manager is apparently not the best source of air pollution advice or technology compared with (2) staff specialist or (3) city agency. Only 1.75 air pollutants per plant—low, under 2.4 average.

SIC 36. Electrical Machinery

The cost of operating per year is only 13 percent of capital plus installation costs.

The greatest portion of yearly operating cost ($24,830) is scored in "power, fuel, and water"—$18,200. The balance was equally split between collected wastes, maintenance labor, and spare parts.

Though over 90 percent stated cost of collected waste disposal is not above expected amount, 86 percent call for better method. Of 10 industries, this one recorded the lowest average cost of collected waste disposal.

Original equipment costs ranged from $10,000 to $200,000 per year; installation costs from $500 to $50,000.

Air pollutants not under control, 18 percent. Yet, half stated they had unsolved air pollution problems. A sampling of problems: paint vapors, plating process fumes, exhausts from chemical cleaning.

Design improvements needed: (1) better corrosive and wear resistant materials, (2) easier preventive maintenance, (3) heavier-duty design, (4) reduce noise in equipment.

Most helpful: (1) staff specialist, (2) state agency, (3) consultant.

Air pollutants were 3.1 per average plant based on 11 states and 13 cities reporting.

This was one of the few industries where breakdown hours exceeded preventive maintenance hours (55 vs. 40). The number of breakdowns were four per year, which indicates few but serious equipment failures.

SIC 38. Professional and Scientific Instruments

In one year, the average operating costs ($11,000) approximately equals the average installation costs ($11,500) of control equipment and is one-third the first cost, $30,000. Nearly all the operating expense is spent on two areas: (1) power fuel water (70 percent) and (2) collected waste disposal (25 percent) An insignificant amount of 5 percent is devoted to maintenance labor or spare parts. For this industry the relatively clean surroundings plus high-quality control equipment keep maintenance to a minimum. However, the equipment requires power and the collected waste must be disposed of.

There were 3.3 air pollutants per average plant (highest of 10 industries). Because of the variety of emission sources, air pollutants not under control is also highest (39 percent). Examples are nitric, chromic, sulfuric acid plating fumes, heat treating fumes, lacquer thinner vapors, caustic salt fumes.

Because of the sophistication and high caliber of employees in this field, respondents stated equipment improvement were (1) less noisy, (2) improved capacity, (3) better waste handling system.

State agencies were most helpful with information and solutions; second, staff specialists, and third, consultants.

Geographic universe: 10 states and 11 major cities.

SIC 39. Aerospace Manufacturing

Average annual operating costs ($47,400) per aerospace plant exceeds combined annual capital ($34,900) plus installation ($5,750) costs. And, again the high items are "power, fuel, and water" (42 percent) and "collected waste disposal" (37 percent).

Unlike "professional and scientific instruments" plants, the aerospace group had only 18 percent of air pollutants not under control and only 24 percent listed as unsolved air pollution problems (one of lowest of 10 industries). Aerospace has an overabundance of technical talent. If problems cannot be solved with present commercially available units, this industry will assign competent talent to devise solutions. Of unsolved problems, aerospace lists oil mist, heat treating smoke, rocket propellent vapors, degreasing and cleaning hydrocarbons.

Most helpful: (1) staff specialist, (2) state agency, (3) consultant.

Most critically needed equipment design improvement: (1) better corrosive and wear resistant materials, (2) improved capacity, (3) improved controls.

On average, 2.9 air pollutants per plant in a universe of 11 states and 13 major cities.

GENERAL SUMMARY

Average plant spent in 1967 (330 reporting):

Capital equipment cost	$ 88,400
Installation costs	$ 53,200
Operating costs	$ 46,550
Total costs	$188,150

Operating costs:		
Power, fuel, water	$19,840	43%
Materials, spare parts	$ 5,100	10–11%
Maintenance labor	$ 7,100	15–16%
Collected waste disposal	$14,510	31%
Total operating costs	$46,550	

Total represents $188,000 per year per plant, of which operating costs average 25 percent approximately. In the United States, there are nearly 50,000 plants of significant size. Thus, the cost of air pollution control as extrapolated for the 50,000 total plants would be 188,000 × 50,000, or $9.4 billion including operating costs of $2.33 billion per year.

Glossary

Part 1. Industrial Air Pollution Control Terms

Abatement Measures taken to reduce or eliminate air pollution which may involve legislative proceedings and technological applications.

Absorber(s) Apparatus used in the process in which one material is employed to retain another. Absorbers are used to remove a gaseous or liquid material selectively from another gas or liquid. Usually the process is performed in cylindrical towers packed with an absorbing material. Devices used for sampling by absorption include the following: scrubber, impinger, packed column. This equipment includes spray chambers, mechanical contractors, bubble cap or sieve plant contractors, or packed towers. A spray chamber is an empty chamber through which a gas stream is passed through curtains of liquid spray. Bubble cap or sieve plate equipment requires that the gas be passed upward through a series of plates on which pools of absorbent are located. Bubble cap trays are used in the former, and porous or perforated plates in the latter to support the liquid layers. Packed towers allow liquid absorbent to flow by gravity downward through a bed of packing material while the gas stream moves either concurrently or countercurrently through the tower. In each case, intimate gas-liquid contact is promoted over large interfacial areas.

Absorption A process in which one material (the absorbent) takes up and retains another (the absorbate) with the formation of a homogeneous solution. The process may involve the physical solution of a gas, liquid, or solid in a liquid, or the chemical reaction of a gas or liquid with a liquid or solid. Absorption may also refer to the process by which molecules of a gas, vapor, liquid, or dissolved substance become attached to a solid surface by physical forces.

Accretion A process that increases the size of particles by external additions; a form of agglomeration.

Acid dew point The dew point of flue gases containing little or no SO_3 is known as the water dew point and is usually in the region of 120°F (49°C). If SO_3 is present in any significant quantity the dew point is raised considerably to about 300°F (149°C). This is known as the acid dew point. Acid condensation begins at this temperature, heavy condensation occurring when the surface temperatures fall below 250°F (121°C).

Activated carbon/active carbon, activated charcoal Any form of carbon characterized by high adsorptive capacity for gases, vapors, and colloidal solids. Carbon must usually be activated to develop adsorptive power; achieved by heating to 800 or 900°C with steam or CO_2 which produces a porous particle structure. This material can be used for clarifying liquids and the purification of solutions (electroplating). The activity of (activated) carbon is the maximum amount of vapor which can be absorbed by a given weight of carbon under specified conditions of temperature, concentration of vapor, and concentration of other vapors.

Additives Agents used to improve and control air pollutant emissions from various types of fuels such as coal, gasoline, petroleum. Additives such as finely powdered solids ($MgCO_3$, Zn metal, ZnO), sprayed liquids (Na_2CO_3), and NH_3 gas, which reacts with SO_3 in flue gases to form solid sulfates, have been used to control acid smut formation.

Adiabatic lapse rate The special process lapse rate of a parcel of dry air as it moves upward in a hydrostatically stable environment and expands slowly to lower environmental pressure without exchange of heat; it is also the rate of increase in temperature for a descending parcel. The lapse rate is g/c_{pd}, where g is the acceleration of gravity and c_{pd} is the specific heat of dry air at constant pressure; equal to 9.767°C/km, or about 5.4°F/1,000 ft.

Adsorbate In the process of adsorption, adsorbates are molecules of gas, dissolved substances, or liquids, which adhere in an extremely thin layer to the surface of a solid body.

Adsorbent In the process of adsorption, an adsorbent is the solid material which collects the adsorbate.

Adsorber(s) Apparatus used for a process in which the molecules of either a gas or liquid are captured by solid material. Usually the capturing solid is activated carbon, which has a large surface area of capillary form. (The carbon can be regenerated by steam, which drives off the adsorbed material in the form of a vapor which can be condensed along with the steam.)

Adsorption A physical or chemical bond process in which molecules of gas, of dissolved substances, or of liquids adhere in an extremely thin layer to the surface of solid bodies with which they are in contact.

Advection To transport an atmospheric property solely by the mass motion (velocity field) of the atmosphere; the term is more particularly applied to the transfer of heat by horizontal motion of air. The transfer of heat from high to low altitudes is the most obvious example of advection. Fog drifts from one place to another by advection. Cold air moves from polar regions southward. Large-scale north-south advection is more prominent in the Northern Hemisphere than in the Southern, but west-to-east advection is prominent on both sides of the equator.

Aeration Bringing about intimate contact between air and a liquid by such methods as spraying the liquid in the air, bubbling air through the liquid, or agitating of the liquid to promote surface absorption of air. In general, any process whereby a substance becomes permeated with air or other gas.

Aerology Study of the atmosphere, including the upper air as well as the more general studies understood by world meteorology. Frequently used as limiting the study to the upper air. The study of the free atmosphere throughout its vertical extent, as distinguished from studies confined to the layer of the atmosphere adjacent to the earth's surface.

Aerometer An instrument used to measure the density of gases.

Aerometric measurements Atmospheric measurements such as (1) temperature measurements made by means of a wide sonde or a temperature sensitive element carried aloft by means of a balloon; (2) air current measurements.

Aerosol(s) A colloidal system in which gas, usually air, is the continuous medium and particles of solid or liquid are dispersed in it. There is no clear-cut upper limit to the particle size of the dispersed phase in an aerosol, but as in all other colloidal systems, it is commonly set at 1 micron. Haze, most smokes, and some fogs and clouds may be regarded as aerosols.

Afterburner(s) In incinerator technology, a burner located so that the combustion gases are made to pass through its flame in order to remove smoke and odors, It may be attached to or be separated from an incinerator proper.

Agglomeration The process by which precipitation particles grow larger by collision or contact with cloud particles or other precipitation particles.

Agglutination The process of uniting solid particles coated with a thin layer of adhesive material or of arresting solid particles by impact on a surface coated with an adhesive.

Air contaminant Any particle matter, gas, or combination thereof, other than water vapor or natural air.

Air ions Small particles in the atmosphere (molecules or microscopic dust particles) which have an induced electrical charge acquired through loss or gain of an electron, or through the adsorption of a molecule which has lost or gained an electron. Sizes are estimated as small, 0.001 to 0.005 micron; intermediate, 0.005 to 0.015 micron; large, 0.015 to 0.1 micron.

Air mass A widespread body of air established while that air was either situated over a particular region of the earth's surface (air mass source region) or undergoing specific modification while in transit away from the source region. Stagnation or long-continued motion of air over a source region permits the vertical temperature and moisture distribution of air to reach relative equilibrium with the underlying surface.

Air monitoring station(s)/continuous air monitoring program System to monitor continuously the concentrations of gaseous pollutants. In conjunction with the Continuous Air Monitoring Program (CAMP), these have been located in or near the principal business district in each city where the program established such a station. Data collected by these stations provide continuous, comparable, concurrent information of significant gaseous air pollutants (SO_2, NO, NO_2, CO, hydrocarbons, O_3, and oxidants) from a number of urban areas.

Air pollutant Substances added in sufficient concentration to produce a measurable effect on man or other animals, vegetation, or material. Pollutants may thus include almost any natural or artifical composition of matter capable of being airborne. They may occur as solid particles, liquid droplets, or gases, or in various admixtures of these forms. It is convenient to consider two general groups: (1) those emitted directly from identifiable sources and (2) those produced in the air by interaction among two or more primary pollutants, or by reaction with normal atmospheric constituents, with or without photoactivation. Exclusive of pollen, fog, and dust, which are of natural origin, about 100 contaminants have been identified and fall into the following categories: solids, sulfur compounds, organic compounds, nitrogen compounds, oxygen compounds, halogen compounds, radioactive compounds, odors. Solids include carbon fly ash, ZnO, PbCl; gaseous and other airborne compounds include SO_2, SO_3, H_2S, mercaptans, aldehydes, hydrocarbons, tars, NO, NO_2, NH_3, O_3, CO, CO_2, HF, HCl; radioactive compounds include radioactive gases and aerosols.

Air pollution control Measures taken to maintain a degree of purity of air resources consistent with (1) the promotion of the public health and welfare; (2) the protection of plant and animal life; (3) the protection of property and other resources; (4) the visibility requirements for safe ground and air transportation; and (5) continued economic development and growth.

Air pollution control district District created and authorized by state legislation to adopt comprehensive control programs and establish the limits of air pollution contaminants which may be emitted from sources within their jurisdiction.

Air pollution index An arbitrarily derived, mathematical combination of air pollutants which give a single number attempting to describe the ambient air quality. The formula $API = 20\ [SO_2 + 1\ (CO) + 2\ (\text{smoke shade})]$ has no scientifically derived basis. Experience indicates that the average value of this index is 12.0. An index reading of 50.0 is considered adverse and a cause for alarm.

Air pollution survey A critical examination of a given geographical area for the purpose of determining the nature, sources, extent, and effects of air pollution.

Air resource management approach The planning and implementation of air pollution control programs with the long-range objectives of the preservation of the health and welfare of man, protection of plant and animal life, preventing damage of physical property, providing visibility for safe air and ground transportation, permitting continued economic growth, and maintaining aesthetically acceptable environment.

Air sampling network A network of air sampling stations to determine the extent and nature of air pollution, the trends in the levels of atmospheric contamination, and relationships between air pollution and socioeconomic, geographic, topographic, meteorological, and other factors.

Alert levels Concentrations of gaseous pollutants indicative of approach to danger to public health. For instance, three such levels are promulaged by the Los Angeles County Air Pollution Control District; the first is an initial warning; the second is a signal for curtailment of certain significant sources; the third is the level at which emergency action must be taken.

Ambient air Surrounding environmental air.

Ambient air quality A general term used to describe the state of the air outside. No qualitative measures are associated with this term. Ambient air quality is usally considered "good" or "bad," depending upon the measurement techniques employed. Some techniques break down and list the actual constituents measured in the air, while others attempt to group all constituents into one arbitrary index number.

Aneroid barometer An instrument for measuring atmospheric pressure.

Ash Mineral content of a product that remains after complete combustion.

Atmosphere (an) A standard unit of pressure.

Atmosphere (the) The whole mass of air surrounding the earth composed largely of oxygen and nitrogen.

Atmospheric assimilation Photochemical reactions whereby chemical substances in the atmosphere are mixed, diffused, and chemically reacted, thereby maintaining concentrations reasonably constant within defined atmospheric regions.

Atomization The dividing of a liquid into extremely minute particles, either by impact with a jet of steam or compressed air, or by passage through some mechanical device.

Atomize To divide a liquid into exptremely minute particles, either by impact with a jet of steam or compressed air, or by passage through some mechanical device.

Background concentration/background level The concentration of air pollutants in a definite area during a fixed period of time prior to the starting up or on the stoppage of a source of emission under control.

Baffle A plate, grating, or refractory wall used especially to block, hinder, or divert a flow or to hinder the passage of a substance.

Baffle chamber In incinerator design, a chamber designed to promote the settling of fly ash and coarse particulate matter by changing the direction and/or reducing the velocity of the gases produced by the combustion of the refuse.

Bag filter Fabric filter used in industrial operations to recover valuable matter as well as to control atmospheric pollution at the source. Such a filter is commonly of a tubular form or similar to an envelope and can be slipped over a wire frame. Applications of this type of filter are used by carbon black, cement, clay, and pharmaceutical plants, as well as by power stations, and where operations involve abrasives and irritating chemicals. Particle laden air is introduced inside the tube usually to allow the larger particles to settle or be projected into the dust hopper before air enters the tubes.

Baghouse filtering Among the many processes which may be used to eliminate intermediate and large (greater than 20 microns in diameter) particles in a "bag filter." This device operates in a way similar to the bag of an electric vacuum cleaner, passing the air and smaller particulate matter, while entrapping the larger particulates.

Bivane A sensitive wind vane used in turbulence studies to obtain a record of the horizontal and vertical components of wind.

Blowdown Hydrocarbons purged during refinery shutdowns and startups which can be manifolded to blowdown systems for recovery, safe venting, or flaring.

Breeching A passage or conduit to conduct products of combustion to the stack or chimney.

Breeze Very fine particles of coke.

Brown fume(s)/brown smoke(s) Smoke emitted by air-blown, superoxygenated air-blown, or oxygen-blown iron or steel converters.

Bubbler Absorption apparatus used in gas sampling. Usually U-tube absorbers filled with a specific amount of reagent and fitted with a glass partition, so that the air or gas led into them passes through the reagent solution in the form of very fine bubbles. Sampling rate is usually about 100 to 150 liters/hr of gas stream.

Carbon black Finely divided forms of carbon made by the incomplete combustion or thermal decomposition of natural gas or liquid hydrocarbons. Principal types, according to the method of production, are channel black, furnace black, and thermal black.

Carbon dioxide (carbonic acid, carbonic anhydride) Colorless, odorless gas or heavy, volatile, colorless liquid, or white, snowlike solid.

Carbon monoxide (CO) Colorless gas, exceedingly faint metallic color and taste; highly poisonous, inducing asphyxiation; 0.2 percent in air is poisonous and 0.43 percent will induce asphyxiation. Burns with a violet flame. It is found in trace quantities in the natural atmosphere but also is produced by the incomplete combustion of carbonaceous fuels.

Cascade Impactor Sampling device in which air is drawn through a series of jets against a series of slides. Air particles then adhere to the microscope slides, which are coated with an adsorbing medium. Jet openings are sized to obtain a size distribution of particles.

Catalysis A process in which the speed of a reaction is altered by the presence of an added substance which remains unchanged at the end of the reaction. Catalysts are usually employed to accelerate reactions and are called "positive" catalysts; in some cases "negative" or retarding catalysts are employed, e.g., glycerol, phenol, and mannite retard the oxidation of sodium sulfite.

Catalyst(s) A substance whose presence changes the rate of a chemical reaction without itself undergoing permanent change in its composition. Catalysts may accelerate or retard the chemical reaction rate, depending on the type used.

Ceilometer An automatic, recording, cloud-height indicator.

Ceramic filter(s) Component of a stack sampling system. These are also known as ceramic thimbles and are suitable for high-temperature (1000°F) use.

Charcoal A product of the destructive distillation of wood used for absorbent, filtering, and decolorizing media. It can absorb various gases and vapors, showing preferential absorption for vapors of liquids that are volatile at ordinary temperatures. One volume of carbon can absorb 170 volumes of NH_3 or 148 volumes of alcohol vapors, as compared with about 16 volumes of O_2 or N_2.

Checkwork A pattern of multiple openings in a refractory wall to promote turbulent mixing of products of combustion.

Chemisorption Chemical adsorption which depends on chemical bond formation between the adsorbent and adsorbate. but which is distinct from a chemical reaction, in that it takes place only in a monolayer on the surface of the adsorbent.

Chimney effect A phenomenon consisting of a vertical movement of a localized mass of air or other gases due to temperature differences.

Chimney/stack Stacks or vertical steel flue for conducting cooled combustion products of a process to the atmosphere, sometimes movement activated by an induced-draft fan.

Chimney height/effective Height of chimney which ensures normal distribution of emissions.

Chimney superelevation As used in dispersion calculations, effective chimney height denotes the maximum height of the center of a plume path above the ground.

Chlorination The process of introducing one or more chlorine atoms into a compound. A common chlorinating agent is gaseous chlorine, using light, heat, or catalyst as reaction promoters.

Chromatographic analysis A method of separation based on selective adsorption. A solution of the substance or substances desired is allowed to flow slowly through a column of adsorbent. Different substances will pass with different speeds down the column and will eventually be separated into zones. The column core can then be pushed out and the zones of material cut apart.

Chromatography (1) Partition chromatography involves selective solution of the desired material between two solvents. The final solvent, usually water, is used to wet the solid material packed in the column, and the first solvent containing the desired material is poured into the column as above. (2) Paper chromatography is a micro method. A drop of the liquid to be investigated is placed near one end of a strip of paper. This end is immersed in solvent, which travels down the paper and distributes the materials present in the original drop selectively. Comparison with known substances allows identification.

Chromatography/gas An analytical technique for separating mixtures of volatile substances. The procedure consists of introducing the sample mixture into a separating column and washing it down with an inert gas. The column is packed with absorbent materials (20 to 200 mesh) which selectively retard the components of the sample. The surface is usually coated with a relatively nonvolatile liquid designated the stationary phase. This gives rise to the term gas-liquid chromatography. If the liquid is not present, the process is gas-solid chromatography. Different components move through the bed of packing at different rates and appear one after another at the effluent end, where they are detected and measured by thermal conductivity changes, density difference, or various type of ionization detectors. (Gas chromatography is advantageous for minute quantity analysis of complete mixtures.)

Cloud chamber (Wilson cloud chamber) A device which illuminates normally invisible paths of high-energy subatomic particles. A supersaturated vapor condition is created in a chamber filled with dust-free air by a sudden adiabatic expansion and cooling. In this environment, the small molecular ions formed along the path of a high-energy particle act as effective condensation nuclei. The line of droplets so formed can be used to mark the path. A device for observing the paths of ionizing particles, based on the principle that supersaturated vapor condenses more readily on ions than on neutral molecules.

Cloud-height indicators The general term for instruments which measure the height of cloud bases. They may be classified according to their principle of operation: (1) One class is based upon height determination by means of triangulation. Examples are the ceilometer and the ceiling light. (2) Another class is based upon pulse techniques. The time required for a pulse of energy to travel from a source located on the ground to the cloud base and back to the ground is measured electrically. Examples are the pulsed-light cloud-height indicator and vertically directed cloud-detection radar.

Coagulation Process of converting a finely divided or colloidally dispersed suspension of a solid into larger-size particles to cause rapid settling or precipitation. Often accomplished by addition of a di- or trivalent metal. Alum, aluminum sulfate, and ferric sulfate are commonly used to clarify water from suspended particles.

Coal desulfurization Process of removing sulfur from coal. Some of the sulfur in coal can be removed by mechanical cleaning processes which depend upon the difference of the specific gravity of coal and the unwanted materials. Thus, iron pyrites, which account for about one-half of the total sulfur content of coal, can be removed mechanically because the specific gravity of iron pyrites is 5 while that of coal is between 1.25 and 1.45.

COH, coefficient of haze Units are based on the transmission of light through a soiled filter.

Coke plant(s) Chemical plants for the purification of gases evolved in the carbonization of coal and where commercial products such as ammonia, tar, benzene, and naphtha are recovered. (1) Sludge coking plant—In the petroleum refining industry, a plant for H_2SO_4 recovery from dry acid sludge. The sludge is thermally decomposed in a slight vacuum to form H_2SO_4, water, hydrocarbons, and coke. (2) By-product coke plant—A plant used in connection with the distillation process to produce coke in which the volatile matter in coal is expelled, collected, and recovered. It consists of coal and coke handling equipment, by-product chemical plant, and other equipment associated with the coking chambers or ovens.

Collection efficiency The percentage of a specified substance retained on passage through a gas cleaning or sampling device.

Collector A device for removing and retaining contaminants from air or other gases. Usually this term is applied to cleaning devices in exhaust systems.

Combustion air (incinerator terminology) Air introduced to the primary chamber of an incinerator through the fuel bed by natural, induced, or forced draft. (Generally referred to as overfire air if supplied above the bed through the sidewalls and/or the bridge wall of the primary chamber.)

Combustion chamber(s) Where combustible solids, vapors, and gases from the primary chamber are burned and settling of fly ash takes place.

Combustion equipment Any furnace, incinerator, fuel burning equipment, refuse burning equipment, boiler, apparatus, device, mechanism, fly ash collector, electrostatic precipitator, smoke arresting or prevention equipment, stack, chimney, breeching, or structure, used for the burning of fuel or other combustible material, or for the emission of products of combustion, or used in connection with any process which generates heat and may emit products of combustion. Included are process furnaces, by-product coke plants, coke baking ovens, mixing kettles, cupolas, blast furnaces, open hearth furnaces, heating and reheating furnaces, puddling furnaces, sintering plants, bessemer converters, electric steel furnaces, ferrous and nonferrous foundries, kilns, stills, driers, roasters, and equipment used in connection therewith; and other equipment used in manufacturing, chemical, metallurgical, or mechanical processing which may cause the emission of smoke, particulate, liquid, gaseous, or other matter.

Combustion gas(es) Gases and vapors produced in furnaces, combustion chambers, or open burning.

Compaction Compressing metallic or nonmetallic powders, as a briquette in a die.

Concentration The amount of a given substance in a stated unit of a mixture, solution, or ore. Common methods of stating concentration are percent by weight or by volume, normality, weight per unit volume.

Condensation The physical process of converting a material from a gaseous phase to a liquid or solid state by decreasing temperature, by increasing pressure, or both. Usually in air sampling only cooling is used.

Condensation nuclei A particle, either liquid or solid, upon which condensation of water vapor begins in the atmosphere. Two general types are (1) "neutral particles" (size range 0.1 to 30 microns and larger) commonly known as dust and fumes; (2) condensation nuclei (size range 0.001 to 0.1 micron), made up of hydroscopic substances.

Condensation sampling Gas sampling technique in which the gas is drawn through trapping tubes immersed in refrigerants. By use of different refrigerant solutions, various fractions of the components of the gas can be measured.

Continuous air monitoring program (CAMP) Program undertaken by the U.S. Public Health Service to increase knowledge about the nature of air pollution by providing information in the form of comparable, continuous, concurrent data on significant gaseous pollutants from a number of urban areas. Monitoring stations have been established in or near the principal business district of such cities as Chicago, Cincinnati, Philadelphia, San Francisco, St. Louis, and Washington to monitor gaseous pollutants, SO_2, NO, NO_2, CO, hydrocarbons, O_3, and oxidants. Measurements comparable with the CAMP operations are also made in cities at stations operated by local groups (such as Detroit, New York, Los Angeles).

Continuous sampling Sampling without interruptions throughout an operation or for a predetermined time.

Control agencies The term "air pollution control agency" means any of the following: (1) a single state agency designated by the governor of that state as the official state air pollution control agency for the purposes of the Clean Air Act; (2) an agency established by two or more states and having substantial powers or duties pertaining to the prevention and control of air pollution; (3) a city, county, or other local government health authority where an agency other than the health authority is charged with responsibility for enforcing ordinances or laws relating to the prevention and control of air pollution; or (4) an agency of two or more municipalities located in the same state or different states and having substantial powers or duties pertaining to the prevention and control of air pollution. The term "interstate air pollution control agency" means (1) an air pollution control agency established by two or more states, or (2) an air pollution control agency of two or more municipalities located in different states.

Control apparatus Any device which prevents or controls the emission of any air contaminant.

Controlled atmosphere(s) Atmospheres which may be used as primary standards for the calibration of analytical techniques or as simulated environments for the study of biological responses or of the resistance of materials. The first requirement of a controlled atmosphere is a source of "pure air"; other components are the test substance or substances; a container, duct, or other boundary for the system; and devices to limit, define, or measure the flux or amount of air and test substances. Atmospheres may be classified as static or dynamic; two intermediate types are quasi-static and quasi-dynamic.

Cooling rate Rate of cooling of a body, substance, or gas environment obtained by plotting time against temperature under constant conditions.

Correlation coefficient(s) A measure of the persistence of eddy velocity as a function of time and space.

Counter(s) Instruments for measuring the size and number of dust particles in the atmosphere as dust counters.

Count median size The median measurement of particle diameters of particulate matter of a gas sample.

Cracking Thermal decomposition processes, such as the cracking of NH_3 to N_2 and H_2, and natural gas hydrocarbons, such as CH_4 to C and H_2, or into other hydrocarbons.

Culm pile Anthracite that has disintegrated into coarse powder (about ⅛-in. screen) underground because of a disturbance of the strata.

Cumulative frequency function A statistical analysis with reference to the cumulative frequency distribution of pollutants.

Cumulative sample A sample obtained over a period of time with (1) the collected atmosphere being retained in a single vessel, or (2) a separated component accumulating into a single whole. Examples are dust sampling in which all the dust separated from the air is accumulated in one mass of fluid; the absorption of acid gas in an alkaline solution; and collection of air in a plastic bag or gasometer. Such a sample does not reflect variations in concentration during the period of sampling.

Cupola A vertical shaft furnace used for melting metals, as distinct from a blast furnace in which ore is melted. Metal, coke, and flux are charged from the top of the furnace onto a bed of hot coke through which air is blown. Hot blast cupolas: the hot blast is derived from a recuperator fed by hot gases drawn from the cupola stack or from a separately fired blast heater. This allows higher melting rates, and grit and dust emissions from the recuperative types tend to be lower than from the conventional plant. The amount of metallurgical fume emitted is higher than that produced in a cold blast cupola.

Curtain wall A partition wall between chambers in an incinerator under which combustion gases pass. Drop arch—A refractory construction or baffle which serves to deflect gases in a downward direction.

Cyclone(s) A cyclone collector is a structure without moving parts in which the velocity of an inlet gas stream is transformed into a confined vortex from which centrifugal forces tend to drive the suspended particles to the wall of the cyclone body. Types in common use may be classified as follows: (1) tangential inlet with axial dust discharge (the common cyclone); (2) tangential inlet with peripheral dust discharge; (3) axial inlet (through swirl vanes) with axial dust discharge; (4) axial inlet (through swirl vanes) with peripheral dust discharge.

Cyclone sampler (cyclone collector) A structural device without moving parts in which the velocity of an inlet gas stream is transformed into a confined vortex from which centrifugal forces tend to drive the suspended particles to the wall of the cyclone body.

Cyclone separator A separator which utilizes centrifugal force to arrest all or part of the particles suspended in a current of gas.

Cyclonic spray tower Liquid scrubbing apparatus where sprays are introduced countercurrent to gases for removal of contaminants. Axial gas velocities can be as high as 3 to 6 fps because drops which might be entrained collect on the cyclone wall. The transfer unit's height is lower than that in a simple spray tower, but the pressure drop is correspondingly higher.

Decay (pollutants in air) Decay rate—In air pollution sampling the gross beta activity of a weighed sample is measured after it is 2 days old to allow decay of the natural alpha and associated beta-gamma activity. The samples may be recounted after 4 to 7 days to establish decay in terms of apparent half-life. Decay values are given as gross beta radioactivity in picocuries per cubic meter of air on the date of sample collection. Decay rate is expressed as half-life beginning on the date of the first count.

Dedusting Operation or process to arrest all or part of the dust present in gases. A distinction is usally implied between dedusting and filtration.

Defoliation Removal of leaves from growing plants due to the effect of liquid or solid chemical agents (herbicides as well as some pesticides).

Degasification Removing the last trace of a gas from wires used in vacuum tubes, from metals to be plated, and from substances to be used in other specialized application.

Demister(s) (collector) Apparatus made of wire mesh or glass fiber and used to eliminate the acid mist as in the manufacture of H_2SO_4.

Deodorizer Equipment for the removal of noxious gases and odors. It may consist of a combustion, an absorption, or an adsorption unit.

Deposit gage General name for air pollution instruments used to measure the amount of material deposited on a given area during a given time.

Deposited matter, dust deposit, deposition The particulate matter from the atmosphere which rapidly settles out of the air.

Desulfurization (1) In coal processing, desulfurization involves elutriation, froth flotation, laundering, magnetic separation. (2) In petroleum refining, removing sulfur or sulfur compounds from a charge stock (oil that is to be treated in a particular unit). (3) Removing sulfur from iron, metal, or an ore.

Detection Establishment of the presence of air pollutants by means of instrumentation, sampling, or visual observation.

Detectors Device which indicates the presence of air pollutants.

Deterioration Degradation of quality. For example, degradation of materials due to the chemical activity of air pollutants, especially evident in cases of corrosion and oxidation.

Diffraction analysis Application of diffraction technique (of x-rays, electrons, neutrons) to study the structure of matter, gases, and solids. This may involve diffraction of electromagnetic radiation or particle beams.

Diffuser A porous plate or tube through which air is forced and divided into minute bubbles for diffusion in liquids. Commonly made of carborundum, alundum, or silica sand.

Diffusion The spreading or scattering of a gaseous or liquid material. (1) Eddy diffusion: the most important mixing process in the atmosphere. It causes a parcel of polluted air to occupy larger and larger volume. (2) Molecular diffusion: a process of spontaneous intermixing of different substances, attributable to molecular motion and tending to produce uniformity of concentration.

Discomfort threshold A value at which a stimulus just produces a discomfort sensation or comes just within the limits of perception.

Dispersion The dilution of a pollutant by diffusion, or turbulent action, etc. Technically a two-phase system of two substances, one of which, the dispersed phase, is uniformly distributed in a finely divided state through the second substance, the dispersion medium. The dispersion medium may be a gas, liquid, or solid, and the dispersed phase may be any one of these.

Dispersion forecasts Prediction of gas plume dispersions.

Dispersion rate Diffusion parameter of a gas plume or stack effluents.

Diurnal Recurring daily. Applied to variations in concentration of air contaminants, diurnal indicates variations which follow a distinctive pattern and which recur from day to day.

Diurnal variation Variations in concentration of air contaminants which follow a distinctive pattern and which recur from day to day.

Dosage Concentration of pollutants in an exposure chamber.

Downdraft A current of air with a marked downward or reversed motion down a chimney or flue.

Downwash Occurs when chimney gases are drawn downward by a system of vortices or eddies which form in the lee of a chimney when wind is blowing. It affects the appearance of a plume and causes blackening of stacks. It may also bring flue gases prematurely to the ground. Downwash may be prevented by discharging gases with a velocity 1½ times that of wind passing the top of the chimney.

Draft A gas flow resulting from the pressure difference between the incinerator, or any component part, and the atmosphere, which moves the products of combustion from the incinerator to the atmosphere. (1) Natural: the negative pressure created by the difference in density between the hot flue gases and the atmosphere. (2) Induced: the negative pressure created by the action of a fan, blower, or ejector which is located between the incinerator and the stack. (3) Forced: the positive pressure created by the action of a fan or blower, which supplies the primary or secondary air.

Drag coefficient A dimensionless ratio of (1) the component of force parallel to the direction of flow (drag) exerted on a body by a fluid to (2) the kinetic energy of the fluid multiplied by a characteristic surface area of the body.

Droplet(s) A small spherical liquid particle that may fall free under still conditions but which may remain suspended in the atmosphere under turbulent conditions. A liquid cloud particle. The size of a droplet is usually less than 0.2 mm in diameter. A drop or liquid precipitation is usually larger than 0.2 mm.

Dry impingement The process that pushes particulate matter carried by a gas stream against a retaining surface. The retaining or collecting surface may be coated with an adhesive film.

Drying agents Substances that remove water, as heat or chemicals or for organic liquids (alcohols); these include metallic sodium, calcium oxide, metallic calcium, barium oxide aluminum amalgam.

Duct(s) A tube or passage that confines and conducts a fluid, as a passage for the flow of air to the compressor of a gas turbine engine or a pipe leading air to a supercharger.

Dust Solid particles projected into the air by natural forces, such as wind, volcanic eruption, or earthquake, and by mechanical or manmade processes such as crushing, grinding, milling, drilling, demolition, shoveling, conveying, screening, bagging, and sweeping. Generally dust particles are about 1 to 100 microns in size. Below 1 micron in size, particles are classed as fumes or smoke. Dusts do not tend to flocculate except under electrostatic forces; they do not diffuse but settle under the influence of gravity.

Dust collector Device for monitoring dust emissions. Equipment to remove and collect dust from process exhaust gases. These may employ the following approaches; sedimentation (dustfall jars, coated slides, papers, settled dust samples); inertial separation (cyclones, impactors, impingers, sticky tapes); precipitation (thermal and electrostatic); filtration.

Dust counter An instrument for counting the dust particles in a known volume of air. In Aitken's dust counter, condensation is made to occur on the nuclei present by adiabatic expansion of air, and the number of drops is counted. In Owen's dust counter, a jet of damp air is forced through a narrow slit in front of a microscope coverglass. The fall of pressure due to expansion of the air after passing the slit causes the formation of a film of moisture on the glass, to which dust adheres, forming a record which can be studied under a microscope.

Dust deposit The quantity of solid matter deposited on the ground from the external air in a given time and over a given area.

Dust fall Particulate matter in the air which falls to the ground under the influence of gravity.

Dust flow Rate at which dust is carried in a gaseous medium.

Dust horizon The top of a dust layer which is confined by a low-level temperature inversion in such a way as to give the appearance of the horizon when viewed from above against the sky. The true horizon is usually obscured by the dust in such instances.

Dust loading The amount of dust in a gas, usually expressed in grains per cubic foot or pounds per thousand pounds of gas. Engineering term for "dust concentration."

Eductors Device for mixing air with water. A liquid pump operating under a jet principle, using liquid under pressure as the operating medium.

Effluent General term denoting any fluid emitted by a source. The various spent liquors that are allowed to flow away as waste from plating shops, pickling tanks, etc. Effluent treatment usually involves (1) neutralization of acid liquids usually with lime; (2) precipitation of the salts of heavy metals as hydroxides by treatment with lime; (3) treatment of cyanides either by removal as hydrocyanic acid or by conversion into cyanates or into prussian blue; (4) reduction of chromates with ferrous sulfate and precipitation with lime. Sewage, water, or other liquid, partially or completely treated, or in its natural state, flowing out of a reservoir, basin, or treatment plant.

Electron charge The smallest possible negative electric charge (4.80298×10^{-10} electrostatic units).

Electrostatic filter(s) Filters where electrostatic charge is applied to the filter element. Fibrous filter material is pleated in the conventional manner between V-shaped supports consisting of metal rods which are insulated from the supporting frame. The rods are electrostatically charged with respect to the frame, and the filter material fibers which have dielectric properties become charged. Collection efficiency of small particles is improved.

Electrostatic precipitation A process to separate particulate matter from air or other gases using an electrostatic field.

Electrostatic precipitator(s)/electrical precipitators Devices that separate particles from a gas stream by passing the carrier gas between the two electrodes across which a unidirectional, high-voltage potential is effected. The particles pass through this field, becoming charged and migrating to the oppositely charged electrode. Single-stage precipitators are those in which gas ionization and particulate collection are combined into a single step. Basic designs for these are plate type and pipe type units. In the pipe type unitts the collecting electrodes are formed by a nest of parallel tubes. In the plate type units the collecting electrodes consist of parallel plates. In the two-stage unit, ionization is achieved by one element of the unit and the collection by the other. Electrostatic precipitators are highly efficient collectors for minute particles.

Elutriation Separating the lighter particles of a powder from the heavier by means of an upward stream of fluid. It is especially useful for very fine particles below the usual screen sizes and is used for pigments, clay dressing, and ore flotation.

Emission The total amount of a solid, liquid, or gaseous pollutant emitted into the atmosphere from a given source in a given time, and indicated in grams per cubic meter of gas or by a relative measure (e.g., smoke density) upon discharge from the source. (1) Primary emissions may be characterized as follows: fine solids (less than 100 microns in diameter); coarse particles (greater than 100 microns in diameter); sulfur compounds, organic compounds, nitrogen compounds, oxygen compounds, halogen compounds, radioactive compounds. (2) Secondary emissions may be characterized as the products of polluted air reactants, such as those which occur in atmospheric photochemical reactions. Secondary pollutants include O_3, formaldehyde, organic hydroperoxides, free radicals, NO, O, etc. Emissions are commonly reported in terms of weight of pollutant per unit of time.

Emission control equipment Equipment to control air pollution by collecting the pollutants. They include gravity settling chambers, inertial separators, cyclonic separators, filters, electrical precipitators, scrubbers (spray towers, jet scrubbers, venturi scrubbers, inertial scrubbers, mechanical scrubbers, and packed scrubbers). Control equipment used for gases and vapors and odors includes combustion, absorption, and adsorption units. Incineration equipment with a single combustion chamber is generally used for the combustion processes. Absorption equipment includes spray chambers, mechanical contactors, bubble cap or sieve plate contactors, and packed towers. Adsorption equipment includes packed beds, and sometimes fluidized beds.

Emission inventories Systematic collections and collations of detailed information concerning the air pollution emissions in a given area. Inventories list the types of sources as well as their contributions in terms of the composition and rates of discharge of the individual pollutants. Supplemental information may include the number and geographical distribution of sources, description of processes, raw materials, and control measures.

Emission monitoring Collection of data on the air quality of the atmosphere. This may be done at a specific emission source as well as for a general area.

Emission point/discharge point The point where a pollutant(s) is discharged temporarily, intermittently, or continuously into the atmosphere from a given source.

Emission source Process, building, furnace, incinerator, or plant releasing foreign matter into the air.

Emittor/emitter Any technical equipment which discharges matter foreign to air into the open air.

Energy absorption Process whereby some or all of the kinetic energy of electromagnetic radiation is lost to another substance.

Energy distribution (energy dispersion) Any disturbance that can be analyzed into two or more harmonic wave trains will undergo dispersion, each component being propagated at its own group velocity, provided this velocity is not zero. The meteorological importance of such dispersion arises from the fact that energy is, in a certain sense, propagated at the group velocity. The synoptic phenomenon of "downstream intensification" following development farther upstream has been attributed to horizontal energy dispersion in the atmosphere.

Enforcement (activation) Concerns the prohibition of discharges into the atmosphere from any source constructed or operated without a permit (if one is required), or in violation of the terms of a permit or rules, regulations, or orders of an agency. This may involve civil and criminal penalties, abatement authority, and judicial review of agency orders.

Entrainment The mist or fog droplets of liquid carried off by the vapor of a boiling liquid, or more frequently from a liquid through which bubbles of gas or vapor are passing rapidly.

Environment An external condition or the sum of such conditions, in which people, living organism, equipment, or a system operates as in temperature environment, vibration environment, or space environment. Environments are usually specified by a range of values, and may be either natural or artificial.

Evaporation A physical process in which a liquid is slowly converted into vapor, at temperatures below the boiling point of the liquid. The rate of evaporation is determined by the temperature and by the pressure of vapor at the liquid surface. (Also called vaporization.)

Exposure Subjecting a person, animal, plant, or material to an environment containing a harmful concentration of an air pollutant or air pollutants.

Exposure chamber A chamber for exposing human, plant, animal, or material to a particular environment. Various parameters, such as pollutants, humidity, temperature, pressure, noise, movement, radiation, and fluid contents, may be controlled.

Exposure dose A measure of radiation exposure based upon its ability to produce ionization. The unit of exposure dose is the roentgen (r). Exposure dose rates are measured in roentgens per unit of time.

Exposure rate Concentration of pollutant or radiation per unit of time to which an animal, plant, or material is exposed.

Eye irritation Caused by air pollutants, such as gases contained in smog and automotive exhausts (SO_2, SO_3, NO_2, O_3, olefins, formic acid, oil, gasoline); secondary pollutants caused by the photochemical reactions of NO or NO_2 and hydrocarbons such as PAN (peroxyacetyl nitrate), formaldehyde, or acrolein; alone, or in combination with aerosols (such as NaCl and silica dust).

Fabric filter A collector designed to remove particles from a carrier gas by filtration of the gas through a porous medium. Two basic types of filter are (1) a fibrous medium utilized as the collecting element; (2) where the medium is utilized as a support for a layer of collected particles, relying on the coat of collected particles to serve as the principal collecting medium. The most common type of fabric collector is the tubular type, consisting of a structure in which cylindrical fabric bags are suspended vertically over a tube sheet with the open end of the bag attached to the sheet. Some of the fabric materials include cotton, wool, nylon, asbestos, Orlon. Particles equal to or greater than 1 micron can be collected.

Fallout/particle fall A measurement of air contamination consisting of the mass rate at which solid particles deposit from the atmosphere. Used in the same sense as the older terms "dust fall" and "soot fall" but without any implication as to the nature and source of the particles. Another example, radioactive fallout from an atomic or thermonuclear explosion.

Fanning In blast furnace operation, the idling period between the blowing periods when the blast pressure is reduced to a minimum. During this period no charging or tapping of slag or metal is carried out.

Fiber glass Generic name for a manufacured nonflammable fiber fabricated from glass. A continuous filament or staple fiber having unusual resistance to heat and chemicals. The strongest fiber known and perfectly elastic up to its ultimate strength. It is attacked by hydrofluoric acid and alkalies but resistant to most other chemicals and solvents.

Filter(s) A porous material on which solid particles are largely caught and retained when a mixture of liquids and solids is passed through it.

Filtration The process of separating suspended solids from a liquid or gas, usually by forcing a carrier gas or liquid through a porous medium.

Flame photometry Emission spectroscopy in the ultraviolet and visible regions which makes use of flame sources. Some seventy-odd elements can be qualitatively or quantitatively determined based on the emission spectra of these elements when they are excited in an arc or high-voltage spark.

Flare (gases) A bright eruption from the sun's chromosphere.

Flash point The lowest temperature at which a substance, such as fuel oil, will give off a vapor that will flash or burn momentarily when ignited.

Flocculation A process of contact and adhesion whereby the particles of a dispersion form larger-size clusters. Synonymous with agglomeration and coagulation. (Coagulation)—The process of converting a finely divided or colloidally dispersed suspension of a solid into particles of such size that reasonably rapid settling occurs. This is usually done by adding the salt of a bi- or trivalent metal, e.g., when alum, aluminum sulfate, and ferric sulfate are added in clarifying water from suspended impurities.

Floccule A small loosely aggregated mass of material suspended in or precipitated from a liquid; a cluster of particles.

Flowmeter An instrument for measuring the rate of flow of a fluid moving through a pipe or duct system. The instrument is calibrated to furnish volume or mass rate of flow.

Flow rate Measure of the volume of fluid per unit of time.

Flow regulator A device used to maintain constant airflow for a given period of time in particulate sampling equipment.

Flue A passage for conducting combustion gases in an incinerator installation. Also, used synonymously with chimney (buildings).

Flue gas Waste gas from combustion processes.

Flue gas scrubber (washer) Equipment for removing fly ash and other objectionable materials from the products of combustion by means of sprays or wet baffles. Also reduces excessive temperatures of effluent.

Fluid flow A stream or movement of air or other fluid, or the rate of fluid movement, in the open or in a duct, pipe, or passage. Various types of fluid flow are; (1) *Uniform* flow is steady in time, or the same at all points in space (2) *Steady* flow is flow of which the velocity at a point fixed with respect to a fixed system of coordinates is independent of time, (3) *Rotational* flow has appreciable vorticity and cannot be described mathematically by a velocity potential function,

(4) *Turbulent* flow is flow in which the fluid velocity at a fixed point fluctuates with time in a nearly random way. The motion is essentially rotational, and is characterized by rates of momentum and mass transfer considerably larger than in the corresponding laminar flow, (5) *Laminar* flow is flow in which the mass of fluid may be considered separate laminae (sheets) with simple shear existing at the surface of contact of laminae should there be any difference in mean speed of the separate laminae. If turbulence exists, its effect is confined to a lamina and there is no exchange of momentum between laminae, (6) *Streamline* flow is flow in which fluid particles move along the streamlines. This motion is characteristic of viscous flow at low Reynolds numbers or of inviscid, irrotational flow, (7) *Secondary* flow is a less rigorously defined term than many of the foregoing types of flow. Flow in pipes and channels is frequently found to possess components at right angles to the axis. These components, which take the form of diffuse vortices with axes parallel to the main flow, form the secondary flow.

Fluorescent particle technique Use of fluorescent particles as visual traces to study diffusion in the atmosphere.

Fluorometer(s) An instrument to measure the intensity and color of fluorescent radiations.

Fly ash Finely divided particles of ash entrained in flue gases resulting from the combustion of fuel. The particles of ash may contain incompletely burned fuel. Predominantly gas-borne ash from boilers with spreader stoker, underfeed stoker, and pulverized fuel (coal) firing. All solids including ash, charred paper, cinders, dust, soot, or other partially incinerated matter, carried in the products of combustion. Three main constituents are unburned coal, shale, and magnetite. The dust consists largely of a mixture of well-defined hollow punctured spheres, or cenospheres of a light shiny appearance, and irregularly shaped particles of ash and carbon of darker appearance (about 10 to 75 microns). Color of the dust varies from light to dark gray; the higher the sulfur content the darker the ash. (Analysis of typical fly ash: silica, 43.2 percent; alumina, 31.6 percent; ferric oxide, 9.8 percent; lime, 6.1 percent; magnesia, 3.4 percent; sulfate, 1.2 percent; alkalies, etc., 0.9 percent; loss on ignition, 3.8 percent.)

Fly ash collector Equipment to remove fly ash from the products of combustion.

Fog chamber A confined space in which air or other gas is supersaturated by reduction of pressure or cooling, to study the movement and interaction of electrified particles by the condensation (fog tracks) they produce.

Fog horizon The top of a fog layer which is confined by a low-level temperature inversion in such a way as to give the appearance of the horizon when viewed from above against the sky. The true horizon is usually obscured by the fog.

Fossil fuels Fuels that are the result of decomposition of deposited vegetation, over a considerable period of time, under the extreme pressure of the overburden of earth which has accumulated since the vegetation was deposited. The hardest and most dense of these deposits is coal. Liquid deposits are oil, and those in gaseous form are found as natural gas.

Fractionation/fractional distillation Separation in successive stages, each removing from a mixture some proportion of one of the substances. The operation may be precipitation, crystallization, or distillation. In such an operation, a part of the vapor is condensed and the resulting liquid contacted with more vapor, usually in a fractionating column with plates or packing. The term fractional distillation is also applied to any distillation in which the product is collected in a series of separate fractions.

Freezing out/combustion A procedure to analyze incinerator flue gas involving a series of traps for the collection of pollutants. Coolants used may become progressively lower-temperature substances, e.g., ice, dry ice, acetone, liquid nitrogen are applied to respective traps in series to bring about separation of the gaseous components. This is a method used for analysis of liquid, vapor, and gaseous organic compounds which cannot be trapped by the filtration procedures used for suspended particulate matter.

Fume Fine solid particle predominantly less than 1 micron in diameter resulting from condensation of vapor from chemical reactions.

Fumigation Fumigation is an atmospheric phenomenon in which pollution, which has been retained by an inversion layer near its level of emission, is brought rapidly to ground level when the inversion breaks up. High concentrations of pollutant can thus be produced at ground level.

Furnace gas Gases used for firing furnaces; these include natural gas, liquefied petroleum, manufactured gas, coke oven gas, blast furnace gas.

Gas black Finely divided carbon produced by incomplete combustion or thermal decomposition of natural gas. Other names used are carbon black, channel black, furnace black.

Gaseous diffusion separator Instrumentation to separate a gas mixture into its components, as when a gas is an isotopic mixture and is allowed to diffuse through another (preferably heavier) gas. Separation of the various isotopes is based on the principle that the lighter isotopic molecules diffuse through the heavier gas more readily than the heavier isotopic molecules.

Gas flow Fluid flow with gas as the fluid.

Gas plume Stack effluent composed of gases alone or gases and particulates. The geometrical form and concentration distribution of gas plumes are dependent on turbulence. Because of distinctly different atmospheric conditions, gas plumes may be classified into five types; (1) *Looping plumes* occur with a high degree of turbulence, especially convective turbulence, typical of a daytime condition with intense solar heating of the earth's surface which causes unstable thermal conditions —similar plume characteristics may occur with large-scale mechanical turbulence caused by large and sharp hills upwind of the stacks, (2) *Coning plumes* occur under more nearly neutral thermal conditions when mechanical turbulence of smaller scale is predominant, (3) *Fanning plumes* occur under stable conditions when mechanical turbulence is suppressed. The vertical component is suppressed more than the horizontal, with the result that plume width is greater than its thickness— this is likely to occur at night when the earth is cooled by outgoing radiation, (4) The *fumigating plumes* is caused by unstable air below and a stable layer above, causing greater downward dispersion of gases and bringing more of the plume to the ground than would occur with unstable conditions above and below. This usually occurs in the morning following a night of marked stability; the rising sun heats the ground causing an unstable layer to develop from the ground upward. When the unstable layer reaches plume level, the compact, highly concentrated plume quickly diffuses downward and a high ground level concentration may occur for a ½ hr, (5) The *lofting plume* is caused by conditions inverse to those causing fumigation—the greater diffusion is upward, and late afternoon and early evening are favorable for this condition. Two or more stack effluents from closely grouped stacks tend to merge into one.

Gob piles Fires which result when air can flow readily through the interstices of large piles of low-combustible refuse from coal mine preparation plants. Slow oxidation occurs, heat is liberated, and higher temperatures develop until ignition occurs.

Gradient(s) Either the rate of change of a quantity (as temperature, pressure, etc.) or a diagram or curve representing this rate.

Grit Solid particles of natural or manmade origin, retained on a 200 mesh British Standard sieve, i.e., 76 microns or greater in diameter.

Ground level concentration The amount of solid, liquid, or gaseous material per unit volume of air, from 0 to 2 meters above the ground.

Ground level inversion/ground inversion Increase in temperature with the increasing height of the atmospheric layers nearest to the ground (normally the temperature drops with increasing height).

Halogenated hydrocarbons Hydrocarbons having one or more hydrogen atoms replaced by a halogen atom.

Halogenation Incorporation of one of the halogen elements, usually chlorine or bromine, into a chemical compound; e.g., benzene is treated with chlorine to form chlorobenzene, and ethylene is treated with bromine to form ethylene dibromide.

Haze A state of atmospheric obscurity due to the presence of fine dust solid particles in suspension; visibility exceeds 1 but is less than 2 km. The particles are so small that they cannot be felt or individually seen with the naked eye. They diminish horizontal visibility and give the atmosphere an opalescent appearance. Many haze formations are caused by the presence of abundant condensation nuclei which may grow in size and become mist, fog, or cloud. (Heat haze is caused by volatilized essential oils of growing vegetation. Constituents of heat haze are examples of naturally occurring particulate materials in the atmosphere.) Distinction may be made between dry (0.1 micron) haze particles and damp (larger) haze particles. The dry haze particles produce a bluish color when haze is viewed against a dark background; this same type of haze when viewed against a light background appears as a yellowish veil.

Haze horizon The top of a haze layer which is confined by a low-level temperature inversion in such a way as to give the appearance of the horizon when viewed from above against the sky. The true horizon is usually obscured by the haze in such instances.

Humidity sensitive elements The transducer of any hygrometer, i.e., that part of a hygrometer that quantitatively "senses" atmospheric water vapor.

Hydrocarbon cracking Decomposition by heat with or without catalysis, of petroleum or heavy petroleum fractions, with production of lower-boiling materials useful as motor fuels, domestic fuel oil, or other needed products. Generally thermal decomposition processes, where ammonia may be cracked into nitrogen and hydrogen, and methane may be cracked into carbon and hydrogen or other hydrocarbons

Hydrocarbons Compounds which consist solely of carbon and hydrogen. (1) *Acetylene hydrocarbons* are those which contain at least one pair of triple-bonded carbon atoms in their structure, satisfying the general formula C_nH_{2n-2}. Also known as alkynes. These are unsaturated hydrocarbons. (2) *Aromatic hydrocarbons* are characterized by a molecular structure involving one or more six-carbon-atom rings and having properties similar to those of benzene, which is the simplest member of this group. (3) *Branch chain hydrocarbons* are nonaromatic hydrocarbons in which not all the carbon atoms of the molecule are in a single chain. The simplest is isobutane. (4) *Olefin hydrocarbons* contain at least one part of double-bonded carbon atoms in their structure, with a single double bond satisfying the formula C_nH_{2n}. These are unsaturated hydrocarbons. (5) *Paraffin hydrocarbons* are those in which the proportion of hydrogen to carbon is such as to satisfy the general formula C_nH_{2n+2}, such as methane, CH_4, or ethane, C_2H_6. These are also called saturated hydrocarbons. (6) *Straight chain hydrocarbons* are those in which all the carbon atoms of the molecule are in a single unbranched chain; they are designated also as normal hydrocarbons. (7) *Unsaturated hydrocarbons* have fewer hydrogen atoms or equivalent groups than the corresponding saturated compounds. (8) *Polycyclic hydrocarbons* are molecules having more than one ring in their structure. These may be aromatic rings, cycloalkane rings, cycloalkene rings, heterocyclic rings (with other atoms besides carbon in the ring); and combinations of such rings.

Hydrocracking unit Process for cracking heavy hydrocarbons to light products in the presence of high partial pressures of hydrogen, and of a special catalyst. Hydrocracked products are considered to be essentially sulfur-free.

Hygrometer(s) An instrument which measures the water vapor content of the atmosphere. There are six basically different means of transduction used in measuring this quantity and an equal number of hygrometer types. These are (1) the psychrometer which utilizes the thermodynamic method; (2) the class of instruments which depends upon a change of physical dimensions due to absorption of moisture (hair hygrometer, torsion hygrometer, goldbeater's skin hygrometer, carbon-film hygrometer element); (3) those which depend upon moisture concentration (dew point hygrometer); (4) the class of instruments which depends upon a change of electrical or chemical properties due to moisture absorption (absorption hygrometer, electrical hygrometer, carbon-film hygrometer element); (5) the class of instruments which depends upon the diffusion of water vapor through a porous membrane (diffusion hygrometer); (6) the class of instruments which depends upon measurements of water vapor absorption spectra.

Immission The occurrence of solid, liquid, and gaseous substances polluting the air primarily close to the ground. Usually equivalent to ground level concentration.

Impaction A forcible contact of particles of matter; synonymously with impingement.

Impactor A sampling device which employs the principle of impingement. The "cascade impactor" refers to a specific instrument which employs several impactions in series to collect successively smaller sizes of particles.

Impingement The act of bringing matter forcibly in contact. As used in air sampling, impingement refers to a process for the collection of particulate matter in which the gas being sampled is directed forcibly against a surface. (1) Dry impingement: the process of impingement in the gas stream where particulate matter is retained upon the surface against which the stream is directed. The collecting surface may be treated with a film of adhesive. (2) Wet impingement: the process of impingement in a liquid body and that liquid retains the particulate matter.

Impinger A sampling instrument employing impingement for the collection of particulate matter. Commonly, this term is applied to specific instruments, the "midget" and "standard" impinger. (1) The midget impinger is a specific instrument employing wet impingement, using a liquid volume of 1 ml and a gas flow of 0.1 cfm. (2) The standard impinger is a specific instrument employing wet impingement, using a liquid volume of 75 ml and a gas flow of 1 cfm. (3) Dry impingers: commercial types usually consist of a series of progressively smaller-size jets impinging on standard microscope slides. This results in progressively higher jet speeds and in the collection of progressively smaller-size particles.

Incinerator(s) Equipment in which solid, semisolid, liquid, or gaseous combustible wastes are ignited and burned, leaving solid residues containing little or no combustible material. (1) Flue-fed incinerator(s): incinerator for multiple-occupancy units in which refuse is charged through openings on each floor into a flue, depositing the refuse into a combustion chamber below. (2) Multiple-chamber incinerator(s): a two-stage combustion mechanism. Drying of the fuel, ignition and combustion of fixed carbon, gasification and partial combustion of the volatile components proceed in the ignition chamber or primary stage. The gas phase combustion is completed in the second stage, which consists of both a mixing chamber and a combustion chamber. Secondary air is admitted to the mixing chamber to aid in combustion. (3) Multiple-chamber incinerator(s): consists of two or more refractory lined chambers, interconnected by gas passage ports or ducts and designed in such a manner as to provide for complete combustion of the material to be burned. Depending upon the arrangement of the chambers, multiple-chamber incinerators are designed as in-line or retort types.

Individual perception threshold (IPT) The lowest concentration of a particular odor at which a subject indicates both an initial positive response and a repeated response.

Inertial separator(s) Includes all dry type collectors utilizing the relatively greater inertia of the particle to effect particulate gas separation. Two types of equipment utilize this principle: cyclonic separators which produce continuous centrifugal force as a means of exerting the greater inertial effects of the particle; and simple inertial or impaction separators which employ incremental changes of direction of the carrier gas stream to exert the greater inertial effects of the particle. Included in this category, besides cyclonic and impingement separators, are gravity settling chambers and high-velocity gas reversal chambers.

Infrared gas analyzer Infrared spectrometer with two equivalent calcium fluoride infrared window sources, one for the comparison cell and the other for the sample cell. It functions on the principle that the sample air will absorb infrared radiation at a different rate than the gas in the comparison cell. Thus, with proper instrumentation, the CO concentration in the sample gas may be measured.

Instantaneous sampling Obtaining a sample of an atmosphere in a very short period of time, such that this sampling time is insignificant in comparison with the duration of the operation or the period being sampled. Also known as grab sampling.

Intermittent sampling Sampling successively for limited periods of time throughout an operation or for a predetermined period of time. The duration of sampling periods and of the intervals between is not ncessarily regular and is not specified.

Inversion Hot gases generated by burning, being warmer than surrounding air, rise into the upper atmosphere. However, since the pressure at higher levels is lower, these gases expand and cool themselves. Under normal circumstances, the upper layers of the atmosphere are colder than those nearer the surface, and the gases continue to rise. Occasionally, however, a warm air mass moves in over a cool one. It is lighter than the cooler air; so this condition is quite stable and may remain for a considerable time. Under these circumstances, hot gases rising through the cold air become heavier than the warm upper air and cannot penetrate the warmer layer but form a layer between the two air masses. This condition is called an inversion.

Inversion height(s) A term commonly used to denote the height of a temperature inversion which is a layer of atmosphere in which the temperature increases with height. This layer may be of any depth above the ground amounting to a discontinuity in the temperature lapse rate. An abbreviation for "inversion of temperature gradient." The temperature of air generally gets lower with increasing height, but occasionally the reverse is the case, and when the temperature increases with height there is said to be an inversion.

Ionization chamber An apparatus used to study the production of small ions in the atmosphere by cosmic ray and radioactive bombardment of air molecules. The chamber is an airtight container usually cylindrical in shape and 25 to 50 l in volume. An insulated electrode is centrally located in the chamber. In operation, a potential is applied between the electrode and the chamber wall. The ions produced in the chamber are collected by the electrode system and measured by an electrometer.

Ion(s) In atmospheric electricity, any of several types of electrically charged submicroscopic particles normally found in the atmosphere. Atmospheric ions are of two principal types, small ions and large ions, and a class of intermediate ions has occasionally been reported. Formation of small ions depends upon cosmic rays and radioactive emanations. Each of these consists of very energetic particles which ionize neutral air molecules by knocking out one or more planetary electrons. The resulting free electron and positively charged molecule (or atom) attach themselves to one or at most a small number of neutral air molecules, forming new small ions. In the presence of Aitken nuclei, some of the small ions will attach themselves to the nuclei, thereby creating new large ions. The two classes of ions differ widely in mobility. Only the highly mobile small ions contribute significantly to the electrical conductivity of the air under most conditions. The intermediate ions and large ions are important in certain space charge effects but are too sluggish to contribute significantly to electrical conductivity. The formation of ions is offset by processes such as recombination (reactions).

Isoconic line A line joining points of equal dust precipitation per unit area and time.

Isokinetic/isotach A line in a given surface connecting points with equal wind speed. (British term for isotach) (also called isovel).

Isokinetic sampling A technique for collecting airborne particulate matter in which the collector is so designed that the airstream entering it has a velocity equal to that of the air passing around and outside the collector. In principle, an isokinetic sampling device has a collection efficiency of unity for all sizes of particulates in sampled air.

Kiln (oven) A furnace in which the heating operations do not involve fusion. Kilns are most frequently used for calcining, and free access of air is usually permitted. The raw materials may be heated by the combustion of solid fuel with which they are mixed, but more usually they are heated by gas or the waste heat from other furnaces.

Konimeter An instrument for determining the dust content of a sample of air (also called conimeter). One form of the instrument consists of a tapered metal tube through which a sample of air is drawn and allowed to impinge upon a glass slide covered with a viscous substance. The particles caught are counted and measured with the aid of a microscope. Because this method does not reveal particles of diameter less than 0.8 micron, and since the dust in steel foundries is generally below this limit, the konimeter is favored for use in mines while Owen's jet test is mainly used in foundries.

Kytoon A captive balloon used to maintain meteorological equipment aloft at approximately a constant height. The kytoon is streamlined, and combines the aerodynamic properties of a balloon and a kite.

Labile A term used synonymously with "unstable" in reference to systems of two phases that undergo spontaneous change as soon as the transition temperature is passed.

Lacrimation/lachrymation The secretion and discharge of tears.

Lapse rate The rate at which temperature decreases or lapses with altitude.

Light attenuation Reduction in the intensity of light due to the action of the transmitting medium (absorption and scattering).

Light scattering Reflection or scattering of light in all directions when it enters a body of matter, however transparent. This is due to the interposition in the light stream of particles of varying size, from microscopic specks down to electrons, and the deflection of light quanta resulting from the encounters with these small obstacles.

Luminescence Any emission of light below temperatures required for incandescence. Frequently a body becomes luminescent because of previous exposure to some form of radiation. If the body is luminous during the time it is being irradiated, the phenomenon is called fluorescence. In phosphorescence the emission of light continues after irradiation.

Macrometeorology Study of the largest-scale aspects of the atmosphere, such as the general circulation, weather types, and the Grosswetterlagen. There is a wide gap between this and the relatively small scale of mesometeorology. The gap is bridged by those atmospheric characteristics referred to as cyclonic scale. (Grosswetterlag is defined as the mean pressure distribution for a time interval during which the positions of the stationary cyclones and anticyclones remain unchanged.)

Manometer An instrument for measuring pressure differences usually consisting of a U-shaped tube containing a liquid. The height of the liquid in one end of the tube moves proportionally with changes in pressure upon the liquid in the other end. Applied to a tube type of differential pressure gage. A mercury barometer is a type of manometer.

Mass concentration Concentration expressed in terms of mass of substance per unit volume of gas or liquid.

Mass median size The median measurement of particle size for samples of particulate matter related to diameter size and thus mass.

Material damage Air pollution damage to materials of inanimate bodies. This may be brought about through (1) abrasion; (2) deposition and removal, where there might not be much damage except for appearance in deposition, but where damage occurs in removal or cleaning operations; (3) direct chemical attack such as the tarnishing of silver by H_2S; (4) indirect chemical attack which is deterioration caused by chemical reactions after absorption of a pollutant; (5) electrochemical corrosion.

Maximum allowable concentration/mac The concentration of a pollutant considered harmless to healthy adults during their working hours, assuming that they breathe uncontaminated air for the remainder of the time.

Maximum emission concentration(s) Standards for maximum concentration of air pollutant emission from stationary or moving sources. These include opacity standards, such as the Ringelmann chart gradations often applied with time specification; gravimetric emission standards (expressed as weight of emitted pollutant per volume or unit weight of carrier gas); volumetric emission standards (percent of gaseous pollutant by volume in the emitted gas or as parts per volume in specific parts of effluent gas, e.g., where SO_2 is limited to 0.2 percent by volume in emitted gases or zinc mineral is limited to 1 part by volume per 1,000 parts of effluent gas).

Membrane filter Controlled pore filters composed of cellulose esters. They can be manufactured with uniformly controlled pore size. Nylon mesh may be used for reinforcement. Types commonly used for air sampling have a pore size of about 0.45 to 0.8 micron. The pores constitute 80 to 85 percent of the filter volume. Because of electrostatic forces and the formation of a precoat of collected particles on the surface, these filters can collect particles down to about 0.1 micron in diameter. They are particularly suited to particle sizing, since most of the collected particles are retained on or near the surface.

Micrometeorology The portion of the science of meteorology which deals with the observation and explanation of the smallest scale physical and dynamic occurrences within the atmosphere. So far, studies within the field are confined to the surface boundary layer of the atmosphere, that is, from the earth's surface to an altitude where the effects of the immediate underlying surface upon air motion and composition become negligible. More confining limits have been suggested. To date, the bulk of the work in this field has centered around evaluation of low-level atmospheric turbulence, diffusion, and heat transfer. Small-scale temperature, evaporation, and radiation studies are included; so far these have been made primarily in the field of microclimatology.

Mist A state of atmospheric obscurity produced by suspended water droplets; invisibility exceeds 1 but is less than 2 km. According to international definition, a hydrometeor consisting of an aggregate of microscopic and more or less hygroscopic water droplets suspended in the atmosphere. It produces, generally, a thin, grayish veil over the landscape. It reduces visibility to a lesser extent than fog. The relative humidity with mist is often less than 95 percent. It is intermediate in all respects between haze (particularly "damp haze") and fog. In popular usage in the United States, the same as drizzle.

Mixing height Height at which stack effluent begins mixing with the atmosphere as it leaves the stack.

Monitoring Instrumentation and procedures for the continuous measurement of air pollutants and the application of regulatory or control measures when an established standard has been exceeded.

Monitoring systems Automatic systems placed in a smokestack to measure and record the amounts of air pollutants being discharged into the atmosphere. The data can either be stored for later examination or be transmitted to a central unit.

National air sampling network (NASN) Network of air sampling stations established in the United States in urban and nonurban communities. The objectives of this network include determination of the extent and nature of air pollution as well as the study of trends in the levels of atmospheric contamination and investigation of relationships between air pollution and socioeconomic, geographic, topographic, meteorological, and other factors. Initially operations of the network included the collection and analysis of suspended particulate matter only. Activities have been expanded to include the collection and analysis of precipitation samples and the determination of atmospheric contamination levels of specific gaseous pollutants. Network established by the Public Health Service in 1953 to measure United States air quality. Orginally, 17 communities participated.

Noxious gases Gases which have been proved to cause ill effects in human beings. In the true sense of the word, CO_2, which is not poisonous, can be noxious in such large concentration as to reduce dangerously the amount of oxygen available. Usually, however, noxious gases are considered to be only those which cause ill effects at low concentrations. Foul-smelling gases, if not harmful, are usually not categorized as noxious.

Nuisance threshold Standard for the concentration of an air pollutant that is considered objectionable. In the case of an odor, that concentration which can be detected by a human being.

Odor threshold Concentration of an odor in air that can be detected by a human being; i.e., in the case of sulfides emitted from a kraft pulping process, an observer can smell these compounds at and above the odor threshold, concentrations of about 1 part in 10^9 parts of air.

Olfactory sensitivity Sensitivity of the olfactory nerves or ability to detect odors.

Opacity rating A measurement of the opacity of emissions, defined as the apparent obscuration of an observer's vision to a degree equal to the apparent obscuration of smoke of a given rating on the Ringelmann chart.

Open burning The occasional burning of outdoor stores of lumber, sawdust, scrapped cars, tires, textiles, and open dumps.

Open hearth furnace(s) Reverberatory furnace, containing a basin-shaped hearth, for melting and refining suitable types of pig iron, iron ore, and scrap for steel production. In all open hearth furnaces, a large amount of dust from ore and other materials and splashings from slag are carried away by the waste gases. This is especially serious in the basic process. To prevent choking of the regenerators, and the fluxing and glazing of the checker-bricks, a supplementary chamber is used for collecting slag and dust (slag pocket or dust catcher).

Overfire Overfire air jets are streams of high-velocity air issuing from nozzles in a furnace enclosure, to provide turbulence and oxygen to aid combustion, or to provide cooling air.

Oxidant Any substance with an oxidation potential greater than oxygen.

Oxides of nitrogen These include nitric anhydride or nitrogen pentoxide (N_2O_5), nitrogen tetroxide (N_2O_4), nitrous anhydride or nitrogen sesquioxide (N_2O_3), nitrogen trioxide (NO_3), nitrogen dioxide (NO_2), nitrous oxide (NO).

Oxides of sulfur These include sulfur dioxide (SO_2), sulfur trioxide (SO_3).

Ozonosphere General stratum of the upper atmosphere in which there is an appreciable ozone concentration and in which ozone plays an important part in the radiative balance of the atmosphere. The region lies between 10 and 50 km, with maximum ozone concentration at about 20 to 25 km (also called ozone layer).

Packed bed Consists of mesh solid support used in gas chromatography. Components of a mixture are separated by volatizing a sample into a carrier gas stream which is passing through and over the bed of packing. The surface of the bed is usually coated with a relatively nonvolatile liquid designated the stationary phase, thus the term gas-liquid chromatography; if the liquid is not present, the process is dry, a gas-solid chromatography. The different components move through the packing at different rates, appearing one after another at the effluent end, where they can be detected and measured separately.

Packed column A vertical column used for distillation, absorption, and extraction, containing packing, e.g., Raschig rings, Berl saddles, or crushed rock, which provides a large contacting surface area between phases. Normally, gas flow is countercurrent to liquid flow in this type of equipment. The separating mechanism is believed to be primarily impingement of the particle on the packing itself with the liquid medium acting to clean the surface of the packing material (also called packed scrubber or packed tower).

Particle A small discrete mass of solid or liquid matter such as aerosols, dusts, fumes, mists, smoke, and sprays. (1) Aerosol—a dispersion of solid or liquid particles of microscopic size in gaseous media, such as smoke, fog, or mist. (2) Dusts—solid particles formed by disintegration processes such as crushing, grinding, demolition. Normally above 2 microns in diameter, but they have been found to be as small as 0.1 micron in diameter. The most noticeable dusts and grits are above 76 micron in size. (3) Fumes—solids generated by condensation of vapors. They may also result from sublimination condensation or foundry processes, and chemical reactions. They are often metals or metal oxides and may differ materially from the substance from which they originated. They are less than 1 micron in diameter. (4) Mists—liquid droplets smaller than 10 microns in diameter and generated by condensation. (5) Smoke—a product of incomplete combustion, consisting of minute carbonaceous particles which remain suspended in the air until removed by precipitation or gravitation. The vast majority of smoke particles are less than 1 micron in size and require the electron microscope for detailed examination. (6) Sprays—liquid droplets larger than 9 microns created by some mechanical disintegration process.

Particle concentration Concentration expressed in terms of number of particles per unit volume of air or other gas.

Particle fall/soot fall A measurement of air contamination consisting of the mass rate at which solid particles deposit from the atmosphere. A term used in the same sense as the older terms "dust fall" and "soot fall" but without any implication as to nature and source of the particles.

Particle size An expression for the size of liquid or solid particles expressed as the average or equivalent diameter.

Particle size distribution The relative percentage by weight or number of each of the different size fractions of particulate matter.

Particulate matter Airborne particulates consisting of solid and liquid particles ranging between particles over 20 microns in diameter and aerosols or suspensions of fine particles extending into the submicron region, down to less than 0.05 micron. More than 20 metallic elements may be identified spectrographically in the inorganic fraction of particulate pollutants. The organic fraction of the particulates is more complex and contains large numbers of aliphatic and aromatic hydrocarbons, acids, bases, phenols, and other compounds. New products are formed by the interaction of contaminants with each other after discharge into the atmosphere. Particles dispersed in a gaseous medium are collectively termed an aerosol.

Peak concentrations During a period of monitoring an air pollutant, or pollutants, the highest concentration of the pollutant or pollutants during that period.

Peak to mean ratio Ratio of the peak concentration to the mean concentration of an air pollutant, or air pollutants, which have been monitored over an established period of time.

Periodic variations Hourly, daily, seasonal, or annual variations in the concentrations of air pollutants due to meteorological, technological, and/or sociological factors. Thus nitric oxide concentrations in the atmosphere have been shown to reflect the same three factors, seasonally and diurnally, namely, the rate of conversion to NO_2, the dilution capacity of the atmosphere, and the strength of combustion sources, the three factors, in turn, being influenced by sunlight patterns.

Permeability The ability to permit penetration or passage. In this sense the term is applied to substances which permit penetration of passage of fluids.

Permissible immission concentration Concentration of matter foreign to air in layers of the atmosphere close to the ground (or deposition of dust) which is tolerable for man, animal, or plant, according to present experience, for a given duration and frequency of effect.

Photochemical reaction(s) A chemical reaction which involves either the absorption or emission of light radiation.

Photochemistry The branch of chemistry that deals with the effect of light in causing or modifying chemical changes. Important examples are in natural photosynthesis, in the production of a photographic image, in reactions of chlorine with hydrocarbons and other organic compounds, in polymerization and cross-linking reactions, and in various degradation processes.

Photoelectric particle counter Instrumentation such as a scintillation counter which consists of several transparent phosphors together with a photomultiplier tube which detects ionizing particles or radiation by means of the light flash emitted when the radiation is absorbed in the phosphors.

Physiological effects In regard to air pollution effects attributed to the direct or indirect action of air pollutants on body functions and health.

Phytotoxicant An agent which produces a toxic effect in vegetation.

Plant damage Damage to living plants caused by air pollutants such as SO_2, fluorides, oxidants such as O_3 and peroxacyl nitrates, nitrogen oxides, ethylene, chlorine, and hydrogen sulfide. Visible symptoms include leaf tissue collapse, chlorosis or other color changes, and growth alterations.

Plume The path taken by the continuous discharges of products from a chimney. The shape of the path and the concentration distribution of gas plumes is dependent on turbulence. Looping, coning, fanning, fumigating, and lofting are among the designations of various shaped plumes.

Precipitation (chemical) The formation of solid particles in a solution. Also, the settling out of small particles in either a liquid or gaseous medium.

Precipitation (electricity) Potential differences of 12,000 to 30,000 (can be as high as 100,000) volts dc are required between spaced electrodes in the electrical precipitation of particles, i.e., where particles suspended in transmitted air are to be ionized for collection at one of the electrodes.

Precipitation (electrostatic) Separating of dust or droplets suspended in a gas or in air by electrical means. In this process, the material to be precipitated is already present in a phase distinct from that of the medium, and it is only necessary to provide the electrical charge by means of ions from the gas to bring about precipitation.

Precipitation (polluting) The air polluting substances separated from the atmosphere in grams per square meter per time.

Precipitation (thermal) Separation which operates on the principle that thermal force (a force greater than that caused by convection) acts on a body suspended in a gas not in thermal equilibrium. Thus, the particles suspended in a gas migrate from a zone of high temperature to one of low temperature because of fluid creep. In general, the thermal force is negligible if the gradient is less than about 750°C/cm. In the "thermal precipitator" (sampling instrument), the air or gas is drawn slowly through a narrow chamber across which extends a heated wire, particulate matter being deposited upon the adjacent collecting surface.

Precipitation (ultrasonic) A process consisting of the separation of particulate matter from air and other gases following agglomeration induced by an ultrasonic field.

Precombustion Preliminary stage of combustion.

Precursor(s) Chemical compounds which are gaseous air pollutants and which react with other substances in the atmosphere to produce different pollutants, such as the case where photochemical reactions producing ozone involve preliminary reactions of NO and NO_2 with the oxygen of the air.

Primary air Air admitted to a furnace during the first part of the firing cycle, i.e., for firing fresh fuel.

Primary emission Pollutants emitted directly into the air from identifiable sources. Primary emissions may be characterized as follows: fine solids (less than 100 microns in diameter), coarse particles (greater than 100 microns in diameter), sulfur compounds, organic compounds, nitrogen compounds, oxygen compounds, halogen compounds, radioactive compounds.

Probe A tube used for sampling or for measuring pressures at a distance from the actual collection or measuring apparatus. It is commonly used for reaching inside stacks and ducts.

Pure air Since the composition of the atmosphere is slightly variable with respect to certain components, the term "pure air" has no precise meaning, but it is commonly air free of nongaseous suspensoids (dust, hydrometeors) and of such gaseous contaminants as industrial effluents. By far the most important gas in air, from the meteorological standpoint, is water vapor. "Dry air" denotes air from which all water vapor has been removed. Pure dry air has a density of 1.2923 g/cu cm at a pressure of 1,013.25 mb and a temperature of 0°C, a specific heat at constant volume of 0.1707 cal/(g)(°C); a specific heat at constant pressure of 0.2396 cal/(g)(°C); and its gas constant (per gram) is 2,870,400 ergs/(g)(°C). The percent by volume of those gases found in relatively constant amount in dry air is very nearly as follows: N_2, 78.084; O_2, 20.946; A, 90.934; CO_2, 0.033; Ne, 0.0018; He, 0.000524 CH_4, 0.0002; Kr, 0.000114; H_2, 0.00005; N_2O, 0.00005; Xe, 0.0000087.

Reinjection Method of disposing of fly ash where it is reinjected in a furnace for further burning.

Reinluft process A dry scrubbing (German) process for the adsorptive separation and recovery of gaseous oxides of sulfur, nitrogen, and phosphorus from industrial gases. This involves bringing the oxide containing gas together with free oxygen into contact with a solid adsorption medium of carbon of low surface activity. The oxide is converted to a higher oxidation stage for which the adsorbing medium, an incxpensive char made from peat, has greater adsorbing power. (Oxidation, and adsorption, consequently, can be accelerated by application of metal catalysts which favor oxidation.) The char travels downward through a large three-stage adsorber, against the rising flow of flue gas. The dioxide (SO_2) is first adsorbed at about 200°F; the trioxide (SO_3) is adsorbed in a smaller lower chamber at about 320°F. In the lowest chamber, N_2 or CO_2 liberates the adsorbed dioxide from the char at 750°F and carries it off in a 50 percent dioxide stream for further processing. The oxides, then being reduced, are separated out, and the medium is cooled and ready for reuse. Simultaneous presence of water vapor increases the efficiency of adsorption, making possible the formation of the corresponding acids of H_2SO_4 and HNO_3.

Residual fuel(s) Liquid or semiliquid products obtained as residues from the distillation of petroleum and used as fuel.

Residual oil When crude oil is extracted from the ground, it is a combination of many organic molecules. In the refining process, the smaller and lighter molecules are evaporated and sold as petroleum gas. The slightly heavier molecules become gasoline and diesel oil. Intermediate molecules are "cracked" under high temperature and pressure to yield additional gasoline. Still heavier oils become lubricants. Paraffin, asphalt, etc., are also extracted from the crude oil. The remaining residual oil, which is not economically usable for other industrial purposes, is sold for burning.

Ringelmann chart A chart used in air pollution evaluation for assigning an arbitrary number, referred to as smoke density, to smoke emanating from any source. This chart is designed specifically for subjective estimating of the density of black smoke and is not applicable to other emissions. The observer compares the grayness of the smoke with a series of shade diagrams formed by horizontal and vertical black lines on a white background.

Robert's smoke chart A disk chart having radial black lines on a white background. When the disk is revolved, solid shades are produced. The chart has an eyelet center and can be spun on a broad awl, nail, or other convenient shaft.

Rotometer A device for measuring rate of fluid flow, based on the principle of Stokes' law. It consists of a tapered vertical tube having a circular cross section. A float inside moves in a vertical path to a height dependent upon the rate of fluid flow upward through the tube.

RUDS Reflectance units of dirt shade units based on the reflection of light.

Sampling network System of installations for sampling air, such as the National Air Sampling Network.

Sanitary landfill(s) Engineered burial of refuse. Refuse is dumped into trenches and compacted by a bulldozer, and microorganisms decompose the organic matter to stable compounds. Moisture is essential for the biological degradation. Groundwater assists this process except when it fills air voids and prevents aerobic metabolism.

Santa Ana wind A hot, dry, desert wind, generally from the northeast or east, especially in the pass and river valley of Santa Ana, Calif., where it is further modified as a mountain gap wind. It blows, sometimes with great force, from the deserts to the east of the Sierra Nevada Mountains and may carry a large amount of dust. It is most frequent in winter.

Screen (sieve) A plate or sheet or a woven cloth, or other device, with regularly spaced apertures of uniform size, mounted in a suitable frame or holder for use in separating material according to size.

Scrubber A type of apparatus used in sampling and in gas cleaning in which the gas is passed through a space containing wetted "packing" or spray. Devices which utilize a liquid to achieve or assist in removal of solid or liquid particles from a carrier gas stream. In general, particles are collected in scrubbers by one or a combination of mechanisms: (1) impingement of particles on liquid medium; (2) diffusion of the particles onto liquid medium; (3) condensation of liquid medium vapors on the particles; (4) partitioning of the gas into extremely small elements to allow collection of the particles by brownian diffusion and gravitational settling on the gas-liquid interface. Designs include spray towers, jet scrubbers, venturi scrubbers, cyclonic scrubbers, inertial scrubbers, mechanical scrubbers, and packed scrubbers. Scrubbers in which liquid contact of the particles is obtained as a result of gas velocity are classified as inertial type scrubbers; these, in turn, consist of the impaction type and the deflection type. Mechanical scrubbers operate by liquid particle contact achieved by simultaneous introduction of liquid medium and gas stream onto rotating disks, blades, or perforated plates. The packed scrubber is a conventional packed tower employing Raschig rings, Berl saddles, fiber glass, or other packing. Normally gas flow is counter to liquid flow. Simple types of scrubbers with low-energy inputs are effective in collecting above 5 to 10-micron particles; more efficient scrubbers will collect particles as small as 1 to 2 microns.

Secondary emission(s) The products of polluted air reactants, such as those which occur in atmospheric photochemical reactions. Secondary pollutants include O_3, formaldehyde, organic hydroperoxides, free radicals, NO, O, etc.

Settling chamber Chamber designed to reduce the velocity of gases in order to permit the settling out of fly ash. It may be either part of, adjacent to, or external to an incinerator.

Settling velocity The terminal rate of fall of a particle through a fluid as induced by gravity or other external force.

Sintering Fly ash accumulated in the smokestack is a salable item for use in the manufacture of cinder blocks and other ceramic products. However, it must first be sintered, or baked, at a very high temperature.

Smog A term to describe a combination of smoke and fog, but widely used in the United States to describe objectionable air pollution. A natural fog contaminated by automobile exhausts and industrial pollutants; a mixture of smoke and fog. This term, coined in 1905 by Des Voeux, has experienced rapid acceptance; but so far it has not been given precise definition. Also used, in the term "Los Angeles smog" to denote a haze produced in the atmosphere by a large-scale photochemical oxidation process, accompanied by eye irritation, plant damage, ozone formation, and characteristic odor.

Smog alert(s) Conditions indicated by meterological forecasts, air sampling, etc., where authorities responsible for air pollution control in a given area recommend or impose reduction of pollutant output by industry and other sources which contribute to smog development. Alert levels of gaseous pollutants (CO, nitrogen oxides, sulfur oxides, O_3) are considered to be concentrations indicative of an approach to danger to public health. There are three such levels promulgated by the Los Angeles Country Air Pollution Control District, the first being an initial warning level, the second a signal for curtailment of certain significant sources, and the third the level at which emergency action must be taken.

Smog index Mathematical correlation between smog and meteorological conditions associated with it which has resulted in a formula, such as the Stanford Research Institute (SRI) smog index, which can ascertain with accuracy whether eye irritating smog is or is not present in the area at the moment for which the index is calculated. Calculation of the SRI smog index S:

$$S = \frac{10(T + 10)}{RW} = \left(\frac{I}{V}\right)^{1/2}$$

where T = °F deviation of 24-hr mean temperature from the mean temperature for that particular day of the year; R = relative humidity at noon; W = total 24-hr wind movement in miles; I = inversion intensity. The five separate aspects of meteorological conditions originally used in the calculations were wind movements for 24 hr; noon visibility in miles; relative humidity at noon; difference between the day's 24-hr mean temperature and the normal mean temperature; height of the temperature inversion level.

Smoke Small gas-borne particles resulting from incomplete combustion, consisting predominantly of carbon and other combustible material, and present in sufficient quantity to be observable independently of the presence of other solids. Used in English in three meanings: (1) An aerosol consisting of all the dispersible particulate products from the incomplete combustion of carbonaceous materials entrained in flue gas as gaseous medium. (2) The persistent aerosol of solid particles and tarry droplets arising from the dilution with air of the above. (3) The particulate material separated from the above aerosol. Black smoke—as dark as or darker than shade 4 on the Ringelmann chart.

Smoke abatement Legal measures which a community may take to enforce laws and regulations concerning smoke emission, i.e., to prevent excessive smoke pollution.

Smoke alarm(s) An instrument which can provide an objective method of continuous measurement and recording of smoke density by means of measuring the amount of light obscured by smoke when a beam of light is shone through the smoke in a flue. Most of the instruments have a scale, graded according to the Ringelmann shades. They can be fitted with an alarm which operates when the smoke is above a preset density. The Ringelmann scale is not applicable to white or colored smokes.

Smoke candle(s) Apparatus used in collecting acid mists. Tubes or candles made from glass or plastic fibers which are pressed into pads with thicknesses up to 2 in. and are mounted in banks. Efficiency is much increased when the glass is treated with silicone oil to repel water, or when normally water-repellent plastic is used.

Smoke density The amount of solid matter contained in smoke and often measured by systems that relate the grayness of the smoke to an established standard.

Smoke horizon The top of a smoke layer which is confined by a low-level temperature inversion in such a way as to give the appearance of the horizon when viewed from above against the sky. The true horizon is usually obscured by the smoke in such instances.

Smoke plume(s) Stack effluent, the geometrical form and concentration of which depend on meteorological factors, such as atmospheric turbulence and wind. Five types due to distinctly different atmospheric conditions are looping, coning, fanning, fumigating, and lofting.

Smoke shade A method for measuring the particulate matter in the air. A clean, white piece of filter paper is placed in an apparatus which forces a known quality of air through it. The particulates which adhere to the paper darken it. A "smoke shade" reading may be obtained by comparing the discolored paper with known standard shades.

Smokestack A chimney which carries the products of combustion away from a fire. A well-designed stack will aid the airflow (draft) to facilitate efficient burning.

Soiling (1) Visible damage to materials by deposition of air pollutants. (2) Soiling may be used as an index for collector efficiency, i.e., by relating the quantities of gas before and after passage through a separator which are necessary to produce equal soiling of a standard filter paper.

Soiling particulates Products of combustion emitted into the atmosphere, such as fly ash, ash, charred paper, cinders, dust, soot, or other partially incinerated matter.

Solar energy Application of sunlight as a special purpose energy source, as in the case of a solar furnace.

Solar radiation The total electromagnetic radiation emitted by the sun. To a first approximation, the sun radiates as a blackbody at a temperature of about 5700°K, 99.9 percent of its energy output falling within the wavelength interval from 0.15 to 4.0 microns, with peak intensity near 0.47 micron. About one-half of the total energy in the solar beam is contained within the visible spectrum from 0.4 to 0.7 micron, and most of the other half lies in the near infrared, a small additional portion lying in the ultraviolet spectrum.

Soot Aggregations of black carbonaceous particles formed during incomplete combustion.

Soot fall During combustion, the high-velocity exhaust gases will draw particulate matter up into the atmosphere. Many of these particles, however, are too large to remain suspended and will be deposited as "soot fall" onto the surrounding terrain.

Specific gravity The ratio between the density of a substance at a given temperature and the density of some substance assumed as standard. For liquids and solids, the standard assumed is either the density of distilled water at 4°C, or the density of distilled water at 60°F. For gases the standards are air, hydrogen, or oxygen at 0°C and a pressure of 760 mm Hg, or distilled water at 4°C. Specific gravity is a relative property that varies with temperature.

Spot test A means of detecting a chemical constituent, or class of constituents, by a color producing chemical reaction visible to the naked eye. The function of spot tests is the rapid, simple qualitative detection of air pollutants.

Stack A vertical passage or chimney, whether of refractory, brick, tile, concrete, metal, or other material, for conducting cooled products of combustion to the atmosphere from a process.

Stack effluent Gaseous and particulate waste products discharged to the atmosphere through stacks of some form.

Stack gas Gaseous waste products discharged to the atmosphere through a stack.

Stack height selection Various limitations prevent the construction of stacks as high as might be desired from the pollution viewpoint. Among these are cost, air traffic considerations and hazards, aesthetic considerations, aerodynamic factors, vibration, materials. Some stacks are now reaching the 700-ft level, and the 1,000-ft level is anticipated.

Stack sampling Collection of representative gaseous and particulate samples of matter flowing through a duct or stack. Acceptable performance should indicate a collection efficiency of 95 ± 5 percent. It may be necessary to use more than one sampling device if partition of the contaminant is desired. Three major requirements must be met in sampling particulate matter: (1) the sample must be taken under isokinetic conditions; (2) there must be a sufficient number of samples; (3) the long axis of the sampling should be parallel to and facing the gas flow direction. Filter paper holders or electrostatic sampling may be used for particulates and scrubbers for gases; the standard impinger may be used for either or both together.

Stagnant inversion During cloudless evenings and nights, particularly in fall and winter, the earth's surface loses heat by radiation and in consequence the air adjacent to the ground is cooled and reduced to some temperature below that of the air at some heights above. Thus, air pollutants are trapped in this layer of air, which being colder than the layer above it, will not rise. During extreme stagnation, pollution reaches such high concentration that a major decrease in solar radiation results over a period of time, coincidentally ensuring the stability of the lower atmosphere.

Stagnation Atmospheric phenomenon responsible for air pollution inversion, where concentrations of air pollutants develop in air layers close to the ground. (1) Stagnation point, a point in a field of flow about a body where the fluid particles have zero velocity with respect to the body. (2) Stagnation pressure, the pressure at a stagnation point. (3) Stagnation region, specifically, the region at the front of a body moving through a fluid where the fluid has negligible relative velocity.

Staining Visible damage to materials caused by chemical reactions between air pollutants and substances contained in materials.

Standard(s) (1) An exact value, or a concept, that has been established by authority or agreement, to serve as model or rule in the measurement of a quantity or in the establishment of a practice or procedure. (2) A document that establishes engineering and technical limitations and applications for items, materials, processes, methods, design, or engineering practices. (Design)—a standard, often set by ordinances, against which plans and specifications are checked before construction is permitted. An air pollution control method often used in lieu of emission or air quality standards. (Fuel)—a form of design standard which limits the kind of fuel that may be burned in equipment of specified design. (Mandatory)—standards which appear in law or regulation.

Stochastic sampling Random sampling.

Stoker A machine for feeding coal into a furnace, and supporting it there during the period of combustion. It may also perform other functions, such as supply of air, control of combustion, distillation of volatile matter. Modern stokers may be classified as overfeed, underfeed, and conveyor. Any mechanical device that feeds fuel uniformly onto a grate or hearth within a furnace.

Sulfur content Both oil and coal are the result of decomposition of vegetation which grew on the earth millions of years ago. These products consist of many complex organic structures, the characteristics of which vary, depending upon the vegetation from which they were formed. If the vegetation contained sulfur, then the resultant oil or coal will also contain sulfur. Such sulfur may be separated from the fuel by either chemical or physical means.

Sulfur dioxide The result of combustion of sulfur and oxygen. It may also be generated by the combustion of some compound which contains sulfur.

Suspended matter/suspended particles A nonuniform distribution of a solid in a liquid without the solid's being dissolved in the dispersion medium.

Temperature gradient Rate of change of temperature or a diagram or curve representing this.

Temperature inversion A layer in which temperature increases with altitude. The principal characteristic of an inversion layer is its marked static stability, so that very little turbulent exchange can occur within it.

Thermosphere Atmospheric shell extending from the top of the mesosphere to outer space. It is a region of more or less steadily increasing temperature with height, starting at 70 or 80 km. It therefore includes the exosphere and most or all of the ionosphere.

Threshold(s) (physiology) That value at which a stimulus just produces a sensation, or comes just within the limits of perception. Also called schwelle. Absolutely the lowest possible limit of stimulation capable of producing sensation.

Threshold limit value (TLV) From AGGIH (documentation of threshold limit values), the concentration of an airborne contaminant to which workers may be exposed repeatedly, day after day, without adverse affect. The TLV recognizes that there are individual variations among workers and that maintaining exposures within the limits may not prevent discomfort or aggravation of a preexisting condition in certain individuals. Most TLVs represent "average" exposures, and some fluctuation during the working day is possible without causing adverse effects.

Toxicity The quality of being poisonous, especially the degree of virulence of a toxic microbe or of a poison. It is expressed by a fraction indicating the ratio between the smallest amount that will cause an animal's death and the weight of that animal.

Tracer A foreign substance mixed with or attached to a given substance to enable the distribution or location of the latter to be determined subsequently. (1) A *physical* tracer is one that is attached by physical means to the object being traced. (2) A *chemical* tracer has chemical properties similar to those of the substance being traced with which it is mixed homogeneously. (3) A *radioactive* tracer is a physical or chemical tracer having radioactivity as its distinctive property. (4) An *isotopic* tracer is a radionuclide or an allobar used as a chemical tracer for the element with which it is isotopic.

Transmission The process by which radiant flux is propagated through a medium or body.

Transmissometer An instrument for measuring the extinction coefficient of the atmosphere and for the determination of visual range. Also called telephotometer, transmittance meter, hazemeter.

Turbulence Atmospheric eddies classified as mechanical or thermal as they are produced primarily by shearing stress or by convection. Turbulence or turbulent energy promotes atmospheric dilution of pollutants. Turbulent fluctuations in stable air are mainly of high frequency with typical periods of the order of seconds (mechanical turbulence). The fluctuations often constitute major deformations of flow and are capable of transporting momentum, energy, and suspended matter at rates far in excess of the rate of transport by diffusion and conduction in a nonturbulent or laminar flow.

Underfeed A stoker consisting of a trough- or pot-shaped retort into which coal is forced by an endless screw or ram. Coal is fed to the fire zone by being pushed up from underneath.

Units ppb = parts per billion; pphm = parts per hundred million; ppm = parts per million.

Units British thermal unit/Btu—the amount of heat required to raise 1 lb water at 60°F, 1°F. This unit is defined for various temperatures, but the general usage seems to be to take the Btu as equal to 252, 15° gram-calories, or 1,055 joules. The amount of heat required to raise the temperature of 1 lb of water through 1°F. Usually the temperature range is from 60 to 61°F. 1 Btu = 252 cal = 778.3 ft-lb = 1,055 joules. Unit of work or energy—energy required to raise the temperature of 1 lb of water through a temperature rise of 1°F without any vaporization. When greater precision is needed, the temperature rise is specified as from 39 to 40°F.

Vapor recovery system System used in petroleum refining for separating a mixed charge of miscellaneous gases and gasolines into desired intermediates for further processing.

Visibility meter The general term for instruments used to make direct measurement of visual range in the atmosphere or of the physical characteristics of the atmosphere. Visibility meters may be classified according to the quantities they measure. Telephotometers and transmissometers measure the transmissivity or, alternatively, the extinction coefficient of the atmosphere. Nephelometers measure the scattering function of the atmospheric suspensoids. A third category of visibility meters makes use of an artificial "haze" of variable density which is used to obscure a marker at a fixed distance from the matter.

Visibility reduction Produced by the scattering of light from the surfaces of airborne particles. The degree of light obstruction is a complex function of particle size, aerosol density, thickness of the affected air mass, and certain more subtle physical factors. Particulates responsible for the phenomenon may be primary pollutants, e.g., coal smoke, or secondary pollution, e.g., Los Angeles photochemical smog.

Visual acuity Acuteness or clearness of vision.

Visual threshold Value of a visual stimulus that just produces a sensation, is just appreciable, or comes just within the limits of perception.

Wind rose Any one of a class of diagrams designed to show the distribution of wind direction experienced at a given location over a considerable period; it thus shows the prevailing wind direction. The most common form consists of a circle from which 8 or 16 lines emanate, one for each compass point. The length of each line is proportional to the frequency of wind from that direction; the frequency of calm conditions is entered in the center. Many variations exist. Some indicate the range of wind speeds from each direction; some relate wind directions with other weather occurrences.

Part 2. Industrial Water Pollution Control Terms

Acid-washed activated carbon Carbon which has been contacted with an acid solution with the purpose of dissolving ash in the activated carbon.

Acidity The capacity of a waste water for neutralizing a base. It is normally associated with the presence of carbon dioxide, mineral and organic acids and salts of strong acids or weak bases.

Acre-foot (1) A term used in measuring the volume of water that is equal to the quantity of water required to cover 1 acre 1 ft deep, or 43,560 cu ft. (2) A term used in sewage treatment in measuring the volume of material in a trickling filter. One acre-foot contains 43,560 cu ft of water.

Activated carbon Carbon which is treated by high-temperature heating with steam or carbon dioxide producing an internal porous particle structure. The internal surface area of granular activated carbon is estimated to be about 3,600 sq ft/gr.

Adsorption The adhesion of an extremely thin layer of molecules (of gas, liquid) to the surfaces of solids (granular activated carbons, for instance) or liquids with which they are in contact.

Adsorption isotherms (activated carbon) A measurement of adsorption determined at a constant temperature by varying the amount of carbon used or the concentration impurity in contact with the carbon.

Advanced waste treatment Any treatment method or process employed following biological treatment (1) to increase the removal of pollution load, (2) to remove substances that may be deleterious to receiving waters or the environment, (3) to produce a high-quality effluent suitable for reuse in any specific manner or for discharge under critical conditions. The term tertiary treatment is commonly used to denote advanced waste treatment methods.

Aeration The bringing about of intimate contact between air and liquid by one of the following methods: spraying the liquid in the air, bubbling air through the liquid (diffused aeration), agitation of the liquid to promote surface absorption of air (mechanical aeration).

Aerobic Living or active only in the presence of oxygen.

Aerobic biological oxidation Any waste treatment or process utilizing aerobic organisms, in the presence of air or oxygen, as the agent for reducing pollution load or oxygen demand of organic substances in waste. The term is used in reference to secondary treatment of wastes.

Alkalinity A term which represents the presence of salts of weak acids. The hydroxides, carbonates, and bicarbonates of calcium, sodium, and magnesium are the common impurities that cause alkalinity. They are expressed in milligrams per liter of calcium carbonate.

Anaerobic Adjective meaning living or active only in the absence of free oxygen.

Anaerobic biological treatment Any treatment method or process utilizing anaerobic or facultative organisms, in the absence of air, for the purpose of reducing the organic matter in wastes or organic solids settled out of wastes; commonly referred to as anaerobic digestion or sludge digestion when applied to the treatment of sludge solids.

Apparent density (activated carbon) The weight per unit volume of activated carbon.

Appurtenances, sewer Structures, devices, and appliances, other than pipe or conduit, which are integral parts of a sewerage system; such as manholes, flush tanks, surface inlets.

Aquifer A subsurface geologic formation which contains water.

Backsiphonage The flowing back of contaminated or polluted water from a plumbing fixture or cross connection, into a water supply line, due to a lowering of the pressure in such line.

Backwashing The process of cleaning a rapid sand or mechanical filter by reversing the flow of water.

Baffles Deflector vanes, guides, grids, gratings, or similar devices constructed or placed in flowing water or sewage, to (1) check or effect a more uniform distribution of velocities; (2) absorb energy; (3) divert, guide, or agitate the liquids; and (4) check eddy currents.

Banks, sludge Accumulations on the bed of a waterway of deposits of solids of sewage or industrial waste origin.

Bed depth (activated carbon) The amount of carbon expressed in length units, which is parallel to the flow of the stream and through which the stream must pass.

Biochemical oxygen demand (BOD) (1) A test which measures the quantity of oxygen utilized in the biochemical oxidation of organic matter in a specified time and at a specified temperature (usually 5 days at 20°C). (2) The amount of oxygen required for the biological oxidation of the organic matter in a liquid.

Blowoff A controlled outlet on a pipeline, tank, or conduit which is used to discharge water or accumulations of material carried by the water.

Branch (1) A special form of vitrified sewer tile and cast iron pipe used for making connections to a sewer or water main. They are called T, Y, T-Y, double Y, and V branches according to their respective shapes. (2) Any part of a piping system other than a main.

Bulking, sludge A phenomenon that occurs in activated sludge plants whereby the sludge occupies excessive volumes and will not concentrate readily.

Cake, sludge The material resulting from air drying or dewatering sludge (usually forkable or spadable).

Carbon column A column filled with granular activated carbon whose primary function is the preferential adsorption of a particular type or types of molecules.

Carbon tetrachloride activity The maximum percentage increase in weight of a bed of activated carbon after air saturated with carbon tetrachloride is passed through it at a given temperature.

Catalyst A substance which accelerates or retards a chemical reaction without undergoing any permanent changes.

Cesspool An underground pit into which raw household sewage or other untreated liquid waste is discharged and from which the liquid seeps into the surrounding soil or is otherwise removed. Sometimes called leaching cesspool.

Chamber detritus A detention chamber larger than a grit chamber, usually with provision for removing sediment without interrupting the flow of liquid. A settling tank of short detention period designed, primarily, to remove heavy settleable solids.

Chamber, flowing-through The upper compartment of a two-story sedimentation tank.

Chamber, grit A small detention chamber or an enlargement of a sewer designed to reduce the velocity of flow of the liquid, to permit the separation of mineral from organic solids by differential sedimentation.

Chemisorption Adsorption where the forces holding the adsorbate to the adsorbent are chemical (valance) instead of physical (van der Waals).

Chemical oxygen demand (COD) (1) A test based on the fact that all organic compounds, with few exceptions, can be oxidized to carbon dioxide and water by the action of strong oxidizing agents under acid conditions. Organic matter is converted to carbon dioxide and water regardless of the biological assimilability of the substances. One of the chief limitations is its inability to differentiate between biologically oxidizable and biologically inert organic matter. The major advantage of this test is the short time required for evaluation (2 hr). (2) The amount of oxygen required for the chemical oxidation of organics in a liquid.

Chlorination The application of chlorine to water, sewage, or industrial wastes, generally for the purpose of disinfection but frequently for accomplishing other biological or chemical results. Prechlorination is chlorination prior to treatment, and postchlorination follows treatment.

Chlorination break point The application of chlorine to water, sewage, or industrial waste containing free ammonia, to the point where free residual chlorine is available.

Chlorination, free residual The application of chlorine to water, sewage, or industrial wastes to produce directly or through the destruction of ammonia, or of certain organic nitrogenous compounds, a free available chlorine residual.

Chlorine, available A term used in rating chlorinated lime and hypochlorites as to their total oxidizing power. Also a term formerly applied to residual chlorine; now obsolete.

Chlorine, combined available residual That portion of the total residual chlorine remaining in water, sewage, or industrial wastes at the end of a specified contact period, which will react chemically and biologically as chloramines, or organic chloramines.

Chlorine demand The quantity of chlorine absorbed by waste water (or water) in a given length of time.

Chlorine, total residual Free residual plus combined residual.

Chlorite, high-test hypo A combination of lime and chlorine consisting largely of calcium hypochloride.

Chlorite, sodium hypo A water solution of sodium hydroxide and chlorine, in which sodium hypochlorite is the essential ingredient.

Clarification Process of removing turbidity and suspended solids by settling. Chemicals can be added to improve and speed up the settling process through coagulation.

Clarifier A sedimentation tank.

Clear well A reservoir containing water which has been previously filtered or purified before going into the standpipes or distribution system.

Coagulation (1) The agglomeration of colloidal or finely divided, suspended matter by the addition to the liquid of an appropriate chemical coagulant, by biological processes or by other means. (2) The process of adding a coagulant and necessary other reacting chemicals.

Coils, digester A system of pipes for hot water or steam installed in a sludge digestion tank for the purpose of heating the sludge being treated.

Collector, grit A device placed in a grit chamber to convey deposited grit to one end of the chamber for removal.

Collector, sludge A mechanical device for scraping the sludge on the bottom of a settling tank to a sump pump, from which it can be drawn by hydrostatic or mechanical action.

Colloids A finely divided dispersion of one material, called the "dispersed phase" (solid) in another material, which is called the "dispersion medium" (liquid). Normally negatively charged.

Color bodies Those complex molecules which impart color (usually undesirable) to a solution.

Concentration, hydrogen ion The weight of hydrogen ions in grams per liter of solution. Commonly expressed as the pH value that represents the logarithms of the reciprocal of the hydrogen ion concentration.

Contamination A general term signifying the introduction into water of microorganisms, chemicals, wastes, or sewage which renders the water unfit for its intended use.

Corporation cock A valve for joining a service pipe to a street water main; it is generally owned and operated by the water utility or department. It cannot be operated from the surface.

Countercurrent efficiency (activated carbon) The unique advantage of a carbon column that permits partially spent activated carbon to adsorb impurities before the semiprocessed stream comes in contact with fresh carbon. This allows the maximum capacity of the activated carbon to be utilized.

Critical bed depth (activated carbon) In a carbon column the critical bed depth is the depth of granular carbon which is partially spent. It lies between the fresh carbon and the spend carbon and is the zone where adsorption takes place. In a single-column system, this is the amount of carbon that is not completely utilized.

Cross-sectional bed area (activated carbon) The area of activated carbon through which the stream flow is perpendicular.

Curb cock A shutoff valve attached to the water service pipe from the water main to the building, installed near the curb, which may be operated by means of a valve key to start or stop flow in the water supply lines of a building.

Curve, oxygen, sag A curve that represents the profile of dissolved oxygen content along the course of a stream, resulting from deoxygenation associated with biochemical oxidation of organic matter, and reoxygenation through the absorption of atmospheric oxygen and through biological photosynthesis.

Degreasing The process of removing greases and oils from sewage, waste, and sludge.

Demand, biochemical oxygen (BOD) The quantity of oxygen utilized in the biochemical oxidation of organic matter in a specified time period and at a specified temperature. It is not related to the oxygen requirements in chemical combustion, being determined entirely by the availability of the material as a biological food and by the amount of oxygen utilized by the microorganisms during the oxidation.

Desorption The opposite of adsorption. A phenomenon where an adsorbed molecule leaves the surface of the adsorbent.

Diffuser A porous plate or tube through which air is forced and divided into minute bubbles for diffusion in liquids. Commonly made of carborundum, alundum, or silica sand.

Diffusion, ridge and furrow air A method of air diffusion in an aeration tank of the activated sludge process where porous tile diffusers are placed in depressions created by the sawtooth construction of the tank bottom, in rows across the tank at right angles to the direction of flow.

Diffusion, spiral flow air A method of diffusing air in an aeration tank of the activated sludge process, where by means of properly designed baffles and the proper location of diffusers, a spiral helical movement is given to both the air and the liquor in the tank.

Digestion The biochemical decomposition of organic matter which results in the formation of mineral and simpler organic compounds.

Disinfection (1) The killing of the larger portion (but not necessarily all) of the harmful and objectionable microorganisms in or on a medium by means of chemicals, heat, ultraviolet light, etc. (2) The use of a chemical additive or other treatment to reduce the number of bacteria, particularly the pathogenic organisms.

Dissolved oxygen (DO) The oxygen dissolved in sewage, water, or other liquid, usually expressed in milligrams per liter or percent of saturation. It is the test used in BOD determination.

Eductor (activated carbon) A device with no moving parts used to force an activated carbon water slurry through pipes to the desired location.

Effluent (1) A liquid which flows out of a containing space. (2) Sewage, water, or other liquid, partially or completely treated, or in its natural state, as the case may be, flowing out of a reservoir, basin, or treatment plant, or part thereof.

Elutriation A process of sludge conditioning in which certain constituents are removed by successive flushing with fresh water or plant effluent, thereby reducing the need for using conditioning chemicals.

Facultative Having the power to live under different conditions (either with or without oxygen).

Feeder, chemical, dry A mechanical device for applying dry chemicals to water or sewage at a rate controlled manually, or automatically, by the rate of flow.

Feeder, chemical, solution A mechanical device for applying chemicals in liquid form to water and sewage at a rate controlled manually, or automatically, by the quantity of flow.

Filter, high-rate A trickling filter operated at a high average daily dosing rate, usually between 10 and 30 mgd/acre, sometimes including recirculation of effluent.

Filter, intermittent A natural or artificial bed of sand or other fine-grained material, to the surface of which sewage is intermittently applied in flooding doses and through which it passes, opportunity being given for filtration and the maintenance of aerobic conditions.

Filter, low-rate A trickling filter designed to receive a small load of BOD per unit volume of filtering material and to have a low dosage rate per unit of surface area (usually 1 to 4 mgd/acre). Also called standard rate filter.

Filter, rapid sand A filter for the purification of water where water which has been previously treated, usually by coagulation and sedimentation, is passed downward through a filtering medium consisting of a layer of sand or prepared anthracite coal or other suitable material, usually from 24 to 30 in. thick and resting on a supporting bed of gravel or a porous medium such as carborundum. The filtrate is removed by an underdrain system. The filter is cleaned periodically by reversing the flow of the water upward through the filtering medium; sometimes supplemented by mechanical or air agitation during backwashing to remove mud and other impurities that are lodged in the sand.

Filter, roughing A sewage filter of relatively coarse material operated at a high rate as a preliminary treatment.

Filter, trickling A filter consisting of an artificial bed of coarse material, such as broken stone, clinkers, slate, slats, or brush, over which sewage is distributed and applied in drops, films, or spray, from troughs, drippers, moving distributors, or fixed nozzles, and through which it trickles to the underdrains, giving opportunity for the formation of zoogleal slimes which clarify and oxidize the sewage.

Filter, vacuum A filter consisting of a cylindrical drum mounted on a horizontal axis, covered with a filter cloth, revolving with a partial submergence in liquid. A vacuum is maintained under the cloth for the larger part of a revolution to extract moisture, and the cake is scraped off continuously.

Filtration, biological The process of passing a liquid through a biological filter containing media on the surfaces of which zoogleal films develop which absorb and adsorb fine suspended, colloidal, and dissolved solids, and which release various biochemical end products.

Floc A very fine, fluffy mass formed by the aggregation of fine suspended particles in a liquid.

Flocculation The formation of flocs.

Flocculator An apparatus designed for the formation of floc in water or sewage.

Flotation A method of raising suspended matter to the surface of the liquid in a tank as scum—by aeration, vacuum, evolution of gas, chemicals, electrolysis, heat, or bacterial decomposition—and the subsequent removal of the scum by skimming.

Grease In sewage, grease includes fats, waxes, free fatty acids, calcium and magnesium soaps, mineral oils, and other nonfatty materials. The type of solvents to be used for its extraction should be stated.

Grit The heavy mineral matter in water or sewage, such as sand, gravel, cinders.

Hardness (1) A measure of the capacity of the water for precipitating soap. (2) A characteristic of water which represents the total concentration of just the calcium and magnesium ions expressed as calcium carbonate.

Heat of adsorption The heat given off when molecules are adsorbed.

Heavy metals A general name given to the ions of metallic elements such as copper, zinc, iron, chromium, and aluminum. They are normally removed from a waste water by forming an insoluble precipitate (usually a metallic hydroxide).

Incineration The combustion (by burning) of organic matter in waste water sludge solids after water evaporation from the solids.

Index, purification A criterion by which may be measured the degree of oxidation, reduction, or nitrification accomplished in sewage treatment.

Index, sludge The volume in milliliters occupied by aerated mixed liquor containing 1 gram of dry solids after settling 30 min. commonly referred to as the Mohlman index. The Donaldson index, which is also commonly used, is obtained by dividing 100 by the Mohlman index.

Influent Sewage, water, or other liquid, either raw or partly treated, flowing into a reservoir, basin, or treatment plant, or any part thereof.

Invert The floor, bottom, or lowest portion of the internal cross section of a closed conduit.

Iodine number (activated carbon) The iodine number is the milligrams of iodine adsorbed by 1 gram of carbon at a filtrate concentration of 0.02*N* iodine.

Ionization the process of the formation of ions by the splitting of molecules of electrolytes in solution.

Irrigation, spray Irrigation by means of nozzles along a pipe on the ground or from perforated overhead pipes.

Lagoon, sludge A relatively shallow basin, or natural depression, used for the storage or digestion of sludge, sometimes for its ultimate detention or dewatering.

Liquor, mixed A mixture of activated sludge and sewage in the aeration tank undergoing activated sludge treatment.

Liquor, supernatant (1) The liquor overlying deposited solids. (2) The liquid in a sludge digestion tank which lies between the sludge at the bottom and the floating scum at the top.

Loss of head gage A gage on a rapid sand filter which indicates the loss of head involved in the filtering operation, whereby the operator is able to ascertain the need for filter backwashing.

Macropore The pores in activated carbon which are larger (diameter) than 1,000 Å.

Makeup carbon Fresh granular activated carbon which must be added to a column system after a regeneration cycle or when deemed necessary to bring the total amount of carbon to specification.

Mesh size (activated carbon) The particle size of granular activated carbon as determined by the U.S. Sieve series. Particle size distribution within a mesh series is given in the specification of the particular carbon.

Methylene blue number (activated carbon) The methylene blue number is the milligrams of methylene blue adsorbed by 1 gram of carbon in equilibrium with a solution of methylene blue having a concentration of 1.0 mg/l.

Micropore The pores in activated carbon which range in size (diameter) from 10 to 1,000 Å.

Milligrams per liter (mg/l) This is a weight per volume designation used in water and waste water analysis.

Molasses number (activated carbon) The molasses number is calculated from the ratio of the optical densities of the filtrate of a molasses solution treated with a standard activated carbon and the activated carbon in question.

Moving bed (activated carbon) A unique application with Filtrasorb granular carbons in which a single carbon column offers the efficiency of several columns in series. This is accomplished by the removal of spent carbon from one end of the carbon bed and the addition of fresh carbon at the other end, with little or no interruption in the process.

Nitrification The conversion of nitrogenous matter into nitrates by bacteria.

Number, indicated (IN) In the testing of bacterial density by the dilution method, the number obtained by taking the reciprocal of the highest possible dilution (smallest quantity of sample) in a decimal series.

Outfall The point or location where sewage or drainage discharges from a sewer, drain, or conduit.

Overflow, storm A weir, orifice, or other device for permitting the discharge from a combined sewer of that part of the flow in excess of that which the sewer is designed to carry.

Oxidation, direct Oxidation of substances in sewage without the benefit of living organisms, by the direct application of air or oxidizing agents such as chlorine.

Oxidation, sewage The process whereby, through the agency of living organisms in the presence of oxygen, the organic matter that is contained in sewage is converted into a more stable or a mineral form.

Oxygen, available The quantity of atmospheric oxygen dissolved in the water of a stream. It is the quantity of dissolved oxygen available for the oxidation of organic matter in sewage.

Oxygen consumed The quantity of oxygen taken up from potassium permanganate in solution by a liquid containing organic matter. Commonly regarded as an index of the carbonaceous matter present. Time and temperature must be specified.

Oxygen dissolved Usually designated as DO. The oxygen dissolved in sewage, water, or other liquid, usually expressed in parts per million or percent of saturation.

Ozone Oxygen in molecular form with three atoms of oxygen forming each molecule. Atmospheric oxygen is in molecular form, but each molecule contains two atoms of oxygen. Ozone is formed by passing high-voltage electric charges through dry air. The third atom of oxygen in each molecule of ozone is loosely bound to it and is easily released. Ozone is sometimes used for the disinfection of water, but more frequently for the oxidation of taste producing substances such as phenol in water and neutralization of odors in gases or air.

pH A symbol denoting the negative logarithm of the hydrogen ion concentration in a solution. pH values run from 0 to 14. The number 7 indicates neutrality, while numbers less than 7 indicate increasing acidity and numbers greater than 7 indicate increasing alkalinity.

Parts per million (ppm) Parts by weight in sewage analysis; ppm by weight is equal to milligrams per liter divided by the specific gravity. It should be noted that in water analysis ppm is always understood to imply weight/weight ratio even though in practice a volume may be measured instead of a weight. By contrast, "percent" may be either a volume/volume or a weight/weight ratio.

Period, aeration (1) The theoretical time, usually expressed in hours, that the mixed liquor is subjected to aeration in an aeration tank undergoing activated sludge treatment; is equal to the volume of the tank divided by the volumetric rate of flow of the sewage and return sludge. (2) The theoretical time that water is subjected to aeration.

Period, detention The theoretical time required to displace the contents of a tank or unit at a given rate of discharge (volume divided by rate of discharge).

Pollution load A measure of the strength of a waste water in terms of its solids or oxygen-demanding characteristics and/or other objectionable physical and chemical characteristics, or in terms of harm to receiving waters.

Pollution, water The introduction into a body of water of substances of such character and of such quantity that its natural quality is so altered as to impair its usefulness or render it offensive to the senses of sight, taste, or smell.

Polyelectrolytes Used as a coagulant or a coagulant aid in water and waste water treatment. (Activated carbon is another coagulant aid.) They are synthetic polymers having a high molecular weight. Anionic: negatively charged. Nonionic: carry both negative and positive charges. [Cationic: positively charged (most popular).]

Pond, sewage oxidation A pond, either natural or artificial, into which partly treated sewage is discharged and in which natural purification processes take place under the influence of sunlight and air.

Pooling, filter The formation of pools of sewage on the surface of filters, caused by surface cloggings.

Pore volume (activated carbon) The pore volume is the difference in the volumetric displacement by granular activated carbon in mercury and in helium at standard conditions.

Preaeration A preparatory treatment of sewage consisting of aeration to remove gases, add oxygen, or promote flotation of grease, and aid coagulation.

Prechlorination (1) Chlorination of water prior to filtration. (2) Chlorination of sewage prior to treatment.

Precipitation, chemical (1) Precipitation induced by addition of chemicals. (2) The process of softening water by the addition of lime and soda ash as the precipitants.

Pretreatment Any waste water treatment process used to reduce pollution load partially before the waste water is introduced into a main sewer system or delivered to a treatment plant for substantial reduction of the pollution load.

Primary treatment A process to remove substantially all floating and settleable solids in waste water and partially to reduce the concentration of suspended solids.

This process can include screening, grit removal, sedimentation, sludge digestion, and sludge disposal.

Process, activated sludge A biological sewage treatment process in which a mixture of sewage and activated sludge is agitated and aerated. The activated sludge is subsequently separated from the treated sewage (mixed liquor) by sedimentation, and wasted or returned to the process as needed. The treated sewage overflows the weir of the settling tank in which separation from the sludge takes place.

Process, biological The process by which the life activities of bacteria, and other microorganisms in the search for food, break down complex organic materials into simple, more stable substances. Self-purification of sewage polluted streams, sludge digestion, and all so-called secondary sewage treatments result from this process. Also called biochemical process.

Process, oxidation Any method of sewage treatment for the oxidation of the putrescible organic matter; the usual methods are biological filtration, and the activated sludge process.

Purification, degree of (1) A measure of the completeness of destruction or removal of objectionable impurities, such as bacteria and hardness, from water by natural means (self-purification) or by treatment. (2) A measure of the removal, oxidation, or destruction of solids, organic matter, bacteria, or other specified substance effected by sewage treatment processes.

Putrefaction Biological decomposition of organic matter accompanied by the production of foul-smelling products associated with anaerobic conditions.

Putrescibility (1) The relative tendency of organic matter to undergo decomposition in the absence of oxygen. (2) The susceptibility of waste waters, sewage, effluent, or sludge to putrefaction. (3) Term used in water or sewage analysis to define stability of polluted water or raw or partially treated sewage.

Rate, oxidation The rate at which the organic matter in sewage is stabilized.

Ratio, dosing The maximum rate of application of sewage to a filter on any unit of area, divided by the average rate of application on that area.

Reactivation (activated carbon) The removal of adsorbates from spent granular activated carbon which will allow the carbon to be reused. This is also called regeneration and revivification.

Reaeration, sludge The continuous aeration of sludge after its initial aeration in the activated sludge process.

Recirculation (1) The refiltration of all or a portion of the effluent in a high-rate trickling filter for the purpose of maintaining a uniform high rate through the filter. (2) The return of effluent to the incoming flow to reduce its strength.

Reduction, overall The percentage reduction in the final effluent as compared with the raw sewage.

Reoxygenation The replenishment of oxygen in a stream from (1) dilution water entering the stream, (2) biological oxygenation through the activities of certain oxygen-producting plants, and (3) atmospheric reaeration.

Reservoir A pond, lake, tank, basin, or other space, either natural in origin or created in whole or in part by building of engineering structures. It is used for storage, regulation, and control of water.

Riparian An adjective pertaining to anything connected with or adjacent to the banks of a stream or other body of water.

Screening The removal of relatively coarse, floating, and suspended solids by straining through racks or screens.

Secondary treatment A process to reduce the amount of dissolved organic matter and further reduce the amount of suspended solids in waste water. The effluent from the primary treatment process is given this additional treatment with processes such as activated sludge or trickling filter.

Sedimentation, final Settling of partly settled, flocculated, or oxidized sewage in a final tank.

Sedimentation, plain The sedimentation of suspended matter in a liquid unaided by chemicals or other special means, and without provision for the decomposition of deposited solids in contact with the sewage.

Seeding, sludge The inoculation of undigested sewage solids with sludge that has undergone decomposition, for the purpose of introducing favorable organisms, thereby accelerating the initial stages of digestion.

Sewage, combined A sewage containing both sanitary sewage and surface or storm water with or without industrial wastes.

Sewage, dilute Sewage containing less than 150 ppm of suspended solids and BOD (weak sewage).

Sewage, industrial Sewage in which industrial wastes predominate.

Sewage, raw Sewage prior to receiving any treatment.

Sewage, settled Sewage from which most of the settleable solids have been removed by sedimentation.

Sewage, storm Liquid flowing in sewers during or following a period of heavy rainfall and resulting therefrom.

Sewer A pipe or conduit, generally closed, but normally not flowing full, for carrying sewage and other waste liquids.

Sewer, intercepting A sewer which receives dry-weather flow from a number of transverse sewers or outlets, and frequently, additional, predetermined quantities of storm water (if from a combined system), and which conducts such waters to a point for treatment or disposal.

Sludge, activated Sludge floc produced in raw or settled sewage by the growth of zoogleal bacteria and other organisms in the presence of dissolved oxygen, and accumulated in sufficient concentration by returning floc previously formed.

Sludge, dewatering The process of removing a part of the water in sludge by any method, such as draining, evaporation, pressing, centrifuging, exhausting, passing between rollers, or acid flotation, with or without heat. It involves reducing from a liquid to spadable condition rather than merely changing the density of the liquid (concentration) on the one hand or kiln drying on the other.

Sludge, digested Sludge digested under anaerobic conditions until the volatile content has been reduced, usually by approximately 50 percent or more.

Sludge, excess activated Excess activated sludge removed from the activated sludge system for ultimate disposal.

Tertiary treatment A process to remove practically all solids and organic matter from waste water. Granular activated carbon filtration is a tertiary treatment process. Phosphate removal by chemical coagulation is also regarded as a step in tertiary treatment.

Total organic carbon (TOC) TOC is a measure of the amount of carbon in a sample originating from organic matter only. The test is run by burning the sample and measuring the CO_2 produced.

Total solids The total amount of solids in a waste water in both solution and suspension.

Turbidity (1) A condition of a liquid due to fine visible material in suspension, which may not be of sufficient size to be seen as individual particles by the naked eye but which prevents the passage of light through the liquid. (2) A measure of fine suspended matter (usually colloidal) in liquids.

Voids The percent by volume of the interstices to total bed volume.

Wavefront The wavefront is the capacity gradient that exists in the critical bed depth. It outlines the gradual transition of the carbon from "fresh" to "spent."

Zeolite softener An ion exchange process used to soften the water which consists of replacing the calcium and magnesium ions with the sodium ion. Salt (sodium chloride) is used for regeneration. When operated on the hydrogen cycle, calcium and magnesium are replaced with the hydrogen, and a mineral acid (usually sulfuric acid) is used for regeneration.

Index*

* References to entries in the Glossary are indicated by the letter G, followed by the page number in the Glossary on which they appear.

UNIVERSITY LIBRARY
NOTTINGHAM